# RAPTORS

## of Western North America

# RAPTORS
## of Western North America

*Text and Photographs by*

BRIAN K. WHEELER

Range maps researched by
John M. Economidy and Brian K. Wheeler
and produced by
John M. Economidy

*Foreword by*
Clayton M. White

PRINCETON UNIVERSITY PRESS
PRINCETON, NEW JERSEY

Most

some

Published by Princeton University Press, 41 William Street, Princeton, New Jersey 08540
In the United Kingdom: Princeton University Press, 3 Market Place, Woodstock, Oxfordshire OX20 1SY



Library of Congress Cataloging-in-Publication Data

Wheeler, Brian K., 1955–
Raptors of western North America : the Wheeler guide / text and photographs by Brian K. Wheeler ; foreword by Clayton M. White.
p. cm.
Includes bibliographical references and index.
ISBN 0-691-11599-0 (cl : alk. paper)
1. Birds of prey—West (U.S.) 2. Birds of prey—Canada, Western. I. Title.

QL677.78 .W54 2003
598.9'0978—dc21 2002042716

This book has been composed in Minion

Printed on acid-free paper. ∞

www.nathist.princeton.edu

Composition by Bytheway Publishing Services, Binghamton, New York

Printed in Italy by Eurografica

10 9 8 7 6 5 4 3 2 1

*To my wife, Lisa, and my son, Garrett, with all my love*

*In memory of my sister,*
*Janet Louise Wheeler Mashue,*
*who enlightened the lives of all who knew her*

# CONTENTS



## SPECIES ACCOUNTS

# FOREWORD

My Australian colleague, Victor Hurley, was seated in the car beside me as we traveled to my family's ranch in the mountains of extreme northwestern Utah. It was early August 2002. We rattled along the twenty-one-mile gravel road, lined with telephone poles, and passed buteo after buteo perched on the poles; three different species of multiple colors and two age classes. Such an array can be confusing. As we passed one bird I asked Victor what species it was. "I dunno, you have too many hawks in the States. But I reckon it was not a bloody wedgie," meaning of course, not an Australian Wedge-tailed Eagle. The bird appeared, from our distance, to have nearly a completely white breast and was obviously an immature Ferruginous Hawk. At these times I have frequently referred guests to the two Clark and Wheeler raptors guides, but at that moment I thought it would be nice to have a more detailed guide I could let Victor use. Identifying raptors can be confusing, and even those whom others refer to as "experts" need help sorting them out at times.

Nearly twenty years earlier, in 1983, Bill Clark told me he had come upon a remarkable new raptor artist and photographer, a fellow by the name of Brian Wheeler, who would be illustrating Bill's forthcoming field guide. Bill asked if I would be willing to review some material for that guide. So along with Dave Mindell, my graduate student at the time, we looked at several color plates Brian had done and also some text. "Holy mackerel," I thought, "This kid is dang good, just like Bill said." Oh, the tail on the plate of the male Rough-legged Hawk needed to be revised a bit, multiple bars rather than just the subterminal band, and we could tell he was still perfecting that part of his craft. But overall a nice job and most useful for a field guide. Now, nearly two decades later, Brian K. Wheeler has produced two companion volumes, *Raptors of Eastern North America* and *Raptors of Western North America*, and in them he provides the descriptive detail I had wished I had for Victor.

Although currently a truck driver by profession, Brian has been wonderfully productive in his avocation as a raptor artist and photographer by providing material for those of us who dabble in the bird of prey world; he is an important part of that world. Traveling as a truck driver has been an excellent opportunity for him to watch for raptors to photograph. But then, his passion for raptors goes back to his childhood days in the mid-1970s in Washington State. When he moved to Connecticut in his early twenties he came under the influence of Fred Sibley, who held the master permit for Brian to band raptors. One can only guess what impact that had on Brian's drive to succeed in the bird world and in his knowledge of raptors. Brian started to take photographs in 1980, but it was not until about 1985 that he perfected his current style. He has photographed raptors throughout the country, has a keen eye for detail, and is excellent at the craft. From what I know, his prize would be photographing a white morph Gyrfalcon on the breeding grounds sitting beside its eyrie. He has participated in the trapping and research efforts of other biologists, such as along the coastal Pacific Northwest, and has a good feeling for what birds look like in the hand. *A Photographic Guide to North American Raptors* used some of his best work, but these two new volumes will outshine that guide, and I think all will agree that the work is better, more detailed. Why would these not be finer guides with 540 photos in the eastern guide and 603 in the western? You will note that he has not used computer enhancements to make things look better than they really are. What you see is what he got.

I am convinced that users of these guides will find the maps more useful than are the broad sweeping maps in most such guidebooks, Brian Wheeler sought help on range maps from Texas attorney John Economidy, who led the Coastal Bend Hawk Watch in Corpus Christi for nine years. Both agreed that the maps should be extremely detailed, show current status, enable readers to envision what range had been lost in previous decades, and serve as a standard for future researchers to gauge distributional gains or declines. You will note in the Acknowledgments that Brian has contacted literally hundreds of people across the U.S., Canada, and Mexico to get the most up-to-date status of a species, with some of it coming in as late as the date of submission to the press.

So, sit back in a comfortable chair and browse through these guides, or better, go into the field and use them. Above all, enjoy what you have in your hands because it will be a long time before something equal or better comes along.

Clayton M. White
Brigham Young University

# PREFACE

Raptors have enthralled me since I was a child. I remember when I was eight and nine years old roaming the parched wheat-field and sagebrush landscape near my house in south-central Washington State looking for hawk nests. In my early teens my family moved east, and there are vivid memories while living in Iowa of seeing Bald Eagles in the winter along the Mississippi River. In my late teens, and living in central Michigan, finding my first Northern Goshawk nest was a truly exhilarating experience (and tough on jackets, as the female would often dive and hit me before I could duck behind a large enough tree!). A 12-year stint living in Connecticut put me in the midst of raptor migration heaven—all within a few minutes or a few hours drive. The 16 years I have now lived in Colorado have enabled me to easily study the vast number of raptors that summer, winter, or migrate through the Great Plains and Rocky Mountains, particularly the variably plumaged western buteos. Besides watching, studying, and photographing raptors in the various regions where I have lived in the U.S., I have traveled widely from Alaska to southern Mexico and from the Pacific to the Atlantic Coast to obtain firsthand experience with raptors. No matter where I am, whether in the lush tropics of Nayarit, Mexico, aiming my lens at a juvenile Common Black-Hawk only 40 feet away or weathering the icy blast of Duluth, Minnesota, photographing Northern Goshawks gliding by at eye level, raptors still illicit a childish exuberance.

The two preceding raptor guides I coauthored with William S. Clark, *A Field Guide to Hawks of North America* (Clark and Wheeler 2001) and particularly *A Photographic Guide to North American Raptors* (Wheeler and Clark 1995) were the foundations for this guide and its eastern counterpart. Having produced standard field guides, I wanted to develop a set of regionally formatted books that went beyond the scope of a field guide and dealt only with species breeding in North America (defined here as the continental U.S. and Canada). I developed the initial format and began working on these two guides in 1996. As data poured in, particularly on status, distribution, and range, the books quickly became even more detailed than originally planned, and they saw several evolutionary changes until the final format was decided.

I have always been keen on regional guides because they allow for more precise and applicable data that cater to a smaller geographic region. In essence, this limits the number of species one has to deal with when trying to solve identification. It also allows for larger and more precise range maps.

The scientifically accepted demarcation of East and West in North America is the 100th meridian that invisibly slices through the eastern Great Plains. This division between East and West varies among authors, however: many adhere to the 100th, some divide along the abrupt eastern edge of the Rocky Mountains, and a few use the Mississippi River. Although the 100th is a reasonably good demarcation on the Plains for breeding Ferruginous Hawks and Prairie Falcons, it is rather meaningless for other western raptors that breed north and south of the Plains. The demarcation line that best befits raptors tends to be the Mississippi River. Only one western species, Swainson's Hawk, has a very small remnant population east of the river, and only a few true western species stray east of the river during migration and winter. Therefore, I have opted to abide by what raptors do and have divided the U.S. at the Mississippi River. Canada has been divided along the Manitoba-Ontario border and along the west coast of Hudson Bay. Hudson Bay separates the arid western Arctic of Nunavut from the damp eastern Arctic of Québec—a demarcation that also defines plumage characteristics of Gyrfalcons and Rough-legged Hawks (and which is in accordance with Gloger's Rule, which states that in many species lightly pigmented individuals occur in dry climates, darker pigmented individuals in humid climates).

The southern cutoff point on range maps in the West was based on trying to show as much of North America as possible. For the most part, western species with full North American maps are cut off at 26° north latitude in Mexico. This latitude is just south of Brownsville, Texas, and extends west to the middle of Baja California, Mexico. However, true geographic delineation of North America extends farther south into the central highlands of Mexico. Golden Eagles, Northern Goshawks, and "American" Peregrine Falcons are considered to be of special concern or endangered in Mexico; thus I felt more detailed maps of their Mexican ranges were needed, so these maps illustrate areas far south of 26° north. Logistics, however, prevented such detailed mapping for all species whose breeding or wintering ranges extend south of this latitude.

The concept of the regional guides was to elaborate on the 34 species of raptors, that breed in North America. For the West this includes 33 species, including 3 New World Cathartidae vultures.

Accidental species from Europe and Mexico are well covered in the two books I coauthored with William S. Clark. Also, there are two great European guides, *A Field Guide to the Raptors of Europe, the Middle East, and North Africa* (Clark 1999) and *The Raptors of Europe and the Middle East* (Forsman 1999), that give additional coverage of the overseas raptors that rarely occur in the U.S. and Canada.

The photographs encompass an expansive range of plumages. Most photographs were taken after the completion of *A Photographic Guide to North American Raptors* in 1994, and the majority after 1995, when I switched over to an autofocus camera system. A few photographs that may qualify for "chance-of-a-lifetime images" date back to the 1980s. Fewer than 10 photographs have been reused from *A Photographic Guide.*

In 1985 I began to perfect, and subsequently turn into an art form of its own, the imagery of flying raptors. I have done this by using good equipment, learning what "naturally" creates a great photograph, and knowing the behavior and habitat of each species and subspecies. Being from the "old school," I am not an advocate of the new rage of computer alteration of photographs. All photographs herein are of wild raptors, and none have been tweaked by computer: what you see is what I saw and photographed. Although many photographs have less-than-aesthetic backgrounds such as wires or utility poles, these are perches that raptors often favor and where one should look for them. The few raptors that sport leg bands were not staged for photography but were captured and released by biologists for scientific study months or years before I or others photographed them. (I am an advocate of all facets of scientific raptor study, which may include banding or telemetry studies; these greatly increase our knowledge and benefit raptors.)

These two volumes are meant to be complete, self-sustaining guides. The detailed data in the species accounts are tailored to moderate- and expert-level bird enthusiasts, and people of all interest levels can enjoy the photographs and short, descriptive captions.

Raptors, like all wildlife, are experiencing severe habitat alterations in our ever-changing world. Some species are adapting and thriving. Some are not adapting and faltering. We must preserve habitat so all species can thrive. It is my hope that these books will help increase our awareness and appreciation of raptors.

Brian K. Wheeler
Longmont, Colorado

# ACKNOWLEDGMENTS

I am most grateful to my wife, Lisa, who assisted me on the prospectus for these guides and who also taught me most of what I needed to know about the computer to create them. Working a full-time job, I diligently created the books in my "spare time," typically for several hours a day, months at a time. Lisa graciously and lovingly supported me during this period.

I apologize to my three-year-old son, Garrett, who had no idea what Daddy was doing. There were many days I could not spend time with him because of work.

I thank all of my family members for their support during this long, arduous project.

This book is unquestionably better and more complete because of the assistance provided by John M. Economidy. I first queried John, after working on the books for one and a half years, for help on Bald Eagle status and range in Texas. Being the Texas regional editor for the Hawk Migration Association of North America at the time, he had access to a considerable amount of information on raptors in his home state. The rest, you could say, is history. Having a deep appreciation for raptors and seeing what I was producing, John quickly became interested in my ever-expanding project, particularly on status and distribution. He soon began assisting in research not only of Texas raptors but of virtually all raptors in North America. For the last four and a half years of the project we worked as a team, gathering the extensive data that went into the Status and Distribution section of the text and the range maps. John spent thousands of hours gathering data. With two years yet to go, he also volunteered to take on the mind-boggling task of creating the range maps on computer. I am most appreciative of his help, diligence, and expertise. Being a journalist as well as an attorney, John also proofread all species accounts for initial "clean up."

My original editor, Andrew Richford, then with Academic Press, was most supportive of my endeavor. Because of a change of publishers, however, he was unable to see my books produced under his editorship. I thank Dr. Richford for all of his help and faith in me. He gave me a chance to produce my "dream books." I also thank my new editor, Robert Kirk. He was also most supportive and, with his expertise, has made my dream come true. Ellen Foos did an incredible job directing production. Elizabeth Pierson and Sandy Sherman did a most superb job copyediting. Dimitri Karetnikov reformulated the range maps to fit the book format.

The following people and museums permitted access to their collections for plumage studies and, in many cases, shipped specimens to the Denver Museum of Nature and Science in Denver, Colorado, for me to study: Bill Alther of the Denver Museum of Nature and Science; Vicki Byre of the Oklahoma Museum of Natural History in Norman; Mary LeCroy of the American Museum of Natural History in New York City; Michael McNall of the Royal British Columbia Museum in Victoria; J. V. Remsen and Steven Cardiff of the Museum of Natural History at Louisiana State University, Baton Rouge; and Robert Zinc and John Klicka of the Bell Museum of Natural History in Minnesota.

Several additional people helped in numerous ways. Dudley and Nancy Edmondson, Dave Gilbertson, and Frank and Kate Nicioletti all encouraged my project and helped cut my travel costs by providing lodging on several of my photographic ventures to Duluth, Minnesota. Ned and Linda Harris were supportive throughout; Ned also joined me on several photographic ventures. Joe and Elaine Harrison encouraged all my efforts; Joe also accompanied and helped me on many trips. Jim Zipp accompanied me on photographic trips, and the use of his super-telephoto lens enhanced my photography on those occasions.

I thank the photographers who provided images that were either better than mine or were images I did not have. Their contributions most certainly enhance the books.

Many people provided data for the various biology headings in the species accounts, especially Status and Distribution. My goal was to "go to the source" to obtain firsthand information. As the books progressed, I could see each one becoming a "book of friends" because of the immense assistance from these people. Without the incredible amount of help I received, these books would not have been possible. My thanks go to all of the following people who helped so much: Paul Adamus, Rory Aikens, Skip Ambrose, Henry Armknecht, John Backlund, John and Helen Baines, Ursula Banasch, Richard Banks, Greg Beatty, Marc Bechard, Chris Benesh, Julia Bent, David Blankinship, Peter Bloom, Tom Bloyton, Ricardo Padilla Borja, Jeff Bouton, Jack Bowling, Michael Bradbury, Seamus Breslin, John Brooks, Jennifer Brookshier, Tim Brush, Kelly Bryan, Charles Burwick, Bill Busby, Jamie Cameron, R. Wayne Campbell, Steve Cardiff, Oscar Carmona, William S. Clark, Keith Corliss, Troy Corman, Jennifer Coulson, Gordon Court, Jerry Craig, Bob Davies, Robert Dickerman, John Dinan, James Dinsmore, Bob Dittrick, Tom Dore, Robert and Jane Dorn, John Driscoll, Jon Dunn, Cory Ellingson, Ernesto Enkerlin-Hoeflich, David Fellows, Elmer Finck, Allen Fish, Craig Flatten, Mark Flippo, Ted Floyd, Lynne Fox-Parrish, Brush Freeman, Rick Fridell, Paul and Cecily Fritz, Gail Garber, Dan Gibson, Joe Grzybowksi, Mary Gustafson, Frank Hein, Steve Hein, Bill Heinrich, Ralph Heuer, Mark Hitchcock, Steve Hoffman, Geoff Holroyd, C. Stuart Houston, Lawrence Igle, Marshall Iliff, Jimmy Jackson, Alan Jenkins, Ron Jurek, Kent Justus, Kansas Department of Wildlife and Parks, John Karger, John Keane, Henry Kendall, Sandra Kinsey, Jack Kirkley, John Koloszar, Leon Lalonde, Dave Lambeth, Mike Lanzone, Kit Larson, Greg Lasley, Laird Law, Tony Leukering, Seymour Levy, Jerry Liguori, Janet Linthicum, Bill Lisowsky, Connie Lyons, Bill Manan, Mark Martell, Ron Martin, Terry Maxwell, Terry McEneaney, Diann McRae, Martin Meyers, Steve Millard, Andrew Miller, Trish Miller, Paul Milotis, Pierre Mineau, Montana Natural Heritage Program, Angel Montoya, Frank Nicoletti, Gary Nielson, Bob Oakleaf, Michael O'Brien, Mark O'Donoghue, Christian Oliva, Chad Olson, Brent Ortega, Joel (Jeep) Pagel, Bill Panell, Max Parker, John Parsons, Rob Parsons, Dennis Paulson, Jim Paxton, Tim Reeves, Martin Reid, J. V. Remsen, Virginia Rettig, Mark Robbins, Robert and Susan Robertson, Karen Rowe, Jeff Rupert, Christopher Rustay, Rex Sallabanks, Gon Sanchez, Rafael Sanchez, Judy Scherpelz, Pat Schlarbaum, John Schmitt, Joe Schmutz, Jay H. Schnell, Hart Schwartz, Larry Semo, Kenneth Seyffert, Cliff Shackelford, Peter Sharpe, Mike Shipman, Steve Shively, Gary Shugart, Joel Simon, Chris Sloan, Helen Snyder, Matt Solensky, Kelly Sorenson, Rick Steenberg, Sylvester Sorola, Karen Steenhof, Glen Stewart, Sharon Stiteler, Brian Sullivan, Glenn Swartz, Ted Swem, Max Thompson, Kelley R. Tucker, Dan Varland, William Vermillion, Mike Wallace, Jim Watson, Roland Wauer, Maure Weigel, Western Resources, Clayton White, David Wiedenfeld, Sartor O. Williams III, Sheri Williamson, John Wittle, Kendall Young, and Jim and Carol Zipp.

The following people provided unpublished telemetry data for "Movements": Michael Bradbury (Swainson's Hawk), Geoff Holroyd (Peregrine Falcon), Mark Martell (Osprey), and Jim Watson and the Washington Department of Fish and Wildlife (Ferruginous Hawk).

# ABBREVIATIONS

*The following abbreviations are used in the text.*

| | |
|---|---|
| AMNH | American Museum of Natural History, New York City |
| AOU | American Ornithologists' Union |
| BLM | Bureau of Land Management |
| Co. (Cos.) | County (Counties) |
| Co. Pk. | County Park |
| CWS | Canadian Wildlife Service |
| EPA | Environmental Protection Agency |
| HMANA | Hawk Migration Association of North America |
| LSU | Louisiana State University, Baton Rouge |
| N.F. | National Forest |
| N.P. | National Park |
| NWR | National Wildlife Refuge |
| Pk. | Park |
| Prov. Pk. | Provincial Park |
| RBCM | Royal British Columbia Museum, Vancouver |
| S.P. | State Park |
| USDA | U.S. Department of Agriculture |
| USFWS | U.S. Fish and Wildlife Service |
| WMA | Wildlife Management Area |

# I. INTRODUCTION

## *Identification: Solving the Puzzle*

Identifying raptors is much like assembling a puzzle: you do it piece by piece. Being mainly shades of brown or black, raptors can initially be difficult to identify. If one knows what to look for and where to look, however, it may take only a few pieces of the puzzle to solve an identification. Many species have distinct traits, and it may take only one plumage marking to make an accurate identification. On some, it may take two and sometimes three markings to make a proper identification.

The major factor inhibiting easy, positive identification is that most raptors are wary and do not afford close viewing. Poor lighting, odd angles of view, and missing or broken feathers also contribute to viewer hardships.

With a good, close look, all raptors can be identified, aged, and sometimes (accipiters and falcons) sexed. Even with a close look, however, not all subspecies can be identified to race. With a distant or poor look, some raptors go unidentified because key features cannot be seen. Skilled observers can identify many distant raptors by silhouette.

**PERCHING:** There are three anatomical areas to consider on perching raptors: the head, wings, and tail. The body is the least important because many species share similar colors or markings. Plumage details are often readily visible because high-powered optics can provide a magnified view.

**Head:** The features of the head can be valuable when determining an identification. Check iris color, lore and forehead color, and bill and cere color. It is also important to know if a pale supercilium and dark eye line or malar mark are present. Most raptors have a pale region on the nape, sometimes with a dark spot inside the white region.

**Wings:** The markings on the greater coverts and secondaries can be important, especially on buteos. The distance from the tips of the primaries to the tail tip (the wingtip-to-tail-tip ratio) is very important on large raptors.

**Tail:** This is the most important area to study on a perched raptor. Study width and amount of dark barring on the dorsal surface. The pattern on the dorsal surface will often clinch an identification. The only drawback is that the dorsal surface of the tail is often partially covered by the folded wings. The ventral surface generally offers little in the way of definitive markings on a perched bird. Tail shape, however, is visible only on the ventral surface and can be of use to clinch identification. In fresh plumage, typically in the fall, many raptors have distinct white terminal bands. By mid-winter and especially spring, the white band wears off (white is a weak feather structure and wears quickly).

**FLYING:** There are two anatomical areas to consider on flying birds: wings and tail. Head markings are often obscured because viewing distances are typically greater than with perching birds and less powerful optics must be used on a moving subject. As on perched raptors, the body markings offer minimal help because of shared markings between species.

**Wings:** Every raptor species has a distinct wing shape. On some raptors, wing shape varies within a species. Juvenile buteos have narrower wings than adults, and juvenile accipiters, eagles, and falcons have wider wings than older ages. However, they all have species-distinct shapes. The only glitch is that wing shape varies with the different flight modes. Learn the wing shape of a common, easy-to-see raptor such as a Red-tailed Hawk and compare other raptors to it to visualize differences.

Wing markings are also important. Even such similar-looking species as Cooper's and Sharp-shinned hawks have subtle but visible differences on the underwing markings (and shape), particularly on juveniles. Compare the extent of barring and other markings on the dorsal and ventral surfaces of the wing, including the secondaries. The dorsal wing surface may also have such vital markings as the small crescent-shaped panel of a Red-shouldered Hawk or the large, pale, rectangular panel of a juvenile Red-tailed Hawk.

The wing attitude—how the wings are held during various flight modes—may assist identification (it is not as reliable as markings, however, because it can change drastically). Look for dihedral wing position and behavioral traits such as hovering or kiting. It is important to watch a raptor as long as possible to get a sense of its attitude and behavior. See *Hawks in Flight* (Dunne et al. 1988) for more discussions on flight attitudes.

**Tail:** This is the second important part of identifying flying raptors. The shape varies considerably according to flight mode and may cause confusion (*see* chapter 6). For instance, the tail tip of a Sharp-shinned Hawk can be quite square when closed while gliding but becomes quite rounded when fanned while soaring. Likewise, the short outer rectrices of a Cooper's Hawk, which greatly help in many identification situations, are often not visible on distant flying birds.

As described above in Perching, use the pattern of dark barring. Most species have distinct patterns of dark markings. In translucent lighting conditions, the dorsal pattern and color, which are the best defined, vividly show on the underside, a factor rarely seen on perching raptors. One thing to keep in mind is that on many species that have dark bars, the ventral surface of the outer one or two rectrix sets often has a different pattern than the other four or five rectrix sets. In strong light, however, the dorsal pattern often overrules the often less distinct ventral pattern.

Enjoy!

## *Book Concept*

The species accounts present detailed information that falls into two broad categories: plumage and biology. The various headings occur in a consistent manner throughout the book.

Information on plumage is written in a formal scientific style and is organized under four main anatomical headings: Head, Body, Wings, and Tail. Much of the information is based on more than 20 years of studies in the field and museum. Information is also based, in part, on my previously published collaborations with William S. Clark. Information procured from others is cited in the text as published data (author's name and date of publication), unpublished data (author's name and "unpubl. data"), or personal communication (author's name and "pers. comm.").

Information on biology is presented in a less formal style, and in-text citations are not used. This style was adopted to ensure an easy reading format. Extensive information is organized under the headings Habitat, Habits (behavior), Feeding, Flight, Voice, Status and Distribution, Nesting, Conservation, and Similar Species.

## *Format for Species Accounts*

**AGES:** Each age is described in detail under its respective plumage heading. I have followed the terminology of Humphrey and Parkes (1959), but instead of attaching the word "basic" to all non-juvenile ages as they did, I often use a combination of "basic" and the more recognizable "subadult"—thus, "basic (subadult)"—for all interim ages between juvenile and adult. I also use "adult" rather than "definitive basic" as preferred by Humphrey and Parkes *See* chapter 4.

Ages are arranged from oldest to youngest. This is because most people recognize the more distinct adult plumages before they do the mundane, often "brown" plumages of younger ages. Major plumage variations that involve two or more anatomical regions are labeled with definitive "type" names (e.g., lightly marked type, moderately marked type, streaked type, etc.). Variations of secondary features on one part of the plumage, such as a distinct tail pattern, may also be addressed as a type.

**MOLT:** The molt sequences on the wings and tail are described in detail for each raptor family in chapter 4, and the species accounts refer readers to this chapter. On many species, however, some information is reiterated in the species account in order to state a particular point. Interesting molt facts of an individual species are also described in this section.

**SUBSPECIES:** Polytypic variation has always interested me because subspecies represent unique adaptations to regional environments. Twenty-three of the 34 North American breeding species of raptors are polytypic. The consensus of the public and the scientific community in the last few decades, however, is that polytypic characters are not important issues. The American Ornithologists' Union (AOU) has not included subspecies in its *Check-list of North American Birds* since 1957 (AOU 1957).

Although I have attached colloquial names to all polytypic variations in the main plumage headings, the trinomial scientific name is adjacent to the colloquial name. On polytypic species, I use only scientific names in the body of the text.

Polytypic species that breed in North America are given as much attention in plumage and biological matters regarding their subspecies in this book as are monotypic species. Subspecies that do not breed in North America are presented in a more basic format, simply because little has changed with the stance of the AOU and because many raptor books have previously published much of this information. When dealing with the magnitude of information incorporated in this book, however, it seemed appropriate to include these references.

I present some new information available in scientific journals but otherwise not widely available. My opinion regarding subspecies plumage and status is also expressed for a few species.

Subspecies are arranged from most common to least common, or adjacent races are arranged next to each other.

**COLOR MORPHS:** Eight North American species, all of which breed in the West and five of which breed in the East, are polymorphic. One western species exhibits polymorphism in non-native subspecies. The East has two additional polymorphic species that are seasonal visitors from the West. Plumages are arranged from lightest to darkest and are described in detail.

**SIZE:** Most data on length (distance from tip of bill to tip of tail) and wingspan were obtained from published sources. Measurements on some subspecies are from both published and unpublished sources, which are cited in the text. My opinion on matters such as separating accipiter sexes by size in the field is also included.

**SPECIES TRAITS:** Anatomical features and markings that are shared by all ages and sexes of a species.

**ADULT TRAITS:** Anatomical features and markings that are shared by both adult sexes of a species.

**BASIC (SUBADULT) TRAITS:** In species with interim ages between juvenile and adult, anatomical features and markings that are shared by both sexes of the respective age category.

**JUVENILE TRAITS:** Anatomical features and markings that are shared by both juvenile sexes of a species. On some kites, this stage may only last a few months; on most raptors, it lasts about a year.

**ABNORMAL PLUMAGES:** Generally describes various forms of albinism but may also include the opposite aberration, melanism.

**HABITAT:** Often variable according to season. For migrant species this section is divided into Summer, Winter, and Migration subheadings. Even species that are resident in a particular area, however, may use different habitat after the nesting season; if so, this section is subdivided accordingly.

**HABITS:** Behavioral traits of a species. A species' general nature is listed first. This varies from being *tame* to *very wary*. *Tame* does not mean "tame as a household pet"; it means that a walking, unconcealed human can approach within 100 feet without a raptor exhibiting alarm. *Fairly tame* means a human cannot approach on foot but if concealed in a vehicle can approach within 100 feet without the raptor exhibiting alarm. *Moderately tame* means a raptor will become alarmed and probably fly if a vehicle approaches within 100 feet. *Wary* means a raptor will fly when approached by a vehicle that is several hundred feet away. *Very wary* means the raptor will fly when a vehicle is sighted.

Other interesting behavioral facets are also covered.

**FEEDING:** The first sentence in all accounts indicates if the species hunts from a perch, in flight, or by both methods. This section also describes what the species preys on and where and how it captures and eats prey.

**FLIGHT:** This section describes the individual variations for the three main flight modes—powered flight (flapping), gliding, and soaring—and the three secondary modes—diving, hovering, and kiting. It also elaborates on migration flight.

**VOICE:** Each species has distinct vocalizations. Not all species regularly vocalize, but if they do, their distinct call notes can assist identification. Nearly all species occasionally vocalize year-round, particularly when agitated. Some species, such as Red-shouldered and Red-tailed hawks, are highly vocal year-round. Most species become fairly vocal during the breeding season, when courting, or when trespassers enter their territory. Even during migration and winter, typically silent raptors such as Sharp-shinned and Cooper's hawks regularly vocalize when agitated and exhibit their distinct, separable call notes. The only species that is notoriously silent is the Golden Eagle.

Interpreting sounds phonetically is difficult, but methods such as sonograms are even more difficult to interpret. All vocalizations are shown in italics.

**STATUS AND DISTRIBUTION:** Below are the status designations used in this book. The numbers that accompany each designation are somewhat arbitrary because they are based on rough estimates. They reflect the most current and accurate data available, however. Some data are from published sources, and some are based on recent information from major hawkwatches (e.g., fall hawkwatch counts in Veracruz, Mexico). For polymorphic species, and for polytypic species whose subspecies are identifiable by sight and/or range, status designations are given for the species as a whole and, in a subheading, for each color morph or subspecies. Many of the status estimates for color morphs are based on my observations of the last 16 years and those of my colleagues.

*Endangered.* Highly threatened and extremely difficult to find; fewer than 100 individuals.

*Very rare.* Threatened and very difficult to find; 100–250 individuals.

*Rare.* Somewhat threatened and quite difficult to find; fewer than 500 individuals.

*Very uncommon.* Fairly low numbers and difficult to find; estimated population ranges from 500 to 10,000 individuals.

*Uncommon.* Moderate numbers and moderately easy to find; estimated population ranges from 10,000 to 100,000 individuals.

*Common.* Large numbers and easy to find. Estimated population ranges from 100,000 to 2 million individuals.

*Very common.* Very large numbers and easy to find. This designation is used only for Turkey Vulture, which has an estimated population of over 3 million individuals.

*Local.* A raptor that can have any of the above designations but is found only in a restricted location and habitat.

*Accidental.* Occurring far out of the normal range and not likely to be seen. Also called *vagrant.*

*Casual.* Occurring out of typical range but infrequently seen.

This section is usually divided into four main subheadings—Summer, Winter, Movements, and Extralimital movements—with other headings added when appropriate.

Summer and winter data correlate with the detailed range maps; if there is a plot on a map, it specifies an exact location. Summer data are mainly from published and unpublished breeding bird atlases. A great deal of information was also gathered on the Internet, by tapping local, state, and regional experts throughout the U.S., Canada, and Mexico. Winter data were primarily taken from many years of Christmas Bird Counts, previously published by the National Audubon Society and now published by the American Birding Association. Data were also obtained from several years of *North American Birds* (previously called *American Birds* and *Field Notes*). Some personal data were also used, based on many years of travel across North America.

"Movements" is further divided into Dispersal (summer or other seasons), Fall migration, and Spring migration. Migration data are based heavily on information in *Hawk Migration*

*Studies*, published by the Hawk Migration Association of North America (HMANA); published banding data; published and unpublished telemetry data; and personal knowledge from more than 20 years of field work. For simplicity and consistency, dates for movements are based on splitting a month into three parts: 1st–10th is "early month," 11th–20th is "mid-month," and 21st–31st is "late month."

The information in "Extralimital movements," which typically lists locations and dates, is based mainly on published data from local, state and provincial, and regional experts. However, a considerable amount of information is based on recent unpublished reports from authoritative persons. A similar method of a tripart month is used here.

**NESTING:** The first segment gives the start and end of the nesting season. This typically spans the period when courtship behavior begins and fledglings become independent. The start of the nesting season is more difficult to determine with raptors that are resident or remain paired year-round. This section also explains when pair formation occurs and the age of first breeding.

Courtship is consistently described, under the subheading Courtship (flight) for nearly all species and Courtship (perched) for the few species that engage in this type of behavior. Rather than describe displays time and again, courting behaviors are described in detail in chapter 5, and this chapter is referenced in the species accounts. Courting behaviors in the species accounts are in italics for easy recognition.

The remainder of this section describes nest sites, nest size, which sex builds the nest and incubates the eggs, incubation period, and fledging time.

**CONSERVATION:** This section elaborates on any measures that have been taken to protect a species. All raptors are covered under the Migratory Bird Treaty Act of 1918. For many years this law was rarely enforced for raptors, however, since virtually all species were considered a nuisance. Endangered and threatened species finally received additional protection under the Endangered Species Protection Act of 1970 and, since 1973, under the Endangered Species Act. Bald Eagles were protected under the Bald Eagle Act of 1940, which was amended in 1962 to include Golden Eagles. Individual states and provinces often have their own protective laws.

The Mortality subheading in this section describes survival problems that raptors encounter. Natural and human-caused mortality issues are discussed.

**SIMILAR SPECIES:** This section gives detailed comparisons of similar-looking raptors. Polytypic and polymorphic variations that cause confusion are also detailed. Comparisons are organized under subheadings of the various ages and/or subspecies and color morphs.

**OTHER NAMES:** North American colloquial names are listed. Since the ranges of some species include Mexico and Canada, Spanish and French names are also given.

**REFERENCES:** This is an abbreviated bibliography: full citations occur in the bibliography at the end of the book.

## *Plates and Captions*

**PLATES:** The photographs depict an extensive array of plumages for all species, ages, sexes, and color morphs of North American breeding raptors. Most subspecies and plumage types are also depicted. However, it is nearly impossible to show all individual variations and types of plumages with photographs (or with illustrations).

Although I considered the aesthetic value of a photograph when making the final selection, the ultimate criterion for every photo was whether it adequately showed particular field markings. More than one photograph is often used for the same type of bird in order to suitably illustrate a field mark or plumage variation. On some, I was fortunate to have perching and flying images of the same individual, which helps in conveying total plumage features.

Individuals of obvious age classes are grouped together. Following the text layout, older ages precede younger ages. This is because adults are generally easier to identify than the often "brown-colored" younger birds. Polymorphic and polytypic species are also grouped according to age class. Polymorphic species are arranged in a light-to-dark format. Perching figures are presented before flying figures for each age class grouping.

**CAPTIONS:** The species name is followed by the bird's age, sex (if obvious), and color morph (where applicable). For polytypic species, all photos are of the "typical," or most widespread, subspecies unless otherwise indicated. Where a subspecies is labeled, both the colloquial and scientific name are given. The month the photograph was taken is listed last in brackets.

Information in the captions is arranged in a "bullet" format for easy, consistent reading. The bullets use the same terminology as the text and are arranged in the same anatomical order: head, body, wings, and tail. *Head* describes the bill, eyes, and important feather markings on the head. It may also include the front and sides of the neck. *Body* includes dorsal and ventral surfaces, leg feathers, tarsi, and undertail coverts. It may also include the front and sides of the neck. *Wings* includes dorsal and ventral surfaces of the remiges and coverts. *Tail* describes the rectrices and uppertail coverts. Only one or two bullets are used if a particular image does not require all four anatomical areas for identification, and to save space, body and wings may be under the same bullet, especially if there are major similarities in color or markings. The *Note* at the end of many captions presents additional information. It also gives credit on images taken by other photographers.

The location where a photograph was taken is typically not listed because it is virtually impossible to take many or most photographs for a regional book within that particular region. I thought it would appear strange to list a large number of photographs that were taken in one region and used them in the other. I photograph where the raptors are easiest to get: accipiters and falcons mainly at hawkwatches; buteos and eagles in the West, where they are much tamer than in the East. This affected the eastern guide more than the western. Case in point: the exceedingly large number of Red-tailed hawk and Swainson's hawk photographs were nearly all taken in the West, where these two species are most common and generally reasonably tame. Under optimal conditions, it would take decades to get such photographs in the East. Photographic locations may be listed if the image was taken outside of the contiguous U.S. or if it was of special interest.

## *Range Maps*

The range maps are meant to show the most exact plotting of distribution possible. Using knowledge of a species' habitat, one can use the maps to more readily locate a particular raptor.

I was most impressed with the range maps in *A Field Guide to Warblers of North America* (Dunn and Garrett 1997). This was the first North American field guide that had thoroughly researched maps that, as much as is possible, abided by geographic regions within a species' range. Inspired by this great book, I decided something of a similar caliber needed to be done for North American raptors.

The distribution maps in this book are the result of a joint effort by my status and distribution coresearcher, John M. Economidy. John was also impressed with Dunn and Garrett's expertise, and also felt that accurate maps were much needed for raptors. We spent nearly five years of daily research tapping experts from virtually every state and province in the U.S., Canada, and Mexico for the latest in known distribution and status of many species. This was done primarily by E-mail. We also used published breeding and wintering data that dated back several years (*see* Status and Distribution, above, in Format of Species Accounts) and every state and provincial breeding bird atlas—including eight unpublished atlases.

Major obstacles that any author faces are the ever-changing habitats and ranges of birds. What might be accurate one week is not accurate the next week! Also counter in the fact that birds have wings—and thus may stray to faraway locations.

The plottings on the maps abide by geographical and topographical areas. If there is a specific location marked on the map, it *is* for a specific location. A unique quality of these maps is the city locations. Plotting cities not only gives more credence to accuracy but also helps plot a species' range. Super-accurate plotting is illustrated on maps that show county delineations. These finely detailed maps truly follow actual habitat zones required for a particular species. Habitat alteration, either by humans or natural causes, may alter the fine detailing of these maps over time, but such detailed plotting gives a substantiated accuracy for this particular period.

For polytypic species whose subspecies are identifiable by sight and/or range, a separate range map was created for each subspecies. These maps show, as do those for monotypic species, detailed, known, typical permanent, summer, and winter ranges. The data are based on published (often state ornithological journals) and unpublished (regional experts tapped by Internet) sources and personal data from hundreds of thousands of miles of travel across North America.

Only on "Krider's" form of Red-tailed Hawk does a color morph have its own map.

Several recent field guides have shown migration routes as colored zones on maps. I chose not to do this. It is fairly easy to see that migrants go from the breeding grounds (point A) to wintering grounds (point B). Also, most raptors migrate in a broad-fronted manner. Thus, all regions between point A and point B would be colored. It would not make any sense to color this region when one can logically figure it out. The captions on the maps may elaborate on heavily used routes and on where a species winters if these are not shown on the map.

Population densities are not shown. Each species' status is labeled on its map. Most raptors are either widespread and reasonably common throughout their typical range or they are local and uncommon to rare in their restricted range.

# KEY TO MAPS

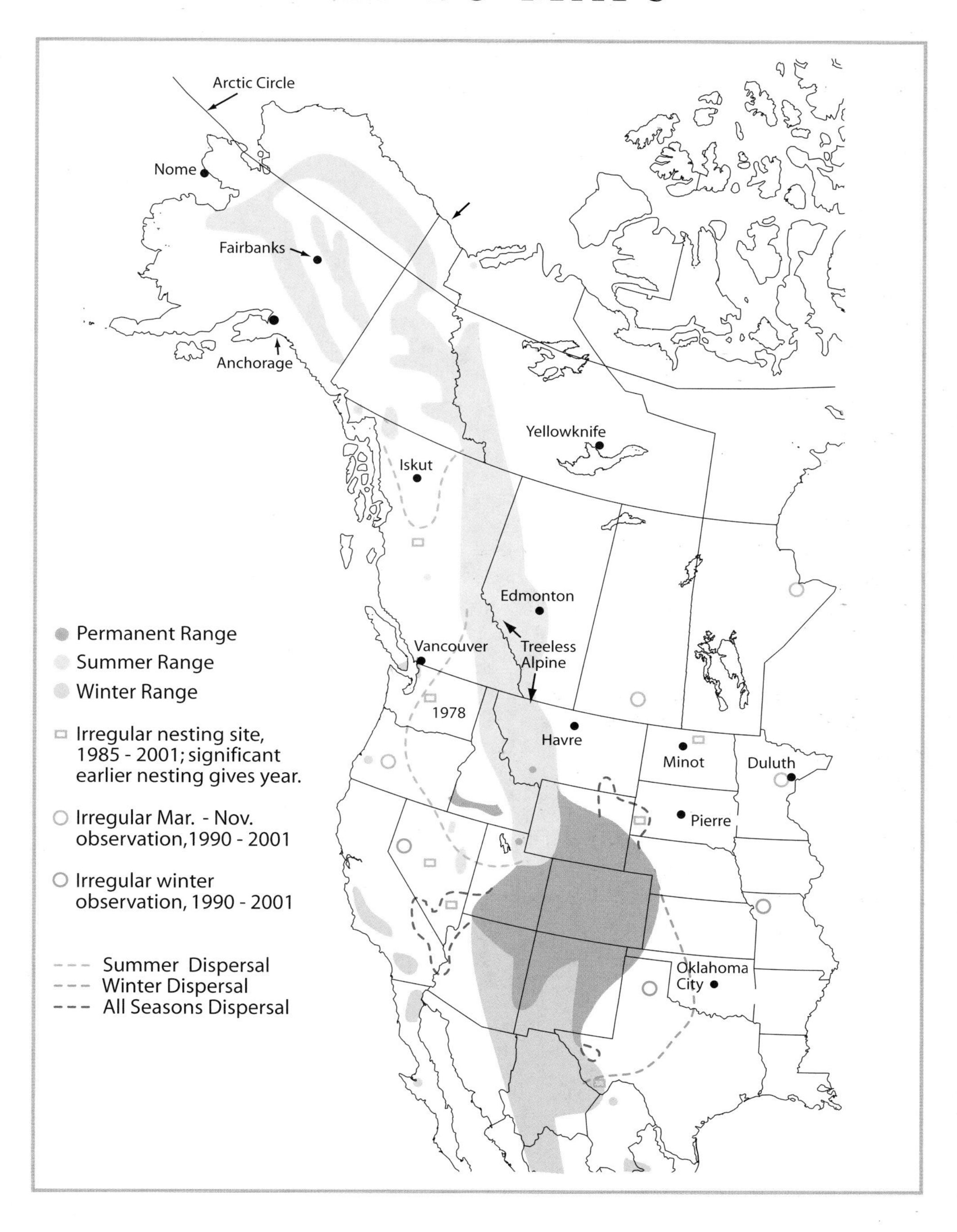

Arctic Circle
Nome
Fairbanks
Anchorage
Yellowknife
Iskut
Edmonton
Vancouver
Treeless Alpine
1978
Havre
Minot
Duluth
Pierre
Oklahoma City
Permanent Range
Summer Range
Winter Range
Irregular nesting site, 1985 - 2001; significant earlier nesting gives year.
Irregular Mar. - Nov. observation,1990 - 2001
Irregular winter observation, 1990 - 2001
Summer Dispersal
Winter Dispersal
All Seasons Dispersal

# II. GENERAL GLOSSARY

**Accipiter.** Name given to the round-winged and long-tailed woodland-dwelling "true" raptors in the genus *Accipiter*. North America has three species in this genus.

**Aerial hunting.** Hunting method in which the raptor is already airborne when it begins pursuit of its intended prey.

**Albinism.** Reduction or absence of pigment in the eyes, skin, or feathers. There are four varieties.

> **Total Albinism.** Complete absence of normal pigmentation in the eyes and skin, with both areas being pink. Feathers are pure white. This is the rarest form and rarely seen in raptors.
>
> **Incomplete Albinism.** Normal pigmentation is completely absent in eyes, skin, or feathers but not in all three areas. Plumage is often pure white or nearly so, but in most cases eyes and skin are normal color. However, talons are often pink colored.
>
> **Imperfect Albinism.** Normal pigmentation is only partially reduced in eyes, skin, or feathers but not totally in any of the three areas. Also known as "dilute plumage." Eye pigmentation is either normal or bluish. Skin areas are normal or slightly paler than normal in color. Most or all of the plumage is a tan color; any typical markings that have a rufous or tawny color are altered to a pale rusty color in this plumage.
>
> **Partial Albinism.** Normal pigmentation is completely lacking on portions of the body. Eye and skin areas are usually of normal coloration. Plumage is often a patchwork of normal and white feathers, or parts of several feathers may be white and other parts normal. Most common type of albinism.

**Allopreening.** Mutual preening between mates or between adult and sibling, but rarely between unrelated individuals.

**AOU.** Abbreviation for American Ornithologists' Union. Governing board of scientists that makes decisions on taxonomy and nomenclature of avian species in North America.

**BLM.** Abbreviation for the Bureau of Land Management. Federal agency within the Department of the Interior that oversees land use on federally owned land. Mainly associated with lands in western U.S.

**Brancher.** *Fledgling*-aged raptor that ventures from the confines of the nest onto nearby branches or to the ground.

**Buteo.** Species of raptor in the genus *Buteo*, which comprises the broad-winged, soaring hawks. There are 12 species in this genus in North America. Buteo species in the Old World are called *buzzards*.

**Butte.** Flat-topped, cone-shaped earth formation that is the rock skeletal remains of a hill.

**Buzzard.** Old World term for *buteo* species of raptors that has become a slang word for vultures in North America. As a reference to New World buteos, the term has never caught on in North America and is rarely used.

**Cline.** Range of variation that is gradual and continuous in scope. Used to define plumages of some species of raptors with *color morphs*, where various individuals show a continuous gradation in plumage characters between two color types.

**Colloquial name.** Regional name given to a *species*, *subspecies*, or *color morph*. It does not have scientific application but makes it easier to express the name in "everyday" language.

**Conifer.** Tree species with needlelike leaves that are retained all year.

**Color morph.** Plumage variation within some raptor species (six buteos and one falcon) that is sometimes geographically designated and occasionally is a climatic adaptation. Darker coloration of these species is the result of increased amounts of *phaleomelanin* and *melanin* (darker feather pigmentation) which results in *erythristic* (rufous morph) and *melanistic* (dark morph) birds, respectively. Many of these species also have an intermediate division between each major color morph. Also called "phase," though "morph" is the more accepted scientific term.

**Continuum.** See *cline.*

**Contour feathers.** General body plumage feathers that cover the downy, next-to-body insulating feathers.

**Crepuscular.** Active during twilight in the early morning and early evening.

**Deciduous.** Tree species that lose their leaves in autumn. Often called "hardwoods."

**Dihedral.** Wing attitude of a flying bird that has wings held in various V angles above the horizontal plane.

**Dihedral, modified.** Wing attitude in which the shoulder-to-wrist angle is held above the horizontal plane and the wrist to primary tips are on a horizontal plane.

**Dihedral, low.** Wing attitude held in a shallow V above the horizontal plane. Angle is less than 30 degrees above the horizontal.

**Dihedral, high.** Wing attitude held in a steep V above the horizontal plane. Angle is 30–60 degrees above the horizontal plane.

**Dihedral, very high.** Wing attitude held in a very steep V above the horizontal plane. Angle is at least 60 degrees above the horizontal plane.

**Dimorphic.** Showing a distinct difference in color or size. Often relates to sexual differences in raptors.

**Diurnal.** Active during daylight.

**Double clutch.** Two egg sets per year.

**Erythrism.** High amount of rufous (*phaeomelanin*) pigmentation in a feather structure. Produces rufous *color morph* in some buteo.

**Escarpment.** Hill formation in which a portion of the hill is eroded away and exposes an inner rock surface that is vertical and clifflike.

**Estivation.** Summer hibernation, used by many western ground squirrel species. When their nutritious food supply disappears in the hot and dry summer, they enter into a "summer sleep" which then usually continues into regular winter *hibernation.* The reproductive cycle of some raptor species is timed to the above-ground season of some ground squirrel species, which are major prey items.

**Eyass.** *Nestling* or *fledgling* falcon.

**Eyrie.** Raptor nest site; particularly used to note the nest area of a falcon. Also spelled *aerie.*

**Falcon.** Species of raptor in the genus *Falco;* there are seven species of falcons in North America. Also, name given to the female of a large falcon species.

**Falconry.** Sport hunting of game birds and animals with a captive raptor (captive reared or taken from the wild). All hunting is done within the prescribed season for the specific game species. A falconer is licensed under state and federal regulations.

**Fall line.** Geographic delineation in the southeastern U.S. in which there is an abrupt difference between upland and lowland elevation. Generally an arch-shaped region from Durham, N.C., to Columbia, S.C., to Columbus, Ga.

**Fledgling.** Young bird that has attained the power of flight but is still under care of parents.

**Form.** Term that defines variations within a species that has variable plumage characters.

**Fratricide.** Older or stronger nestling and/or fledgling that kills (and may eat) a weaker sibling; due primarily to food competition. Also known as "siblicide."

**Genus.** Taxonomic grouping that represents one or more closely related species. The genus name is the first part of a binomial or trinomial designation of a species' or subspecies' scientific name and is always capitalized.

**Gloger's Rule.** Ecological "rule" coined in C. W. L. Gloger's 1833 book *The Variation of Birds under the Influence of Climate.* The rule states that species inhabiting humid climate regions have darker pigmentation (increased amount of *melanin*) than those inhabiting drier regions. In raptors this pertains primarily to a climatic adaptation that often results in a darker colored *subspecies* and occasionally produces a higher percentage of darker *color morphs* in a more humid region.

**Greenery.** Green leaves or twigs with green leaves on them that are used by some nesting raptors to line or decorate the top of the nest.

**Hacking.** Human-assisted rearing of young raptors, using captive fledglings or relocating fledglings from areas that have a stable population to areas where the natural population has been diminished or eradicated by habitat alteration or pesticides. Hacked fledglings do not have contact with their human helpers so are not imprinted and thus retain a natural fear of humans.

**Hibernation.** "Winter sleep" by many rodents and other animals.

**Intergrade (plumage).** Occurs when an individual of one *subspecies* has plumage traits of an adjacent-breeding-area subspecies.

**Intermediate (plumage).** Occurs in *polymorphic* species when an individual has plumage traits intermediate in color between the two *color morphs.*

**Jack.** Name given to a male Merlin.

**Jerkin.** Name given to a male Gyrfalcon.

**Kettle.** Term used to describe a "flock" of raptors, usually during migration.

**Melanin.** Dark brown or black pigmentation in the structure of a feather.

**Melanistic.** Showing a high degree of *melanin* in feather structure. The added melanin produces a dark *color morph* in six buteo and one falcon species. Increased melanin also creates a darker colored subspecies in three species of raptors. Rarely, an abnormal amount of melanin produces an abnormally dark-colored individual in a species that does not typically have color morphs or subspecies.

**Mixed woodland/forest.** Wooded area that has a combination of *coniferous* and *deciduous* trees.

**Monotypic.** Species that does not have *subspecies.*

**Neotropical.** Referring to the region that extends from south of the central Mexican Highlands through Central America, South America, and the West Indies.

**Nominate.** In species with *subspecies,* the race selected as the most "typical." In the trinomial scientific name, the subspecific designation is the same as the specific designation.

**Olfactory.** Pertaining to the sense of smell.

**Perch hunting.** Detecting prey from a perch and then becoming airborne to catch it.

**Pesticides.** Organochlorine pesticides: Deadly and long-lived chemicals that persist for decades in the environment. DDT (dichlorodiphenyltrichloroethane) was the famous pesticide that caused the population decline of many North American raptors, most notably Osprey, Bald Eagle, and Peregrine Falcon. When metabolized, DDT becomes DDE (dichlorodiphenylethylene) and accumulates in the fat deposits of mammals and birds. In birds, DDE most commonly reduces eggshell thickness, causing eggs to break under an incubating bird. Rarely it causes direct mortality by ingestion of contaminated prey.

DDT was first used in 1946 to control insect infestations on crops and forests in Canada and the U.S. It was banned for most uses in Canada and the U.S. in 1972 and 1973, respectively. Canada and the U.S. had limited public health use of DDT, however, until 1985 and 1989, respectively. Mexico ceased its limited public health use of DDT for malaria control in 2002 and plans a total ban in 2006. Although 122 countries have banned DDT, some countries in Central and South America still use it. Other deadly organochlorine pesticides that were used in the 1950s and 1960s included dieldrin, heptachlor epoxide, and polychlorinated biphenyls (PCBs).

**Organophosphate pesticides.** Deadly chemicals but not as long-lived in the environment as organochlorine pesticides. Even when properly used, such chemicals are known to kill animal life. Improper and illegal use causes even greater mortality to predators and scavengers. Some organophosphate pesticides have been known to kill large numbers of Bald Eagles, Golden Eagles, Red-tailed Hawks, and other birds of prey in several cattle-ranching and dairy-farming states. Some types of organophosphate pesticides used in eastern Canada killed songbirds as well as friendly insects.

**Carbamate pesticides.** Less harmful to the environment than organophosphate chemicals.

**Pyrethroid pesticides.** Newly developed pesticides that are harmless to birds and mammals but are harmful to aquatic life.

**Phaeomelanin.** Rufous pigmentation in the structure of a feather. A high degree of it produces an *erythristic*, or rufous *color morph* in some species.

**Phase.** Alternate name for *morph.*

**Plucking post.** Favorite elevated perch used by a raptor to pluck feathers or other inedible portions off prey before eating or delivering it to the nest to feed a mate or youngsters.

**Polygamy.** Having more than one mate. There are two varieties: polygyny is one male with multiple females; polyandry is one female with multiple males.

**Polymorphic.** Referring to species that have *color morphs.*

**Polytypic.** Referring to species that have *subspecies.*

**Preening.** Grooming action of a bird in which individual feathers are drawn through the bill to restore their neatness and waterproofness. The bill is rubbed across the oil (uropygial) gland, which is located on the top basal region of the tail and supplies waterproofing oil for the feathers.

**Race.** Alternate name for *subspecies.*

**Riparian.** Near water.

**Rousing.** Cleansing action of a bird in which all *contour feathers* are elevated, wings and tail are held loosely, and the bird vigorously shakes itself. It is done either while perched or flying.

**Scientific name.** Two or (if referring to a subspecies) three-part name accepted worldwide that identifies each organism. The first word in the name denotes the *genus* level, the second the *species* level, and the third the *subspecies* level (if the species has one or more geographic variations). Scientific names are italicized.

**Shelterbelt.** Trees planted in linear formation to serve as windbreaks; planted throughout much of the West to reduce soil erosion. Shelterbelts are popular roosting and nesting areas for many raptors, especially in regions naturally void of the safe-haven elevation of trees.

**Species.** Population of similar individuals that are reproductively isolated.

**Specimen (museum).** Bird that is collected (shot or found dead) and preserved in a scientific collection. It is preserved by being "stuffed" and laid on its back in an airtight drawer in a large case. Each individual has a data tag attached to its feet that signifies the species, age, sex, collection locale, and any other pertinent details. Museum specimens are valuable tools for scientific studies. Properly cared for, they may last for centuries.

**Status.** Indication of how abundant a species is, how easily it is found, and how at risk it is. Estimated sizes of a species' population are given in numerical figures or in general terms. Several terms are used to define status.

**Subspecies.** Distinct variation within a breeding population of a *species* that is partially or totally isolated by a geographic barrier. Seventy-five percent of the population share similar plumage and/or size traits. Subspecies are adaptations to regional environments. Plumage adaptations are due mostly to climate but may also be evolutionary adaptations to selective feeding, hunting, and migratory factors. *Intergrades* occur on the periphery ranges of adjacent subspecies where individuals of the two subspecies interbreed.

**Superspecies.** Two or more *species* that are reproductively isolated but are otherwise very similar and probably came from similar ancestry.

**Thornbush.** Low to moderately high tree covered with thorns. In dense growths, areas of thornbush are virtually impenetrable.

**Translucent.** Allowing the passage of diffused light through a surface. Used here to describe wing and tail feathers of flying raptors when viewed from underneath with sunlight shining through. The underparts are in shadow, and diffused sunlight is transmitted through the paler areas of these feathers.

**Type.** Term that describes a plumage variation, often within a *species* that has *color morphs* or *subspecies.*

**Urohydrosis.** Cooling mechanism by which vultures (and storks) regulate body temperature by excreting onto their legs and feet. Normally pinkish and/or reddish colored legs turn chalky white.

**Vermiculation.** Irregular, narrow zigzag formation of markings on feathers.

**Wash.** In arid regions, an area between hills that during periods of heavy rain absorbs moisture and allows for more substantial vegetation growth, especially of trees. Washes are dry during non-rainy periods. Also called "dry wash."

**Whitewash.** White-colored excrement from raptors. Often seen as white stains below a nest or favorite perch. Especially visible on cliff-area nest locales used by falcons.

**Wing chord.** Distance on a folded wing from the wrist to the tip of the longest primary; it is the most accurate measurement of a bird's size.

# III. ANATOMY AND FEATHER GLOSSARY

This glossary is divided into four sections: Head, Body, Wings, and Tail. This is the same order in which plumage and caption data are arranged in the species accounts.

## *Head*

**Auriculars.** Short, stiff feathers that cover the ear hole. This area merges with the *cheek* to the front and is often referred to as the auricular/cheek area.

**Bill.** Sheath of hardened tissue over two skeletal bone sections, the upper and lower *mandibles*. The upper mandible of all raptors has a distinct, long hook.

**Bill notch.** On falcons, a jagged notch near the tip of the upper *mandible*, used for severing the vertebrae of prey.

**Cere.** Fleshy region at the base of the upper *mandible* of the bill that surrounds the *nostrils*.

**Cheek.** Area directly behind and under the eye. The cheek merges onto the *auriculars*, often without any separation of markings. Often referred as the auricular/cheek area.

**Crown.** Top of the head.

**Ear.** Small hole covered by the *auriculars* (except on vultures, which show the ear hole because of their bare head).

**Eyeline.** Dark linear mark directly behind each eye.

**Facial disk.** *Auricular/cheek* area that appears owl-like with distinct semi-circle formation to accentuate sound reception.

**Forehead.** Front, top area of head, directly behind the *cere*.

**Gape.** Fleshy "lip" surrounding the mouth and connecting the *mandibles*.

**Hackles.** Feathers on the *nape*, on the rear of the head. They are elevated or lowered in response to mood/temperature, being raised when a bird is agitated, alarmed, or cold and lowered most other times.

**Hindneck.** Area below the *nape* and above the *back*.

**Lores.** Area between the *cere* and front of the eyes. Often covered with stiff, hairlike feathers.

**Malar mark.** Dark mark on the lower jaw area. The mark may connect to the bottom of the eye or be isolated below the *gape*.

**Mandibles.** Two sections of bone that are covered with hardened tissue sheaths. The upper mandible is mostly stationary and in raptors has a long hook on the tip; the lower mandible is attached to the hinged lower jaw and is movable.

**Mustache mark.** Same as *malar mark*, but a name given to falcons which have a similar dark mark on the lower jaw area; on many falcons, this mark may extend up under the eye, more so than in most buteos. It lessens glare.

**Nape.** Rear area of the head. These feathers are erectible: See *hackles*.

**Nostril.** Hole on each side of the fleshy *cere* for air intake. Also called "nares" (naris, sing.).

**Nostril baffle.** On some raptors, especially falcons, a postlike structure in the center of each nostril to permit breathing, by slotting air into the nostrils, during high-speed aerial maneuvers. The baffles can be seen at close range.

**Ocelli.** Dark or light spots on the *nape* that appear as "fake" eyes, possibly to deter predators; commonly found on falcons.

**Orbital ring.** Fleshy eyelid surrounding the entire eye on falcons; the region in front of the eyes on the inner *lores* has a broad fleshy area. There is usually a smaller fleshy area behind the eyes.

**Sclera.** Membrane surrounding the eyeball. A small part is visible on the edge of the eye on some birds if it is brightly colored (e.g., red on older California Condors).

**Supercilium.** Pale linear area over the top of the eyes. Also called "superciliary line" or "eyebrow."

**Supraorbital ridge.** Bony projection over the top of the eyes on most raptors that shields and shades the eyes; covered by a thin strip of bare skin. This feature gives raptors their "fierce" look.

**Throat.** Area under the lower *mandible.*

## *Body*

**Back.** Area covered by V-shaped feather tract that merges with the hindneck and is between the two *scapular* tracts. The back tract overlaps inner portion of each scapular tract. Also called the "mantle."

**Belly.** Mid-section of the *underparts;* centered over the central pectoral muscle area. Feathers that cover this region are short to moderate in length.

**Belly band.** Dark band on the *belly* and *flank* region.

**Breast.** Area below the neck and above the belly. Covered by two feather tracts running from the neck onto the *belly* and *flank* tracts.

**Crop.** An elastic "holding pouch" on the *breast* where food first enters the body after swallowing. In raptors that have recently eaten, the crop shows as a distended bulge. Caracaras and vultures show a large exposed fleshy area when the crop is full because the two feather tracts covering the breast separate. Also called the "craw."

**Flags.** Elongated, quite visible feathers that attach to the tibia region on raptors. Often called "thigh feathers."

**Flanks.** Elongated feathers attaching on the sides of the large pectoral muscles. They streamline the sides of the body.

**Heel.** Upper part of the *tarsus.*

**Hindneck.** Area below the *nape* and above the *back.* It is covered by a single feather tract that runs down the rear vertebrae onto the *back.*

**Leg.** General term for *tarsus* and *tibia.*

**Leg feathers.** Long, fluffy feathers that attach on the *tibia.*

**Lower back.** Area below the *back* and above the *rump.* Usually hidden by the *scapulars,* except in flight when this area may be visible.

**Lower belly.** Area covered by elongated feathers that are an extension from the *belly;* they often cover the feet when a bird is perched. There are two feather tracts which are sometimes separated on perched birds to accommodate a raised foot when resting.

**Rump.** Area below the *lower back.* Elongated feathers on the rump cover the *uppertail coverts* and also the oil (uropygial) gland.

**Scapulars.** Two distinct, large feather tracts that cover the inner area of the wings on the dorsal surface. They streamline the junction of wings and body in flight.

**Talon.** Long, sharply curved, needlelike toenails.

**Tarsus.** Visible bare flesh or scaled part of the leg: except on three species, the tarsus is mostly feathered. *Tarsi* is plural.

**Tibia.** Muscular "drumstick" area of the leg, above the *tarsus.*

**Toes.** Four in number. Toe length and exterior construction are adaptations to the type and size of prey a raptor usually feeds on.

**Underparts.** General term that collectively applies to *breast, belly, flanks,* and *undertail coverts.*

**Upperparts.** General term that collectively applies to *back, scapulars, lower back,* and *rump.* Occasionally includes the upperwing.

## *Wings*

**Alula.** Stiff, knife-shaped feather group comprising three or four feathers on top of the wing, near the wrist. Controls airflow over the wing during slow flight; also allows control when wings are folded close to the body during dives.

**Axillaries.** Group of feathers that attach on the underside of the wing, in the "armpit" area. They streamline the junction between the wings and flanks when a bird is in flight.

**Fingers.** Outermost *primaries.* There are usually five to seven, and they are distinctly longer than inner primaries. Visible on many buteos and eagles.

**Flight feathers.** Group of long, stiff feathers comprising the *primaries* and *secondaries.* Collectively known as *remiges.*

**Patagium.** Elastic membrane on the front edge of the inner wing that unites the shoulder, elbow, and wrist areas. Patagial is the adjective.

**Primaries.** Ten stiff feathers that attach on the "hand" area of the outer wing. Part of the *flight feathers/remiges* (along with *secondaries*). They are numbered from inner to outer, with p1 being the innermost and p10 the outermost.

**Primary coverts.** Fairly short, stiff feathers attaching on the dorsal side of the wing at the base of each of the ten *primaries.*

**Primary projection.** The distance from the rear of the *tertials* to the tips of the *primaries* on a perched raptor. In the text, referred to as "short projection" (short distance between tertials and wingtips) and "long projection" (long distance between tertials and wingtips).

**Primary underwing coverts.** Row of short feathers attaching on the basal area of the underside of the *primaries.*

**Remiges.** Name given to the *flight feathers:* the *primaries* and *secondaries.* Remix is singular.

**Secondaries.** Several stiff feathers (the number varies somewhat) that attach on the rear of the forearm area and are part of the *flight feathers/remiges* (along with the *primaries*). They are often numbered for quick reference, especially when describing molt. Outermost is 1 or s1.

**Secondary underwing coverts.** Partially visible row of feathers attaching on the underside of the *secondaries.*

**Tertials.** Innermost three or so feathers of the *secondaries,* attaching on the elbow region. Visible mostly when perched; in flight, they are usually covered by the *scapulars.*

**Trailing edge.** Rear edge of all the *flight feathers* (*remiges*).

**Underwing coverts.** Several rows of feathers, on some species rather indistinct, that attach on the forearm and *patagial* area. Also called "wing lining."

> **Greater underwing coverts.** Attach on rear of forearm and form a visible, well-developed row overlapping the inner *secondaries*.
>
> **Median underwing coverts.** Attach on forearm and are rather long in length. They comprise a major feather row and are visible on most species.
>
> **Lesser underwing coverts.** Attach on *patagium* as an elastic group and for the most part do not have defined rows.
>
> **Upperwing coverts.** Comprise several well-defined, overlapping rows of feathers.
>
> **Greater upperwing coverts.** Distinct row of fairly long feathers that overlap the inner *secondaries*. There is one greater upperwing covert for each secondary.
>
> **Median upperwing coverts.** Fairly distinct row of medium-length feathers attaching above the *greater coverts*. There is one median covert for every greater covert.
>
> **Lesser upperwing coverts.** Small feathers comprising several usually indistinct rows on the *patagium*. First row attaches above the median coverts. Also called "shoulder area," which technically is a misnomer.

**Window.** Pale underside of the *wing panel* through which light shines on pale-colored *primaries*.

**Wing panel.** Pale upperside surface on a portion or most of the *primaries*.

**Wingtip-to-tail-tip ratio.** Distance on a perched raptor from the tips of the *primaries* to the tail tip. This "measurement" is very useful in the field when separating confusing raptor species.

## *Tail*

**Deck set.** Center pair of *rectrices*, which on dorsal surface overlap all others when tail is closed. Only one *rectrix* of the set is totally visible at all times. It is labeled as the r1 set.

**Inner web.** Portion of *rectrix* that is visible on underside of tail.

**Outer web.** Portion of *rectrix* that is visible on dorsal side of tail. Exception is on the top most of the *deck set*, which has both sides of the feather visible.

**Rectrices.** Six symmetrical sets of long, stiff feathers that form the tail. On the dorsal side, the inner feather sets overlap the outer sets. On the ventral side, the outer feather sets overlap the inner sets. The sets are numbered from innermost (*deck set*) to outermost, from r1 to r6. Rectrix is singular.

**Subterminal band.** Darker band just inside the tail tip.

**Tail band.** Horizontal band across the dorsal of the tail.

**Undertail coverts.** Long, often rather fluffy, contour feathers attaching on the underside of the tail bone that cover the base of the undertail. On the outer area, there is one covert for every *rectrix*.

**Uppertail coverts.** Short, fairly stiff feathers, one attaching at the base of each *rectrix*, overlapped on the inside by the *rump* feathers. Mistakenly called "rump" on some species.

fingers
primaries
remiges
secondaries
greater primary coverts
scapulars
lower back
back
rectrices (tail)
greater upperwing coverts
alula
uppertail coverts
rump
trailing edge
wing panel
supraorbital ridge
lores
orbital ring
nostril
cere
bill
bill notch
gape
cheek
auriculars
breast
crown
ocelli
forehead
throat
malar mark
neck
breast
eyeline
supercilium
nape
hindneck
malar mark
back
wrist
scapulars
lesser upperwing coverts
median upperwing coverts
greater upperwing coverts
tertials
secondaries
tarsi
toes
talon
primary projection
primaries
deck feathers
wingtip-to-tail tip ratio
leading edge
window
patagium
remiges (flight feathers)
lesser underwing coverts
median underwing coverts
axillaries
greater underwing coverts
breast
flanks
patagial mark
leg feathers
underwing coverts
undertail coverts
subterminal band
primary underwing coverts
belly
lower belly
secondary underwing coverts

# IV. PLUMAGE, MOLT, AND AGE GLOSSARY

## Plumages

Confusion abounds in labeling the various plumages and ages of raptors. The uncertainty stems from the fact that there is currently no consistently used terminology in the ornithological world. The terminology used in this book is described below and I hope will ease some of the confusion. To better understand the age progression in raptors, a basic knowledge of molt is useful (*see below*).

The descriptions here are based primarily on the widely accepted plumage and molt terminology of Humphrey and Parkes (1959). The plumage changes acquired by each age class are based on molt cycles. In most raptors, a molt cycle corresponds near the hatching date. In some cases, however, it may happen when the raptor is only a few months old (i.e., kites). The plumage descriptions for each age class below follow such molt cycles, and each plumage change that is due to molt signifies advancement into the next older age class.

The terminology used by the U.S. Department of the Interior Bird Banding Laboratory and Canadian Bird Banding Office is not used here. This terminology uses calendar age cycles rather than molt cycles. With this system, age of a bird automatically changes on January 1 of each year, not by a plumage change.

Humphrey and Parkes divided bird plumages into two categories. Basic plumage is the primary, or winter, plumage for species that have one year-round plumage. This plumage is acquired by a single annual molt that occurs during and after the breeding season; on most species and/or subspecies, it is completed prior to winter. Alternate plumage is the temporary breeding, or summer, plumage that some species, mainly songbirds, acquire by molting prior to the breeding season. They molt again after the breeding season into a basic plumage for the winter.

Diurnal raptors undergo one annual molt and for the most part have one year-round plumage. They are therefore classified as birds that have a basic plumage. On some species, the molt is not complete or is prolonged and is often temporarily suspended; however, such species still typically have only one molt per year.

Since diurnal raptors adhere to the basic plumage concept, each molt is called a prebasic molt, meaning "the molt before the basic plumage." For birds that take more than one molt to acquire adult plumage, these molts are often labeled numerically, i.e., "first prebasic molt," "second prebasic molt," and so forth. The molt into the adult plumage is sometimes called the definitive prebasic molt by some authors because it is the final molt into adult plumage. Even definitive prebasic molts are often numerically designated, i.e., "first definitive prebasic molt."

In this book, the first molt is labeled simply the "first prebasic molt." Additional numerical designation precedes the "prebasic" term if more than one plumage is needed to attain adult plumage. In adult plumages, the "definitive" numerical designation is not used (*see* Ages, below).

## Molt

**Molt.** Replacement of old, worn body feathers (contour feathers) and wing and tail feathers with new feathers. Degree and timing of molt vary considerably. On average, some or all of the plumage is replaced on an annual basis. Age, sex, latitude, elevation, migration, and food can play a role in molt timing and duration. It usually takes several months to replace all the plumage (generally during summer and fall). In nesting adults, females begin molt prior to males. Males molt later, after their young are mostly grown or have fledged. This permits the male—the major food supplier to the female and nestlings—to have his plumage in prime condition in order to hunt effectively, and also not expend nutrients on growing his own feathers. In large raptors such as eagles, it takes 4 years to replace all remiges and rectrices.

**Molt center.** Beginning point on the wing and tail where molt begins. Molt progresses in a particular direction, generally symmetrically, on each wing and the tail from the molt center.

**Molt unit.** Group of feathers on the wings that molt as a unit. Molt units can advance in variable directions from the *molt center.*

**Molt wave.** Partial or complete molt of a *molt unit.* If a partial molt of the molt unit occurs, then molt begins during the next molt at the location where it ceased (see *suspended molt*). Also called "molt front."

**Serially descendant molt.** Simultaneous molt at more than one *molt center* and *molt unit.*

**Suspended molt.** Temporary suspension of molt. This normally occurs during winter and migration in order to save energy growing feathers when food is in short supply or when energy is needed for migration. Molt resumes from the location of the stopping point at a later date.

## MOLT PATTERNS ACCORDING TO FAMILY CLASSIFICATION

**Accipitridae Wing and Tail Molt Pattern—Accipiters, Buteos, and Harriers.** (Data based in part on Edelstam 1984 and Forsman 1999.)

**Wing (Remiges).** Primaries comprise 10 feathers with a molt center on the innermost primary (p1). Molt advances outwardly to the outermost (p10) in a single, complete molt unit. On larger species, however, molt may be partial and may not extend all the way to p10 in one molt season. In a partial molt, p8–10 are often retained old feathers. Primary greater coverts molt in unison with the remiges.

Secondaries and tertials comprise 13 feathers with 4 molt centers and molt units. The outermost secondary is s1, the innermost s13. There is a molt center at s1 and molt advances inwardly to s4 (molted in order of s1, 2, 3, 4). On the mid-wing there is a molt center at s5 that advances inwardly to s8 (molted in order of s5, 6, 7, 8). On the inner secondaries (tertials), a molt center is at s13 and molt advances outwardly to s11 (molted in order of s13, 12, 11). From s10, molt advances to s9.

On most hawks, this is a complete molt each year. On some large species, however, the outer two to four primaries and mid-wing secondaries may not be replaced during the first molt, and subsequent molts may not be complete either. In general, s4 and 8 are the last to molt and are the remiges most likely to be retained in a partial molt. Secondary greater coverts molt in unison with the remiges. Tertial molt is completed prior to the rest of secondary molt.

**Tail (rectrices).** The tail has 12 rectrices with 2 molt units. Molt begins on the deck set (r1) and goes to the outermost set (r6). Molt on r6 may begin before r1 is fully grown. The typical sequence is for r3 and 2, and r4 and 5, to molt after r1 and 6 are grown. Rectix molt may be serially descendant, with r3 and 5, and 4 and 2, molting simultaneously.

**Accipitridae Wing and Tail Molt Pattern—Eagles.** (Data based on Edelstam 1984 and Forsman 1999.)

**Wing (Remiges).** Primaries molt as a single molt unit; however, it takes up to 4 years, with annual molt waves, to replace the entire 10 feathers. Advancement of the annual molt waves depends on latitude, health, and diet. Molt begins with a molt center at p1 and advances outwardly to p10. Inner primaries may molt two or three times prior to the replacement of the outer primaries (p9 and 10). Primary greater coverts molt in unison with the remiges.

Secondaries and tertials comprise 17 feathers with 4 molt centers and molt units. Molt is serially descendant. On the outer secondaries, there is a molt center at s1 and it advances inwardly as a molt unit to s4 (molted in order of s1, 2, 3, 4). On the mid-secondaries, there is a molt center at s5 and it advances inwardly as a molt unit to s9 (molted in order of s5, 6, 7, 8, 9). On the inner secondaries, a molt center is at s14 and advances outwardly as a molt unit to s10 (molted in order of s14, 13, 12, 11, 10). On the tertials, a molt center is at s17 and advances outwardly as a molt unit to s15 (molted in order of s17, 16, 15). Secondary greater coverts usually molt in unison with the remiges. See individual eagle accounts for variables in molt of the remiges.

**Tail (rectrices).** There are 12 rectrices, and molt sequence is similar to that of smaller Accipitridae, but molt timing varies with the two eagle species. See individual species accounts for species variation.

**Pandioninae Wing and Tail Molt Pattern.** (Data based on Edelstam 1984 and Palmer 1988.)

**Wing (Remiges).** Ten primaries, molted in a similar pattern as in other Accipitridae (p1–10). However, only part of the primaries are molted before a new wave begins on the inner primaries. Most inner primaries may have molted twice before p10 is molted for the first time (p10 is the last to be replaced). Therefore, the primaries will have three ages of remiges once molt has progressed far enough along.

Secondary molt is similar to that of eagles, with 17 feathers.

**Tail (rectrices).** There are 12 rectrices. Molt begins after wings have started their molt. It begins from the deck feathers (r1) and proceeds outward (with some variation).

**Cathartidae Wing and Tail Molt Pattern.**

**Wing (Remiges).** Molt is similar to that of eagles (Accipitridae; *see above*), with 10 primaries and 17 secondaries. California Condors molt their remiges in annual waves and may take 4 years to complete all feather replacement.

**Tail (rectrices).** There are 12 rectrices. Molt is symmetrical and begins on the deck set (r1), then the outermost set (r6), then r2–5. Rectrix molt may be serially descendant as noted in Acciptridae.

**Falconidae Wing and Tail Molt Pattern.** (Data based on Edelstam 1984, Forsman 1999, and Palmer 1988.)

**Wing (Remiges).** Primaries comprise 10 feathers, and molt begins on the mid-region (p4), then usually proceeds to p5; it sometimes begins on p5, however. From p4 (or 5), it advances both inwardly and outwardly in an irregular sequence that varies with each species. Regardless, p10 is the last primary to molt. Primary greater coverts molt in unison with the remiges.

Secondaries and tertials comprise 13 feathers with 2 molt centers and 3 molt units. Molt advances outwardly and inwardly from the mid-secondaries (s5). From s5, it extends outwardly in the order of s4, 3, 2, 1. Also from s5, it extends inwardly in the order of s6, 7, 8, 9, 10. The outermost secondary, s1, is the last to molt. Tertial molt begins on s13, goes to s12, and ends at s11. Tertial molt is finished prior to completion of molt of the outer secondaries. Secondary greater coverts molt in unison with the remiges.

**Tail (rectrices).** Molt is identical to that of Accipitridae, with 12 rectrices.

## *Ages (from Youngest to Oldest)*

**Nestling.** Young raptor confined to the nest. Early stage: adorned in a downy feather coat (usually with two downy stages). Mid-stage: contour feather coat is acquired. Late stage: contour feather coat is quite grown but downy feathers retained on head; wings and tail not fully grown.

**Fledgling.** Young raptor that has left the nest but is still fed by the parents. Adorned in the early stage in a contour feather coat that still is not fully grown, especially remiges and rectrices. Often labeled as a brancher at this stage. Birds are not capable of flight. In mid-stage, plumage is mostly grown but wings and tail are still not fully developed. Capable of reckless flights. In late stage, the contour plumage has been aquired for a while and wings and tail are fully grown. Capable of controlled, sustained flights.

**Juvenile.** Young raptor that is fully grown and totally self-sufficient. Juvenile plumage may be retained for only a few months on a few species, but usually for 1 year on most species. Some large raptors may retain portions of this first plumage for several years. "Juvenal" also designates any less-than-adult age. A prebasic molt precedes any change into another plumage and/or age class. See *immature. Note:* Some authors use "juvenal" as an adjectivial term to describe plumage features of this age class.

**Immature.** General term for any plumage and/or age that is less than adult.

**Subadult.** General term for any plumage and/or age that is older than juvenile but not yet adult. *Note:* In this book, "subadult" is used synonymously with "basic" plumage and/or age. On species with one interim subadult plumage and/or age, it is labeled "basic I (subadult I)." On species with more than one subadult plumage and/or age, additional designations have a numerical labeling, i.e., basic I (subadult I), basic II (subadult II), etc.

**First adult.** Adult plumage with some immature traits. Under field conditions, however, these characteristics may not always be visible and the bird may look like a typical adult.

**Young adult.** Full adult plumage and/or age that has one or possibly more minor traits that are considered to be adult, but differs somewhat from older-aged adults. Often pertains only to eye color.

**Adult.** Final stage of plumage development. Minor plumage changes may continue to occur with advancing age—mainly of the plumage becoming less marked and paler. Technically known as "definitive basic plumage," or the final stage of basic plumage and/or age development.

# V. FLYING AND PERCHING DISPLAYS GLOSSARY

## *Flying Displays*

This section describes many of the aerial displays used by raptors. Most are used for courting, though some may be used at any time of year for social interaction. Most species engage in at least some type of aerial antics in order to entice a mate, establish or defend territories, or warn of danger. Most raptor species perform one or more of these displays, although a few do not engage in any type of aerial display. Aerial displays are italicized in the species accounts for easy recognition. Courtship displays are listed in the Nesting section of each account; other aerial displays may be listed in Habits or Flight.

**Aerial food transfer.** Male gives prey item to female, either by talons or bill, or prey may be dropped by male and caught by lower-flying female. Moderate- or high-altitude display.

**Aerial-kissing.** Closely flying pair of birds touch bills while airborne. Moderate- or high-altitude display.

**Begging-flight.** Level flight by female with wingbeats kept below the horizontal plane and tail fanned. Accompanied by whining calls. Also called "cuckoo-flight." Performed at any altitude.

**Cartwheeling.** Closely flying birds grasp and lock talons, then engage in a free-fall tumble. They often come very close to ground level before separating. Display may be performed at any time of year by some species. It begins at a high altitude and ends at a low altitude.

**Diving swoop.** One sex of a pair or prospective pair swoops down from a higher altitude in a moderate-angled dive toward lower bird, then overshoots lower bird at same altitude and glides past.

**Drop-and-catch.** Playful antics of a bird dropping and catching objects in mid-air. Moderate- or high-altitude display.

**Escort flight.** Territorial flight in which individual of an occupied area flies alongside an intruding conspecific to boundaries of territory. See also *leg-lowering.*

**Exaggerated-flapping.** Wings are raised very high on upstroke and lowered very low on downstroke. Flight is on a level course or slightly undulating. Moderate- or high-altitude display. This is a more accentuated version of *slow-flapping.*

**Eyrie fly-by.** Purposeful level flight in front of prospective nest site. Moderate- or high-altitude display.

**Figure eight.** Performed usually by males of larger falcons in which they loop around in a figure-eight pattern.

**Flash roll.** Male of a large falcon rolls 90 degrees to one side and then the other side in level flight. See also *roll* and *undulating roll.*

**Flight-play.** One individual dives toward another, often engaging in brief sideways rollover maneuvers and, when passing each other, often extending talons. Moderate- to high-altitude dislay.

**Flutter-flight.** Level flight in which wings are flapped in very shallow strokes. Generally performed by females of all falcons. Also called "flutter-glide."

**Flutter-flying.** Level flight in circles or in figure-eight loops with rapid, shallow wingbeats. Performed at any altitude.

**Fly-around.** Seen mainly in Bald Eagles, where male intensely flies around its presumed territory.

**Follow-flight.** Generally in mated pairs of Turkey Vultures. A bird follows closely behind lead bird, twisting and turning, often for extended periods. A gentle variation of *Tail-chasing.*

**High-circling.** High-altitude soaring over territorial area by an individual of a prospective or mated pair. Often called "mutual high-circling" when two birds engage in this display. If performed by a pair, they may be in the same thermal but not necessarily near each other. High-altitude display.

**Hover-flight.** Courtship flight performed by male Osprey in which bird angles its body upward with slow, labored, nearly stalled flight. A fish is carried in talons during the display.

**Leg-dangling (leg-hanging).** Legs are lowered downward while one or both birds engage in *high-circling, flutter-flight,* or *V-flutter.* High-altitude display.

**Leg-lowering.** Territorial interaction in which two individuals fly close together along territorial borders and repeatedly glide downward and land with stiffly held, lowered legs. Moderate- and low-altitude display. Often accompanies *escort flight.*

**Mutual floating display.** In large falcons, male is positioned slightly above female while soaring, then both slowly descend with wings partially closed and held a bit above horizontal, tail spread, and legs lowered. Both birds maintain same distance as they descend.

**Pair-flight.** Both sexes glide and soar close together in synchronized formation. High-altitude display.

**Parachuting.** Male (usually) glides down to female with his wings held high above body and fluttering and his legs lowered. Female rolls over and presents her extended talons to male when he approaches her. Moderate- or high-altitude display.

**Passing-and-leading display.** Male of a large falcon flies past a female, then begins to weave back and forth in front of her. Moderate- or high-altitude display.

**Pothooks.** Series of short, steep vertical dives interspersed with short upswings. High-altitude display.

**Power-diving.** High-speed vertical dive accompanied by sideways twists. High-altitude display.

**Power-flight.** High-speed direct flapping interspersed with rollover maneuvers. Performed at any altitude.

**Pursuit-flight.** Male closely chasing a female.

**Rocking-glide.** Sideways rollover maneuvers while gliding. Performed at any altitude.

**Roll.** Male of a large falcon rolls back and forth while in a steep dive.

**Sky-dancing.** Succession of shallow or moderately deep undulations consisting of closed-winged dives. Dives can be fairly long and end with moderate upswings. Undulations may be interspersed with short distances of gliding or flapping. Often performed at high altitudes.

**Sky-dancing (high intensity).** Succession of deep undulations consisting of closed-winged dives ending with sharp upswings that may include an upside-down rollover. Moderate- or high-altitude display, occasional at lower altitudes.

**Slow-flapping.** Accentuated, slow, and deliberate wing flapping with wings raised high above body on upstroke in a very accentuated manner on a level flight course. Moderate- or high-altitude display. See also *exaggerated-flapping.*

**Slow-landing.** Wings are held high, with slow flaps from wrist area. Legs are extended forward with tail fanned and lowered as raptor approaches a perch. Moderate- or low-altitude display.

**Tail-chasing.** Typically a male pursuing a female, often in gentle swooping downward maneuvers. Performed at any altitude.

**Talon-grappling.** Upper individual dives toward or is flying above a lower individual while lower bird rolls upside-down and both birds grasp each other's talons. High-altitude display.

**Undulating roll.** Male of a large falcon rolls sideways while in level flight, then begins a steep dive, then swoops up and repeats maneuver. A roll variation of *sky-dancing.*

**V-flight.** Long, steep, angled dive that ends with a sharp upswing. High-altitude display.

**V-flutter.** Level flight with wings held in a pronounced dihedral high over body and fluttered just enough to remain airborne; legs extended downward. Low- to moderate-altitude display.

**Z-flight.** Series of long, steep dives interspersed with short-distance level flight. Display begins at a high altitude and ends at a low altitude.

## *Perching Displays*

Except for falcons, most raptors do not engage in displays while perched. Many falcons, however, have quite elaborate sexual courtship displays while perched. Most of the displays below correspond to falcons unless otherwise noted. Other types of raptors may engage in one or two perching displays. Perching displays are shown in italics in the species accounts. Displays associated with courtship are in the Nesting section; others are in Habits and sometimes Voice.

**Allopreening.** Mutual preening within a pair. Also seen among related individuals and occasionally among unrelated individuals.

**Billing.** Pair members nibble at each other's bill; seen in large falcons.

**Circle display.** Performed by vultures. Male struts around female, his neck bowed and arched forward and his wings partially or fully spread, and circles female. Male usually faces female, but sometimes his back is toward her. Male may inflate throat and neck air pouches and emit hissing noises.

**Copulation-solicitation.** Female assumes a horizontal stance with wings held loosely and tail lowered (*Accipiter*), or wings drooped and tail raised (*Falco*).

**Curved-neck display.** Prior to copulation, males of large falcons stand erect and arch neck forward with bill pointing downward. No vocalization is made.

**Female ledge display.** Female of a large falcon walks to nest scrape with exaggerated high stepping motion with body in a horizontal posture. She utters call notes but does not look at male. See also *male ledge display.*

**Food-begging.** Female vocalizes in a continuous whine, usually with her plumage fluffed out, or sometimes perched in a squatted position.

**Food transfer.** Bill-to-bill transfer of prey item from male to female. Performed by virtually all raptor species.

**Head-bow posture.** Head is pointed downward, often accompainied by vocalization and up-and-down head bowing. Can be performed by both sexes.

**Head-throwback display.** Performed only by Crested Caracara. Performed when there is agonistic confrontation with conspecifics, particularly in social feeding situations. Head is quickly snapped onto back and bill is pointed skyward; usually accompanied by vocalization.

**High-perching.** Solitary individual or a pair perching on a high exposed perch to advertise territory. If a pair, both birds may perch near each other.

**Hitched-wing display.** After landing in a *slow-landing* display, male raises wrists of folded wings, lowers his head, and compresses body plumage.

**Horizontal head-low bow.** Either sex of a large falcon has its body at a horizontal posture with all feathers compressed and head bowed 90 degrees from body angle.

**Male ledge display.** Male of a large falcon walks toward nest scrape with exaggerated high steps, with head bowed low and body at a horizontal position, and looks at female. He utters call notes. See also *female ledge display*.

**Mutual ledge display.** Both sexes of a large falcon walk to nest scrape performing high stepping actions with bodies in a horizontal posture. See also *male/female ledge displays*.

**Mutual plumage stroking.** Bald Eagle pairs gently stroke each other's plumage.

**Nest displays.** Concerns Merlins: both birds are at potential nest site and make *tic* calls. Male lies on nest, arches back, droops and shakes wings, and fans tail.

**Perch-and-fly.** One or both individuals of a large falcon pair land at prospective nest site or other nearby perch, sit or walk around momentarily, then quickly take off.

**Tiptoe-walk.** Male approaches nest site in an exaggerated high stepping motion.

**Vertical head-low bow.** Either sex of a large falcon perches in an upright posture with body feathers compressed and head lowered, often pointing away from mate.

**Wail-pluck display.** Male of a large falcon emits a "wail" call as he plucks prey.

# VI. PERCHING AND FLYING ATTITUDES

Raptors exhibit numerous behavioral attitudes when perching and flying. The position of a perching bird indicates if the individual is relaxed, reconditioning feathers, alarmed, feeding, or ready to take flight. The position of the wings and tail on a flying bird indicates if the bird is soaring, flying fast or slow, hunting, or feeding. Major body positions that a viewer will encounter are shown and described below.

## *Perching*

**A. Relaxed or in light wind.** Body is postured at a vertical angle. A bird may stand on two feet, but when it is very relaxed, one of the legs and feet are raised and tucked up, often hidden underneath the belly feathers. Or one foot may be clenched and its knuckles rested on top of the perch. Feathers are fluffed and a bird appears larger in cool or cold temperatures (fluffed feathers allow the downy feathers next to the body to trap more air and thus provide more insulation). *Note:* Juvenile "Eastern" Red-tailed Hawk.

**B. Forward leg-extension.** A very relaxed posture in which one leg is extended forward and rests on the top or forward portion of the perch. The rear of the tarsus rests on the perch, or the entire leg is extended forward so the tibia rests on the front of the perch and the tarsus and foot are in a straightened, stiff-legged position. *Note:* Juvenile Mississippi Kite.

**C. Preening.** Body is postured at any angle but is in a vertical angle in light winds. Raptors only preen when they feel safe and relaxed. Feathers are slipped through the bill, which is typically oiled by the uropygial gland: Birds typically close their eyelids when preening to avoid injury to their eyes. *Note:* Adult Rough-legged Hawk.

**D. Rousing.** Between preening sessions and often prior to taking flight, birds fluff all their feathers and vigorously shake their bodies to rid themselves of dirt and loose feathers. *Note:* Juvenile White-tailed Hawk.

**E. Sunning.** Kites and vultures are especially inclined to sun themselves. The dorsal area is situated to face the sun, the wings may be partially or fully extended, and the tail is typically spread. All raptors spread their wings and tail to dry out after getting wet. *Note:* Juvenile Mississippi Kite.

**F. Mantling.** Wings are partially or fully spread and the tail is spread: This posture is used in three behavioral attitudes: (1) when a raptor that has prey on the ground attempts to shield it from the view of other raptors that may pirate the food; (2) as a threat posture to ward off approaching or nearby raptors from taking captured or pirated prey; and (3) as a threat posture if the raptor is injured and is being approached by a human or predator. In the last two cases, the nape feathers (hackles) are erected to exhibit defiance and to make the bird appear larger and more dominant. *Note:* Adult Red-tailed Hawk.

**G. Feeding.** Raptors typically stand on prey to hold it in place. If perched on a branch, falcons and kites hold prey with a lowered leg that braces the rear of the tarsus (heel) on the perch. Avian prey is plucked of feathers in the region that will be eaten, mainly the breast. Mammalian prey is plucked of its fur in areas that will be eaten, mainly the inner, thick leg muscles. Small bones, and sometimes larger leg bones, are devoured. Ospreys eat the forward part of the fish first. *Note:* Adult Osprey eating a fish.

**H. Wing-leg stretch.** Prior to taking flight, raptors typically stand on one leg, then simultaneously raise and stretch the other leg and same-side wing; first one side, then the other. The tail is often simultaneously spread with each wing-leg stretch. *Note:* Adult Swainson's Hawk.

**I. Wing flex.** Prior to taking flight, and usually after *defecating* and performing the *wing-leg stretch*, raptors lean over at a horizontal angle and raise both partially opened wings above their body. The head and neck are bowed downward and extended outward. *Note:* Basic II (subadult II) Bald Eagle.

**J. Defecating.** Prior to taking flight, all raptors lean over at a horizontal angle, lower their lower belly feathers, raise their tail, then shoot out a stream of white excrement. They may defecate and then assume a regular perching position; however, flight is usually imminent within seconds or minutes after defecating. *Note:* Juvenile Sharp-shinned Hawk.

**K. Alarmed or in strong wind.** Alarmed birds that are ready to take flight lean over in a more horizontal position. In strong winds, even relaxed raptors lean over and face into the wind to be more aerodynamic and to reduce feather ruffling caused by wind. Feathers are compressed and birds appear thin and sleek in warm and hot temperatures (feather compression reduces the insulating factor of the downy feathers next to the body). *Note:* Juvenile "Eastern" Red-tailed Hawk.

## *Flying*

**L. Powered flight.** Wings are flapped. Tail is usually closed but is sometimes opened for stability. Flight mode is used to gain speed quickly or maintain a fast speed. *Note:* Adult "Harlan's" Red-tailed Hawk.

**M. Soaring.** An energy-efficient flight mode in which a bird maximizes a rising air mass. The rising air provides excellent conditions in which a bird rotates in a circular motion with wings fully extended and tail typically fanned. Falcons have a slight backward bend at the wrist, even when the wing is fully extended; all other raptors have a straight or forward thrusting leading edge of the wing. There are two types of soaring: (1) **thermal soaring**, when a bird rises on a warm, rising air column created by a warm ground surface, and (2) **dynamic soaring**, when a bird rises on strong winds deflecting off a land form. *Note:* Juvenile Sharp-shinned Hawk.

**N. Slow glide.** Wings are slightly closed and all raptors show a slight bend at the wrist. Wingtips are partially closed and become somewhat pointed. Tail is closed to reduce drag. Some raptors may also use this position when *soaring*, particularly when *dynamic soaring. Note:* Juvenile Sharp-shinned Hawk.

**O. Moderate glide.** Wings are nearly half closed and all raptors show a definitive bend at the wrist. Wingtips are fairly pointed. Tail is closed to reduce drag. *Note:* Juvenile Sharp-shinned Hawk.

**P. Fast glide.** Wings are more than half closed and all raptors show a definitive bend at the wrist. Wingtips become pointed. Tail is closed to reduce drag. Some raptors may also use this position when *kiting. Note:* Juvenile Sharp-shinned Hawk.

**Q. Banking.** Wings are fully extended and even falcons may exhibit a straight leading edge of the wings as they push against the air. Alula are extended for stabilization. Tail is widely fanned to act as a brake and rudder. *Note:* Juvenile Sharp-shinned Hawk.

**R. Diving.** Wings are partially opened in most dives but are nearly closed during the highest-speed dives. Tail is partially open for stability but can be closed. *Note:* Adult Peregrine Falcon.

**S. Hovering.** A stationary altitude maintained by flapping the wings, typically when hunting. Head points downward searching for prey. Tail is widely fanned to maintain or gain altitude. Legs are sometimes lowered to assist balance. *Note:* Juvenile light morph Rough-legged Hawk.

**T. Kiting.** A stationary altitude is maintained by gliding on a brisk wind current. Wings are partially closed and tail is closed. *Note:* Adult dark morph Short-tailed Hawk.

**U. Aerial feeding.** Falcons, kites, and a few buteos may feed on small prey while in flight. Feeding typically occurs when *soaring* or in a *slow glide*. The foot holding the prey is extended forward and the head is lowered to reach the prey. *Note:* Adult Mississippi Kite.

**V. Distended crop.** Raptors typically gorge themselves when a large amount of food is available. After an extensive feeding session, they often exhibit an enlarged breast because of an overly stuffed crop. On many raptors, the distended breast notably affects the bird's appearance. Also, the added weight of food in the crop may affect how a bird flies. Particularly on smaller raptors, wingbeats become labored and slower than normal, and the alteration may well make identification more difficult at moderate and long distances. *Note:* Juvenile Golden Eagle.

A. J-Eastern Red-Tail
B. J-Mississippi Kite
C. A-Rough-legged Hawk
D. J-White-tailed Hawk
E. J-Mississippi Kite
F. A-Red-tail Hawk

G. A - OSPREY

H. A - Swainsons Hawk

I. SAll - Bald Eagle

J. J - Sharp - Shinned

K. J - Eastern Red - Tail

L. A - HARLANS Red - Tail

M. J - Sharp - Shinned

N. J - Sharp - Shinned

O. J - Sharp-shinned

P. J - Sharp-shinned

Q. J - Sharp-shinned

R. A - Peregrine Falcon

S. J lite morph Rough-Legged

T. A dark morph Short-tailed

U. A - Mississippi Kite

V. J - Golden Eagle

# VII. PHOTOGRAPHY

I was a painter of gallery-type renditions of life-sized birds and an illustrator of raptor books long before I began to pursue raptor photography. I began to dabble with photography in the early 1970s, but mainly as a reference tool for my paintings. During this time, I used photography to catalog images of fresh foliage and captive birds. Throughout the 1970s and into the early 1980s, I was not keen on using photography for anything but obtaining reference material. I did not like the limitations imposed with wildlife photography and could not afford better, more expensive equipment that can help overcome some of the limitations.

I began intense research and illustration on the original *A Field Guide to Hawks of North America* in 1983. My photography, now mainly of wild and captive raptors, began to be more intense, but it was still relegated to reference use. This slowly began to change, however. A few quality images began to emerge from my large stock of mostly poor-quality reference images.

By 1984, the "bug" was beginning to infect me; I desperately wanted to obtain higher quality photographs of raptors, particularly flying raptors. By this time, I realized there are three criteria for obtaining great raptor photos, particularly of the difficult-to-get underside flight images. First, one needs to have a quality lens that lets in a maximum amount of light, to help "overexpose" the typically shadowed, dark underside of flying raptors. Second, one has to photograph flying shadow-side raptors over a pale-colored ground surface, in order to assist the overexposure by reflecting light up underneath raptors. And third, a blue or dark gray sky is advantageous in creating an aesthetically pleasing background that also contrasts with the pale feather edgings of the wings and tail.

August 1985 was the turning point in my photography. After trading and selling my photographic equipment over a period of years in order to gradually get better optics, I finally scraped together enough money to purchase my dream lens: a Nikkor 400 mm f/3.5. This is the lens I used for the next 10 years. The large aperture let in a maximum amount of light to help in my overexposure method of flight photography. Although the lens was heavy, I was still able to hand-hold it for easy maneuverability (I have never used gunstock mounts and rarely use tripods). From this point on, I tried to coordinate the benefits of this extraodinary lens and situating myself near photographically conducive backgrounds. The images in *A Photographic Guide to North American Raptors* (including the images by W. S. Clark) were photographed exclusively with this lens. For additional magnification on skittish perched raptors, I attached a Nikon 1.4X extender, which created a respectable 560-mm telephoto lens (at the time, all brands of 2X extenders were of poor quality). During my 10-year use of the Nikkor 400 mm, I first used Nikon FM2 bodies, then graduated to workhorse F3s, and finally to superb F4s. I used all camera bodies with motor-drive units, either attached or integral (F4).

The photographic world, however, was coming of age: autofocus equipment was entering mainstream professional use. I was a disbeliever until famed bird artist Arthur Morris, who had joined me on a photography stint in Duluth, Minnesota in October 1995, showed me the astounding results of autofocus. I was quickly converted! By November of that year, I made the switch and sold my beloved Nikkor 400 mm (and Nikon F4 bodies) and bought Canon autofocus equipment, which at the time was the leader in high-speed, silent-focusing autofocus technology.

The exquisitely sharp Canon 300 mm f/2.8 autofocus lens has been my primary lens ever since. For additional magnification of more distant flying raptors, I use the incredibly sharp Canon 1.4X extender; for perched birds, I used a Canon 2X extender. At first, I used Canon 1N bodies with motor drives and a superquiet Canon A2 body for close-perching birds (raptors are very sensitive to the sound of a camera shutter, even when 100 feet away). I now use the EOS 3 body and still use an A2 for perching birds (Canon has since discontinued making the A2).

I rely on my knowledge of raptor behavior to get close to the birds and have rarely used the large supertelephoto lenses. The Canon 600-mm autofocus lens (now with the amazing IS [Image

Stabilizer] system), especially with a 1.4X or 2X extender attached, produces a coveted 840- or 1,200-mm focal length of impecable sharpness.

Autofocus is not infallible, however, and it still takes skill and some luck to obtain superb results. Regardless, one is still able to garner more sharp, in-focus images than with manual focus equipment.

My film usage has also varied over time. I used Kodachrome 64 until 1992. This fine-grained, sharp film is slow and does not reproduce warm colors very well. I tried the old Fujichrome 100 but was not impressed. In 1993 and 1994, I used Fujichrome Velvia (ISO 50) pushed one stop (100 ISO). I obtained many great images with this finely grained film, but the reds were often too overbearing. I began using Fujichrome Sensia in 1995 and Sensia II (both ISO 100) when it came out a few years later. These are superbly color-balanced, natural-looking films. For the last two years I have been using the professional-grade Sensia II, Provia F (100 ISO), and in the last year I have been pushing Provia F one stop (200 ISO) with superb results.

Digital camera bodies are now being used by many professionals. Although they have some advantages over film cameras (e.g., no film, quiet), I do not see myself jumping into the digital game just yet.

## *Raptor Music*

Feathered batons slicing the air,

Composing the stirring notes of flight.

Twisting, tumbling through thermal melodies,

Wind played like a drum.

Wings pounding, rolling,

Now lilting skyward harp-like,

Then sliding down-scale,

Gliding over medleys of unseen currents,

Exposing chords of ruffled delight.

A symphonic aerial ballet

Rendered with instruments of perfection.

  Encore!

*by Steve Millard*

# Species Accounts

# TAXONOMY OF VULTURES

Black Vulture, Turkey Vulture, and California Condor are no longer considered true diurnal raptors in the order Falconiformes. These three species are part of the New World vulture family, Cathartidae. These vultures have many of their own unique traits; however, they also share several anatomical and behavioral traits (e.g., inability to grasp objects with their feet, urohydrosis, voicelessness, courtship displays, and laying down on the ground or perch) with the New World stork family, Ciconiidae.

The taxonomic niche for Cathartidae, however, is still in question. DNA research verifies that these vultures are not related to true raptors, but recent studies are inconclusive as to whether they are taxonomically closer to storks or raptors (Seibold and Helbig 1995). Recent morphology (syringeal) research still places these vultures in Falconiformes (Griffiths 1994).

Despite the ongoing taxonomic dilemma, the AOU (1998) has reassigned the New World vultures from Falconiformes to Ciconiiformes, the storklike birds. The vultures still remain in their own family, Cathartidae, but are now placed immediately after Ciconiidae and are separated from Falconiformes by Phoenicopteriformes (flamingos, ibises, and spoonbills) and Anseriformes (geese and ducks).

Regardless of taxonomy, these vultures are unquestionably very raptorlike, particularly in flight. Considering their historical taxonomic association with true raptors, and the fact that they pose identification problems with raptors, they are therefore treated here.

---

## BLACK VULTURE
### *(Coragyps atratus)*

**AGES:** Adult and juvenile, with a transitional subadult stage with development of head and upper neck features and molt. Adult plumage and head and neck characters are acquired during the second year when 1 year old. Juvenile plumage is worn the first year but with a gradual change to more adultlike head and bill toward the end of the period. Adults and juveniles have minor plumage differences and different head features. Sexes are identical.

**MOLT:** Cathartidae wing and tail molt pattern (*see* chapter 4). First prebasic molt from juvenile to adult begins when about 1 year of age in spring or early summer and extends until fall. Bare skin region of the upper neck and bill features of juveniles, however, begin to gradually change to partial adultlike characters when about 6–8 months of age. Little is known of subsequent prebasic body molts.

**SUBSPECIES:** Polytypic with two races in North America. Nominate *C. a. atratus* is found in Texas, Oklahoma, and eastward; also in portions of Mexico. *C. a brasiliensis*, which is smaller, inhabits s. Arizona and tropical portions of Mexico and south into South America. *C. a. foetens* is a large race inhabiting parts of South America. Subspecies are based primarily on subtle differences in size. *C. a. atratus* and *brasiliensis* are not separable in the field.

**COLOR MORPHS:** None.

**SIZE:** Large raptorlike bird. Length: 23–28 in. (58–71 cm); wingspan: 55–63 in. (140–160 cm).

**SPECIES TRAITS:** HEAD.—**Gray or black featherless head and upper neck.** Neck skin is highly adjustable and can be retracted or distended at will. Bare skin portion of upper neck can look long or short depending on adjustment of feathered portion of the lower neck (ruff). Lower neck conceals bare upper neck and nape when fully raised. **Very narrow bill.** BODY.—**Black. Long tarsi and toes are gray, but urohydrosis turns them white by the chalky accumulation of excrement.** WINGS.—**When perched, primaries extend a short distance beyond tertials. In flight, outer six primaries (p5–10) have white quills and whitish area on each feather, forming large white panel.** Panel is visible on both upper and lower wing surfaces, but is most pronounced on underside of the wings. TAIL.—**Black, short, and square-tipped.** *Note:* Dorsal plumage is often splattered with white excrement. Since Black Vultures often roost communally (*see* Habits), higher perched birds may excrete onto lower perched birds.

**ADULT TRAITS:** HEAD.—Pale yellowish brown tip on bill. **Very wrinkled, tubercle-covered, pale gray or medium gray skin on head and upper half of neck.** Lower neck feathers are glossy iridescent greenish black. BODY.—**Glossy iridescent greenish black.**

**BASIC (SUBADULT) TRAITS (FIRST PREBASIC MOLT STAGE):** This is a transitional stage from juvenile to adult head, neck, and plumage features and occurs during the second year when 1 year old. It begins in late spring or early summer and extends into fall. HEAD.—Essentially like late-stage juvenile in the early part of this period. Bill tip may also be dusky on the central part of the upper mandible as in late-stage juvenile, but many subadults quickly acquire the all-yellow adultlike bill tip. The lower part of the bare neck skin gradually acquires more adultlike, enlarged, pale gray wrinkles. In the later stages, adultlike wrinkles finally appear on the nape, crown, and face. BODY.—First prebasic molt replaces rather faded brownish black juvenile feathering with darker, glossy iridescent greenish black adult feathering. WINGS.—In the early stages, new darker adult feathers are first noticeable on the tertials of perched birds. As molt continues, glossy blackish green adult feathers gradually replace the worn and faded brownish juvenile feathering.

**JUVENILE TRAITS (LATE STAGE):** HEAD.—**Bill tip partially changes to pale yellowish brown on portions of both mandibles. Small grayish wrinkles form on lower front region of bare neck.** BODY.—Except for fading, plumage features do not alter from early/mid-stages until the first prebasic molt. Blackish plumage turns more brownish because of wearing and fading.

**JUVENILE TRAITS (EARLY/MID-STAGES):** HEAD.—**Black tip on upper and lower mandibles of bill. Smooth black skin on head and upper half of neck.** Neck feathers are brownish black and when ruff is not elevated, extend up hindneck a bit farther than on adults, but similar to adults when feathered neck is retracted. *Note:* For a few weeks following fledging, a few remnant buff-colored, wispy downy feathers are still attached to the very upper part of lower feathered portion of the neck and on bare skin of the nape and crown. There is often a rather large patch of wispy downy feathers on the nape. BODY.—**Black with little iridescent quality.**

**ABNORMAL PLUMAGES: Total/incomplete albino.**—Several records of mainly white individuals (Palmer 1988; H. Kendall pers. comm.). **Imperfect albino.**—Several records of pale whitish brown individuals (Palmer 1988). *Note:* Both aberrant plumages rare and not depicted.

**HABITAT:** Semi-open humid lowland regions from south coastal Texas, Oklahoma, and eastward. Semi-open arid regions in southern inland areas of Texas and Arizona. Inhabits lowland subtropical and lower mid-latitude elevations. Absent from high-elevation montane areas but occupies mid-level mountains in s. Arizona. In w. Texas, found only along the Rio Grande in Presidio and Brewster Cos. *Mexico* (mapped area).—Lowland tropical and arid regions below 4,900 ft. (1,500 m). Smaller numbers inhabit higher elevations. In Sonora, mainly inhabits agricultural areas, towns and villages, riparian zones, and coastal regions. In e. Chihuahua and Coahuila, favors areas along the Rio Grande, villages and towns, and lower elevation riparian zones. Absent in arid areas away from riparian zones and human-settled areas, and generally absent or rare at montane elevations. Regular in all of Tamaulipas, particularly around farmlands, cities, towns, villages, riparian zones, and coastal areas.

**HABITS:** Tame in southern areas but somewhat wary in northern regions. Gregarious. Flocks may comprise hundreds of individuals, especially in the nonbreeding season. Exposed, elevated natural or human-made objects of any type used for perches. Also a very terrestrial species. Black Vultures are fond of clean, fresh water and regularly drink and bathe. Large congregations often form at favorite watering areas.

Allopreening occurs between pairs and between parents and offspring; also occasionally with juvenile and subadult Crested Caracaras. Caracaras are the recipient of the preening action. Caracaras bow their heads to have their napes preened or straighten heads to have their chin preened by vultures. Caracaras initiate the preening behavior by moving closer to a nearby vulture and making head gestures.

When walking or galloping, cocks tail upwards and assumes a horizontal posture. Black

Vultures regularly "sun" themselves. Body temperature is regulated by urohydrosis. In cool temperatures, bare neck skin is retracted and feathered portion of lower neck is raised so all of neck appears feathered, including nape. Black Vultures regularly sit upright on their haunches (heels) or lie down when on a perch or ground.

**FEEDING:** An aerial scavenger. Feeds primarily on any type and size of dead creature. Prefers fresh carrion but eats decaying carrion and occasionally eats vegetable matter. On rare occasions, kills newborn animals such as piglets and calves by pecking out their eyes. Also preys on eggs and live nestlings of colonial nesting birds such as herons and cormorants and is known to kill skunks and opossums. In Mexico has been seen feeding on human excrement. Regularly forages in open-pit garbage dumps. This species is aggressive when feeding, with constant squabbling among individuals, and may drive off other scavengers, including Turkey Vultures, from communal feeding sites. Black Vultures cannot smell and must rely on visual acuity to locate food sources. Food is located by flying at high altitudes and observing the behavior of other predators and scavengers.

**FLIGHT:** Powered flight consists of short bursts of snappy, stiff-winged flaps interspersed with variable-length glide sequences. Wings are held in low dihedral when soaring and on a flat plane when gliding. Black Vultures often hop several times on the ground to gain momentum for flight but if necessary are capable of launching into direct flight. When landing, they also may hop or bounce several times before coming to a stop. Black Vultures do not hover or kite. They "dip" wings frequently to maintain speed and stability.

**VOICE:** Technically voiceless. Grunts and hisses when food-begging or agitated. An airy, low-toned *wuff* is emitted when alarmed.

**STATUS AND DISTRIBUTION: Permanent resident.**—Estimated population is unknown but is probably stable in the U.S. *Common* in typical range in the U.S. Range extension first occurred into s. Arizona from Sonora, Mexico, in the 1920s. In Presidio and Brewster Cos., Tex., mainly found along the Rio Grande. Rare north of the river. *Mexico:* Fairly common to common in appropriate habitat. Numbers have possibly decreased in w. and cen. Mexico in recent years. Much of proper habitat in Sonora is inhabited. Sparsely distributed in e. Chihuahua and Coahuila in appropriate but isolated habitat zones.

Black Vulture population is probably much reduced from former times when refuse disposal was less regulated and different farming and ranching practices allowed vast numbers of this species to thrive. Texas ranchers shot and poisoned untold thousands in the early to mid-1900s because of possible predation on newborn livestock, transmission of cattle diseases, and contamination of cattle water supplies. Suitable safe-haven breeding sites have also been greatly reduced with land management changes since the mid-1900s.

**Winter.**—Primarily forms large groups comprising hundreds of individuals. Many remain in their general breeding range, even in very northern areas in sw. Missouri. Christmas Bird Count data, however, show some possible withdrawal from breeding areas in s.-cen. and se. Missouri. Considerable localized wandering occurs during this season in quest of ample food and to retreat from excessively cold weather.

**Movements.**—Regional dispersal is most common. Local and regional dispersal occurs in any region depending on food supply in fall and winter. Actual migration may be undertaken in some northern areas. Southern populations and possibly some northern individuals are sedentary. Seasonal movements are not well documented in the West. It is often difficult to separate migrants from local foraging populations. Both migrants and foraging birds may accompany flocks of migrating Turkey Vultures, Broad-winged Hawks, and Swainson's Hawks. However, hawkwatches in coastal and s. Texas document birds that appear to be actually migrating.

*Fall migration:* Sep. to early Nov. with no distinct peak. Passage through s. Texas, as seen at Hazel Bazemore Co. Pk., Corpus Christi, is from early Sep. to early Nov. Peak numbers occur from late Oct. to early Nov. Over 200 have been seen at Hazel Bazemore in a single day and 1,400 in a season.

*Spring migration:* Jan.–Apr. Prolonged period of movement. Black Vultures regularly disperse to n.-cen. and sw. Oklahoma.

**Extralimital movements.**—Casual in w. and

nw. Texas from spring through fall. Records exist for the following counties: El Paso, Culbertson, Pecos, Oldham, Potter, Gray, Armstrong, Bailey, Cottle, and Foard. Rarely seen away from the Rio Grande in Presidio and Brewster Cos. A sighting in Aug. 2001 in Big Bend Ranch State Natural Area, Presidio Co., was an unusual record north of the Rio Grande.

Accidental in the following states and provinces: *California.*—Arcata, Humbolt Co., from mid-Sep. 1993 to early Feb. 1994, is the only accepted state record. A 1972 individual at Chico was considered "of questionable origin" and is currently not accepted. *New Mexico.*—First state record was at Rodeo, Hidalgo Co., in late Jul. 1996. *Colorado.*—First state record in mid-Aug. 2002, Bent Co. *Iowa.*—(1) Dallas Co. in mid-Sep. 1933; (2) Winnebago Co. in late Aug. 1959; (3) Pottawattomie Co. in late Sep. 2002. *Minnesota.*—Only state record is from Duluth in late Aug. 2001. *Kansas.*—Several unconfirmed records spanning several years in Elk, Cowley, and Chautauqua Cos. from Jul. to Sep. *Canada.*—(1) A sight record at Fort Qu' Appelle, Sask., in late May 1992; (2) a sight record in Okanagan Falls, B.C., in late Jun. 1981; (3) a photographed individual at Kluane Lake, Yukon, in early Jul. 1982.

**NESTING:** Jan.–Aug.

**Courtship (flight).**—*Pair-flight* and *tail-chasing* (*see* chapter 5).

**Courtship (perched).**—*Circle display* and *allopreening* (*see* chapter 5). Regulates inflation of neck sacs during courtship. Pair formation may last until a mate dies.

No nest is built. Nest sites are in dark protected locales, usually on the ground under dense thickets or in hollow logs or caves. Occasional nests in tree cavities or abandoned buildings. Egg-laying begins mid- to late Jan. in s. Texas; early Feb. along the Gulf Coast of Texas and Louisiana; Mar.–Apr. in Oklahoma, Arkansas, and Missouri. Two eggs, rarely 1–3, are incubated by both sexes for 37–41 days. Downy feather coat of nestlings is buff-colored. Nestlings and fledglings are fed regurgitated food by both parents. Fledge in 75–80 days and depend on parents an additional 56–84 days.

**CONSERVATION:** There are no measures taken.

**Mortality.**—Illegal shooting. Mass poisoning that occurred in the past is not legal now; however, second-hand poisoning for "varmints" probably kills many. Collision with vehicles causes some mortality.

**SIMILAR SPECIES: (1) Turkey Vulture.**—PERCHED (only juveniles are confused with Black Vultures).—Grayish head is pinkish around nares. Brownish black plumage. Long primary projection beyond tertials. FLIGHT.—Remiges are uniformly pale gray on underside, and much paler than dark underwing coverts. Wings held in a more discernible dihedral when soaring, gliding; wingbeats are slow, labored. Long tail extends far beyond feet. Like Black Vulture, also "dips" wings. **(2) Bald and Golden eagles, immatures.**—FLIGHT.—Similar dark color, but any white in wings is on inner primaries, not tip. Very slow wingbeats.

**OTHER NAMES:** BV. *Spanish:* Zopilote Negro. *French:* Urubu noir.

---

REFERENCES: AOU 1997, 1998; Arizona Breeding Bird Atlas 1993–1997; Baumgartner and Baumgartner 1992; Bylan 1998, 1999; Clark and Wheeler 2001; Dodge 1988–1997; Godfrey 1986; Griffiths 1994; Howell and Webb 1995; Jacobs and Wilson 1997; James and Neal 1986; Kaufman 1996; Kellogg 2000; Kent and Dinsmore 1996; McCaskie and San Miguel 1999; Morrison 1996; Oberholser 1974; Palmer 1988; Robbins and Easterla 1992; Russell and Monson 1998; Seibold and Helbig 1995; Smith 1996; Sutton 1967; Thompson and Ely 1989; Wood and Schnell 1984.

**BLACK VULTURE,** *Coragyps atratus:* Uncommon, local in AZ. Common in core range of U.S. and Mexico. Regularly disperses beyond mapped range in TX. Accidental in CA, CO, IA, KS, BC, and YK.

**Plate 1. Black Vulture, adult** [May] ▪ Yellow bill tip. Pale gray wrinkled skin on head and upper neck. ▪ Black plumage. White tarsi and feet. ▪ *Note:* In bright light, plumage is iridescent greenish black.

**Plate 2. Black Vulture, subadult (1st prebasic molt stage)** [May] ▪ Yellow bill tip. Juvenile-like black skin on head and upper neck with some adultlike pale gray wrinkles. ▪ Black plumage with a few old brownish black juvenile feathers. ▪ *Note:* Head and body changing from juvenile to adult features. Worn juvenile feathers among new black adult feathers.

**Plate 3. Black Vulture, juvenile (late stage)** [Nov.] ▪ Mostly yellow bill tip. Smooth black skin on head and upper neck with fuzzy brown feathers. Adultlike pale gray wrinkles on front neck. ▪ Black plumage. White tarsi and feet. ▪ *Note:* Adultlike features appear on bill and lower part of bare neck when 6–8 months old. Plumage fades to brownish black.

**Plate 4. Black Vulture, juvenile (early/mid-stages)** [May] ▪ Black bill tip. Smooth black skin on head and upper neck with fuzzy brown feathers. ▪ Black plumage. White tarsi and feet. ▪ *Note:* Recently fledged birds have wispy, buff-colored feathers on top of neck feathers and head. Downy neck feathers are lost soon after fledging.

**Plate 5. Black Vulture, adult [Feb.]** ▪ Gray head. ▪ Large white panel on tip of broad, black underwings. ▪ Short black tail. ▪ *Note:* White panel on outer 6 primaries. Soaring and banking birds fan tail widely.

**Plate 6. Black Vulture, juvenile [Nov.]** ▪ Black head. ▪ Large white panel on tip of broad, black underwings. ▪ Short black tail. ▪ *Note:* Juveniles similar to adults but have darker heads, neater plumages. Gliding birds close tail.

**Plate 7. Black Vulture, adult [Feb.]** ▪ Gray head with yellow bill tip. ▪ Large white panel on tip of upper wings.

## TURKEY VULTURE
### *(Cathartes aura)*

**AGES:** Adult, subadult, and juvenile. Adult characters are probably attained by the end of the second year. Subadult is a transitional stage that occurs during the second year. It acquires full adult plumage and gradually loses juvenile traits on the head and bill. Juvenile plumage is worn the first year. However, there is a gradual but noticeable change to more adultlike head and bill coloration that begins when about 6–8 months old.

**MOLT:** Cathartidae wing and tail molt pattern (Accipitridae; *see* chapter 4). First prebasic molt begins on the innermost primary, lower neck feathers, and back. Much of the neck molts prior to molt on other body regions. Molt can be detected on some southern-latitude-hatched juveniles in Nov. and on mid- and northern-latitude juveniles in early spring (head and bill color also slowly changes to more adultlike character during this time; *see* Juvenile Traits). Several primaries are replaced prior to the start of molt on the secondaries. Rectrix molt begins after remix molt is well under way. In mid-latitude birds, p1–5 may be replaced and rectrix molt just beginning by early May.

Subsequent prebasic molts on remiges and rectrices may occur in any month of the year, but only a minimal amount of molt occurs in winter months (Kirk and Mossman 1998). Remix molt is often serially descendent, with two or three molt centers molting at the same time along the molt waves. This molt strategy is typical of large birds that cannot replace all flight feathers in 1 year. In early May, mid- and northern-latitude adults may have replaced p1–5 but few if any secondaries; by mid-Sep., p10 may still be molting. Body molt is complete in the first prebasic molt and probably fairly complete in subsequent molts.

**SUBSPECIES:** Polytypic with three races in North America; however, races are often difficult to distinguish in the field. Subspecies ranges and adult plumage distinctions are based on Kirk and Mossman (1998). Nominate race, *C. a. aura*, breeds in the Southwest and Greater Antilles. *C. a. meridionalis* breeds north of *aura* and west of *C. a. septentrionalis*. Some taxonomists consider *meridionalis* part of *aura*. The two subspecies have similar plumage and head features and are often distinguishable only by the length of in-hand-measured wing bones (Rea 1998). *Septentrionalis* is separable from *aura/meridionalis* only in adults.

Three additional subspecies reside from Central America to s. South America: (1) *C. a. ruficollis* inhabits the lowlands from s. Costa Rica to n. Argentina, including areas east of the Andes in South America; (2) *C. a. jota* is in the highlands of cen. and s. Colombia and south to s. Argentina; and (3) *C. a. falklandica* is found in the Pacific Coast region from Ecuador south to s. Chile, including the Falkland Islands.

**COLOR MORPHS:** None.

**SIZE:** A large raptorlike bird. Males average slightly larger than females, but sexes are not distinguishable in the field. *C. a. septentrionalis* averages the largest, but there is some overlap with mid-sized *meridionalis. C. a. aura* averages smallest but overlaps with *meridionalis* (there is no overlap in wing bone lengths between these two races; Rea 1998). The following measurements encompass ranges for the three subspecies. Length: 24–28 in. (61–71 cm); wingspan: 63–71 in. (160–180 cm).

**SPECIES TRAITS:** HEAD.—Bare skin on the head and upper neck. Neck skin can be distended or retracted depending on temperature and mood. Bare skin area of upper neck is long when distended but concealed by lower neck (ruff) feathers when retracted. Nape skin is smooth when the upper neck skin is distended but is wrinkled when retracted. **Short, thick bill. Large, see-through naris opening.** Medium brown irises. BODY.—Blackish plumage. **Tarsi are reddish but often variably pinkish or whitish from urohydrosis:** (1) **only the upper front portion of the tarsi is reddish and rest of tarsi and feet is white,** (2) **all of tarsi is reddish and only feet are white, or** (3) **all of tarsi and feet is clean and reddish (mainly after being cleansed by walking in moist areas).** Bare-skinned crop becomes distended after extensive feeding. In flight, toes extend to the rear border of the undertail coverts. WINGS.—**In flight, wings are long with nearly parallel front and trailing edges.**

**There are six distinct "fingers" on the outer primaries. Underside of remiges is uniformly pale gray and contrasts sharply with the dark gray greater coverts and blackish brown underwing coverts.** Upperwing coverts have broad, pale brownish edges. Primaries have yellowish white quills on both ventral and dorsal surfaces. Long primary projection beyond the tertials when perched. TAIL.—**Fairly long and wedge-shaped. The outer rectrices are sequentially shorter than the deck set. Ventral surface is uniformly gray and darker than the underside of the remiges.** *Note:* Turkey Vultures regularly have white excrement splattered on dorsal areas of the body. Since this species often roosts communally (*see* Habits), higher perched birds may excrete onto lower perched birds.

**ADULT TRAITS:** HEAD.—**White (ivory) bill. Medium or bright red with a purplish hindneck.** Variably dense, short black hairlike feathers adorn the crown, nape, and hindneck: some birds are sparsely covered, others quite fuzzy. **Lower neck is covered with iridescent, glossy black feathers. White tubercles may be extensive, moderate, or lacking depending on subspecies and geographic region (*see below*).** *Note on tubercles.*—As a rule, the farther west, the fewer the tubercles (also generally noted by Kirk and Mossman 1998). (1) *Very extensive tubercles:* Large mass of white bumps on the lores, under and above the eyes, and on the crown. This type has not been documented on *septentrionalis* in western areas of the East but is occasionally seen on more eastern birds; not found on other races. (2) *Extensive tubercles:* A large cluster of white bumps on the lores and under the eyes. A few black hairlike tufts may grow between the white bumps. (3) *Moderate tubercles:* A small linear group of small white bumps on the lores and under the eyes. A moderate amount of black hairlike feathers on lores surrounding tubercles. (4) *No tubercles:* The lores have a large black hairlike feather patch and lack white tubercle bumps; no tubercles under the eyes. BODY.—**Blackish brown. The body is more brownish than the glossy black of the lower neck.** Some blackish iridescence on the central portion of the back and scapular feathers. Scapular are variably edged with pale brown. Non-iridescent blackish brown breast, belly, flanks, leg feathers, and undertail coverts. WINGS (dorsal).—Wing coverts are edged with pale brown. Secondaries and tertials are narrowly edged with pale brown. The inner portion of some median and lesser coverts and most greater coverts has an iridescent quality, particularly in fresh plumage. In fresh plumage, new secondaries are also partially iridescent. WINGS (ventral).—Remiges are pale gray. TAIL.—Blackish brown on the dorsal surface and medium gray on the ventral surface.

**ADULT "EASTERN" (*C. a. septentrionalis*):** HEAD.—Has mainly extensive tubercles. A few may have very extensive tubercles. Feather edging on scapulars is very pale and quite broad. WINGS.—Broad, very pale brown edges on all lesser upperwing coverts and secondaries.

**ADULT "WESTERN/SOUTHWESTERN" (*C. a. meridionalis/C. a. aura*):** HEAD.—Extensive tubercles on many individuals in eastern regions. Birds of far western regions typically lack tubercles or have moderate tubercles. Very rarely do far western birds have extensive tubercle, and very extensive tubercle types do not occur. Moderate or extensive tubercle are fairly common on *meridionalis* of eastern areas of the West (e.g., Minnesota), on the Great Plains, and in the Rocky Mts. of Colorado and Wyoming. Only a few Plains and Rocky Mts. adults lack tubercles. *C. a. aura* of s. Arizona typically lack tubercles. Of 186 spring migrant and resident adult vultures in Kern Co., Calif., 40 had moderate tubercles, 1 individual had extensive tubercles, and 145 had none (J. Schmitt unpubl. data). Resident *aura* of s. Texas vary in the amount of tubercles: some lack them whereas others have an extensive amount (as much as any *septentrionalis;* plumage is also similar). BODY.—As in Adult Traits, but brown feather edgings are not as broad or as pale as on *septentrionalis.* WINGS.—Pale brown edging on upperwing coverts is not as defined or as pale as on *septentrionalis;* however, this distinction is not readily visible. *Note:* Plumage and size differences between *meridionalis/aura* and *septentrionalis* are indeed very subtle and often not field-separable, even by skilled observers.

**BASIC (SUBADULT) TRAITS (LATE STAGE [all races]):** HEAD.—**Small dusky area on tip of white (ivory) bill on the upper mandible; the lower mandible is white.** The dusky bill tip probably disappears by the end of the second year; it is the remnant dark area of the largely dark bill of juvenile age class. **The head gradually acquires the bright red as on adults.** White tubercles are partially or fully formed on individuals and races that possess them as adults. The lower neck feathers are glossy, iridescent black as on adults. BODY, WINGS, and TAIL.—As on adults. *Note:* This age class constitutes birds that are 1.5 to 2 years old that have lost most juvenile bill markings and gained full or nearly full adult head features and full adult plumage.

**BASIC (SUBADULT) TRAITS (EARLY STAGE [all races]):** This is basically a late-stage juvenile that has begun to show adult feathering on the neck and exhibits some wing molt. As bill and head color continues to change and body continues to molt, it gradually changes into a late-stage subadult. HEAD.—**Dark tip on the distal area of the bill is reduced to less than one-half or one-third of the upper mandible; the lower mandible is white. Medium or dark pink head gradually turns more reddish but still retains a varying amount of fuzzy juvenile feathering on the nape and hindneck.** Lower neck molts into iridescent black feathering as on adults. BODY.—Faded brownish body contrasts with the new dark black lower neck feathers. Additional molt produces the darker adult plumage, with broad pale brown edges on all feathers (but still retains the darker black neck). The crop is pale grayish or grayish pink. *Note:* This is the transitional molt stage and change in head features from juvenile to subadult that occurs late in the first year and into the early part of the second year.

**JUVENILE TRAITS (LATE STAGE [all races]):** HEAD.—**Bill is blackish or dark brownish on the distal half and whitish on basal half of both mandibles. The head is uniformly medium pink and retains the thick growth of a grayish brown downy feather coat on the crown, nape, and hindneck from early stage.** *Note:* There is a gradual transition from early-stage head color to late stage (*see below*). Lower neck feathers are blackish brown with little or no iridescent quality. BODY.—Much of the iridescence on the back and scapulars wears off and the upperparts appear more brownish and not as blackish as in early stage. All of the body, including lower neck, is uniformly blackish brown. Crop is pale grayish. WINGS.—The early stage narrow white edgings on the greater, median, and first row of lesser upperwing coverts has worn off. However, the neat, same-age, pale brown edged appearance on the upperwing coverts is retained. Underside of remiges fades to pale gray and is similar to the color on adults. TAIL.—Underside fades to medium gray. *Note:* This age distinction pertains to birds older than 6–8 months, when more adultlike head and bill colors and markings become noticeable.

**JUVENILE TRAITS (EARLY STAGE [all races]):** HEAD.—**Bill is blackish on the distal two-thirds and whitish on the basal one-third on both mandibles. The naris opening and distal one-half of the lower mandible are pale pink or pale pinkish gray and gradually turn medium pink with increasing age. The rest of head is smooth-skinned and medium gray. The crown, nape, and hindneck are covered with thick, fuzzy, grayish brown downy feathers. The lores have a black hairlike feather patch. A pinkish color gradually develops around the eye and ear regions with increasing age. Lower neck feathers are brownish black with only a minimal or moderate iridescent quality.** *Note:* In recently fledged birds, wispy, white downy feathers adorn the top of the lower neck feathers and, sparsely, the nape and crown. BODY.—Uniformly brownish black, including the lower neck, with neat, pale brown edges on all back and scapular feathers. Dorsal feathers have minimal or moderate amount of greenish, bluish, or purplish iridescent quality. Ventral areas of the body are non-iridescent brownish black. WINGS (dorsal).—All coverts are neatly edged with pale brown. The greater, median, and first row of lesser coverts have thin white edges. WINGS (ventral).—**Remiges are medium pale gray but, being same age and new, are slightly darker than the irregular-age remiges of adults.** TAIL.—Medium dark gray on ventral surface and slightly darker than on adults. *Note:* This age distinction is from fledging to about 6–8 months.

**ABNORMAL PLUMAGES:** Although rare, various degrees of albinism are more common in Turkey Vultures than in any other raptorlike bird except Red-tailed Hawks (Kirk and Mossman 1998; BKW pers. obs.). Numerous records exist of all-white birds. Incomplete or partial albinos, with scattered white feathers, are reported regularly. Imperfect albinos are also fairly regular: head can be the normal red, but irises may be whitish or bluish; body feathers may be a moderate dark brownish with very pale edgings; remiges and rectrices are often tan or whitish. *Note:* An imperfect albino is depicted in plate 23; other types are not shown.

**HABITAT: Summer.**—Inhabits a vast array of geographic and topographic regions. Range varies from subtropical to northern temperate zones in both humid and arid climates. Inhabits rural and undisturbed areas that may be densely wooded, semi-open, or open. Extensive agricultural regions are rarely inhabited. Topography can be flat, hilly, or moderate montane elevations. High montane elevations are occasionally inhabited. Areas may have dense undergrowths, hollow logs, hollow stumps, rocky outcrops, or cliffs for suitable nest sites.

**Winter.**—Habitat can be similar to summer areas, but at lower elevations and more southern latitudes. There is no requirement for dense vegetation, rock slides, or cliffs in this season. Rural and extensive agricultural areas are inhabited more extensively than in the summer. In the winter, Turkey Vultures inhabit southern temperate and tropical regions, generally at elevations and latitudes below regular snowfall.

**Migration.**—Similar to summer and winter habitat.

**HABITS:** Tame to fairly tame in the South, especially in s. Texas; moderately wary or very wary at mid- and northern latitudes. Very wary when feeding. Solitary in breeding season. Solitary or gregarious in other seasons. Night roosts may number in the hundreds or, in southern latitudes, in the thousands. Where ranges seasonally overlap, often roosts with Black Vultures. Exposed branches, buildings, utility poles (but not wires), towers, and other elevated structures are used for perches. A very terrestrial species. Sunning occurs regularly, especially in mornings prior to daily flight and after rain. Body temperature is regulated by urohydrosis. Vultures commonly sit upright on haunches (heels) or lie down when perched on thick branches or on the ground.

**FEEDING:** An aerial scavenger. Food sources are detected by sight or smell. Has a highly developed olfactory system and is the only North American vulture species that can smell. Prefers eating fresh carrion but if a fresh source is not available will eat decaying carrion. Vegetable matter is rarely eaten, but may eat decaying pumpkins. Feeds on an array of carrion, including small and large mammals, birds, reptiles, amphibians, and stranded fish. More likely to feed on smaller carrion than Black Vultures. Turkey Vultures approach carrion warily. Vultures land far away from the food source, then timidly walk up to feed, but are always poised to fly if danger arises. Small prey objects may be carried short distances in the bill. Open-pit garbage dumps are common feeding areas. Turkey Vultures are submissive to the smaller, more aggressive, dominant Black Vulture. Turkey Vultures rarely kill small prey.

**FLIGHT:** A highly aerial species. Wings are held in a high dihedral when soaring and in a high or modified dihedral when gliding. In moderate or strong winds, the energy-efficient back-and-forth rocking motion is the most common flight mode. Flight is steady in calm winds. Powered flight is used to gain altitude quickly in inclement weather or to launch into flight. Flight may be at low altitudes, barely skimming the ground or treetops, or, particularly when migrating, at very high altitudes. Turkey Vultures do not hover or kite. Wings are frequently "dipped" when flying to maintain speed and stability. On windy days, migrants may begin daily flights before dawn.

**VOICE:** Technically voiceless, but grunts and hisses when agitated.

**STATUS AND DISTRIBUTION:** *Very common.* Possibly the most common raptor-type bird. Fall counts of migrants passing through Veracruz, Mexico, tally extraordinary numbers: 1.7 million in 1998, 1.5 million in 1999, 1.8 million in 2000, 2 million in 2001, and 2.6 million in 2002. These migrants come mainly from the w. and cen. U.S. and s. Canada. Some undoubtedly come from western portions of the East (a juvenile banded in Wisconsin was found in Belize in Central America). There are thousands of vultures that winter in the southern areas of the w. U.S. and n. Mexico. Large numbers also breed and winter in the East and winter in southern regions of the East. Total population may well exceed 3 million birds. *Note:* Since the three subspecies are difficult, if not impossible, to separate under field conditions, the range map does not reflect each race's distribution.

**Summer.**—In the West, *C. a. septentrionalis* breeds from Minnesota south to e. Texas. *C. a. meridionalis* breeds west of *septentrionalis* from British Columbia east to Manitoba, south to s.-cen. Texas to n. New Mexico and west through Arizona, and from s. Nevada to cen. California. *C. a. aura* breeds south of *meridionalis* to Costa Rica. *C. a. aura* is also found on various islands of the Caribbean: the Bahama Islands (Grand Bahama, Great Abaco, Andros), Cuba, Jamaica, Isla de la Juventud, Hispaniola, and sw. Puerto Rico. *Aura* was introduced on some of these Islands.

Range in Washington, n. Idaho, n. Montana, n. and e. North Dakota, e. South Dakota, e. Colorado, w. Kansas, s. Minnesota, and n. Iowa, is irregular and disjunct. Turkey Vultures are absent or rare from extensive portions in many of these states because of montane elevations, arid plains, or extensive agriculture. Turkey Vultures were naturally absent from many prairie regions because of lack of suitable nesting habitat. The agricultural conversion of many historically inhabited prairie regions, where breeding habitat was available, has reduced the species' range even more.

Range reduction from historical times has taken place in Alberta (formerly about 100 pairs), Saskatchewan, and North Dakota. Range has expanded, however, in Washington, e. Colorado, Nebraska, Kansas, n. Iowa, South Dakota, and s. Minnesota. Vultures are still rare in the Columbia Basin in cen. Washington because of the lack of ample food and nesting habitat. Turkey Vultures are found throughout all of Nebraska in the summer but are most common in western, northern, and eastern regions of the state. In Minnesota, expansion has occurred in recent decades in several counties: Big Stone, Chisago, Kandiyohi, Lincoln, Pine, Pipestone, Pope, Stearns, and Traverse. Areas along the Minnesota River have seen a dramatic population increase. An increase along riparian corridors due to the amount of mature wooded areas, which creates fallen logs for nest sites, and less intensive agriculture have assisted range expansion.

Generally absent from high montane elevations. In sw. Colorado, however, breeds up to 12,000 ft. (3,700 m) in Montezuma and La Plata Cos. Occasionally seen in summer at similar elevations in Huerfano Co., Colo.

**Winter.**—*C. a. aura* winters south of the U.S. from Mexico to Panama. *C. a. meridionalis* winters from w. and cen. California, s. Arizona, n. Mexico, cen. and e. Texas, and e. Oklahoma southward to s. Brazil and Paraguay. A few may winter from w.-cen. to s. Vancouver, Island, B.C. *C. a. septentrionalis* winters from s. Missouri and e. Oklahoma south through s. Texas and into Central America. Rare in mid-winter in sw. and n.-cen. Colorado, Iowa, and sw. Oregon. Extremely rare in winter in s. Saskatchewan.

**Movements.**—Northern populations are quite migratory and depart Canada and most of the northern and central latitudes of the w. U.S. in winter. Turkey Vultures migrate in small or large flocks. By the time they reach southern latitudes, flocks can comprise hundreds or thousands of individuals.

*Fall migration:* In s. Canada and n. U.S., begins in mid-Aug., peaks in late Sep. or early Oct., and ends abruptly in late Oct. In the Great Basin, begins in late Aug., peaks in late Sep., and ends in mid-Oct. In the Pacific Northwest, occurs from mid-Sep. to mid-Oct., and peaks in late Sep. or early Oct. In s. Texas, migrants first arrive in mid-Sep., peak in late Oct., and continue moving until late Nov. and possibly Dec. Migration is very prolonged at southern latitudes. In Veracruz, Mexico, migrants arrive in mid-Sep., peak in mid- to late Oct., and continue moving until late Nov. and probably into Dec. Migrants often join flocks of Broad-winged and Swainson's hawks. Peak migratory periods of Turkey Vulture flights are nearly synonymous with Swainson's Hawk peak flights.

*Spring migration:* In s. Texas, begins in early Mar., possibly peaks in late Mar., and continues through early May. In the cen. Rocky Mts., begins in early to mid-Mar., peaks in early Apr., and continues at least to mid-May and possibly through May.

**NESTING:** Begins in Feb. in southern latitudes and Apr.–May in northern regions. Nesting is completed as late as early Sep. in northern ar-

**TURKEY VULTURE,** *Cathartes aura:* Very common. Expanding northward in all areas. Subspecies' ranges not depicted.

eas. Pairs stay together as long as mate survives. Age of first breeding is unknown but probably not until a few years old.

**Courtship (flight).**—*Follow-flight,* which may include stints of *tail-chasing,* may last up to 3 hours.

**Courtship (perched).**—*Circle display* with inflated throat and neck sacs. Birds may utter low-pitched groaning sounds.

No nest is built. Eggs are laid on the ground. Vultures do not prepare a nest site. Nest sites are in dark protected areas away from human disturbance. Primary sites are in rocky outcrops with caves, in crevices, under rock ledges, or fallen stumps, in hollow stumps, under thick brush piles, or dense undergrowths, on the floor of abandoned buildings, and rarely in elevated nests of other large birds. Nest sites are often within dense woodlands. Sites may be used annually for several years.

Two eggs, rarely 1–3, are incubated by both sexes for 38–40 days. Youngsters fledge in about 77 days but depend on parents for several more weeks. Nestlings are fed regurgitated, digested foods by both parents. Nestlings are covered with white downy feathers.

**CONSERVATION:** No measures are taken. A stable, adaptive, and thriving species.

**Mortality.**—Illegal shooting, lead poisoning from ingested bullet fragments and shot pellets in carcasses of game mammals and game birds that are not retrieved by hunters, ingested poisons meant for varmints, collisions with power lines and vehicles, and electrocution from wires on utility poles.

**SIMILAR SPECIES:** **(1) Black Vulture.**—PERCHED (confused only with juvenile Turkey Vulture).—Adults have pale or medium gray, very wrinkled fleshy heads. Juveniles have a smooth black head and upper neck. Tarsi may be a similar whitish but are never pinkish. Primary projection is short. FLIGHT.—White panels on the outer primaries. White toes extend to the tail tip. Short, square black tail. **(2) Rough-legged Hawk, dark morph juveniles.**—FLIGHT.—Undersides of remiges and rectrices can appear uniformly pale with a two-toned effect; however, the trailing edge has a dusky band. **(3) Golden Eagle.**—PERCHED.—Yellow cere. Pale tawny nape. FLIGHT (ages/individuals with all-dark underwings).—Adults and subadults have a fairly distinct dark trailing edge on the underside of the remiges. Underwings of juveniles are a similar uniform medium gray and are also two-toned. Both species have six "fingers" on outer primaries and a wedge-shaped tail. Head and bill are moderately large on eagles. Eagles may rock back and forth in strong winds but are not as tipsy as most vultures. Trailing edge of wing bows discernibly outward. Dihedral of wing attitude is not as pronounced.

**OTHER NAMES:** TV. *Spanish:* Aura Cabecirroja, Zopilote Aura. *French*: Urubu à Tête Rouge.

---

REFERENCES: Alcorn 1988; Andrews and Righter 1992; AOU 1997; Baicich and Harrison 1997; Baumgartner and Baumgartner 1992; Busby and Zimmerman 2001; Bylan 1998, 1999; Dinsmore 1996, 1997; Dodge 1988–1997; Gilligan et al. 1994; Godfrey 1986; Jacobs and Wilson 1997; Janssen 1987; Kellogg 2000; Kent and Dinsmore 1996; Kingery 1998; Kirk and Mossman 1998; Montana Bird Distribution Committee 1996; North American Bird Information web site 2003; Palmer 1988; Peterson 1995; Rea 1998; Seibold and Helbig 1995; Semenchuk 1992; Sharpe et al. 2001; Smith 1996; Smith et al. 2002; Stewart 1975; Wheeler and Clark 1995.

**Plate 8. Turkey Vulture, adult (extensive tubercle type) "Eastern" (*C. a. septentrionalis*) [Feb.]** ▪ White bill. Bare red head and upper neck; large cluster of white tubercles in front of and below eyes. ▪ Glossy black neck feathers. ▪ Distinct pale edges on brownish black wings. ▪ Reddish tarsi. ▪ *Note:* Average amount of white tubercles.

**Plate 9. Turkey Vulture, adult (moderate tubercle type) "Western" or "Eastern" (*C. a. meridionalis*/or *C. a. septentrionalis*) [Jun.]** ▪ White bill. Bare red head and upper neck; small cluster of white tubercles in front of and below eyes. ▪ Glossy black neck feathers. ▪ Distinct pale feather edges on brownish black wings. ▪ Mostly white tarsi. ▪ *Note:* Large amount of tubercles for *meridionalis* and small amount for *septentrionalis.*

**Plate 10. Turkey Vulture, adult (no tubercle type), "Western/Southwestern" (*C. a. meridionalis*/ *C. a. aura*) [Feb.]** ▪ White bill. Bare red head and upper neck; lacks white tubercles. ▪ Glossy black neck feathers. ▪ Indistinct feather edges on brownish black wings. ▪ *Note:* W. North American migrant wintering in Nayarit, Mexico.

**Plate 11. Turkey Vulture, adult "Eastern" (*C. a. septentrionalis*) [Dec.]** ▪ Sunning posture: wings spread with back facing sun. Common posture in morning and after rain.

**Plate 12. Turkey Vulture, adult "Eastern" (*C. a. septentrionalis*) [Dec.]** ▪ White bill. Bare red head. ▪ Uniformly pale gray remiges contrast with black underwing coverts and body. ▪ Uniformly medium gray tail.

**Plate 13. Turkey Vulture, adult "Eastern" (*C. a. septentrionalis*) [Dec.]** ▪ White bill. Bare red head. ▪ Brownish upperwing coverts.

**Plate 14. Turkey Vulture, adult "Eastern" (*C. a. septentrionalis*) [Dec.]** ▪ Wings held in high dihedral when gliding and soaring.

**Plate 15. Turkey Vulture, subadult (late stage) [Dec.]** ▪ White bill with dark tip. Bare red head and neck. ▪ Glossy black neck feathers. ▪ Distinct pale edges on brownish black wings. ▪ *Note:* Identical to adult except bill tip darker. All races similar.

**Plate 16. Turkey Vulture, subadult (early stage)** [**Dec.**] ▪ White bill with dark tip. Bare pink head. ▪ Black adult feathers growing on neck and back. Brownish plumage. ▪ Molting (missing) inner two primaries (p1 and 2). ▪ *Note:* As on late-stage juvenile except molting adult feathers. Much of body and wings molts before head color changes to more adultlike character. All races similar.

**Plate 17. Turkey Vulture, subadult (early stage)** [**Dec.**] ▪ White bill with dark tip. Bare pink head. ▪ Brownish black plumage with new black adult feathers on neck. ▪ Uniformly pale gray remiges. Molting (missing) inner two primaries (p1 and 2). ▪ *Note:* All races similar.

**Plate 18. Turkey Vulture, subadult (early stage)** [**Dec.**] ▪ Engaging in "wing dip" motion. ▪ *Note:* Wings often dipped in downward motion when gliding to increase speed and maintain stability.

**Plate 19. Turkey Vulture, juvenile (late stage)** [**Dec.**] ▪ Outer half of bill dark, inner half white. Medium pink head and neck; fuzzy brown downy feathers on crown and hindneck. ▪ Uniformly brownish black plumage. ▪ *Note:* 6–8 months old with pinkish head; plumage full juvenile except white tips on upperwing coverts worn off. All races similar.

**Plate 20. Turkey Vulture, juvenile** (early stage) [**Dec.**] ▪ Outer 2/3 of bill dark, inner 1/3 white. Fuzzy brown downy feathers on pale pinkish gray head and neck. ▪ Uniformly brownish black plumage with narrow white tips on wing coverts. ▪ *Note:* This bird is younger than bird on plate 19 but older than bird on plate 21. All races similar.

**Plate 21. Turkey Vulture, juvenile** (early stage) [**Aug.**] ▪ Outer 2/3 of bill dark, inner 1/3 white. Gray fuzzy downy feathers on head and neck with pale pink nares and outer half of lower mandible. ▪ Uniformly brownish black plumage with narrow white tips on wing coverts. ▪ *Note:* Recently fledged and about 4 months old. All races similar.

**Plate 22. Turkey Vulture, juvenile** (early stage) [**Oct.**] ▪ Dark bill and gray head. ▪ Brownish black plumage. ▪ Uniformly medium gray remiges contrast with dark underwing coverts. ▪ *Note:* Underside of remiges is darker than on older birds. All races similar.

**Plate 23. Turkey Vulture, adult (imperfect albino)** [**Aug.**] ▪ White bill and reddish head. Pale iris. ▪ Blackish neck feathers. Washed out color on rest of body. ▪ *Note:* Turkey Vultures rarely exhibit albinism.

# CALIFORNIA CONDOR
## *(Gymnogyps californianus)*

**AGES:** Adult, juvenile, and five slowly changing interim subadult plumages. Juvenile plumage is retained for the first year.

*Note:* Age and molt data are based on 18 known-aged individuals studied over a 7-year period (J. Schmitt unpubl. data) and personal study of photographs of 14 known-aged released Condors. Head and neck color and amount of white on the axillaries and underwing coverts vary considerably within each subadult age class. Gradual color change occurs in fleshy areas on the head and neck with advancing age. Age progression is also noticeable with molt on remiges and, to a lesser extent, as condor attains more white on the axillaries and underwing coverts.

Subsequent age class characters are attained at different rates. This slow and highly variable rate of individual transition into older age classes is typical of species that take several years to achieve adult plumage.

All photographs are of known-aged individuals. The hatching date and photograph date of depicted individuals are given at the end of each age class description.

**MOLT:** Cathartidae wing and tail molt pattern (*see* chapter 4). Prebasic molts occur on a partial basis each year beginning at about 1 year of age. All ages beyond the juvenile plumage have multi-aged feathering on the body, remiges, and rectrices. Molt occurs Feb.–Dec. When Condor is in flight, age can often be determined by sequence of remix molt for at least the first 3 years. Newly acquired secondaries, tertials, and respective greater coverts have a pale gray sheen on the dorsal surface. Over time and before the next annual molt, the pale sheen gradually wears off and the feathers become dark gray or blackish.

**SUBSPECIES:** Monotypic.

**COLOR MORPHS:** None.

**SIZE:** An extremely large raptorlike bird. Length: 43–50 in. (109–127 cm); wingspan: 98–118 in. (249–300 cm).

**SPECIES TRAITS:** HEAD.—Throat and neck regions of all ages can be voluntarily inflated to be very puffy. Lower neck feathers create a "ruff" of spikelike feathers. The ruff can be raised or lowered at will; if raised, it conceals all of the bare upper neck skin and even the nape. In flight, the loose neck skin often overlaps the nape as a fold. BODY.—Blackish. Whitish tarsi, which are actually pink or reddish, but turn whitish or grayish from urohydrosis. Fleshy, pale skin of the crop shows on the breast and typically becomes distended after extensive feeding. WINGS.—Very long with all-black remiges on the underside. **Whitish axillaries and underwing coverts form a long, narrow triangle-shaped region on the underwing.** TAIL.—**Black, short, and square-edged.**

*Note:* All birds currently in the wild have highly visible large plastic wing markers with bold numerical and/or alphabetical designations attached on dorsal and ventral surfaces of the patagial area of one or both wings. Patagial markers are visible at great distances. On perched birds they are visible only on the dorsal surface. On flying birds, they are visible on both dorsal and ventral surfaces. A telemetry antenna is attached to the dorsal side of each patagial marker. The patagial markers are white with black code numbers, black with white numbers, or colored and have white numbers. Some birds may have a combination of colored markers, but most have same-marker tag. Those released by Ventana Wilderness Society: blue in Jan. 1998, yellow in Jan. 1999, orange in Feb. 2000, white in Apr. 2001.

**ADULT TRAITS:** HEAD.—**Pale bluish or grayish bill is black on the basal region. Yellowish orange head has a black hairlike feathered patch on the forehead, lores, and below the eyes.** There is a reddish area on the lower front area of the pink neck skin. **Red irises and sclera.** BODY.—Black. Reddish crop. WINGS.—**Immaculate white axillaries and underwing coverts form a long, narrow, white triangle extending onto the carpal region of the underwing. Broad white bar on the upperwing's inner portion of the secondary greater coverts.** Inner secondaries and tertials have a very defined white line on the outer edge of each feather. Upperwing surface of tertials, secondaries, and greater coverts is very pale gray with multi-ages of feathers: some very pale gray, some dark gray.

*Note:* Six years and older.

**Photographs:** Condor #11, 7.5 years old (hatched May 12, 1994, photographed Dec. 12, 2001) is a classic example of a full adult with pure white underwing region and orange head. Condor #2, 6 years old (hatched Mar. 28, 1994, photographed Apr. 7, 2000), is not yet showing the sharply defined, white-edged secondaries and greater coverts. This individual is also an example of the raised neck ruff concealing bare skin of the upper neck. Condor #5, at 5 years and 5 months old (*see* Basic V), is also adultlike.

**BASIC V (SUBADULT V) TRAITS:** HEAD AND BODY.—Identical to adult in most head and body characters. Generally not separable from an adult at many angles of view or at long distances. HEAD.–**Yellowish orange. Red irises and sclera.** BODY.—Black. Reddish crop. WINGS.—**Axillaries and underwing coverts can be immaculate white or retain a small amount of dusky color on the shaft area of each feather. The white feathers on the carpal area also have some dark centers.** Pale bar on the upperwing's inner portion of the greater secondary coverts is fairly narrow and dusky-white. Newly molted feathers on the greater secondary coverts attain the wide white bar of a full adult. Tertials, secondaries, and greater coverts are multi-aged, and upper surfaces are a mix of very pale gray and darker gray, including the outer edges of the inner secondaries and tertials.

*Note:* Five to 6 years old; can be labeled as an adult.

**Photographs:** Condor #5, 5 years and 5 months old (hatched Apr. 18, 1994; photographed Sep. 19, 1999), is a classic example of this age class, but could also pass for a full 6-year-old adult (*see above*).

**BASIC IV (SUBADULT IV) TRAITS:** HEAD.—Grayish bill is streaked with black or can be mostly dark. *Late stage:* **Orangish yellow head is similar to adult's but duller.** Some individuals have sparse, dark feather tufts on the crown and nape. Others have pronounced dark freckling on the crown and around the eyes that creates a partial dark mask. There is a considerable amount of individual variation in head and neck color in this age class. Hindneck and upper front area of the neck area often have dark brownish feather tufts that create a dark spot on the throat junction. *Early stage:* Slow-changing individuals may retain yellow-gray freckling around the gape, on the sides of the head, and on the nape. The neck is bright pink and resembles late-stage Basic III. Brownish red or red irises with a red sclera. BODY.—Black. Pinkish crop. WINGS.—**Axillaries and underwing coverts can be mostly white with small amount of dark only on the central shaft area of each feather.** Early stage individuals may have a rather extensive amount of dusky on the central area of each feather, as on juveniles and young subadults. Underwing coverts have dusky, dark-centered feathers. The pale bar on the inner portion of upperwing's greater secondary coverts is dusky and fairly wide on newly molted feathers, but eventually wears off to a narrow linear strip. The tertials, secondaries, and greater coverts are multi-aged pale and dark gray, including edges of inner secondaries and tertials. There is considerable variation in amount of white on the underwing region in this age class.

*Note:* Four to 5 years old.

**Photographs:** Condor #14, 4 years and 3 months old (hatched Apr. 7, 1995; photographed Jul. 1999), is an early-stage; Condor #7, 4 years and 10 months old (hatched Apr. 29, 1994; photographed Feb. 27, 1999), and Condor #8, 4 years and 11 months old (hatched Mar. 28, 1994; photographed Feb. 27, 1999), are typical of late-stage types of this age class.

**BASIC III (SUBADULT III) TRAITS:** HEAD.—Gray or dark bluish gray bill with some black streaking. Reddish brown irises with a red sclera. *Late stage:* **Distinct transition to adult-like character with a dull yellow or orangish yellow head and a dark pink neck. Extensive black freckling surrounds the eyes and is on the cheeks, creating a dark masked appearance. Brownish downy feathers on the hindneck and front area of the upper neck create a dark area at the throat junction.** *Early stage:* **Dark grayish with dull yellow or orangish yellow freckles around the gape, cere, and cheeks. Sometimes all of the face is dark grayish and lacks yellowish freckles. Pinkish or pinkish gray neck has brownish, fuzzy downy feathers on the head, hindneck, and throat. Pinkish region is hidden when the neck ruff is raised.** BODY.—Black. Pinkish gray crop.

WINGS.—**Whitish axillaries and underwing coverts have a considerable amount of dusky color on the central portion of each feather.** Pale bar on upperwing's inner portion of greater secondary coverts is fairly wide on newly molted feathers, but becomes narrower with feather wear. Multi-aged remiges. Upper surface of new tertials, secondaries, and greater coverts has a pale gray sheen, including edges of the inner secondaries and tertials. Only a few juvenile remiges are retained. Dorsal surface of remiges is dark gray because the pale grayish sheen is worn off.

*Note:* Three to 4 years old.

**Photographs:** Condor #25, 3 years and 9 months old (hatched Jun. 2, 1995; photographed Feb. 27, 1999), depicts a late-stage category. Condor #56, 3 years old (hatched Apr. 7, 1997; photographed Apr. 7, 2000), and Condor #55, 3 years old (hatched Apr. 8, 1997; photographed Apr. 7, 2000), are examples of early-stage types.

**BASIC II (SUBADULT II) TRAITS:** HEAD.—**Dark gray bill has irregular black streaking. Head is dark gray with an extensive amount of brownish downy feathering on throat, nape, hindneck, and front and sides of the neck. Yellow speckling adorns the gape and nares. The crown, sides of head, and throat have very sparse yellow speckling. Neck is dull pink; however, it is often concealed by the raised neck ruff and contracted upper neck skin when perched and in flight.** Brown irises with a red sclera. BODY.—Blackish brown. Grayish or pinkish gray crop. WINGS.—**Axillaries typically have a large dusky area on the central region of each feather with whitish on the outer edges. Underwing coverts are dusky with dark-centered feathers.** Numerous, multi-aged remiges. This is the first age class with significant wing molt. A large number of juvenile remiges have been molted. Trailing edge of remiges has an irregular, serrated border because of molt that occurs from spring through fall. Pale dusky bar on the upperwing's inner portion of greater secondary coverts is fairly wide, but becomes narrower with feather wear. Upper surface of the new tertials, secondaries, and greater upperwing coverts has a pale gray sheen (paler than the retained juvenile-aged feathers); outer edges of the inner secondaries and tertials are similarly colored.

*Note:* Two to 3 years old. Often called "ring-necked age" because of the distinction of the dark head and pale, pinkish neck.

**Photographs:** Not depicted. Condors #55 and #56 (*see* Basic III), which have just turned 3 years old, are classic examples of this age class.

**BASIC I (SUBADULT I) TRAITS:** HEAD.—**Blackish bill. Head and neck are similar to a juvenile. Head and neck are uniformly dark gray or blackish and densely covered with brownish, fuzzy downy feathers except on the forehead and sides of head from the nares and jaws to the ear holes. Downy feathering is sparse on the front and sides of the neck. Brown irises have a red sclera.** BODY.—Blackish brown. Grayish crop. WINGS.—**On underwings, axillaries have distinct dusky centers. There is a mottled grayish or whitish linear formation on the underwing coverts, but the carpal region is fairly dark.** Axillaries can appear quite dark on some, with whitish only on the very outer edge of each feather. Molt first appears in this age class. First prebasic molt begins on the primaries (p1–5 or p1–6), secondaries (s1, sometimes s5), and one or two tertials on each wing. Molt sequence easily defines this age class. Irregular, serrated trailing edge of remiges is apparent with a few molting, rounded-tipped feathers among pointed-tipped juvenile remiges during a span of spring through fall. With the beginning stage of multi-aged remiges, the upper surface of the new secondaries has a pale gray sheen and is paler than the retained juvenile-age feathers. The outer edges of the inner secondaries and tertials are similarly colored. Tips of the inner portion of the greater secondary coverts have a narrow, pale dusky bar, which is narrow because of feather wear.

*Note:* One to 2 years old.

**Photographs:** Condor #85, 1 year and 5 months old (hatched Apr. 13, 1998; photographed Sep. 19, 1999), is a classic example of this age class. In-flight photograph illustrates molt on p1–5 with p5 still growing.

**JUVENILE TRAITS:** HEAD.—**Blackish bill. Head, neck, and throat, are uniformly dark gray and/or blackish. Head and neck are covered with a dense coat of brownish, fuzzy downy feathers, except on the forehead and sides of the head.** Grayish brown irises with a pinkish sclera. BODY.—Blackish brown. Dark gray crop. WINGS.—Whitish axillaries typically have a large dusky area on the central region and may appear quite dark, with whitish only on the outer edge of each feather. Whitish underwing coverts are mottled dusky, especially on the carpal region. Considerable variation exists in amount of white on the axillaries and underwing coverts: Some are quite dusky, others very white. Tips of the inner portion of the greater secondary coverts form a fairly wide, neat, pale dusky bar. Same-age, pointed-tipped remiges produce a uniform neatly, serrated trailing edge of the wing. Secondaries, including outer edges of the inner secondaries, tertials, and respective greater coverts, have a slight grayish sheen on the upper surface, which is paler than the rest of the upperparts.

*Note:* Less than 1 year old.

**Photographs:** Not depicted. Similar to Condor #85 (1 year and 5 months old; *see* Basic I), but would lack wing and body molt. In flight, the trailing edge of wings would be neat since there would be no molt.

**ABNORMAL PLUMAGES:** None documented.

**HABITAT: California** (Monterey, San Luis Obispo, Ventura, Kern, Tulare Cos.).—Rugged, hilly, and montane regions, from sea level to 9,000 ft. (2,700 m). Moderately arid to very arid and seasonably warm to hot climate. Generally breeds (pre-capture/release-era population) at elevations from 1,500 ft. (460 m) to 4,500 ft. (1,300 m), but has occurred up to 6,500 ft. (2,000 m). **Arizona** (Coconino Co.).—Similar to California, but base elevation is much higher, generally around 4,000 ft. (1,200 m). Arid climate. Condors historically inhabited this state and now do so again (*see* Status). In dispersal areas outside of Arizona, similar arid and rugged topography is also inhabited.

**HABITS:** Tame and very inquisitive. Released birds are conditioned to distance themselves from humans; however this is not always successfully accomplished. Gregarious except when nesting. Elevated, exposed objects are used for perch sites. A very terrestrial species. Their feet cannot grasp objects. Condors often sun during early morning. Depending on weather conditions, condors may perch or be airborne for considerable lengths of time each day. Drinking and bathing occur often. Allopreening occurs between pairs. They regularly sit upright on their haunches (heels) or lay down on perches or on the ground.

**FEEDING:** An aerial scavenger. Large dead mammals, particularly herbivores, are primary food sources. Condors also feed on very large marine mammals such as beached whales and sea lions. Stranded fish form a small portion of their diet. Food is located by sight. Condors observe behavior of other scavengers, including Golden Eagles and Common Ravens, to assist in locating prey, but relinquish feeding status to the smaller, more aggressive Golden Eagle. Foraging often occurs long distances from roosting and nesting areas. Historically, nesting pairs foraged within 40 miles (64 km) from nest sites, but traveled up to 112 miles (180 km) in one day. Immatures and nonbreeders have traveled 124 miles (200 km) in 1 day. During periods of inclement weather, California Condors are sedentary at roost sites and will fast for several days; after gorging themselves, they may not eat again for several days.

*Note:* "Subsidy" feeding is regularly done for all released populations to ensure an ample, safe food supply. Since lead poisoning is a major problem (*see* Mortality), supplying prey that is free of possible lead poisoning is vital for the small wild populations. Carcasses, mainly of stillborn livestock, are placed in select, isolated areas in California. Once acclimated to life in the wild, released birds also readily feed on wild, natural carrion.

**FLIGHT:** Virtually all airborne time is spent in an energy-efficient soaring mode. Wings are held in a slight dihedral with the outer primaries flexing substantially upwards when soaring and gliding. Soaring stints may last for over an hour without resorting to powered flight. Very stable mannerisms when soaring and slowly completes each soaring revolution. Powered flapping is used sparingly to gain altitude quickly when taking off from nest sites or perches; when first becoming airborne after eating or bathing; or during aerial interactions with other California Condors, Golden Eagles, or Common Ravens. Flapping is sometimes used at higher altitudes when atmospheric lift conditions are poor due to inclement weather. Like all vultures, California Condors regularly "dip" their wings in order to regain momentum and aerial stability. California Condors may occasionally kite in very strong winds along canyon updrafts, but do not hover.

**VOICE:** Rarely heard. Hissing and grunting noises are made when agitated during communal feeding confrontations; sometimes when courting. Technically voiceless.

**STATUS AND DISTRIBUTION: Prehistoric.**—Fossils have been recovered in Arizona, California, Florida, Nevada, New Mexico, New York, Oregon, Texas, and n. Mexico. Population decline probably began with the loss of the large prehistoric mammals as a readily available prey base.

**Historical (to mid-1960s).**—Ranged from s. British Columbia into Baja California. Condor population decline continued and unquestionably accelerated in the mid-1800s with increased human colonization of the West. *Oregon:* Regularly recorded from 1805 to 1854 along the Columbia River (Lewis and Clark). Last sightings were in Douglas Co. in 1903 and 1904 (3 and 4 birds, respectively). *Arizona:* Remnant population sparingly inhabited the western part of the state through the late 1800s; most recent record was in 1924. All are sight records. *California:* Human alteration of the historical environment is major factor for decline in the California Condor's last stronghold. Prior to 1939, up to 200 birds were thought to exist. By the late 1960s, possibly 60 birds remained.

**Recent (late 1960s to early 1987).**—*Mexico:* A single bird was seen in 1971 in the Sierra San Pedro Mártir, Baja California. *California:* The condor population declined precipitously from the 1960s to the early 1980s. There was a growing concern at this time on the California Condor's fate and threat of extinction. The first chick was brought in for a captive breeding program in 1982 to try to boost the population and ensure survival of the species. The population reached its lowest level in late 1982 with 22 birds. In 1983, researchers discovered that when they removed the first egg (for artificial incubation), adults would lay more than 1 egg annually, thus greatly increasing population growth (*see* Nesting). The last California Condor egg hatched from a wild pair in 1984. During the winter of 1984–1985, four of the five known wild pairs lost mates. The decision was made in 1985 to capture the remaining six wild California Condors to save the species from extinction and boost captive breeding stock. On Apr. 19, 1987, the last wild California Condor was captured.

**Current (mid-1987 to present).**—*Endangered.* Until recently, on the verge of extinction. Federally listed as an Endangered Species since 1967.

*California:* Listed as a State Endangered Species since 1971. From mid-Apr. 1987 to early Jan. 1992, all condors were held in captivity for propagation and to rebuild their fragile numbers. Andean Condors, from South America, were experimentally released into former California Condor range from 1988 to 1990 to study behavior, foraging, and movement patterns. This experiment was done in hopes of fine-tuning eventual release of native California Condors. The first two California Condors (juveniles), from captive breeders, were released into the wild in Jan. 1992. The 1992 release also included two "companion" Andean Condors. The Andean Condors were brought back into captivity in Sep. 1992 and subsequently re-released in South America.

In early Apr. 2000, one of the last two California Condors from natural wild origins from the pre-captive-release era that were captured in 1987, AC8 (adult condor #8), was released back into the wild in s. California. After 14 years in captivity and no longer able to lay eggs for the captive-release program, she was released back into the wild. In May 2002, the last pre-captive release-era adult male condor, AC9 (adult condor #9), was also released back into the wild in s. California.

*Arizona:* The first California Condors were released in the state in mid-Dec. 1996. Releases have occurred annually since then.

*Total population:* As of Jul. 2002, 205 captive and wild condors. Seventy-four are in the wild in three populations: 32 in Arizona, 18 in cen. California (Big Sur region), and 24 in s. California; 118 are in the three breeding facilities; 13 are in holding pens to be released.

**Winter.**—Basically sedentary within normal range, but they have a tendency to wander extensively. Breeding adults with fledglings are often bound by parental duties during this season and do not wander far from nesting and foraging areas.

**Movements.**—Seasonal shifting. All released individuals in California and Arizona are tracked via telemetry.

*California:* Individuals wander within historical regions in the state mainly during spring through fall. Wanderlust individuals travel up to 160 miles (257 km) in a day. The southern population from Santa Barbara and Ventura Cos. have gone as far north as s. Fresno Co. in the Sierra Nevada and w.-cen. Monterey Co. in the Coast Range Mts. In spring of 2000, the northern population in the Big Sur region wandered south and visited individuals from the southern population for the first time. Since then, the Big Sur population made regular, if not weekly excursions to visit the s. California population. The two California populations have not embarked on the long-distance journeys as seen in the Arizona population (*see below*).

*Arizona:* The captive-released population undertakes rather extensive spring-through-fall dispersal in all directions. In the summer of 1998, some birds began to take extraordinary northward excursions from n. Arizona release sites. In late Aug. 1998, three California Condors dispersed northward near Grand Junction, Mesa Co., Colo. One embarked on a 13-day, nearly 700-mile (1,100 km) excursion north to Flaming Gorge Reservoir, Sweetwater Co., Wyo. (most distant movement noted to date).

In 1999, dispersal movements from Arizona were not as far-ranging as in 1998, but occurred in more areas. Movements extended from late Apr. to late Sep. *Utah:* Cedar City, Milford, Zion N.P., Bryce Canyon N.P., and Lake Powell region. *Nevada:* Westward dispersal into the Virgin Mts., south of Mesquite.

*Colorado:* Movements occurred northeast into Mesa Verde N.P., Montezuma Co. in late Jun. *Arizona:* Southward movements within the state extended to the San Francisco Mts. north of Flagstaff. Currently, dispersing individuals have returned to areas near their release sites in n. Arizona after exploratory periods.

**NESTING:** Begins late Jan.–May. Courtship may begin in mid-Dec. Eggs are normally laid from early Feb. through mid-Mar.; captive breeders may begin laying in early Jan. Eggs hatch in Apr. and May. Breeding age is typically attained from 6 to 8 years old; however, breeding may occur when 5 years old.

**Courtship (flight).**—*Pair-flight.*

**Courtship (perched).**—*Circle display*, with neck skin hanging loosely and *allopreening* (*See* chapter 5).

Nests are in large cavelike structures on cliffs or rocky hillsides and rarely in tree cavities. *Note:* The 2001 and 2002 pairs nested in caves. Nests are a collection of small rocks and gravel or other debris from the immediate cavity area. The single egg is incubated 56 days by both sexes, with each parent taking 1- to 5-day shifts. Condors fledge in 5–6 months and are independent in 11–12 months. Due to long parental duties, breeding occurs irregularly. Condors may breed every other year or breed 2 consecutive years, then skip a year due to very late breeding and fledging the previous year. *Note:* Under controlled conditions, captive condors may lay up to 3 eggs per year.

**Current Nesting Data.**—*Arizona:* The first egg-laying of a released pair was discovered in late Mar. 2001 at Grand Canyon N.P. This was a historical benchmark of a captive-released individual laying an egg under natural conditions. (The last pre-restoration-era egg-laying occurred in 1984 in California.) The egg, however, was found broken, a common occurrence in a pair's first nesting attempt. The female egg-laying California Condor was 6 years old and released in the spring of 1997 at the Vermilion Cliffs in Coconino Co., Ariz. In the spring of 2002, two 7-year-old pairs exhibited nesting behavior at Grand Canyon N.P. but did not nest.

*California:* Two pairs consisting of 8-year-old birds and one pair of an 8-year-old male and a 5-year old female successfully hatched eggs in the spring of 2002. The first captive-release-era pair to successfully hatch an egg in the wild occurred in mid-Apr. 2002 in Ventura Co. Two other pairs also successfully hatched their single eggs in Ventura Co. in early and late May 2002. As of Jul. 2002, all three youngsters were doing well in their natural environment.

In mid-May 2001, the first successful egg-laying of a captive-released bird occurred in the wild at Lion Canyon region in Santa Barbara Co. The nest actually had two eggs from two females that were 7 years old. One egg contained a dead embryo when collected, and the other egg was retrieved for artificial incubation to ensure safe hatching. The real eggs were replaced with artificial eggs. The viable egg hatched in mid-Jun. at the Los Angeles Zoo. The nestling was subsequently returned to the nest in the wild but was found dead 2.5 days later. One of the two incubating female condors killed the nestling. Also, this is the first documentation that two females attempted nesting in the same cavity. This created an incubation problem and death of the nestling.

**CONSERVATION:** Recovery effort is administered by the USFWS (California Condor Recovery Program) in cooperation with California Department of Fish and Game; Ventana Wilderness Society; The Peregrine Fund's World Center for Birds of Prey in Boise, Idaho; BLM; Arizona Game and Fish Department; and the National Park Service. Financial support is also greatly aided by the public. All wild populations are from captive-released birds. Condors are bred and raised at three facilities: Los Angeles Zoo, San Diego Wild Animal Park, and The World Center for Birds of Prey.

Condors had been released at least at 7 months old, but now are not released until 1 year old since older birds adapt better. Many older birds have been brought back into captivity for various reasons and rereleased–sometimes numerous times.

**Release sites.**—*California:* Released birds are transferred to Hopper Mountain NWR, Ventura Co., prior to release. California Condors are released at four sites: Sespe Condor Refuge, Ventura Co.; Lion Canyon, Santa Barbara Co.; Castle Crags, San Luis Obispo Co.; and Ventana Wilderness Sanctuary (Big Sur region), Monterey Co.

*Arizona:* Two release sites: (1) Vermilion Cliffs in n.-cen. Coconino Co., where releases began in late Oct. 1996, and (2) Hurricane Cliffs in nw. Coconino Co., first used as a release site in the fall of 1998. Arizona historically had California Condors, and the state is now being used as a release site because its isolated habitat may help reduce human-induced mortality. Mortality of released birds, however, has been quite high since reintroduction began.

*Mexico:* The Zoological Society of San Diego/San Diego Zoo requested permits in Jun. 2002 to export condors for release in the Sierra San Pedro Mártir N.P. in Baja California.

In mid-Aug. 2002, six condors, five juvenile and one adult, were sent to Mexico to undergo quarantine before being released in the fall in n. Baja California in the n. Sierra San Pedro Mártir N.P. The adult will be recaptured later and returned to the U.S.

**Recovery goal.**—The recovery program's goal is to have 300 California Condors in the wild: 150 in two populations in California and Arizona with at least 15 breeding pairs in each state. The captive breeding program has been an ongoing success; however, high mortality has plagued efforts. Successful breeding and hatching of wild pairs in the spring of 2002 is an encouraging sign that the goals will eventually be attained.

**Mortality.**—A combination of natural and human-induced mortality factors has greatly affected the California Condor population. High mortality continues to plague current reintroduction efforts.

*Human-induced mortality:* (1) *Scientific collecting.*—Nearly 300 California Condors were collected from the late 1700s to 1976. (2) *Lead poisoning.*—A major cause of mortality since the settlement of California. Lead poisoning still plagues captive-released birds in California and Arizona. Ingestion of lead bullets, fragments of lead bullets, and lead pellets embedded in prey animals, especially big-game mammals, causes deadly lead poisoning. This is particularly the case with wounded big-game mammals that are not recovered by hunters during the fall hunting season. Lead poisoning is possible year-round from birds that might feed on Wild Boar in California, which are hunted all year and may not be retrieved by hunters. *Note:* Nonlead, nontoxic bullets are currently being tested that are equal to the ballistics of lead ammunition for mammal hunting, but such ammunition is not yet commercially available. If approved, use of nontoxic ammunition by hunters will greatly benefit Condor survival. (3) *Illegal shooting.*—An ongoing problem, but not as major a threat as once thought it might be. (4) *Varmint poisoning.*—Grizzly Bears and Coyotes in the past; but in recent times, Coyote poisoning has contributed to some California Condor loss.
(5) Electrocution and/or collisions with power lines has been a modern-day loss factor. Even though released birds undergo conditioning to avoid utility poles and power lines, mortality of this fashion continues.
(6) *Antifreeze.*—The green liquid has killed at

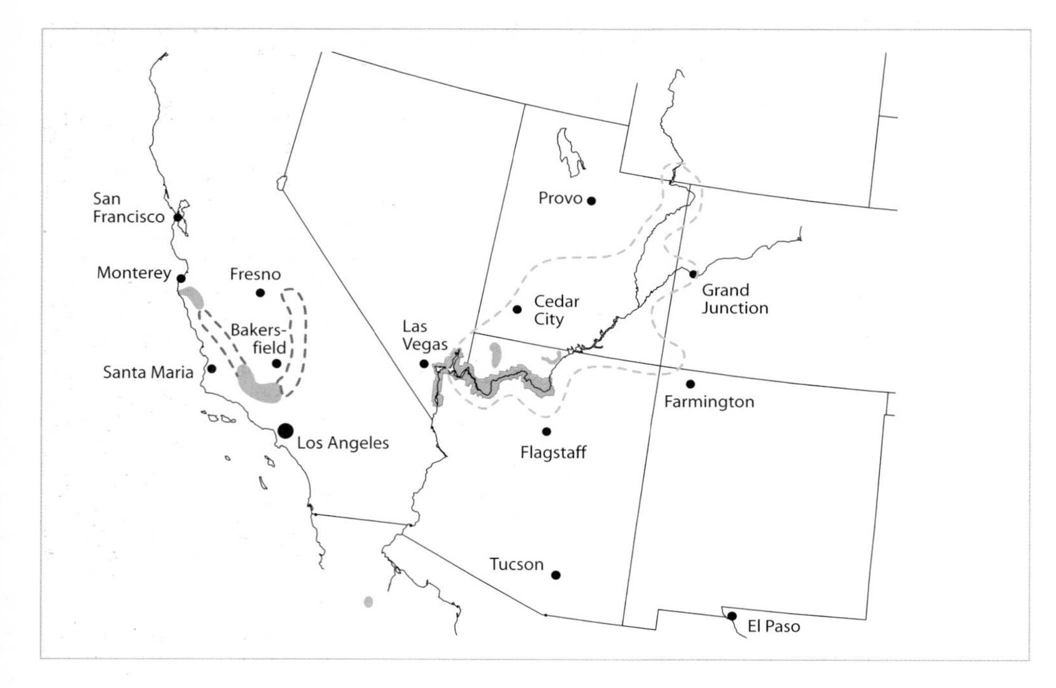


**CALIFORNIA CONDOR,** *Gymnogyps californianus:* Condors in wild as of Jul. 2002: CA: 42 + 8 in pens, AZ: 32 + 5 in pens.

least one condor when the bird drank it from a puddle.

*Natural mortality:* (1) *Common Ravens.*—Wreak havoc on nesting birds by depredation on eggs. This may also become a major threat with current breeding pairs. (2) *Golden Eagles.*—Kill nestlings and even immatures intruding into their territories. (3) *Drowning.*—Birds have drowned in natural, steep-sided pools of water. (4) *Botulism.*—One bird contracted the disease but was captured, treated, and recovered. (5) *Coyotes.*—Kill young released birds when feeding at carcasses. An adult pair (of 2 adult pairs released in late 2000) was killed by Coyotes in late 2000 in Arizona. Having lived in captivity, even adult birds are not acclimated to the dangers of life in the wild. Intense conditioning for all ages of the perils of life in the wild is vital for survival. (6) *Unknown.*—A few have perished of undetermined causes.

**SIMILAR SPECIES: (1) Turkey Vulture.**—FLIGHT.—Uniformly dark upperwing. Dark axillaries and underwing coverts; pale gray remiges. Long, wedge-shaped tail. Unstable, tilting flight. Much smaller size. **(2) Bald Eagle, immatures.**—FLIGHT.—Share whitish axillaries and wing linings. Very large head, bill. Wings held on horizontal plane when soaring. Uppersides of remiges often darker than coverts. **(3) Golden Eagle.**—FLIGHT.—All-dark color. Older birds have pale tawny bar on upperwing: use caution with pale bar on coverts on most condors. Immatures may have white on inner primaries on underwing and at times on upperwing. Immatures have white on basal region of tail. Tawny nape area. In flight, wings held in a dihedral.

**OTHER NAMES:** Condor. *Spanish:* Condor Californiano.

---

REFERENCES: AOU 1997; Bent 1961; Bridges 2002; Davey 1999a, 1999b; Gilligan et al. 1994; Griffiths 1994; Koford 1953; Kricher 1999; Meretsky and Snyder 1992; Meretsky et al. 2000. Palmer 1988; Snyder and Rea 1998; Snyder and Snyder 1991; The Peregrine Fund 1999b, 1999c, 2001; USFWS 1997a, 1998a, 2001, 2002; Ventana Wilderness Society 2000b; Wauer 1992.

**Plate 24. California Condor, adult (6 yrs. old) [hatched 3-28-94, photographed 4-7-00]** ▪ Bluish bill. Orange head with black feather patch on forehead. ▪ *Note:* Condor #2. Acquiring white bar tips of greater coverts. Raised neck ruff conceals bare upper neck skin. Photograph by Ned Harris.

**Plate 25. California Condor, basic (subadult) V (5 yrs. and 5 mos. old) [hatched 4-18-94, photographed 9-19-99]** ▪ Bluish bill. Orange head with black feather patch on forehead. ▪ *Note:* Condor #5. Similar to adult, but lacks broad white bar on greater coverts. Downy feather is stuck on forehead from preening. Photograph by Ned Harris.

**Plate 26. California Condor, adult (6 yrs. and 6 mos. old) [hatched 5-12-94, photographed 12-12-01]** ▪ Orange head. ▪ Black body. ▪ Large, immaculate white region on black underwing. ▪ Square-shaped tail. ▪ *Note:* Condor #11. Classic full adult with very white underwing region. Photograph by Ned Harris.

**Plate 27. California Condor, basic (subadult) V (4 yrs. and 11 mos. old) [hatched 3-28-94, photographed 2-27-99]** ▪ Orange head like adult. ▪ Large white region on black underwing; dark speckling on white area. ▪ *Note:* Condor #8. Similar to an adult but has dark markings on white underwing region. Photograph by Ned Harris.

**Plate 28. California Condor, basic (subadult) IV (4 yrs. and 3 mos. old) [hatched 4-7-95, photographed 7-99]** ▪ Dusky head is freckled with yellow; neck is reddish. Reddish iris. ▪ *Note:* Condor #14. Head beginning to acquire yellowish-orangish adult color. Photograph by William S. Clark.

**Plate 29. California Condor, basic (subadult) IV (4 yrs. and 10 mos. old) [hatched 4-29-94, photographed 2-27-99]** ▪ Yellowish orange head with dark freckling. ▪ White region on the underwing has extensive dark areas. ▪ *Note:* Condor #7. Has younger head and plumage features than Condor #8. Classic example for age class. Photograph by Ned Harris.

**Plate 30. California Condor, basic (subadult) III (early stage) (3 yrs. old) [hatched 4-8-97, photographed 4-8-00]** ▪ Dark bill with bluish freckling. Dark gray head with fuzzy brown feathers on nape and hindneck. Neck is pale pink. ▪ *Note:* Condor #55. Head and neck color is identical to subadult II (not depicted). New World Vultures often lay down on perches. Photograph by Ned Harris.

**Plate 31. California Condor, basic (subadult) III (early stage) (3 yrs. old) [hatched 4-7-97, photographed 4-7-00]** ▪ Dark bill; some bluish freckling. Dark gray head with fuzzy brown feathers on nape and hindneck. Neck ruff conceals pink neck. ▪ Molt on wings and body. ▪ *Note:* Condor #56. Head is similar to a subadult II (not depicted). Photograph by Ned Harris.

**Plate 32. California Condor, basic (subadult) III (late stage) (3 yrs. and 9 mos. old) [hatched 6-2-95, photographed 2-27-99]** ▪ Yellowish head. ▪ Whitish under wing region has extensive dark markings. ▪ *Note:* Condor #25. Amount of dark markings on underwing is highly variable. Photograph by Ned Harris.

**Plate 33. California Condor, basic (subadult) I (1 yr. and 5 mos. old) [hatched 4-13-98, photographed 9-19-99]** ▪ Dark bill. Dark gray head and neck with fuzzy brown feathers. ▪ *Note:* Condor #85. Exhibiting molt. Juvenile (not shown) is similar but lacks molt. Photograph by Ned Harris.

**Plate 34. California Condor, basic (subadult) I (1 yr. and 5 mos. old) [hatched 4-13-98, photographed 9-19-99]** ▪ Gray head. ▪ Whitish underwing has extensive amount of dusky markings. Inner 5 primaries have molted. ▪ *Note:* Condor #85. Same bird as plate 33. Juvenile (not shown) is similar but lacks molt. Photograph by Ned Harris.

# OSPREY
## *(Pandion haliaetus)*

**AGES:** Adult and juvenile. Sexes of both ages are somewhat dimorphic in the amount of breast markings, but there is overlap. Those breeding in northern latitudes tend to be more extensively marked on the head and breast than southern birds (Palmer 1988). Except for minor variation on breast markings, there are virtually no other plumage variations in either age group.

Substantial feather wear occurs on the dorsal feathers of juveniles by mid-fall. By late Oct., the white scalloped dorsal feather edgings may be virtually worn off on some individuals. Iris color of juveniles gradually changes to more adultlike by autumn.

**MOLT:** Pandioninae wing and tail molt pattern. First prebasic molt on the primaries begins in the winter when juveniles are about 6 months old (Forsman 1999). Outermost primary (p10), which becomes excessively worn, is retained for nearly 2 years (Edelstam 1984). Serially descendant molt waves replace the inner primaries at least two times before p10 is replaced for the first time. Molt on the secondaries begins 1–3 months after primaries have begun molting (Forsman 1999). Rectrix molt also begins after primaries have been molting for some time. No data on body molt.

Subsequent prebasic remix molt of adults is continuous except during periods of migration and, for males, during the breeding season, when molt generally ceases. Remiges may have two to four molt waves in progress at the same time (Forsman 1999). Body feathers may be replaced annually since the plumage of many adults can be nearly uniformly worn and faded.

**SUBSPECIES:** Polytypic. One primary race, *P. h. carolinensis*, is in the U.S. and Canada. This race inhabits the contiguous U.S. south to the Florida Keys; portions of coastal Baja California and coastal Sonora, Mexico; and the boreal forest of Canada and Alaska.

Unless otherwise noted, all data herein pertains to *carolinensis*.

"Ridgway's" race, *P. h. ridgwayi*, is resident mainly in the Caribbean region. *P. h. ridgwayi* or intergrades with *carolinensis* can also be found in very small numbers among the more numerous *carolinensis* in s. Florida, particularly on the Keys (Palmer 1988; BKW pers. obs. and photographs on Big Pine Key, Fla.).

Subspecies elsewhere in the world: (1) Nominate *P. h. haliaetus*, which is fairly similar in plumage to *carolinensis*, but is consistently darker and more heavily marked (del Hoyo et al. 1994, Palmer 1988). It breeds across Eurasia, Japan, Taiwan, and the Canary and Cape Verde Islands. *Haliaetus* winters to s. Africa, India, w. Indonesia, and the Philippines. (2) *P. h. cristatus* is resident in Australia, east to New Caledonia, north to New Guinea and the Philippines, and west to Sumatra. This small race lacks dark crown and nape markings, and the dark auricular mark does not connect to the hindneck. Breast markings may be present on both sexes, but are more prevalent on females. Plumage data are based on photographs in Olsen (1998) and Debus (1998).

**COLOR MORPHS:** None.

**SIZE:** A large raptor. Males average somewhat smaller than females. Length: 20–25 in. (51–64 cm); wingspan: 59–67 in. (150–170 cm).

**SPECIES TRAITS:** HEAD.—Black bill except the basal part of the lower mandible, which is pale blue. Pale blue cere. Black lores with a black crescent above the eyes. **Black auricular stripe extends onto the side of the head and connects to the brown hindneck.** Long, bushy nape feathers (hackles) can be compressed or erected at will. Small black spot on the central crown may connect with the black crescent over the eyes. Small black triangular-shaped mark on the top rear area of the hackles. BODY.—Dark brown upperparts. White underparts. Variably marked breast: (1) unmarked (most males, very few females), (2) moderately marked (some males, many females), (3) heavily marked (very few males, most females). Legs are long and the tarsi are pale bluish gray. Rear area of the tarsi and toes are covered with short, sharp spicules for grasping slippery fish. WINGS.—Long, fairly narrow with parallel front and trailing edges. Wingtips are rounded. **A large black rectangular patch is on the carpal region of the underwing. The black carpal patch merges with the**

**black secondary greater underwing coverts, which form a diagonal black line to the body.** Inner primaries are whitish, barred, and form a whitish panel against the darker gray secondaries. Perched birds exhibit a white linear stripe from the wrist and shoulder area to the elbow along the inner edge of the patagial region. Wingtips of perched birds extend far beyond the tail tip. TAIL.—Brown on the dorsal surface and white on the ventral surface and banded with black.

**ADULT TRAITS:** HEAD.—Top of the head is white except for the black crown and top-of-nape spots. Iris color is medium yellow, but varies in either being a cool yellow or a bit warmer and slightly orangish. BODY.—All dorsal areas are uniformly dark brown. However, worn-plumaged birds can be medium or even pale brown, in patches or on extensive areas. WINGS (dorsal).—Uniformly dark brown upperwing. WINGS (ventral).—Underwing coverts are white. **Carpal patch on the underwing is generally solid black.** Secondaries are dark gray with narrow, pale barring on the outer feathers and solid dark gray on the inner feathers. TAIL.—White terminal band is irregularly worn and fairly narrow.

**JUVENILE TRAITS:** HEAD.—Iris color is orange as nestlings, fledglings, and younger juveniles. Some retain the orangish iris color until late fall. Iris color gradually changes to more adultlike yellow. By mid-Oct., some may possess near adultlike yellow iris color. Crown is extensively streaked with black and the nape region is washed with tawny. BODY.—**Dark brown upperparts have broad white tipped feathers on all back, scapular, and uppertail covert feathers on old nestlings, fledglings, and younger juveniles.** WINGS (dorsal).—**Dark brown with broad white tips on all covert feathers through fall as described in Body.** By Oct., white feather tips on older juveniles may show signs of wear and become narrower. By late fall or winter, much of the white tips on much older juveniles wears off and upperparts become more uniformly brown. WINGS (ventral).—**Black carpal patch is often mottled with tawny.** Dark gray secondaries have distinct whitish bands on the outer feathers and gray or indistinct pale bands on the inner feathers. Underside of the secondaries appear paler and more banded than on adults. White underwing coverts and axillaries are washed with tawny. TAIL.—Neatly formed broad white terminal band.

**ABNORMAL PLUMAGES:** None known for the West. A melanistic adult was in Monroe Co., Fla., in the mid- to late 1990s (Clark 1998). This bird was dark brown throughout, including the head. Undersides of the remiges were uniformly gray, but the undertail was pale and banded.

**HABITAT: Summer.**—Found along fresh, brackish, or saltwater areas in temperate and subtropical regions. Water areas vary from artificial impoundments, natural lakes and streams, and seashores. Ospreys readily accesses human-used waterways. Breeding occurs from sea level to montane elevations. In Arizona, they breed above 5,000 ft. (1,500 m). Elevation may extend up near treeline in mountain areas of other western states.

**Winter.**—In the U.S., mainly found in subtropical areas, but some occupy the southern temperate zone in Oregon, n. California, and Oklahoma. Most winter in subtropical and tropical areas in Central and South America.

**Migration.**—Ospreys typically stay near waterways; however, inland migrants are often found far from water in desert areas and on the Great Plains for portions of the migratory journey.

**HABITS:** Individuals vary from being wary to tame. Those around human-accessed areas can be quite tame. Ospreys perch on any type of elevated structure. Migrants crossing the Great Plains may perch on the ground. Basically a solitary species, but they may nest in loose colonies in some locations. Outer front toe can be rotated to face to the rear in order to grasp fish more firmly. Bushy nape feathers are compressed in hot weather and erected in cool weather and when alarmed. Ospreys regularly swoop down and skim their clenched feet in the water.

**FEEDING:** An aerial hunter. Ospreys feed almost exclusively on live fish. Dead fish are rarely taken. Also rarely feeds on other types of prey: birds (mainly small waterfowl) and small amphibians and reptiles. Ospreys dive in a headfirst plunge from as high as 100 ft. (30 m). Feet are lowered during the dive and extended forward of the head just upon impact with the water and fish. The Osprey may snatch the fish by skimming the water's surface in a shallow-angle dive, if the fish is very near the surface, or make a steep dive into the water and be momentarily partially submerged, then fly out of the water, grasping the fish and vigorously shaking off water. Prior to transport, bird immediately manipulates fish to point head first and belly down; clutches most fish with both feet taking prey to a perch. However, partially eaten, small-size fish remnants and small intact fish may be transported in flight with one foot. Prey is eaten while standing on a perch.

**FLIGHT:** Wings held in an arched position when gliding and soaring. Wings are folded at a sharp angle at the wrist when gliding. Gliding birds create an "M"-shaped appearance with their sharply angled wings. Wings are extended in a nearly straight, perpendicular angle from the body when soaring. Powered flight is with slow, deep wingbeats with irregular gliding and flapping sequences. Ospreys regularly hover when hunting. In certain areas and wind conditions, they may briefly kite.

**VOICE:** A vocal species in all seasons. Vocalizations occur between pairs, unrelated conspecifics, and towards intruders of any kind crossing into nesting territories. The most commonly heard call is a loud, sharp, high-pitched whistled *cheeurp*, either as a single note or as a long, rapid series of notes. When agitated, the *cheeurp* may be quite harsh and uttered for extended periods. A drawn-out, loud, equally high-pitched *eeeep* is emitted at various times. Highly vocal during courting activities.

**STATUS AND DISTRIBUTION: Summer.**—Locally *fairly common.* Estimated population in the U.S. in 1981 was 8,000 pairs; estimated population in 1994 was nearly 14,200 nesting pairs. Current status is undoubtedly considerably greater since their population continues to grow (*see below*). The substantial number of Ospreys that nest in w. Canada and Alaska is not represented in published material. The Osprey is still listed as a Threatened Species in several states, but it is not a Federally listed species. The Osprey was on the Blue List in Canada from 1972 to 1981, but was delisted due to a promising recovery following the ban on DDT.

Osprey populations were greatly harmed by the adverse effects of the organochlorine era of the late 1940s to mid-1970s. In states with formerly low nesting numbers, former pesticide poisoning extirpated the breeding population. Populations in many western states, however, have greatly increased if not exploded since the late 1970s due to the ban of DDT in Canada and the U.S. and the implementation of human-assisted programs to boost breeding potential. Midwestern states, which historically lacked breeding populations or had only a few pairs, have experienced slower-paced growth since the organochlorine era. Overall, pairs are expanding into formerly inhabited and newly colonized regions.

Number of breeding pairs in most western states and provinces continues to grow. The number of breeding pairs listed are from 1994 data from Houghton and Rymon (1997); if more up-to-date estimates are known, they are noted in parentheses: Arizona: 25–35 (over 50), Arkansas: 0 (2 in 1997, 3 in 1999), California: 500–700, Colorado: 17 (over 20), Idaho: 400–425, Iowa: 0 (1 in 2000; 3 hacking programs currently in progress), Kansas: 0 (2 hack sites beginning in 1996), Louisiana: 10, Minnesota: 350–450 (increasing), Missouri: 0 (2 in 1999), Montana: 500–600, Nebraska: 0 (no change), Nevada: 4 (no change), New Mexico: 2 (6), North Dakota: 0 (2), Oklahoma: 0 (no change), Oregon: 675–700, South Dakota: 2 (1 or 2), Texas: 3 (no change), Utah: 30, Washington: 350–400, and Wyoming: 150. Additionally, 810 pairs were estimated for Baja California and the Gulf of California in 1977.

In British Columbia, the Osprey is listed as an uncommon to fairly common summer resident, but rare in the northern one-third of the province (north of 56°N). Ospreys occasionally occur on the Queen Charlotte Islands, British Columbia, from spring through fall but do not breed. The Osprey's exact status in Alberta is unknown, but it is considered relatively common in the southern boreal forest, but is a sparse breeder in the northern and northwestern parts of the province. Saskatchewan lists it as being fairly common but local in primary range in the southern boreal forest, but is rare in the northern boreal forest. Its range has extended southward in the prairie parkland region in Saskatchewan mainly because breeders are utilizing utility poles. The species is widespread in much of Manitoba.

The Osprey is a fairly common permanent resident in Sonora, Mexico. Fairly common to common in Baja California and Sinaloa.

Two-year-olds return from wintering areas to the U.S. and Canada, but only one-quarter to one-half return to locales near their natal areas; however, they do not breed. Nearly all 3-year-olds return to their original natal areas to breed. Nonbreeding individuals are irregularly recorded during summer in areas not mapped as typical breeding range.

**Winter.**—*Adults:* Northern adult breeding populations winter from s. coastal Oregon, the Central Valley of California, major waterways in w. and s. Arizona, and s. and e. Texas southward to s. Brazil, Bolivia, and Peru in South America. Isolated regular wintering has occurred for a few years near Tulsa, Okla. Several breeding adults from Oregon that were tracked with telemetry wintered from coastal areas of cen. and s. Mexico to Central America. Adults may reach wintering areas as far south as n. South America by mid-Sep. Movements south of the U.S. may span until Nov. or even Dec.

*Juveniles:* Winter from Cuba and Belize southward to areas in South America, as described for adults, and remain on the winter grounds until the spring of their second year.

**Movements.**—Highly migratory except perhaps for Mexican populations. Migration routes may take individuals over inland routes often far from major water sources and often far from any water sources.

Migration data below are from preliminary telemetry studies by The Raptor Center of the University of Minnesota.

*Fall migration: Adults.*—Adult migration begins early, although not as early as in the e. U.S. Fall migration begins in late Aug. and Sep. for most western adults. An adult female tracked with telemetry from Saskatchewan left her breeding area in early Sep. Migration is a protracted event that spans 3–4 months for all the North American population to move to wintering areas. On average, females tend to migrate before males. Once migration has started, movement to the winter grounds is rapid and often accomplished in 2–8 weeks. The Saskatchewan-tracked female covered an amazing 1,274 miles (2,052 km) in 2 days. Based on telemetry data, adults from Saskatchewan, Oregon, and Minnesota may reach winter grounds in Mexico, Central America, and n. South America 2–3 weeks after leaving breeding areas. Those nesting in southern latitudes begin migration at an earlier date than later-nesting northern birds.

*Juveniles.*—Most recently fledged juveniles begin migration in Sep. Movement may also be rapid and may reach wintering areas as quickly as adults.

*All ages.*—Based on telemetry and banding data, adults and recently fledged juveniles from Minnesota, Oregon, and Saskatchewan may (1) take a southeasterly diagonal course in the fall to the East Coast to Florida, then onto the Caribbean islands for winter, or continue to South America; (2) head more or less due south of breeding and natal areas to the Gulf Coast then make an over-water crossing of the Gulf of Mexico to the Yucatán Peninsula or to South America; (3) head south through the Midwest or southeasterly to the Pacific Coast from breeding and natal areas, then into Mexico and follow the east or west coast of Mexico and Central America; or (4) take an inland route south or southeasterly into interior Mexico, then eventually onto coastal areas in Central America.

Along the w. Great Lakes, peak movements occur in mid-Sep. with migration ending by mid-Oct. In the central latitudes of the U.S., peak flights are in late Sep. with movements extending through Oct., and stragglers occurring until late Nov. In Veracruz, Mexico, migrants are seen from late Aug. to late Nov., with a peak in early Oct.

*Spring migration: All ages.*—Breeding adults leave the winter grounds in Feb. In inland mid-latitude areas, the first adults are seen in late Mar., but may arrive as early as late Feb. Peak movements of adults are in mid-Apr. Along the w. Great Lakes, the first adults arrive in mid-Apr. (rarely in early Apr.), peak in late Apr. or early May, and with stragglers occurring into Jun. Arrivals in interior s. Canada are seen in mid- to late Apr. Migration extends to at least mid-Jun. for returning nonbreeders.

**Extralimital movements.**—Ospreys have been recorded in spring on Kodiak and Gambell Islands, Alaska.

**NESTING:** Feb.–May in Sonora, Mexico, but courtship and copulation may occur in late Dec. in this region. Depending on latitude and elevation, nesting generally spans from late Mar. and Apr. to late Aug. in the U.S. and Canada. Ospreys do not attempt to breed until at least 3 years old. Pairs often reunite to breed each year unless one dies. However, unlike eagles, most pairs do not remain together year-round. (Telemetry has shown nesting pairs take far different migration courses and utilize different wintering areas.) Polygyny may occur in areas of high Osprey density.

**Courtship (flight).**—*High-circling* by both members of the pair and *sky-dancing* and *hover-flight* by males (*see* chapter 5). A great deal of vocalization occurs among courting birds.

Both sexes bring nest materials to the nest site, but females do most of the actual construction. New nests may be only a mere shallow layer of sticks. Nests, however, are regularly reused, and refurbished old nests may be up to 6 ft. (2 m) deep; in Sonora, Mexico, nests have been up to 13 ft. (4 m) deep. Nests are placed up to 100 ft. (30 m) high in trees, on rocks, and in Sonora, on large cacti. Nests may also be placed on the ground where terrestrial predators and human disturbance are absent, particularly on islands. Tree nests are typically located in the uppermost part of trees. Dead trees are often used; if live trees are used, nests are built in those with a barren top section above the foliage. Various artificial structures such as buoys, metal towers, buildings, barrels, and other unusual human-made implements are used. Ospreys also readily use human-made platforms erected on poles. The poles supporting the platforms have metal shields that prevent depredation by terrestrial predators. Nests are placed near or over the water, but have been up to 6 miles (10 km) from water.

Both sexes incubate, although females perform the majority of the 38-day task. Males mainly provide food during incubation and after hatching. Typical clutch consists of 3 eggs; although, clutches sometimes contain 2 eggs, but rarely 4. A single clutch is laid per season. Nestlings fledge in 44–59 days. Fledglings are cared for by their parents until they are 93–103 days old.

**CONSERVATION: Reintroduction programs.**—Several states have developed hacking programs to create new or boost existing populations. *Colorado:* Colorado Division of Wildlife sponsored a hacking program in several areas of the state, including Larimer Co. Forty-five birds were released from 1991 to 1993. *Iowa:* A small hacking program began in 1997 at Coralville Reservoir (Lake McBride) in Johnson Co. with four birds taken from Wisconsin nests. Since then, hacking has also taken place at Saylorville Reservoir in Polk Co. and near Cedar Falls, Black Hawk Co. *Kansas:* Eight birds were released in 1996 and seven in 1997 at El Dorado Reservoir and Wolf Creek Lake, Coffey Co. *Minnesota:* Limited hacking near Minneapolis-St. Paul.

**Artificial nest construction programs.**—Numerous states and provinces also assisted naturally existing and hacked populations by erecting artificial nesting platforms on moderate-height poles. These artificial structures not only increase nest site availability but also produce sturdy and safe nest sites. Slippery-surfaced shields are attached to the poles of the nest platforms to prevent terrestrial predators from climbing up to the platform and raiding eggs and nestlings. Substantial number of Osprey pairs in many states and provinces utilize artificial nest structures.

**Pesticide bans.**—The ban on organochlorine pesticide use, particularly DDT, greatly assisted Osprey populations to recover in the last three decades. This lethal pesticide was first used in 1946 in Canada and the U.S.

Canada took a series of steps to discontinue the sale and use of DDT that began in 1968 with a ban on spraying forests in national parks. The major Canadian ban came on Jan. 1, 1970 (announced Nov. 3, 1969), when DDT use was permitted for insecticide use on only 12 of the 62 previously sprayed food crops. However, all registration for insecticide use on food crops was stopped by 1978. Canada, however, permitted DDT use for bat control and medicinal purposes until 1985. Canadian users and distributors were also allowed to use existing supplies of DDT until Dec. 31, 1990.

The U.S. also had a series of steps to ban DDT, but halted the overall sale and use more quickly. In 1969, the USDA stopped the spraying on shade trees, tobacco crops, aquatic locations, and in-home use. The USDA placed further bans on its use on crops, commercial plants, and for building purposes in 1970. The Environmental Protection Agency banned all DDT sale and use on Dec. 31, 1972. However, limited use for military and medicinal purposes was permitted until Oct. 1989. In 1974, the U.S. banned the use of Aldrin and Dieldrin, both deadly chemicals that may have affected wildlife as much as did DDT.

Mexico was expected to discontinue government-sponsored DDT use for malaria control in 2002 and has planned a total ban of DDT by 2006.

There are 122 countries, including the U.S. and Canada, that have signed a United Nations-sponsored treaty banning eight deadly organochlorine pesticides: Aldrin, Chlordane, DDT, Dieldrin, Endrin, Heptachlor, Mirex, and Toxaphene. There are also two industrial chemicals, Hexachlorobenzene (also a pesticide) and PCBs, and two by-products of industrial processes, dioxins and furans, that have been banned.

Organophosphate pesticides, although not as persistent in the environment as organochlorine pesticides, can be deadly to some aquatic life that fish eat and, therefore, affect Osprey.

Carbamate pesticides were less harmful to the environment than organophosphate chemicals and were used to control insect infestations in e. Canadian forests.

Pyrethroid pesticides are harmless to birds and mammals and are currently used to control insect infestations in e. Canadian forests. However, some of these may be deadly to aquatic life and affect the food that fish eat, thereby affecting Osprey.

**Mortality.**—Ospreys probably still suffer in a limited manner in local areas from organophosphate and carbamate pesticide poisoning. Acid rain affects fish populations and reduces Osprey food supply in some regions. Illegal shooting undoubtedly occurs on both the breeding and wintering grounds. Excessive human disturbance of nest sites may cause nest abandonment. Avian and mammalian predators abound. Gulls and Raccoons cause mortality on unattended eggs and nestlings. Coyotes kill youngsters on the ground.

**SIMILAR SPECIES: (1) Bald Eagle, lightly marked type and moderately marked type 1- to 3-year-olds.**—PERCHED.—All have a dark auricular and side of head stripe, which is very Ospreylike. One-year-olds are similar in that bills and ceres are black or gray; 2- and 3-year-olds have brownish or yellowish bills and ceres. All have a small dark breast patch contrasting to a white belly, and dark leg feathers. Wingtips are shorter than tail tip. FLIGHT.—All ages exhibit considerable white on the underside, but do not have a black carpal patch. Tails are often whitish but unbanded. **(2) Herring and Western gulls.**—FLIGHT.—Dark wingtips; lack dark carpal patch on the underwing.

**OTHER NAMES:** Fish Hawk. *Spanish:* Gavilan Pescador. *French:* Balbuzard Pecheur.

---

REFERENCES: Adamus et al. 2001; Arizona Breeding Bird Atlas 1993–1999; Bylan 1998; Campbell et al. 1990; Clark 1998; Debus 1998; del Hoyo et al. 1994; Dinsmore 1997, 2000; Dodd and Vahle 1998; Dodge 1988–1997; Dorn and Dorn 1990; Edelstam 1984; Ewins 1995; Environment Canada 2001; Ewins and Houston 1992; Forsman 1999; Gilligan et al. 1994; Godfrey 1986; Henny and Anderson 1979; Houghton and Rymon 1997; Houston 1983; Houston and Scott 2001; Howell and Webb 1995; Kellogg 1999; Kent and Dinsmore 1996; Kingery 1998; Montana Bird Distribution Committee 1996; Olsen 1998; Palmer 1988; Peterson 1995; Russell and Monson 1998; Scott and Houston 1983; Semenchuk 1992; Small 1994; Smith 1996; Smith et al. 1997; The Raptor Center 1999a; Wheeler and Clark 1995.

**OSPREY,** *Pandion haliaetus carolinensis:* Fairly common. Local high densities. Nonbreeders along coastal Texas in summer. Migrants often found far from water. Winters to cen. South America. Irregular in winter north to Vancouver Island, B.C.

**Plate 35. Osprey, adult** [**Feb.**] ▪ Yellow iris. Black eyeline connects to hindneck. ▪ Type with unmarked breast. Uniformly dark brown upperparts. ▪ *Note:* Holding a captured fish.

**Plate 36. Osprey, adult** [**Dec.**] ▪ Yellow iris. Black eyeline connects to hindneck. Bushy nape. ▪ Type with heavily marked breast. Uniformly dark brown upperparts.

**Plate 37. Osprey, adult** [**Mar.**] ▪ Black eyeline. ▪ Type with moderately marked breast. ▪ Large, solid black carpal patch on wrist of wing; black diagonal line on greater coverts. Dark secondaries and pale primaries. Wings held in "M" shape when gliding.

**Plate 38. Osprey, adult** [**May**] ▪ Head-on view with wings held in an arch when gliding.

**Plate 39. Osprey, adult** [**Mar.**] ▪ Black eyeline. ▪ Large, solid black carpal patch on wrist of wing. ▪ *Note:* Fish carried with head pointing forward, usually with both feet (fish is partially eaten).

**Plate 40. Osprey, juvenile** (**older**) [**Oct.**] ▪ Black eyeline. ▪ Dark brown upperparts with very narrow white feather tips. ▪ *Note:* White tips on dorsal feathers often wear off by mid-fall.

**Plate 41. Osprey, juvenile** (**fledgling**) [**May**] ▪ Orange iris. Crown is heavily marked; Black eyeline connects to hindneck. ▪ Dark brown upperparts tipped with white. ▪ *Note:* This bird could fly well but returned to the nest to feed and rest.

**Plate 42. Osprey, juvenile** (**fledgling**) [**May**] ▪ Black eye line. ▪ Dark brown upperparts tipped with white.

**Plate 43. Osprey, juvenile (younger)** [**Oct.**] ▪ Black eyeline. ▪ Type with moderately marked breast. ▪ Large, black carpal patch is mottled with tawny. Tawny wash on underwing coverts. Very dark secondaries and pale primaries. Wings held in "M" shape when gliding. ▪ *Note:* Younger than juvenile on Plate 41.

## HOOK-BILLED KITE
## *(Chondrohierax uncinatus)*

**AGES:** Adult and juvenile. Adult plumage is acquired during the second year. Juvenile plumage is worn for much of the first year. Adult light morphs have very dimorphic plumages, but adult dark morphs have fairly similar plumages. A considerable amount of plumage variation occurs on juvenile light morphs, but there is very little variation in juvenile dark morphs. Juveniles may have slightly longer tails than adults.

**MOLT:** Accipitridae wing and tail molt pattern (*see* chapter 4). Data on first prebasic molt is based on the molt sequence of museum specimens (AMNH, LSU) from Mexico and Central America and field photographs. There are no known-age birds for definitive data on molt timing.

There is a moderate amount of wearing and fading on the plumage of birds entering the first prebasic molt. This degree of feather wear indicates that they are probably older than 6 months of age, and older than most kite species entering the first prebasic molt. First prebasic molt from juvenile to adult probably begins when they are 8–10 months old. Molt first appears on the crown of the head and hindneck (as early as Jan. or Feb. in s. and cen. Mexico). Molt then appears on the back, forward scapulars, and breast, and then continues to the rear of the body. The rear portion of the flanks, leg feathers, and undertail coverts are the last body areas to molt. Most of the ventral areas of the body are molted before dorsal areas complete their molt. Upperwing coverts and underwing coverts molt after much of the new adult body plumage is attained. Molt begins on the remiges on the innermost primary (p1) when nearly 1 year old, and after much of the body has attained adult feathering. Rectrices begin molting the central set (r1) after primaries have undergone a fair amount of molt.

In early Oct., when about 4 months old, Texas juveniles are in fresh plumage and do not exhibit signs of molt. By early Apr., when Texas juveniles are about 10–11 months old, body plumage is still fairly similar to autumn-plumaged birds. Also in early Apr., juvenile rectrices and remiges are retained and do not exhibit molt, but show a considerable amount of wear on the nearly worn-off white tips.

Subsequent prebasic molts on adults begin in late spring and continue until late fall. Adults in Texas and Mexico are involved in a substantial amount of feather replacement during summer and fall. Molt is not evident in winter and early spring.

**SUBSPECIES:** Polytypic. *C. u. aquilonis* is found from s. Texas to Tabasco and Chiapas, Mexico. Nominate *C. u. uncinatus* is found south of *aquilonis* to n. Argentina.

Two island subspecies: (1) *C. u. mirus* is an endangered race in Grenada, and (2) *C. u. wilsonii* is in e. Cuba. *Note: C. u. wilsonii* is considered a separate species by some authorities.

**COLOR MORPHS:** Polymorphic. *C. u. aquilonis* has two color morphs: light and dark with no intermediate morphs. Dark morphs are prevalent farther south in this species' core range. It was only recently discovered in the U.S. (*see* Status). Dark morph plumage data is based on specimens from the AMNH ($n = 4$); specimens from the Museum of Natural Science, LSU ($n = 4$); and field-photographs of two individuals by J. and P. Culbertson and two birds by W. S. Clark in Hidalgo Co., Texas.

**SIZE:** A medium-sized raptor. Males average slightly smaller than females, although size difference is not apparent in the field. Length: 16–20 in. (41–51 cm); wingspan: 34–38 in. (86–97 cm).

**SPECIES TRAITS:** HEAD.—**Thick, hooked bill. Greenish or yellowish fleshy cere and lores, and a pale orangish or yellowish horizontal teardrop-shaped mark above the eyes. A supraorbital ridge is absent.** BODY.—Very short, thick tarsi and toes. WINGS.—**In flight, the moderately long, broad wings have rounded wingtips and become narrower at the body junction.** TAIL.—Long. Outer rectrices are slightly shorter than the inner rectrix sets. Ventral surface tail patterns have identical formations of dark and light bands on all rectrices for the respective age and color morph.

**ADULT TRAITS:** HEAD.—**White irises. Greenish upper eyelids and lores, yellow to yellowish orange teardrop mark above the eyes, and a yellowish patch under the nares.** TAIL.—**Black ventral surface has two broad white bands; however, the inner band is partially covered by undertail coverts.**

**ADULT LIGHT MORPH:** There are no shared plumage traits between male and female adult light morphs.

**Adult male light morph:** HEAD.—Dark gray head. BODY.—Dark gray upperparts. Underparts are dark gray with variable-width white barring: white barring may be (1) very narrow or (2) moderate in width and up to 50 percent of width of the gray barring. Breast may have a lesser amount of white barring than rest of the underparts and may be solid gray and create a hooded effect in conjunction with the head. White undertail coverts may be finely patterned with sparse, gray barring. WINGS.—**Underwing has four rows of black and white bars on the outer six or seven primaries, but only three rows show well in translucent lighting. The secondaries are uniformly dark gray on the underwing, but may show faint pale bars on outer area. Finely barred, whitish underwing coverts.** TAIL.—**Black dorsal surface has one broad pale gray band on the mid-section; rarely, one additional, narrower gray or whitish basal band.** Ventral surface is black or gray with one broad white band on the mid-section. Males may show an additional narrow white band on the very basal area if tail is widely fanned when soaring.

**Atypical adult male light morph:** A femalelike variation of adult male is in LSU collection (BKW pers. obs.) and described by Howell and Webb (1995). HEAD.—A partial rufous hindneck collar. Head is gray as in typical males. BODY.—Ventral barring is rufous-brown and grayish. *Note:* An unusual plumage and not depicted.

**Adult female light morph:** HEAD.—**Black cap. Gray forehead with a gray area above and under the eyes. Auriculars and hindneck collar are rufous.** BODY.—White or tawny underparts are coarsely barred with rufous or rufous-brown. Dark grayish brown upperparts. WINGS.—**Dark grayish brown upperwings. Inner four primaries (p1–4) have a rufous wash on both the inner and outer webs. Rufous-colored primaries are visible only in flight, especially when viewed from underneath in translucent light. Whitish underside of all remiges are distinctly barred with dark brown and have a wide dark band on the trail-**

ing edge. Underwing coverts are finely barred with rufous. TAIL.—**Black dorsal side of the tail has a wide gray band on the mid-section and usually a narrower basal gray band. The basal gray band may have a considerable amount of white on the very inner region. Ventral side of the tail is dark gray or black with two or three fairly broad white bands.**

**ADULT DARK MORPH:** HEAD.—Black. BODY.—Black or grayish black upperparts and underparts, including the undertail coverts. TAIL.—Black with at least one broad white band on the mid-section on ventral surface. Sexes may not always be separable under field conditions because of subtle differences in body color. Although not detected in specimens from AMNH and LSU, there may be some sexual overlap in tail patterns. Howell and Webb (1995) state that there is overlap in plumage color and tail pattern between sexes. According to AMNH and LSU specimens, sexual variations were noticeable but subtle, especially on ventral areas.

**Adult male dark morph:** Specimens ($n = 4$); photographs ($n = 1$). HEAD.—Uniformly grayish black, including crown of head. BODY.—Uniformly grayish black, including the undertail coverts. WINGS.—Undersides of remiges are uniformly black. TAIL.—**One broad all-white band on the mid-section of both dorsal and ventral surfaces.**

**Adult Female dark morph:** Specimens ($n = 4$); photographs ($n = 1$). HEAD.—**Grayish black, sometimes with the crown being subtly darker and forming a black cap or all of the head is uniformly blackish. On individuals with caps, this feature is visible only at close range and in good light.** BODY.—Body, including undertail coverts, is grayish black, but is more blackish than on males; sometimes has a slight brownish tone. WINGS.—Undersides of remiges are uniformly black or one or more primaries have large, irregular white spotting. TAIL (dorsal).—**Black with one broad white band on the mid-section of the dorsal side. Grayish tinge on the distal portion of the broad white dorsal band of each rectrice.** TAIL (ventral).—**A second narrow white band on the very basal region may show on the ventral side if the tail is widely fanned when soaring.**

**JUVENILE TRAITS:** HEAD.—Iris color is usually pale brown in summer and fall. Irises may be medium brown or darker, depending on lighting conditions during this period (W. S. Clark pers. comm.). Irises turn white by spring. **Yellow facial skin, with a distinct yellow "tear drop" mark over the eyes.** TAIL.—**Four black and three pale brown equal-width bands on the dorsal tail. Four dark and three white equal-width bands on the undertail.** Light morph plumage data based on Howell and Webb 1995, museum specimens, and in-field photographs.

**JUVENILE LIGHT MORPH (HEAVILY BARRED TYPE):** HEAD.—**Black or dark brown cap on the crown and nape and a narrow white collar on the hindneck, except #4 below: hindneck sparsely mottled with white.** Some lack a defined dark cap area (*see below*). Color of cheeks, auriculars, and forehead is quite variable: (1) Cheeks and auriculars are whitish; (2) cheeks are gray, including area over the eyes and forehead, and auriculars are whitish; (3) cheeks, auriculars, and area over the eyes and forehead are gray; and (4) cheeks and auriculars are dark brown and blend with the dark brown cap. Throat is (1) whitish, (2) covered with dark barring, or (3) nearly solid grayish. BODY.—Upperparts are dark brown and narrowly edged with rufous in fresh plumage but pale edgings wear off by late winter and the upperparts become uniformly brown. Underparts are cream-colored or pale tawny when recently fledged but fade to white by fall. **Underparts have widely spaced, moderately broad dark brown barring on the breast, belly, and flanks. Thin barring on the leg feathers, axillaries, and underwing coverts. Barring may be grayish or rufous on some feathers, even in fresh plumage. Underparts may be interspersed with an irregular, blotchy array of very broadly barred feathers.** WINGS.—Dark brown upperwing coverts have narrow rufous edges on each feather. Undersides of the remiges are whitish, prominently barred, and have a narrow dark gray trailing edge. Solid brown upperside of primaries may have a slight tawny-rufous cast on the inner primaries, but is not as intensely colored as on adult females. When bird is flying overhead, inner primaries are typically grayish or whitish when viewed in translucent light.

**JUVENILE LIGHT MORPH (LIGHTLY BARRED**

**TYPE):** HEAD.—As on "heavily barred type." Those with gray heads and throats appear very hooded. BODY.—**Underparts are mostly unmarked and cream-colored or white. Flanks and leg feathers have very thin brown barring.** The pale underparts contrast sharply to the dark throats and heads of some individuals. WINGS.—Underwing coverts are white and unmarked. Axillaries may have thin dark barring. Remiges marked as on heavily marked type.

*Note:* When in first prebasic molt, the dark adult head and ventral feathering on the breast region contrasts very sharply to the virtually unmarked belly and flanks.

**JUVENILE DARK MORPH:** Specimens ($n = 2$). HEAD.—Brownish black. **At close range, crown of head may be a deeper black and form a subtly darker cap or all of head is uniformly blackish.** BODY.—All of body, including undertail coverts, is brownish black. Plumage is subtly more brownish than on adult female dark morphs and discernibly more brownish black than on adult male dark morphs. Scapulars have narrow tawny-rufous edgings in fresh plumage. Underparts may have very thin rufous edges on the flanks and lower belly. Some have whitish mottling on the breast and some white speckling on the belly. WINGS.—**Undersides of remiges are black with three narrow white bands.** Outermost primary may be unmarked. TAIL.—**Three moderately wide black bands and two moderately wide, pale grayish brown bands on the dorsal side. A thin or broad white border edges the upper portion of each gray band on dorsal surface of each rectrice.** Ventral surface has three black bands and two wide white bands. In soaring flight, a third narrow white band may sometimes show on the very basal portion of the under side of tail when widely fanned.

**ABNORMAL PLUMAGES:** Some juveniles in fresh, presumably non-molting plumages have rather unusual-looking markings that are very adultlike, with grayish heads and gray or rufous barring on the ventral areas.

**HABITAT:** Inhabits either semi-open or densely wooded deciduous thornbush tracts of medium-sized and tall trees. Climate is humid and seasonably hot.

**HABITS:** Moderately tame to tame raptor. A reclusive raptor. Their movements are very sluggish. Exposed and concealed branches are used for perches. Hook-billed Kites do not perch on poles, posts, or wires. Generally solitary, but kites may be found in small, loose groups in nonbreeding season. Groups are typically family members that may stay intact long past the breeding season, but also may be comprised of unrelated individuals. They often soar together and regularly engage in mutual, playful airborne antics.

**FEEDING:** A perch hunter. Very arboreal habits when foraging. Feeds by climbing on limbs and trunks in quest of tree snails. Snail shells are partially broken by the kite's thick bill, extracted, then eaten. Hunting occurs within or beneath tree canopies. Favorite perches may have a large collection of snail shells littering the ground below, particularly at nest sites. Kites have bills that vary in size, even within the same locale. Variability in bill size is possibly an adaptation to minimize competition of preying on same size snail. Differences in bill size have not been detected in Texas population, and all appear large-billed.

**FLIGHT:** Powered by slow, rather deliberate, floppy wingbeats interspersed by irregular gliding sequences. Wings are held in an arched position when gliding. Wings are held on a flat plane when soaring, with the wingtips flexing somewhat upwards. Flight is often low over treetops. Soaring flight occurs at various times of the day and regularly to high altitudes. Diving is a common aerial antic, especially when two or more kites are playing. Hook-billed Kites do not hover or kite.

**VOICE:** A staccato *keh-keh-eh-eh-eh-eh*. Sounds much like the highly vocal Golden-fronted Woodpecker, which is abundant within the kite's range. Vocalizes only when agitated in nesting territories.

**STATUS AND DISTRIBUTION: Light morph.**—*Very uncommon* and only found along the lower Rio Grande in s. Texas. Perhaps up to 50 pairs breeding along the U.S. side of the Rio Grande. Hook-billed Kite is a relative newcomer to the U.S. The species was first seen in May 1964 at Santa Ana NWR, Hidalgo Co., Texas. Kites were not seen again in the U.S. until Dec. 1975 when two pairs were observed at

Santa Ana N.W.R. Nesting was first documented in the U.S. in the spring of 1976 at Santa Ana NWR. Hook-billed Kites have since resided and expanded into available thorn forest woodland patches along the Rio Grande from Falcon Dam to somewhat east of Brownsville.

*Mexico:* Fairly common. Hook-billed Kites have been nesting in Nuevo León since the mid-1970s. Kites are in remnant, very fragmented habitat in n. Tamaulipas, but become more prevalent in the southern part of the state where habitat is more contiguous.

**Dark morph.**—Very rare in the U.S. The first U.S. record (photograph herein on plate 45), of an adult male, was in early Dec. 1998 at Bentsen-Rio Grande Valley S.P. in Hidalgo Co., Texas. Sightings of dark morphs have irregularly occurred since then at Bentsen-Rio Grande Valley S.P. An adult female dark morph was photographed in the state park in Dec. 2001. In Jun. 2002, an adult female dark morph successfully nested with a light morph adult at the state park (raised 2 light morph youngsters). A juvenile dark morph was photographed in the state park in Oct. 2002. Dark morphs are *fairly common* south of the U.S.

**Winter.**—Little is known about this species' winter biology. Some individuals are probably found year-round in typical Texas range. Hook-billed Kites were assumed to be fairly sedentary, but a few U.S. birds undoubtedly winter in Mexico.

**Movements.**—There are no data concerning movements of the Texas population. Formerly presumed to be sedentary, permanent residents in breeding areas. Hook-billed Kites first reached the lower Rio Grande valley in Texas, however, by dispersing north of their previously known Mexican range. In recent years, a significant migration has been documented at fall hawkwatch sites in Veracruz, Mexico. Kites are probably migrating from Tamaulipas and Nuevo León, Mexico, and undoubtedly Texas. There were 139 counted in the fall of 1998, 166 in 1999, 302 in 2000, 256 in 2001, and 154 in 2002. In Veracruz, Mexico, movements began in early Sep., peak in mid- to late Oct., and end by mid-Nov.

**CONSERVATION:** Since 1979, the U.S.F.W.S., State of Texas, and many local and private conservation organizations have been steadily acquiring acreage to increase wilderness habitat in the lower Rio Grande valley that forms a conglomerate of land parcels that forms the Lower Rio Grande NWR. The increased habitat may assist Hook-billed Kites and other wildlife species inhabiting the valley. The valley's wilderness habitat is currently 5 percent of the original 4,000 square miles (1 million ha) prior to massive agricultural conversion that began in 1920.

There are large expanses in n. Tamaulipas, Mexico, that lack suitable habitat, including most areas along the Mexican side of the Rio Grande. Habitat tracts do exist between Falcon Dam and Salineño on the Mexican side of the river, and possibly in a few other small areas.

**Mortality.**—Unknown. Susceptible to illegal shooting; however, most areas inhabited in Texas are on protected lands. Possible shooting occurs south of the U.S. as lazily soaring birds make easy targets. Large-scale habitat destruction occurs south of the U.S. border.

**NESTING:** Mar.–Aug.

**Courtship (flight).**—*High-circling*, but may involve other antics such as *tail-chasing* (see chapter 5).

Nest trees are within a group of dominant-sized trees. Nests are flimsy, see-through shallow platforms of thin twigs that are built by both sexes and average 12 in. (31 cm) in diameter. Nests resemble large dove nests. Nests are placed about 20 ft. (6 m) high on horizontal branches of the lower canopy about 10 ft. (3 m) from the main trunk. Hook-billed Kites breed relatively late in the season for such a southern latitude. Two eggs, laid Apr. through early Jun. Incubation and fledging periods are unknown, but are probably similar to other same-sized raptors (about 30 days and 30–40 days, respectively). Fledglings stay with parents for an extended period, possibly for several months after fledging.

**SIMILAR SPECIES:** COMPARED TO ADULT MALE LIGHT MORPH.—(1) **Gray Hawk, adults.**—PERCHED.—Supraorbital ridge, dark irises, whitish lores. Long tarsi. White band on mid-section of dorsal side of tail. Dorsal and ventral color and markings are virtually identical in the two species. FLIGHT.—Underside of wings uniformly whitish. "Finger" definition

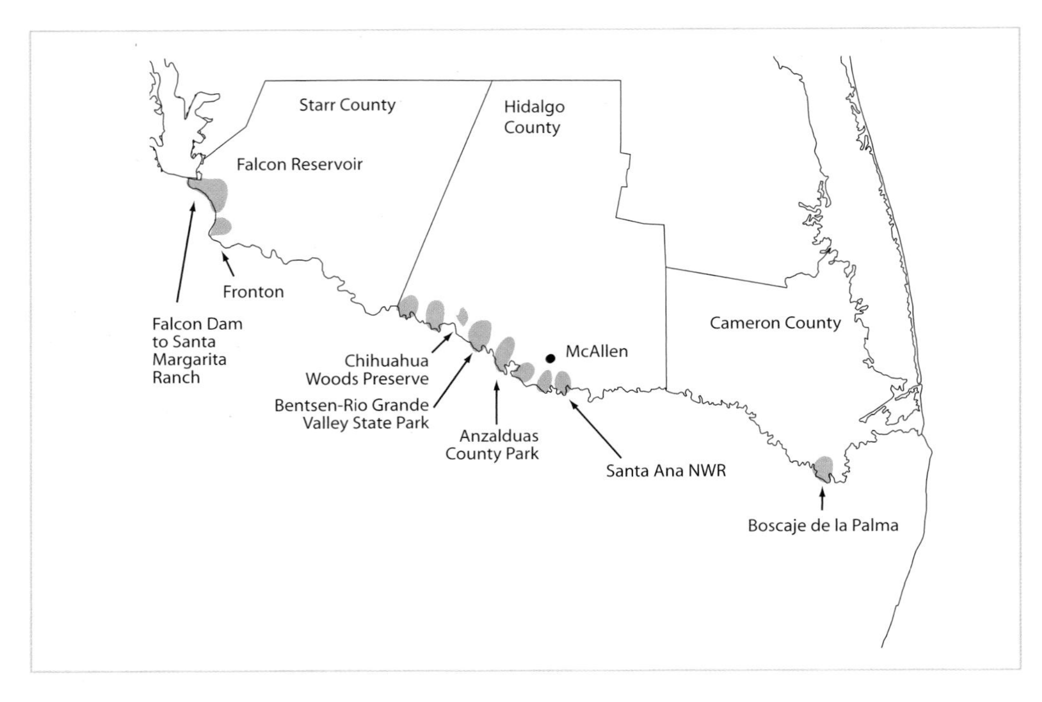


**HOOK-BILLED KITE,** *Chondrohierax uncinatus:* Very uncommon. 20–50 pairs in U.S.

on outer five primaries. Snappy, stiff, and rapid wingbeats. (2) **Cooper's Hawk, subadult/adult females.**—PERCHED.—Supraorbital ridge, orange or red irises, whitish lores. Long tarsi. Three pale, equal-width bands on dorsal surface of tail. Shares similar dark cap, rufous grayish nape; rufous barred underparts, grayish brown upperparts. Underside of outer rectrix set has narrow black bands that do not align with inner rectrices. FLIGHT.—Uniformly colored upper and lower surfaces of wing; nondescript gray trailing edge of remiges. "Finger" definition on outer six primaries. Rapid, stiff wingbeats; glides with wings on a flat plane. Tail as described in Perched. (3) **Red-shouldered Hawk, adults.**—PERCHED.—Supraorbital ridge; dark irises; whitish lores. Black and white checkered wings. Long tarsi. Narrow white tail bands. Rufous barring on underparts generally narrow and brightly colored. FLIGHT.—Underwing similar when not backlit. When backlit, has white crescent-shaped windows on primaries. "Finger" definition on outer five primaries. Wings held in similar arched manner when gliding. (4) **Broad-winged Hawk, adult.**—PERCHED.—Supraorbital ridge; whitish lores; orangish eyes. Similar rufous barred underparts and tail pattern. FLIGHT.—Pointed wingtips. Wings held on flat plane when soaring; quick wingbeats. Underparts and tail patterns as described in Perched. COMPARED TO JUVENILE LIGHT MORPH.—(5) **Cooper's Hawk, juveniles.**—Shape and size are similar when perched and flying. PERCHED.—Supraorbital ridge, whitish lores, uniformly brown head and nape. Streaked underparts. Moderately long tarsi. Tail length and equal-width pattern of bands similar on dorsal surface; but on ventral surface, outer rectrix set has narrow black bands that do not align with rest of rectrices. FLYING.—Outer six primaries have "finger" definition. Fairly quick wingbeats; glides with wings on flat plane. Streaked underparts. Tail as in Perched. COMPARED TO DARK MORPH.—(6) **Broad-winged Hawk, adult dark morphs.**—PERCHED.—Whitish lores, orangish brown irises. Gray band on dorsal side of tail. Use caution on similar white band on underside of tail. FLIGHT.—Pointed wingtips; white underside of remiges. One or two white tail bands on ventral surface of tail, similar to Hook-billed Kite. **(7) Common Black Hawk.**—Much larger. PERCHED.—Dark irises. Long yellow tarsi. Similar white tail band on dorsal side of tail. FLIGHT.—Remiges are paler than underwing coverts. Similar white tail band on dorsal and ventral surfaces of tail. Long yellow tarsi extend the feet into the white mid-tail band. **(8) Zone-tailed Hawk.**—PERCHED.—Dark irises. Gray band on dorsal side of tail. FLIGHT.—Underside of remiges are gray and heavily barred. Similar white tail band(s) on ventral side of tail. COMPARED TO ALL SEXES/AGES.—**(9) Harris's Hawk.**—FLIGHT. Identical wing and body proportions, especially when silhouetted; try to see plumage markings. Adults have solid colored underparts and remiges. Juveniles have large, square-shaped, pale-colored panel on undersides of primaries. Streaked belly. Very long tarsi. Flight mannerisms similar, but wingtips are flat on soaring Harris's.

**OTHER NAMES:** Hookbill. *Spanish:* Milan Piquiganchudo. *French:* Unknown.

---

REFERENCES: Clark 2002; del Hoyo et al. 1994; Hiller 1976; Howell and Webb 1995; Kaufman 1996; Kellogg 2000; Montiel de la Garza and Contreras-Balderas 1990; Palmer 1988; Parvin 1988; Snyder and Snyder 1991.

**Plate 44. Hook-billed Kite, adult female light morph** [May] ▪ Large bill. Fleshy green lores, Yellow teardrop over eyes. White iris. Black cap. Rufous auriculars and hindneck collar. ▪ Underparts are barred rufous and white. Short tarsi and toes.

**Plate 45. Hook-billed Kite, adult female light morph** [May] ▪ White iris. Black cap. Rufous auriculars and hindneck. ▪ Two, wide, pale brown dorsal tail bands.

**Plate 46. Hook-billed Kite, adult female light morph** [**May**] ▪ White underparts are barred with rufous. ▪ Broad rounded wings are narrower at junction of body. Rufous window on primaries. Remiges are narrowly barred with black. ▪ *Note:* This bird has damaged secondaries.

**Plate 47. Hook-billed Kite, adult male light morph (top), adult female dark morph (bottom)** [**Dec.**] ▪ *Both:* Large bills. Fleshy green lores, pale yellow "teardrop" over eyes. White irises. ▪ *Light morph:* White-and-gray barred underparts. *Dark morph:* Brownish blackish body. ▪ *Dark morph:* 1 white tail band (second white band is hidden). ▪ *Note:* Photo: Hidalgo Co., Texas, Jim & Pat Culbertson.

**Plate 48. Hook-billed Kite, adult male light morph** [**May**] ▪ Dark gray head. ▪ Dark gray underparts barred with white. White undertail coverts. ▪ Broad rounded wings are narrower at junction of body. Black-and-white barred primaries and solid black secondaries; barred coverts.

**Plate 49. Hook-billed Kite, adult male dark morph** [**Dec.**] ▪ Large bill. Fleshy green lores, pale orange "teardrop" over eyes. White iris. Blackish gray head and body. ▪ One broad white dorsal tail band. ▪ *Note:* In underside flight (not shown), remiges are solid black and tail has one broad white band. Photo: Hidalgo Co., Texas, 1998, Jim & Pat Culbertson (first U.S. record of a dark morph).

**Plate 50. Hook-billed Kite, adult female dark morph** [**Jun.**] ▪ Blackish head with white iris. Yellow teardrop over eyes. ▪ Brownish black or grayish black body. ▪ Underside of the broad, rounded wings are either uniformly black or have irregular white spotting on 1 or more primaries. ▪ Two broad white tail bands (inner band often partially or mostly covered by undertail coverts). ▪ *Note:* Photo: Hidalgo Co., Texas, William S. Clark.

**Plate 51. Hook-billed Kite, juvenile light morph (heavily marked type)** [**Oct.**] ▪ Large bill. Yellowish fleshy lores and teardrop over eyes. Pale to medium brown iris. Dark cap contrasts with gray or white auriculars. ▪ White underparts heavily barred with brown. ▪ Broad, rounded wings become narrower at junction of body. White remiges are broadly barred. ▪ *Note:* Some have lightly barred or unmarked white underparts.

**Plate 52. Hook-billed Kite, juvenile dark morph** [**Oct.**] ▪ Yellowish fleshy lores and teardrop over eyes. Pale brown iris. ▪ Brownish black underparts. ▪ Broad, rounded wings. Remiges are black with neat rows of large white spots. (White underwing coverts are disarranged feathers caused by air turbulence.) ▪ Black tail has 2 broad white bands. ▪ *Note:* Photo: Hidalgo Co., Texas, 2002 William S. Clark (1st U.S. record of a juvenile dark morph).

# SWALLOW-TAILED KITE
## *(Elanoides forficatus)*

**AGES:** Adult and juvenile. Sexes are similar. Ages are superficially alike, but are separable by tail length and, to a lesser degree, by dorsal color. Juvenile age class is held for nearly 1 year.
**MOLT:** Accipitridae wing molt pattern (*see* chapter 4). Primaries molt a few remiges prior to beginning molt on the secondaries. Based on close in-field photography (BKW pers. obs.), rectrix molt appears to begin, as on most raptors, on the central deck set (r1). Based on in-field observations, Meyer (1995; K. Meyer pers. comm.) suggests that some may begin rectrix molt on the outermost set (r6).

Prebasic molt begins on juveniles and non-breeding adults in mid- to late Apr. with the innermost primary. By May, 1-year-olds may exhibit several white blotches on the dorsal surface of the upperwing coverts caused by missing feathers. Because of the missing upper-wing coverts, small translucent "windows" may be visible on the ventral surface of the white underwing coverts when seen in shadowed overhead flight. Rectrix molt begins in May or Jun. after remix molt is under way. Body, remix, and rectrix molt is complete each year.

*Note:* One-year-olds photographed in Florida in late May had replaced p1 and 2, dropped p3 and 4, and replaced the innermost rectrix set, but had not yet molted the outer-most rectrix set.

Adult molt data are based on Meyer (1995). On breeding adults, and typical of most nesting raptors, remix molt begins in late May to early Jun. on females and late Jun. to mid-Jul. on males. All remiges, rectrices, and contour feathers are completely molted annually. Remix molt is rapid and is mainly complete by mid-Aug. to early Sep.
**SUBSPECIES:** Polytypic with two races. *E. f. forficatus* inhabits the s. U.S. and *E. f. yetapa* is found from s. Mexico through the northern two-thirds of South America.

Adults of *yetapa* are similar to the dorsal color of adult *forficatus*, but with more of a greenish cast on the inner portion of each feather. Juveniles of *yetapa* are overall more greenish on the dorsal areas than *forficatus*.

**COLOR MORPHS:** None.
**SIZE:** A medium-sized raptor. Length: 20–25 in. (51–64 cm). Tails of adults are a minimum of 2 in. (5 cm) longer than those of juveniles and generally fairly distinct from the shorter tailed juveniles. Recently fledged birds will appear very short tailed since it takes an additional few weeks after fledging to complete the growth of the long feathers of the outer rectrix set. Wingspan: 47–54 in. (119–137 cm).
**SPECIES TRAITS:** HEAD.—Small black bill and pale bluish cere. Iris color is dark brown. Lacks a supraorbital ridge (presence of the supraorbital ridge gives raptors the fierce look). White head. BODY.—**White underparts. Glossy iridescent, medium purplish blue upperparts with a non-iridescent black area on the back and forward one-third of the scapulars.** Feet are small and pale bluish gray to gray; sometimes flesh colored on portions of the feet. WINGS.—**Long, narrow, and pointed. White underwing coverts and black undersides of the remiges. Dorsal surface of the remiges and distal upperwing coverts are an iridescent medium purplish blue. The forward portion of the lesser coverts is noniridescent black and forms a distinct dark black band across the upperwing coverts. This black band blends with the black region on the forward scapulars and back. Inner tertials and inner greater secondary coverts are mainly white. When kites are perched, this region shows as a large white patch behind the scapulars. All of the white area on the tertials and inner coverts are concealed by the scapulars when in flight.** TAIL.—**Black and deeply forked. Innermost rectrix set (r1) is very short. Each sequential outer rectrix set gets progressively longer and creates the forked shape.**
**ADULT TRAITS:** HEAD.—White. BODY.—Ventral areas are immaculate white. Iridescence on upperparts is a medium purplish blue. WINGS.—Iridescence on upperwing is also a medium purplish blue. TAIL.—Very long and very deeply forked. The outer rectrix set (r6) is much longer than the preceding set (r5).
**JUVENILE TRAITS:** HEAD.—At close range, nestlings and fledglings have a tawny wash and

tawny shaft streaking on the head and neck. The tawny areas fade quickly and appear adult-like white soon after fledging. BODY.—Breast may also have a tawny wash or slight tawny shaft streaking through the fledgling period; however, this also wears and fades off and is rarely seen on fledged birds. Iridescent gloss on upperparts is a medium purplish blue with a greenish cast. Dorsal color is subtly different than adults'. The iridescent greenish dorsal cast is sometimes difficult to detect in the field. The iridescence on the rear scapulars contrasts with the noniridescent darker black band on the back and forward scapulars. WINGS.—Iridescence on the greater upperwing coverts is also the same color as the rear scapulars and contrasts with the darker black back, forward scapulars, and lesser upperwing coverts. In very fresh plumage, greater upperwing coverts, including primary coverts and tips of the primaries, have very narrow white tips. The white tips are not readily visible at long distances on recently fledged birds. The white tips quickly wear off by late summer. TAIL.—Moderately long and deeply forked. Outermost rectrix set (r6) is moderately longer than the preceding set (r5). The r6 set of juveniles is noticeably shorter than the r6 set on adults. When recently fledged with still-growing feathers on the r6 rectrix set, the tail is fairly short and only moderately forked. In fresh plumage, a very narrow white edge adorns each rectrix tip; however, this feature is practically invisible under field conditions. On the dorsal surface, the rectrices may exhibit a slight greenish cast.

**ABNORMAL PLUMAGES:** None reported.

**HABITAT: Summer.**—Semi-open areas with small or extensive pine and/or deciduous tracts. Wooded tracts contain variable amounts of second growth and mature, tall trees. Generally inhabits riparian areas such as marshes, bayous, rivers, lakes, and deciduous and/or coniferous swamps. Semi-open dry upland areas, including meadows, agricultural fields (soybean), and suburban areas, are accessed for foraging; however, these regions are typically in the vicinity of a riparian zone. Climate is hot and humid. Formerly inhabited temperate, moderately humid latitudes.

**Migration.**—Found in above-mentioned habitats.

**Winter.**—Similar to summer habitat but in tropical latitudes in South America.

**HABITS:** Tame. Gregarious. Kites often nest, forage, roost, and migrate in variable-size groups. Swallow-tailed Kites exhibit little fear of humans. A highly aerial species and one of the most aerial North American raptors. Perching occurs mainly for night roosting, during inclement weather, and for certain nesting activities. Sunning behavior occurs in the morning prior to flight and also after periods of rain in order to dry feathers.

Swallow-tailed Kites associate with Mississippi Kites in breeding season in all areas of typical range in the West. They are often seen with Mississippi Kites when migrating along the Coastal Bend region of Texas. Out-of-range dispersing Swallow-tailed Kites are sometimes in the company of Mississippi Kites.

**FEEDING:** Strictly an aerial hunter. Small and large insects, frogs, lizards, nestling birds, and small snakes are primary prey items. Bats roosting in trees are eaten during the breeding season. Kites regularly bring wasp nests, including stinging species, back to nests sites and feed on the larvae. Insects are mainly captured in flight, and Swallow-tailed Kites rarely engage in anything more than very short, but often acrobatic pursuits. Insects and other types of prey are masterfully snatched from outer branches and foliage of trees and bushes and from other types of short or tall vegetation. Captures prey with feet. Prey is eaten while soaring or gliding except when feeding young or transferring prey to another parent to feed young. Aerial feeding is accomplished by extending the foot clutching the prey forward and bowing the head down to the extended foot. Low-level aerial hunting occurs most often in the morning and late afternoon when insect prey is also at low altitudes. Foraging altitudes may increase substantially during midday; however, low-altitude hunting may also occur at this time. Swallow-tailed Kites primarily feed on insects except during the breeding season, when vertebrates form an extensive dietary component. Carrion is not eaten.

**FLIGHT:** Supremely buoyant and elegant aerial maneuvers. Most aerial activities involve energy efficient soaring and gliding, with the long forked tail used as a constantly adjusting rudder. Powered flight is used in short pursuits of prey, when chasing intruders at nest sites, and for stabilization during low-altitude foraging. Wingbeats are slow and methodical. Wings are held on a flat plane with wingtips gently bending upwards when soaring and held in a similar flat-winged fashion or bowed slightly downward when gliding. When pursuing prey, very acrobatic maneuvers are often made with exquisite tail fluctuations. When soaring, the tail is widely spread, exhibiting the deeply forked shape when soaring. When gliding, the tail may be widely fanned, moderately fanned, or nearly closed.

Long over-water flights occur during migration. Migrating birds may travel at very high altitudes. Drink and bathe by skimming the surface of the water. Swallow-tailed Kites do not hover or kite.

**VOICE:** Three main call notes: (1) A clear, sweet, sharp whistled *klee, klee, klee* repeated two to four times is the common call when agitated, by males after copulating, and during food transfers. Emitted by solitary birds, pairs, or by individuals of a group flying overhead. (2) *Tew-whee* is similar to *klee* call, but slurs upward at the end of the note and repeated two to six times. Emitted during courtship flights, by females after copulating, and by individuals on nests when mates are approaching. *Eeep* is uttered by both sexes and given as a single note or in a short series of notes. Call is given by males or females when passing food (while perched), by females soliciting copulation or food, and by nestlings and fledglings as a food-begging call. (3) *Chitter* is a very soft call at roost sites when birds take off and land before settling down or before morning flight.

**STATUS AND DISTRIBUTION: Summer.**—*Very uncommon* summer resident. Most of the species' population is in the e. U.S., and is estimated to be 800–1,150 pairs or approximately 3,200–4,600 birds at end of the breeding season. Population is low, but probably overall stable at the present time.

*Texas:* Only a few nesting pairs (perhaps 20). Regular sightings have occurred in e. Texas since the 1940s. Sizable numbers were surveyed in the early 1990s, but nesting was not confirmed. In 1994, the first nest in 80 years was located. Subsequently, additional pairs have been found. A reward was offered by the state for confirmed nesting in the late 1990s. Summer residence and probable breeding occurs in the following counties: Chambers, Jefferson, Liberty, Newton, Orange, and Tyler. Primarily found along the lower Neches, Sabine, and Trinity River basins. The range of the Swallow-tailed Kite is gradually increasing in e. Texas. Since they have a tendency to wander, extralimital sightings occur regularly just north of their typical range. Short-distance extralimital occurrences are known for the following counties: Bexar, Comal, Robertson, Rusk, and Washington. Moderate extralimital movements have extended to Hagerman NWR in Grayson Co. *Louisiana:* Probably 100 pairs. Confirmed breeding occurs in the following parishes: Beauregard, Iberville, and St. Landry. Probable breeding occurs in the following parishes: Calcasieu, Evangeline, Fourche, St. Charles, St. Martin, St. Tammany, Tangipahoa, and Washington. Adults have been seen in summer at Tensas River NWR in Tensas Parish, but breeding is not confirmed. *Arkansas:* Documented summer sightings since at least 1998, and through 2001, with suspected nesting in the White River NWR in Arkansas Co.

**Historical:**—Breeding range formerly extended to n.-cen. Minnesota. A drastic population decline took place from the late 1800s to mid-1900s. By 1910, the kite's breeding range had shrunk substantially to the states it currently inhabits due to massive habitat alteration of mature woodland riparian areas and human persecution. *Minnesota:* Formerly bred as far north as Becker and Hennipin Cos. until 1907. Nesting pairs remained in this state longer than any other state north of their present range during the decline period. *Iowa:* Nested in isolated areas in much of the state, west to Pottawattamie Co., until at least the late 1870s. A few summer records existed through 1917, a specimen record exists for 1931, then they were not seen again in the state until 1992. *Missouri:* Last nested in the state in 1912 in Jackson Co. Birds were seen in 1916, but not seen again in the state until the 1970s. *Nebraska:* Little data exists on actual breeding but probably occurred along the Missouri River prior to 1900 in Cass, Dixon, Douglas, and Washington Cos. Extirpated in the state around 1910. *Kansas:* Last known breeding was in 1876 in Woodson Co. There are few sight records since the late 1800s. *Oklahoma:* Last recorded nesting was in 1902 and last regular sighting was in 1910. *Arkansas:* Extirpated as breeder in the late 1800s or early 1900s. Summer sightings occurred in 1913 in the western part of state. There were no records from 1949 to 1990. A single bird was seen in Van Buren Co. from late Aug. to early Sep. 1991 (*see* Summer, Arkansas). *Texas:* Last documented nesting occurred from 1911 to 1914 in Harris Co. (*see* Summer, Texas).

**Winter.**—The species' gregarious nature of foraging and roosting at communal sites is the same as during other seasons. Most of the North American population winters in a fairly small region in w.-cen. and sw. Brazil. *E. f. Forficatus* winters in the breeding range of Brazil's population of *yetapa.*

**Movements.**—Migrates in small flocks or singly.

*Fall migration:* An early season migrant. After nesting, birds may disperse in any direction prior to actually migrating. On the s. Pearl River in s. Louisiana, fairly large premigration roosts of over 100 birds form by late Jul. Actual migration may begin in mid-Jul. and extend into Sep. In the U.S., the majority of birds move from late Jul. through Aug. Stragglers occur into Oct. in the s. U.S. South of the U.S., migration may last until late Oct., when most kites arrive on the winter grounds. Adults precede most juveniles. Movements may span 3 months.

Only small numbers of Swallow-tailed Kites are seen during fall in the West. Three different migratory routes may be taken in the fall between the U.S. and s. Mexico: (1) An easterly over-land coastal route from Louisiana to n. Florida, then south through the Florida Peninsula. As seen with telemetry data by K. Meyer (unpubl.), an over-water route is taken between s. Florida and the Yucatán Peninsula in Mexico. Some birds may stop in or fly over w. Cuba en route from Florida to the Yucatán. (2) Observers on off-shore oil rigs in the Gulf of Mexico have seen kites making a trans-Gulf of Mexico crossing between the n. Gulf Coast of the U.S. and the Yucatán Peninsula or other areas of s. Mexico. (3) Hawkwatches in s. Louisiana, e. and s. Texas, and e. Mexico record many kites migrating along coastal areas. Movement through Mexico occurs east of the Sierra Madre Oriental.

From s. Mexico, Swallow-tailed Kites wend their way on mainly a land-based course through e. Central America. They enter South America at w. Colombia and cross the Cordilleras, then angle southeast to the winter grounds in Brazil.

The peak movement, as in the East, occurs prior to mid-Aug. and before Texas hawkwatch sites at Smith Point in Chambers Co. and at Hazel Bazemore County Park in Nueces Co. are staffed. Hawkwatch sites along coastal Texas are seeing migrants when they officially begin counting in mid-Aug. Migration through Texas trickles during Sep. with stragglers occurring into early Oct.

In Veracruz, Mexico, official counts begin in mid-Aug., and kites are also being seen at this time. Peak migratory period in Veracruz varies from late Aug. to early Sep., with stragglers occurring until early Oct.

South of the U.S., hawkwatches observe considerably more Swallow-tailed Kites. The 2001 Veracruz count logged 286 birds with a peak in late Aug. of 112 birds. The new Talamanca, Costa Rica, Hawkwatch tallied 1,319 in the fall of 2001. This later count gives credence to the belief that the majority of kites take an over-water route and bypass land-based hawkwatch sites.

*Spring migration:* First adults typically appear in Texas and Louisiana in early to mid-Mar. The earliest adult arrivals in Louisiana have occurred in late Feb. in St. Tammany and Plaquemines parishes. Nonbreeding adults and 1-year-olds migrate later. Movements extend at least until mid-May on the upper Texas coast as seen at the hawkwatch at LaPorte and at sites along coastal Louisiana in Cameron and Plaquemines parishes. A fair number of 1-year-olds and nonbreeders peak on the upper Texas coast in mid- to late Apr., with lesser numbers straggling through until mid-May.

It is assumed that most migrants retrace their route they took in the fall. A small to moderate number of migrants are seen along coastal areas of Texas and Louisiana in the spring. Migrants have also been observed making a trans-gulf crossing from oil rigs. Migrants have been seen leaving the Yucatán in Feb. and heading north over water at very high altitudes.

**Extralimital movements.**—Irregular northward dispersal spring through fall. Considered *acci-*

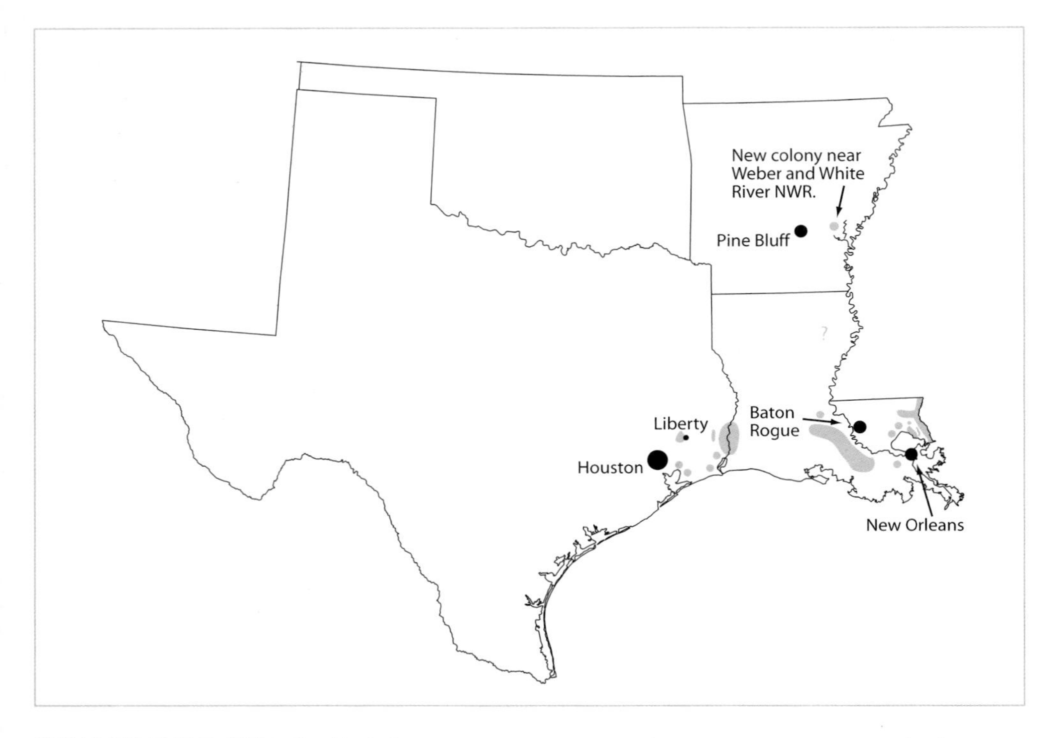


**SWALLOW-TAILED KITE,** *Elanoides forficatus:* Very uncommon. Local. Increasing in se. Texas and Arkansas. 850–1,150 pairs, mainly in e. U.S. Irregularly disperses Apr.–Sept. northwest to CO and north to MN. Winters in w.-cen. and sw. Brazil.

*dental* north of current breeding states. Sightings have occurred northward to historical breeding locales in the Midwest. *Minnesota:* One individual was seen for several days in Le Sueur Co. in mid-May 1999. This is the first state sighting in nearly 40 years. Historically, this state was the northernmost breeding area. *Iowa:* Ten records spanning from 1867 to 1931. Recent sightings: (1) Black Hawk Co. in May 1992, and (2) Cerro Gordo Co. for 2 days in mid-Sep. 2000. *Missouri:* (1) Ozark Co. in Jul. 1972, and (2) Greene Co. in Aug. 1975. *Nebraska:* There are no records since about 1910. *Kansas:* Shawnee Co. in early Sep. 1972. *Oklahoma:* (1) Texas Co. in Aug. 1988, (2) McCurtain Co. in Jun. 1993, and (3) Le Flore Co. in Apr. 1998. *Arkansas:* (1) Van Buren Co. from late Aug. to early Sep. 1991, and (2) Little River Co. in late Apr. 1999. *Colorado:* Prowers Co. in Jul. 1993.

**NESTING:** Mar.–Jul. Breeds singly or in small, loosely formed colonies. One or two nonbreeding birds often accompany breeding pairs but rarely assist in nesting duties.

**Courtship (flight).**—*Diving-swoop* (*see* chapter 5). In *diving-swoop*, any pair member may swoop down on the other gliding bird; often two or more pairs may be involved. Courtship flights are accompanied by considerable vocalization (mainly *tew-whee* but also *klee* notes).

Breeding does not occur until birds are at least 2 years old. Nests are built by both members of a pair. Old nests may be refurbished and used again. Nests are built in live trees and placed near the top of one of the tallest trees in the tract. Nests are often at least 100 ft. (30 m) high. Nests may be somewhat oblong in shape and average 20 in. (52 cm) long and 17 in. (42 cm) wide. Nest structure comprises small sticks, which are broken off from branches while in flight, and lichens and Spanish Moss. Material is continually added to nests during the course of use. Both sexes incubate, but incubation stints average longer for females.

Two eggs are typical, but may only have 1, and occasionally 3 eggs. The eggs are incubated for about 28 days. Youngsters fledge in 36–42 days. Only one clutch is laid.

**CONSERVATION:** No measures are currently exercised. This species is not "listed" in Texas, Louisiana, or Arkansas. Many pairs nest on private lands and enjoy protection from concerned landowners. Ongoing telemetry studies will further assist knowledge of migration routes, roosting areas, and wintering areas.

The Kansas Department of Wildlife and Parks attempted a reintroduction in the summer of 1982 by cross-fostering chicks into Mississippi Kite nests. Although one or two chicks fledged, the program was discontinued.

**Mortality.**—Historically, shooting was a factor in decline of the population; illegal shooting still occurs but appears to be a minor threat.

Natural mortality occurs with loss of eggs and nestlings due to inclement weather, especially being blown out of nests and nest trees being toppled over. A minimal amount of mortality results from mammalian predation. Some mortality occurs from avian predators, including from Bald Eagles. Parasites cause nestling mortality.

**SIMILAR SPECIES:** None concerning raptors. At a distance might be confused with the deeply forked-tailed Magnificent Frigatebird, particularly juveniles, which have white heads and underparts.

**OTHER NAMES:** Swallowtail. *Spanish:* Milan Tijereta. *French:* Milan à Queue Fourchue.

---

REFERENCES: Baicich and Harrison 1997; Baumgartner and Baumgartner 1992; Brown et al. 1997; Bylan 1998, 1999; Cardiff 1999; Dinsmore 2002; Ducey 1988; Eberly 1999; Granlund 1999; James and Neal 1986; Kellogg 2000; Kent and Dinsmore 1996; Meyer 1995; Meyer et al. 1997; Purrington 2000; Ripple 2000; Robbins and Esterla 1992; Thompson and Ely 1989.

**Plate 53. Swallow-tailed Kite, adult** [May] ▪ White head. ▪ White undersides. ▪ Very long, deeply forked black tail.

**Plate 54. Swallow-tailed Kite, adult** [May] ▪ White head. ▪ Black iridescent dorsal upperparts. ▪ Large white patch on base of tertials and inner greater upperwing coverts. ▪ Very long, deeply forked black tail.

**Plate 55. Swallow-tailed Kite, adult** [**May**] ▪ White ventral body. ▪ White underwing coverts and black remiges. ▪ Very long, deeply forked black tail. ▪ *Note:* Inner 2 primaries (p1 and 2) are molting and missing.

**Plate 56. Swallow-tailed Kite, adult** [**May**] ▪ Iridescent purplish blue upperparts have a non-iridescent black band across the back and upperwing coverts. ▪ Very long, deeply forked black tail.

**Plate 57. Swallow-tailed Kite, juvenile** (**1-year-old**) [**May**] ▪ White ventral body. ▪ White underwing coverts and black remiges. ▪ Long, deeply forked black tail. ▪ *Note:* Molting the inner 4 primaries (p1–4), upperwing coverts, and 2 central rectrices (r1 set). Molting upperwing coverts create translucent white spots on underwing coverts.

**Plate 58. Swallow-tailed Kite, juvenile** (**1-year-old**) [**May**] ▪ Purplish-blue upperparts may show a greenish cast. Noniridescent black band on back and upperwing coverts. White spots on upperparts are molting, missing feathers. ▪ *Note:* Fresh-plumaged juveniles in summer and fall are not in molt.

# WHITE-TAILED KITE
## *(Elanus leucurus)*

**AGES:** Adult, subadult (basic I), and juvenile. Adult plumage is acquired when 1 year old. Subadult plumage is a transitional molt stage from juvenile to adult; however, there is often a short period with minimal molt or lack of molt. Juvenile plumage is retained for only a few months. However, since the West has a prolonged breeding season, it is possible to see individuals in juvenile plumage May–Feb. in s. California and s. Texas.

**MOLT:** Accipitridae wing and tail molt pattern (*see* chapter 4). First prebasic molt from juvenile to subadult begins when juveniles have been fledged for about 2 months and are about 3.5–4 months old (Palmer 1988, Dunk 1995). Contour body molt is completed when about 6 months old, but may take longer. Molt begins on the crown of the head, then appears on the back, scapulars, and breast. Rectrices begin molting when kites are about 7–8 months old after contour body molt is nearly complete or is complete. Wing covert molt begins prior to remix molt, but is completed during the span of the remix molt. Remiges do not molt until approximately 1 year of age. At this time, body molt has been completed for several months, and at least some or most rectrices have also been molted. There is a short time span of about 1–2 months when contour feather molt is basically complete and rectrix molt has not started. When the first prebasic molt is finally completed on the wings, full adult plumage is attained.

First prebasic body molt mainly occurs in summer through fall. Juveniles hatched in late summer and fall may be in body molt in fall and winter.

Subsequent prebasic molts are annual and complete. Molt is primarily spring through fall. Contour feathers, rectrices, and remiges molt in unison.

**SUBSPECIES:** Polytypic with two races. *E. l. majusculus* inhabits the U.S. and Central America and *E. l. leucurus* inhabits South America. *E. l. leucurus* is smaller and has a shorter tail but is otherwise identical in plumage to *majusculus*.

The White-tailed Kite was considered a subspecies of the Black-shouldered Kite (*Elanus caeruleus*) of Africa, s. Europe, and Asia; however, studies on plumage, proportion, and behavioral differences illustrated that they were different species (Clark and Banks 1992). In 1994, the AOU formally split them.

**COLOR MORPHS:** None.

**SIZE:** A medium-sized raptor. Males average smaller than females, but there is considerable overlap. Length: 14–16 in. (36–41 cm); wingspan: 37–40 in. (94–102 cm).

**SPECIES TRAITS:** HEAD.—**Narrow black bill. Yellow cere and gape.** Black inner lores. Large eyes. BODY.—**Short yellow or orangish tarsi and feet.** WINGS.—**In flight, long and narrow with parallel front and trailing edges. Wingtips are moderately pointed.** When perched, wingtips are just short of the tail tip. WINGS (ventral).—**Large black spot or rectangle shape on the carpal region. Primaries are dark gray, but graduate to a paler gray on the inner primaries.** WINGS (dorsal).—**Large black shoulder patch on the median coverts and all lesser upperwing coverts.** TAIL.—**Moderately long. The deck rectrix set is a bit shorter than all other rectrices and creates a notched or somewhat forked shape when closed. Deck rectrix set is pale gray, and all other sets are white.**

**ADULT TRAITS:** HEAD.—Iris color varies from reddish orange to red. **Crown and nape are pale gray, rest of head is white.** BODY.—Immaculate white ventral surface. **Back and scapulars are uniformly pale gray. Yellowish orange tarsi and feet.** WINGS.—**Ventral surface of the secondaries is very pale gray. Uniformly black shoulder patch on the upperwing coverts.** TAIL.—**The gray deck rectrix set and white outer rectrix sets are unmarked.**

**BASIC I/SUBADULT I (LATE STAGE):** HEAD.—As on adults, but iris color is often more orangish. Dark streaking behind the eyes of early stage is absent. BODY.—As on adults on dorsal and ventral areas. WINGS.—All remiges are retained juvenile feathers. White tips of the juvenile plumage on the tertials, secondaries, and primary greater coverts are completely worn off; however, narrow white tips may be retained on the inner primaries. Upperwing coverts may be retained juvenile feathers, with all white edgings worn off, or possess some non-white-tipped adult feathering. TAIL.—Retained juvenile rectrices with the dusky subterminal band or may have molted in a few unmarked adult rectrices. Molting rectrices are often partially grown.

*Note:* There may be a span of 2 or more months in which birds may be in a nonmolting plumage.

**BASIC I/SUBADULT I (EARLY STAGE):** HEAD.—Newly molted gray crown is similar to an adult's; however, the crown and the area immediately behind the eyes may retain some sparse dark brown juvenile streaking. BODY.—A moderate amount of new, pale gray adult feathers are among juvenile feathers on all areas of the back and scapulars. Retained juvenile feathers are a faded pale brownish gray with worn, narrow white tips. At a distance, the dorsal area can appear rather uniformly grayish. Tarsi and feet are yellow. WINGS.—Retained juvenile remiges. White tips of the tertials, secondaries, and greater upperwing coverts are either worn and very narrow or are completely worn off. The white tips of median upperwing coverts are worn off and all brownish-tipped lesser upperwing coverts are also worn off. TAIL.—Retained juvenile rectrices with the dusky subterminal band.

*Note:* A continual change into adult plumage occurs with ongoing molt.

**BASIC I/SUBADULT I (VERY EARLY STAGE):** HEAD.—Crown is mainly adultlike pale gray with possibly some retained juvenile dark streaking. The area behind the eyes also has retained juvenile dark streaking. Iris color is pale orangish brown, but can be quite reddish. BODY.—As on juveniles, but the rufous-tawny "necklace" wash on the breast is often reduced by molt and sun bleaching. Back and scapulars are medium brownish gray with somewhat worn, moderately wide white tips. Tarsi and feet are yellow. WINGS and TAIL.—As on juveniles.

*Note:* This stage is the earliest prebasic molt stage in the molt sequence beyond the juvenile plumage. It essentially appears like a juvenile with the exception of the grayish crown and narrower, worn white tips on dorsal feathers.

**JUVENILE TRAITS:** HEAD.—Crown is white or washed with pale tawny and narrowly streaked with dark brown. The area immediately behind the eyes is also streaked with dark brown. Iris color is pale or medium brown, but may be orangish brown. BODY.—Back and scapulars are medium brownish gray with a broad white tip on each feather. Breast is washed with a variable-sized, rich tawny "necklace." Short tawny streaks may extend below the necklace onto the belly and forward flanks. Tarsi and feet are yellow. WINGS.—All remiges are broadly tipped with white. Greater upperwing coverts are broadly tipped with white. The feathers of the median covert tract on the black shoulder patch have narrow white tips; many lesser coverts are narrowly tipped with pale brown. Gray upper surface of the remiges and greater upperwing coverts is slightly darker than on adults. Ventral surface of the secondaries is pale gray (and slightly darker than on adults). TAIL.—Narrow dusky subterminal band.

*Note:* This pumage is retained for about 3.5 months.

**ABNORMAL PLUMAGES:** None.

**HABITAT:** Open and semi-open wet or dry human-altered or natural grasslands, pasturelands, meadows, idle fields, grassy coastal dunes, marshes, and agricultural areas with ample nonfarmed edge strips. In arid climates, kites are particularly fond of irrigated low intensity farmlands such as alfalfa and grass-hay. Widely scattered bushes or trees dot the open areas and moderately scattered small tracts or single bushes or trees are in semi-open regions. Climate is seasonally hot and varies from humid or arid, depending on geographic region. Topography varies from flat or moderately hilly. In California, the elevation may reach 2,000 ft. (600 m) in the western foothills of the Sierra Nevada and nearly 4,000 ft. (1,200 m) on the flat grasslands of Arizona. Kites are found up to 3,400 ft. (1,030 m) in Sonora, Mexico. Dispersing birds may be found at higher elevation flat plains (e.g., over 6,000 ft. [1,800 m] in Laramie Co., Wyo.).

**HABITS:** White-tailed Kites vary from being a tame to wary species. Exposed branches, posts, and wires are favored perches. They do not perch on utility poles. The highest perch on a tree is typically used. In very windy conditions, kites may seek shelter in protected, heavily foliaged, lower portions of a tree.

The tail is "bobbed" repeatedly: it is cocked

upward from the angle of the body, then slowly lowered. The tail-bobbing action is a display against conspecific intruders into territory. *Talon-grappling* (*see* chapter 5) occurs as an agonistic confrontation between conspecifics.

White-tailed Kites may be gregarious and form large communal night roosts in late fall and winter. Roosts of 100 or more kites occur in California and Texas. Roosts are primarily in bushes and trees but sugar cane fields may be used in s. Texas. Kites may perch on the ground as a staging area prior to going to the roost site.

**FEEDING:** An aerial hunter. Hunting flights are at altitudes of 15–80 ft. (5–25 m). White-tailed Kites prey almost exclusively on small rodents. Kites occasionally prey on small birds, lizards, and insects that are on the ground.

**FLIGHT:** Wings are held in high dihedral when soaring and gliding. Powered flight is with moderately fast wingbeats. Kites hover with shallow-beating wings when hunting. Legs are sometimes lowered when hovering. If legs are lowered, they are often fully extended, with toes widely spread. In strong winds, hunting White-tailed Kites may kite for short stints. Prey is captured by an awkward-looking dive, with the bird angled headfirst and wings fully extended high above the body. Landing birds daintily glide to or gently drop onto a perch with feet fully extended downward, often with quivering wingbeats.

**VOICE:** A fairly vocal species, particularly when disturbed. A soft *cherp* is emitted when mildly agitated. A grating, raspy *kree-aak* is emitted when very agitated, when bringing prey to a nest, and when nestlings and/or fledglings are food-begging (sounds much like a Barn Owl's call). A guttural, grating *grrkkk* is called when chasing intruders.

**STATUS AND DISTRIBUTION:** Historically threatened with possible extirpation in much of its U.S. range in the early 1900s due to habitat destruction, shooting, and egg-collecting. The species has rebounded on its own in recent decades, and its range in Mexico and the U.S. is more extensive now than during historical times. White-tailed Kites benefit in many regions from human-altered, but ecologically friendly environments.

*California:* Nearly extinct in the state by the 1930s. The kite population increased from the 1940s to the early 1970s. However, their numbers decreased in the mid-1970s due to drought and apparent movement into Oregon. Numbers have increased since the 1970s. Currently fairly common and widespread. They occupy a vast portion of suitable lower elevations in the state. Somewhat extralimital distribution for the state has occurred to Honey Lake area in Lassen Co. Isolated breeding occurs in Imperial Co. The population in n. California is tied to the cyclic fluctuations of the California Vole. *Oregon:* Uncommon and local. Kites were first seen in the state in the mid-1920s, but were not seen again until 1933. In the 1960s and 1970s, kite numbers increased dramatically from birds dispersing from California. The population leveled off by 1990, then decreased until 1993, when it began increasing. Formerly a late summer through spring visitor; however, year-round territories and breeding now occur. Kites first bred in the state in 1976 near Medford, Jackson Co., and currently breed in Benton, Coos, Curry, Douglas, and Tillamook Cos. Regular sightings with possible breeding occurs in Josephine and Polk Cos. *Washington:* Rare and very local. White-tailed Kites were first recorded in the state in Jul. 1975. Expansion into the state subsequently followed expansion into Oregon from California. The first state nesting record was in 1988 in Pacific Co. A few pairs currently nest in nw. Pacific, cen. Thurston, and cen. Wahkiakum Cos. They are regularly seen, with possible breeding, in Lewis Co. There are recent spring sight records for nw. King and w. Skagit Cos. *Arizona:* Very uncommon and very local. Kites utilize human-altered, irrigated farmlands in many areas of the state. White-tailed Kites were first seen in the state in 1972 in Cochise Co. The first breeding record was in 1983 in Pinal Co. Confirmed breeding has occurred in ne. Cochise, se. Graham, nw. Maricopa, and s. Pima Cos. Kites are regularly seen and possibly breed in e.-cen. La Paz Co. *Texas:* Fairly common to locally common. Massive agricultural conversion of s. Texas in the early and mid-1900s wiped out an extensive amount of habitat. However, they have adapted to remnant, small-habitat plots and now thrive in this region. The Texas population increased after 1960 and rapidly increased after 1980. Their primary range is from w. Jefferson Co. west to se. Harris and s. Waller Cos., then south along the Coastal Bend region through s. Goliad Co., to se. Duval, e. and s. Hidalgo, and s. Starr Cos. Except along the immediate coastal areas, kites are absent from much of the extensively farmed Nueces Co. Kites are isolated residents in irrigated farmlands in Dimmit, La Salle, Maverick, Webb, and Zavala Cos. Very small numbers are resident in the very isolated northern counties of Navarro and Throckmorton. *Louisiana:* Breeds in Calcasieu and Cameron Parishes and probably breeds in Beauregard Parish. *Mexico:* First discovered in Sonora in late Mar. 1979. Sightings increased in the following decades. Breeding is likely but not verified.

**Extralimital breeding.**—*Oklahoma:* Kites first nested in 1860 in Murray Co. They did not nest again until 1982, when two nests were found in Latimer Co. Most recent nesting occurred in Atoka Co. in 1994 and 1997. *Kansas:* Only state nesting record was in Pottawatomie Co. in 1989 (nest was blown out of the tree). *North Dakota:* Only state nesting was discovered in McKenzie Co. in late Sep. 1987 with an adult pair and five fledged young. The nest was not located.

**Movements.**—Basically a sedentary species. White-tailed Kites may embark on variable-distance seasonal dispersal or nomadic wanderings from fall through early summer. Movements take them north and east of typical range. Minor shifting occurs within typical range in order to locate suitable nesting and feeding areas during nonbreeding periods. No. California birds embark in northward dispersal when populations of California Voles drop. In Oregon, dispersing birds arrive from California in early Aug., increase in Sep. and Oct., and leave in late Mar. and Apr.

**Dispersal movements.**—Highly irregular long-distance dispersal and nomadic excursions may take birds far north and east of typical range from spring through fall. *Oklahoma* (other than nesting records): Tillman Co. in 1982; Commanche Co. in 1983; Marshall Co. in 1984; and Tulsa, Osage Co., in 1984. *Missouri:* (1) Old River, Miller Co., in mid-Dec. 1976; (2) Russellville, Pope Co., in mid-Apr. 1978 (2 birds); (3) Arkadelphia in Clark Co. in early Mar. 1981. *Missouri:* Single record for Springfield, Greene Co., in mid-May 1983. *Wyoming:* (1) Moose, Teton Co., in mid-Aug. 1982; (2) west of Cheyenne, Laramie Co., in mid-Nov. 1984; (3) Casper, Natrona Co., from mid-Jun. to late Jul. 1989; (4) Yellowstone NP in late Aug. 1991; (5) Casper, Natrona Co., in mid-Apr. 1992. *North Dakota:* Dickinson, Stark Co., in early Jun. 1996. *See also* Extralimital Breeding. *Minnesota:* Single record for Washington Co. in mid-May 2000. *Nevada:* Numerous but not well documented sightings in the 1970s: Carson Lake near Fallon; Mason Valley area and Wabuska, Lyon Co.; Walker Lake, Mineral Co.; near Reno (multiple sightings in the 1970s); Pyramid Lake area, Washoe Co. In the 1980s: (1) Mormon Farm area, Clark Co., in mid-Sep. (two birds for a 2-week period); (2) Las Vegas for 5 days in late Apr. 1986. *British Columbia:* Becoming regular in the southwestern part of the province. First sightings occurred in 1990 with three records near Vancouver that pertained to separate individuals: (1) Westham Island near Vancouver in late Apr., (2) Fraser River delta on Jun. 5, and (3) Pitt Meadows on Jun. 7. There were four sightings in the Vancouver vicinity from mid-Apr. to early May 1999. The most extralimital record is near Nanaimo on Vancouver Island in late Apr. 1999.

**NESTING:** Jan. through fall. Very prolonged breeding season. Pairs may remain intact year-round. Newly formed pairs unite Dec.–Aug. Kites regularly double clutch, and a few may attempt a third clutch (locally in California, possibly occurs in Texas). Nest-building occurs Jan.–Aug.

**Courtship flight.**—*Flutter-flight* and *V-flutter flight*, with *leg-dangling* (*see* chapter 5), are performed by males. In *leg-dangling*, legs are fully extended downwards and toes widely spread. In *V-flutter flight*, the quivering wings may be held at such a high angle that they may nearly touch each other.

Nests are built by females and placed in the upper portion of bushes or trees and are 9–60 ft. (3–18 m) high. Nest structures are 20 in. (51 cm) in diameter and 8 in. (21 cm) deep, very compact, and made of twigs and lined with grasses. If the first nest fails or a second brood is started, a new nest is built in another location within the territory. Clutches range from 3 to 6 eggs, with 4 being most typical. Clutches of 6 eggs are rare. The 30- to 32-day incubation is performed only by females. Only

**WHITE-TAILED KITE,** *Elanus leucurus:* Common. Local. Expanding range in AZ and northward in BC, WA, and OR. Mainly resident but disperses widely. Forms large winter roosts in CA.

females feed the nestlings. Males hunt and guard territories. Fratricide (siblicide) does not occur. Youngsters fledge when 28–35 days old. Fledglings are fed by aerial prey transfer, while on nests, or on perches. Pairs may start a second nest while still feeding fledglings from the first brood. However, parents cease caring for the first brood once the second brood hatches.
**CONSERVATION:** No measures are taken. The White-tailed Kite is one of few species of raptors that has benefited from human alteration of the environment. As long as sufficient buffer habitat exists for foraging, kites readily adapt to cultivated and irrigated farming communities. Crops such as alfalfa and other hay types harbor prey species; thus, White-tailed Kites can take advantage of extensive agricultural use of the land. Brush clearing and burning have created habitat and assisted range expansion.
**Mortality.**—Little known other than natural depredation by larger diurnal raptors and owls. Nest depredation by grackles, crows, mammals, and snakes. As with any raptor, they are susceptible to being shot.
**SIMILAR SPECIES:** **(1) Mississippi Kite.**—No range overlap during breeding season. **(1A) Adults.**—PERCHED.—Uniformly dark gray dorsum, including upperwing coverts. Black tail. FLIGHT.—Gray ventral surface of wings. Black tail. Does not hover. **(1B) Juveniles.**—PERCHED.—Lightly streaked individuals are similar to recently fledged White-tailed Kites that are heavily marked with rufous-tawny on ventral region. Uniformly dark brown upperwing coverts. Tail is blackish (and may not be banded). FLIGHT.—Uniformly mottled rufous-brown underwing coverts. Dark tail. **(2) Northern Harrier.**—Range overlap Sep.–Apr. **(2A) Adult males.**—PERCHED.—Pale lores. Long tarsi. Wingtips are shorter than the tail

tip. FLIGHT.—Head and neck are gray and appear hooded. Black area on outer half of the primaries; black bar on trailing edge of secondaries. White patch on the uppertail coverts. **(2B) Adult females.**—PERCHED.—Same as for adult males. FLIGHT.—Hooded look with brown head and neck. Barred underside of remiges; dark bar on trailing edge of underwing. In dorsal view, exhibits white uppertail covert patch. **(3) Nonraptors (gulls).**—Several small and medium-sized gulls may superficially appear similar and cause a double-take; however, wings are held in an arch when gliding or soaring; all may hover like a kite. Similar gulls: Bonaparte's Gull (winter), Laughing Gull (nonbreeding plumages; Texas and Louisiana), Mew Gull (winter; along Pacific Coast), and Ring-billed Gull (winter).

**OTHER NAMES:** None used. *Spanish:* Milano coliblanco, Milano Cola Blanco. *French:* Élanion à Queue Blanche.

REFERENCES: Adamus et al. 2001, Alcorn 1988, Arizona Breeding Bird Atlas 1993–1997, Baicich and Harrison 1997, Baumgartner and Baumgartner 1992, Blanchard 1999. Clark and Banks 1992, Clark and Wheeler 1989, Dorn and Dorn 1990, Dunk 1995, Gatz 1998, Gilligan et al. 1994, James and Neal 1986, Oberholser 1974, Palmer 1988, Robbins and Esterla 1992, Russell and Monson 1998, Shepard 1999, Small 1994, Smith and Ireland 1992, Smith et al. 1997.

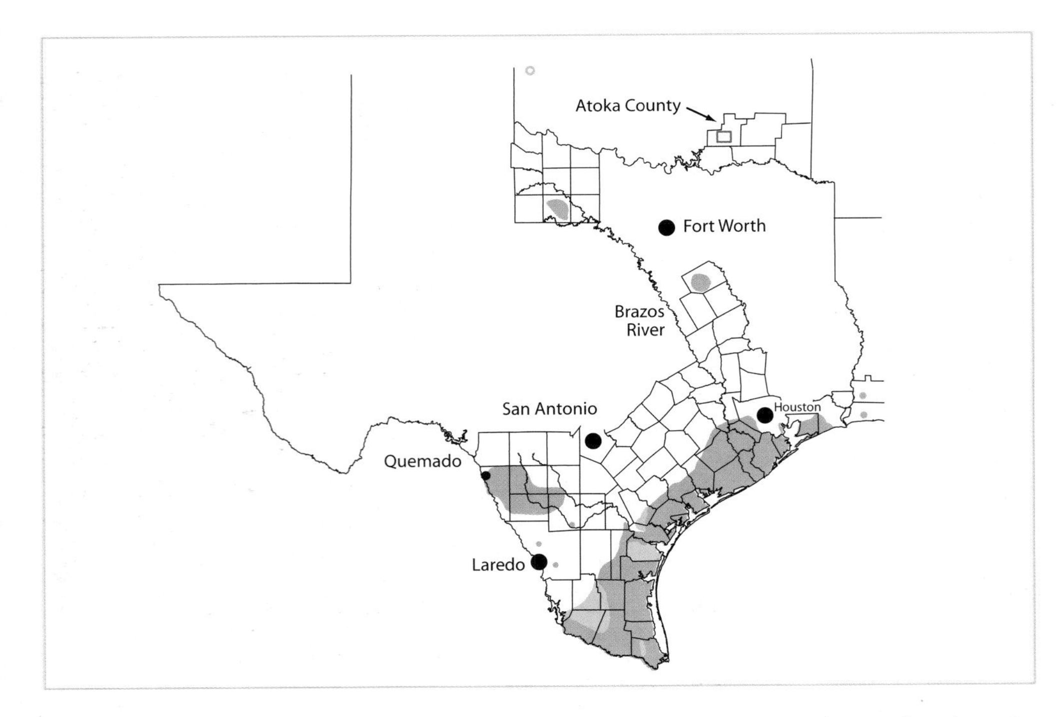


**WHITE-TAILED KITE,** *Elanus leucurus:* Fairly common. Local. Mainly resident. Expanding north and east. Irregular dispersal north to MT, ND, MN, and WY. Nested McKenzie Co., ND, 1987; Pottawatomie Co., KS, 1989; & Atoka Co., OK, 1994 and 1997. Forms large roosts in winter in Texas.

**Plate 59. White-tailed Kite, adult** [**Oct.**] ▪ Red iris, yellow cere, and black bill. ▪ Gray back and scapulars. ▪ Large black shoulder patch.

**Plate 60. White-tailed Kite, adult** [**Oct.**] ▪ Red iris, yellow cere, and black bill. ▪ Mainly white head and white underparts. ▪ Large black shoulder patch. ▪ Underside of tail is plain white.

**Plate 61. White-tailed Kite, adult** [**Feb.**] ▪ White underparts. ▪ Small black carpal spot and dark gray primaries; very pale gray secondaries. ▪ Underside of tail is plain white.

**Plate 62 White-tailed Kite, adult** [**Dec.**] ▪ Gray remiges with a large black shoulder patch. ▪ Upperside of tail is white with 2 gray mid-rectrices.

**Plate 63. White-tailed Kite, adult** [**May**] ▪ When hunting, kites hover, often with lowered legs.

**Plate 64. White-tailed Kite, adult** [**Sep.**] ▪ When prey is sighted, kites dive head-first to the ground with wings raised above their bodies and legs lowered.

**Plate 65. White-tailed Kite, subadult** (**late stage**) [**Feb.**] ▪ White underparts and gray upperparts are adult feathers. ▪ Large black shoulder patch. Retained juvenile remiges have narrow white tips on primaries. ▪ Retained juvenile tail with dusky subterminal band. ▪ *Note:* Adult body plumage, but retain very worn juvenile remiges and rectrices. Tail is cocked upwards.

**Plate 66. White-tailed Kite, subadult** (**late stage**) [**Dec.**] ▪ Gray upperparts. ▪ Large black shoulder patch; gray upper wing. ▪ Retained juvenile tail with dusky subterminal band; gray mid-rectrices. ▪ *Note:* Retained juvenile remiges and rectrices. White tips on primary coverts.

**Plate 67. White-tailed Kite, subadult (early stage) [Jan.]** ▪ Orangish iris. Most brownish colored juvenile crown feathers have molted. ▪ Faded juvenile upperparts with a few gray adult feathers. White underparts. ▪ Large, black shoulder patch. Narrow white tips on greater coverts and primaries. ▪ *Note:* Worn juvenile feathers with new adult feathers on head, underparts, and some scapulars.

**Plate 68. White-tailed Kite, subadult (very early stage) [Feb.]** ▪ Orangish iris. Most brownish colored juvenile crown feathers have molted. ▪ Brown scapulars with white tips; orangish breast. ▪ Large black shoulder patch. Greater coverts and remiges tipped with white. ▪ Dusky subterminal tail band. ▪ *Note:* Mainly juvenile plumage with adultlike head.

**Plate 69. White-tailed Kite, juvenile [May]** ▪ Brownish iris. Brown streaking on crown. ▪ Grayish brown upperparts with broad white tips. ▪ Large black shoulder patch. Broad white tips on greater coverts and remiges. ▪ *Note:* Recently fledged.

**Plate 70. White-tailed Kite, juvenile [May]** ▪ White underparts have tawny breast markings. ▪ Small black carpal patch. Dark gray primaries; pale gray secondaries. ▪ Underside of tail is white with dusky subterminal band. ▪ *Note:* Recently fledged. Hovering flight.

## MISSISSIPPI KITE
### *(Ictinia mississippiensis)*

**AGES:** Adult, basic I (subadult I), and juvenile. Adults have sexually dimorphic plumages that are fairly diagnostic at field distances.

Basic I (Subadult) birds also have sexually dimorphic plumages and adhere to the same sexual traits as seen in the respective adult sex. Subadult age is not a separate plumage, but is a transitional molt stage consisting of incoming adult and retained juvenile feathers. Subadults also have individual variation because of variable stages of molt progression.

Juveniles have a considerable amount of individual variation on body, wing, and tail markings, but do not have sexually dimorphic plumages. Plumage is retained only a few months.

**MOLT:** Accipitridae wing and tail molt pattern (*see* chapter 4).

**Adult.**—Acquired when 1.5 years old. Females begin molt in Jun. or Jul. and begins on males in Jul. or Aug. Molt begins with loss of the innermost primary (p1). Some females molt several inner remiges; males generally molt only one or two remiges while in the U.S. Some females may also molt the innermost (r1) and outermost (r6) rectrices. Many females—but no males—begin rectrix molt while in the U.S. Most molt, including body molt, occurs on the winter grounds. Any wing and tail molt that began in Jun. through Aug. is probably suspended during fall migration, then resumed once on the winter grounds. An adult specimen from the winter grounds in n. Argentina in mid-Jan. was just completing its annual wing molt with the replacement of the outermost primary set (p10).

**Basic I (Subadult).**—This transitional stage consists of a very prolonged, continuously molting, first prebasic molt from juvenile to adult. The metamorphic plumage change begins when juveniles are on the winter grounds. Exact timing is unknown, but probably begins when about 4–6 months old, which is typical of many kite species. *Note:* Juveniles in early and mid-Oct. do not show signs of prebasic molt, and are up to 4.5 months old at this time ($n = 2$ from Honduras; LSU). Molt continues when the kites return to the U.S. and is completed when once again on the winter grounds. Molt is not completed until the birds are 1.5 years old. Head and breast are molted prior to other feather regions. *Note:* All new plumage acquired by molt is adult feathers.

Eye, cere, and foot colors change to adult colors prior to returning to the U.S. in spring. A considerable amount of body molt generally takes place on the winter grounds, but only a moderate amount occurs in summer. Body plumage is fairly adultlike when the kites return to the U.S. Further body molt progression into adult plumage is usually, but not always, completed by the end of summer.

Wings and tail molt on a partial basis during summer. Wing molt begins within a span from mid-May to mid-Jun. on the innermost primary (p1), then sequentially advances outwardly throughout the summer and fall, but rarely past p7 while in the U.S. As in most raptors, one or two old primaries are dropped ahead of the incoming new feathers and produce only a small gap of missing feathers. In mid-Jun., some individuals, however, may lose up to four inner primaries in quick succession before new adult replacements come in, thus creating a huge gap in the mid-wing. Only the outermost secondary (s1) and the three tertials are molted while in the U.S. Molt on primaries and secondaries is completed on the winter grounds.

Most greater upperwing coverts are gradually replaced during the summer. Underwing coverts are also gradually replaced during the summer, except the greater underwing covert tract, which is not molted until on the winter grounds. Axillaries molt prior to the underwing coverts.

Tail molt begins from Jul. to Sep. (or later). Rectrix molt begins on deck rectrix set (r1), then proceeds to the outermost set (r6). However, odd birds may have other sequences. Molt into adult rectrices rarely surpasses the innermost and outermost sets while in the U.S., and is completed on the winter grounds. Rarely, tail molt may begin as early as May.

Plumage transition is divided into three headings: very early stage, early stage, and late stage. However, advanced or delayed molt in various anatomical areas makes it difficult to accurately label plumage stages in some individuals. Very early and early stages correspond to a span from spring to early summer and late stage aligns with mid-summer to early fall periods.

**SUBSPECIES:** Monotypic. Closely related to and forms a superspecies with Plumbeous Kite of Central America (*see* Adult Female Traits: Band-tailed Type).

**COLOR MORPHS:** None.

**SIZE:** A medium-sized raptor. Males average somewhat smaller than females, and size is often visibly apparent in the field. Length: 12–15 in. (30–38 cm); wingspan: 29–33 in. (74–84 cm).

**SPECIES TRAITS:** HEAD.—Large black lore spot in front of the eyes. BODY.—Short, thick tarsi and toes. WINGS.—**When perched, wingtips extend somewhat beyond the tail tip. In flight, wings are very long and narrow, with parallel front and trailing edges on the secondaries, and taper to pointed wingtips. In soaring flight, outermost primary (p10) is distinctly shorter than longest primaries (p8 and 9).** However, when gliding, p10 is usually concealed by the overlapping longer primaries. When perched, secondaries are sometimes nearly covered by the greater coverts. TAIL.—**Square-tipped. When perched, tail is often notched, and on males, it is often forked. Outer rectrix set flares slightly outwards on the outer feather edge.**

**ADULT TRAITS:** HEAD.—**Pale grayish. Dark gray bill and cere, but cere can be pale gray.** Red irises. BODY.—**Tarsi and toes are typically gray on the upperside and yellowish orange or orange on underside, but may be bright orange on the upper tarsi. Toes and tarsi are occasionally orangish on the upperside. Medium gray underparts and dark bluish gray upperparts.** WINGS.—**Upper surface of secondaries are very pale gray, but appear white at a distance. White bar on the trailing edge of secondaries is apparent on the underwing when in flight.** Pale upper surface of secondaries is most visible in dorsal view in flight. **Secondaries appear as a narrow pale gray bar when perched; however, they are often covered by the greater coverts and barely visible. Primaries are black on the upper surface and dark gray on the undersides, and may have a variable amount of rufous on both the inner and outer webs.** TAIL.—**All black on upper surface.**

**ADULT MALE:** HEAD.—**Pale gray or very pale gray and often appears white at a distance; palest on crown.** BODY.—**Uniformly pale gray underparts.** Lower belly and undertail coverts are the same color as the breast, belly, and flanks. **Upperparts are uniformly dark bluish gray.** Scapulars occasionally have a small white spot on the basal region of some feathers. WINGS.—Pale gray dorsal surface of the secondaries remains unchanged throughout the year. Underwing coverts are uniformly gray. TAIL.—**Uniformly black on the undersides.** Quills on the outermost set of rectrices (r6) are either black or white only on the very basal region.

**ADULT FEMALE:** HEAD.—**Medium gray, but sometimes pale gray. Medium gray-headed individuals often have a discernibly pale, whitish forehead, supercilium, and throat.** BODY.—**Medium gray or sometimes pale gray underparts.** Lower belly is either (1) the same gray color as the rest of the underparts or (2) white and appears as a fairly large white patch, which is most visible in flight as a white patch ahead of the tarsi and feet. Underparts may have small white spots on basal region of many feathers on the belly and flanks (and appear much like a subadult). **Upperparts are very dark bluish gray with a brownish cast.** Scapulars often have white spots on basal region of some feathers. Undertail coverts (1) are white with a gray mark on the tip of each feather and appear spotted, blotched, or barred; (2) may be all white; or (3) are solid gray like the rest of the underparts. WINGS.—The pale grayish sheen may wear off on the dorsal surface of the secondaries and become medium grayish or brownish gray. Underwing coverts often have pale speckling. TAIL.—Quills of the outermost rectrix set (r6) are white their entire length. There are two main variations of undertail patterns: (1) *Pale type:* **Pale gray on the inner three-fourths or four-fifths of each feather of the outer rectrix set with dusky or black tips on the distal one-fourth or one-fifth; all other rectrices are solid black. Tail appears to be pale gray with a moderately wide darker terminal band when**

**closed.** *Note:* Common pattern. (2) *Black type*: **Undertail is uniformly black (the quills of outermost rectrix set are still all white).** *Note:* Fairly common pattern.

**ADULT FEMALE (BAND-TAILED TYPE):** Two adult females with partial narrow white tail bands and one with a fully banded tail (with extensive rufous in the primaries) were photographed in Lamar, Colo., in 1999 and 2000; a fully banded type was also carefully noted in Jun. 2001 and one was photographed in Aug. 2002 at the same location. Plumages of these birds were otherwise like typical adult females. It is unknown if this unusual pattern exists throughout the entire population. *Note:* A rare plumage and easily confused with subadults and easily overlooked as an adult plumage variation.

**BASIC I (SUBADULT I) TRAITS: Bill, cere, eyes, and feet are like adult's. Back and scapulars are primarily adultlike. Head, body, and undertail coverts are also like the respective adult sex.** During typical residence in the U.S. (May–Sep.), transformation to adultlike plumage occurs in replacing most retained old rufous or brown juvenile feathering on the breast, belly, flanks, and upperwing coverts. Transition to adult plumage is partial on underwing coverts, primaries, and tail while in the U.S. Transition of scapulars and upperwing coverts to adult plumage is mainly complete. Secondaries are primarily retained, worn brown juvenile feathers. Old, tattered juvenile secondaries are uniformly brown at all times because the grayish sheen on upper surface and white trailing edge of fresh juvenile plumage are worn off. Subadults have the same four underwing and three undertail patterns described in Juvenile Traits until partially altered by molt. *See* Adult Traits and Juvenile Traits for full descriptions of plumage that subadults retain from each of these age classes.

**BASIC (SUBADULT; LATE STAGE):** HEAD.—As on respective sex of adult. BODY.—Underparts acquires the adultlike gray color; however, some retain a few remnant juvenile rufous or brown markings on the underparts. Some females exhibit white spotting on underparts and both sexes can have scattered white spots on the scapulars. WINGS.—Upperwing coverts are primarily adultlike dark bluish gray. Greater secondary coverts are new adult feathers and, like adults, are often a slightly paler gray than the median and lesser coverts. Underwing coverts are mostly or totally adultlike gray, except the greater coverts, which are retained mottled or spotted brown or rufous juvenile feathers. New adult inner primaries are steadily acquired during this stage. New dark grayish or blackish primaries have crisply delineated, thin white tips and contrast sharply to the faded, worn, brown outer juvenile primaries. Several new primaries project beyond the tertials when perched and are quite visible when perched and in flight. Since newly molted primaries are adult feathers, and many adults exhibit rufous markings on several primaries, many subadults in this stage will exhibit a rufous tinge on the newly acquired primaries. Secondaries are retained brown-colored juvenile, except the outermost (s1), as well as the three tertials, which molt into adult pale gray feathers. TAIL.—Deck rectrix set (r1) and outermost rectrix set (r6) often molt to adult blackish feathers while in the U.S. A few acquire only adult rectrices on the deck set (r1) while in the U.S. *Note:* When perched and viewed from the front, late-stage females with partially banded or unbanded types look similar to adult females.

**BASIC (SUBADULT; EARLY STAGE):** HEAD.—As on respective sex of adult. BODY.—(1) Underparts regularly retain up to 50 percent rufous or brown streaked and barred, pale-edged, remnant juvenile ventral feathers. Many have white streaks and blotches on the breast, belly, and flanks. These white areas are caused by pale edges of the retained juvenile feathers, gaps of missing feathers, and, on some females, by white basal areas of new adult feathers. Mottled juvenile-mixed underparts may be retained into the fall. (2) Underparts may also be solid, adultlike gray on some individuals, even in May. WINGS.—(1) Upperwing coverts may be nearly full adult; (2) most upperwing coverts are adultlike with a few retained brown juvenile feathers intermixed; or (3) upperwing coverts may be mainly retained brown juvenile feathers with a few grayish adult feathers intermixed. Greater upper covert tract is primarily brown juvenile feathering; however, the inner one-half of the tract can be replaced with adult gray in the early part of this stage. Large white blotches are often present on the innermost greater coverts and tertials because previously overlapping feathers are temporarily lost due to molt. Underwing coverts are retained rufous or brown juvenile feathers; however, they are sometimes mixed with adult gray feathering. Axillaries obtain the adultlike gray before the rest of the underwing coverts. Newly acquired inner primaries are dark grayish or blackish adult feathers with crisply edged white tips. New adult primaries are not readily visible when perched, as there are only a few of them and they are covered by the long tertials; however, the new dark feathers are highly visible in flight. A variable-sized gap often occurs on the inner primaries with dropped, but not quickly replaced, new primaries from mid-Jun. to mid-Jul. Up to four primaries may be dropped before new adult replacements come in. Since newly molted primaries are adult feathers, and many adults exhibit rufous markings on several primaries, many subadults in this stage will exhibit a rufous tinge on the few newly acquired primaries that have grown in. TAIL.—Fully retained juvenile.

**BASIC (SUBADULT; VERY EARLY STAGE):** HEAD.—As on respective sex of adult. BODY.—Underparts are mainly adultlike gray on the breast, but the belly and flanks can be 50–100 percent retained juvenile streaking. The grayish head (especially on females) and breast, contrasting with the pale belly and flank area on birds that were paler type juveniles, may create a hooded appearance. Upperparts are primarily retained, worn and faded brown juvenile feathers on the scapulars; the back region is partially adultlike gray. WINGS.—Upperwing coverts are mostly retained brown-colored juvenile with few, if any, gray adult feathers. Underwing coverts are fully retained juvenile feathers, including the axillaries. There is some molt on the inner primaries, but with few adult replacements. TAIL.—Retained juvenile of any of the three variations.

*Note:* Only small number of birds are in this barely molted first prebasic molt stage in May through Jun.

**JUVENILE TRAITS:** HEAD.—**Grayish-streaked head (lacking a dark malar mark); short, broad, white supercilium.** Black lore area. Iris color is medium brown. Yellow cere. BODY (ventral).—Variable-width rufous or brown streaking on white or tawny breast and belly; flanks are typically barred. There are three variations of underpart markings on the breast, belly, and flanks. (1) *Narrowly streaked type:* Narrow streaking with dark areas being less than 50 percent of feather width. *Note:* Common pattern. (2) *Moderately streaked type:* Moderately wide dark streaking being 50 percent of feather width. *Note:* Common pattern. (3) *Broadly streaked type:* Each feather is mainly dark with a very narrow pale outer edge. At a distance, underparts may appear nearly uniformly rufous or brown. *Note:* Uncommon pattern. BODY (dorsal).—Dark brown upperparts have very narrow white or tawny edgings on the scapulars. Scapulars also have large white area on the basal region of most feathers, creating a spotted or blotched appearance when fluffed. Tarsi and toes are yellow throughout, but may be brownish or grayish on upperside of the toes (as in older ages). WINGS.—**Prominent white bar on the trailing edge of the secondaries. Dorsal surface of secondaries has a pale grayish sheen.** Underwing coverts vary from being solid rufous or brown to mottled rufous or brown and white markings on undersides of the remiges are highly variable; there are four major variations. (1) *All-dark type:* All remiges are solid dark gray. *Note:* Common pattern. (2) *Moderate white type:* Some white on inner primaries (p1–3) and basal area of the outermost primary (p10). *Note:* Common pattern. (3) *Extensive white type:* **Basal region of all primaries is white and forms a large white panel.** *Note:* Common pattern. (4) *Very extensive white type:* **Basal region of all primaries and a narrow band of basal region of all secondaries are white.** The extensive amount of white forms a very large white panel on the primaries and a broad white linear area on the basal region of the secondaries. *Note:* Uncommon pattern. TAIL.—White terminal band is very narrow or absent. Tail is mostly solid brownish black on the dorsal surface; when fanned, partial white banding, if present, may show on inner webs of the rectrices. Ventral surface of tail is dark gray with a moderately wide darker terminal band. Undertail pattern has three major variations: (1) *Banded type:* Three or four complete, narrow white bands on all rectrices. White bands are visible at all times. Darker terminal band is sometimes distinct. *Note:* Common pattern. (2) *Partially banded type:* One or two narrow white bands on the basal area of inner rectrices. Outer one or two rectrix sets are often unbanded. Since inner rectrices are overlapped and hidden by outer rectrices when the tail is closed, undertail appears unbanded on a closed tail (*see below*). White bands are typically visible only when the tail is fanned. However, sometimes a faint, narrow white band may be present on the inner web of the basal region of outer rectrix set and visible when tail is closed. Wide dark terminal band is very distinct. *Note:* Fairly common pattern. (3) *Unbanded type:* Undertail lacks any definition of white bands, and the wide dark terminal band is distinct. *Note:* Uncommon to rare pattern.

**JUVENILE (HEAVILY MARKED TYPE):** HEAD.—As on typical juveniles. BODY.—Upperparts as on typical juveniles. Underparts are the broadly streaked type. WINGS.—Undersides of remiges are consistently the all-dark type. TAIL.—Undertail patterns are either the partially banded type or unbanded type. *Note:* Uncommon plumage.

**ABNORMAL PLUMAGES:** None; however, the band-tailed type adult female is a rather unusual occurrence (BKW pers. obs.).

**HABITAT: Summer.**—Inhabits a variety of geographically hot climate areas of the West. In humid eastern portions, found in lowland riparian floodplains, lakes, and swamps that are in semi-open tracts of tall, old-growth deciduous trees or mixed deciduous and coniferous. Kites are rarely found in pure mature coniferous woodlands. In semi-arid and arid western areas of the s. and cen. Great Plains and the Southwest, tall, old-growth deciduous trees in lowland riparian areas along rivers and lakes are inhabited. Tree growth in the Plains and desert regions are often limited to narrow strips along immediate riparian corridors. Surrounding landscape is generally quite open. Dense, artificially planted, deciduous shelterbelt groves, mainly in double-row plantings, have been favored breeding areas since the 1930s in s. Kansas, Oklahoma, and n. Texas. Open regions with solitary trees may also be occupied. Locales with very small trees or bushes are inhabited in undisturbed regions. In human-occupied regions, areas with medium-height and tall trees are used. Mississippi Kites are readily adapting to urban locales with ample deciduous tree growth, even areas with extensive human activity such as residential areas, golf courses, parks, campuses, and cemeteries. In eastern and southern areas, breeds at sea level; in northern areas, breeding occurs in wooded portions of cities and towns, and in riparian regions on high-elevation plains up to 4,600 ft. (1,400 m) in s.-cen. Colorado.

**HABITS:** Tame. A tolerant species that is highly adaptable to human activity. Where acclimated to humans, kites can be approached closely while on foot. Nesting birds are sometimes very aggressive towards human intruders. A gregarious species; retains such feeding and roosting habits in winter. Sunning is a regular activity, especially during the morning. Exposed branches, wires, and poles are favored perches. Leisure perching occurs periodically throughout the day, often for extended periods. All ages, but especially fledglings, may perch within concealed branches or shaded, exposed branches for long periods to escape the hot sun. Kites regularly communally bathe and drink from small puddles and pools and along river shorelines. Mississippi Kites are quite playful and intently chase each other. Kites also engage in harmless but intense, playful pursuits of larger songbirds too large to kill (e.g., Common Grackles and European Starlings). An awkward stance is assumed upon landing—tipping forward—almost as if losing balance.

**FEEDING:** Aerial and perch hunter. Kites mainly feed on large flying insects, particularly cicadas. Mississippi Kites occasionally feed on small bats and birds, particularly nestlings and inexperienced fledglings. They will prey on species as large as recently fledged Purple Martins and Blue Jays. Kites may also prey on small terrestrial mammals, amphibians, and reptiles. When feeding on insects, they will discard inedible parts such as legs and wings prior to eating. Carrion is rarely eaten. Virtually all aerial species of prey are captured while airborne. Aerial-hunting takes place at moderate to high altitudes. Perch-hunting is at low to very low altitudes, often just above ground level. Perch-hunting occurs mainly in the morning and late afternoon when insect forms are also at low altitude or when atmospheric conditions are not favorable for high-altitude flight.

Prey is captured with the feet. If engaged in aerial feeding, prey is held in the forward-extended foot and the head is lowered down. When transporting prey to nests or fledglings, prey is transferred to the bill and carried in the bill.

**FLIGHT:** Considerable amount of time is spent lithely gliding and soaring at various altitudes. *Wings are held on a flat plane and primary tips flex slightly upward or in a low dihedral when soaring.* Wings are held on a flat plane when gliding. Powered flight is accessed a moderate amount of the time. Kites may flap for extended periods when atmospheric lift conditions are poor, when rising from the ground, or when in pursuit of prey. Wingbeats are fairly deep, loose, and moderately fast. Uses very acrobatic aerial dives and twists when pursuing prey or in playful antics. Kites regularly dive from high altitudes to a chosen perch, or, with adults, to feed nestlings and fledglings. Mississippi Kites do not hover or kite.

**VOICE:** A loud, high-pitched, whistled *phee-toooooo:* the last syllable is drawn out and often a decrescendo. A piercing, monotone staccato whistled *phee-too-too-too-too* is a greeting call. Sometimes an abrupt, high-pitched, short whistled *phee-too* is emitted, which may be repeated several times; it also is a greeting call. During courting encounters and when whining for food, adult females and fledglings, respectively, emit a very soft, drawn-out whistled *pheer.* Quite vocal during nesting season, especially fledglings, which call incessantly. *Note:* Calls are imitated exceptionally well by Northern Mockingbirds and European Starlings.

**STATUS AND DISTRIBUTION: Summer.**—*Common* in very localized areas of the West. Estimated overall population for the U.S. is based on full-season fall migration counts in Veracruz, Mexico, where virtually all of the U.S. population passes during migration. Fall season counts at Veracruz: 186,000 in 1998, 127,800 in 1999, 101,800 in 2000, 214,000 in 2001, and 308,500 in 2002. Full-season counts are currently too few to detect a definitive trend; however, the most recent Veracruz tally shows that a healthy population exists.

*Arizona:* Listed as a Threatened Species with 30–50 pairs. Kites were first discovered in the state in Jun. 1970. Kites are sporadically distributed along portions of the San Pedro River, with small colonies at Dudleyville-Winkelman and from Fairbanks to the Mexican border. Mississippi Kites inhabit isolated areas along the eastern portion of the Gila River near Fort Thomas, Dripping Springs Wash, and Three Way. Kites also inhabit various stretches along the upper and lower Verde River, along the Hassayampa River near Wickenburg, along the Big Sandy River near Wickieup, and in pecan groves on the lower stretches of the Santa Cruz River. Virtually all populations are centered in riparian cottonwood zones. *New Mexico:* State listed as an Endangered Species with about 100 pairs. Found mainly along the Rio Grande in riparian cottonwood tracts and are irregularly and patchily distributed from Bernalillo Co. to Valencia Co. They are also irregularly found at Socorro, nesting occurred at Bosque del Apache NWR in the summer of 2000. Kites possibly breed at Hatch and in the Mesilla Valley from Las Cruces to El Paso. All populations in the Pecos River region and the eastern Staked Plains are centered in wooded portions of larger towns and cities: Tucumcari, Fort Sumner, Clovis, Portales, Roswell, Artesia, Carlsbad, Hobbs, and possibly a few other small towns in the southeastern region. *Colorado:* Most nesting pairs are in towns along the Arkansas River from Pueblo (50–100 pairs) and eastward. Small colonies are in Rocky Ford, La Junta (where nesting was first discovered in the state in 1971), Las Animas, Lamar, Granada, and Holly. They also nest in cottonwood riparian canyon locales in n. Animas Co., se. Animas Co., and sw. Baca Co. Isolated, local expansion has taken place in Holyoke, Phillips Co. (nesting first occurred in 1992). Canyon locations have only a few pairs. *Nebraska:* A small colony that was established in 1991 at Ogallala, Keith Co., is still active. Probable nesting occurred in Polk Co., with birds seen irregularly in summer from 1983 to 1990. *Kansas:* The southern counties of this state make up the northern fringe of the densely inhabited s. Great Plains population. In the arid western plains, kites breed only in towns and cities. In the central region, they breed in wooded riparian tracts, sheltered wooded tracts, shelterbelts, and towns and cities. Confirmed breeding occurs in the following counties: sw. Morton in the town of Elkhart, cen. Grant, cen. Stevens, sw. Seward, e. Meade, s. Clark, Commanche, Kiowa, e. Pawnee, Stratford, w. Pratt, Barber, Reno, Kingman, Sumner, w. Cowley, Sedgwick, cen. Harvey, and, isolated in the northeastern part of the state, in ne. Johnson. Possible breeding occurs in the following counties: cen. Sherman, s.-cen. Wallace, s. Gray, e.-cen. Hodgeman, se. Russell, sw. Ellsworth, Saline, s.-cen. Marion, n. Douglas, and ne. Wilson. *Oklahoma:* This state probably harbors the highest and densest nesting population in the species' entire U.S. range in the western half of the state. In the panhandle region, pairs are confined to riparian areas and towns. In all other areas, kites are found in wooded groves, shelterbelts, and towns and cites. In the southeast, they are found only along the Red River. *Texas:* Kites densely populate the north-central part of the state and southward to San Angelo. Kites are found as far south as Midland and Odessa; north of here and in the panhandle, found only in towns, cities, and wooded riverine stretches, including the Red and Canadian Rivers. Isolated breeding colonies exist in El Paso, which is part of the Rio Grande population extending north into New Mexico. In n. Texas, primarily breeds along the Red River to Bowie Co. and probably south of the Red River to Marion Co. Mineral Wells, Parker Co., and in w. Fort Worth and in Keller, Tarrant Co., have irregularly nesting pairs. In se. Texas, breeding occurs in Newton, Trinity, and Chambers Cos. Isolated breeding occurs in at least three locations in Washington Co. In the summer of 2000, there were also isolated pairs in Richmond, Fort Bend Co.; Eagle Lake, Colorado Co.; Edna, Jackson Co.; and in Victoria, Victoria Co. *Iowa:* Summer resident in the Des Moines suburb of Clive, Polk Co., since 1991. Breeding was confirmed in Clive in 1995 for the first state breeding record. Kites have been regular in areas surrounding Des Moines since 1996, including Windsor Heights and Urbandale (both near Clive), the Wakonda Golf Course in s. Des Moines, and in West Des Moines. The second confirmed breeding record for the state was in 2000 in Urbandale. Kites have also been seen fairly regularly at Ledges S.P., Boone Co., since 1996; a pair built a nest in 1997, but it was destroyed by high winds. *Missouri:* Breeding occurs in Pike Co.; St. Louis, St. Louis Co.; New Madrid Co.; and Pemiscot Co. Recent breeding expansion has extended to Kansas City, Jackson Co.; Joplin, Jasper Co.; and in areas of Henry Co. Possibly breeds in Cape Girardeau and Mississippi Cos. *Arkansas:* Widespread in the eastern one-third of the state in a diagonal line from Little River Co. in the southwest to Mississippi Co. in the northeast in lowland areas. *Louisiana:* Largely found statewide but is lacking as a summer resident or breeder from upland pine forest areas and extensive swamp regions. Breeds locally in the following parishes: s. Bossier, s. Webster, cen. Claiborne, n. and e. Bienville, cen. Jackson, e. Sabine, sw. Vernon, n. Beauregard, w.-cen. Evangeline, n. Terrebonne, n. Plaquemines, w. St. Bernard, and se. Washington.

In many areas, the Mississippi Kite is truly becoming the "urban raptor" because it so readily nests in towns and cities. *Mexico.*—Documented breeding in Mexico at Casas Grandes in nw. Chihuahua in the late 1990s. Recently nested along the Rio Grande southeast of Ciudad Juarez, Chihuahua.

**Winter.**—South America, east of the Andes from n. Colombia south to Paraguay and n. Argentina. Based on recent data from a Bolivia hawkwatch (*see below*), much of the population may winter in the southern part of the winter range.

**Movements.**—A long-distant migrant. Migrants typically travel in flocks, but they may also migrate singly. North of Texas, flocks are generally small; however, in s. Texas, flock size may comprise of thousands of individuals. The largest flock at Hazel Bazemore Co. Pk., Corpus Christi, Texas, was 5,000 on Aug. 20, 1991. Large night roosts often form during migration. Also in Texas, large numbers are tallied in the fall at Smith Point, Chambers Co., and inland at Falfurrias, Brooks Co.

From all regions, birds move to and from the U.S. via s. Texas, with many moving along the Coastal Bend region, but most appear to stay inland from coastal hawkwatch sites. From the U.S., the kites move through e. Mexico (east of the Sierra Madre Oriental), through Central America, and to and from South America via w. Colombia. They cross the Cordilleras in nw. Colombia when entering and leaving Colombia.

*Fall migration*: Very punctual early season migrant. Migration extends from early Aug. to early Oct., peaks in s. Texas from late Aug. to early Sep., and with stragglers thereafter. All ages move simultaneously. Late-nesting adults and late-fledged juveniles probably make up most late Sep. and Oct. stragglers. Those nesting at the northern periphery of their range in s.-cen. Iowa and se. Colorado may not leave breeding and natal areas until mid-Sep.

In the State of Veracruz, Mexico, at Cardél and Chichicaxtle, the migration season spans mid-Aug. to mid-Oct. (largest numbers are seen at Chichicaxtle). There is a major first peak from late Aug. to early Sep. (for all ages), which lasts several days, and a secondary peak in mid-Sep. with a short, but abrupt upswing in numbers (probably juveniles and late-nesting adults). In 1999, peak day on Aug. 31 brought over 15,000 kites at Chichicaxtle; the secondary peak day, on Sep. 15, brought nearly 11,000 kites. In 2000, the Sep. 5 peak day had over 15,500 kites; the secondary peak occurred on Sep. 18, with over 13,500 kites. In 2002, 170,000 Kites passed between Aug. 31–Sep. 2, with a peak of 96,000 on Sep. 1. Most migrants have passed by late Sep. with only a trickle of movement in Oct.

Large numbers are counted in cen. Bolivia. In the fall of 2000, nearly 38,000 were seen during partial-season coverage. In the fall of 2001, with extended full-season coverage, over 118,000 were tallied: 6,600 in Sep. (begining in mid-Sep.), nearly 71,000 in Oct., and nearly 41,000 in Nov. Peak numbers are seen in mid-Oct. This substantial number shows that many or most winter in the southern part of their winter range.

*Spring migration*: Apr. to mid-May for adults. Peak movement in s. Texas of adults is in early Apr. The northernmost breeding grounds in se. Colorado, cen. Kansas, and s.-cen. Iowa are reached by adults in early to mid-May. Subadult migration extends from May to mid-Jun., and for some, throughout the summer. Subadults are seen moving through s. coastal Texas in decent numbers in mid-May.

**Extralimital movements.**—Subadults and a few adults regularly overshoot typical breeding range from late Mar. to early Oct., with bulk of extralimital sightings occurring in May and Jun. being subadults. Juveniles also engage in excursions far north of usual natal areas from Aug. to Oct. *California:* Twenty-six accepted records for the state from 1933 to 1992, with sightings from May to Sep.; two additional unconfirmed sightings occurred in 1994. The following counties have recorded sightings, and some counties have multiple records: Humboldt, Marin, Santa Barbara, Inyo, Kern, Mono, Los Angeles, San Diego, and Riverside. *Nevada:* (1) Corn Creek, near Las Vegas, Clark Co., in 1979 (no date given); (2) Alamo, Lincoln Co., in late Mar. 1982; (3) Las Vegas, Clark Co., in early Jun. 1986. *Minnesota:* Nineteen records as of the spring of 2001 that span from mid-May to late Oct. Sightings are regular and sometimes annual, and often more than one individual is seen in a year. Counties with sightings (sightings are of single birds): (1) Lincoln in late Aug. 1973 for the first state record, (2) Olmsted in late May 1975, (3) Traverse in late May 1980, (4) Olmsted in late Oct. 1982, (5) Fillmore in mid-May 1986, (6) Ramsey in mid-May 1986, (7) Fillmore in mid-Aug. 1986, (8) St. Louis (Duluth) in late Aug. 1991, (9) St. Louis in early Sep. 1992 (Duluth), (10) Wabasha in mid-Sep. 1992, (11) St. Louis (Duluth) in early Sep. 1993, (12) Clay in late May 1994, (13) Roseau in early Aug. 1994, (14) Anoka in mid-Sep. 1994, (15) St. Louis (Duluth) in early Aug. 1996, (16) Marshall in mid-May 1998, (17) St. Louis (Solway) in late Jun. 1998, (18) Ramsey in May 2000, and (19) St. Louis (Duluth) in mid-May 2001. *Iowa:* Sightings of single birds, except as noted. Records outside of the typical Des Moines areas: (1) Winneshiek Co. in late May 1978; (2) Waubonsie S.P., Fremont Co., in mid-May 1980; (3) Dudgeon Lake, Benton Co., for 5 days in early Jun. 1989; (4) Coralville Reservoir, Johnson Co., in early May 1990; (5) Rock Creek Park, Jasper Co., in mid-May 1996; (6) Minbury, Dallas Co., in mid-Jun. 1997; (7) Lake Ahquabi, Warren Co., in late Jun. 1997; (8) Shimek State Forest, Lee Co., in mid-May 1999; (9) Lime Creek Nature Area, Cerro Gordo Co., in mid-May 1999; (10) Hickory Hill Park, Johnson Co., in mid-May 2000; (11) Saylorville Reservoir, Polk Co., in mid-Sep. 2000; (12) Hitchcock Wildlife Area (hawkwatch), Pottawattamie Co., with 38 individuals seen from mid-Aug. to early Oct. (an exceptionally large number for this latitude; peak of 14 occurred on Sep. 24); and (13) Algona, Kossuth Co., in early Oct. 2000. *Wyoming:* (1) Sheridan, Big Horn Co., in mid-Jun. 1987; and (2) Cheyenne, Laramie Co., in late May 1998, with two individuals. *Nebraska:* (1) Hall Co. in early Jul. 1983 in Hall Co. (adult); (2) Hall Co. in mid-Jun. 1988 (adult); (3) Kearney, Kearney Co., in early Jul. 1992; (4) Hastings, Adams Co., in mid-Aug. 2000 (adult); (5) Lancaster Co. in early September 1978 (subadult); (6) Lancaster Co. in mid-Sep. 1988; (7) Lincoln, Lancaster Co., in early Sep. 1990; (8) Lincoln, Lancaster Co., in late Aug. 1998; (9) Lincoln, Lancaster Co., in late Jul. 1999; (10) Mitchell, Scotts Bluff Co., for 5 days from late May to early Jun. 1998, (11) Dawes Co. in late Jul. 1990; (12) Oliver Reservoir in Kimball Co. in late Aug. 2000. Records for Douglas and Sarpy Cos.: Late May 1965, mid- and late May 1975, mid-May 1977, early Sep. 1980, early Sep. 1983, mid-Aug. 1984, late Sep. 1986, late Jul. 1988, and mid-Sep. 2000. *Saskatchewan*: (1) A bird spent 7 days in Regina in mid-Aug. 1985, and (2) one individual also at the same location in Regina for 1 day in late Jun. 1992. These are the northernmost records for North America. *South Dakota:* Pending acceptance by state committee, a possible first state record of a subadult photographed at Hartford Beach S.P., Roberts Co., in mid-May 2001.

**NESTING:** May to mid-Sep., mainly completed mid-Jul. to late Aug. Late-nesting pairs or renesting attempts may not begin until early Jul. in northern latitudes. Nests singly or in small or large, loosely formed colonies of up to 20 pairs. In colonies, nesting pairs are typically widely spaced and occupy separate trees. In towns and cities, pairs cluster in various parts of the urban area. No more than one or two nesting pairs typically occupy a city block. Adults may display territorial aggression towards other kites in the immediate nest area.

**Courtship (flight).**—*Sky-dancing* has occasionally been seen (*see* page chapter 5). Kites typically exhibit little if any aerial displays.

**Courtship (perched).**—Males offer prey to females. Females, which may be flying, are enticed to the perched males by the *phee-too-too-too-too* call. Males may offer prey to females prior to copulation. Most do not breed until 2 years old.

Nests are small, often oval-shaped, and about 11 x 14 in. (28 x 36 cm) in diameter and 5 in. (13 cm) deep. However, some new nests are very small and may measure 6 x 8 in. (15 x 20 cm). In these small nests, the tails of the incubating or brooding adults will extend beyond the edge of the nests. The nest structures may be either poorly constructed or compactly made of small, thin twigs and branches. The nests are decorated throughout the entire period with a considerable amount of greenery. Nests are well concealed in the very upper portion of secondary branch clusters in densely foliaged deciduous trees. Nests are rarely located in pine trees; if so, most often in Loblolly Pines. Nest materials are broken off with their bills and generally carried in bills, but large pieces are sometimes carried with their feet. When breaking off nest twigs, kites often engage in rather awkward antics by regularly hanging upside down and flapping their wings for balance while vigorously trying to snap off twigs. In undisturbed arid regions, nests are as low as 4 ft. (1 m). In humid eastern regions, nests are typically in tall trees and nests may be 135 ft. (40 m) high. Old nests are often reused. Both sexes build nests and share nest duties, including feeding. The 2 eggs are incubated for 30 days. Youngsters fledge in about 34 days. Most typically fledge mid-Jul. to late Aug., but can be as late as early Sep. in northern areas, especially if it is a renest attempt. Fledglings are fed primarily by males for another 21–28 days. Young typically gain independence from late Jul. to mid-Sep. Recently fledged youngsters form groups and are often still food-begging and being fed by their parents. Adults feed only their offspring.

Subadults form groups in larger colonies. Some assist nest-building and regularly carry nest materials during the nest-building period, although not always to actual nest sites. A few may assist adults in feeding nestlings. Very rarely, subadults will mate with adults.

**CONSERVATION:** No measures are taken in the West since the population is stable. Kites are currently expanding their breeding range in Colorado, Nebraska, Iowa, and Missouri.

**Mortality.**—Illegal nest destruction has occurred in many towns and cities where pairs are aggressive towards humans intruding into territories. Illegal shooting has been documented in Texas, especially by dove hunters (dove season occurs, in part, during the late summer and early fall migration of kites). Since Mississippi Kites are tame, many are probably shot elsewhere south of the U.S. during migration or on winter grounds. Since nests are rather frail and are placed in the tops of tall trees, the nests, eggs, and nestlings are

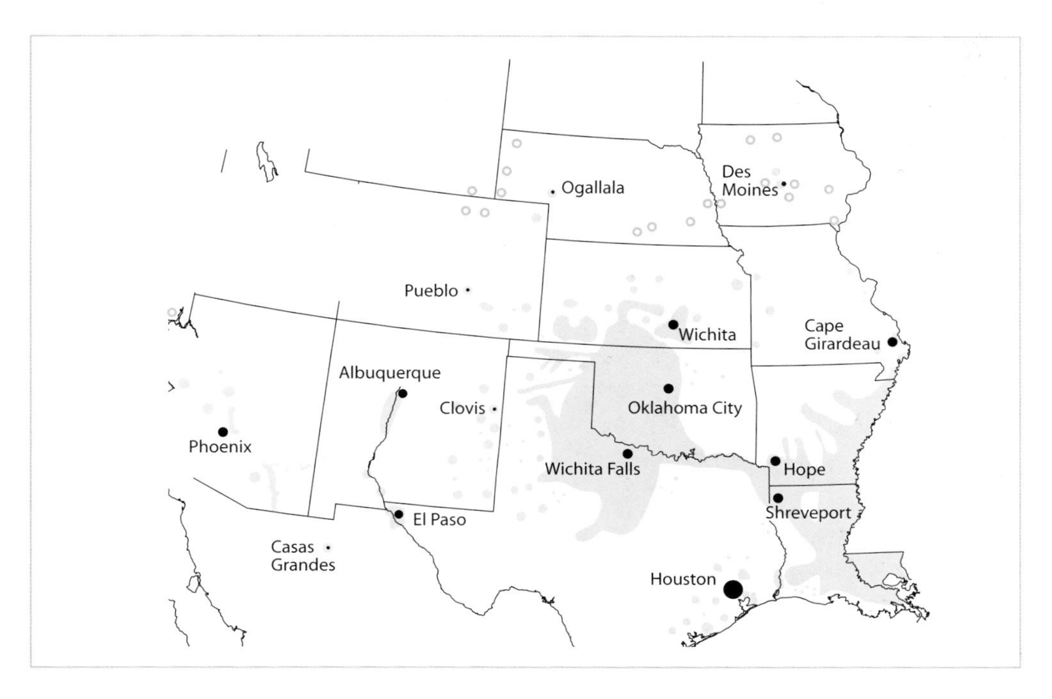


**MISSISSIPPI KITE,** *Ictinia mississippiensis:* Locally common. Expanding to north and west. Disperses in spring and fall. Winters in South America.

susceptible to being blown out of trees during severe summer storms.

**SIMILAR SPECIES: (1) White-tailed Kite.**—PERCHED.—Identical black lores. Large black "shoulder" on upperwing coverts. Tail whitish in all plumages. FLIGHT.—Large black spot on carpal region of underwings. Upperwings as in Perched. Frequently hovers. **(1A) Adults.**—Identical red irises. White underparts, pale gray upperparts. Bright yellow feet. **(1B) Juveniles.**—Similar when perched. Orangish brown irises. Orangish rufous markings on underparts similar, but they are more restricted to breast region. **(2) Sharp-shinned Hawk.**—PERCHED.—Yellow irises; pale area in front of eyes. Similar rufous or brown streaked underparts. Similar dark brown upperparts, including white spotting. Long, thin tarsi. Wingtips are much shorter than the tail tip. Tail is distinctly banded on dorsal side. **(3) Merlin, adult females/juveniles** of ***F. c. columbarius.***—PERCHED.—Yellow fleshy orbital area. Dark irises are similar. Brown upperparts similar; underparts similarly streaked. Wingtips are much shorter than tail tip. Tail pattern similar, but usually partial/full narrow pale bands on dorsal side and a broad white terminal band. FLIGHT.—Similar shape. Outermost primary almost same length as longest, requiring caution in autumn: Adult females often still molting p10, and it will be much shorter than longest primaries (p8 and 9) and appear similar to a kite's proportion. Underwings distinctly spotted on all remiges. Rapid wingbeats. Tail pattern as in Perched. **(4) Peregrine Falcon.**—PERCHED.—Dark "mustache." Pale fleshy orbital region. Wingtip-to-tail-tip ratio is similar. FLIGHT.—Similar shape; use caution. Outermost primary almost as long as wingtip. Wingbeats powerful, may flap for considerable distances. **(4A) Adults.**—Black cap and "mustache." Yellow orbital area. Fine barring on flanks. Upperparts similar bluish to adult kite. Underwing very barred or spotted. **(4B) Juveniles.**—Dark eyeline connects with dark "mustache." Pale blue orbital area. Underparts similar to juvenile Mississippi Kite, but streaking always dark brown. Upperparts similar;

lack white spotting. Underwing similar to all-dark type underwing of juvenile kite, but always has tawny spotting on remiges. Tail has broad white terminal band.

**OTHER NAMES:** None regularly used. *Spanish:* Milano Migratorio, Milano de Mississippi. *French:* Milan du Mississippi.

---

REFERENCES: Alcorn 1988; Arizona Breeding Bird Atlas 1993–1997; Baicich and Harrison 1997; Bent 1961; Bolen and Flores 1993; Busby and Zimmerman 2001; Bylan 1998, 1999; Conrads et al. 1989; del Hoyo et al. 1994; Dinsmore 1997, 2000; Dorn and Dorn 1990; Fuller 2001; Glinski 1998b; Jacobs and Wilson 1997; James and Neal 1986; Justus 1997; Kellogg 2000; Kent and Dinsmore 1996; Kingery 1998; Oberholser 1974; Olivo 2001, 2002; Palmer 1998; Parker 1999; Pulich 1988; Robbins and Easterla 1992; Rottenborn and Morlan 2000; Sharpe et al. 2001; Small 1994; Smith 1996; Snyder and Snyder 1991; Sutton 1967; Walsh 1996; Williams 1998.

**Plate 71. Mississippi Kite, adult male [Jun.]** ▪ Very pale gray head. Black bill, gray cere. Red iris. ▪ Dark bluish gray upperparts. ▪ Whitish secondaries form a narrow whitish bar. ▪ Black dorsal surface of tail. Tail notched when closed.

**Plate 72. Mississippi Kite, adult male [Aug.]** ▪ Pale or very pale gray head. Black bill, gray cere. Red iris. ▪ Medium gray underparts. ▪ Wingtips extend beyond tail tip. ▪ Black tail; quills are black.

**Plate 73. Mississippi Kite, adult male [Aug.]** ▪ Pale or very pale gray head. Black bill, gray cere. Red iris. ▪ Dark gray upperparts. ▪ Whitish secondaries; black primaries, often with rufous markings. ▪ Black tail. ▪ *Note:* Sunning posture: Wings may be fully extended. Males begin annual molt in Aug.–Sep., starting with inner primaries: 2 inner primaries are new (grayish).

**Plate 74. Mississippi Kite, adult male [Sep.]** ▪ Pale or very pale gray head. ▪ Uniformly medium gray underparts. ▪ Long, pointed wings with outermost primary (p10) much shorter than wingtips. ▪ Black, squared-edged tail; quills are black. ▪ *Note:* Holding a cicada in left foot.

**Plate 75. Mississippi Kite, adult male [Sep.]** ▪ Pale or very pale gray head. ▪ Uniformly medium gray underparts. ▪ Long, pointed wings with outermost primary (p10) much shorter than wingtips. ▪ Black, squared-edged tail; quills are black. ▪ *Note:* Aerial feeding posture when gliding and soaring. Feeding on a cicada held in its left foot.

**Plate 76. Mississippi Kite, adult male [Aug.]** ▪ Pale or very pale gray head. ▪ Uniformly medium gray underparts. ▪ Long, pointed wings with outermost primary (p10) much shorter than wingtips. Inner 3 primaries lost due to molt, creating a gap in wing. ▪ Black, squared-edged tail; quills are black. ▪ *Note:* Males begin annual molt starting with inner primaries in Aug.–Sep.

**Plate 77. Mississippi Kite, adult female [Jun.]** ▪ Pale to medium gray head. Black bill, gray cere. Red iris. ▪ Very dark gray upperparts. ▪ Gray secondaries. ▪ *Note:* Most females are slightly darker than males. Dorsal surface of secondaries are typically darker than on males.

**Plate 78. Mississippi Kite, adult female [Aug.]** ▪ Pale to medium gray head. Black bill, gray cere. Red iris. ▪ Medium gray underparts. Some have white on the undertail coverts. ▪ Gray secondaries. ▪ Pale type tail has dark terminal band on pale gray outer rectrix set; quills are white.

**Plate 79. Mississippi Kite, adult female [Aug.]** ▪ Pale to medium gray head. Black bill, gray cere. Red iris. ▪ Medium gray underparts. Some have gray undertail coverts. ▪ Gray secondaries. ▪ Black type tail has uniformly black rectrices; quills are white.

**Plate 80. Mississippi Kite, adult female [Jun.]** ▪ Pale or medium gray head. ▪ Medium gray underparts with whitish lower belly. ▪ Long, pointed wings with outermost primary (p10) much shorter than wingtips. ▪ Pale type tail has a dark band on outer rectrix set; quills are white on outer set. ▪ *Note:* Amount of rufous on primaries is not a sexual trait.

**Plate 81. Mississippi Kite, adult female (band-tailed type) [Aug.]** ▪ Outermost primary (p10) much shorter than wingtips. Rufous on some primaries. Inner 3 primaries are new; p4 is dropped (typical molt stage for females in late summer). ▪ Pale type tail has partial white and gray bands. ▪ *Note:* Rare tail pattern on adult females. Photographed in Prowers Co., Colo.

**Plate 82. Mississippi Kite, adult female [Jun.]** ▪ Pale or medium gray head. ▪ Very dark gray upperparts. ▪ Whitish secondaries contrast with rest of the upperparts. ▪ *Note:* Females may not show extensive white on secondaries; males nearly always have distinct white secondaries.

**Plate 83. Mississippi Kite, subadult female (late stage) [Aug.]** ▪ Pale or medium gray head. Black bill, gray cere. Red iris. ▪ Very dark gray upperparts. ▪ Very dark gray upperwing coverts except new, paler gray greater coverts. New pale gray tertials. Brownish, retained juvenile secondaries. New (gray) inner primaries. ▪ Black central rectrix is partially grown on the tail.

**Plate 84. Mississippi Kite, subadult female (early stage) [Jun.]** ▪ Pale or medium gray head. Black bill, gray cere. Red iris. ▪ Very dark gray upperparts; some white spotting. ▪ Wing coverts are mainly brown juvenile with incoming new, gray feathers. Old, brown juvenile remiges. Often white blotches on inner wing. ▪ *Note:* Less-advanced molt than on many subadults by early Jun.

**Plate 85. Mississippi Kite, subadult female (early stage) [Jun.]** ▪ Pale or medium gray head. Black bill, gray cere. Red iris. ▪ Medium gray underparts are often speckled with white and have a few retained, brownish juvenile feathers. ▪ Banded type retained juvenile tail has thin white bands.

**Plate 86. Mississippi Kite, subadult male (early stage) [Jun.]** ▪ Very pale or pale gray head. Black bill, gray cere. Red iris. ▪ Medium gray underparts may be fully adult on some birds, even in late spring. ▪ Mix of old brown juvenile and new gray adult wing coverts. Brown juvenile secondaries. ▪ Banded type retained juvenile tail has thin white bands.

**Plate 87. Mississippi Kite, subadult female (late stage) [Aug.]** ▪ Medium gray underparts with whitish lower belly and undertail coverts. ▪ Mainly gray underwing coverts. ▪ Banded type juvenile tail has thin white bands, r6 set is partially grown black adult feathers. ▪ *Note:* Kites regularly dive to a chosen perch.

**Plate 88. Mississippi Kite, subadult female (late stage) [Aug.]** ▪ Medium gray underparts with whitish lower belly and undertail coverts. ▪ Mainly gray underwing coverts. Extensive-white type wing has white juvenile primaries. P1–5 are new, dark feathers. ▪ Banded type juvenile tail has thin white bands; black, new adult middle (r1) set.

**Plate 89. Mississippi Kite, subadult female (late stage) [Jul.]** ▪ Uniformly medium gray underparts. ▪ All-dark type wing. P1–3 are new/growing dark feathers, p4 is missing. ▪ Partially banded type/all-dark type juvenile tail is all-dark when tail is closed. Partially banded type shows thin white bands when widely fanned.

**Plate 90. Mississippi Kite, subadult female (early stage) [Jun.]** ▪ Pale or medium gray head. ▪ Gray underparts are mix of juvenile streaking and white speckling. White lower belly. ▪ Moderate-white type wing with white on outer primaries. P1–3 new/growing dark feathers. ▪ Banded type juvenile tail has thin white bands.

**Plate 91. Mississippi Kite, subadult female (early stage) [Jun.]** ▪ Gray underparts have some white areas; white lower belly. White, spotted undertail coverts. ▪ Extensive-white type wing with large white patch on primaries. Missing p1–4, creating large gap in wing. Outermost primary (p10) is much shorter than wingtip. ▪ Banded type juvenile tail has thin white bands.

**Plate 92. Mississippi Kite, subadult female (early stage) [Jun.]** ▪ Gray underparts with moderate amount of white speckling and retained juvenile streaking. Whitish lower belly. ▪ Very extensive type of wing with white patch on primaries and base of all secondaries. Missing p1 and 2. ▪ Banded type juvenile tail has thin white bands. ▪ *Note:* Entering into first stage of remix molt.

**Plate 93. Mississippi Kite, subadult male** (**early stage**) [**Jun.**] ▪ Gray underparts with only a few white specks. ▪ Moderate type wing with whitish areas on inner primaries and outer 2 primaries. Coverts are mainly juvenile with some gray; axillaries have molted into gray feathers. No remix molt. ▪ Banded type juvenile tail has thin white bands.

**Plate 94. Mississippi Kite, juvenile** (**either sex**) [**Sep.**] ▪ Black bill, yellow cere. Dark brown iris. Gray head with a short, white supercilium and white throat; black lore spot. ▪ Moderately wide rufous or brown streaking. ▪ Wingtips reach or extend beyond tail tip. ▪ Banded type tail has 3 or 4 thin white bands.

**Plate 95. Mississippi Kite, juvenile** (**either sex**) [**Aug.**] ▪ Black bill, yellow cere. Dark brown iris. Gray head with a short, white supercilium and white throat; black lore spot. ▪ Moderately wide rufous or brown streaking. ▪ Wingtips reach or extend beyond tail tip. ▪ Partially banded type tail: thin, partial white bands are hidden when tail is closed. *Note:* Recently fledged youngsters are highly vocal.

**Plate 96. Mississippi Kite, juvenile** (**heavily marked type; either sex**) [**Sep.**] ▪ Black bill, yellow cere. Dark brown iris. Gray head with a short, white supercilium and white throat; black lore spot. ▪ Broadly streaked underparts appear solid rufous or brown. ▪ Wingtips equal to or extend beyond tail. ▪ Partially banded type tail: thin, partial white bands are hidden when tail is closed.

**Plate 97. Mississippi Kite, juvenile (either sex)** [**Sep.**] ▪ Black bill, yellow cere. Dark brown iris. Gray head with a short, white supercilium and white throat; black lore spot. ▪ Dark brown upperparts often have large white spots on scapulars. ▪ Wingtips equal to or extend beyond tail. ▪ Tail is uniformly black on dorsal side.

**Plate 98. Mississippi Kite, juvenile (either sex)** [**Sep.**] ▪ Underparts are moderately streaked with rufous or brown. ▪ Extensive-white type wing with large white patch on base of primaries. Rufous brown markings on coverts. ▪ Banded type tail has thin white bands. ▪ *Note:* Juveniles have somewhat broader secondaries than older birds.

**Plate 99. Mississippi Kite, juvenile (either sex)** [**Sep.**] ▪ Underparts are streaked with rufous or brown. ▪ Moderate-white type wing with small white areas on inner and outer primaries. Rufous brown markings on coverts. ▪ Banded type tail has thin white bands. ▪ *Note:* Juveniles have somewhat broader secondaries than older birds.

**Plate 100. Mississippi Kite, juvenile (heavily marked type; either sex)** [**Aug.**] ▪ Broadly streaked underparts appear solid rufous or brown. ▪ All-dark type wing lacks white markings. Coverts are heavily marked. ▪ Partially banded type tail shows partial, thin white bands when widely fanned. ▪ *Note:* Juveniles have somewhat broader secondaries than older birds.

# BALD EAGLE
## *(Haliaeetus leucocephalus)*

**AGES:** Adult (basic V and older; sometimes basic IV), four interim basic/subadult ages (basic/subadult I–IV) that correspond to 1- to 4-year-old birds in their second to fifth years of life, and juvenile during the first year of life. There is a minimal amount of variation in adults, a considerable amount of variation in subadults, and a moderate amount of variation in juveniles. Sexes are similar.

Adult plumage is typically acquired as 5-year-old when a basic/subadult V is in its sixth year of life, but it can be attained as a 4-year-old when a basic/subadult IV is in its fifth year of life (McCollough 1989; *see* Figure 2 in Clark and Wheeler 2001). Some adults may retain traces of younger-age characters on the head until at least 8 years old (McCollough 1989).

Basic ages/subadults slowly change from juvenile to adult plumage through partial annual molts. Each age class is held for 1 year. Basic ages/subadults best exhibit their respective age characters when in flight and when viewed at a dorsal angle when perched. With a considerable amount of individual variation, the three older subadult ages can often be difficult and sometimes impossible to separate from adjacent ages. Basic/subadult IV and basic/subadult V and older ages can be identical in all aspects. Basic/subadult IV and III can share similar bill, head, body, and tail markings. Basic/subadult III and II can also share similar markings as previously described. Basic/subadult I is distinct and is easily separable from older subadults. Juvenile feathering is retained in the first two subadult plumages and sometimes in the third plumage.

Juvenile plumage is held for the first year. Juveniles are easily separable from older birds. Juveniles have longer, more pointed secondaries and rectrices than older ages. This creates the appearance of a broader wing and a longer tail.

**MOLT:** Accipitridae wing and tail molt pattern (*see* chapter 4). Wing and tail molts are complex and variable, and it is important to understand the various molt centers and molt waves in order to age subadult birds. Latitude, diet, and other factors affect timing and intensity of molt. Birds that are born in Florida begin molting in Nov.; birds from n. U.S. and Canada begin molting in Mar. or Apr. Eagles from interim latitudes begin molting at interim times.

*Note:* Eagles have the typical 10 primaries as most raptors, but have 17 secondaries, four more than smaller raptors; the secondary molt is an important segment in the sequence of aging.

The annual molt is an incomplete feather replacement on most anatomical areas. A large portion of body and wing covert feathers are replaced, a small portion of remiges are replaced, and all or virtually all rectrices are replaced. Younger subadults may have one or two molt waves in progress at the same time, and older subadults and adults may have two or three molt waves in progress. Birds from southern latitudes have a longer period in which to molt than do birds from northern latitudes; thus they typically replace more feathers during the annual molt.

Molt is a similar sequence to that of White-tailed Eagles; however, in most cases, molt and the progression of molt waves is slightly more accelerated for the respective age of Bald Eagles.

Data on molt are based on Edelstam (1984) and Forsman (1999) on White-tailed Eagles, Clark and Wheeler (2001), Wheeler and Clark (1999), W. S. Clark pers. comm., and BKW pers. obs. of wild birds and museum specimens. McCollough (1989) had exceptional plumage study data on 135 known-aged birds but did not correlate it with molt sequences, particularly on the remiges.

On the remiges and upperwing coverts, retained juvenile feathers are longer and more pointed than new, older-age subadult feathers. The longer the juvenile feathers are retained, the more they become faded brown, frayed, or broken. Subadult remiges can also be aged by the amount of fading and fraying: new feathers are dark and neat, older feathers are more brownish and ragged on the their rounded tips. Molt is serially descendent, with two or three molt waves in progress at the same time.

**First prebasic molt.**—Molt out of juvenile into subadult I (1-year-old) plumage. Molt is complete on the head and tail. Typically, molt is fairly complete on the contour feathers of the dorsal and ventral body; however, only a partial molt is undertaken by some eagles, and they may retain a large amount of old juvenile feathering. Molt is always incomplete on the upper- and underwing coverts, which retain some juvenile feathering. The greater upperwing coverts molt with the respective remiges; all other upperwing coverts molt in an irregular pattern. Remix molt is incomplete: *Slow molt.*—On the primaries, p1–4 are replaced (in that order); p5–10 are retained juvenile. On the secondaries, s1 is replaced and the three tertials, s17–15, are replaced (in that order); all other 13 secondaries are retained juvenile. Molt may occur on s5. *Moderate molt.*—On the primaries, p1–5 are replaced; p6–10 are retained juvenile; however, p6 may molt on southern-breeding birds. On the secondaries, s1 and 2, s5 and 6, s14–12 (in that order), and the tertials, s17–15 (in that order), are replaced. The other seven secondaries, s3 and 4 and s7–11, on the middle section of the wing are retained juvenile. *Fast molt.*—Mainly seen on birds born at southern latitudes. On the secondaries, s7 and 11 may be replaced.

*Note:* Key factor on perched and flying birds are the large number of retained faded, long, and pointed juvenile secondaries; on perched birds, the tertials (s15–17) are new and dark colored subadult I and contrast sharply with the retained juvenile secondaries.

**Second prebasic molt.**—Molt out of subadult I into subadult II (2-year-old) plumage. Molt is fairly complete on the head and complete on the tail. Molt is incomplete on the body, which retains a small amount of subadult I feathering. Remix molt is incomplete: *Slow molt.*—On the primaries, p1 is replaced for the second time; p5–8 or, less commonly, p5–9 are replaced for the first time. P2–4 are retained subadult I feathers. P9 and 10 or p10 are retained juvenile. On the secondaries, s4 and 9 are retained juvenile, s1 is an old subadult I, s2 and 3 are new subadult II, s5–8 are new subadult II or s5 is an older subadult I, s10–14 are new subadult II, and s15–17 are replaced for the second time and are new subadult II. *Moderate molt.*—On the primaries, p1 and sometimes p2 are replaced for the second time, p6–9 are replaced for the first time, p10 and sometimes p9 are retained old juvenile feathers, and p2–5 or p3–5 are retained subadult I feathers. On the secondaries, s4 is a retained old juvenile (and is the last juvenile secondary to be lost), s1 and 2 are old subadult I, s3 is replaced for the first time and is a new subadult II, s5 and 6 are old subadult I, s7–9 are replaced for the first time and are new subadult II, s10 and 11 are replaced for the first time and are new subadult II, s12–14 are old subadult I feathers, and s15–17 are replaced for the second time and are new subadult II feathers. *Fast molt.*—On the primaries, p1 and 2 are replaced for the second time, p6–9 are replaced for the first time and are new subadult II, p10 is a very old juvenile, and p2–5 or p3–5 are retained subadult I feathers. On the secondaries, all juvenile feathers have been replaced. S1 and 2 are old subadult I feathers, s3 and 4 are new subadult II, s5 and 6 are old subadult I feathers, s7–9 are new subadult II feathers, s10 and 11 are replaced for the first time and are new subadult II, s12–14 are old subadult I, and s15–17 are replaced for the second time and are new subadult II feathers.

*Note:* Key factor mainly on perched birds viewed from the rear is the new, dark tertials (s15–17) that contrast with the older subadult I s12–14 and new, dark subadult II s10 and 11.

**Third prebasic molt.**—Molt out of subadult II into subadult III (3-year-old) plumage. Molt is fairly complete on the head and complete on the tail. Body molt is incomplete but retains only some feathering from the previous year or two. Wing covert molt is incomplete. Remix molt is incomplete: *Slow molt.*—On the primaries, p1 is usually a retained subadult II (replaced annually in the first two molts; rarely replaced in this molt), p2–4 are replaced for the second time, p5–9 are retained subadult II feathers, and p10 is a retained juvenile. On the secondaries, s4 and 9 may be the newest and darkest feathers. S14 and 13 (molted in that order) are also new and dark; s11 and 10 were molted in the second prebasic molt and are older than the adjacent outer secondaries; and s12, which is a very old subadult I feather and a "key" feather for aging this subadult class, is one of the oldest subadult type of feathers of the secondaries (sometimes along with s1 and 2 and s5 and 6). S15–17 are not molted (but were molted in the each of the prior two molts) and somewhat faded. Secondaries that were molted in the first prebasic molt, such as s1 and 2 and s5 and 6, may also molt for the second time. *Fast molt.*—On the primaries, p1 and 2 have been replaced twice and are retained subadult II; p3–5 are also replaced for the second time, but during the subadult III period; p6–9 are old subadult II feathers; and p10 is replaced for the first time as a new subadult III feather. On the secondaries, the molt pattern is similar as described for slow molt; however, some birds molt the inner secondaries more rapidly and, rather than molting only s14 and 13 (in that order), they may extend molt into s12 and sometimes s11—which leaves s10 as an old subadult II feather and the oldest of that portion of the wing.

*Note:* Key factor of this age class is the very old, faded, and frayed retained subadult I s12. This very old feather is visible in flight and at a dorsal angle when perched. Also, the fairly old retained subadult II tertials (s15–17) are important feathers to see for proper aging.

**Fourth prebasic molt.**—Molt out of subadult III into subadult IV (4-year-old) plumage; adult plumage is sometimes attained. Molt on the head is incomplete. Tail molt may be incomplete; if so, retains only one or two rectrices from the third prebasic molt. Body molt is incomplete and may retain a small amount of feathering from the previous one or two ages. Remix molt is incomplete and becomes even more irregular than in the previous three molts: On the primaries, p1–5 or 1–6 are newer; p1, or p1 and 2, and p4–6 are the last to molt and the newest feathers. Birds that were in slow molt in the previous prebasic molt may have a new p10; otherwise, p10 is moderate in age and fade character. On the secondaries, replacement of several feathers occurs in the approximate sequence of earlier molts; however, sequence may become irregular.

**Subsequent prebasic molts.**—Molts within the adult age class. Subsequent molts may gradually lessen any retained subadult traits on the head and sometimes tail. Molt is incomplete on virtually all parts of the body, wing, and tail. Remix molt is often an irregular sequence of feather replacement.

**SUBSPECIES:** Polytypic. They are considered to have two subspecies based solely on size, that spans from small to large birds. The demarcation between sizes is extremely arbitrary since they vary on a clinal north-south trend. *H. l. alascanus* breeds mainly north of 40°N and *H. l. leucocephalus* breeds south of this latitude. *See* Size for more information.

*Note:* As with Harris's Hawk and Northern Goshawk, the author does not consider subspecies that are based only on size, particularly if there is a clinal trend.

**COLOR MORPHS:** None.

**SIZE:** A large raptor. Within regional populations, there is no overlap between sexes: Males are smaller than females. There is a clinal trend in size, with southern-latitude birds being the smallest and northern-latitude birds, particularly from Alaska, being the largest. Southern females average slightly smaller than northern males; southern birds can be up to 7 in. (18 cm) shorter than northern birds of the same sex (Stalmaster 1987). Juveniles, with their longer tails, are 1.25–1.5 in. (3–4 cm) longer than adults of the same sex and similar latitude (Stalmaster 1987). The measurements are for the range of sizes of southern and northern birds. Length: 27–37 in. (69–94 cm); wingspan: 71–96 in. (180–244 cm).

*Note:* Due to restoration programs that released hundreds of nestling eagles from various origins, the size and subspecies factor is distorted (*see* Conservation). Large-sized eagles from Alaska and small-sized eagles from Florida were released into areas of the U.S., including Arkansas, California, and Oklahoma, that hosted either smaller or larger "natural" sizes. However, states tried to keep larger eagles in northern latitudes and smaller eagles in southern latitudes to conform with trend of larger birds in the north and smaller birds in the south.

**SPECIES TRAITS:** HEAD.—**Large, deep bill.** Forehead, nape, and hindneck feathers are regularly elevated in cool temperatures and during certain moods. BODY.—The yellow to orangish yellow tarsi are unfeathered on the very lower part. WINGS.—**There are six "fingers" on the outer primaries.**

**ADULT TRAITS:** HEAD.—**Uniformly medium yellow or orangish yellow bill, cere, and gape; gape and sometimes cere are paler than the bill. Iris is pale lemon yellow. Head and neck are white with a sharp demarcation line between the white lower neck and the brown back and breast.** Some full adults, even birds that are at least 8 years old, may have a small amount of dark speckling on the auriculars, nape, and hindneck (McCollough 1989). BODY (dorsal).—**Dark brown with thin pale tawny edges on most feathers. Upper half of the lower back is dark brown and the lower half of the back and rump are white.** BODY (ventral).—**Dark brown with thin pale tawny tips on most feathers, creating a scalloped appearance. The undertail coverts are white.** WINGS (dorsal and ventral).—**Dark brown coverts are edged with pale tawny and appear scalloped.** *Note:* Molting birds may show patches of white on the underwing coverts due to missing feathers. **Remiges are blackish. On the ventral surface, the basal region on the outer three or four primaries is pale gray.** TAIL (dorsal and ventral).—**Rectrices and uppertail coverts are white and form a white unit with the white lower back and undertail coverts.** Occasionally, older adults still retain some black speckling on the tips of some rectrices.

*Note:* This definitive basic plumage may be found on a few basic/subadult IV (4-year-olds), but is most typical of birds that are basic V (5-year-olds) or older. *Additional note:* Basic/subadult IV that appear totally adultlike on the bill, head, and body when perched may still exhibit subadult characters on the underwing (*see below*).

**BASIC IV (SUBADULT IV) TRAITS:** There are two main variations. Eagles that are more advanced in molt are similar to full adults and those that retain one or more younger subadult traits are fairly similar to adults. HEAD.—Both types have a sharp demarcation line between the white head and the brown back and breast. WINGS.—All remiges have been replaced for at least the first time, which includes the outermost primary (p10) on slow-molting birds; some remiges have been replaced two or three times. Five or six inner primaries (p1–5 or 6) will be newer, darker, and less frayed then the outer primaries (p6 or 7–10). The three tertials (s15–17) may be new feathers since they were not molted the previous year.

*Note:* Since subadult III and IV both may have new p10, the larger number of newer, darker inner primaries helps separate the two ages in flight.

**BASIC IV (SUBADULT IV; ADVANCED/ADULT TYPE):** HEAD.—**Uniformly medium yellow or orangish yellow bill and cere.** Gape is pale yellow. **Iris is pale lemon yellow.** Head and neck have immaculate white feathers as on most adults. BODY (dorsal).—Dark brown with adultlike feathering, including the white lower half of the back and rump. BODY (ventral).—Dark brown and adultlike on the belly and flanks. The undertail coverts are all white. WINGS (dorsal).—Adultlike dark brown with thin pale edges on all coverts. WINGS (ventral).—A few very advanced individuals may have all-dark underwing coverts and remiges and are identical to full adults. However, as described in Adults Traits, even birds with very adultlike heads and bodies can have a considerable amount of white on the under wing region. TAIL.—Immaculate white uppertail coverts and rectrices.

**BASIC IV (SUBADULT IV; TYPICAL TYPE):** HEAD.—**Medium yellow or orangish yellow bill with a brownish or grayish smudge on the lower mid-part of the upper mandible. Cere is often gray or brown with a yellow fringe around the nares and on the front edge. Iris is either pale lemon yellow or pale gray.** Head has a very narrow brown line or patch on the top of the auriculars that often extends down the sides of the neck. Some may also have brown feathering on the white forehead and sometimes on the crown. BODY (dorsal).—As in adult type, but often has a few white specks on the back and dark blotches on the white portion of the lower back and rump. BODY (ventral).—Some are all dark as adult type, but many also retain some white blotching on the belly and flanks. The white undertail coverts my be all white or have a few dark blotches. WINGS (dorsal).—Adultlike dark brown with pale edges on all coverts. WINGS (ventral).—Most have a small to moderate amount of white subadult markings on the axillaries, underwing coverts, and secondaries. TAIL.—Most have white rectrices with irregular dark tips on a few or most feathers, which may form a partial or complete dark terminal band. Sometimes there is a dark strip on the outer web of the outermost rectrices and scattered black mottling on the inner portion of some or many feathers. The uppertail coverts are white, but may have dark blotches.

**BASIC III (SUBADULT III) TRAITS:** This is the

first age class with fairly adultlike characters, but all birds still retain a fair amount of younger subadult traits. There are two main plumage types, which reflect the amount of retained younger subadult characters. Advanced types have acquired more adultlike characters with a darker plumage and paler head, and those with younger subadult traits have more white blotches on the body and wings. HEAD.—Bill and cere color is variable and does not correlate with the plumage stage; gape is always pale yellow: (1) Advanced birds have mostly medium yellow or orangish yellow bills, including the cere. (2) Less-advanced birds may have yellowish brown or grayish bills with a more yellowish region on the lower basal part of the upper mandible. The central part of the bill is streaked with dark brownish or grayish that extends from the cere to the tip of the bill. The lower mandible is yellowish but still has some brownish or grayish areas. The cere is also brownish or grayish with yellow fringing the front edge, as well as the nares. The head and neck are whitish, but the demarcation line between the brown back and breast is only moderately defined. Dark brown feathers form streaks that create an ill-defined border between the lower neck and back and breast. This is especially apparent on the front and rear of the neck. The whitish nape and hindneck, in particular, are often extensively streaked with dark brown. Iris color varies: pale brown, pale gray, or pale yellow. BODY.—*See* Types below. WINGS.—Slow-molting individuals may retain a very old juvenile outermost primary (p10), which by this time, is very faded brown and frayed. Many have replaced p10, which would then be a neat, dark colored feather (and similar to a subadult IV eagle). All other primaries have been replaced one or two times and are new or fairly new, and are dark and neatly formed feathers. Of note are the newer, darker p1–4, which are more new feathers than on a subadult II and fewer than on a subadult IV. All secondaries have been replaced one or two times by annual molts, and the ventral surface may still exhibit white patterns as seen on younger birds. Virtually all have an old subadult I feather on the inner region of the secondaries at the s12 position, which is older and more faded brown than any nearby feather. S13 and 14 are newly grown subadult III, dark, and neatly formed. Adjacent secondaries, s10 and 11, are moderately old and moderately faded subadult II feathers. The tertials, s15–17, are also older subadult II feathers and are more faded and brown than s13 and 14, but are not as pale as the subadult I feather, s12. On rapidly molting individuals, s10 may be the oldest feather of the inner secondaries. On the upperwing coverts, most eagles retain a very small number of old subadult II median and lesser coverts on the inner part of the wing; some may still have one or more very old subadult I coverts. TAIL.—Highly variable in the amount of black-and-white pattern. Dorsal surface varies from being (1) all black with some white mottling, (2) mainly white on the central rectrices with an irregular black terminal band and black on the outer feathers, or (3) very white on most rectrices with an irregular black terminal band. The ventral surface is mainly white with a black strip on the outer web on the outer rectrix set and an irregular black terminal band. Some have irregular black edges on several rectrices; a few lack the black outer edge. The uppertail coverts are often white.

**BASIC III (SUBADULT III; ADVANCED/HEAVILY MARKED TYPE):** HEAD.—Head is fairly white with a narrow dark patch on the upper half of the auriculars that does not attach to the gape. Forehead, crown, nape, and hindneck have a moderate amount of dark brown streaking on the white feathers. BODY (dorsal).—Adultlike, but there are usually blotches on the white part of the lower back and rump. The back may have a few white specks. BODY (ventral).—Adultlike, but the white undertail coverts have some dark blotching and mottling. WINGS (dorsal).—Adultlike. WINGS (ventral).—There is a small amount of white speckling on the underwing coverts; axillaries may be all dark brown or have a very small amount of white speckling. The secondaries may be all dark and adultlike or have only a small amount of white on a couple of feathers. TAIL.—Any of the three patterns described in Basic III (Subadult III) Traits, but more likely to have a more whitish dorsal surface.

**BASIC III (SUBADULT III; MODERATELY MARKED TYPE):** HEAD.—Head is moderately white with a moderately wide dark patch on all of the auriculars that extends from the gape on the lower edge of the dark mark (dark patch is as wide as on most subadult II eagles). Forehead, crown, nape, and hindneck have an extensive amount of dark streaking on pale tawny or white feathers. A fairly distinct whitish supercilium is formed by the dark auricular patch and the dark forehead and crown. The front of the neck and hindneck are quite dark and form an ill-defined border from the back and breast. BODY (dorsal).—The back has a moderate or extensive amount of white speckling that forms an inverted white triangle. *Note:* The white triangle can be paler and more distinct than on heavily marked types of subadult I and II. The lower back and rump are dark with some white mottling. BODY (ventral).—A moderate amount of white speckling on the dark brown flanks and belly. The white undertail coverts have a considerable amount of dark markings. WINGS (dorsal).—Adultlike, but most birds have remnant subadult II feathering that may have distinct white edges on one or more greater and median upperwing coverts on the inner part of the wing. WINGS (ventral).—There is a moderate or sometimes an extensive amount of white markings on the underwing coverts, axillaries, and some secondaries. Any secondaries with white markings have the white contained inside a broad black terminal band. TAIL.—Any of the three variations described in Basic/Subadult III Traits; however, they are more likely to have darker tail variations.

*Note:* Not separable from heavily marked type of subadult II unless wing molt sequence is well observed.

**BASIC II (SUBADULT II) TRAITS:** There are two main plumage types that have a variable amount of white on the different anatomical areas. As a rule, they are darker with a lesser amount of white than all types of subadult I eagles except lightly marked type. HEAD.—Bill and cere color is highly variable and does not correlate with the different plumage types. Gape is always yellow, but there are three main variations: (1) Brownish or grayish with a yellowish area on the lower basal region of the upper mandible; lower mandible is mostly yellowish. Cere is also a similar grayish with yellow surrounding the nares. *Note:* Common type. (2) Medium yellow with a wide brownish or grayish strip from the mid-cere to the tip of the bill; lower mandible is yellow. Cere is brownish or grayish with yellow fringing the nares. *Note:* Fairly common type. (3) All-yellow upper and lower mandibles and similar to color of adults. *Note:* Uncommon to very uncommon type. Iris color is usually pale brown or pale gray, but is occasionally pale yellow. BODY (dorsal and ventral).—Highly variable. *See* Types below. WINGS (dorsal and ventral).—*See* Types. Key features of this age class, which are most visible in flight: (1) The faded, retained subadult I secondaries s12–14; (2) birds in slow and moderate stages of remix molt will show long, pointed, and faded retained juvenile secondaries in the s4, or s4 and 9 positions; (3) the outer one or two primaries, p9 and 10, will also be worn and frayed retained juvenile feathers; and (4) p1 and 2 are new and dark feathers. On perched birds, the freshly molted tertials (s15–17) are darker and less worn than the adjacent s12–14 feathers. White markings on the upperwing coverts are less obvious than on subadult I eagles. A few scattered, very bleached tan-colored juvenile feathers are retained on the lesser upperwing coverts, particularly on the inner part of the wing. TAIL.—Two main types: (1) Dorsal surface, which is on the outer web of each feather, is primarily black with some white mottling, and (2) dorsal surface is white on the central one or two sets of rectrices and all other sets have black outer webs; tips of all rectrices have a wide black terminal band. In both types, the inner feather web is white or white sprinkled with black; when the tail is widely spread, it will appear very white with narrow dark edges and a black terminal band. Since the inner webs are white, the ventral surface is white with a variable amount of black mottling. The outer edge of the tail is black, and there is a moderately wide black terminal band.

**BASIC II (SUBADULT II; MODERATELY MARKED TYPE):** HEAD.—The demarcation line on the base of the whitish neck and the dark brown back and breast is moderately defined. Two types of head patterns: (1) *Moderately white type.*—Moderately wide dark brown auricular patch extends from the gape and connects with the moderately dark streaking on the nape and hindneck. The forehead and crown are brownish and often create a broad supercilium. The white areas of the head often have narrow dark streaking. *Note:* Common type. (2) *White* *type.*—A narrow brown patch on the upper half of the auriculars that is similar to subadult III and connects under the eyes but does not extend to the gape. The patch connects with the lightly streaked nape and hindneck. The white areas of the head are often unmarked or very lightly streaked. *Note:* Uncommon type. BODY (dorsal).—Whitish hindneck merges with the white inverted triangle-shaped area on the back. The white back has a small to moderate amount of dark blotches. The lower back and rump are dark brown and mottled with white. BODY (ventral).—Dark brown breast forms a large, distinct bib that contrasts with the white belly and flanks. The dark bib is sometimes mottled with white but is still distinct. The white belly and flanks are lightly to moderately spotted with dark brown (but never unmarked, as in some eagles of subadult I age). Leg feathers are dark brown; in flight, they form a dark "V" against the white belly and flanks. The undertail coverts are a mix of white and dark brown markings. WINGS (dorsal).—Moderately broad white edges on most median and first row of lesser upperwing coverts. WINGS (ventral).—There is an extensive amount of white on the underwing coverts and axillaries. There is often a distinct white bar on the median underwing coverts. The inner one to three primaries may show some white, and several middle and inner secondaries will also exhibit white patches that are bordered by a wide black terminal band on the trailing edge.

**BASIC II (SUBADULT II; HEAVILY MARKED TYPE):** HEAD.—Rarely has the white type and regularly has the moderately white type. A large number of eagles of this plumage type, however, have a darker head pattern. *Heavily marked type.*—Head is tawny with a moderately wide dark brown auricular patch that extends from the gape and connects with the dark brown nape and hindneck; crown is a darker tawny and is often darker on the forehead. Demarcation of the neck to the back and breast is very ill-defined. The throat is whitish. BODY (dorsal).—Dark brown and adultlike with only a few white specks on the back; some may lack white speckling. *Note:* Dorsal region is as dark as many subadult III and some subadult IV birds. Lower back and rump are dark brown with some white speckling. BODY (ventral).—Dark brown with a very small amount of white speckling on the belly and flanks; sometimes lacks speckling. Undertail coverts can be dark with some light areas or moderately white with some dark areas. WINGS (dorsal).—Mainly dark brown with little if any white edges on the coverts. WINGS (ventral).—Mainly dark brown coverts with a small amount of white mottling on the axillaries, lesser underwing coverts, and median coverts (lacks a white bar on median coverts). Underside of the remiges is dark or has only a small amount of white markings.

*Note:* This plumage variation is very similar and often inseparable from moderately marked type of subadult III. It is necessary to see wing molt in order to categorize this age, as iris and bill colors overlap with subadult III.

**BASIC I (SUBADULT I) TRAITS:** There are three main plumage types that are labeled according to the amount of white feathering. Unlike older subadults, variations do not appear to be correlated to a more rapid advancement towards adult plumage. HEAD.—Unlike older subadults, there is little variation on bill and cere color: (1) Bill is dark gray with a paler grayish or yellowish area on the lower basal region of the upper mandible and a yellow fringe on front of the cere and around the nares, or (2) uniformly gray on bill and cere. Gape is pale yellow. Iris color is pale brown, pale gray, or rarely pale yellow. BODY (dorsal and ventral).—Variable in the amount of white markings; *see* types. WINGS (dorsal and ventral).—Variable in the amount of white markings; *see* Types below. The stage of remix molt easily identifies this age class. Even though there are minor variations, they are always separable from older subadults and juveniles. On perched birds viewed from behind, the three new tertials (s15–17) contrast sharply to the more faded, worn, and longer brownish juvenile feathers on most of the rest of the secondaries. In flight, the jagged trailing edge created by the large number of retained juvenile secondaries contrasts with the shorter, more rounded subadult I feathers. White markings on retained juvenile feathers often extend to the tip of each feather. White markings on new subadult I feathers are bordered on the trailing edge of each feather by a wide black band. The outer half of the wing has worn and frayed juvenile primaries. TAIL.—Two main variations that can be found on any plumage type: (1) *All-black type.*—All black or lightly sprinkled with white on the dorsal surface. (2) *White type.*—white on the central one or two rectrix sets and black on all other sets. On both types, the inner feather web of each rectrix is white or sprinkled with black. When widely fanned, the tail can appear fairly white on the black type and very white on the white type. On the ventral surface, the tail is mainly white and lightly or moderately sprinkled with black, with a moderately defined black terminal band.

**BASIC I (SUBADULT I; LIGHTLY MARKED TYPE):** A very lightly marked plumage type that can appear to be albinistic. However, all-white feather markings are simply more extensive than in the two darker types and are not irregular white areas, as found on albinistic birds. HEAD.—*White type.*—Moderately wide, isolated dark brown patch that extends from the gape onto the auriculars (same width of auricular patch as on most subadult II eagles). The rest of the head and neck, including the crown, nape, and hindneck, are white. BODY (dorsal).—The back is white with a very small amount of dark spotting and merges with the white hindneck. The white back forms a distinct inverted white triangle that blends with the white hindneck. Many of the scapulars have white edges. The lower back is white with some dark spotting. BODY (ventral).—The breast has a small to moderate-sized dark brown bib. On some, the breast is lightly mottled with dark brown and the bib is poorly defined. The white lower neck contrasts sharply with the top of the brown bib. The belly and flanks are either white with small dark spots or unmarked and pure white. The leg feathers are dark brown with some white mottling and form a dark "V" against the rest of the white underparts when in flight. The undertail coverts are dark with some white areas. WINGS (dorsal).—The median and first rows of lesser upperwing coverts are white with narrow dark center streaks. The tips of many greater upperwing coverts are broadly edged with white. The rest of the coverts are dark brown. WINGS (ventral).—All of the underwing coverts and axillaries form an extensive white area. The retained juvenile secondaries are largely white. The new subadult I secondaries, typically s1 and 2, s5 and 6, and s12–14, may be covered with extensive white markings but they are bordered on the trailing edge by the wide black terminal band. TAIL.—Mainly the white type.

**BASIC I (SUBADULT I; MODERATELY MARKED TYPE):** HEAD.—Two variations of head patterns: *Moderate type.*—Moderately wide dark patch extends from the gape and onto the auriculars. The head and neck, including the crown, nape, and hindneck, are tawny brown and streaked with dark brown. *Dark type.*—A wide dark patch extends from the lower mandible and onto and below the auriculars as a large dark side-of-head patch. The head and neck are tawny brown or dark tawny brown and streaked with dark brown and is only slightly paler than the breast. The throat is whitish. BODY (dorsal).—The back is white with moderate to large dark spots and forms a distinct inverted white triangle. Scapulars have narrow white or tawny edges on many feathers. The lower back has some white spotting. BODY (ventral).—The breast is dark brown and forms a distinct bib. The dark bib is often mottled with white, but is still very distinct. The belly and flanks are white and covered with small to moderate-sized dark spots; occasionally nearly unmarked. The leg feathers are dark brown and, in flight, contrast against the white belly and flanks. The undertail coverts are dark brown. WINGS (dorsal).—The median and first rows of lesser upperwing coverts may have moderate or wide white edges; sometimes rather white with narrow dark center streaks on each feather. The inner greater upperwing coverts may have white or tawny mottling or edging on most feathers. WINGS (ventral).—The axillaries are white, the lesser underwing coverts are mottled brown and white, and the median coverts are often white and form a distinct white bar along the midwing. The remiges may have a moderate or extensive amount of white markings. TAIL.—Either of the two types described in Subadult Traits.

**BASIC I (SUBADULT I; HEAVILY MARKED TYPE):** HEAD.—Either the moderate type or dark type, but more likely to be the latter. The throat is whitish. BODY (dorsal).—The back is dark brown on the upper two-thirds and speckled with white on the distal one-third (a white triangle is not apparent, as on the other types). The back is sometimes all dark. Scapulars lack pale edges on the feathers. Lower back is dark brown. BODY (ventral).—Large dark brown bib on the breast. The belly and flanks are white and covered with moderate to large dark brown spots. The leg feathers are dark brown and form a dark "V" when the bird is flying. The undertail coverts are dark brown. WINGS (dorsal).—The upperwing coverts are either uniformly dark brown or the inner median and greater coverts may have narrow pale tips or edges. WINGS (ventral).—Axillaries are mainly white, the lesser coverts are darker and mottled with brown and white, and a fairly distinct white bar is on the median covert tract. Remiges have a moderate amount of white on each feather. TAIL.—Mainly the dark type.

**JUVENILE TRAITS:** HEAD.—Uniformly dark brown with a whitish throat. On some birds, small tawny tips form on the nape and hindneck feathers. Most have a few white streaks on the sides of the neck. **Bill and cere are black; gape is pale yellow.** Iris is dark brown. BODY (dorsal).—A paler brown than the head. BODY (ventral).—**Breast is darker brown than the tawny flanks and lower belly and forms a bib.** Many eagles exhibit a small amount of white streaking on the lower part of the bib, and many have few dark brown blotches on the flanks and upper belly. **Leg feathers are dark brown and contrast with the paler belly and flanks.** The undertail coverts vary from being dark brown to white with dark streaks down the center of each feather. WINGS (dorsal).—Upperwing coverts are the same brown color as the back and scapulars and are paler than the head and remiges. The inner one-third of the greater and first rows of lesser coverts are pale tawny and forms a pale brown patch on the inner wing that is visible while perching or flying. The tips and outer edges of the inner six secondaries (s12–17 are pale tawny or pale brown and much paler than the rest of the remiges. The greater upperwing coverts are medium brown and paler than the rest of the upperwing coverts. **In flight, the black remiges contrast with the paler brown upperwing coverts and form a two-toned effect.** WINGS (ventral).—There are two types of underwing patterns: (1) *White type.*—**Underwing coverts and axillaries are nearly all white, including all of the patagial region. There is a narrow dark bar along much of the first row of lesser coverts and on part of the inner portion of greater covert feathers. The greater coverts have an extensive amount of white on a large part of each feather. The remiges have an extensive amount of white on the inner two or three primaries, on the outermost secondary, and on the inner 12 secondaries. Only three or four secondaries are dark (s2–4), and they divide the white regions on the inner primaries and secondaries. On the inner secondaries, the white extends to the tips of the feathers; on the middle secondaries the white usually ends before reaching the tips.** (2) *Dark type.*—**The axillaries are white, there is some white mottling on patagial coverts, a white bar extends out on the median coverts, and the secondary greater coverts have white tips. The remiges have a minimum to moderate amount of white on the outer one to three primaries and inner few secondaries.** TAIL.—Three variations on dorsal and ventral surfaces: (1) *All-black type.*—Dorsal and ventral surfaces are completely black. (2) *Moderately white type.*—The dorsal surface is black with a sprinkling of white, particularly on the central one or two rectrix sets. On the ventral surface, the tail is also a mix of black and white with a moderately wide or wide black terminal band and a narrow black band on the outer edge. The ventral surface can be mainly white with the black terminal band and narrow black outer edge. (3) *White type.*—Tail is white except the outer web of each feather is black, and the moderately wide or wide terminal band is black on both surfaces.

**JUVENILE (WORN PLUMAGE):** BODY (dorsal and ventral).—A few months after fledging, and especially towards the end of the age cycle, the body plumage fades considerably due to exposure to the sun. The back, scapulars, and upperwing coverts bleach to pale brown. The belly and flanks bleach to pale tawny or very pale tawny and contrast sharply with the dark brown breast, neck, head, and leg feathers.

**JUVENILE (FRESH PLUMAGE):** BODY (dorsal and ventral).—When recently fledged and lacking extensive exposure to the sun, the plumage is darker than on older juveniles. The back, scapulars, and upperwing coverts are medium brown, but still a bit paler than the dark brown head. The belly and flanks are medium to dark tawny, but still contrast a bit against the dark brown breast, neck, head, and leg feathers.

**JUVENILE (STREAKED TYPE):** As typical juvenile in either fresh or worn plumage except for the flanks and portions of the belly. BODY.—The flanks may be white instead of tawny and streaked with dark brown. The belly is mostly tawny but may be somewhat whitish on the upper area near the flanks. The belly is variably streaked and spotted with dark brown. The leg feathers are dark brown as on typical birds.

**ABNORMAL PLUMAGES:** There are numerous records of imperfect albino (dilute plumage) juveniles and adults (Clark and Wheeler 2001; BKW pers. obs.). See photograph of an imperfect albino adult in Clark and Wheeler (2001). Imperfect albino juveniles have similar color patterns as normal juveniles but have nearly white bellies and flanks and light brown head, neck, breast, and upperparts. There is an extensive amount of white on the underwing and undertail (BKW pers. obs. and photos). There are records of partial albino juveniles in the West (Clark and Wheeler 2001).

**HABITAT: Summer.**—Bald Eagles nest in a wide variety of habitats, including lowland deserts, lowland forests, montane forests, prairies, and seashores. Virtually all eagles breed and summer near bays, creeks, lakes, marshes, rivers, sea coasts, or swamps. Most regions have tall trees for nesting sites. On the Great Plains, open prairies may have narrow riparian strips along small creeks. In Arizona, some pairs nest in desert lowlands, using cliffs, rock pinnacles, or tall trees for nest sites. In Sonora, Mexico, eagles inhabit deserts with cliffs, large or small trees, or cacti. If there are no terrestrial predators, treeless islands, as on the Aleutian Islands in Alaska, are used.

Climate varies: hot and arid in the deserts, warm to hot and moist south and east of the Great Plains, cool and moderately moist in montane areas and the boreal forest, and cool and very moist in the Pacific Northwest and coastal Alaska.

Bald Eagles typically nest in isolated regions, but in the Pacific Northwest, some pairs nest in human-inhabited areas.

**Winter.**—Similar habitats as in summer where there is open water. Montane and northern-latitude breeders winter at lower elevations and more southern latitudes. If there is sufficient prey, eagles are also found at middle and southern latitudes in vast open expanses that lack water. Large numbers seasonally gather at salmon spawning areas along rivers in s. Alaska and British Columbia. Lowland marshes and lakes with an abundance of waterfowl are also favored wintering habitats.

**Migration.**—Found in any type of habitat, including montane areas, but the largest numbers are seen near lowland riparian locations.

**HABITS:** Typically a wary species; however, some individuals are fairly tame, particularly in southern regions. Nesting pairs are quite wary and become highly agitated by humans intruding into nesting territories.

Bald Eagles may perch in one location for several hours, especially during winter, to conserve energy or wait for a feeding opportunity. Large, exposed branches are preferred perches. Eagles also readily perch on the ground, ice, and many artificial objects. Except from n. California to Washington, Bald Eagles in the West rarely perch on utility poles. During the hot summer months, southern birds will perch in sheltered, shaded areas of trees.

They are solitary during the breeding season, but become gregarious in the nonbreeding season, and large numbers may congregate to feed and roost. Prime winter feeding and roosting locations may host several hundred or even a few thousand eagles. Pairs remain together year-round, even if they migrate. Pairs often perch side by side.

When perched, two animated body positions exhibit social and sexual behavior and are accompanied by vocalization: (1) *Straight-necked.*—Neck is fully extended in a straight line ahead of the body, with the head and bill also held in the same line. (2) *Head-toss.*—Head and neck are tossed upwards at a vertical angle, with the neck fully extended.

**FEEDING:** Perch and aerial hunter. Bald Eagles are opportunistic feeders; however, they are adept hunters and mainly feed on live prey during the breeding season. During the nonbreeding season, eagles become more opportunistic and lazy, pirate food from conspecifics and other raptors, and scavenge for carrion of all types and sizes. Pairs will cooperatively hunt.

Perch hunters fly from elevated perches to capture most prey. Bald Eagles also walk or stand in shallow water or along the edge of a shoreline or ice to catch fish.

Aerial hunters make random, low- to high-altitude forays over potential feeding areas. Prey may be captured on the ground or water and occasionally in the air (*see* Flight).

Eagles capture most prey by grabbing it with their talons, but when standing or walking in shallow water or along shorelines, may also use their bills.

Diet varies regionally and seasonally, but fish are a primary prey in most regions and seasons. Live fish are caught in all seasons, but mainly in the breeding season. Dead, dying, or stunned fish become important food in the nonbreeding season. In spring, feeds on winter-kill fish from thawing lakes, ponds, rivers, and streams.

Live and dead ducks, geese, and other water-type birds, especially American Coots, are preyed upon in all seasons, but probably more so in the winter.

Smaller mammals form a small part of its diet in much of the eagle's range, especially Muskrats. In parts of Colorado and Wyoming and possibly other parts of the West, nesting eagles specialize in preying on prairie dogs and ground squirrels. Jack rabbits and Black-tailed Prairie Dogs, whether hunted, pirated, or scavenged, become important prey for eagles wintering in arid regions.

Wild and domestic ungulates killed by hunters, vehicles, or natural causes form an integral part of the diet of scavenging eagles that do not associate with water. Eagles also scavenge at open-pit garbage dumps.

**FLIGHT:** Soars and glides with wings held on a flat plane. May soar with the wings held in a very low dihedral. Powered flight consists of unique wingbeats that separate them from other raptors and large birds, even when observed at great distances. The upstroke is high and the downstroke is low, which creates an exaggerated upwards motion on the upstroke. When entering a glide or soar, Bald Eagles nearly always end a wingbeat on a downstroke (F. Nicoletti, unpubl. data). Powered flight is an irregular series of flapping and gliding.

Also unique to Bald Eagles is the "flare-up" display. Eagles of all ages will flare up to signal danger to nearby flying eagles by banking sharply upwards at a vertical angle with their wings flapping and legs either tucked up or lowered.

**VOICE:** A highly vocal raptor during all seasons. The calls of males are higher-pitched than females. Call notes are high-pitched, crisp, and metallic-sounding. (1) *Ca-ack* or *kah* call is emitted during alarm, especially around nest sites, and may be repeated numerous times. (2) *Whee-he-he-he* is a staccato-sounding call given in a decrescendo and slows towards the end. It is uttered when eagles are annoyed by conspecifics or other bird species (crows, gulls, ravens), during pair formation (accompanied by *head-toss*), when a female begs for food from a male, and when copulating. (3) *Yaap* is a wailing, gull-like call that is repeated numerous times and is often intermixed with the #2 call. The call is given at the nest by displaying adults and hungry chicks, and also away from the nest by fledged, food-begging young. (4) *Chatter* or *chitter* is a series of rapidly uttered short *ca-ack* notes given by adults at nest sites and with eagles engaging in conflicts with conspecifics at feeding and roosting sites.

**STATUS AND DISTRIBUTION:** Currently listed as a Threatened Species in the lower 48 states. Alaska and Canada have always had healthy breeding populations, and the Bald Eagle was never listed as endangered or threatened.

Laws prohibiting killing Bald Eagles, organochlorine pesticide bans in Canada and the U.S., state programs that assisted population growth, and retention of critical habitat have helped the Bald Eagle gain a new foothold and re-establish itself. With this assistance, eagle populations have been steadily growing, often significantly, in many areas.

**Summer.**—CANADA: Large numbers breed in Canada except for much of the southern one-third of Alberta and Saskatchewan and treeless Nunavut. N. Saskatchewan has the largest inland population in North America. They are absent from high montane regions of British Columbia, the Northwest Territories, and the Yukon Territory.

UNITED STATES: Known pairs are given for 1982/1998 for comparison of population increase. More current data, if known, are in parentheses. In 1999, the lower 48 states had an estimated 5,748 pairs of Bald Eagles.

*Alaska:* 30,000 pairs. Admiralty Island, south of Juneau, has the highest nesting concentration of Bald Eagles in North America. *Arizona:* 15/36 pairs (43 pairs in 2002). Many are desert dwelling. Only two pairs remained in 1970. Nests are on Alamo Lake on the Bill Williams River, Lake Pleasant, on the Verde River from Mesa to Prescott, on the Salt River east of Mesa through Theodore Roosevelt Lake and east to the split of the river into the Black River and the White River. Also nests on Tonto Creek below Gisela. *Arkansas:* 1/29 pairs (36 pairs in 2000). The first successful nesting since the 1930s occurred in 1982. The state reintroduced hacked eagles. Nests on Greers Ferry Lake, Cleburne Co., and Lake Conway, Faulkner Co. Nesting also occurs in the following counties: Ashley, Arkansas, Benton, Crittenden, Desha, Drew, Franklin, Fulton, Grant, Hempstead, Jackson, Lafayette, Little River, Logan, Mississippi, Monroe, and Van Buren. *California:* 43/143 pairs (167 pairs in 1999). The population had dropped to around 20 pairs by the late 1960s due to pesticide contamination. Nests in 27 counties, most in the northern counties. After an absence of 60 years, a released population of about five pairs now nests within the central coast near Livermore in Alameda Co., in Santa Barbara Co., at San Antonio Reservoir in Monterey Co., and at Nacimiento and Santa Margarita reservoirs in San Luis Obispo Co. Eagles originally nested on Santa Catalina Island, but were wiped out by pesticides; however, after reintroduction, they have nested there since 1992. Four pairs now nest on the island. However, as of 2002, the Santa Catalina Island birds are still not producing viable eggs due to pesticide contamination. They are also resident near Gregory Lake in the San Bernardino Mts. and at Tinemaha Reservoir in the Owens Valley north of Independence. Eagles have recently nested on the California side of Lake Tahoe. *Colorado:* 4/27 pairs (51 pairs in 2001, 45 pairs in 2002). One nest site has been used since 1974. Nesting pairs are mainly located in the mountains; however, a few pairs reside along rivers and reservoirs on the flatlands adjacent to the foothills of the Front Range: A nest at Barr Lake in Adams Co. has been active since 1986; Boulder Co. is a newly established area. *Idaho:* 15/92 pairs (113 in 2001). Nesting occurs in the following counties: Benewah, Bingham, Blaine, Boise, Boundary, Caribou, Clark, Custer, Elmore, Fremont, Gem, Jefferson, Jerome, Kootenai, Lemhi, Madison, Teton, and Valley. *Iowa:* 1/83 pairs (130 pairs in 2001). This state has seen an explosive natural recolonization of Bald Eagles since the mid-1970s. Most nests are in large cottonwood trees along riparian corridors. Eagles were absent as a nesting species from 1906 to 1976. Eagles successfully nested in 1977 at New Albin, Allamakee Co. They are now resident in the following counties: Allamakee, Appanoose, Black Hawk, Buena Vista, Clayton, Clinton, Delaware, Des Moines, Fayette, Fremont, Howard, Iowa, Jackson, Jefferson, Jones, Louisa, Muscatine, Sac, and Winneshiek. *Kansas:* 0/8 pairs (10 pairs in 2000). Nests are located at John Redmond Reservoir in Coffey Co., Clinton Reservoir in Douglas Co., Perry Reservoir and on the Kansas River in Jefferson Co., Hillsdale Reservoir in Miami Co., Glen Elder Reservoir in Mitchell Co., and Norton Reservoir in Norton Co. A pair is also present on Tuttle Creek near Olsburg. There were active nests in the 1990s in Hodgeman and Stafford Cos. *Louisiana:* 18/135 pairs. Nesting is concentrated in the southeastern part of the state. Nesting pairs in the northern part of the state are found in Cypress Bayou Reservoir north of Shreveport, Lake D'Arbonne in the north-central part, and in Morehouse Parish. Nests in the west-central region in Sabine Parish and in the southeast region in Calcasieu Parish. *Minnesota:* 207/618 (in 1995) pairs (680 pairs in 2000). Common in breeding range. The most southwestern site is in Lac qui Parle Lake, Lac qui Parle Co. Pairs recently began nesting along the Red River in Polk Co. Nesting has also recently expanded to Rice Co. south of Minneapolis; several pairs are around Minneapolis-St. Paul. *Missouri:* 1/45 pairs (47 pairs in 2001). There were no nests in the state from 1965 to 1982. Hacking programs helped rebuild the population. Nesting occurs on Harry Truman Reservoir and Lake of the Ozarks and their adjoining waterways. Pairs also nest at Mingo NWR in the southeastern part of the state and along the Mississippi River. *Montana:* 37/212 pairs. Most nesting pairs are located in the western part of the state and along the Yellowstone River, which crosses the southern part of the state. *Nebraska:* 0/14 pairs (23 pairs in 2001). In 1991, Douglas Co. had the first nest in the state since the late

1800s. Since then, eagles have nested at Lake Alice in Scotts Bluff Co., near Arnold in Custer Co., near Beatrice in Gage Co., DeSoto NWR in Washington Co., near Guide Rock in Webster Co., Hackberry Lake in Garden Co., Calamus Reservoir in Loup Co., on the Niobrara River in Knox Co., near Ord in Valley Co., in Pawnee Co., in Nemaha Co., Sherman Reservoir in Sherman Co., and Swan Lake in Cherry Co. Nesting also occurs on the Platte River in Buffalo and Dodge Cos. and along the Missouri River in Boyd, Buffalo, Cedar, Dixon, Holt, and Nemaha Cos. *Nevada:* 0/1 pair (2 pairs in 2001). Both current nesting pairs are on the Carson River. *New Mexico:* 0/4 pairs. Nesting occurs in Colfax and Sierra Cos. *North Dakota:* 0/9 pairs. Nesting pairs are mainly along the Missouri River and Lake Oahe north and south of Bismarck. The most recent pair is near McKenzie in Burleigh Co. In the northeast, nests are at Fordville Dam, near Rolette, southeast of Lake Alice, Minnewaukan Flats, and in the vicinity of Benson, Nelson, and Ramsey Cos. *Oklahoma:* 0/26 pairs (39 pairs in 2002). Northeast of Oklahoma City: Nesting pairs are on the Arkansas River in Pawnee Co., Kaw Lake in Osage Co., and Bear Creek in Noble Co. Northeast of Tulsa: On or near Grand Lake of the Cherokees in Ottawa Co. Tulsa area: Two nests on the Arkansas River west of Tulsa and several nests on the river southeast of Tulsa through Robert S. Kerr Reservoir. Many nests are also on or near Eufala Lake. Three pairs nest on or near the Poteau River from Poteau, Okla., to Fort Smith, Ark. The state had a hacking program to rebuild its population. *Oregon:* 100/324 pairs. Urban nesting occurs in Portland. Most pairs are found west of a line from Portland to the junction of Oregon, California, and Nevada. *South Dakota:* 0/13 (15–20 pairs in 2002). Pairs are located on the Missouri River in Bon Homme, Gregory, and Yankton Cos. Pairs are also present on the Belle Fourche River in Meade Co. and the reservoirs in Brown Co. *Texas:* 13/62 pairs (98 in 2001). Only four nests were active in 1974. The population recovered by natural recolonization. Nests are mostly northeast and southwest of Houston at Toledo Bend Reservoir and on the following rivers: Angelina, Brazos, Colorado, Guadalupe, Navasota, Neches, Sabine, San Marcos, and Trinity. There has been a recent westward extension into Bosque and Llano Cos. *Utah:* 0/4 pairs. Nesting pairs are on the Jordan River near Salt Lake City and near Castledale. The other two nest sites are on the Colorado River between Westwater Creek and Cisco. *Washington:* 137/630 pairs (658 pairs in 2001). Pairs nest in urban settings around Puget Sound. Large numbers are found along all coastal areas. Many pairs nest in the Cascade Mts. and a few pairs are scattered in the central and northeastern part of the state. *Wyoming:* 23/78 pairs (over 100 in 2002). Majority of pairs are in the northwestern part of the state in and around Yellowstone N.P. Pairs also nest at Pathfinder and Seminoe reservoirs in Carbon Co., at the headwaters of the North Platte River in Carbon Co., and along Rock Creek in Albany and Carbon Cos.

MEXICO: The Baja has three pairs at Magdalena Bay, and Sonora has three pairs on the Rio Aros and Rio Yaqui. The first Bald Eagles recorded in Sonora were seen in 1976. Nesting was documented in 1986–1990 with five nesting pairs.

**Winter.**—Thousands of Bald Eagles remain on the Aleutian Islands and southern coastal areas of Alaska and along the coastal regions of British Columbia. The famed Chilkat River in se. Alaska brings in several thousand eagles each autumn and early winter because of the late salmon-run. A similarly large gathering occurs for a late salmon-run near Brackendale in se. British Columbia.

South of Canada, the largest congregations occur in areas of open water with an abundant supply of waterfowl or fish. Large winter concentrations form at Lower Klamath NWR and Tule Lake NWR in ne. Siskiyou Co., in n. California. In the Midwest, about 2,500–4,000 Bald Eagles winter in various locations along the Mississippi River from Minneapolis, Minn., to near St. Louis, Mo., mainly near the numerous locks and dams that provide open water and abundant fish.

Small numbers winter throughout the Great Basin, Great Plains, and into the boreal forest of Minnesota. Many of these birds wintering in these arid regions are found far away from water sources. Very small numbers winter in the interior regions of Canada. Isolated wintering occurs in interior Alaska and the Yukon Territory.

Eagles winter as far north as open water or ample food permits. Frigid winter weather is not a deterrent if there is adequate food.

Some adults breeding from the Northwest Territories and Saskatchewan cross the Rocky Mts. and winter in cen. and s. California. Based on banding data, birds from cen. Canada appear to winter mainly east of the Rocky Mts. and along the Mississippi River.

**Movements.**—Resident or short- to moderate-distance migrant. Pairs in interior Canada and Alaska migrate due to a lack of food during winter. Pairs on coastal Alaska and British Columbia are mainly resident. Most lower elevation pairs in the U.S. are resident. Pairs nesting in most montane elevations must leave during winter.

Only two hawkwatches in the West see sizable numbers of Bald Eagles in the spring and fall: Windy Point Hawkwatch southwest of Calgary, Alberta, and Hawk Ridge Nature Reserve (fall) and Enger Tower (spring) in Duluth, Minn. Cape Flattery, near Neah Bay on the Olympic Peninsula in Washington, sees fairly large numbers in the spring.

*Northward spring and summer dispersal:* Juveniles born at southern and middle latitudes of the w. U.S. disperse north from natal areas after becoming independent. Subadults and some adults from similar latitudes also disperse north from former natal or soon-to-be breeding areas for the spring and summer.

Northward dispersal begins in spring; however, some n. California juveniles do not leave natal areas until Jul. to mid-Aug. California juveniles' northernmost dispersal area is generally reached by late Aug. to early Sep. Northward movement may be rapid. Once reaching their northernmost point of travel, they may begin heading back south; linger through Sep. and Oct., then head south; or, in mild climates of the Pacific Northwest, may remain at their northernmost point throughout winter and head south in Feb. (when northern-breeding eagles are heading north).

Arizona youngsters that have been tracked with radiotelemetry returned to natal regions by fall.

After dispersing, California early-returning birds remain in the general region of their natal areas from late fall to early summer, then return north for late summer and fall. California birds that remained north all winter may return to natal areas in Apr. and remain until late Jun. or early Jul., then return north, at least for the fall and possibly the winter.

Arizona juveniles and subadults may regularly disperse to British Columbia, Montana, or Wyoming. One tracked juvenile went as far as Manitoba. California-tracked juveniles and subadults mainly headed up to British Columbia and may remain there for winter, but one bird went up to the Northwest Territories and stayed until late Aug.

Most youngsters from the n. U.S. and Canada do not fledge early enough to disperse.

*Fall migration:* As seen in Duluth, adults make up a fair percentage of early migrants in Sep., which lends credence to some adults also dispersing north like most younger birds from southern locations.

There is overlap between the southward movement of southern birds and northern birds. There is a distinct mid-Sep. peak at Hawk Ridge Nature Reserve in Duluth, probably of returning southern birds. Then there is a lull before the major peak of northern birds that occurs from late Oct. to mid-Nov. Movements continue at Duluth through late Nov. and into Dec. At Windy Point in the Rocky Mts. of sw. Alberta, there is a peak of probably mainly southern eagles from late Sep. to early Oct. The second Windy Point peak in late Oct. or early Nov. is probably made up of some southern birds (as seen with California juveniles remaining north until late Oct.) and most northern birds.

Based on telemetry, adults breeding in w. Saskatchewan and the Northwest Territories may leave nesting areas from early Sep. to early Nov. and head southwest to California for the winter.

Fall migration, however, is an unpredictable event for many northern birds because it correlates with the severity of weather. If the autumn is mild, migration may occur late and be protracted. If winterlike weather comes early, a mass exodus from the northland occurs early and quickly.

*Spring migration:* According to telemetry data in s. and cen. California, adults may begin leaving the winter grounds from early Feb. to early Mar. There is a noticeable staging of migrants on the cen. Great Plains and along the

Mississippi River in late Feb. and early Mar. The first migrants are seen at hawkwatches in late Feb. or early Mar., even in s. Alberta. The peak movement of all ages at Duluth is from late Mar. to early Apr. The peak in s. Alberta at Windy Point is typically in late Mar.

**NESTING:** Nesting period varies with latitude and elevation. Arizona birds may begin laying eggs in late Dec., but most do not start until Jan. or Feb. Nesting begins in Jan. and Feb. in most low-elevation southern and central regions. In northern latitudes and high elevations, nesting may begin in Mar. and Apr. Nesting ends from late spring in the south and central regions to early fall in northern areas.

Breeding may occur when 3 years old, but it usually does not occur until birds are at least 4–5 years old. Breeding occurs at younger age in areas where the population is growing, especially in the last few decades as eagles are reestablishing or expanding into new territories with the population boom. Where the population is established and saturated, only older birds may breed.

Pairs typically remain together until one dies. Polyandry and polygyny infrequently occur. Males tend to return closer to natal/hack sites to breed than do females, a trait seen in many raptor species.

**Courtship (flight).**—Both sexes perform *mutual soaring, pursuit-flight,* and *talon-grappling* that turns into *cartwheeling.* Males perform *fly-around* and *sky-dancing.*

**Courtship (perched).**—*Billing* and *mutual plumage-stroking.* The *head-toss* is used at times. Pairs often perch side-by-side.

Nest locations are typically near water. Nests are often used for many years and gradually become larger with repeated use. Alternate nests are often in a territory. Nests are mainly in large, live trees, and placed below the canopy. Bald Eagles favor large conifers in montane and boreal forest regions and large cottonwoods in lowland regions. Small trees and cacti are used in Mexico. Nests are also placed on rock out-crops, cliffs, and pinnacles in areas of rugged terrain. Ground nests are used on islands where there are no terrestrial predators, particularly on the Aleutian Islands, Alaska.

New nests may be 1–3 ft. (0.3–1 m) deep and 3–6 ft. (1–2 m) in diameter. Old nests can be up to 12 ft. (4 m) deep and 8.5 ft. (3 m) in diameter. The nest is comprised of large sticks, and the nest bowl is regularly lined with greenery consisting of grasses, stems, and often pine needles. Both sexes build or refurbish nests.

Eggs are laid at several-day intervals. Both sexes incubate the typical clutch of 2 eggs for 34–36 days. Three eggs are sometime laid (fairly common in British Columbia), and 1-egg clutches are rare. Fratricide (siblicide) is common, with the larger, stronger nestling overpowering the smaller, weaker nestling. Youngsters fledge in 70–77 days, but occasionally take up to 88 days to fly. Fledglings stay with their parent for up to 2 additional months.

**CONSERVATION:** The Bald Eagle, along with the Osprey and Peregrine Falcon, suffered from the widespread use of organochlorine chemicals from the late 1940s to the mid-1970s. The eagle population in some states was either decimated or greatly reduced during this period. The bans on deadly pesticide use allowed eagles in most areas to rebound. As with all raptors, Bald Eagles suffered tremendously from shooting until public awareness turned the tide to produce legal protection. A few states assisted population recovery with reintroduction programs.

**Pesticide bans.**—The first step that was taken, not only to help Bald Eagles, but to benefit all wildlife and humankind, was to ban organochlorine pesticides, particularly DDT. This lethal pesticide was first used in 1946 in Canada and the U.S. This ban, however, was nearly too late to help the decimated population of Bald Eagles in many states.

Canada took a series of steps to discontinue the sale and use of DDT that began in 1968 with a ban on spraying forests in national parks. The major Canadian ban came on Jan. 1, 1970 (announced Nov. 3, 1969) when DDT use was permitted for insecticide use on only 12 of the 62 previously sprayed food crops. However, all registration for insecticide use on food crops was stopped by 1978. Canada, however, permitted DDT use for bat control and medicinal purposes until 1985. Canadian users and distributors were also allowed to use existing supplies of DDT until Dec. 31, 1990.

The U.S. also had a series of steps to ban DDT, but halted the overall sale and use more quickly. In 1969, the USDA stopped the spraying on shade trees, tobacco crops, aquatic locations, and in-home use. The USDA placed further bans on its use on crops, commercial plants, and for building purposes in 1970. The Environmental Protection Agency banned all DDT sale and use on Dec. 31, 1972. However, limited use for military and medicinal purposes were permitted until Oct. 1989. In 1974, the U.S. banned the use of Aldrin and Dieldrin, both deadly chemicals that may have affected wildlife as much as DDT.

Mexico was expected to discontinue government-sponsored DDT use for malaria control in 2002, and has planned a total ban of DDT by 2006.

As of 2000, possibly five other Latin American countries still use DDT and other organochlorine chemicals without restrictions.

The U.S. and Canada, along with 120 other countries, have signed a United Nations-sponsored treaty banning eight deadly organochlorine pesticides: Aldrin, Chlordane, DDT, Dieldrin, Endrin, Heptachlor, Mirex, and Toxaphene. There are also two industrial chemicals, Hexachlorobenzene (also a pesticide) PCBs, and two by-products of industrial processes, dioxins and furans, that have also been banned.

Organophosphate pesticides, although not as persistent in the environment as organochlorine pesticides, can be deadly to some aquatic life that fish eat and therefore affect eagles.

Carbamate pesticides were less harmful to the environment than organophosphate chemicals and were used to control insect infestations in some forests.

Pyrethroid pesticides are harmless to birds and mammals and are currently used to control insect infestations in some forests. However, some may be deadly to aquatic life and affect the food that fish eat, therefore affecting eagles.

**Protective laws.**—Even though the Bald Eagle has been the national symbol of the U.S. for over 2 centuries, wanton killing still occurred through the 1930s. Bounties were paid in many states and provinces for dead eagles until legal protection was enacted in 1940, when the Bald Eagle Protection Act was passed (it was amended in 1962 to include the Golden Eagle). The 1940 act prohibited killing, taking, or selling body parts and feathers. To counteract the decimating effects of organochlorine pesticides on the Bald Eagle population in the contiguous 48 states, the USFWS on Mar. 11, 1967, listed the Bald Eagle as endangered south of the 40th parallel under the Endangered Species Preservation Act of 1966. With the enactment of the more powerful Endangered Species Act in 1973, the USFWS listed the Bald Eagle as an Endangered Species on Feb. 14, 1978 in the lower 48 states except in Michigan, Minnesota, Oregon, Washington, and Wisconsin, where the eagle was listed as a Threatened Species. With the Bald Eagle's surprisingly strong comeback from near decimation from pesticides, on Jul. 12, 1995, the USFWS reclassified it from an Endangered Species to a Threatened Species in all lower 48 states. On Jul. 6, 1999, the USFWS *proposed* delisting the eagle. Many states still classify the Bald Eagle as endangered.

In 1986, the U.S. Supreme Court held that Native Americans must abide by mandated protective laws. Since the 1970s, the USFWS has collected eagle body parts and feathers from eagles that have found dead in the wild, illegally killed and confiscated eagles, and eagles in rehabilitation facilities and zoos; the parts have been stored at its National Eagle Repository in Commerce City, Colo. The USFWS distributes the parts and feathers to Native American tribes for ceremonial use. The repository annually receives about 900 Bald and Golden eagle carcasses for distribution.

**Restoration programs.**—Four western states, Arkansas, California, Montana, and Oklahoma, implemented programs to re-establish breeding populations or assist population growth.

Three methods were used to rebuild populations: (1) *Egg transplant.*—Putting healthy eggs taken from wild nests and putting them into nests of remnant pairs that were experiencing thin-shelled eggs due to pesticides. (In 1970, Maine was the first state to use this method.) (2) *Fostering.*—Released birds were obtained by three methods: (A) Importing wild nestlings from other states and provinces with healthy populations; (B) eggs taken from wild nests from other states and provinces with healthy populations were hatched and raised in captivity until release; (C) nestlings from captive origins, including captive breeding facilities at the USFWS Patuxent Research Laboratory in Laurel, Md.; zoos; and rehabilitators.

The nestlings were put in nests and fostered by extant wild pairs. (3) *Hacking/releasing.*—Young eagles were obtained by the same three methods as described for fostering. Once old enough, they were put into large, sheltered elevated wooden structures (usually tall towers) in suitable natural environments. The fledglings were fed by concealed caretakers until they could fend for themselves.

Alaska supplied most of the eaglets that were released in various states. Manitoba, Minnesota, N.S., Saskatchewan, and Wisconsin each supplied a large number; California, Ontario, and Washington each contributed moderate numbers; Florida, Maryland, Michigan, Montana, and Virginia supplied a few eaglets.

From 1984 to 1992, Florida supplied 393 eggs to restoration programs in the U.S. The eggs were artificially incubated and the nestlings raised and distributed to foster nests and hack sites in Oklahoma (and four eastern states).

*Arkansas:* Began releasing eagles in 1982. *California:* Bald Eagles formerly bred along the central and s. California coast until being wiped out by the 1960s by pesticide contamination. Santa Catalina Island.—A restoration project by the Institute for Wildlife Studies that began in 1980 will continue until a viable breeding population is established. Thirty-three eagles were hacked from 1980 to 1986. Unsuccessful breeding occurred in 1987 and 1988 from hacked stock. Fostering has occurred since 1989. As of 2002, 31 eagles have been fostered into the current four nesting pairs and an additional 20 eagles have been hacked. Because of broken eggs caused by local DDT residues, successful breeding still does not occur. Santa Cruz Island.—A 5-year restoration project by the Institute for Wildlife Studies began in Jun. 2002. Plans are to hack 12 eagles each year. The first four eagles were released in late Jun. 2002. Monterey Co.—The Ventana Wilderness Society conducted a hacking program in the Big Sur region 1986–2000, with 72 eagles being released. Eaglets were taken from healthy populations in se. Alaska, Canada, and n. California. The first release-era nesting occurred in 1993. There are currently five pairs that have formed from the hacked stock. *Missouri:* Hacked eagles 1981–1990 at Mingo NWR, Wayne Co., and the Schell-Osage Conservation Area, Vernon Co. Eighteen eagles were released through 1985. *Oklahoma:* 90 eagles released 1985–1990 by the Sutton Avian Research Center, which raised and distributed eggs supplied by Florida taken from nests there. From 1985 to 1992, the center raised and distributed 275 eagles not only for Oklahoma but also for Alabama, Georgia, Mississippi, and North Carolina.

**Mortality.**—The Bald Eagle has few natural enemies once it has attained fledging age. Nestlings may be killed by Bobcats, Great Horned Owls, and Raccoons while in the nest. Fledglings that are on the ground may also fall prey to Coyotes or Timber Wolves. Severe weather can blow nestlings out of nests or destroy nests and occupants. As with all nestling raptors, parasites cause some deaths.

Human-induced mortality takes its toll on the eagle. Illegal shooting still occurs, even with the passage of the Bald Eagle Act of 1940. Pesticide contamination is now minimal but still occurs, with organophosphate and carbamate pesticides. Eagles are also victims of electrocution, collisions with vehicles, entanglement in fishing lines, and leg-hold traps Lead poisoning from eagles ingesting shot pellets in dead or wounded waterfowl was a morality factor until the 1991 nationwide ban on lead shot.

Bald Eagles have recently succumbed to a new disease, avian vacuolar myelinopathy. It was first called the Coot and Eagle Brain Lesion Syndrome since eagles first contracted it from eating American Coots, a major prey in some winter areas; however, the disease has now been discovered in other species of waterfowl. The disease creates lesions on the myelin sheath that insulates nerve fibers in the brain and causes disorientation, motor problems, and death. Dead Bald Eagles were first discovered with the disease in 1994 on DeGray Lake in Clark Co., Ark. Since then, diseased eagles also have been found in Georgia, North Carolina, and South Carolina. The exact cause of the disease is unknown.

**SIMILAR SPECIES:** COMPARED TO JUVENILE AND SUBADULT.—**(1) Golden Eagle, adult.**—PERCHED.—Bill is blackish on outer half and bluish on inner half; cere is yellow. Distinct pale tawny nape and hindneck. Feathered tarsi are often not visible. FLIGHT.—

All-dark underwing generally separates from all younger Bald Eagles. However, in extensive molt, white blotches, even on the axillaries, caused by missing feathers on the underwing coverts may appear very Bald Eagle-like. Typically glides and soars with wings in dihedral. **(2) Golden Eagle, juvenile and subadult.**— PERCHED.—As in adult. FLIGHT.—Individuals with a small amount of white on the underside of the remiges are similar but typically lack white on the axillaries. Use caution on summer and fall molting birds as they will have varying amounts of white blotching on the underwing coverts and even axillaries. On juveniles, the outer web of the outer tail feather is white; on subadults, it may be dark and similar to that of Bald Eagles. **(3) Osprey.**—Subadults I–III can have a very distinct dark patch or line on the auriculars and appear Osprey-like. Lightly marked type subadult I and moderately marked type subadult II Bald Eagles have an extensive amount of white and can appear as pale on the ventral areas as an Osprey. COMPARED TO NON-RAPTORS.—**(4) Great Blue Heron.**—FLIGHT.—At moderate to long distances with the neck retracted, and nearly invisible long legs, the heron can be eaglelike. Wingbeats, however, are steady, shallow, with an even up-and-down stroke.

**OTHER NAMES:** American Eagle, White-headed Eagle. *Spanish:* Áquila Cabeza Blanca, Áquila Cabeciblanca. *French:* Pygargue à tête blanche.

---

REFERENCES: Adamus et al. 2001; Arnold 2001; Baicich and Harrison 1997; Brown 1988; Brown et al. 1987; Bylan 1998, 1999; Clark and Wheeler 2001; Commission for Environmental Cooperation 1997; Dinsmore 2000; Dodge 1988–1997; Edelstam 1984; Ehresman 1999; Environment Canada 2001; Forsman 1999; Godfrey 1986; Henny et al. 1993; Howell and Webb 1995; Hunt 1998; Idaho Conservation Data Center 1999; Jacobs and Wilson 1997; Janssen 1987; Jeffers 2000; Johnsgard 1990; Kellogg 2000; Kent and Dinsmore 1996; Kingery 1988; McCollough 1989; Millar 2002; Montana Bird Distribution Committee 1996; Nye 1988; Oberholser 1974; Palmer 1988; Peterson 1995; Pierce 1998; Robbins and Esterla 1992; Russell and Monson 1997; Semenchuk 1992; Small 1994; Smith 1996; Smith et al. 1997; Stalmaster 1987; Stevenson and Anderson 1994; Santa Cruz Predatory Bird Research Group 1999a; USFWS 1995, 1999b; Ventana Wilderness Society 2000a; Wheeler and Clark 1995, 1999.

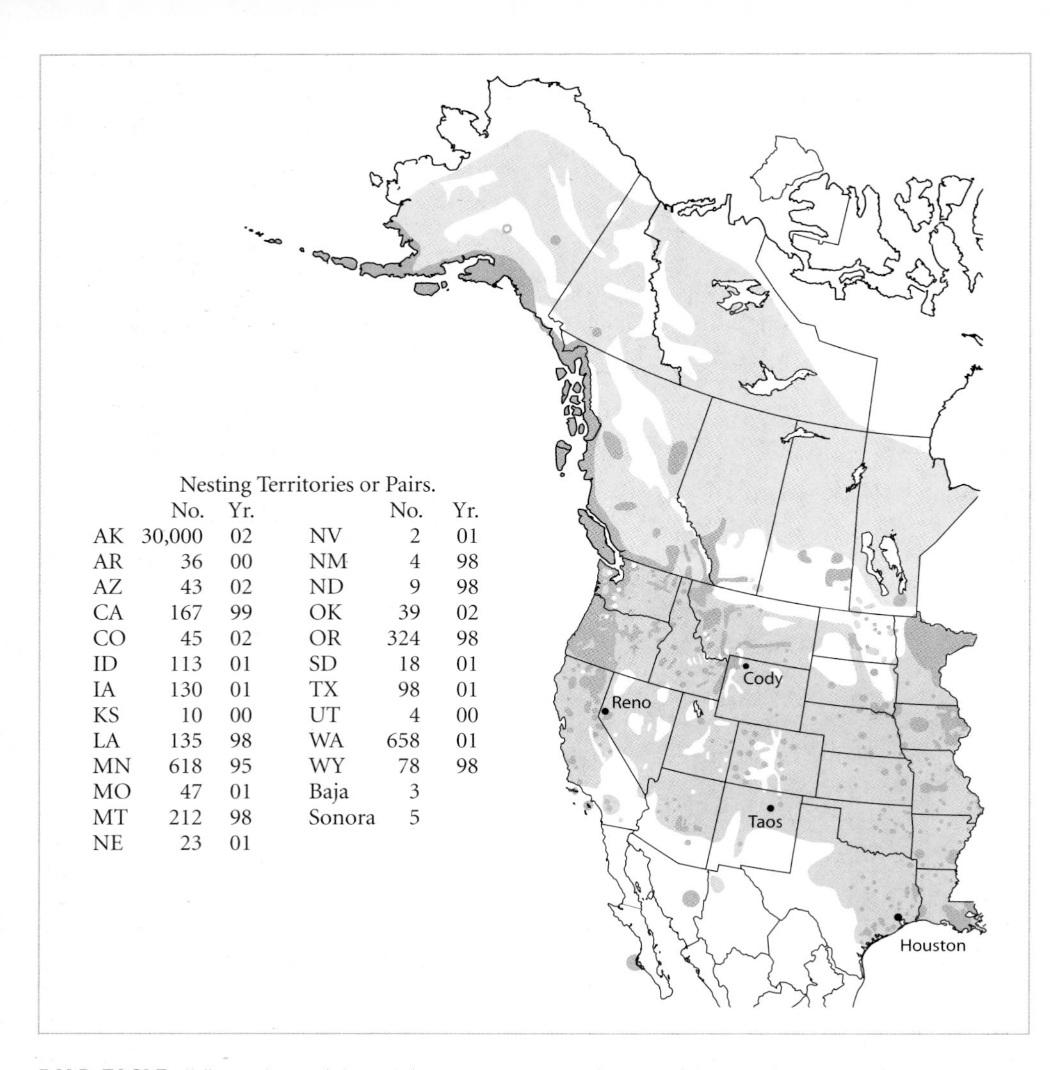


Nesting Territories or Pairs.

| | No. | Yr. | | No. | Yr. |
|---|---|---|---|---|---|
| AK | 30,000 | 02 | NV | 2 | 01 |
| AR | 36 | 00 | NM | 4 | 98 |
| AZ | 43 | 02 | ND | 9 | 98 |
| CA | 167 | 99 | OK | 39 | 02 |
| CO | 45 | 02 | OR | 324 | 98 |
| ID | 113 | 01 | SD | 18 | 01 |
| IA | 130 | 01 | TX | 98 | 01 |
| KS | 10 | 00 | UT | 4 | 00 |
| LA | 135 | 98 | WA | 658 | 01 |
| MN | 618 | 95 | WY | 78 | 98 |
| MO | 47 | 01 | Baja | 3 | |
| MT | 212 | 98 | Sonora | 5 | |
| NE | 23 | 01 | | | |

**BALD EAGLE,** *Haliaeetus leucocephalus:* Southern immatures and some adults spend spring and summer in n. U.S. and Canada. Common in AK and Canada. Listed as endangered in U.S. except for MI, MN, OR, WA, and WI, 1967–95. Then, redesignated threatened. Delisting proposed in 1999. Winters over broad areas, often far from water. Subspecies not delineated.

**Plate 101. Bald Eagle, adult** [**Dec.**] ▪ Yellow bill and cere. Pale yellow iris. White head and neck. ▪ Brown body and white undertail coverts. ▪ White tail.

**Plate 102. Bald Eagle, adult** [**May**] ▪ White head and neck. ▪ Brown body. ▪ Dark underside of wings. ▪ White tail.

**Plate 103. Bald Eagle, adult** [**May**] ▪ White head and neck. ▪ White lower back. ▪ White uppertail coverts and tail.

**Plate 104. Bald Eagle, basic/subadult IV** [**Mar.**] ▪ Yellow bill. Pale yellow iris. White head often has dark markings on auriculars and crown. ▪ Brown body. ▪ Brown upperwings. ▪ White tail may retain dark markings on portions of rectrices (retained dark subadult III rectrix). ▪ *Note:* May attain full adult plumage at this age.

**Plate 105. Bald Eagle, basic/subadult IV** [**Feb.**] ▪ Brownish areas on yellow bill. White head and neck often have dark markings on auriculars and crown. ▪ Brown body. ▪ Most have white areas on coverts, sometimes on secondaries. ▪ White tail may have dark areas on tips of most rectrices. White undertail coverts. ▪ *Note:* Photograph by Ned Harris.

**Plate 106. Bald Eagle, basic/subadult III (moderately marked type)** [**Mar.**] ▪ Yellowish brown bill, yellow cere. Pale brown iris. Whitish head and neck; dark auricular patch. ▪ Many have white speckling on back. ▪ Old tan-colored subadult I secondary (s12) and respective greater covert identifies this age. ▪ *Note:* Told from heavily marked type subadult II by wing molt.

**Plate 107. Bald Eagle, basic/subadult III (moderately marked type)** [**Jan.**] ▪ Yellowish brown bill and cere. Whitish head and neck; dark auricular patch; ill-defined junction of whitish neck and brown body. ▪ White markings on coverts and secondaries. Very old, faded brown juvenile primary (p10); 4 new inner primaries. ▪ White tail with dark border.

**Plate 108. Bald Eagle, basic/subadult II (moderately marked type)** [**Feb.**] ▪ Yellowish brown bill and cere. Pale brown iris. Whitish head; dark auricular patch. ▪ Dark spotting on white back and belly. Dark bib. ▪ A few worn, remnant white-edged subadult I wing coverts. ▪ *Note:* Photograph by Ned Harris.

**Plate 109. Bald Eagle, basic/subadult II (heavily marked type) [Jan.]** ▪ Yellowish brown bill, yellow cere. Pale brown iris. Dark auricular patch; tawny head and neck. ▪ Mainly dark brown underparts. ▪ Whitish tail with dark border. ▪ *Note:* Similar to subadult III; separable by remiges. This eagle retains old juvenile remiges (p9 and 10; s4) [not shown].

**Plate 110. Bald Eagle, basic/subadult II (moderately marked type) [Feb.]** ▪ Yellowish brown bill and cere. Dark auricular patch; tawny head and neck. ▪ Dark bib and white-spotted belly; dark leg feathers. ▪ Whitish axillaries and some wing coverts. Retains frayed juvenile outermost primaries (p9 and 10). Neatly edged trailing edge of wings. ▪ *Note:* Some subadult II birds will retain only p10.

**Plate 111. Bald Eagle, basic/subadult II (heavily marked type) [Sep.]** ▪ Dark auricular patch; tawny head and neck. ▪ Dark body with a few white specks. ▪ Whitish axillaries. Outermost primaries are retained, frayed juvenile (p9 and 10); retained, long juvenile secondary (s4). *Note:* Similar to subadult III, but separable by retained juvenile remiges.

**Plate 112. Bald Eagle, basic/subadult II (moderately marked type) [Dec.]** ▪ Pale head, dark auricular patch. ▪ Dark bib, white-speckled belly. ▪ Mainly whitish axillaries and wing coverts. ▪ Retains juvenile outermost primaries (p9 and 10), 2 retained, long, faded brown juvenile secondaries (s4 and 9). ▪ *Note:* Slow molt type for age class.

**Plate 113. Bald Eagle, basic/subadult II (moderately marked type) [Feb.]** ▪ Yellow bill and cere. Dark auricular patch; tawny head and neck; ▪ White inverted triangle on back; mottled white on lower back. ▪ Outer primaries retain frayed juvenile feathers (p9 and 10). ▪ Upper surface of tail varies from blackish to whitish. ▪ *Note:* Photograph by Ned Harris.

**Plate 114. Bald Eagle, basic/subadult I (lightly marked type) [Feb.]** ▪ Gray or black bill. Pale gray iris. Wide dark auricular patch; whitish head and neck. ▪ Small, mottled brown bib. Very white belly. ▪ Extensive white band on upperwing coverts; white tips on greater coverts. A few bleached, tan-colored, retained juvenile upperwing coverts. ▪ *Note:* Photograph by Ned Harris.

**Plate 115. Bald Eagle, basic/subadult I (moderately marked type) [Nov.]** ▪ Black bill with a pale spot on base of upper mandible; yellow fringe on cere. Pale yellow iris. Wide dark auricular patch; tawny head and neck. ▪ Large, brown bib. White belly with brown spotting. ▪ White band on some upperwing coverts.

**Plate 116. Bald Eagle, basic/subadult I (moderately marked type) [Mar.]** ▪ Gray bill with a pale spot on base of upper mandible. Pale gray iris. wide dark auricular patch; tawny head and neck. ▪ White inverted triangle on back. ▪ Retained, bleached, tan-colored juvenile feathers on inner part of upperwing coverts.

**Plate 117. Bald Eagle, basic/subadult I (heavily marked type) [Dec.]** ▪ Black bill; yellow fringe on cere. Pale brown iris. Very wide dark auricular patch extends from the lower mandible on the tawny-brown head. ▪ Dark back. ▪ Dark upperwing coverts. Most secondaries are faded brown juvenile feathers; new dark tertials and inner secondaries.

**Plate 118. Bald Eagle, basic/subadult I (moderately marked type) [Dec.]** ▪ Dark bib, white-spotted belly; dark leg feathers. ▪ White axillaries and white-mottled coverts. Jagged rear edge of secondaries have two groups of long, pointed juvenile feathers. Juvenile outer primaries (p6–10).

**Plate 119. Bald Eagle, basic/subadult I (lightly marked type) [Dec.]** ▪ Very wide dark auricular patch; white head and neck. ▪ Small, brown bib; white belly; dark leg feathers. ▪ White axillaries and coverts. Extensive white on secondaries and inner primaries. Jagged rear edge of secondaries have two groups of long, pointed juvenile feathers. Juvenile outer primaries (p5–10).

**Plate 120. Bald Eagle, basic/subadult I (moderately marked type) [Mar.]** ▪ Tawny head and neck. ▪ Dark bib; white belly is heavily spotted; dark leg feathers. ▪ White axillaries; white mottled coverts. Slow molt with only outermost and a few inner secondaries replaced, all others are retained juvenile. Outer 6 primaries (p5–10) are juvenile feathers.

**Plate 121. Bald Eagle, basic/subadult I (moderately marked type)** [**Feb.**] ▪ White back and belly are spotted with brown. ▪ A few faded tan-colored juvenile feathers on the inner coverts. Jagged rear edge of secondaries retains two groups of long, pointed, juvenile feathers (s2–4, s7–11). ▪ Tails are whitish on some birds. ▪ *Note:* Photograph by Ned Harris.

**Plate 122. Bald Eagle, juvenile (worn plumage)** [**Mar.**] ▪ Black bill and cere. Dark brown iris. Dark brown head and neck; crown sometimes paler. ▪ Dark brown bib; faded, pale tawny belly. ▪ Moderately white type tail with some white mottling.

**Plate 123. Bald Eagle, juvenile (worn plumage)** [**Mar.**] ▪ Black bill and cere. Dark brown iris. Dark brown head and neck; white streaking on neck. ▪ Faded pale brown upperparts. ▪ Faded pale brown wing coverts; palest on inner greater coverts. Blackish secondaries contrast with rest of dorsal wing. Tertials have pale tips.

**Plate 124. Bald Eagle, juvenile (fresh plumage)** [**Jul.**] ▪ Dark brown head with tawny tips on nape feathers. ▪ Medium dark brown upperparts. ▪ Medium dark brown wing coverts; outer greater coverts are often dark; secondaries are blackish. ▪ White type tail with extensive white on inner rectrix sets. Black terminal band and outer rectrix sets.

**Plate 125. Bald Eagle, juvenile (fresh plumage; fledgling) [Jul.]** ▪ Black bill and cere. Dark brown iris. Very dark brown head and neck. ▪ Very dark brown bib; dark tawny belly is slightly paler than breast.

**Plate 126. Bald Eagle, juvenile (fresh plumage; fledgling) [Jul.]** ▪ Black bill and cere. Dark brown iris. Very dark brown head and neck. ▪ Dark brown upperparts. ▪ Dark brown wing coverts; blackish secondaries. ▪ All-black type tail.

**Plate 127. Bald Eagle, juvenile (worn plumage) [Mar.]** ▪ Dark brown head and neck; pale throat. ▪ Dark brown bib; pale tawny belly; dark brown leg feathers. ▪ White type underwing with nearly all-white axillaries and coverts. Extensive amount of white on inner secondaries and inner primaries. Neat, serrated rear edge of secondaries.

**Plate 128. Bald Eagle, juvenile (fresh plumage) [Sep.]** ▪ Dark brown head and neck; pale throat. ▪ Dark brown bib; dark tawny belly; dark brown leg feathers. ▪ Dark type underwing with white axillaries and mottled white coverts; white bar on median coverts. Small amount of white on inner secondaries and inner primaries. Neat, serrated rear edge of secondaries.

**Plate 129. Bald Eagle, juvenile (streaked type; worn plumage) [Dec.]** ▪ Dark brown head and neck. ▪ Dark brown bib. Pale tawny or whitish belly is spotted and streaked with dark brown. Dark brown leg feathers. ▪ White type underwing. Axillaries and coverts are white. Extensive amount of white on inner secondaries and inner primaries.

**Plate 130. Bald Eagle, juvenile (fresh plumage) [Sep.]** ▪ Dark brown head and neck. ▪ Medium brown upperparts. ▪ Wing coverts are medium brown with a pale brown area on the inner greater- and median coverts. Remiges are blackish and contrast with rest of wings and body. ▪ All-black type of dorsal tail.

**Plate 131. Bald Eagle, juvenile (worn plumage) [Feb.]** ▪ Dark brown head and neck. ▪ Pale brown upperparts. ▪ Pale brown wing coverts. Remiges and most greater coverts are blackish and contrast with rest of wings and body. ▪ White type tail with white on middle rectrix set and dark outer webs on all other sets. *Note:* Photograph by Ned Harris.

# NORTHERN HARRIER
## *(Circus cyaneus)*

**AGES:** Adult, basic I/subadult (first-adult), and juvenile. Adults and subadults have sexually dimorphic plumages; however, age distinction within each sex class is not readily apparent at field distances. Some adults may alter subadult plumage features gradually over a few molts (Macwhirter and Bildstein 1996). Subadult age is a separate plumage, but may retain a very small amount of juvenile feathering. Juvenile sexes are separable at close range by iris color and size (*see* Size). Juvenile plumage held for one year, but iris color gradually alters during the course of this time.

**MOLT:** Accipitridae wing and tail molt pattern (*see* chapter 4). First prebasic molt begins in a span from late Apr. to May and is completed by late Oct. or Nov. This is a complete molt except some birds retain one or more juvenile greater underwing coverts (J. Liguori unpubl. data). This plumage feature is very difficult to see under field conditions and, if present, is more obvious on males. Birds are quite advanced in molt by late Sep.

Subsequent prebasic molts are complete. Molt may begin in early to mid-May on females and about 2 weeks later on males (Macwhirter and Bildstein 1996). However, a substantial number of birds in adult-to-adult prebasic molts are still in exceedingly heavy molt during Sep. By early Sep., some have molted a majority of the primaries, but only a few secondaries have molted (s1 and s5) and body molt is less than 50 percent complete. Molt is completed within a span from late Oct. to late Nov.

**SUBSPECIES:** Polytypic with two races. *C. c. hudsonius* inhabits North America. Nominate race, *C. c. cyaneus* (Hen Harrier), breeds in Europe and Asia and winters south to n. Africa and tropical regions of Asia.

*Note:* The two races are considered by some authors to be separate species. Adult males are quite different from each other. The Hen Harrier is uniformly solid pale gray on head, breast, back, scapulars, upperwing, and upper tail and has more extensive black on ventral surface of outer primaries than does the North American race. Adult females are similar to one another. Juveniles are also similar, but Hen Harriers tend to be much more heavily streaked on the underparts and paler orangish.

**COLOR MORPHS:** None regularly occurring (*see* Abnormal Plumages).

**SIZE:** Sexually dimorphic with little or no overlap. Males are smaller. With practice, size is often visibly apparent in the field. MALE.—Length: 16–18 in. (41–46 cm); wingspan: 38–43 in. (97–109 cm). FEMALE.—Length: 18–20 in. (41–51 cm); wingspan 43–48 in. (109–122 cm).

**SPECIES TRAITS:** HEAD.—Greenish yellow or yellow cere. **Defined auricular disks. Whitish, short supercilium and slash mark under each eye forms a pale spectacle that surrounds the eye.** Supraorbital ridge is present, but lacks bare skin area on the ridge. Small, flat-topped head. **Head and neck are darker than ventral areas and appear hooded.** BODY.—Slim body. Long yellow tarsi. **White uppertail coverts, sometimes with narrow dark streak on some feathers.** WINGS.—**Long and moderately narrow wings with moderately rounded wing tips. Five very distinct "fingers" on the outer primaries.** Wingtips can appear rather pointed when gliding. **Wide black band on trailing edge of underside of the secondaries.** TAIL.—Long and moderately rounded.

**ADULT AND SUBADULT TRAITS:** HEAD.—Adults have yellow irises that are generally somewhat brighter on males. There are no other shared traits between the adult sexes or between subadults.

**ADULT MALE (FRESH PLUMAGE):** HEAD.—Pale or bright orangish yellow irises. **Some individuals exhibit a distinct whitish spectacle area of the supercilium and slash mark under the eye, and others have a uniformly medium gray head. Rear edge of auriculars lacks an outline of pale fringed feathers. Front of neck and upper breast are medium gray, and along with the gray head, contribute to the hooded appearance.** BODY (ventral).—Lower breast, belly, flanks, leg feathers, and undertail coverts are white with variable amount of mainly rufous but sometimes gray markings. Lower breast and belly are lightly or moderately spotted or are sometimes thinly streaked on the breast area. Some have a rufous wash on the lower breast. Flanks are also lightly or moderately marked and are likely to be somewhat barred. Leg feathers are generally unmarked, but may be lightly spotted. Undertail coverts are marked with a dark arrowhead-shaped design on each feather and often have a tawny or rufous wash on a portion of each feather. BODY (dorsal).—Neck, back, and scapulars are a medium gray, dark brownish gray, or grayish brown. Some harriers can be quite dark-backed and brownish in fresh autumn plumage. WINGS (ventral).—**White with an extensive black patch on the outer five primaries ("fingers"), a grayish trailing edge of inner five primaries, and a broad black band on the trailing edge of all secondaries.** Inner portion of all remiges is either unmarked or partially marked with irregular, fine barring. Axillaries are always finely spotted or barred with rufous or gray. Underwing coverts, including greater primary coverts, are either unmarked or partially spotted or barred with fine rufous or gray markings. WINGS (dorsal).—Medium brownish gray lesser upperwing coverts have a whitish or sometimes tawny mottled area on the first row of coverts; greater coverts have a pale gray band on the inner portion of the tract. **A large black area covers the outer five primaries, with the wide black band on the trailing edge of secondaries. The inner primaries and inner portion of secondaries are pale gray and generally much paler than the more brownish wing coverts, scapulars, and back.** TAIL.—Pale gray on the dorsal surface with a moderately wide black subterminal band and four or five narrow dark bands on the inner tail. Deck rectrix sets and sometimes other inner rectrix sets may be unmarked. Outer dark bands on the basal region of tail are often rufous, but can only be distinguished at close range.

*Note:* Plumage spans from mid-summer to early winter when adorned in freshly molted feathers. Older males may gradually become paler with subsequent molts on dorsal areas and may have fewer ventral markings. There undoubtedly is also individual variation, with some birds being paler and less heavily marked and others being darker and more heavily marked.

**ADULT MALE (WORN PLUMAGE):** BODY and WINGS (dorsal).—Back, scapulars, and upperwing coverts fade to a medium brownish gray or a pale brownish gray; however, most have a discernibly brownish cast. A few individuals can appear almost uniformly gray on dorsal areas; however, head and very pale gray dorsal surface of the remiges are still slightly paler than rest of the upperparts.

*Note:* Plumage span is from late winter to early summer, when it has faded due to sun bleaching and wear.

**BASIC I (SUBADULT I) MALE (FRESH PLUMAGE):** HEAD.—Iris color is lemon yellow, but occasionally orangish yellow, as on adult males. **Head color is variable: (1) Crown and auriculars are often brown or rufous brown and contrast rather sharply with the brownish gray nape and neck; (2) uniformly brownish gray on the crown and auriculars and forms a uniform color unit with head, nape, and neck; or (3) uniformly grayish and similar to older males, and paler than the rest of the upperparts. Auriculars lack a pale border fringe and are not distinct. Front of neck is brownish gray or whitish and often has a rufous wash on the lower part and, along with the head, forms a dark hood. Whitish spectacle of the supercilium and patch area below each eye is very distinct on all birds.** BODY (ventral).—Breast is washed with tawny or rufous and often streaked with dark rufous or brown; it is occasionally white with rufous or brown streaking. Belly and lower belly are variably marked with rufous spots or arrowhead shapes and vary from being rather heavily marked to lightly marked. Leg feathers are generally lightly spotted with rufous. BODY (dorsal).—Dark grayish brown or dark brown but without a grayish tinge on the back and scapulars. WINGS (ventral).—**As on adult males and exhibit the identical broad black area on outer primaries and the wide black band on trailing edge of secondaries.** Axillaries are marked with rufous and are either narrowly barred or thickly barred. Highly variable in the amount of markings on the white underwing coverts: (1) All underwing coverts are extensively marked with rufous spots on the lesser coverts (patagial area) and rufous-brown or brown barring on median coverts. The greater primary coverts are barred. (2) Underwing coverts are lightly marked or unmarked. If marked, they are concentrated on the median coverts and greater primary coverts. Remiges vary from being unbarred to moderately barred, mainly at the base of the primaries and some secondaries. WINGS (dorsal).—All coverts are dark grayish brown or dark brown with a pale tawny or whitish mottled area on inner lesser coverts. **Except for the black outer primaries and wide bar on the trailing edge of secondaries, remiges are pale brownish gray.** TAIL.—**Pale gray on the dorsal surface with a moderately wide, dark subterminal band and about five narrower dark inner bands. The outermost dark bands are often rufous.** Undertail is pale with similar pattern as on dorsal surface.

*Note:* Plumage spans from fall to early winter.

**BASIC I (SUBADULT I) MALE (WORN PLUMAGE):** BODY and WINGS (dorsal).—Upperparts fade to medium grayish brown or medium brown. All other features are as in fresh plumage.

*Note:* Plumage span is from late winter to early summer, when extreme fading from sun bleaching and wearing occurs.

**ADULT FEMALE (FRESH PLUMAGE):** HEAD.—Crown and auriculars are either solid dark brown or brown with tawny streaking. **Neck and hindneck are brown with distinct pale tawny streaking and are darker than rest of the ventral region and, along with the fairly dark head, form a hood. Pale tawny spectacles surround the eyes. Facial disk is well defined with a pale white or tawny fringe surrounding outer edge of the auriculars.** Iris color varies with age but takes a minimum of 3 years to attain the all-yellow color of older birds (Hamerstrom 1968). Iris color varies: (1) pale brownish yellow, (2) lemon yellow, or (3) pale orangish yellow. BODY (ventral).—Base color of underparts is a rich tawny or rufous-tawny in fresh autumn plumage, but fades within a span of late fall through winter (*see below*). Underparts are variably streaked on the belly and lower belly: (1) occasionally heavily streaked, (2) moderately marked, or (3) have only sparse streaking or dash markings on the belly and lower belly and may appear virtually unmarked at field distances. Flanks are streaked on the forward half and have a broad, arrowhead-shaped dark brown mark on the rear half. Some feathers on the rear flank area may have a dark brown cross bar on basal portion of the arrowhead-shaped marked feathers. *Note:* The arrowhead-shaped and cross-bar markings on rear flank feathers are not always readily visible at many distances or angles, particularly if overhead. Undertail coverts have a short, dark streak on each feather. Leg feathers are lightly covered with small, brown diamond-shaped or arrowhead-shaped markings. BODY (dorsal).—Upperparts are dark brown. Scapulars may have some tawny edgings and a few tawny blotches. WINGS (ventral).—**Overall appearance of underwing covert region is pale with a broad dark band on the axillaries and median coverts. At close range, the dark brown median coverts have a narrow tawny fringe on each feather and some pale spotting; all other coverts are tawny with narrow dark markings. Dark brown axillaries have a minimal amount of pale spotting and edging, thus appear very dark. The remiges are uniformly pale grayish, but are sometimes a slightly darker gray on the secondaries. There is a broad dark band on the trailing edge on the secondaries, two narrow dark bands on the inner secondaries, and narrow barring on all primaries.** WINGS (dorsal).—Secondaries and primaries are medium brownish gray or sometimes pale grayish with distinct dark bands (Macwhirter and Bildstein 1996). *Note:* Pale, barred dorsal surface of the remiges separates adult ages of females from juveniles, which have dark, unbarred secondaries. Those with pale grayish remiges are typically separable from most subadult females. Dark brown median coverts have a pale gray bar on the basal area of each feather. All coverts have pale tawny feather edgings, and tawny blotches occur on the distal rows of lesser coverts. TAIL (ventral).—**Pale with two, broad, nearly equal-width dark bands. The outer rectrix set is pale and unmarked except for the dark subterminal band or faint bands on the basal region.** TAIL (dorsal).—**Deck rectrix set and adjacent inner one to three sets (r1–4 sets) are medium gray with four or five nearly equal-width dark bands. The subterminal band is**

**usually somewhat wider, and the outer rectrix sets are a rich tawny with three or four dark bands.**

*Note:* Overall plumage fades by late winter through early summer.

**ADULT FEMALE (WORN PLUMAGE):** BODY (ventral).—Base color of ventral areas fades to pale tawny or white. BODY and WINGS (dorsal).—Back, scapulars, and upperwing coverts fade to medium brown. Remiges fade to grayish with very discernible dark bands.

*Note:* Plumage span occurs from late winter to early summer.

**BASIC I (SUBADULT I) FEMALE (FRESH PLUMAGE):** HEAD.—**Medium brown crown and auriculars with little if any tawny streaking. A moderate amount of tawny streaking is on the brown neck and hindneck and forms a hood along with the head. A well-defined whitish or tawny rim surrounds the auriculars. Sharply defined pale tawny or whitish spectacles around the eyes.** Iris color is variable, but lightens and turns more yellowish with age. Iris color based on Hamerstrom (1968). Iris colors: (1) medium brown, (2) pale brown, or (3) pale brownish yellow. On some younger birds, the medium brown iris color may not change from the springtime color of most juvenile females. BODY (ventral).—Base color of ventral areas is a rich tawny or rufous-tawny. At times, freshly molted ventral color can be nearly as intense as on some juveniles. A variable amount of adult femalelike streaking is on the belly, lower belly, and forward flanks; broad arrowhead-shaped markings are on the distal flanks. As with older females, ventral markings range from heavily marked to very sparsely marked individuals. *Note:* Age does not seem to be a factor in the amount of ventral markings. Data are based on in-hand live birds and close field photographs correlating eye color and molt. Leg feathers are lightly marked with small, brown diamond shapes, arrowhead shapes, or spots. BODY (dorsal).—Dark brown with minimal pale edgings or blotches. WINGS (ventral).—As in adult females, including all remiges being either uniformly pale or secondaries being slightly darker gray. WINGS (dorsal).—Remiges are either a medium brownish gray or pale brownish gray and are distinctly barred with black and similar to many adult females (dorsal color of the remiges of some adult females are quite gray). Median and lesser coverts are often not as pale tawny-edged as on older females. Basal area of greater coverts has a pale gray bar. At least through Oct., faded, retained juvenile secondaries may be apparent (mainly s4 and s7 and 8). *Note:* The paler grayish and barred dorsal surface of the secondaries usually separates subadult and adult females from juveniles at field distances. TAIL.—As on adult females.

**BASIC I (SUBADULT I) FEMALE (WORN PLUMAGE):** BODY and WINGS (ventral).—As on adult females in worn plumage. BODY (dorsal).—As on adult females in worn plumage.

**JUVENILE TRAITS (FRESH PLUMAGE):** HEAD.—**Dark brown crown and auriculars with a distinct pale outer rim on the auricular disks; pale spectacles are also very distinct. Dark brown neck and hindneck have a minimal or moderate amount of rich rufous-tawny streaking and form a hood along with the head.** BODY (ventral).—**Bright, rich, rufous-tawny (orange) belly, lower belly, and flanks.** Breast and flanks are narrowly streaked with dark brown, the belly and lower belly are unmarked. Leg feathers are unmarked. Undertail coverts are unmarked. BODY (dorsal).—Solid dark brown back and scapulars. WINGS (ventral).—Rufous-tawny (orange) wing coverts, but the carpal region is often paler and more whitish. **Median coverts and axillaries are dark brown with narrow tawny edges and appear as a broad dark band. Pale gray primaries are fully barred; secondaries are medium gray and discernibly darker than the primaries, with the innermost secondaries being even darker. Secondaries have the wide black band on the trailing edge and two narrower dark inner bands that may be obscured by the fairly dark gray inner secondaries.** WINGS (dorsal).—Uniformly dark brown on the secondaries. *Note:* The uniformly dark brown dorsal surface of secondaries separates juveniles from subadult and adult females, which have paler remiges and exhibit noticeable barring. Primaries may be somewhat paler than the secondaries and exhibit faint barring. A broad, pale tawny bar on the lesser coverts. TAIL (dorsal).—**Deck rectrix set is medium gray and the outer five sets are pale tawny. A**

**wide dark subterminal band with four or five moderately wide, dark inner bands.** TAIL (ventral).—Two wide dark bands are visible; the outer rectrix set often does not show dark banding, if it does, dark bands are faint.

*Note:* Plumage color and markings are similar for both sexes.

**JUVENILE TRAITS (WORN PLUMAGE):** BODY (ventral).—Underparts fade to a pale tawny and often nearly white by late spring. Fading may be noticeable as early as late Nov. and is usually quite obvious by Feb. Extreme fading occurs in Apr. and May. However, there are a few individuals that are still a fairly rich tawny color even in May. BODY and WINGS (dorsal).—Brown upperparts typically fade and wear to medium brown; however, less faded individuals may still retain a fairly dark brown color in spring.

**JUVENILE MALE:** Iris color from summer to early fall and sometimes until mid-winter, is pale gray or pale brown. Iris color gradually changes to pale lemon yellow midway through juvenile plumage. On early-changing birds, iris color may become quite yellowish by late Sep., but color generally does not change substantially until Oct. or Nov. By Dec., many have attained lemon-colored irises. By Feb., most have changed to the lemon coloration. However, slow-changing individuals may still have pale grayish or brownish irises even in mid-winter. Virtually all springtime birds have attained the lemon-colored irises. Iris color may undergo little additional change and remain a pale lemon into subadult age. *Note:* Very rarely, fall juvenile males will have dark brown irises (B. Sullivan pers. comm.).

**JUVENILE FEMALE:** Iris color from summer to early winter is dark brown. By mid- to late winter, iris color gradually changes to medium brown and may have little if any additional change as harrier enters into the subadult age class.

**ABNORMAL PLUMAGES:** Two records in the w. U.S. of melanistic individuals. An adult male from Glenn Co., Calif., in Jan. 1991 (Howell et al. 1992); an adult female seen in Oct. 1998 in the Mission Valley area in Lake Co., Mon. (C. Olson pers. comm.). Both birds lacked the white uppertail coverts of typical Northern Harriers.

**HABITAT: Summer.**—Moist, open areas: freshwater, saltwater, and brackish marshes; also wet meadows, lightly grazed pastures, abandoned fields, bogs, moorlands, tundra, and alpine meadows. Also commonly breeds in dry areas: meadows, fields, tundra, alpine meadows, and prairies (including dry sage scrub). Prairie regions planted with non-native, tall grasses are favored foraging and nesting areas.

**Winter.**—Similar areas as in summer; but harriers commonly occupy dry locales, including harvested agricultural fields. Harriers vacate northern and high-elevation areas of the Arctic and alpine meadows. Regularly found in rural areas.

**Migration.**—Similar to all above-mentioned zones. Also seen along high mountain ridges but may seek lower-elevation open areas for roosting and foraging each day.

**HABITS:** A wary species. Harriers rarely remain perched around humans, even when being viewed from the shelter of a distant vehicle. Gregarious in all seasons, but especially in the nonbreeding season. It is common to see up to four individuals, mainly juveniles, migrating together; however, birds of mixed ages may also occur in these small groups. Large communal night roosts often form in winter on the ground. Winter territories may be defended or shared by several individuals. Northern Harriers commonly stand on the ground; however, they seem to prefer to perch on slightly elevated clumps of earth or low vegetation, fence posts, or, to a lesser extent, tops of low bushes. In very windy conditions, they stand on the lee side of tall vegetation or clumps of earth.

**FEEDING:** An aerial hunter. Harriers forage at low altitudes and rely on their acute hearing and sight.

Prey is captured on the ground or, with avian prey, sometimes while they are airborne up to about 20 ft. (6 m) in altitude. Harriers mainly feed on small prey: rodents, rabbits, hares, birds, reptiles, amphibians, and insects. Harriers are dependent on *Microtus* voles in many regions. Males feed more on small avian prey than do females. Females often capture avian prey that are somewhat heavier than themselves by grabbing them and holding them down. Females may drown small ducks on the water in this manner. Unusual prey of females include Ring-billed Gulls that are captured in flight. Prey is typically devoured at the point of capture or is taken to a slightly elevated perch or to a safer location a short distance away. Drowned prey is dragged to shore and eaten. Hunting is done during all daylight hours but is often suspended during hot periods of day. A crepuscular species, and hunting often takes place early and late in the day. Carrion is often consumed, particularly species that are too large to kill.

**FLIGHT:** Wings are held in a high dihedral, low dihedral, or modified dihedral when gliding. Wings are held in a low dihedral when soaring. At low-altitude flight, the glide mode is often in a side-to-side tilting manner, especially in windy conditions; in light winds, glides are level and steady. Powered flight is with mechanical, shallow wingbeats. Being smaller, males have quicker beats than females. Three main flight mannerisms: hunting, migrating, and courtship.

*Hunting:* Slow, methodical coursing at low altitudes over the ground or low vegetation with head pointing downward and looking and listening. Flight is at low altitudes and is either at a stable level or in an undulating style, skimming the ground or vegetation tops, then swinging upward in a low arch, then back down to ground-skimming height. A zigzag, erratic path may be taken over large habitat areas or may follow a straight course down leading lines of tall vegetation along fence lines, canals, and ditches. Hunting birds may intermittently flap, glide, or hover. Northern Harriers may occasionally kite for very short stints in strong winds. When prey is detected, harriers quickly bank—flipping sideways, tail fanned, then drop to the ground with fully extended legs. Harriers may dive repeatedly from low altitudes in vole-infested areas, especially at vole nest sites.

*Migrating:* Flight is at high altitudes in light or moderate winds and at low altitudes in strong winds. Flight is a combination of glides interspersed with a series of methodical wingbeats. Glides are stable and birds do not tilt side to side. Sometimes harriers flap their wings for considerable distances before gliding. Northern Harriers occasionally migrate nocturnally, including at least 15-mile (25 km) over-water crossings. *Note:* Appears falconlike in high-altitude flight.

*Courtship:* See Nesting.

**VOICE:** When agitated, emits a variable-length series of rapid chatterlike, moderately pitched, squeaky notes: *cheh-cheh, cheh-cheh.* When mildly agitated or curious, produces a squeeze-toy-like, high-pitched: *squee-aah*; may be a single note or repeated numerous times.

**STATUS AND DISTRIBUTION:** *Fairly common.* Regular in very small numbers on Galveston Island in Galveston Co., Texas. Breeds irregularly and sparingly in isolated locations in Montana, Iowa, w. Oklahoma., n. Texas, and ne. New Mexico. Very irregular breeding has occurred in Arizona: in 1985 in w.-cen. Yavapai Co. and se. Cochise Co.; only recent nesting was in 1998 near Joseph City, sw. Navajo Co. Restricted nesting takes place in n. Baja California and coastal and cen. California. Widespread elsewhere in its range to the northern limit of the treeline in Alaska and Canada. Absent from high montane areas.

Extralimital summer sightings in New Mexico in 2000 in Bitter Lake NWR, Chaves Co. in early Jun. (two birds; possibly breeding); and a single bird at Fort Sumner, N.M. in early Jun.

*Note:* Most extralimital and periphery-of-range nesting occurs very infrequently and is based on variable prey populations within isolated, proper habitat.

**Winter.**—Locally dense in the southern portion of winter range in Kansas, w. Oklahoma, e. Colorado, s. Arizona, and California. Also fairly densely populated in isolated areas as far north as s. British Columbia. Winters regularly on Kodiak Island, Alaska; irregularly winters along se. Alaska and coastal British Columbia, including the Queen Charlotte Islands, B.C. Winter range extends to s. Central America. In the East, wintering birds occur as far south as Cuba, Bahamas, and the Dominican Republic.

**Movements.**—Protracted migration in both fall and spring.

*Fall migration:* Mid-Aug. to early Nov., but extends into Dec. in s. Texas. No definitive peaks. There is a fairly steady movement throughout the period. Juveniles predominate in the early portion of the season and adults in the latter part, especially males. All ages may move throughout the entire migration period.

*Spring migration:* Apr. to mid-May with

stragglers occurring into mid-Jun. Movement begins in late Feb. in the s. U.S. and in late Mar. in s. Canada. Adult males move first, followed by adult females, and then juveniles. Some adult females and juveniles move throughout the entire migration period. Juveniles peak in late Apr. to early May, but stragglers may continue moving into mid-Jun.

**NESTING:** Begins in mid-Mar. in southern latitudes and mid-May or early Jun. in northern regions. The nesting cycle ends in Jul. or Aug.

Pairs may be monogamous or males may practice polygyny and have up to five females in a harem. Harriers often nest in loosely formed colonies. A few 1-year-olds, in juvenile plumage, may breed. This is more likely to occur with females than with males. Most do not breed until 2 years old.

**Courtship (flight).**—*Sky-dancing* (high intensity) with very elaborate maneuvers by males and to a lesser extent by a few females. Males may perform sky-dancing during spring migration at various nonbreeding locales. *High-circling* and *aerial food transfer* are performed by mated pairs. Once territories are established, males exhibit *leg-lowering* to other males. Females perform *leg-lowering, escort-flight,* and *talon-grappling* to conspecific intruders (mainly females). (*See* chapter 5 for display descriptions.) *Cheh-cheh-cheh* chatter call accompanies many types of displays.

Nests are built on the ground surrounded by protective vegetation. Some open-country sites are not well concealed or may not be concealed. Also nests in tall clumps of vegetation over water. Nests are 15–25 in. (39–63 cm) in diameter. Generally 1–7 eggs, but up to 10 are known. Large clutches may be a sign of two females laying in the same nest. Eggs are incubated 30–32 days by females. Youngsters fledge in 30–35 days, with females, being larger, taking longer than males. Females feed the young; males hunt and deliver prey to the females. If females are not present at the nest, males will drop the prey but do not feed the nestlings.

**CONSERVATION:** Listed as state Endangered Species in Iowa, and in the Yukon Territory, Canada, listed as a Sensitive Species. All other Canadian provinces list them as a Secure Species, except in Labrador, where status is undetermined. There is no designation for other states or provinces.

**Mortality.**—Breeding is undoubtedly affected in many areas by agricultural practices and declining marsh and grasslands. Northern Harriers suffered from organochlorine era of late 1940s to early 1970s, but rebounded quickly after the ban of DDT in the 1970s. Organophosphate poisoning for rodent control is a possible threat. Illegal shooting occurs in certain areas. Harriers are susceptible to illegal shooting by errant waterfowl hunters. Lead poisoning from eating dead and wounded waterfowl was a mortality factor until the 1991 nationwide ban on lead shot.

**SIMILAR SPECIES:** **(1) Northern Goshawk, juveniles.**—PERCHED.—Similar to subadult and adult females in many aspects in color and markings. Pale supercilium, but lacks pale area under eyes; auricular disk often well defined: use caution. Pale tawny bar on greater upperwing coverts. Short tarsi. Primaries much shorter than tail tip. FLIGHT.—Six "fingers" on outer primaries. Uniformly pale underwing coverts. Dark uppertail coverts. Wings held on a flat plane when gliding and soaring. **(2) Swainson's Hawk, light morphs.**—PERCHED.—Confusion possible between juvenile Swainson's Hawk and subadult and adult female harriers. Wingtips equal to tail tip. Pale mottling on scapulars. FLIGHT.—Wings held in similar dihedral at all times; use caution. Narrow white "U" on uppertail coverts; narrowly banded tail. Narrow dark bars on underside of remiges; solid dark outermost primaries. **(3) Rough-legged Hawk.**—FLIGHT.—Very similar flight attitude (jizz) and easily confused at long- and moderate distances, especially to heavier, bulkier female harriers. If uncertain of identification at a distance, wait for closer views for plumage markings. Harriers share white uppertail coverts of light morph; look for dark carpal area on underwing of Rough-legged Hawk. **(4) Peregrine Falcon.**—FLIGHT.—Deceptively similar in high-altitude silhouette view. Falcon's body-to-wrist distance is very short (long, angular on harrier); tail is shorter. Wingbeats steady and powerful; soars and glides with wings on a flat plane.

**OTHER NAMES:** Harrier, Gray Ghost (adult male); formerly called Marsh Hawk. *Spanish:* Gavilan Pantanero, Gavilan Rastrero. *French:* Busard Saint-Martin.

REFERENCES: Adamus et al. 2001; Andrews and Righter 1992; Arizona Breeding Bird Atlas 1993–1999; Baumgartner and Baumgartner 1992; Bildstein and Collopy 1985; Bylan 1998, 1999; Call 1978; Clark and Wheeler 2001; Craig et al. 1982; Dechant et al. 1999; Dinsmore 1999; Dodge 1988–1997; Environment Canada 2001; Forsman 1999; Godfrey 1986; Hamerstrom 1968; Herron et al. 1985; Howell et al. 1992; Jacobs and Wilson 1997; Kent and Dinsmore 1996; Kingery 1998; Macwhirter and Bildstein 1996; Montana Bird Distribution Committee 1996; Montaperto 1988; Palmer 1988; Peterson 1995; Pierce 1998; Semenchuk 1992; Small 1994; Smith 1996; Smith et al. 1997; Snyder 1998a; Stewart 1975; Williams 2000.

**NORTHERN HARRIER,** *Circus cyaneus hudsonius:* Fairly common. Found in wet or dry habitat. Winters to s. Central America.

**Plate 132. Northern Harrier, adult male (fresh plumage) [Nov.]** ▪ Defined auricular disk; all-gray head and neck; lack whitish spectacles around eyes. Orangish yellow iris. Flat-topped head. ▪ Dark grayish brown back and scapulars. ▪ Grayish upperwing coverts. ▪ *Note:* Harriers regularly stand on the ground.

**Plate 133. Northern Harrier, adult male (fresh plumage) [Nov.]** ▪ Defined white spectacles around eyes. ▪ Dark grayish brown back and scapulars. ▪ Dark grayish brown upperwing coverts. Pale gray remiges with a wide black bar on rear edge of secondaries; large black patch on outer primaries. ▪ White uppertail coverts.

**Plate 134. Northern Harrier, adult male (worn plumage) [Apr.]** ▪ White spectacles around eyes. ▪ Medium grayish brown back and scapulars. ▪ Medium grayish brown upperwing coverts. Pale gray remiges with a wide black bar on rear edge of secondaries; large black patch on outer primaries. ▪ White uppertail coverts. ▪ *Note:* Dorsal color fades considerably by spring.

**Plate 135. Northern Harrier, adult male (worn plumage) [Apr.]** ▪ Uniformly gray head and neck. ▪ Medium pale brownish gray back and scapulars. ▪ Medium pale brownish gray upperwing coverts. Pale gray remiges with a large black patch on outer primaries. ▪ White uppertail coverts. ▪ *Note:* Palest type of male; possibly an older bird.

**Plate 136. Northern Harrier, adult male (fresh plumage) [Nov.]** ▪ Gray head and neck form a hood; slight spectacle definition around eyes. ▪ White underparts spotted with gray or rufous. ▪ White underwing has a wide black bar on rear edge of the secondaries; large black patch on outer primaries.

**Plate 137. Northern Harrier, subadult male (fresh plumage) [Jan.]** ▪ Rufous-brown auriculars and crown with distinct auriculars; gray nape. Pale yellow iris. ▪ White underparts with rufous markings. ▪ *Note:* Harriers often perch on low, open objects.

**Plate 138. Northern Harrier, subadult male (fresh plumage) [Dec.]** ▪ Rufous-brown auriculars. Grayish brown neck. Tawny wash on breast. Belly and flanks are heavily marked with rufous. Underwing coverts are heavily marked with rufous. Wide black bar on rear edge of the secondaries; large black patch on the outer primaries. ▪ *Note:* Some are more lightly marked.

**Plate 139. Northern Harrier, adult female (fresh plumage) [Nov.]** ▪ Streaked tawny head and neck with distinct pale spectacles around eyes and a pale rim on auricular disks. Yellow iris. ▪ Whitish underparts are narrowly streaked. ▪ Wing coverts are edged with tawny and have white blotches on the median coverts. Pale gray primaries are distinctly barred.

**Plate 140. Northern Harrier, adult female (worn plumage) [Apr.]** ▪ Distinct spectacles around eyes and pale rim on auricular disk. ▪ Medium brown upperwing coverts with pale tawny feather edges. ▪ Pale gray remiges with very distinct barring, including a wide black bar on rear edge of secondaries. ▪ White uppertail coverts.

**Plate 141. Northern Harrier, adult female (fresh plumage) [Jan.]** ▪ Brownish head and neck form a hood. Defined auricular disk. Yellow iris. ▪ Rear flanks are barred. ▪ Barred, pale gray remiges, including a wide dark bar on rear edge of secondaries. Secondaries sometimes slightly darker than primaries. Dark axillaries and greater coverts.

**Plate 142. Northern Harrier, subadult female (fresh plumage) [Jan.]** ▪ Dark auricular disk with pale rim. Dark tawny head and neck form a hood. Brown iris. ▪ Pale tawny underparts are barred on rear flanks and moderately streaked on forward flanks and belly. ▪ *Note:* Iris color changes from medium or pale brown to yellow in about 3 years.

**Plate 143. Northern Harrier, subadult female (fresh plumage) [Sep.]** ▪ Brown iris. ▪ Dark brown back and scapulars. Medium gray remiges with distinct black barring. Retains a few pale brown juvenile secondaries, which will be replaced in a few weeks. ▪ White uppertail coverts. Banded tail with gray center and tawny outer rectrices.

**Plate 144. Northern Harrier, subadult female (worn plumage) [Feb.]** ▪ Brownish head and neck creates a hood. ▪ Lightly streaked/spotted underparts; rear flanks have brown arrowhead-shaped markings. ▪ Pale remiges; sometimes secondaries slightly darker; wide black bar on secondaries.

**Plate 145. Northern Harrier, juvenile female (fresh plumage) [Dec.]** ▪ Dark brown iris. Dark brown head and neck. Pale spectacles around eyes and pale auricular disk rim. ▪ Dark brown back and scapulars. ▪ Dark brown wing coverts with tawny patch. Solid brown remiges. ▪ *Note:* Dark brown irises may lighten to medium brown by spring.

**Plate 146. Northern Harrier, juvenile male (fresh plumage) [Jan.]** ▪ Pale yellow iris. Dark brown head and neck. Pale spectacles around eyes and pale auricular disk rim. ▪ Medium tawny-orange underparts with brown streaking on breast and flanks. ▪ *Note:* Iris color changes from pale brown or pale gray to pale yellow by late fall to mid-winter.

**Plate 147. Northern Harrier, juvenile male/female (fresh plumage) [Jan.]** ▪ Dark brown head and neck. ▪ Tawny-orange underparts. ▪ Wings held in a high dihedral. Dark secondaries. ▪ White uppertail coverts. ▪ *Note:* Classic hunting strategy by coursing low over the ground with head pointing down.

**Plate 148. Northern Harrier, juvenile male (worn plumage) [Mar.]** ▪ White spectacles around eyes; pale yellow iris. ▪ Medium brown back and scapulars. ▪ Remiges have a minimal amount of dark barring (less than in subadult and adult females). ▪ White uppertail coverts. Banded tail with gray center and pale tawny outer rectrices. ▪ *Note:* Dorsal color of both sexes fades by spring.

**Plate 149. Northern Harrier, juvenile female (fresh plumage) [Sep.]** ▪ Dark brown head and neck form a hood. Pale spectacles around eyes, pale rim on auricular disk. Dark brown iris. ▪ Bright tawny orange underparts. ▪ Dark brown axillaries and median coverts. Secondaries are darker than primaries. Barred remiges with a wide black bar on rear edge of secondaries.

**Plate 150. Northern Harrier, juvenile male (fresh plumage) [Oct.]** ▪ Dark brown head and neck. White spectacles around eyes and pale rim on auricular disk. Pale gray iris. ▪ Tawny-orange underparts with brown streaking on the flanks. ▪ Dark brown axillaries and median coverts. Secondaries darker than primaries.

**Plate 151. Northern Harrier, juvenile female (worn plumage) [Apr.]** ▪ Brown head and neck form a distinct hood. ▪ Underparts fade from the bright tawny orange of autumn birds to very pale tawny or whitish on spring birds. ▪ Dark brown axillaries and median coverts. Dark secondaries and pale primaries. ▪ *Note:* This is the palest extreme, which occurs from sun bleaching.

# SHARP-SHINNED HAWK
## *(Accipiter striatus)*

**AGES:** Adult, basic I (subadult I), and juvenile. Adult plumage is usually attained when 2 years old, but some males may acquire full adult plumage when 1 year old. Adult dorsal color is sexually dimorphic, rarely overlaps, and is visible at field distances. If there are similarities in dorsal color, it is generally females assuming malelike color.

Basic I (subadult I) plumage is acquired when 1 year old. Plumage is identical to respective adult sex but may retain a small amount of juvenile feathering. Juvenile feathering, however, is not always visible in the field. As previously described, some males may molt entirely out of juvenile feathering and appear as adults. Iris color of subadults (orange) overlaps with the color of some adults.

Juvenile plumage is worn the first year. Sexes are basically identical but show considerable individual variation, especially on the underparts. Iris color brightens and plumage color fades from mid-winter through spring.

**MOLT:** Accipitridae wing and tail molt pattern (*see* chapter 4). First prebasic molt from juvenile to subadult or adult occurs from mid-May to Nov. Molt begins on the innermost primaries, then proceeds to the tertials, back, and scapulars. By mid-Jun., only the three innermost primaries and tertials may be molted and only a few new dorsal feathers may have grown in. The first prebasic molt may be incomplete and retain a few juvenile feathers on the upperwing coverts and rump. Retained rump feathers are not readily visible under field conditions. First prebasic molt may be complete on some males. During fall migration, subadults may be molting the outermost primaries, secondaries (s4 and 8), and rectrices (r2 and 5, which are possibly the last to molt). All remiges and rectrices are fully replaced by the end of the molt cycle each fall.

Subadult to adult and subsequent adult annual prebasic molts are fairly complete and extend from late May through Nov. Males begin molt later than females. Adults occasionally retain a few old adult feathers on the upperwing coverts, secondaries, rump, uppertail coverts, and tail (r2 and 5). All primaries are replaced by the end of the molt in Oct. or Nov. During autumn migration, adults are molting the outer primaries, secondaries, and rectrices (r2 and 5, but often r6 are the last to molt).

**SUBSPECIES:** Three races occur in the West. Widespread *A. s. velox* occurs throughout much of mainland North America.

"Queen Charlotte" race, *A. s. perobscurus*, is an example of Gloger's Rule. Increased melanin in the feather structure produces the darker, more saturated plumage of this race. Range and plumage data are based, in part, on Friedmann (1950). Museum specimens were also examined from British Columbia island locales from the RBCM and AMNH, ($n = 14$). Notes and sketches on field-studied individuals were supplied by N. J. Schmitt ($n = 2$). Photographs of probable *perobscurus* and/or *perobscurus-velox* types from museum specimens were supplied by N. J. Schmitt ($n = 17$). These are winter specimens from coastal locations in California. The specimens are from the Santa Barbara Museum of Systematic and Ecology and Vertebrate Collection (UCSBM), University of California, Goleta, and the Santa Barbara Museum of Natural History (SBMNH), Santa Barbara.

*A. s. perobscurus* breeds in British Columbia on the Queen Charlotte Islands and near-coastal islands and in coastal mainland regions from Yakutat Bay, AK, south to Vancouver Island, B.C., and possibly on the Olympic Peninsula, Wash. *Perobscurus* and *velox* appear to intermix in all seasons in much of their range (e.g., recently fledged, classic *velox* specimens from Ketchikan, AK, and Vancouver Island, B.C.).

"Sutton's" race, *A. s. suttoni*, is a Mexican subspecies that irregularly extends its range into extreme se. Arizona and possibly sw. New Mexico and s. Texas (*see* Status and Distribution). *Suttoni* typically occupies montane elevations in the Sierra Madre Occidental and Sierra Madre Oriental in Mexico. *Suttoni* plumage data are based on Mexican specimens examined at the AMNH ($n = 3$); Mexican specimens loaned by LSU ($n = 10$); in-field photographs of one individual by H. Snyder; and Friedmann (1950).

Four additional races: (1) *A. s. madrensis* is in s. Mexico; (2) *A. s. fringilloides* is a very rare race on Cuba; (3) nominate *A. s. striatus* is on Hispaniola; and (4) *A. s. venator* is on Puerto Rico and has been on the U.S. Endangered Species List since 1994.

Three species listed below are often considered subspecies of *A. striatus* by some taxonomists. del Hoyo et al. (1994) treat them as separate species that form a superspecies with *A. striatus*: (1) White-breasted Hawk (*A. chionogaster*) in the highland areas from s. Mexico to n.-cen. Nicaragua; (2) Plain-breasted Hawk (*A. ventralis*) in the mountains of n. and se. Venezuela, Colombia, Ecuador, Peru, and w. Bolivia; and (3) Rufous-thighed Hawk (*A. erythronemius*) of s. Brazil, Uruguay, and se. Bolivia to n. Argentina.

**COLOR MORPHS:** None.

**SIZE:** A small raptor and smallest of the three accipiters. Sexually dimorphic with no overlap. Proportionately, the most sexually dimorphic North American raptor. With practice, sexes are separable in the field. Males are considerably smaller than females. Measurements are based on *velox;* length and wingspan are not available for *perobscurus* or *suttoni.* MALE.—Length: 9–11 in. (23–28 cm); wingspan: 20–22 in. (51–56 cm). FEMALE.—Length: 11–13 in. (28–33 cm); wingspan: 23–26 in. (58–66 cm). *Note:* Wing chord measurements based on Friedmann (1950) and studies of LSU specimens: females of *perobscurus* ($n = 1$, Friedmann 1950) and *suttoni* ($n = 1$, Friedmann 1950; $n = 2$, LSU) are similar in length or longer than *velox* females; wing chords of males of *perobscurus* ($n = 1$, Friedmann 1950) and *suttoni* ($n = 2$, Friedmann 1950; $n = 6$, LSU) are longer than most *velox* males.

**SPECIES TRAITS:** HEAD.—Petite bill. Small, round-shaped head, even when nape feathers are erected. **Yellow skin on supraorbital ridge, if exposed, is visible only at close range.** In flight, appears short-necked and small-headed. BODY.—**Long, thin tarsi and toes are thinner than a pencil. Tarsi are yellowish orange or yellow on males and yellow on females.** WINGS.—When perched, wingtips are considerably shorter than tail tip. **In flight, moderately short and broad with very rounded wingtips.** Undersides of all remiges are distinctly barred. Accipiters have six "fingers" on the outer primaries. TAIL.—Long. Three or four equal-width dark bands on the dorsal side. On the ventral side, the outer rectrix set has narrow dark bands that do not align with the wide dark inner bands. Four distinct tail-tip shapes are visible on the ventral surface of a closed tail. (1) *Notched type* (males only): **Outer rectrix sets are longer than central set and form a notch on the center of the tail.** *Note:* Some females exhibit notched tails, mainly because of abnormal feather overlap. (2) *Squared type* (both sexes but mainly males): **All rectrices are the same length.** (3) *Rounded type* (both sexes): Outer rectrix sets are somewhat shorter than inner sets. (4) *Very rounded type* (females only): A few have very rounded or wedged-shaped rectrices. Central set (r1) is longest, and each outer set is sequentially shorter. *Note:* All tail patterns appear more rounded on fanned tails, particularly when soaring. In soaring flight, widely fanned tail seen in certain angles of intense light can also make the paler gray area behind the dark subterminal band appear as a fairly wide white terminal band.

**ADULT TRAITS:** HEAD—Iris color varies from orange to dark red. Iris color of males attains a deeper red and darkens more quickly than on females. The change in iris color does not occur at a rate that can be used as an exact aging criteria. Birds with red irises are older than 1 year. **A black mask may surround under and behind the eyes and may extend as a black eyeline. Pale rufous auriculars often fade to pale grayish by spring. Crown and nape are medium bluish or grayish and typically the same color as the upperparts. The crown and sometimes nape may be somewhat darker gray.** Some birds have a partial or full narrow white supercilium. BODY.—**Uniformly medium bluish or grayish upperparts.** Breast, belly, flanks, and leg feathers have sharply delineated rufous and white barring. Undertail coverts are white and unmarked. WINGS.—**Upperwing coverts are medium bluish or grayish and are same color as rest of the upperparts. Underwing coverts are mainly tawny but sometimes whitish.** Underwing coverts may have a combination of dark brown or rufous streaking, spotting, or barring. The axillaries are rufous barred. TAIL.—Bluish or grayish pale uppertail bands with equal-width, wide dark bands. *Note:* In late summer through late fall (mainly Sep.–Oct.), adults are in extensive tail molt and many rectrices are still growing. Irregular feather lengths may alter shape of tail tip.

**ADULTS OF "TYPICAL" (*A. s. velox*):** BODY.—Breast, belly, flanks, and leg feathers are white with sharply delineated narrow rufous barring. Undertail coverts are white and unmarked.

**Adult male *velox:*** HEAD.—**Medium grayish blue crown and nape.** Rarely has pale supercilium. BODY.—**Medium grayish blue upperparts. Rarely has bluish gray upperparts.** Leg feathers are occasionally nearly solid rufous with narrow white tips. WINGS.—**Medium grayish blue upperwing coverts.** TAIL.—In fresh plumage in fall and winter, regularly has a distinct, fairly wide white terminal band. **Grayish blue pale bands on uppertail.**

**Adult female *velox:*** HEAD.—**Medium gray crown and nape. Regularly has whitish supercilium.** BODY and WINGS.—**Medium gray. On some, upperparts fade to grayish brown by spring.** TAIL.—**Tip is edged pale gray or has a very narrow white band.** Medium gray pale uppertail bands.

**Adult female *velox* (blue-backed type): (1) A few females are malelike grayish blue on the head, nape, back, and forward part of scapulars during all seasons. Lower scapulars, rump, and upperwing coverts are the typical female gray color. (2) Some are a uniform, malelike bluish gray on all upperparts, but generally a bit more grayish than most males. On both types, the pale grayish tail bands are as on typical adult females.**

**ADULTS OF "QUEEN CHARLOTTE" (*A. s. perobscurus*):** Rufous ventral barring is thicker and more dense than on *velox*. Many males are not readily separable from *velox* when seen in the field, but some females are separable at long distances.

**Adult male *perobscurus:*** Based on Friedmann 1950 ($n = 1$) and RBCM collection ($n = 4$). BODY.—Friedmann specimen was "slightly darker, especially below." Dorsal color is identical to respective sex of *velox* on RBCM specimens. Breast and sides of the neck can be nearly solid rufous. Belly and flanks are barred similarly to *velox*, but rufous barring is often broader on flanks. Leg feathers are virtually solid rufous with only narrow white on tips of most feathers. (Some *velox* males also have fairly solid rufous leg feathers.) Undertail coverts are white and unmarked on all specimens.

**Adult female *perobscurus:*** Based on a Friedmann (1950) specimen (location not given); museum specimens from SBNMH in winter/early spring ($n = 3$); and sight records by N. J. Schmitt from Riverside and Los Angeles Cos., Calif., in winter ($n = 2$). HEAD.—As *velox*. BODY.—Friedmann specimen noted as "slightly darker above." Two of the three museum specimens were slightly darker and decidedly more bluish on the dorsal compared to the same-season *velox*. On the two sight records, dorsum color was not recorded. Underparts, including leg feathers, as described for males on all individuals. *Note:* Solid rufous leg feathers are very rare on *velox* females. Undertail covert pattern was not mentioned on the Friedmann specimen. What may be unique with this subspecies are the field-visible, fairly large dark markings on the undertail coverts (also visible on many juveniles of this race; *see below*). SBNMH museum specimens and the two field-observed individuals had small to moderate-sized, field-visible, rufous streaked, arrowhead- or diamond-shaped marks on some or all of the undertail covert feathers. (*A. s. velox* females do not have markings on undertail coverts.) *Note:* An adult female from Mendocino Co., Calif., from mid-Jul. (UCSBM) is a fairly rich bluish gray on the dorsum and very *perobscurus*-like on the ventral areas. Underparts are broadly barred with rufous, with virtually solid rufous leg feathers and a few scattered small grayish marks on some undertail coverts.

**ADULTS OF "SUTTON'S" (*A. s. suttoni*):** HEAD.—In fresh plumage, cheeks and auriculars are medium rufous or tawny-rufous. In worn and faded spring and summer condition, cheeks and auriculars become faded and pale and can be pale tawny, whitish, or even pale grayish. BODY.—Subspecies trait of the diagnostic, uniformly colored rufous flanks and leg feathers is very apparent on classic individuals of both sexes. Breast is either finely barred or solid rufous and blends as a uniformly rufous area with the all-rufous flanks. Belly is always barred; however, the contrast of the white and rufous barring is softly edged and not harshly delineated as on *velox*. Breast area becomes quite faded on summer birds and may be pale tawny-rufous or even quite whitish. *Note:* Separable from *velox* at moderate distance.

**Adult male *suttoni*:** Based on plumage data from the AMNH collection from Jalisco, Mexico ($n = 1$ in Dec.); LSU collection from Coahuila, Jalisco, San Luis Potosí, and Sonora, Mexico ($n = 6$ from late May, Jun., Jul., and Oct.); and Friedmann 1950 ($n = 2$). HEAD.—Auriculars may be rufous as in *velox* or paler and quite grayish. BODY.—Dorsal color is identical to the same-season bluish plumage color of *velox* on all specimens, but the LSU specimens were on the darker end of the spectrum. All individuals were classic *suttoni* on ventral areas.

**Adult female *suttoni*:** Based on plumage data from the AMNH collection from Jalisco, Mexico ($n = 1$ in Sep.); LSU collection from Michoacán and San Luis Potosí, Mexico ($n = 2$ in Nov., Feb.); H. Snyder photograph of a nesting individual from the Huachuca Mts., Cochise Co., Ariz., from 1971; and Friedmann 1950 ($n = 1$). BODY.—Dorsal color is identical to same-season *velox* on all specimens. Two specimens were decidedly brownish gray and one was quite bluish (blue-backed type). Dorsal color on the photographed individual was a darker bluish and also similar to a blue-backed type of adult female *velox*. *Note:* The female from Michoacán (Nov.) was primarily *velox*-like on ventral areas except for the solid rufous leg feathers.

**BASIC I (SUBADULT I) TRAITS:** As previously described, plumage is basically identical to respective adult sex and often not separable from adults under field conditions. Iris color is always orange. (Little if any color change occurs from springtime juveniles.) Many birds have a partial or full thin whitish supercilium. Subadults may retain worn, pale brown juvenile feathering on the upperwing coverts and rump until the next molt in late spring and summer. Retained juvenile feathers on the upperwing coverts are uniformly brown; those on the rump have narrow tawny-rufous tips on each brown feather but are not visible under field conditions. *Note:* In the fall, old, worn upperwing coverts on adults may also be very brownish and similar to worn juvenile feathers of subadults.

**JUVENILE TRAITS:** HEAD.—**Iris color is medium orangish yellow or yellow ochre from fledging until mid-winter; rarely pale lemon yellow.** BODY.—Rufous or brown streaked breast and belly. Flanks are barred. WINGS.—*Underwing* (male): **(1) Trailing edge of inner primaries and secondaries may have a narrow or moderately wide, sharply defined dark band; (2) a nondescript gray band.** *Underwing* (female): **(1) Trailing edge of inner primaries and secondaries has a nondescript, moderately wide gray band; (2) very rarely, a sharply defined dark band, which is more prevalent on heavily marked type females. Axillaries are typically barred with rurous; some males may be barred with dark brown.** TAIL.—Pale brown upper surface with the three or four equal-width, wide dark bands. Tail tip has a narrow pale gray band or sometimes a narrow white band.

**JUVENILES OF "TYPICAL" (*A. s. velox*):** HEAD.—**Crown, nape, and hindneck are dark brown or tawny brown with pale tawny streaking, especially on the nape. Auriculars are tawny-rufous. Long, thin, white supercilium. Supercilium, however, may be partial or lacking. Iris color is medium orangish yellow or yellow ochre from fledging until mid-winter; rarely pale lemon yellow.** BODY (ventral).—White or tawny underparts are variably streaked with rufous or brown on the breast, belly, and lower belly. The flanks are distinctly barred with rufous or brown. There are three variations of ventral markings. (1) *Narrowly streaked type* (many males; a few females).—Breast, belly, and lower belly are narrowly streaked. The lower belly generally has less streaking or occasionally is unmarked. Flanks may be partially barred. Both rufous and brown markings are common on this variation, but females are more likely to have rufous markings. (2) *Moderately streaked type* (both sexes).—**Fairly broad uniform streaking on breast, belly, and lower belly. The lower belly often has less streaking. Streaking often broadens into partial "lobes" on the breast and belly feathers and forms a partially barred pattern. Flanks are broadly barred.** Rufous coloration predominates with this variation. (3) *Broadly streaked type* (some males; many females).—**Breast, belly, and lower belly are uniformly marked with broad, dense streaking. Streaking typically broadens into "lobes" or partial bars on the breast and belly and makes the underparts appear quite**

**barred (and adultlike). Distinct, broad barring on the flanks.** Virtually all markings are rufous in this variation. Leg feathers are either streaked or are variably barred with rufous or dark brown. BODY (dorsal).—Dark brown upperparts have narrow rufous or tawny edges, most pronounced on the upperwing coverts. Males are more likely to have pronounced pale feather edges on scapulars and back, but there is considerable variation. Many females also exhibit distinct pale feather edges, and some males have only a minimal amount of pale edging. Basal region of each scapular has a large white area, making the scapulars appear very "blotched" when fluffed in cold temperatures or when rousing. White, unmarked undertail coverts. WINGS (dorsal).—Upperwing coverts have pale tawny-rufous edges. WINGS (ventral).—*Male:* **(1) Trailing edge of the inner primaries and secondaries may have a narrow or moderately wide, sharply defined dark band or (2) a nondescript gray band.** *Female:* **(1) Trailing edge of inner primaries and secondaries has a nondescript gray band** or **(2) rarely, a well-defined dark grayish or black band, which is most prevalent on heavily marked type females.** Underwing coverts are tawny, rarely whitish.

**JUVENILES OF "TYPICAL" (LIGHTLY MARKED TYPE):** HEAD.—Crown, nape, and hindneck are brown and streaked with tawny. Supercilium is white and wide. Cheeks and most of the auriculars are white except for a small brownish area on the rear part of the auriculars. BODY.—White underparts are narrowly streaked as in narrowly streaked type. Flanks may have partial barring or be narrowly streaked. Narrow brown or rufous streaking on leg feathers. Overall appearance of the underparts is very pale. Upperparts have pronounced tawny edges on all back and scapular feathers. WINGS.—Upperwing coverts are distinctly edged with pale tawny. White underwing coverts are finely marked with rufous or brown. TAIL.—As on typical *velox*. *Note:* Plumage type is on some males and a few females.

**JUVENILES OF "TYPICAL" (HEAVILY MARKED TYPE):** HEAD.—Crown, auriculars, nape, and hindneck are dark brown and lack tawny streaking. Pale supercilium is absent. BODY.—Dark brown upperparts generally lack paler tawny-rufous feather edgings. Underpart streaking varies as described in the three variations under Juveniles of "Typical." Females typically have broadly streaked type underparts, but males often have narrowly or moderately streaked type underparts. WINGS.—Upperwing coverts are dark brown and lack pale feather edgings. Tawny underwing coverts are heavily marked with dark brown. TAIL.—As on typical *velox*. *Note:* Plumage type is similar to *perobscurus* that lack distinct undertail covert streaking and may be difficult to separate in the field, especially in males.

**JUVENILES OF "QUEEN CHARLOTTE" (*A. s. perobscurus*):** Based on description by Friedmann 1950 (unknown number of specimens) and specimens from the AMNH and RBCM. Specimens were from Graham Island in the Queen Charlotte Islands, B.C. ($n = 8$ recently fledged/winter); Salt Spring Island, B.C. ($n = 1$ recently fledged); and Banks Island, B.C. ($n = 1$ recently fledged). Also, specimens from SBMNH from Santa Barbara Co., Calif., with *perobscurus*-like traits ($n = 2$ females from winter). HEAD.—Crown, nape, and hindneck are solid blackish brown. Auriculars are uniformly dark brown. Supercilium is typically absent but may be partial and faintly delineated. BODY.—Uniformly blackish brown upperparts. Some birds may have faint, narrow, dark rufous edging on a few upperwing coverts and distal scapular. White spots on the basal region of each scapular are reduced. Distinct rufous or brown barring on flanks. Leg feathers can be uniformly rufous, barred, or variably streaked. Either sex may have distinct dark brown or rufous streaks on the undertail coverts. *Note:* A few other SBMNH winter-season specimens from coastal California (e.g., Santa Barbara Co.) had moderately field-visible streaking on the undertail coverts, but upperparts were similar to typical *velox*. Natal origins of these birds are unknown. They are possible intergrades. *Note:* The undertail coverts were white and unmarked on the few breeding-season specimens from Vancouver Island, B.C., and adjacent islands of British Columbia and se. Alaska; however, the possibility exists of having markings. WINGS.—Upperwing coverts have little if any rufous edging and appear uniformly dark brown. Underwing coverts are dusky or tawny and heavily marked with dark brown. Trailing edge of the underside of the remiges as on respective sex of *velox*. TAIL.—As on *velox* but darker brown on the pale dorsal bands and lacks a pale terminal band. *Note:* Separable from velox at moderate distances, especially if a bird has distinct undertail covert markings. On average, females are darker and more extensively marked on the ventral region and more separable from *velox* than are males. Not depicted.

**Juvenile male *perobscurus:*** BODY.—Dark brown or rufous streaking on the underparts can be similar to *velox* (narrowly, moderately, or broadly streaked types).

**Juvenile female *perobscurus:*** BODY.—Ventral streaking is predominantly rufous on females and may be moderately, broadly, or very broadly streaked type. The dense streaking of very broadly streaked type makes underparts appear nearly solid rufous.

**JUVENILES OF "SUTTON'S" (*A. s. suttoni*):** HEAD.—Like *velox*. BODY.—Subspecies' trait of the uniformly rufous flanks and leg feathers is apparent on most individuals. Dorsal color and markings as on *velox*, including large white spots on the basal area of the scapular and tawny feather edgings. Undertail coverts are white and unmarked as on *velox*. *Note:* Separable from *velox* at moderate distances.

**Juvenile male *suttoni:***—Data based on specimens from LSU collection from San Luis Potosí, Mexico (*n* = 3 from May, Jun., and Aug.). Only one bird had classic subspecies traits on the ventral surface. HEAD.—As on *velox*. BODY.—Streaking on the breast and belly is similar to the variations described for *velox*. The specimens had narrowly and moderately streaked types. Streaking is either tawny-rufous and blends onto the flanks or is a slightly darker brown and separable as more brownish than the rufous flanks and legs. *Note:* As in all subspecies, individual variation and probable intergrades occur with *velox*. A recently fledged male from San Luis Potosí in mid-Aug. appears very much like many *velox*, including barred, not rufous, legs. It is unlikely that this bird was an early migrant this far south at this season. A 1-year-old male from mid-Jun. also was very *velox*-like, and also unlikely to have been a late spring migrant at this latitude.

**Juvenile female *suttoni:*** Data based on specimen from the AMNH collection from Jalisco, Mexico (*n* = 1 in Aug.). BODY.—Moderately streaked type ventral streaking is rufous and blends into uniformly rufous flanks.

**JUVENILE (LATE WINTER/SPRING: ALL TYPES/RACES):** HEAD.—Iris color changes to orange (and remains this color into the subadult age class). BODY and WINGS (dorsal).—All pale edgings wear off and the upperparts fade and wear to a uniform medium brown. *Note: Perobscurus* may still be quite dark with minimal fading.

**ABNORMAL PLUMAGES: Imperfect albinism.**—Occurs primarily in juvenile females (sightings and specimens have been females). Bill, irises, cere, and legs can be normally colored, or some or all anatomical regions may be paler, including pale bluish irises. Most of plumage is white, including the upperparts. As is typical in this type of albinism, rufous-pigmented anatomical regions found on normal individuals are also exhibited in this plumage. Head and neck are pale rufous-brown. Pale rufous streaking is present on the breast, belly, and flanks. Scapulars and upperwing coverts have pale rufous edges on most feathers. Remiges and rectrices are white and lack barring. Data based on Clark and Wheeler 2001; F. Nicoletti sighting at Duluth, Minn. (pers. comm.); museum specimen. *Note:* Very rare plumage type and not depicted.

**HABITAT: *A. s. velox.* Summer.**—Breeds in dense stands of thickly foliaged young and mid-aged, succession-stage, coniferous, mixed coniferous-deciduous, and rarely pure deciduous wooded areas with dense understories. Wooded areas may be shelterbelts, small woodlots, or extensive forests. Artificial conifer plantations of moderate-aged trees may also be occupied. Virtually all areas have a water source; if it is a stream or river, it is quiet-running water.

Dense conifer are the main component of suitable nesting territories, and most nests are placed in conifers. Conifers may be in the following settings: (1) same-age, contiguous coniferous forests; (2) a small group of conifers surrounded by shorter deciduous trees or bushes; (3) a small group of conifers surrounded by taller deciduous trees; (4) occasionally a single conifer that is shorter than the surrounding densely growing deciduous trees. Very dense pure deciduous areas are occasionally used for nesting territories. Nest sites are in wooded areas with small or moderate-sized trees that are covered with dense branches and foliage or have abnormally dense, diseased growths of branches and dense foliage.

Elevation ranges from sea level to timberline. Sharp-shinned Hawks primarily inhabit montane elevations in much of the w. U.S. In Alaska and Canada they are in the montane and low-elevation boreal forest. In montane areas, nesting territories are often on the base of a slope. In montane areas of Oregon and possibly many other states, territories are often on the cooler, more densely grown north slopes. Developed areas are generally avoided, but Sharp-shinned Hawks have been known to nest in conifer in suburban settings.

East of the Rocky Mts. and south of the boreal forest of Canada, isolated montane or highland areas are sparsely inhabited. The species is sparingly found in low-elevation areas in coniferous or dense deciduous habitat. On the s. Great Plains (e.g., s. Kansas), rare breeding has occurred in dense groves of deciduous Osage Orange trees. In Arkansas, breeding has been confirmed only in high-elevation pine woodlands. However, summer sightings (and possible nesting) have occurred in low-elevation Shortleaf and Loblolly pine stands in Arkansas. In e. Texas and n. Louisiana, found in low-elevation, mature Shortleaf and Loblolly pine woodlands with moderately dense understories. These areas are often national forests or privately owned forests that are managed for lumber use. Many forests are managed for the endangered Red-cockaded Woodpecker. *Note:* Low-elevation pine woodlands in Arkansas that support Red-cockaded Woodpeckers may also harbor breeding Sharp-shinned Hawks.

***A. s. velox.* Winter.**—May be identical to breeding habitat and at identical elevations (e.g., adult males seen at 10,000 ft. [3000 m] in Colorado). More commonly, wooded and semi-open areas at more southern latitudes and lower elevations without regard to tree species or size. Sharp-shinned Hawks are regularly found in rural and urban areas with moderate tree growth. Small numbers are found on the arid Great Plains and vegetated desert regions void of trees. However, these arid regions possess localized growths of short woody vegetation or tall grasses or weeds that provide haven for prey. On the shortgrass habitat of the Great Plains, tall grass and weed habitats are often found along irrigation ditches, canals, and "living" snow fences of small trees.

***A. s. velox.* Migration.**—Similar to summer and winter habitat. Regularly found in open prairies and deserts void of trees.

***A. s. perobscurus.* Summer.**—Breeds in densely wooded rainforest tracts of younger, succession-aged trees. Coniferous and mixed coniferous-deciduous woodlands from sea level up to montane elevations are inhabited. Climate is cool and wet.

***A. s. perobscurus.* Winter/migration.**—Habitat is probably the same as in summer, but many birds use more open areas, including rural and urban locales. Urban areas on the Queen Charlotte Islands (Massett) are used. Individuals wintering to coastal s. California are often in rural or urban areas. Climate is fairly arid in California.

***A. s. suttoni.* Summer.**—In the U.S., this race has bred in desert "sky island" montane-elevation forests in the Huachuca and Chiricahua Mts. in Cochise Co., Ariz. The two nests that have been found were in dense stands of mixed deciduous and coniferous woodlands. Elevation of the Huachuca Mts. nest was 5,200 ft. (1,600 m); the Chiricahua Mts. nest was at 6,200 ft. (1,900 m). Habitat, however, extends to higher elevations. Montane climate is seasonally warm or cool and somewhat moist. In Mexico, breeds in "sky island" mountains in n. Sonora and Chihuahua but in more contiguous mountain formations in southern regions. Breeding and summer sightings in Sonora occur in montane elevations at 4,900–6,500 ft. (1,500–2,000 m). Breeding-season specimens from Coahuila, San Luis Potosí, and Jalisco were collected at elevations of 6,000–7,000 ft. (1,800–2,100 m). The species is resident up to 9,800 ft. (3,000 m) in portions of Mexico.

***A. s. suttoni.* Winter/migration.**—Habitat is not known in the U.S., but in s. Arizona, Sharp-shinned Hawks (race not determined) usually winter below 6,000 ft. (1,800 m). In Mexico, *suttoni* may remain in moist montane forests or inhabit moist or dry climate areas at lower elevations in the foothills.

**HABITS:** Relatively tame to tame. Possesses an extremely hyper and antagonistic temperament. Constantly harasses large passerines too large to kill and larger raptors. Perches mainly on concealed branches but sometimes on exposed branches. Occasionally perches on wires and posts. *Sharp-shinned Hawks do not perch on telephone poles.* A very reclusive species, especially in the breeding season.

**FEEDING:** Perch and aerial hunter. Perches for short stints when perch hunting. Aerial hunting consists of random flight excursions. Small avian species form bulk of diet. Males generally feed on sparrow- and warbler-size prey; females up to American Robin and even Steller's Jay or quail-sized prey. Sharp-shinned Hawks rarely feed on small rodents, bats, reptiles, and insects (mainly large moths). Hunting is done in the upper canopy of tall trees, in or beneath the canopy in small trees, along edges of wooded and brushy areas, and in open areas with minimal vegetation. May occur at high altitude above treetops. Prey is captured while it is airborne or perched. In the breeding season, plucking posts are used. In other seasons, prey is devoured at point of capture. Sharp-shinned Hawks often sit on the ground where prey is plucked and then eaten. Prey is sometimes taken to a more concealed area on the ground under low overhanging foliage or high among densely foliaged branches. Prey is captured with lightning-fast grasps with the long, spindly toes. Sharp-shinned Hawks do not feed on carrion.

**FLIGHT:** Powered flight consists of a regular cadence of several rapid wingbeats interspersed with short glides. Soars often. Rarely completes a full revolution without resorting to a stint of flapping. When soaring, wings are held on a flat plane and front edge of wings is forward of a perpendicular angle to the body. Wings are held either on a flat plane or bowed slightly downward when gliding. With their short, broad wings and long tail, Sharp-shinned Hawks fly through densely wooded tangles with incredible agility. Flight is at very high altitudes when migrating in light winds but at treetop or near ground level in strong winds. Impressive short or moderate-length vertical dives are often performed to pursue airborne prey or to perch in tree canopies. When hunting in open areas, Sharp-shinned Hawks often skim the ground or low vegetation and sometimes use an undulating, songbirdlike flight mode. The songbirdlike flight consists of a few quick flaps interspersed with a glide in which the wings are closed next to the body. Sharp-shinned Hawks do not hover or kite.

**VOICE:** Generally silent except during courtship, when agitated in nesting territory, and when harassing other predators. Emits a very soft, single-note, songbirdlike chirp when mildly agitated or curious. When very agitated, a high-pitched, rapid *kee-kee-kee* or soft, high-pitched *kyew, kyew, kyew*; usually a three-note call, but can be four or more notes in a series. Nestlings and fledglings emit an equally high-pitched *kree*: single notes or a series of notes. During food transfers, females utter a single, clear *kek*. Voice is easily separable from nasal call notes of Cooper's Hawks.

**STATUS AND DISTRIBUTION:** ***A. s. velox.*** **Summer.**—*Common.* Estimated population is unknown, but this is one of the most common raptors. Population is probably stable. Breeding density is highest in coniferous woodland areas of the western states and especially in the boreal forest of Canada and Alaska. Very small numbers breed or summer south of the boreal forest and east of the Rocky Mts. Nonbreeding 1-year-olds may sporadically inhabit local riparian zones in summer on the Great Plains.

*North Dakota:* Breeds in the Turtle Mts. in Bottineau Co.; n. Pembina Co.; Slope Co.; near Bismark along the Missouri River; and possibly along isolated portions of the Red River. *South Dakota:* Breeds mainly in the Black Hills. *Nebraska:* Very local mainly in isolated Ponderosa Pine woodlands in the western part of the state. Possible irregular breeding and summering in se. Nebraska in isolated conifer groves and dense deciduous tracts. Documented breeding in Brown and Sherman Cos. in previous decades. Recent breeding records in Saunders Co. Probably breeds in Cherry Co. near the Niobrara River, in Sioux Co., and in Carter Canyon in Scott's Bluff Co. Also possibly breeds in isolated Ponderosa Pine tracts in Dawes Co. Although not documented, the Ponderosa Pine woodlands of the Nebraska N.F. in Thomas Co. and the Samuel R. McKelvie N.F. in Cherry Co. may have suitable breeding habitat. Birds were seen in the summer in Dixon and Lincoln Cos. in the 1990s. Single birds have also been seen in Garden, Johnson, and Lancaster Cos. *Kansas:* Breeding documented in Douglas Co. in 1993 and in

Cowley Co. in 1995. There are several records of single birds from late spring and summer in s.-cen. and ne. Kansas. These, however, may be late migrant juveniles or summering, non-breeding 1-year-olds. *Iowa:* Two documented breeding records for Hardin and Monona Cos. There are a few summer records from the southeastern area. *Missouri:* Sparse nesting mainly in wooded, higher elevations in the Ozark Mts. and the southern half of the state. *Oklahoma:* Breeding does not occur. *Arkansas:* Recently discovered nesting in northern and northwestern parts of the state in high-elevation pine woodlands in Newton and Pope Cos. in the Ozark Mts. Nesting may occur in the Quachita Mts. in Cleburne and Polk Cos. Listed as breeding at Felsenthal NWR in Ashley, Bradley, and Union Cos. An adult was seen in summer in Dewitt, Arkansas Co. Breeds in very low density throughout additional lower elevations with isolated Shortleaf and Loblolly pine tracts in the following counties: Calhoun, Dallas, Drew, Lafayette, Nevada, and Quachita. These six counties have populations of Red-cockaded Woodpeckers, and as seen in Texas and Louisiana, Sharp-shinned Hawks are confirmed breeders in many Red-cockaded Woodpecker inhabited forests. *Texas:* Recently found nesting in pine woodlands in Jasper and San Augustine Cos. *Louisiana:* Nesting occurs in low-elevation pine woodlands. Confirmed nesting in Bienville and Vernon Parishes, probable nesting in Lincoln and Natchitoches Parishes, and possible nesting in Catahoula Parish. Many areas are inhabited by Red-cockaded Woodpeckers.

At Rocky Mt. and Great Basin hawkwatches, numbers barely surpass those of Cooper's Hawk. Along coastal areas of w.-cen. California, Sharp-shinned far outnumber Cooper's in migration.

***A. s. perobscurus.* Summer.**—Status is unknown.

***A. s. suttoni.* Summer.**—*Arizona:* Accidental in Cochise Co. Two documented nesting records: (1) 1971 in the Huachuca Mts. (photograph on plate 162 of adult female) and (2) 1993 in the Chiricahua Mts. *Texas:* A nesting pair discovered in Tobe Canyon in the Davis Mts. in Jeff Davis Co. in 2000 and 2001 may perhaps be this race. An adult carrying food in the Cat Tank area of the Davis Mts. was described by observers as being "very red below." As many as five pairs of Sharp-shinned Hawks may nest in the Davis Mts. *New Mexico:* This race may occur in areas that support adequate habitat: Coronado N.F., Animas Mts. and Alamo Hueco Mts. in Hidalgo Co., and the Cedar Mts. in Luna Co. Sharp-shinned Hawks (as a species) are very secretive in the breeding season, and this easy-to-overlook subspecies may breed more frequently than is currently documented in the U.S. *Mexico:* Fairly common in the Sierra Madre Occidental and Sierra Madre Oriental. Found in Sonora, Chihuahua, and south to Puebla. Also nests in the Burro Mts. in n. Coahuila.

***A. s. velox.* Winter.**—Found farther south and at lower elevations than breeding and natal areas. Adult males have been seen at elevations to at least 10,000 ft. (3,000 m) in the Rocky Mts. of Colorado. Juvenile females winter farthest south: to Costa Rica. Solitary and quite nomadic. Territories may be established in areas with ample, stable food supply. Common in rural and urban locales with birdfeeders. Recently published studies based on stable-isotope research by a biologist in New Mexico indicate that juveniles born at northern latitudes in Alaska and nw. Canada winter at more northern latitudes than juveniles born at more southern latitudes.

***A. s. perobscurus.* Winter.**—Mainly winters in breeding range on coastal islands and mainland areas of se. Alaska and British Columbia south to n. Oregon. Some juveniles and adult females winter along the Pacific Coast as far south as s. California in Los Angeles, Riverside, and Santa Barbara Cos. Based on museum data and sight records, and similar to *velox*, juvenile and adult females appear to winter farther south than adult and juvenile males. This race also haunts rural and urban areas with birdfeeders.

***A. s. suttoni.* Winter.**—Wintering individuals have not been documented in the U.S. Those in Mexico remain sedentary in montane breeding areas or move to lower elevations.

***A. s. velox.* Movements.**—Short- to long-distance migrant. Large concentrations of all ages occur along geographic barriers of coastlines and mountain ridges. Large numbers pass along west shore of Lake Superior and along the cen. California coast in fall. Concentrations occur in spring and fall along mountain ridges in the Rocky Mts. and especially the Great Basin. Although travel is in a solitary manner, large numbers that get congested on optimal migrating days appear to be traveling in small flocks.

*Fall migration:* Mid-Aug. to Nov., with stragglers into Nov. Movements are age and sex coordinated. Juveniles migrate first with females peaking in mid-Sep. and males in mid- to late Sep. Based on stable-isotope research done by a biologist in New Mexico, southern-latitude juveniles pass through New Mexico before birds from more northern latitudes. Adults and subadults migrate later. In adults, females precede males, and subadults of each sex precede the respective adult. Adult females peak late Sep., adult males in early to mid-Oct. A few juvenile females move with adult males.

*Spring migration:* Mar. to early Jun. Adults precede juveniles. Adults peak from late Apr. to early May with males moving before most females. Juveniles peak in early to mid-May with stragglers continuing into Jun.

***A. s. perobscurus.* Movements.**—Little is known of this subspecies' movements. Some may remain sedentary, others may move short distances between islands and coastal areas of British Columbia and se. Alaska. Fairly long-distance movements are undertaken by some juvenile females and a few adult females along the Pacific Coast south to s. California.

***A. s. suttoni.* Movements.**—Unknown.

**NESTING:** May to mid-Aug. Males arrive on the breeding grounds before females and establish territories. One-year-olds sometimes breed while in juvenile plumage. These are primarily juvenile females mated to adult males, but sometimes two juvenile-aged birds mate. Most do not breed until 2 years old.

**Courtship (flight).**—*High-circling, slow-flapping,* and *sky-dancing* (*see* chapter 5). Undertail coverts are flared and retracted outwards beyond the closed tail during most aerial courting activities. Vocalization may accompany high-circling.

Sharp-shinned Hawks may return to the same territory but rarely reuse the same nest. Old nests may occasionally be used for 2 consecutive years, very rarely for 3 years. They typically build a new nest each year, often on top of old American Crow nests, leafy squirrel nests, or mistletoe growths. Both sexes bring nest material, but females build the structure. Nests are lined with bark chips but not greenery. They are constructed of thin sticks and are 24 in. (60 cm) in diameter and about 6 in. (15.2 cm) deep. Nests are placed 16–70 ft. (5–21 m) high and are typically hidden next to the trunks of densely foliaged conifer, rarely in densely foliaged or branched deciduous trees. Females assume all nest duties, and males hunt. Males are not known to incubate. Clutches vary from 2 to 6 eggs but typically have 4 or 5. The eggs are white or bluish and speckled with brown and The eggs are incubated 34–35 days. Youngsters branch in 21–24 days, fledge in 24–27 days, are fully developed at 38–40 days, and become independent in about 49 days. Nesting pairs are highly secretive but may become agitated and vociferous with human presence at nest sites, but they are rarely aggressive.

*A. s. perobscurus:* Information on only one nest that was located in a densely foliaged alder tree near a stream.

*A. s. suttoni:* The 1971 nest in the Huachuca Mts. in Cochise Co., Ariz., was located in an Emory Oak in mixed Chihuahuan Pine stand.

**CONSERVATION:** Sharp-shinned Hawks suffered, as did all avian-feeding raptors (which are near the top of the food chain), during the organochlorine era. Numbers currently appear healthy and stable. The ban on organochlorine pesticide use, particularly DDT, no doubt helped Sharp-shinned Hawk populations recover in the last three decades.

**Mortality.**—*A. s. velox* is still undoubtedly affected by pesticides. Carbamate pesticides may have contributed to some mortality. Newer generation pyrethroid pesticides are harmless to birds and mammals (but affect aquatic life). Collisions with windows and vehicles injure or kill many. Illegal shooting occurs in the U.S. and south of the U.S. in winter. In Mexico, shooting affects juvenile females more than juvenile males and adults because they are more common at that latitude; it may have decreased in recent years, however, as shown by fewer band recoveries of shot birds.

*A. s. perobscurus* mortality is poorly known. Window-killed specimens exist in museums. Heavy logging could temporarily reduce habitat.

*A. s. suttoni* mortality is also unknown. Susceptible to shooting in its primary range in Mexico.

**SIMILAR SPECIES:** (1) **Cooper's Hawk, males.**—An age-old problematic species for both adult and juvenile as concerns female Sharp-shinned Hawks. Cooper's Hawks are always *longer in length*, on average 3 in. (8 cm), and *longer in wingspan*, on average 4 in. (10 cm). Cooper's Hawks west of the Great Plains, especially in the Pacific Northwest, are smaller than other populations of Cooper's Hawks but still longer in length and somewhat longer in wingspan than female Sharp-shinned Hawks (some overlap in wing chord but not in wingspan). With practice, both species can be separated on proportional differences without ever seeing plumage marks. Vocalizations of the two species very different: nasal *kek* of a Cooper's Hawk is separable from the high-pitched *kee* of Sharp-shinned Hawk. PERCHED.—Hackle may be raised, which makes the head appear very large; however, in warm temperatures head feathers may be tightly compressed and head may appear quite small. Gray supraorbital ridge skin, if exposed, is visible at only close range. Tarsi and toes are moderately thick (near pencil thickness). On underside of tail, outer rectrices are sequentially shorter than inner ones: use caution when comparing very rounded type tail of female Sharp-shinned Hawks as they can be very like Cooper's Hawks. White terminal band on tail is wide in fall but may be mostly worn off by spring. FLIGHT.—When soaring, front edge of wings is perpendicular to body angle. Head often projects quite a distance beyond front of wings. Often soars with wings in a low dihedral. Longer tailed; shape as for Perched. **(1A) Adult males.**—Crown always darker than upperparts. Upperparts more bluish and rarely overlap with gray color of female Sharp-shinned Hawks. Auriculars and nape are pale gray or rufous (nape is always medium bluish gray on Sharp-shinned Hawks). Underside of inner primaries and secondaries often has minimal amount of dark barring (Sharp-shinned Hawks always have prominent barring). **(1B) Subadult males.**—Similar rufous auriculars, but nape is either pale grayish or pale rufous (nape is always medium bluish gray on Sharp-shinned Hawks). Bluish gray upperparts (gray on adult female/subadult Sharp-shinned Hawks). **(1C) Juvenile males.**—Iris color is generally duller at all seasons: pale gray or lemon yellow in fall and medium orangish yellow in spring. Underparts are similar, especially compared to lightly marked type Sharp-shinned Hawks: use caution on thickness and extent of streaking. Streaking on a Cooper's Hawk is almost always dark brown, rarely rufous. Crisp, dark band on trailing edge of underwing. Axillaries are barred with dark brown (rufous on most Sharp-shinned Hawks). Upperparts usually have more tawny feather edges. **(2) Merlin, all races.**—PERCHED.—Dark brown irises. Throat feathers often puffed. Wingtips nearly reach tail tip. Tail has narrow whitish or tan bands; wide white terminal band (except *F. c. suckleyi*). FLIGHT.—Pointed wingtips. Uniformly dark underwing. Constant, rapid wingbeats. Tail as described in Perched. **(3) Kestrel.**—PERCHED.—Dark irises. Dark facial stripes. Bobs tail down and up. Wingtips near tail tip. FLIGHT.—Pointed wingtips. Erratic flapping and gliding sequences. Rufous tail. **(4) Mississippi Kite, juveniles.**—PERCHED.—Dark irises. Dark lores in front of eyes. Underpart streaking similar. Short, thick tarsi and toes. Wingtips equal or extend beyond tail tip. Dark tail has narrow white bands on underside or lacks bands.

**OTHER NAMES:** Sharp-shin, Sharpie, Shin. *Spanish:* Gavilán Estriado, Gavilán Pajarero. *French:* Épervier Brun.

---

REFERENCES: Adamus et al. 2001; Andrews and Righter 1992; Arizona Breeding Bird Atlas 1993–1999; Burnside 1983; Busby and Zimmerman 2001; Bylan 1998, 1999; Campbell et al. 1990; Clark and Wheeler 1998, 2001; del Hoyo et al. 1994; Dodge 1988–1997; Dorn and Dorn 1990; Ducey 1988; Friedmann 1950; Garner 1999; Herron et al. 1985; Islands Protection Society 1984; Jacobs and Wilson 1997; James and Neal 1986; Johnsgard 1990; Kent and Dinsmore 1996; Meehan et al. 1997; Montana Bird Distribution Committee 1996; Moore and Henny 1983; Palmer 1988; Peterson 1995; Quinn 1991; Reynolds 1983; Reynolds and Meslow 1984; Reynolds and Wight 1978; Robbins and Easterla 1992; Rosenfield and Evans 1980; Russell and Monson 1998; Semenchuk 1992; Shackleford et al. 1996; Sharpe et al. 2001; Small 1994; Smith 2002; Smith et al. 1997; Snyder and Snyder 1991, 1998a; Snyder Stewart 1975; USFWS 1994.

**SHARP-SHINNED HAWK,** *Accipiter striatus velox:* Common. Secretive in summer but conspicuous in migration and winter. Very local in summer south of Canadian boreal forests and east of Rocky Mts. Winters south to Costa Rica.

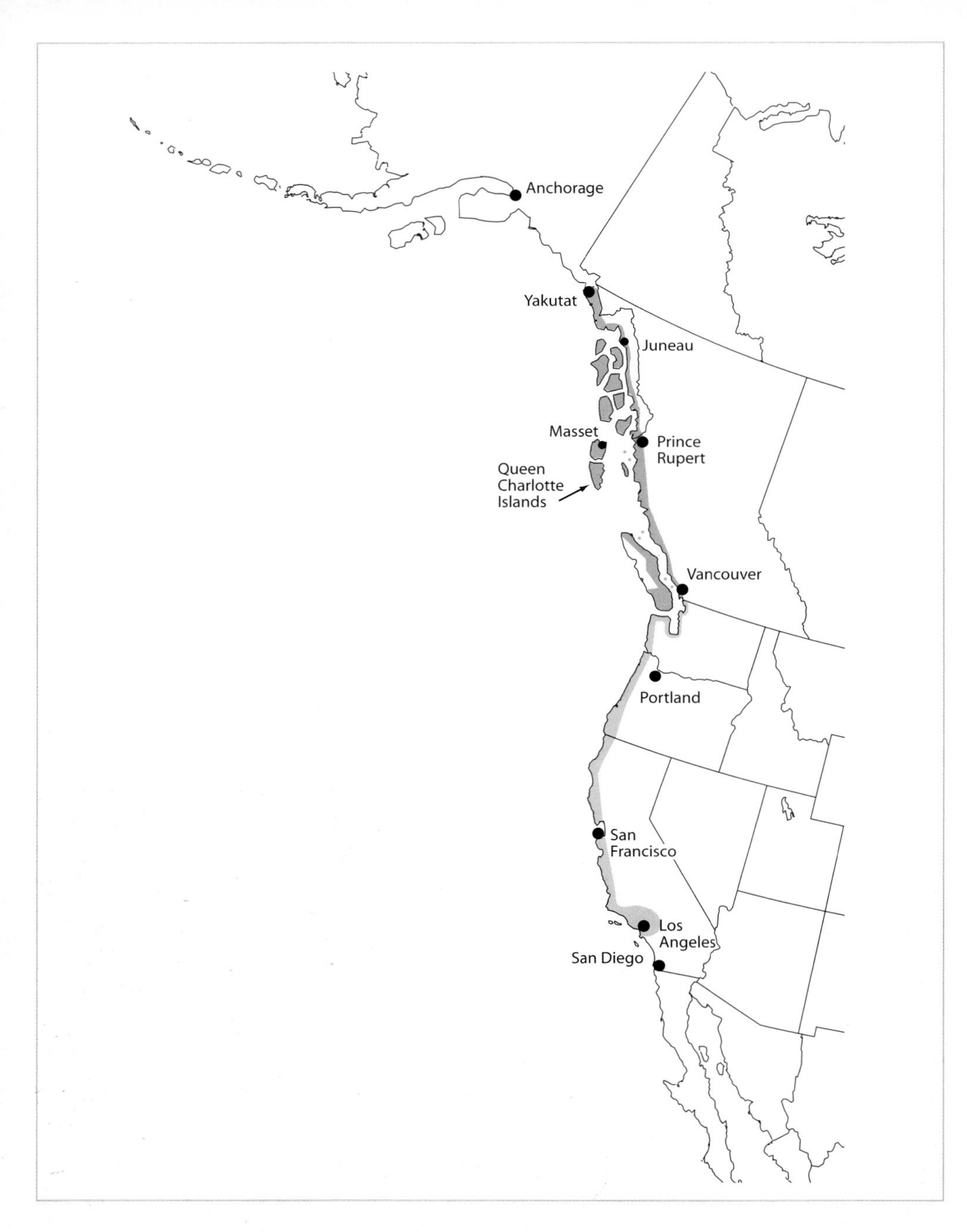


**"QUEEN CHARLOTTE" SHARP-SHINNED HAWK,** *Accipiter striatus perobscurus:* Primarily on Queen Charlotte Islands. Intermixes on islands and coastal mainland with *A. s. velox*. A few adult females and juveniles winter south of breeding areas.

**"SUTTON'S" SHARP-SHINNED HAWK,** *Accipiter striatus suttoni:* Status and seasonal occurrence poorly known. Nested in AZ in 1971 and 1993. Nesting in TX in 2000 and 2001 may have been this race. Possibly breeds in sw. NM. Primary range is in the mountains of Mexico.

**Plate 152. Sharp-shinned Hawk, adult male [Sep.]** ▪ Medium grayish blue crown and nape. Black area encircling eyes. Rufous auriculars. Red iris. ▪ Medium grayish blue back and scapulars. Rufous-barred white underparts. Long, spindly tarsi and toes. ▪ Medium bluish gray wing coverts.

**Plate 153. Sharp-shinned Hawk, adult male [Sep.]** ▪ Medium grayish blue crown and nape. Black area encircling eyes. Rufous auriculars. Reddish orange iris. ▪ Medium grayish blue upperparts. Long, spindly tarsi and toes. ▪ Distinct white terminal band.

**Plate 154. Sharp-shinned Hawk, adult/subadult male [Sep.]** ▪ Medium grayish blue crown and nape. Black area encircling eyes. Rufous auriculars. Orange iris. ▪ Medium grayish blue upperparts. Long, spindly tarsi and toes. ▪ Long, grayish blue tail has 3 or 4 equal-width black bands; distinct white terminal band. ▪ *Note:* Subadults and some younger adults have orange irises.

**Plate 155. Sharp-shinned Hawk, adult male [Sep.]** ▪ Grayish blue crown and nape; rufous auriculars. ▪ Grayish blue upperparts. ▪ Long, grayish blue tail has 3 or 4 equal-width black bands; square-tipped tail; distinct white terminal band.

**Plate 156. Sharp-shinned Hawk, adult female [Sep.]** ▪ Medium gray crown and nape, often with thin whitish supercilium; black encircling eyes; rufous auriculars. Reddish orange iris. ▪ Medium gray upperparts. Rufous-barred white underparts, including leg feathers. Long, spindly tarsi and toes.

**Plate 157. Sharp-shinned Hawk, adult female (blue-backed type) [Sep.]** ▪ Medium bluish gray crown and nape, often with thin whitish supercilium; black encircling eyes; rufous auriculars. Orangish red iris. ▪ Medium bluish gray back and scapulars. Long, spindly tarsi and toes. ▪ Medium bluish gray wing coverts. ▪ *Note:* Dorsal color on some females nearly as bluish on males.

**Plate 158. Sharp-shinned Hawk, subadult female [Sep.]** ▪ Medium gray crown and nape. Black encircling eyes; rufous auriculars. Orange iris. ▪ Medium gray back and scapulars. ▪ A few retained brown juvenile feathers on upperwing coverts. ▪ *Note:* Females often retain some brown juvenile feathering on dorsal areas.

**Plate 159. Sharp-shinned Hawk, adult female (blue backed type) [Sep.]** ▪ Medium bluish gray crown and nape; rufous auriculars. ▪ Medium bluish gray upperparts and upperwing coverts. ▪ Very rounded type tail tip on some females has outer rectrices much shorter than inner feathers. Distinct white terminal band. ▪ *Note:* Tail tip is rounded as on Cooper's Hawk.

**Plate 160. Sharp-shinned Hawk, adult female [Sep.]** ▪ Rufous-barred white underparts. ▪ Remiges always distinctly barred with black. Tawny underwing coverts spotted and barred with brown or black. Six "fingers" on outer primaries. ▪ Square-type tail tip on some females (and many males) has rectrices that are same length.

**Plate 161. Sharp-shinned Hawk, adult female [Sep.]** ▪ Rufous-barred white underparts. ▪ Remiges always distinctly barred with black. Tawny underwing coverts spotted and barred with brown or black. Six "fingers" on outer primaries. ▪ Square-type tail tip on some females (and many males) has rectrices that are same length.

**Plate 162. Sharp-shinned Hawk, adult female "Sutton's" (*A. s. suttoni*) [Jul.]** ▪ Medium gray crown and nape. Grayish auriculars (more so than *A. s. velox*). ▪ Medium gray upperparts. Solid rufous breast, flanks, and leg feathers (unique to this race). ▪ *Note:* Huachuca Mts., Cochise Co., Ariz., in 1971 (first U.S. nesting record). A Mexican subspecies. Photograph by Helen Snyder.

**Plate 163. Sharp-shinned Hawk, juvenile female [Sep.]** ▪ Medium yellow iris. Brown head and auriculars with narrow whitish supercilium. ▪ Dark brown upperparts often uniformly colored. Long, thin, spindly tarsi. ▪ Indistinct black barring on secondaries. Coverts often uniformly brown. ▪ Long, gray tail has 3 or 4 equal-width black bands; thin pale terminal band.

**Plate 164. Sharp-shinned Hawk, juvenile male [Sep.]** ▪ Medium yellow iris. Brown head and auriculars with narrow white supercilium. ▪ Dark brown upperparts often narrowly edged with tawny-rufous. Long, thin, spindly tarsi. ▪ Coverts often edged with tawny-rufous. ▪ Long, gray tail has 3 or 4 equal-width black bands; thin pale terminal band.

**Plate 165. Sharp-shinned Hawk, juvenile female [Sep.]** ▪ Medium yellow iris. Brown head and auriculars with narrow white supercilium. ▪ Moderately streaked type underparts have moderately wide rufous-brown streaks. Long, thin, spindly tarsi. ▪ Rounded-type tail tip on both sexes with outer rectrices somewhat shorter than inner rectrices.

**Plate 166. Sharp-shinned Hawk, juvenile female (lightly marked type) [Oct.]** ▪ Medium yellow iris. Pale head with distinct white supercilium. ▪ Narrowly streaked type underparts have thin rufous-brown or dark brown streaks. ▪ Rounded-type tail tip with outer rectrices somewhat shorter than inner rectrices. ▪ *Note:* Palest extreme. Photographed Cape May, N.J.

**Plate 167. Sharp-shinned Hawk, juvenile male (heavily marked type) [Oct.]** ▪ Medium yellow iris. Uniformly dark brown head. ▪ Uniformly dark brown upperparts. ▪ Narrow grayish terminal band. ▪ *Note:* Darkest extreme of *A. s. velox.* Identical to juvenile male *A. s. perobscurus.* Photographed Cape May, N.J.

**Plate 168. Sharp-shinned Hawk, juvenile female (springtime) [May]** ▪ Iris changes from medium yellow to orange by spring. ▪ Dorsal color fades to medium brown.

**Plate 169. Sharp-shinned Hawk juvenile male [Oct.]** ▪ Moderately streaked type underparts uniformly marked with moderately wide brown/rufous-brown streaks. ▪ Remiges barred with black; males have wide black or gray band on rear edge of wings. Six "fingers" on outer primaries. Tawny underwing coverts. ▪ Square-type tail tip has rectrices that are same length. Long tail.

**Plate 170. Sharp-shinned Hawk, juvenile female** **[Oct.]** ▪ Moderately streaked type underparts uniformly marked with moderately wide rufous-brown streaks. ▪ Barred remiges; females typically have wide gray band on rear edge of wings. Six "fingers" on outer primaries. ▪ Round-type tail tip with outer rectrices somewhat shorter than inner rectrices. Long tail.

**Plate 171. Sharp-shinned Hawk, juvenile female** **[Oct].** ▪ Broadly streaked type underparts are marked with wide rufous-brown streaks. ▪ Barred remiges; females typically have wide gray band on rear edge of wings. Six "fingers" on outer primaries. Rufous-barred axillaries. ▪ Very rounded type tail tip with outer rectices much shorter than inner rectrices. Tail often wedge-shaped.

**Plate 172. Sharp-shinned Hawk, juvenile female** **[Oct.]** ▪ Moderately streaked type underparts uniformly marked with moderately wide rufous-brown streaks. ▪ Barred remiges; females typically have wide gray band on rear edge of wings. Six "fingers" on outer primaries. ▪ Square-type tail tip with all rectrices the same length. Long tail.

**Plate 173. Sharp-shinned Hawk, juvenile male** **[Oct.]** ▪ Narrowly streaked type underparts are narrowly streaked with brown/rufous-brown; often less marked on lower belly. ▪ Wings angled forward of perpendicular line to body when soaring. Barred remiges; males have wide black or gray band on rear edge of wings. ▪ Square-type tail tip with all rectrices same length.

# COOPER'S HAWK
## *(Accipiter cooperii)*

**AGES:** Adult, basic I (subadult I), and juvenile. Adult plumage is acquired when 2 years old. Adults are sexually dimorphic on the head and dorsal color and are separable in the field.

Basic I (subadult I) plumage is acquired when 1 year old. It is similar to respective adult sex but has visible color differences on the auriculars and nape. Juvenile feathering may be retained on the upperwing coverts and rump. The rump is not readily visible under field conditions. *Note:* Head and nape characters of adult and subadult sexes are based primarily on studies by J. Liguori (unpubl. data).

Juvenile plumage is retained the first year with minor individual variation but no sexual variation. Juveniles have slightly longer tails than adults and subadults.

**MOLT:** Accipitridae wing and tail molt pattern (*see* chapter 4). First prebasic molt from juvenile to subadult is often an incomplete molt that occurs from mid-Apr. through Nov. Juvenile feathering may be retained on the upperwing coverts and rump until the next prebasic molt. Outermost primary (p10), usually a secondary (s4), and some rectrice sets (mainly r2 and 5) are the last feathers to molt. A fairly complete molt from subadult to adult and subsequent annual adult prebasic molts occur May–Nov. Molt begins later on males. Some adults may retain a few old worn adult feathers on the upperwing coverts, rump, uppertail coverts, outer primaries (p9 and 10), secondaries (s4 and 8), and occasionally (r2 and 5).

**SUBSPECIES:** Monotypic. However, the population from the Pacific Coast to the Rocky Mts. was formerly considered a separate subspecies, *A. c. mexicanus,* based on smaller size and more intense color (Whaley and White 1994). Many western individuals have visible plumage and size differences in both adult and juvenile ages from those in the East.

**COLOR MORPHS:** None.

**SIZE:** Medium-sized raptor and mid-sized of the three accipiters. Data on geographic size variation are based on Whaley and White (1994). Sexually dimorphic in most breeding regions, with little or no overlap, and separable in the field. Males are smaller than females. However, in s. Arizona overlap has occasionally been noted in wing chord (and possibly wingspan and length) between the largest males and smallest females. There is a marked geographic cline in sizes in the West. Smallest individuals inhabit the Pacific Northwest, particularly sw. British Columbia. Sizes clinally increase south and east from the Pacific Northwest. Cooper's Hawks of the Southwest, mainly from s. California to s. New Mexico, are considerably larger than birds of more northern latitudes of the West. Southwestern birds are similar in size to those inhabiting the eastern portions of the West on the Great Plains and to most populations in the East. Large males from the Southwest may be comparable in size or larger than Pacific Northwest females. Sizes listed are for populations of large individuals; Pacific Northwest individuals are smaller. MALE.—Length: 14–16 in. (36–41 cm); wingspan: 28–30 in. (71–76 cm). FEMALE.—Length: 16–19 in. (41–48 cm); wingspan: 31–34 in. (79–86 cm).

**SPECIES TRAITS:** HEAD.—Long hackles can be elevated to produce a large, square-headed appearance. Gray supraorbital ridge skin, if exposed, is visible only at close range. BODY.—**Moderately long, thick tarsi and toes are about pencil thickness.** WINGS.—**In flight, short and broad with very rounded wingtips.** Wings are proportionately shorter and more round-tipped than in other accipiters. Accipiters have six "fingers" on the outermost primaries. Wingtips extend halfway down the tail on perched birds. TAIL.—**Very long and proportionately the longest of the accipiters. Central rectrix set is longest with each outer set sequentially shorter than the previous set. Sequential feather length distinction is visible only on the underside of the closed tail. Tail may appear somewhat square-shaped when closed and quite rounded when fanned. Each rectrix tip is very rounded with a broad, white terminal band.** White terminal band on the two deck rectrices often wears off by spring. Uppertail has three or four neatly formed, equal-width black bands. On the ventral side, the outermost rectrix set has narrow black bands that do not align with the equal-width bands.

**ADULT TRAITS:** HEAD.—Iris color varies from orange to red and generally darkens with age. Color change advances more quickly and gets darker in males. Eye color cannot be used for aging criteria because the rate of change varies individually and sexually. Eye color data are based on Rosenfield and Bielefeldt (1997). **Dark gray or black crown forms a cap that varies from being marginally darker to much darker than the upperparts. Auriculars and nape region are always paler than rest of the upperparts and form a distinct separation between the cap and upperparts (pale nape is visible only when birds are perched).** BODY.—**Medium grayish or bluish upperparts.** White breast, belly, flanks, and leg feathers are barred with rufous. Some individuals from far western regions have broad rufous barring on the underparts and virtually solid rufous leg feathers with white only on the tip of each feather. WINGS.—White axillaries are barred with rufous. White underwing coverts are distinctly spotted and barred with rufous. Undersides of the remiges are prominently barred on the outer primaries and variably barred on the inner primaries and secondaries.

**Adult male:** HEAD.—Iris color varies from orange to dark red and may take 4–7 years to darken to the deepest red. **Auriculars and nape are uniformly pale gray; however, they sometimes have a slight pale rufous tinge (but never as rufous as on subadult males). Black cap is typically much darker than dorsum.** BODY.—**Back, scapulars, rump, and uppertail coverts are medium grayish blue and remain this color in all seasons.** WINGS.—**Medium grayish blue upperwing coverts. Undersides of the outer six or so primaries are strongly barred, but the four inner primaries and secondaries are pale gray with a minimal amount of dark barring** (F. Nicoletti unpubl. data).

**Adult female:** HEAD.—Iris color varies in being orange, orangish red, or medium red. **Rear of nape is always pale gray, but the auriculars and sides of nape are variably colored: (1) pale rufous, (2) pale rufous with grayish tinge,** or **(3) pale gray (uncommon variation).** Some have a pale grayish forehead and short supercilium lines: if present, they reduce prominence of the darker cap. **Cap is either dark gray and only marginally darker than the upperparts or much darker blackish.** BODY.—**Medium gray back, scapulars, rump, and uppertail coverts; may retain this color all seasons.** WINGS.—**Medium gray upperwing coverts.**

**Adult female (brown-backed type):** HEAD:—**Head features as on typical adult females except cap is dark brown.** BODY and WINGS.—**Upperparts fade and wear to medium brown during a span from mid-winter to mid-summer.** Medium gray plumage is attained in summer through fall molt; however, dorsal plumage may still be more brownish than on typical adult females. TAIL.—Pale dorsal bands are pale brown. *Note:* Uncommon fade and wear pattern.

**BASIC I (SUBADULT I) TRAITS:** Identical in most aspects to respective sex of adult. Separable from respective adult only at fairly close range. HEAD.—Separable from respective adult by auricular and nape color. **As in adults, pale auriculars and nape are distinct. Cap may be slightly darker or much darker than upperparts.** Regularly has a pale rufous forehead and supercilium; if present, reduces prominence of the darker cap. Iris color varies and is either yellowish orange or orange. BODY.—A large patch or scatter of juvenile feathers may be retained on the rump. The rump is rarely visible in the field except at close-range when viewed from above. WINGS.—May retain scattered brown juvenile feathering on the upperwing coverts. Undersides of all remiges are distinctly barred on both sexes.

**Subadult male:** HEAD.—Iris color is mainly orange but may be yellowish orange. **Auriculars are pale rufous, but nape color is variable: (1) pale gray on all of nape, (2) pale rufous sides of nape and pale gray on rear of nape,** or **(3) all-rufous nape (and auriculars; like subadult female).** BODY.—**Medium grayish blue back, scapulars, rump, and uppertail coverts and retains this color in all seasons.** WINGS.—**Medium grayish blue upperwing coverts except any retained brown juvenile feathers.**

**Subadult female:** HEAD:—Iris color is orangish yellow in fall and turns orange by spring. **Uniformly pale rufous auriculars and nape (with no variations).** BODY.—**Medium**

**gray back, scapulars, rump, and uppertail coverts and may retain this color in all seasons.** WINGS.—**Medium gray upperwing coverts except any retained brown juvenile feathers.**
**Subadult female (brown-backed type):** HEAD.—**As on typical subadult females but cap is dark brown.** BODY and WINGS.—As on brown-backed type adult females. TAIL.—Dorsal bands are pale brown.
**JUVENILE TRAITS:** HEAD.—**Iris color is (1) pale gray, (2) pale grayish green,** or **(3) pale lemon yellow, from fledging until early winter; rarely medium orangish yellow.** Tawny or brown head and often appears hooded. Some birds have a short, pale tawny or white supercilium. BODY (ventral).—Variably streaked underparts are either (1) all white or (2) have a tawny wash on the breast and rest of the ventral regions are white. Ventral markings: (1) breast is adorned with narrow brown streaking, then tapers to very narrow streaking on the belly and extremely narrow streaking or a lack of streaking on the lower belly; (2) on the palest individuals, breast is narrowly streaked and belly and lower belly are unmarked. Flanks are (1) narrowly streaked, (2) have narrow diamond-shaped markings, or (3) may have a partial dark brown bar inside the streak or diamond-shaped mark on the inner portion of the longest feathers. Dark brown markings on the leg feathers are (1) narrowly streaked, (2) diamond-shaped, or (3) partially barred. Unmarked white undertail coverts. Overall effect is a very pale-bellied appearance. White undertail coverts. *Note:* Undertail coverts of some individuals west of the Rocky Mts. have a field-visible, narrow, dark brown streak along the shaft area of each feather, particularly the distal feathers. This has been seen on several specimens of both sexes from Arizona, California, and British Columbia (AMNH); undoubtedly, also on individuals in Oregon and Washington, and possibly sparingly other states. This is an uncommon plumage trait. BODY (dorsal).—Dark brown with a tawny edge on each feather. Most scapular have a large white basal area that creates a blotched look on the dorsal region, especially when the feathers are fluffed in cold temperatures and when rousing. WINGS.—Upperwing coverts, including greater coverts, are dark brown and edged with tawny. Undersides of all remiges are prominently barred. *Both sexes have a sharply defined, crisp dark gray or black band on the trailing edge of the remiges of underwing.* Axillaries are barred with dark brown. TAIL.—On the dorsal side, brown with three to four equal-width dark bands; rarely, has narrow white linear borders along some dark bands.
**JUVENILE (HEAVILY STREAKED TYPE):** Head, upperparts, wings, and tail as on typical juveniles. BODY.—Underparts are much more densely and heavily streaked than on typical juveniles. Breast, belly, and lower belly can be uniformly marked with dense brown streaking, or lower belly may have somewhat fewer markings. Markings are often heart- or diamond-shaped on the belly, lower belly, and flanks. Flanks typically have a discernible dark brown bar on the inner portion of each feather. Leg feathers are densely covered with large (1) heart-shaped, (2) diamond-shaped, or (3) thick bars and create a rather dark-legged appearance. Undertail coverts may be unmarked white but are likely to have field-visible streaking, especially on the distal feathers. *Note:* Not depicted.
**JUVENILE (RUFOUS STREAKED TYPE):** HEAD.—As on typical juveniles. BODY.—Ventral areas are streaked with rufous-brown and are similar to juvenile Sharp-shinned Hawk. *Note:* Rufous-brown streaking is documented only on a heavily streaked type male specimen from California (J. Schmitt unpubl. data/photos).
**JUVENILE (LATE WINTER/SPRING; ALL TYPES):** HEAD.—Iris color changes to orangish yellow and may remain this color into subadult age class on some females. BODY and WINGS (dorsal).—Pale feather edgings wear off and upperparts fade and wear to a uniform medium brown.
**ABNORMAL PLUMAGES: Atypically barred subadult.**—Upper breast is streaked with rufous, and lower breast and belly have rufous heart-shaped markings; flanks are broadly barred with rufous. Head, dorsal plumage, and tail features are normal. *Note:* Uncommon plumage variation for both subadult sexes (BKW pers. obs.). Not depicted. **Unbarred adult.**—Underparts are white with only a hint of faint, partial rufous barring on the upper breast. White, unmarked underwing coverts. All other plumage features, including dorsal areas, are

normal (BKW pers. obs.). *Note:* Rare plumage type. Female specimen AMNH (from New York). **Imperfect albino.**—Primarily seen in juveniles. Iris, bill, and foot can be normal, or, some or all of these areas may be paler. Iris may be pale blue. Head and neck are pale rufous or whitish. White upperparts have pale rufous edges on many feathers. Remiges and retrices are white and lack dark banding. White underparts may have faint rufous streaking (Clark and Wheeler 1987). *Note:* Very rare plumage type and not depicted.

**HABITAT: Summer.**—Dry upland regions and moist lowland or upland riparian areas with moderately old coniferous or mixed coniferous-deciduous trees (prefers larger trees than Sharp-shinned Hawks but smaller trees than Northern Goshawks). Large and small woodlands are typically inhabited in the more wooded, coniferous regions of the West. Moderately old coniferous plantations are also inhabited in the southeastern areas of the West. In virtually all areas, there are nearby semi-open expanses and a quiet water source. In the n. Great Plains and Southwest, narrow riparian strips with moderate-age and old deciduous trees and deciduous shelterbelt groves are readily used. In Texas, Arizona, California, British Columbia, and North Dakota, Cooper's Hawks are found in suburban and urban settings, including city parks, campuses, business parks, cemeteries, and residential areas—including those with actively used yards. Densely foliaged deciduous and particularly coniferous trees are integral parts of suburban and urban nesting territories. Most nesting locales have densely foliaged canopies and understories; however, many suburban and urban nesting areas often lack understory growth. In Mexico, higher elevation woodlands and major tree-lined riparian areas are used. Habitats encompass humid to arid climates from sea level to near timberline.

**Winter.**—Can be identical to summer habitat, but usually at lower elevations and at more southern latitudes and without regard to a particular tree species. Regularly found in semi-open and open areas with minimal tree growth, especially desert regions with low brush or scrub vegetation, provided elevated perches are available. Inhabits rural, suburban, and urban locales with moderate tree growth.

**Migration.**—Found in the above habitats. Common at high elevations.

**HABITS:** Moderately tame, but urban individuals can be quite tame. Fairly hyper temperament. Long hackles are often elevated, particularly when threatened or in cool temperatures, giving the large, square-headed look. Concealed and exposed perches are used. Fairly reclusive, but the most likely accipiter to perch in the open on fence posts and wires. *Cooper's Hawk is the only accipiter to perch on telephone poles.* Solitary except when breeding.

**FEEDING:** Perch and aerial hunter. Perch hunting consists of short or moderate-length stints of perching. Commonly perches at low heights in dense foliage or branches in order to surprise potential prey. Aerial hunting forays are random excursions. A variety of small to medium-sized songbirds and game birds and small to medium-sized rodents form diet. In deserts and other arid regions, lizards are also taken. Hunts in open sections of forest understories, along edges of woodlands, clearings, and open expanses. Prey is readily pursued on foot into dense vegetation. Plucking posts on larger branches, logs, or stumps are used in nesting territories. In all other seasons, prey is often consumed at the point of capture, including on the ground. Avian prey is plucked prior to being eaten.

**FLIGHT:** Powered flight is a regular cadence of several fast wingbeats interspersed with short glides. Glides with wings held on a flat plane or bowed slightly downward. Soars readily, with wings held on flat plane or in a low dihedral with the front edges held rigidly perpendicular to the body. Several soaring revolutions may be completed without resorting to powered flight. Head appears to project far beyond front of wings when gliding or soaring (compared to similar Sharp-shinned Hawk)—an optical illusion created by the perpendicularly held wings. With their short, broad, rounded wings and long tail, Cooper's Hawks navigate through densely wooded vegetation with incredible agility. Most hunting is at low to moderate altitudes. In open areas, they may skim the ground or low vegetation at very low altitudes and may dive from high altitudes. Migratory flights are at high altitudes in light winds and at treetop or ground level in strong winds. Cooper's Hawks hover and kite only for short stints in strong winds during antagonistic encounters and when searching for hidden prey.

**VOICE:** Very nasal *kek-kek-kek-kek* when alarmed or agitated. A single, nasal *kek* by males delivering food to females or when approaching the nest. When curious or mildly agitated, a soft, single, nasal *kyew*. The nasal quality of all vocalizations is readily separable from the calls of Sharp-shinned Hawk and Northern Goshawk.

**STATUS AND DISTRIBUTION:** Overall *uncommon* to *fairly common* in West. Estimated population unknown. Numbers are probably stable. Undoubtedly affected during the organochlorine pesticide era in North America.

**Summer.**—A highly adaptable species that is quickly acclimating—and thriving—in human-altered environments. Becoming increasingly common in towns and cities in various regions and truly becoming an "urban raptor" (e.g., San Antonio, Tex.; Tucson, Ariz.; Los Angeles metro and San Francisco Bay area, Calif.; Victoria, B.C.; and several cities in North Dakota). Urban nesting Cooper's are regionally adapting to large densities of House Sparrows, Starlings, Common Grackles, and various dove species.

Absent as a breeder in much of the cen. and s. Great Plains. A common breeder on the n. Plains, particularly in North Dakota. Has not yet adapted to riparian and urban regions on the cen. and s. Plains as it has in the Southwest and along the Pacific Coast. However, two adult males were seen in the summer of 2000 in Sedgewick and Yuma Cos, Colo. in suitable riparian habitat, but breeding was not confirmed. In Nebraska, recent breeding has occurred in Platte River S.P., Cass Co.; Lake Werhspann, Douglas Co.; Lancaster Co.; Schramm Pk., Sarpy Co.; and Saunders Co. Omaha, Neb., had an urban nesting attempt in a city park in 1997. Recent breeding in the western Plains region is also seen in Frontier, Scott's Bluff (Carter Canyon), and Thomas Cos. Although no documentation, possible summering or breeding may occur in Ponderosa Pine tracts and riparian wooded areas in Cherry, Dawes, and Sioux Cos.

In Mexico (mapped area), a fairly common to uncommon breeder in n. Baja, California, w. Chihuahua, Durango, and n.-cen. and e. Sonora.

**Winter.**—Southern-latitude birds are generally fairly sedentary. Northern-latitude birds typically winter farther south and at lower elevations than breeding and natal areas. Northern individuals may winter at the same latitudes as southern resident populations or leapfrog them and winter farther south. Northern birds are often nomadic, but some may establish territories in areas with abundant and stable prey. Many haunt areas with birdfeeders in rural and urban locales. Juveniles, especially females, winter farther south than other ages. Large numbers of juveniles winter in Mexico. Winter range may extend to Costa Rica and, rarely, Colombia.

**Movements.**—Short- and moderate-distance migrant. Some Pacific Coast and s. U.S. populations are sedentary. Cooper's Hawks become quite conspicuous during migration, especially in regions with geographic barriers such as mountain ridges and shorelines.

*Fall migration:* Mid-Aug. to late Oct. Juveniles precede adults, and females precede males of both ages. Juveniles peak in mid- to late Sep., adults in late Sep. to early Oct.

*Spring migration:* Mid-Mar. to mid-May. Adults precede juveniles and peak in early to mid-Apr. Males, who establish breeding territories, precede females. Juveniles peak in late Apr.

At hawkwatch sites in the w. Rocky Mts. and Great Basin, Cooper's Hawk numbers almost rival those of Sharp-shinned Hawk. Along coastal California, however, Cooper's Hawks are considerably less common than Sharp-shinned Hawks during migration.

**NESTING:** Mar.–Jun. in southern latitudes and Apr.–Jul. in northern latitudes. One-year-old females in juvenile plumage often nest with adult males. Rarely, two juvenile-aged birds pair and nest. Nesting pairs are quite secretive.

**Courtship (flight).**—*Sky-dancing, slow-flapping,* and *high-circling* (*see* chapter 5). Undertail coverts are flared and retracted outward beyond the closed tail in the last two courtship flights.

In wooded areas, nest trees are located among similar-aged trees. In suburban and urban areas, they may also be in same-age trees but may be in taller dominant trees. Most nesting areas have trees with dense canopies and understories and are near some sort of a clearing and a quiet water source. Some suburban and urban nests have little or no ground foliage surrounding nest sites. Nests are well hidden and typically placed in the very upper portion of a tree. Suburban and urban nests are often located in conifers, mainly pines, for added secrecy.

A new nest is often built each year in a different tree in the same territory, but pairs may reuse an old nest for up to 3 years. Males often do the bulk of the nest-building. Nests are 24–28 in. (61–71 cm) in diameter and may become quite deep and bulky with reuse. Stick nests are lined with bark chips and some greenery. Leafy squirrel nests or mistletoe are often used as platforms. Nests are usually 20–60 ft. (6–18 m) high; up to 135 ft. (41 m) high in tall conifers in the Pacific Northwest. In deciduous trees, nests are placed primarily in main crotches but can be in minor, outer crotches and sometimes lower-canopy horizontal branches. In conifers, nests are placed next to trunks.

Normally 4 (white) eggs, but up to 7 are laid. The eggs are incubated for 34–36 days. Youngsters fledge in 30–34 days and are independent in about 56 days. Female tends to nesting duties; male hunts, but incubates while female feeds. Male rarely delivers prey to the nest: it is transferred to female at a point away from the nest and the female feeds the nestlings. Extra food is cached on branches of nearby trees for later use. Pairs are occasionally aggressive toward humans intruding into nesting territories.

**CONSERVATION:** Though uncommon, this highly adaptable species seems to be thriving in our constantly changing, human-altered environment. Being a predator mainly of birds, it has benefitted from the ban on organochlorine pesticide use, particularly DDT, in the last few decades.

**Mortality.**—Carbamate pesticides may have contributed to some mortality. Newer generation pyrethroid pesticides are harmless to birds and mammals (but affect aquatic life). Illegal shooting occurs in North America and is prevalent south of the U.S. in winter. Many birds fly into windows. Trichomoniasis, caused by eating doves, affects fledgling mortality in Tucson, Ariz. Pesticide contamination of prey species that winter in South America may still affect Cooper's Hawks.

**SIMILAR SPECIES: (1) Sharp-shinned Hawk, females.**—A long-standing identification dilemma for adults and juveniles when compared to male Cooper's Hawks. There is no overlap in size between species. Cooper's Hawks breeding north of the Southwest, especially in the Pacific Northwest, are small. However, large female Sharp-shinned Hawks are still shorter in length and have a shorter wingspan than male Cooper's Hawks. Numerous differences separate the two species. Proportional differences can separate them even when actual markings are not seen. PERCHED.—Small, round-shaped head at all times; however, in warm temperatures Cooper's Hawks regularly compress head feathers so head also appears quite small. Supraorbital ridge skin, if visible, is yellow. Tarsi are long and thin. Rectrix length on underside of tail helpful; however, a very round-tipped type tail on female Sharp-shinned Hawks is identical to that of Cooper's Hawks. Voice is very different from Cooper's Hawk, with soft *chirp* and high-pitched *kee* notes. *Note:* The only real confusion when perched should be with subadult male Cooper's Hawks, but their nape is always darker rufous/gray. FLIGHT.—When soaring, front edge of wings is held forward of perpendicular angle to body. Tail shape as in Perched. **(1A) Adult females.**—Bluish/grayish crown of head and nape are the same color as rest of dorsal areas and always darker than rufous auriculars. Females are very gray on dorsal areas, not bluish as on male Cooper's Hawks. Less distinct rufous markings on underwing coverts. In flight, use wing attitude and proportions. **(1B) Juvenile females.**—Irises are medium orangish yellow or yellow ochre in fall; orange in late winter/spring. Females generally have rufous-streaked underparts; a few have narrow dark brown streaking. Beware of lightly marked type, they rival lightly streaked underparts of male Cooper's Hawks. Underwing in flight has grayish band on trailing edge, not crisp, dark band of Cooper's Hawks. Axillaries are typically barred with rufous (dark brown on Cooper's Hawk). COMPARED TO ADULT.—
**(2) Hook-billed Kite, adult females.**—South Texas only. Size and coloration similar to adult female Cooper's Hawks. PERCHED.—Dark crown, rufous nape, dorsal color similar. White

irises; large bill; short tarsi. FLIGHT.—Rufous on inner primaries. COMPARED TO JUVENILE.—**(3) Northern Goshawk, juveniles.**—PERCHED.—Irises are a brighter yellowish orange, except in recently fledged birds, in which they are pale grayish: use caution in mid- to late summer. Short, thick tarsi. Greater upperwing coverts have a pale tawny bar. Tail shape can be similar with sequentially shorter outer rectrices: each rectrix is more pointed on goshawk. Tail of both species has a broad white terminal band in fall. FLIGHT.—Trailing edge of remiges on underwing has similar dark band. Use caution with lightly marked type: underparts on these pale individuals are as narrowly streaked as on Cooper's Hawks. Many have little or no streaking on the undertail coverts (many western Cooper's Hawks have prominent streaking on undertail coverts). Tail as in Perched. **(4) Gray Hawk, juveniles.**—S. Arizona and Texas only. PERCHED and FLIGHT.—Dark brown irises. Dark eyeline and malar stripe. Size and flight mannerism are deceptively similar; wings held very perpendicularly to body. Undersides of all remiges are narrowly barred; trailing edge has a grayish band. Numerous variable-width dark tail bands. Long, thick tarsi. **(5) Red-shouldered Hawk, juveniles.**—PERCHED.—Medium brownish irises. Dark malar stripe. Wingtips extend to near tail tip when perched. Long tarsi. FLIGHT.—Tawny or white panel or window on upper and lower primaries. *B. l. elegans* and some *alleni* may have fairly distinct barring on underside of remiges: use caution. Tail has narrow gray or white bands. **(6) Broad-winged Hawk, juveniles.**—PERCHED.—Medium brownish irises. Dark malar stripe. Wingtips extend to near tail tip. Short, thick tarsi. FLIGHT.—Pointed wingtips. Undersides of remiges have little if any barring. All rectrices are same length on underside of tail. Use caution with wide-banded type tail: very accipiter-like with three nearly equal-width bands, including non-aligning outer set.

**OTHER NAMES:** Cooper's, Coop. *Spanish:* Gavilán Pollero, Gavilán de Cooper. *French:* Épervier de Cooper.

---

REFERENCES: Adamus et al. 2001; Andrews and Righter 1992; Arizona Breeding Bird Atlas 1993–1999; Busby and Zimmerman 2001; Bylan 1998, 1999; Call 1978; Campbell et al. 1990; Clark and Wheeler 2001; Conrads 1997; del Hoyo et al. 1994; Dodge 1988–1997; Dorn and Dorn 1990; Ducey 1988; Garner 1999; Harrison 1979; Herron et al. 1985; Howell and Webb 1995; Jacobs and Wilson 1997; Kellogg 2000; Kent and Dinsmore 1996; Kingery 1998; Montana Bird Distribution Committee 1996; Moore and Henny 1983; Oberholser 1974; Peterson 1995; Reynolds 1983; Reynolds and Meslow 1984; Reynolds and Wight 1978; Rosenfield and Bielefeldt 1993, 1997; Russell and Monson 1998; Semenchuk 1992; Sharpe et al. 2001; Small 1994; Smith et al. 1997; Snyder and Snyder 1991; Stewart 1975; Sutton 1967; Wood and Schnell 1984; Whaley and White 1994.

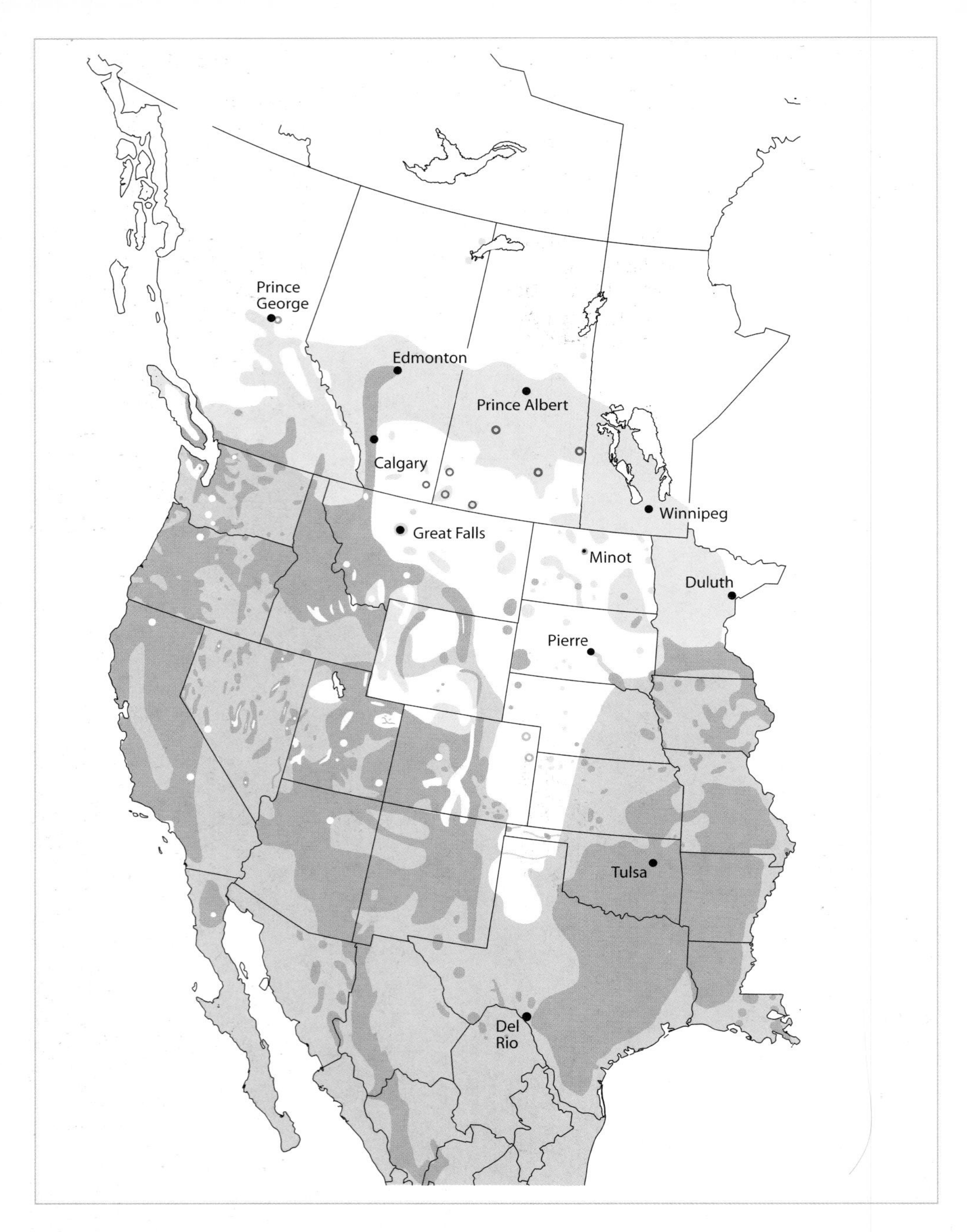


**COOPER'S HAWK,** *Accipiter cooperii:* Uncommon to fairly common. Increasing rapidly as an urban nesting species. Very rare on Great Plains in winter. Winters south to Costa Rica, rarely to Colombia.

**Plate 174. Cooper's Hawk, adult male** [Apr.] ▪ Black crown (cap); pale gray auriculars and nape. Red iris. ▪ Grayish blue dorsal areas. White underparts finely barred with rufous. ▪ Very long tail with much shorter outer rectrix sets.

**Plate 175. Cooper's Hawk, adult male** [Apr.] ▪ Black crown (cap); pale gray auriculars and nape. Red iris. ▪ Grayish blue dorsal areas.

**Plate 176. Cooper's Hawk, adult male** [Sep.] ▪ Black crown (cap); pale gray auriculars and nape. Red iris. ▪ Grayish blue dorsal areas. ▪ Very long tail has 3 or 4 equal-width black bands; broad white tail tip; much shorter outer rectrix sets. ▪ *Note:* Still molting new secondaries; dark feathers are new adult feathers. Photograph by Jerry Liguori.

**Plate 177. Cooper's Hawk, adult male** [Sep.] ▪ Gray auriculars and nape. ▪ White underparts finely barred with rufous. ▪ Primaries distinctly barred, secondaries finely barred and often nearly lack barring. Coverts spotted and barred with rufous. ▪ Very long tail with much shorter outer rectrix sets; broad white tail tip.

**Plate 178. Cooper's Hawk, adult female [Apr.]** ▪ Black crown (cap), pale rufous auriculars, pale gray nape. Red iris. ▪ White underparts finely barred with rufous. ▪ Gray upperwing coverts and back. ▪ Very long tail with much shorter outer rectrix sets ▪ *Note:* Upperparts distinctly more grayish than on adult or subadult males.

**Plate 179. Cooper's Hawk, adult female (brown-backed type) [Apr.]** ▪ Dark brown crown (cap), pale rufous auriculars, pale gray nape. Reddish orange iris. ▪ White underparts finely barred with rufous. Back and scapulars medium brown. ▪ Upperwing coverts medium brown. ▪ *Note:* Uncommon variant for adult and subadult females.

**Plate 180. Cooper's Hawk, adult female [Sep.]** ▪ Black crown (cap), pale rufous auriculars, gray nape. ▪ Gray upperparts. ▪ Very long, pale gray tail has 3 or 4 equal-width black bands. Broad white tail tip; much shorter outer rectrix sets. ▪ *Note:* Photograph by Jerry Liguori.

**Plate 181. Cooper's Hawk, adult female [Oct.]** ▪ Black crown (cap), rufous auriculars, pale gray nape. ▪ White underparts finely barred with rufous. ▪ Remiges fully barred. May have dark gray band on rear edge of remiges. Rufous markings on underwing coverts. ▪ Very long tail with much shorter outer rectrix sets.

**Plate 182. Cooper's Hawk, adult female (left), subadult male (right)** [**Apr.**] Female: Dark gray crown (cap), pale supercilium, gray auriculars and nape. Gray upperparts ▪ Male: Black crown (cap), pale rufous supercilium, rufous auriculars and nape. Grayish blue upperparts. ▪ *Note:* Mated pair. Typical size difference between sexes.

**Plate 183. Cooper's Hawk, subadult male** [Oct.] ▪ Wings held at perpendicular angle to body when soaring and banking. Rufous markings on coverts. Barred remiges. May have dark gray band on rear edge of remiges. ▪ White underparts finely barred with rufous. ▪ Very long tail with much shorter outer rectrix sets.

**Plate 184. Cooper's Hawk, juvenile male** [Oct.] ▪ Pale gray iris. Tawny-brown head; sometimes with short pale supercilium. ▪ Long, moderately thick tarsi and toes. ▪ Very long, pale gray tail with 3 or 4 equal-width black bands and broad white tip. ▪ *Note:* Sexes similar.

**Plate 185. Cooper's Hawk, juvenile male** [Oct.] ▪ Pale gray iris. Tawny-brown head, sometimes with short pale supercilium. ▪ White underparts narrowly streaked with dark brown; lower belly often less heavily streaked. Distal flank feathers may be partially barred. Long, moderately thick tarsi and toes. ▪ Very long tail with much shorter outer rectrix sets. ▪ *Note:* Sexes are similar.

**Plate 186. Cooper's Hawk, juvenile male [Oct.]** ▪ Pale gray iris. Pale tawny-brown head, often with short supercilium. ▪ White underparts narrowly streaked on breast, belly, and flanks and unmarked on lower belly. ▪ Very long tail with much shorter outer rectrix sets. ▪ *Note:* Sexes similar. Type with lightly streaked underparts. Palest variant.

**Plate 187. Cooper's Hawk, juvenile female (springtime) [Apr.]** ▪ Orangish yellow iris. ▪ Upperparts fade to medium brown. ▪ White tail tip often wears off. ▪ *Note:* Iris color is much more yellowish than in juvenile Sharp-shinned Hawk or Northern Goshawk, which have orange irises in spring. Prey is Great-tailed Grackle (photographed in s. Texas).

**Plate 188. Cooper's Hawk, juvenile female [Oct.]** ▪ Tawny-brown head. ▪ Dark brown upperparts with pale tips on some upperwing coverts. ▪ Very long, gray tail with 3 or 4 equal-width black bands; broad white tail tip. Tail rounded when fanned.

**Plate 189. Cooper's Hawk, juvenile male [Oct.]** ▪ Long neck. ▪ Underparts narrowly streaked, often less heavily streaked on lower belly. ▪ Barred remiges with crisp black rear edge (darker, more defined than on juvenile female Sharp-shinned Hawk). Brown-barred axillaries (rufous on Sharp-shinned Hawk). ▪ Very long tail with much shorter outer rectrix sets. Broad white tail tip.

**Plate 190. Cooper's Hawk, juvenile male** [Oct.] ▪ White underparts narrowly streaked, often less heavily streaked on lower belly. ▪ Remiges have crisp black rear edge (darker, more defined than on juvenile female Sharp-shinned Hawk). Wings held at perpendicular angle to body when soaring. Brown-barred axillaries (rufous on Sharp-shinned Hawk). ▪ Very long rounded tail with shorter outer rectrix sets. White tail tip not always distinct.

## NORTHERN GOSHAWK
### *(Accipiter gentilis)*

**AGES:** Adult, basic I (subadult I), and juvenile. Adult and subadult I exhibit moderately sexually dimorphic plumages. Adult plumage is acquired when 2 years old. Subadult plumage is acquired when 1 year old. Juvenile plumage is worn the first year. Juveniles have longer secondaries, which create a broader wing, and have longer tails than older ages.

**MOLT:** Accipitridae wing and tail molt pattern (*see* chapter 4). First prebasic molt from juvenile to subadult occurs mid-Apr. through Nov. This is often an incomplete molt and may retain a variable amount of juvenile feathering. The secondaries often retain old, worn juvenile feathers on s4, s4 and 8, or s7 and 8 and 9. Some males molt fully into new subadult feathering on all the remiges. Almost all subadults have scattered juvenile feathering on the upperwing coverts. The rump tract is partially or totally retained juvenile feathers. The majority of the tail (r2–5) may be retained juvenile rectrices, or some males molt fully into new subadult rectrices.

Subadult to adult molt and subsequent annual adult prebasic molts begin May–Jul., with males beginning later than females, and are finished in Nov. Annual molts are not complete and retain some old adult feathering on the body, secondaries, and tail. Some adults retain a few old juvenile feathers into the second prebasic molt.

**SUBSPECIES:** Two North American races are currently recognized by most taxonomists: *A. g. atricapillus* and *A. g. laingi*. Both are in the West. *A. g. atricapillus* is found throughout mainland U.S. and Canada. It also probably extends southward, at montane elevations, to s. Jalisco and e. Guerrero, Mexico (*see A. g. apache*). *A. g. atricapillus*-like plumaged birds also inhabit virtually all islands of se. Alaska and British Columbia (*see* "Queen Charlotte" race and Size).

The "Queen Charlotte" race, *A. g. laingi*, is an example of Gloger's Rule. Increased melanin in the feather structure produces a darker plumage in all ages. *A. g. laingi* distribution and plumage information are based in part on Taverner (1940) and on specimens from the RBCM ($n = 9$) and AMNH ($n = 2$). Specimens were from the Queen Charlotte Islands (Graham Island) and Vancouver Island (Comox and Port Hardy), B.C. Data are also based on in-hand photographs of live, breeding season *laingi* from se. Alaska ($n = 7$ adults, 1 juvenile) from Prince of Wales Island, Kuiu Island, Kupreanof Island, and Cape Fanshaw.

In British Columbia, consistently dark-plumaged *laingi*-like birds occur year-round on the fairly isolated Queen Charlotte Islands, which perhaps are this subspecies' core range. *A. g. laingi*-like dark-plumaged birds are sparingly found on near-coastal islands and immediate coastal mainland areas of British Columbia west of the Coast Mts. and including Vancouver Island. Dark-plumaged birds are also sparingly found in se. Alaska from Prince of Wales Island north to Chichagof Island and on near-coastal mainland regions from Juneau to Ketchikan.

Individuals inhabiting near-coastal islands in British Columbia and Alaska that are adjacent to the Queen Charlotte Islands have minimal geographic isolation from the mainland. Both *laingi* and *atricapillus* types breed in this region. There is much intergrading, with *atricapillus*-type plumaged birds predominating. Individuals of the more isolated Queen Charlotte Islands, based on museum specimen data, appear to be a consistently dark-plumaged subgroup. *See* Note on "Queen Charlotte" race, Size, and Movements. *A. g. laingi* traits are rarely seen on Northern Goshawks of the Olympic Peninsula, Wash. (*see* Live specimens).

*Note on "Queen Charlotte" race* (Taverner 1940): Based on adults and subadults ($n = 5$) and juveniles ($n = 5$) from the Queen Charlotte Islands. Two specimens were from the breeding season, one in winter, and two did not have collection dates. Taverner noted that there were no *atricapillus* types or individuals with intergrade characters from the Queen Charlotte Islands. All individuals, regardless of season, were dark plumaged. He studied 16 adults and/or subadults and 19 juveniles from Vancouver Island from various seasons. At least seven adults and/or subadults were very *laingi*-like and "many juveniles were said to be intermediate between the two races." Taverner's synopsis also included individuals that could be assigned to either race. A few specimens from se. Alaska (undisclosed number and season) were also said to be paler than individuals of the Queen Charlotte Islands.

*Live specimens (photos supplied by C. Flatten).—Alaska:* 29 ($n = 20$ adults, 9 juveniles) breeding-season Northern Goshawks from se. Alaska were examined. Nesting pairs and fledglings from the following locations: Prince of Wales Island, Kuiu Island, Kupreanof Island, Cape Fanshaw, Chichagof Island, Mitkoff Island, Douglas Island, and Juneau. Only seven adults (35%) and one juvenile (9%) had dark *laingi* plumage traits. A few other adults were possible intergrades; however, the majority were classic pale *atricapillus*. Eight of the nine juveniles from various locations were unquestionably mainland-like pale *atricapillus*. No subadults were in the series. More northerly islands and mainland regions had fewer, if any, dark *laingi*-like birds. Of particular interest is the low percentage of dark *laingi*-plumaged juveniles (juveniles are the easiest age class in which to judge subspecies characters). *Washington:* Although in close proximity to Vancouver Island, B.C., which has some known breeding *laingi*-plumaged birds, nesting birds on the Olympic Peninsula have minimal geographic isolation from other mainland populations and appear to have predominantly *atricapillus* traits. Photographs of 11 live, in-hand, breeding-season individuals (9 adults, 2 juveniles) from various nest sites on the peninsula were studied. From this sample, there was only one adult male from the northern part of the peninsula that had somewhat *laingi*-like head features (reduced supercilium); all other individuals were classic *atricapillus*.

*Museum specimens (RBCM and AMNH).—Breeding season:* Few *laingi*-type plumage breeding-season specimens exist in museum collections from the Queen Charlotte Islands and Vancouver Island. *Queen Charlotte Islands.*—From randomly selected specimens (*n* = 4), one subadult and three juveniles were examined. All were classic dark *laingi. Vancouver Island.*—RBCM total breeding-season sample was small (*n* = 6). One *laingi* adult female (randomly selected); however, five specimens were not available and subspecies could not be determined. *Nonbreeding season: Queen Charlotte Islands.*—Two RBCM nonbreeding individuals were randomly selected. Both were classic dark juvenile *laingi.* Additional nonbreeding specimens are in the collection from the Queen Charlotte Islands and Vancouver Island but were unaccessable at the time. *Vancouver Island.*—Five randomly selected RCBM nonbreeding-season specimens were examined. Four were *laingi* (2 subadults, 2 juveniles [1 subadult and 1 juvenile had intergrade traits]) and one was a typical *atricapillus*.

Birds inhabiting se. Arizona and n. Mexico are assigned by some authors to the race *apache.* Individuals average slightly larger-sized and longer-winged than individuals from n. North America (*see* Size). The subspecies was initially proposed by van Rossem (1938) based on slightly longer wing-chords of three specimens. Snyder and Snyder (1991) and Whaley and White (1994) support this subspecies designation. Whaley and White based their study on measurements of 6 specimens from the breeding season and 17 from the nonbreeding season assigned to this race. *Apache* is not recognized by the AOU and is treated here as part of *atricapillus. Note:* Specimens examined in the AMNH from Jalisco, Mexico and Cochise Co., Ariz. (*n* = 1 adult, 1 juvenile, respectively) and a nestling from Sonora, Mexico, from LSU are identical in plumage to the respective age and sex of *atricapillus. Additional Note:* In North America, only the Bald Eagle and Harris's Hawk are divided into subspecies based solely on size and not on plumage differences (i.e., Harris's Hawk is not accepted by some authorities as having separable races).

There are six additional races of Northern Goshawk: (1) nominate subspecies *A. g. gentilis* of Europe and extreme nw. Africa, (2) *A. g. arrigonii* of Corsica and Sardinia, (3) *A. g. buteoides* of n. Eurasia, (4) *A. g. albidus* is in ne. Siberia and Kamchatka, (5) *A. g. schvedowi* of Asia to s. and cen. China, and (6) *A. g. fujiyamae* of Japan.

**COLOR MORPHS:** None.

**SIZE:** A large raptor and largest of the three accipiters. All data based on Whaley and White (1994). Sexually dimorphic without overlap; however, size difference is not always separable in the field. Proportionately, Northern Goshawk is the least sexually dimorphic of the three accipiters. Males are smaller than females.

Northern Goshawks inhabiting coastal islands of British Columbia and Alaska, which includes dark *laingi* types and, in part, moderately dark *laingi-atricapillus* intergrades and some pale *atricapillus*, are the smallest in North America. Relative to wing chord and body size, the island populations have longer tails. Pacific Northwest island populations average smaller than those on adjacent coastal mainland areas and much smaller than *atricapillus* of mainland North America. There is also a clinal increase in size between island and coastal mainland individuals. Sizes also increase clinally on the mainland north, east, and south of coastal Alaska and British Columbia. Northern Goshawks of the sw. U.S. and n. Mexico average largest in North America. Sizes listed below are average-sized, mainland *atricapillus* (e.g., interior Alaska, Canada, and the U.S. north of s. Arizona). Northern Goshawks inhabiting the islands of se. Alaska and British Columbia are much smaller; those of the Southwest are, at best, marginally larger and not field separable (based on wing chord; total length and wingspan are not available). MALE.—Length: 18–20 in. (46–51 cm); wingspan: 38–41 in. (97–104 cm). FEMALE.—Length: 21–24 in. (53–61 cm); wingspan: 41–45 in. (104–114 cm).

**SPECIES TRAITS:** BODY.—**Fairly short, thick tarsi and toes.** WINGS.—**In flight, moderately short with broad secondaries that taper to narrower, rounded wingtips.** Wingtips can become quite pointed when gliding. Accipiters have six "fingers" on the outermost primaries, which are most visible when soaring. When perched, wingtips extend halfway down the tail. TAIL.—**Long and wedge-shaped. On the underside, each outer rectrix set is sequentially shorter than the preceding inner set. Deck set (r1) is much longer than the outermost set (r6).**

**ADULT TRAITS:** HEAD.—**Reddish irises. Dark gray or black crown, whitish supercilium, and black auriculars.** BODY.—**Gray vermiculated or barred underparts. Bluish gray upperparts. Underwing appears uniformly pale gray with**

**distinct, wide dark barring on the outer primaries and a minimal amount of barring or no barring on the grayish secondaries.** *Note:* Sexes may share similar color and markings of both upperparts and underparts; however, females are more likely than males to have traits of the opposite sex. TAIL.—Rectrices have rounded or square-shaped tips. Bluish gray base color of the uppertail is variably barred.

**ADULTS OF "NORTHERN" (*A. g. atricapillus*):** HEAD.—**Dark gray or black crown and nape, long, broad white supercilium, and black auricular region. The black auriculars, sandwiched between the supercilium and pale cheek, are often the most distinctive feature from many angles of view.** The white supercilium has a minimal amount of dark streaking and often extends as a mottled white area on the nape. Iris color varies from orange to red, typically darkens with age, and is generally a brighter red on males. BODY.—**Gray vermiculated or barred underparts.** Immaculate, often fluffy, long white undertail coverts. Bluish gray upperparts, including the hindneck, contrast sharply with the black nape and auriculars. WINGS.—**Dark uppersides of the remiges contrast sharply with the paler grayish upperwing coverts and other dorsal areas.**

**Adult male *atricapillus:*** HEAD.—Iris color varies from reddish orange to bright red. BODY.—**Underparts are finely vermiculated with gray with little if any dark shaft streaking. Underparts appear uniformly pale gray at field distances.** Underparts are occasionally more coarsely vermiculated and partially barred and appear adult-female-like. Pale bluish gray upperparts. TAIL.—Uppertail pattern is variable and may be: (1) **solid gray,** (2) **have two or three partial dark bands on the central rectrix set,** or (3) **occasionally have partial dark bands on all rectrices.** Undertail is also variable and may be: (1) **solid gray,** (2) **have a darker subterminal band,** or (3) **have two or three partial or complete wide dark bands.** If banded on the undertail, outer rectrix set may be unbanded or have narrow, non-aligning dark bands. Uppertail coverts are generally uniformly bluish gray.

**Adult female *atricapillus:*** HEAD.—Iris color varies from orangish to dark red. Iris color is rarely reddish brown and very rarely dark brown. Iris color typically is not as bright red as on males. BODY.—**Underparts are variably covered with gray vermiculated or barred markings, and each feather has a fairly distinct black shaft streak: (1) moderately barred, (2) finely or coarsely patterned with vermiculation and barring,** or (3) **occasionally covered with malelike fine vermiculation and may lack distinct shaft streaking.** Medium bluish gray or medium gray upperparts often have a slightly more grayish and/or brownish tone than on adult males. The upperparts are more bluish and adult-male-like on some individuals. TAIL.—**Uppertail has three or four distinct equal-width wide dark bands.** Undertail also has distinct equal-width dark bands, except the outer rectrix set which has narrow, non-aligning dark bands. Uppertail coverts may be tipped with white.

**ADULTS OF "QUEEN CHARLOTTE" (*A. g. laingi*):** Darker than the respective sex of *atricapillus*, particularly on the head and upperparts. HEAD.—Extensive amount of black on the crown, auriculars, nape, and hindneck. White supercilium is restricted, heavily streaked with black, and rarely extends onto the nape. The head may appear as a black hood with the black crown, nape, auriculars, and hindneck. There may be only a sprinkling of white on the supercilium or a small white patch above or behind the eyes. The white supercilium may extend onto nape as on virtually all *atricapillus* (probable *atricapillus* intergrades). Blackish or dark gray hindneck blends with the dark dorsal color. BODY and WINGS.—Ventral markings are similar to those of respective sex of *atricapillus.* Back, scapulars, and upperwing coverts are darker than on *atricapillus* and only slightly paler than the black crown, auriculars, nape, and hindneck. Dark grayish upperwing coverts, scapulars, and back are nearly as dark as the dark remiges and lack the two-tone upperwing effect exhibited on *atricapillus.* Underparts are generally similar to the respective sex of *atricapillus.* TAIL.—As on respective sex of *atricapillus*, but darker gray on the dorsal surface and has a narrow terminal white band. *Note:* Neither sex is depicted.

**Adult male *laingi:*** BODY.—Medium bluish gray upperparts.

**Adult female *laingi:*** BODY.—Dark bluish gray or dark gray upperparts have a distinct brownish cast.

**BASIC I (SUBADULT I) TRAITS:** Superficially as on respective sex of adult but with more pronounced ventral markings and darker dorsal coloration. Iris color varies from yellowish orange to pale red and averages brighter on males. BODY.—Lower back and rump retain either (1) scattered worn and faded brown juvenile feathers or (2) a large brown patch of remnant juvenile feathers. A few old brown juvenile feathers are often retained on the scapulars and uppertail coverts. WINGS.—Scattered brown juvenile feathers are on some upperwing coverts. Secondaries may retain some juvenile feathers (s4, s7 and 8, sometimes also s9) or molt totally into new subadult feathering. Primaries generally molt into new feathers, but many retain juvenile outer primaries (p8–10). New subadult remiges are distinctly barred on the underside. **As in adults, paler blue-gray upperwing coverts, scapulars, and back contrast with the darker flight feathers.** TAIL.—Rectrices may either (1) molt totally into bluish gray subadult feathers or (2) retain several brown juvenile rectrices. If old juvenile rectrices are retained, the deck set (r1) and outermost set (r6) are always new subadult feathers. White terminal band on old juvenile rectrices is always worn off. New subadult rectrices have a fairly broad white terminal band. The uppertail coverts may be tipped with white. *Note:* Many males advance almost totally out of any retained juvenile feathering.

**BASIC I (SUBADULT I) OF "NORTHERN" (*A. g. atricapillus*):** HEAD.—Broad white supercilium often extends onto the black nape and is distinctly streaked with black.

**Basic I (Subadult I) male *atricapillus:*** HEAD.—Iris color varies from orangish yellow to pale red. BODY.—Medium bluish gray or medium gray upperparts, including the hindneck and back. Upperparts are typically the same color as in most adult females, rarely paler and adult-male-like. **Underparts are virtually identical to most adult females, but a few birds have a more vermiculated pattern and are more adult-male-like. Unique to subadult males: many have a variably distinct "bib" on the breast, consisting of large, dark arrowhead-shaped marks or thick barring.** TAIL.—**Uppertail has three or four equal-width wide dark bands, rarely has partial black bands on the central rectrices. A few birds lack dark tail banding and appear much like many adult males.** Undertail has equal-width dark bands, except the outer rectrix set which has narrow, non-aligning dark bands. *Note:* Often not readily separable from adult males at field distances.

**Basic I (subadult I) female *atricapillus:*** HEAD.—Iris color varies from orangish yellow to orange. White supercilium is heavily streaked with black. BODY.—Dark gray upperparts have a distinct brownish cast and are darker than on adults and subadult males. **Underparts are prominently marked with wide dark gray or black barring throughout, and each feather has a wide black shaft streak.** TAIL.—**Uppertail has three or four prominent equal-width dark bands.** Undertail has equal-width dark bands, except the outer rectrix set which has non-aligning, narrow dark bands.

**BASIC I (SUBADULT I) OF "QUEEN CHARLOTTE" (*A. g. laingi*):** More heavily marked on the undersides and darker on the uppersides than the respective sex of adult *laingi.* HEAD.—Darker than subadult *atricapillus.* White supercilium is generally ill defined but can be moderately wide but heavily streaked with black. Black hooded appearance of head and dark hindneck as on adult *laingi.* BODY and WINGS.—As on adult *laingi,* upperwing coverts and rest of dorsal areas are nearly the same grayish color as the remiges and rarely show contrast. TAIL.—Terminal white band is narrow. *Note:* Neither sex depicted.

**Basic I (subadult I) male *laingi:*** HEAD.—Dark bluish gray or dark gray upperparts with distinct brownish cast are similar to *laingi* adult females. BODY.—**Underparts are similar to *laingi* adult females and most *atricapillus* adult females and subadult males. As in *atricapillus,* some birds have a variably distinct "bib" on the breast consisting of large, dark, arrowhead or barred markings.** TAIL.—Pattern as in *atricapillus* subadult males.

**Basic I (subadult I) female *laingi:*** BODY.—Dark gray upperparts have distinct brownish cast. Underparts are heavily marked with wide black barring throughout and have pronounced, wide black shaft streaking. Dorsal and ventral coloration is darker and more heavily marked than on any adult types. TAIL.—Pattern is distinctly barred as in *atricapillus* subadult females.

**JUVENILE TRAITS:** HEAD.—**A facial-disk pattern often borders the rear of the brown auricular patch. Most birds exhibit a partial dark malar mark. Orangish yellow iris.** *Note:* Nestlings and recently fledged birds have pale gray irises (Jun.–Aug.) that typically begin turning yellowish orange soon after fledging. Irises are rarely medium brown. **By spring, iris color typically changes to orange.** WINGS.—In flight, broad secondaries have an indentation at the narrower inner primary junction. Underside of all remiges often has irregular or wavy, dark barring, especially on the outer primaries and secondaries. Remiges have a moderately wide dark band on the trailing edge. TAIL.—**Long and wedge-shaped. Each rectrix tip is pointed. On the upperside of the tail, each rectrix has three or four wide, often offset, equal-width dark bands bordered by thin white lines. On a fanned tail, dark bands may exhibit an irregular or wavy pattern.**

**JUVENILE NORTHERN (*A. g. atricapillus*):** HEAD.—**Most have a long, broad white supercilium; however, a few have a nondescript supercilium. Most have a partial, dark malar mark.** BODY.—Whitish or tawny underparts have moderately wide to very wide dark brown streaking from the breast to lower belly. Dark brown leg feather markings are (1) streaked, (2) diamond shaped, or (3) partially barred. **Undertail coverts are also variably marked: (1) partially streaked on only the distal feathers, (2) totally streaked,** or **(3) with large diamond or arrowhead-shaped dark brown markings.** Medium to dark brown upperparts, including rump and uppertail coverts, are mottled and edged with tawny and make the upperparts appear pale and blotchy. Crown, nape, and all of hindneck and neck are streaked with tawny. WINGS.—**Greater upperwing coverts and sometimes median upperwing coverts have a broad tawny or white bar; however, some birds lack a discernible pale bar.** TAIL.—Broad white terminal band extends to the distal edge of the dark subterminal band; by spring, however, the white band on the central rectrix set often wears off. *Note:* Recently fledged pale gray, brown, and springtime eye colors are not depicted.

**JUVENILE (LIGHTLY MARKED TYPE):** HEAD.—Similar to typical juveniles but always has a broad white supercilium. BODY.—Upperparts are consistently paler brown and more extensively mottled with tawny than on typical juveniles. Pale tawny or whitish underparts are either uniformly marked on breast, flanks, and lower belly with narrow dark brown streaking or have less streaking on lower belly. Leg feathers are lightly streaked or unmarked. White undertail coverts have thin dark shaft streaks on only the longest distal coverts or are unmarked and pure white. WINGS.—**Distinct tawny or white bar on greater upperwing coverts and sometimes on the median upperwing coverts.** TAIL.—As on typical juveniles.

*Note:* Classic specimens are in the Bell Museum of Natural History, University of Minnesota, Minneapolis, Minn. ($n = 2$ from autumn in Minnesota) and the AMNH ($n = 2$ autumn birds from N.J. and N.Y.).

**JUVENILES OF "QUEEN CHARLOTTE" (*A. g. laingi*):** Considerably darker than *atricapillus*, especially on the head and dorsal areas. HEAD.—Crown and auriculars are solid blackish brown; however, a few birds are somewhat streaked with tawny on the auriculars. Pale supercilium is very reduced or nearly absent. Solid blackish brown nape and neck regions have little if any pale tawny streaking. BODY.—Back and scapulars are a solid blackish brown. Uniformly blackish brown rump. Uppertail coverts are blackish brown with narrow tawny tips. WINGS.—Upperwing coverts have narrow tawny edges. Greater and median coverts are uniformly blackish brown (and lack the pale tawny bar of most *atricapillus*). Tawny underparts are densely marked with broad, dark brown streaks. Undertail coverts have large, dark brown, diamond or arrowhead-shaped markings. TAIL.—Dorsal surface has thin, faint, pale linear borders along the equal-width dark bands. White terminal band is very narrow and does not extend to distal edge of the dark subterminal band. *Note:* Not depicted.

**ABNORMAL PLUMAGES: Partial albinism.**—An adult with some white feathers (Clark and Wheeler 2001). *Note:* Very rare plumage and not depicted.

**HABITAT: *A. g. atricapillus.* Summer.**—Breeds in remote tracts of old-growth or older second-growth coniferous, deciduous, or mixed coniferous-deciduous forests (uses larger trees than Cooper's Hawks). Forest canopies are moderately to densely foliaged. Understories have a minimal amount of foliage and an assortment of scattered dead stumps, trees, and logs. Wooded tracts are moderate to extensive in size and generally at least 30 acres (12 ha) in size. Mature forest tracts may be fragmented and adjacent to younger second-growth woodlands and open expanses, often of sagebrush. Deciduous and mixed habitat typically consists of Quaking Aspen and, in northern latitudes, also Paper Birch (primary nesting habitat in interior Alaska). In montane areas, also found in small, narrow tracts of Quaking Aspen (aspen stringers) amidst open expanses. In Nevada, aspen stringers are generally at least 75 ft. (23 m) wide and 600 ft. (183 m) long. Inhabits elevations up to timberline. Found above 6,000 ft. (1,800 m) in the cen. and s. Rocky Mts. In the Great Basin, at similar elevation but also rarely found as low as 5,500 ft. (1,700 m) in narrow cottonwood riparian tracts. Montane elevations in n. Rocky Mts., Sierra Nevada, and Cascade Mts. are inhabited. In coastal mainland and island areas of Washington, British Columbia, and Alaska, primarily found in coniferous rain forest from sea level to high montane elevations. The low-elevation boreal forest of Canada, Alaska, and Minnesota is prime habitat. In Minnesota, nesting pairs have been located in mature Red Pine plantations that are adjacent to mature aspen (and nest in the aspen). Foraging habitat can be in the understory of old-growth woodlands; in second growth woodlands; and in open, low vegetation expanses, often consisting of sagebrush. In Mexico, in Sonora, breeds in oak woodlands as low as 3,900 ft. (1,200 m). In Sonora and Chihuahua, mainly found in high-elevation pine forests up to 6,900 ft. (2,100 m).

***A. g. atricapillus.* Winter.**—Essentially the same as breeding habitat for all ages. All ages may also inhabit semi-open and open montane scrub expanses; occasionally, rural and suburban areas; and periodically, open, arid prairie regions.

***A. g. laingi.* All seasons.**—Occupies temperate, humid, low-elevation islands and near-coastal mainlands with extensive tracts of coniferous-deciduous old-growth rain forests. Forests comprise dense growths or semi-open stands of Douglas Fir, Western Hemlock, Sitka Spruce, and Red Alder.

**HABITS:** Varies from a fairly tame to moderately wary species. Temperament is fairly high strung. A reclusive species that has minimal contact with humans. Breeding birds can be very aggressive towards humans intruding into nesting territories. Concealed branches are used for perches. *Northern Goshawks do not perch on posts and telephone poles along well-traveled roadways.* They are solitary away from the breeding grounds. Highly tolerant of cold weather as long as sufficient food is available. Undertail coverts are fluffed when agitated and during some courting displays (*see* Nesting).

**FEEDING:** Perch and aerial hunter. Sits for moderate lengths of time when perch hunting. Aerial hunting consists of random excursion flights. An extremely aggressive and relentless species when pursuing prey. Prey is often pursued on foot into thick brush. In parts of West, goshawks chase ground squirrels, particularly Richardson's Ground Squirrels, on foot in mountain meadows. Hunting typically occurs while flying: under the tree canopy; near forest floors; in small woodland openings; along woodland edges; and over large, open expanses (for ground squirrels, rabbits, and hares). Various species of rodents, rabbits, hares, large songbirds, and small to medium-sized game birds comprise diet. Avian prey is captured while it is on the ground or, rarely, when airborne at low altitude and is plucked prior to being eaten. Adults have plucking posts in territories. Prey may also be consumed at point of capture on the ground. Nesting birds cache extra food for later use.

*A. g. atricapillus* feeds on an assortment of mammals and birds but specializes in species that are dominant in a particular geographic region and season. In some areas, mammalian prey forms the major prey base; in a few locales, especially the Pacific Northwest, where mammalian species are in low numbers, avian species form the bulk of the diet.

Northern Goshawks are regionally and seasonally dependent on the following species: Douglas' Squirrel, Red Squirrel, Golden-mantled Ground Squirrel, Belding's Ground Squirrel, Richardson's Ground Squirrel, Columbia Ground Squirrel, cottontail rabbit (*Sylvilagus* spp.), Blue Grouse, Spruce Grouse, Willow Ptarmigan, Montezuma Quail, Band-tailed Pigeon (particularly in Mexico), American Robin, Varied Thrush, Blue Jay, Steller's Jay, and Northern Flicker. Across the spruce-aspen woods of the boreal forest of Minnesota, Canada, and Alaska, goshawks are highly dependent on, and subsequently quite affected, by populations cycles of Snowshoe Hare and Ruffed Grouse.

*A. g. laingi* feeds extensively on birds because of the lack of mammalian prey in its island range. Varied Thrush, Northwestern Crow, Blue Grouse, and particularly Steller's Jay are major prey species in *laingi*'s entire island range, including the Queen Charlotte Islands. Woodpeckers are minor prey species. Blue and Spruce grouse and ptarmigan species comprise a fair portion of the diet in se. Alaska. Band-tailed Pigeons is a major prey species on Vancouver Island and sw. coastal British Columbia, but it rarely extends its range to the more isolated Queen Charlotte Islands. Ruffed Grouse is found throughout Vancouver Island and se. British Columbia but is not a major prey species. The patchily distributed Red Squirrel is found on the coastal mainland and only some islands. It is absent from the isolated Queen Charlotte Islands. *Note: A. g. laingi's* smaller size and proportionately longer tail provide superior maneuverability and are probably adaptations that enables it to pursue avian prey through dense rain-forest habitat.

**FLIGHT:** Powered by a regular cadence of moderately fast, stiff wingbeats interspersed with short to medium-length glides. Wings held on a flat plane when gliding and soaring. When soaring, front edge of wings is held at a slight angle forward of a perpendicular line to body. Hunting occurs at low altitudes. Flight may be at high altitudes at other times, especially during migration and courtship. Quite agile for its large size. Rarely kites, and only for short periods in strong winds.

**VOICE:** Generally silent except around the nest and during courtship. A repetitive, harsh, loud, and resonant *cack, cack, cack* is a typical agitated call. The beginning of each *cack* is emphatic. Males have a higher-pitched call. A drawn-out, high-pitched, wailing *kree* is common courtship and food-begging call. Adult males emit a soft, throaty, staccato *kuk* when delivering prey to females and nestlings (C. Flatten data). Fledglings are highly vocal.

**STATUS AND DISTRIBUTION (*A. g. atricapillus*):** Estimated population is unknown. *Uncommon* and distributed in low density. A considerable amount of breeding habitat in the West includes forests of old-growth, commercially favored conifers such as Douglas Fir, Ponderosa Pine, and Lodgepole Pine. Because goshawk habitat includes these forests, the population in the w. U.S. has been the focus of much concern. Environmentalists have repeatedly petitioned to have Northern Goshawks west of the 100th meridian in the U.S. listed as threatened or endangered in recent years. The USFWS released a finding in Jun. 1998 that the estimated 3,200 territories in 11 western states still appear to broadly occupy their historical range and that the species is not necessarily threatened or endangered. It is not known, however, if the population in this region is stable or changing. A court ruling in Oregon in Jun. 2001 upheld the 1998 finding.

Estimated territories, by state, according to 1998 USFWS data: Arizona 377, California 816, Colorado 83, Idaho 212, Montana 173, Nevada 194, New Mexico 125, Oregon 484, South Dakota 80, Utah 210, Washington 267, and Wyoming 222. Additionally, a minimum of 41 pairs are found in n. Minnesota. Northern Goshawk is listed by the USFWS as a Sensitive Species in some regions and as a Species of Special Concern in Arizona, Idaho, and Montana. Combined population in the boreal forest of Canada and interior Alaska may exceed that in the Lower 48 but is highly variable depending on prey cycles. The Canadian population is estimated at 10,000–50,000 individuals. In Canada, Northern Goshawk is listed as a Sensitive Species in the Yukon Territory, Nunavut, Alberta, and Saskatchewan and a Secure Species in the Northwest Territories, British Columbia, and Manitoba.

Large acreage heavy-cut logging, which involves clear-cutting and other methods that remove large numbers of older trees from a stand, was practiced extensively in previous decades on federal and private lands and is still done on many private lands. Heavy-cut logging may be detrimental to nesting and foraging habitat, which may be destroyed for decades, possibly for 100 years. Heavy-cut logging has decreased in recent years, especially on federal lands in the U.S. Eighty percent of Northern Goshawk habitat in the West is thought to be on federally owned and regulated lands.

Heavy-cut logging is prevalent in Alberta and mainland British Columbia.

In Mexico, Northern Goshawk has been considered a Threatened Species since 1994. Main range is in the Sierra Madre Occidental from n. Sonora south to n. Jalisco, then from s. Jalisco east through Michoacán. Goshawks are also found in isolated locations in Hidalgo (Los Marmoles N. P. and El Chico N. P.) and in Querétaro. The southernmost nesting occurs in cen. Guerrero. The southern habitat loss is thought to be the main cause for low numbers. Only remnant stands of old-growth forests remain in the Sierra Madre Occidental. Over 98% of this mountain range has been logged. However, old-growth mountain habitat designated to assist the endangered Thick-billed Parrot may provide haven for goshawk populations (the hawk, however, is a major predator of the parrot). Only a few goshawk nest sites have been documented in Mexico.

In the spruce-aspen boreal forest of n. Minnesota, Manitoba, Saskatchewan, Alberta, British Columbia, Yukon Territory, Northwest Territories, and Alaska, goshawks are greatly affected by the natural phenomenon of punctual, widespread, and widely fluctuating cyclic population trends of Snowshoe Hare and, in part, Ruffed Grouse. Both species thrive in this low-diversity, near monotypic habitat but in turn are susceptible to great population fluctuations. Hare and grouse populations revolve on about 10-year cycles, varying from 8 to 11 years in Snowshoe Hares. Hare populations, in particular, reach epidemic highs and then plunge to devastating lows, throwing the boreal forest prey-predator ecosystem into chaos.

As with several predators in the boreal forest, particularly the Lynx but also the Coyote and Great Horned Owl, goshawk status in this region is closely tied to cyclic fluctuations of Snowshoe Hare. When hare numbers are high, predators, including Northern Goshawks, also increase with the abundant food supply. When hare populations decline rapidly on the downside of the cycle, however, predators scramble for ample prey to feed their increased numbers. Ruffed Grouse is a secondary but probably crucial factor affecting goshawk status. Grouse come into the scenario when Snowshoe Hare numbers sharply decline and predators feed more extensively on Ruffed Grouse and subsequently cause a rapid population decline of grouse. On the low end of the hare-grouse population cycle, Northern Goshawk winter survival and reproduction are greatly affected by the loss of these two primary species in this region. Low population cycles of these two major prey species also prompt adult Northern Goshawks, which typically have minimal amount of seasonal movements, to depart food-void regions en masse in a southward "irruption" migration (*see* Movements). The irruption phenomenon is accentuated when year-round secondary prey species cannot supply an ample food source for the hawk, especially in winter when seasonal prey species have migrated or hibernated.

In the w. U.S., the Snowshoe Hare occurs from the Rocky Mts. south to n. New Mexico and s.-cen. Utah and in the Cascade Mts. and Sierra Nevada. The hare population is stable in this more diverse southern montane ecosystem. Likewise, Ruffed Grouse in the w. U.S. are found in the Cascade Mts. and n. Rocky Mts., from Idaho and Montana south to w. Wyoming and n. Utah. Grouse exhibit little or no cyclic patterns in this region. Northern Goshawk populations in these areas are stable because primary prey consists of non-cyclic species. In the very southern portion of goshawk range, neither the hare nor grouse are present and goshawks feed on other, also stable prey.

*Additional Note on Snowshoe Hare and Ruffed Grouse: Snowshoe Hare.*—In w. Canada, and particularly in the Yukon Territory, the greatest degree of fluctuations in Snowshoe Hare populations occur. E. Canada has a weaker cyclic trend for the hare, and the Rocky Mts. the least amount of cyclic activity. The Yukon produces the largest and most broods per season. Up to four broods with up to eight young are produced in the Yukon, verses two broods with two to four young in Colorado. Populations also increase more rapidly in far northern latitudes because of long summer days. Cyclic downward trends also occur more quickly in the Yukon than in the boreal forests of Alberta or n. Minnesota. Cycles may occur in a short time span over hundreds or even thousands of miles across the boreal forest. All of the boreal forest of Alaska and Canada may be subjected to a widespread Snowshoe Hare cyclic crash in a span of 1–2 years. Declines in the Yukon have occurred within 2 years of reaching peak numbers and in 3–4 years elsewhere.

In the 1970s to 1990s (and probably 1960s) Snowshoe Hare declines have triggered neatly patterned 10-year cyles of declines in goshawk (and Lynx, Coyote, Great Horned Owl) in n. Canada and subsequent southward irruptions of adult goshawks into the U.S. These irruptions occurred in the fall of 1972, 1982, and 1992. The 1982 and 1992 irruptions (*see* Movements) in the n. U.S. mirrored the population crashes of the well-studied Snowshoe Hare in w. Canada during the same time period. In the Yukon, Snowshoe Hare numbers peaked from 1976 to 1981, then quickly plummeted and sparked the 1982 goshawk irruption. Hare populations then began to increase again in 1984, peaked in 1989 and 1990, then declined rapidly again in the winter of 1990–1991, and again sparked a goshawk irruption in 1992. Snowshoe Hares peaked most recently in sw. Yukon Territory in 1999, and a goshawk irruption seemed imminent with the increasing number of adults seen in key migration locations such as Duluth, Minn. (*see* Movements). However, the latest irruption, which occurred in 2001, was much smaller than previously recorded adult exoduses from the boreal forest.

With an overall 8- to 11-year cycle, Snowshoe Hare population crashes and southward adult goshawk irruptions have not always occurred in the early part of each decade as they have in recent history. In the latter 19th century and 20th century, Snowshoe Hare declines and goshawk irruptions occurred in the middle and latter parts of the decades.

There are several reasons for Snowshoe Hare population fluctuations. Rapid reproduction leads to severe complications because of a burgeoning population. Hare population may increase 200 times over the low end of the cycle, followed by starvation, rampant disease, and extremely high depredation by an increasingly high number of predators, particularly in winter. Some biologists also suggest a possible link to the regular cyclic solar flares of the sun affecting the low-diversity ecosystem of the boreal forest, which is correlated with the study of growth rings and heavy feeding cycles when hares are at the top of the cycle.

*Ruffed Grouse.*—This species is also on approximate 10-year population cycles featuring highs and lows. Ruffed Grouse share the identical range of the Snowshoe Hare throughout w. Canada and Alaska. Population declines of Ruffed Grouse, as noted previously, are considered by biologists to be attributed more to excessively heavy depredation when Snowshoe Hare populations quickly dwindle after reaching their peak. A Ruffed Grouse population crash usually follows a year or so after a Snowshoe Hare population crash. However, long periods of inclement weather and poor food availability in winter may also accentuate rapid grouse declines. Ruffed Grouse reach their highest population density in aspen habitat, which, when available, is also extensively used by goshawks. Aspen trees themselves have cyclic production of male catkins, and the catkins form an extensive portion of the grouse's winter diet. Although Ruffed Grouse may not trigger a cyclic trend of some predator species in the boreal forest zone as Snowshoe Hares do, high or low numbers of grouse may reflect the severity of a major boreal forest food shortage for predators.

*Additional status data:* Some regions have localized Northern Goshawk population cycles that are due to fluctuations of other major prey species. From the Cascade Mts. in Washington south to the Sierra Nevada in California, goshawks feed extensively on Douglas' Squirrels which undergo irregular population cycles that are due to variances in coniferous cone-crop production. Low squirrel cycles temporarily affect goshawk breeding success and status but do not produce an irruption. In Nevada, seasonal and cyclic Belding's Ground Squirrel populations may temporarily affect goshawk breeding success. In se. Arizona, Northern Goshawks are not affected by cyclic prey but have experienced a major population decline because of several factors.

**Winter.**—Many adults remain on or near breeding territories unless forced to vacate because of food shortages, mainly with sharp declines of Snowshoe Hare and Ruffed Grouse. Adults in some areas and juveniles in virtually all areas winter at latitudes that are farther south and at lower elevations than breeding and natal areas. Individuals that leave breeding and natal territories are solitary and often highly nomadic. In sw. Montana, however, some adults winter in the same away-from-breeding areas year after year. Juveniles and occasionally even adults may roam far out of typical range and habitat in any winter and are regularly seen in Iowa and occasionally in Missouri, Kansas, cen. Oklahoma, and w. Texas. Irruption years may force all ages far south of typical range to s.-cen. Texas and s. Louisiana.

**Movements.**—Depending on age, sex, and location, birds are (1) sedentary, (2) engage in short seasonal dispersal or shifting movements in any direction, or (3) perform short-distance southward migrations. Adult females are likely to embark in short-distance post-breeding dispersal movements in any direction to new breeding territories. Males are more likely to remain somewhat sedentary within the general breeding area. However, as verified with telemetry in sw. Montana and possibly in other regions, both adult sexes may disperse or shift in any direction and at varying distances, elevations, and habitats away from breeding areas in fall and winter. Some individuals may perform a variable-length shifting movement to the same winter territory for consecutive years.

During migration, juveniles outnumber adults except during irruptions. Because of low breeding success, adults far outnumber juveniles during irruption-year movements. In Duluth, Minn., in the 1992 irruption, 88% were adult and 12% were juveniles, which is the opposite of the age percent of nonirruption years. A large number of subadults and adults and fewer juveniles were seen in fall of 2000 in Duluth, thus showing definitive signs of a forthcoming irruption. In the fall of 2000, approximately 70% were adults and subadults and 30% juveniles. In the fall of 2001, which constituted the latest irruption, of the 1,154 goshawks an impressive number of adults were tallied. Adult ages tallied 90%, juveniles tallied 10% of the migrants in the fall of 2001. Based on the meager number of adults in the fall of 2002 (of 653 goshawks tallied), 2001 was the adult-irruption-year of the decade. It broke the 10-year increment order that has stood for many decades. *Note:* Even though 2000 had slightly higher number of goshawks (1,236) than in 2001, adults comprised a lower number of the total than in 2001.

*Fall migration* (irruption): Adult irruptions in the 1970s–1990s (and probably 1960s) have occurred in punctual 10-year increments in the West: in the fall and winter of 1972, 1982, and 1992. High numbers of northward-returning birds were detected the subsequent spring of each respective fall and winter irruption. The number of adults departing breeding regions is lowest in the middle of the 10-year phenomenon and remains low until 1–2 years prior to an irruption. An "echo" movement occurs the subsequent autumn with fairly high adult numbers. *Note:* Based on data from Hawk Ridge Nature Reserve, Duluth, Minn., numbers of adults affected by prey cycles have been considerably lower in recent decades for unknown reasons (5,382 in 1972 and 5,819 in 1982, then dropped to 2,247 in 1992 and to only 1,154 in 2001). This is possibly correlated with the intensity of Snowshoe Hare population cycles and possible retention of higher numbers of Ruffed Grouse and secondary prey species in the boreal forest. This does not necessarily mean fewer adults, just fewer adults that have departed favored, northern haunts (because there are ample food supplies).

Irruption occur as a mass exodus from breeding and natal regions from mid-Aug. to Jan. (and later). The first adult push comes in mid-Oct. with the first peak in late Oct. and a secondary peak in mid-Nov. (fall 1992 Duluth, Minn., data). Continual southward movement, however, may occur throughout winter as the prey base in northern regions wanes. The number of dispersing adults seen at hawkwatch sites increases in the Rocky Mts. and Great Basin surrounding irruptions, but migration timing remains the same. Virtually all birds noted in this region undoubtedly come from more northern regions in the Canadian boreal forest. A northwest to southeast movement of some Canadian adults surrounding irruption years is verified by adults banded in the summer in ne. British Columbia and captured in the fall at Duluth, Minn. Subadults may form a large percentage of the "adult" class in pre-irruption years.

*Fall migration* (non-irruption): Most juveniles and some subadults and adults migrate southward each year. Movements span from mid-Aug. to Dec. Juveniles peak from late Oct. to early Nov., adults in mid-Nov. Subadults may peak before adults, in late Oct. to early Nov.

In the s. and cen. Rocky Mts. and the Great

Basin, movements during all years occur from mid-Aug. to mid-Nov. The highest numbers are seen in Oct. There are no distinct peaks, but juveniles predominant Aug.–Sep.

Adults breeding in the Great Basin mountains may move to lower elevations or short distances south each winter because of lack of prey, although some may not move unless forced by severe weather. Belding's Ground Squirrel, a staple summer prey species, estivates and hibernates, and many avian prey species migrate; thus, goshawks lose many food sources. In the Sierra Nevada, Douglas' Squirrel population crashes do not prompt adult goshawks to disperse or shift out of breeding areas. In Carbon and Albany Cos., Wyo., all ages migrate short distances south of breeding and natal locales, probably because of lack of prey and locally severe winter weather.

*Spring migration* (all years): Feb.–May. Adults peak in mid-Mar. to mid-Apr. Juveniles move throughout the entire period but peak in Apr., with stragglers occurring into May.

**STATUS AND DISTRIBUTION (*A. g. laingi*):** *Queen Charlotte Islands.—Rare* to *very uncommon.* This subspecies may be in grave danger (only three known nest sites in the 1990s). It has been deemed a Special Concern Species since 1995 by the Canadian Government and Critically Imperiled by the British Columbia Conservation Data Centre. Intense, large-acreage heavy-cut logging that occurred in previous decades may have had a negative affect on goshawk habitat on these islands. However, heavy-cut logging is no longer practiced. Logging is still done on Graham and n. Moresby Islands but in a "light-cut" or "patch-work" approach consisting of small-acreage clear-cuts, which retain adequate habitat. Several areas on the islands are also now protected from logging. *Vancouver Island.*—Numerous pairs nest on the island. These include a few *laingi* but also *atricapillus* and intergrade types (*see* Note on "Queen Charlotte" race under Subspecies). *Alaska.*—Currently recognizes *laingi* as a Species of Special Concern. Breeding numbers and densities are very low. Since 1991, 50 nesting territories have been documented in se. Alaska. As on Vancouver Island, they include a few *laingi* but also *atricapillus* and intergrade plumage types. There may be 120 pairs breeding in se. Alaska.

*A. g. laingi* is generally not affected by cyclic fluctuations of prey because its diet consists of stable, non-cyclic prey. Snowshoe Hare are absent from virtually the entire range since the hare does not occupy islands and the only area that Ruffed Grouse occupies is on Vancouver Island and south-coastal mainland British Columbia. Ruffed Grouse is not considered a major prey species in this region. In British Columbia, goshawks may be affected by possible decline of Steller's Jays.

**Winter.**—Remains in typical year-round range and habitat.

**Movements.**—Movements are confined mainly to dispersal activities by adult females and juveniles as noted in se. Alaska. Some adult females on coastal islands of se. Alaska move short distances to new territories after the breeding season (up to 100 miles [160 km]). Juveniles disperse short distances out of natal areas. Movement occurs between nearby islands and between near-coastal islands and the coastal mainland. There is probably minimal movement to and from the Queen Charlotte Islands to other islands of British Columbia and se. Alaska because of considerable overwater distances. Museum data currently indicate only dark *laingi* have been collected on the Queen Charlotte Islands. Over 32 miles (51 km) of open water separate the Queen Charlotte Islands from Dall and Prince of Wales Islands, AK, and from Dundas, Stephens, and Porcher Islands, B.C. *Note:* As a rule, the two larger accipiter species, the Northern Goshawk and Cooper's Hawk, are more reluctant to cross large expanses of water than the smaller Sharp-shinned Hawk.

**NESTING (all races):** Late Mar. to mid-Sep., depending on latitude and elevation.

**Courtship (flight).**—*High-circling, slow-flapping,* and *sky-dancing* (*see* chapter 5). Undertail coverts are flared and retracted regularly with tail closed or partially spread, especially in high-circling.

Nests are up to 36 in. (91 cm) in diameter, composed of large sticks, and decorated with greenery. They are 15–75 ft. (5–23 m) high, but up to 90 ft. (27 m) high in the Pacific Northwest. Nests are typically in the largest trees in the tract and located at the base of the canopy level. In deciduous trees, nests are placed in primary or secondary forks; in conifers, usually next to trunks, but sometimes away from trunks on horizontal branches. Up to five nests can be located within a territory, and pairs regularly alternate among nest sites every 1–5 years. Northern Goshawks normally build their own nest but will use unoccupied nests of Common Ravens or nests of another large raptor species. Nest trees are in mature wooded tracts of at least 30 acres (12 ha). Younger-aged second-growth tracts or large open expanses may be adjacent to the nest tree stand. Sometimes dead trees are used as nest sites, with nearby live trees offering canopy cover. Rarely, nests on metal power-line poles or on ledges in abandoned cabins. Nest trees are near small clearings consisting of meadows, an area of fallen trees, paths, lightly used roads (especially logging roads), and in tracts of more widely spaced trees. A quiet water source such as a lake, stream, or ground seepage is nearby. In hilly or mountainous terrain, nesting locales are often near the base of a slope. In many parts of the U.S., nesting areas are on north slopes; in Alaska, they tend to be on south slopes.

Females assume nest duties while males hunt. One to 4 eggs are incubated for 34–35 days. Youngsters fledge in 36–42 days and are independent in 84–94 days. Because of low prey abundance, *atricapillus* nesting success may be poor or birds may not nest, often over extensive regions, surrounding irruption years. Because of ample prey abundance, nesting success is high between irruption years. One-year-old females in juvenile plumage may successfully nest. These younger-aged females mate with adult males, most often in regions that are affected by prey declines and in years following sharp prey declines. Most birds do not breed until 2 years old. Nesting pairs are highly agitated by human presence in territories.

**CONSERVATION:** Logging methods, such as light-cut, which includes various means of thinning but retains ample quantity of old-growth trees, are now being practiced in many areas, particularly on federal lands. Light-cut logging retains adequate habitat and, in many cases, improves the quality of habitat. Northern Goshawks also occupy forests comprising tree species that have minimal or moderate commercial value (e.g., Quaking Aspen, Paper Birch). Populations in the Southwest, and in the Pacific Northwest to southern mainland British Columbia, may benefit from old-growth forestry management techniques implemented for the northern race of the endangered Spotted Owl.

For *laingi*, the southern two-thirds of Moresby Island's Gwaii Haanas N.P. and most smaller islands of the Queen Charlotte archipelago are designated national park and protected from logging. In logging areas, a 30-acre (12-ha) buffer is left around known nest sites until additional protective measures are possibly implemented in the future. In the U.S., environmentalists have repeatedly petitioned to have *laingi* occupying se. Alaska and coastal British Columbia listed Threatened or Endangered because of low numbers and threat of extensive heavy-cut logging, particularly in the Tongass N.F. in se. Alaska.

**Mortality (all races).**—Loss of breeding habitat and prey because of heavy-cut logging of old-growth forests. Great Horned Owls prey on all ages. Common Ravens steal eggs. Goshawks occupying open areas during winter may suffer mortality from Golden Eagles. Illegal shooting in areas where there is human contact. Pesticide contamination has not affected goshawks. Trichomoniasis has been a mortality factor in fledglings in s. Arizona.

**SIMILAR SPECIES:** COMPARED TO JUVENILE *atricapillus.*—**(1) Northern Harrier, adult females.**—PERCHED.—Facial disk, pale brown or yellow irises, streaked underparts, mottled brown upperparts, tail length and pattern similar to goshawk: use caution. Long, thin tarsi. FLIGHT.—Dark axillaries and median underwing coverts. Wings held in dihedral when soaring and gliding. Five "fingers" on the outer primaries. White uppertail coverts. **(2) Cooper's Hawk, juveniles.**—PERCHED.—May have a short pale supercilium; rarely a hint of dark malar stripe or brown auricular patch. Iris color is pale grayish to pale lemon yellow in fall, medium orangish yellow in spring, but always paler and not as orangish as seen in respective-season goshawk. Iris color is never orange or brown: use caution with nestling/recently fledged goshawks as they have pale gray irises (Jun.–Aug.). Moderately long tarsi. Upperwing coverts are uniformly dark. Tail has rounded tips on rectrices (both species have broad white terminal band in fall).

Tail rarely has thin white borders along dark bands on dorsal surface, and dark bands are generally neat and well aligned. Exercise caution on underpart markings: heavily marked type Cooper's Hawks are as heavily streaked as lightly marked type goshawks. Some western Cooper's Hawks have obvious dark streaking on undertail coverts and are similar to goshawk's. Cooper's Hawks perch on posts and poles along roadways. FLIGHT.—When soaring, wingtips very rounded and wings held perpendicular to body. Both species have distinct dark band on trailing edge of underwing. Voice is a nasal *kek, kek* or soft, nasal *kyew*. **(3) Red-shouldered Hawk, juveniles.**—PERCHED.—Pale supercilium; wide dark malar mark. Iris color is medium gray-brown: use caution with "brown-eyed" goshawks. Long, thin tarsi. FLIGHT.—Distinct tawny/white crescent-shaped panel on underside and topside of outer primaries. Five "fingers" on the outer primaries. *B. l. lineatus* has little if any barring on undersides of remiges and narrow gray trailing edge on underwing. Numerous narrow pale tail bands. *B. l. elegans* has considerable barring on underside of remiges; white window or panel on primaries; tail has a few narrow white or gray bands. Voice: *keyair*. **(4) Broad-winged Hawk, light morph juveniles.**—PERCHED.—Pale supercilium; wide dark malar mark. Iris color is medium gray or brown: use caution with "brown-eyed" goshawks. Uniformly dark upperwing coverts. Wingtips almost reach tail tip. Short and thick tarsi and toes. Perch along roadways on poles and wires. FLIGHT.—Underside of remiges has little if any barring and narrow gray trailing edge on underwing. Three to four "fingers" on outer primaries. Wingtips are very pointed at all times. Voice: high-pitched *pee-heeee*. **(5) Gyrfalcon.**—PERCHED.—Dark eyes; bare orbital skin. FLIGHT.—No "finger" definition on outer primaries. When soaring, front edge of wings always bent at wrist, does not form a straight line. **(5A) Gray morph juveniles.**—Faint dark malar distinction. Underparts similarly streaked: use caution. Underside of wing in flight has faint, narrow dark barring; often darker axillaries, medium coverts. Bluish tarsi and toes. COMPARED TO JUVENILE *laingi*.—**(5B) Dark morph adults/juveniles.**—Head (except dark irises), underparts, and upperparts similar. Underwing has unmarked gray remiges and contrasting dark coverts. Tail all dark on top, all gray on undersides; may have faint, narrow pale bands. Juveniles have bluish tarsi and toes. COMPARED TO ADULTS (BOTH RACES).—**(5C) Gray morph adults.**—Dark irises. Minimal to moderate dark malar distinction; shares white supercilium but has gray (not black) auriculars. White underparts have short streaks and spots; barred only on flanks and leg feathers. Underside of primaries has numerous narrow faint bars and primaries; paler than secondaries. Undertail coverts have some barring. Numerous narrow, dark and light tail bands.

**OTHER NAMES:** Goshawk, Gos, Golden Gos. *Spanish:* Gavilán Azor. *French:* Autour des Palombes.

---

REFERENCES: Adamus et al. 2001; Alaska Natural Heritage Program 1998; Andrews and Righter 1992; AOU 1957; Baumgartner and Baumgartner 1992; Bent 1961; Block et al. 1994; Boal et al. 2001; Brockman 1968; Bylan 1998, 1999; Call 1978; Cecil 1997, 1999; Clark and Wheeler 2001; Dodge 1988–1997; Dorn and Dorn 1990; Duncan and Kirk 1994; Environment Canada 1999; Finn et al. 1998; Gibson and Kessel 1997; Gilligan et al. 1994; Godfrey 1986; Herron 1999, Heron et al. 1985; Howell and Webb 1995; Ingle 1999; Islands Protection Society 1984; James and Neal 1986; Johnsgard 1973; Kellogg 2000; Kent 1996, 1997, 1998; Kent and Dinsmore 1996; Kingery 1998; Lammertink et al. 1996; Lowery 1974; Meehan et al. 1997; Montana Bird Distribution Committee 1996; Moore and Henny 1983; Oberholser 1974; Palmer 1988; Peterson 1995; Reynolds 1983; Reynolds and Meslow 1984; Reynolds and Wight 1978; Robbins and Easterla 1992; Russell and Monson 1998; Shefferly 1996; Shipman and Bechard 1998; Small 1994; Smith 1996; Smith et al. 1997; Snyder and Snyder 1991, 1998a; Southwest Center for Biological Diversity 1999; Taverner 1940; Thompson and Ely 1989; USFWS 1997b, 1998b; Van Rossem 1938; Watson et al. 1998; Whaley and White 1994; White and Kiff 1998; White 1965; Yukon Dept. Renewable Resources 1990.

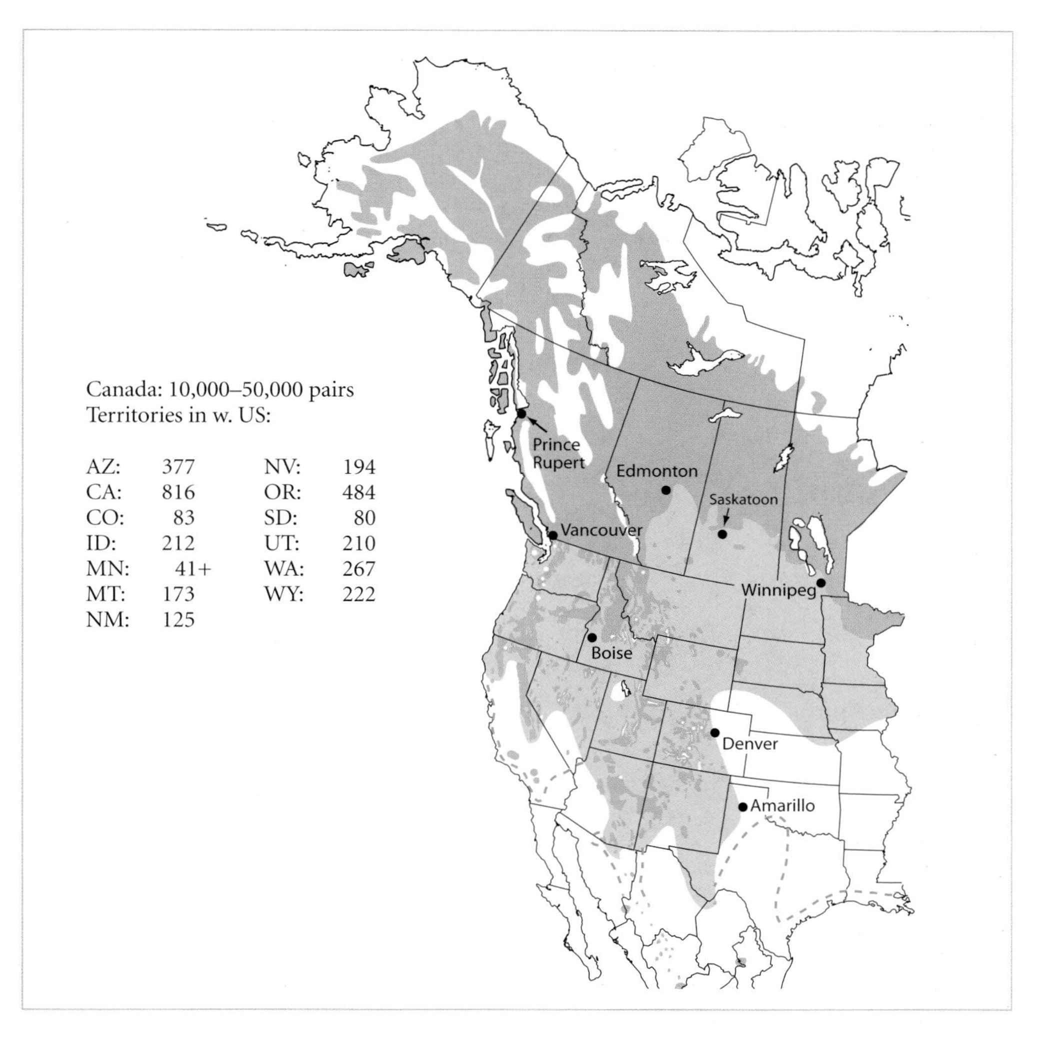


**NORTHERN GOSHAWK,** *Accipiter gentilis atricapillus:* Uncommon. Juveniles, some subadults, and a few adults occur in winter-only range. All ages may extend to dashed lines in irruption years.

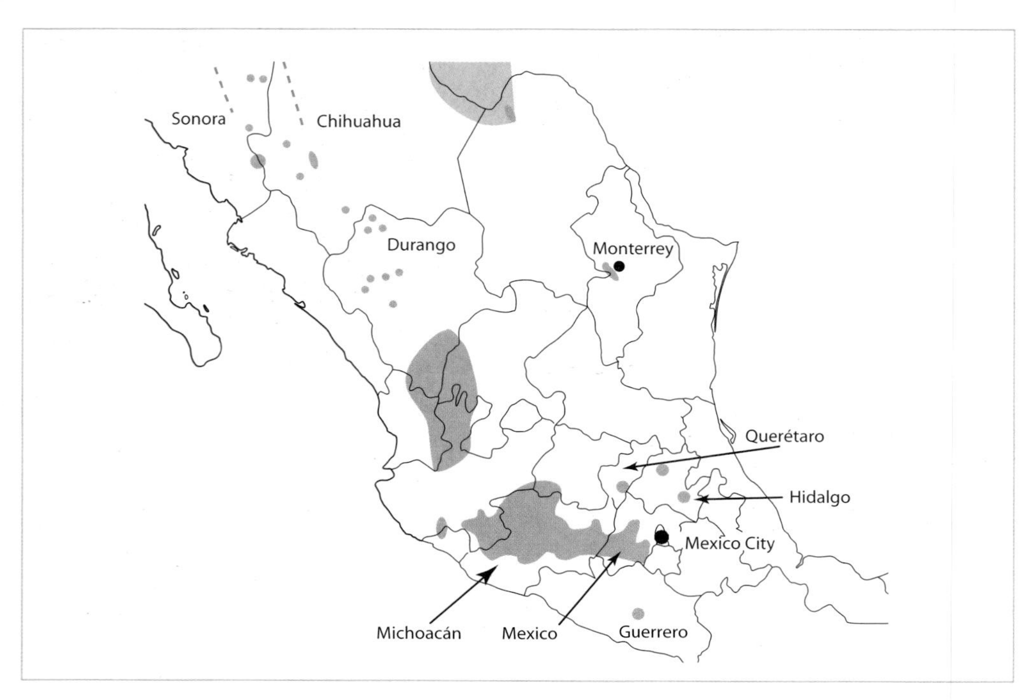


**NORTHERN GOSHAWK,** *Accipiter gentilis atricapillus:* Mexico classifies as threatened.

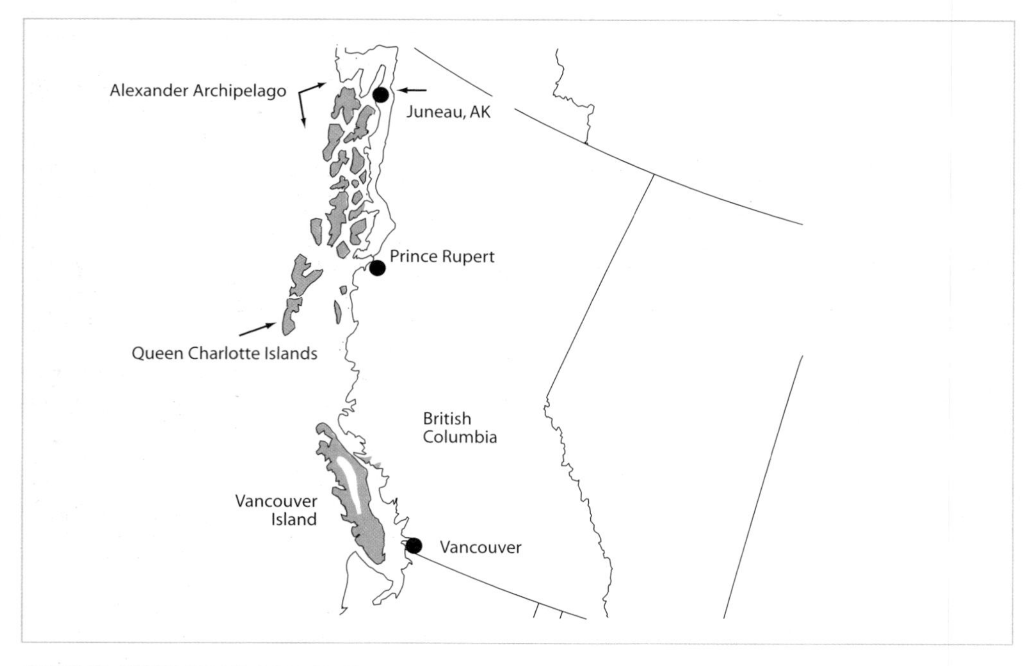


**"QUEEN CHARLOTTE" GOSHAWK,** *Accipiter gentilis laingi:* Rare to very uncommon. Core range is probably on Queen Charlotte Is. Elsewhere mixes and intergrades with *A. g. atricapillus.*

**Plate 191. Northern Goshawk, adult female** [Oct.] ▪ Black auricular patch contrasts with white supercilium and cheeks. Orangish red iris. ▪ Bluish gray upperparts. Underparts finely barred with gray; dark. White, often fluffy, undertail coverts. ▪ *Note:* Adult males similar but have very finely marked underparts. *A. g. laingi* darker on upperparts.

**Plate 192. Northern Goshawk, adult male** [Oct.] ▪ Black auricular patch contrasts with white supercilium and cheeks. Red iris. ▪ Pale bluish gray upperparts. ▪ Dark brownish black remiges contrast with bluish wing coverts. ▪ Long grayish tail typically unbanded. ▪ *Note: A. g. laingi* darker, more grayish on upperparts.

**Plate 193. Northern Goshawk, adult female** [Oct.] ▪ Black auricular patch contrasts with white supercilium and cheeks. Orangish red iris. ▪ Bluish gray or gray upperparts. ▪ Remiges darker than grayish wing coverts. ▪ Black tail bands. ▪ *Note:* Upperparts of adult females often more grayish than on adult males. *A. g. laingi* darker on upperparts.

**Plate 194. Northern Goshawk, adult female** [Oct.] ▪ Black auricular patch contrasts with white supercilium and cheeks. Reddish iris. ▪ Underparts finely barred with gray. ▪ Gray remiges barred (males often barred only on primaries). ▪ Broad black tail bands. ▪ *Note:* Sexes often inseparable when seen from below in flight. Males often appear solid gray on underparts.

**Plate 195. Northern Goshawk, subadult male** [**Oct.**] ▪ Black auricular patch contrasts with white supercilium and gray cheeks. Orange iris. ▪ Bluish gray back and scapulars. Finely barred underparts. ▪ Retained brownish juvenile feathers on bluish gray upperwing coverts. Retained juvenile secondaries (s4 and 8). ▪ *A. g. laingi* darker on upperparts.

**Plate 196. Northern Goshawk, subadult female** [**Oct.**] ▪ Black auricular patch contrasts with white supercilium and gray cheeks. Yellowish orange iris. ▪ White underparts coarsely barred with black; each feather has distinct black shaft streak. Upperparts (not shown) dark grayish brown. ▪ *Note: A. g. laingi* blackish on hindneck and more prominent dark barring on underparts.

**Plate 197. Northern Goshawk, subadult male** [**Oct.**] ▪ White supercilium, black auricular patch. ▪ Bluish gray upperparts with retained brown juvenile feathers on rump. ▪ Retained brown juvenile feathers on wing coverts and secondaries (s4 and 8). ▪ Gray banded tail fully molted into subadult age class. ▪ *Note:* Many birds replace all secondaries into subadult feathers.

**Plate 198. Northern Goshawk, subadult male** [**Oct.**] ▪ Black auriculars contrast with white supercilium and grayish cheeks. Orangish red iris. ▪ Underparts barred with gray. Partial bib on breast consisting of larger black markings; some subadult males have extensive black markings on breast. ▪ Barred, gray remiges are fully subadult feathers. ▪ Wedge-shaped tail.

**Plate 199. Northern Goshawk, juvenile [Oct.]** ▪ Tawny head and hindneck; brown auricular patch; white supercilium; partial, dark malar mark; orangish yellow iris. ▪ Brown upperparts edged with tawny or white and blotched with white. ▪ White or tawny bar on greater upperwing coverts. ▪ Thin white lines border some of irregular black tail bands. Pointed rectrix tips.

**Plate 200. Northern Goshawk, juvenile [Sep.]** ▪ Tawny head and neck; brown auricular patch; partial, dark malar mark; orangish yellow iris. Supercilium sometimes indistinct. ▪ Moderately streaked tawny underparts. Large, brown, diamond-shaped marks on undertail coverts. ▪ Wedge-shaped tail. Pointed retrix tips.

**Plate 201. Northern Goshawk, juvenile [Oct.]** ▪ White supercilium; brown auricular patch; partial, dark malar mark. ▪ White bar on greater and some median upperwing coverts. *A. g. laingi* much darker, lacks pale wing bars.

**Plate 202. Northern Goshawk, juvenile [Oct.]** ▪ Heavily streaked underparts. Arrowhead-shaped dark marks on all undertail coverts. ▪ Dark wavy bands on outer primaries and secondaries. Wide dusky band on trailing edge of remiges. ▪ Pointed rectrix tips. ▪ *Note:* Heavily marked type of typical juvenile *A. g. atricapillus. A. g. laingi* more heavily marked.

**Plate 203. Northern Goshawk, juvenile [Oct.]** ▪ Dark malar mark. ▪ Moderately streaked underparts. Dark marks on undertail coverts. ▪ Wavy bands on outer primaries and secondaries. Wide black band on trailing edge of remiges. ▪ Wedge-shaped tail with pointed rectrices. ▪ *Note:* Moderately marked type of typical juvenile *A. g. atricapillus.*

**Plate 204. Northern Goshawk, juvenile (lightly marked type) [Oct.]** ▪ White supercilium and partial, dark malar mark. ▪ White underparts narrowly streaked on flanks, breast, and belly; unmarked lower belly and leg feathers. Unmarked, white undertail coverts. ▪ Irregular dark tail bands. Pointed rectrices. ▪ *Note:* Palest plumage extreme of juvenile *A. g. atricapillus.* Ventral markings, including pale underwing coverts, rival pale-marked juvenile Cooper's Hawks. Sexed as female by size.

# COMMON BLACK-HAWK

## *(Buteogallus anthracinus)*

**AGES:** Adult, subadult, and juvenile. Full adult plumage is not acquired until birds are 2-year-olds in their third year of life. Subadults are 1-year-olds in their second year of life. Subdults have a small amount of remnant juvenile feathering on the remiges and body but are otherwise in adult plumage. Juvenile plumage is worn the first year. Juveniles have somewhat narrower wings and distinctly longer tails than adults.

**MOLT:** Accipitridae wing and tail molt pattern (*see* chapter 4). First prebasic molt from juvenile to subdult begins in May or early Jun. and ends in late fall. Molt is generally incomplete. The outermost primaries may be retained juvenile feathers (p8–10), as may a few secondaries (s4 and 8 or 9), and a few underwing coverts. The rectrices are fully molted into adult feathers. Subsequent prebasic molts are also incomplete each year. Molt begins in Jun. or Jul. and ends in late fall. In adults, females begin their molt before males.

**SUBSPECIES:** Polytypic. *B. a. anthracinus* is the only race found in the U.S. It extends its range south into e. Sonora and w. Chihuahua, Mexico, and along the Pacific Slope of Mexico. This race is also found on the Atlantic Slope of Mexico from cen. Tamaulipas south to n. South America in n. Peru, n. Colombia, and e. Guyana; also inhabits Trinidad and some of the Windward Islands. One additional race, *B. a. gundlacchii,* is on Cuba and Isla de la Juventud.

**COLOR MORPHS:** None.

**SIZE:** A large raptor. Males average smaller than females, but sizes overlap. Length: 20–22 in. (51–56 cm); wingspan: 40–50 in. (102–127 cm).

**SPECIES TRAITS:** HEAD.—**Fairly long, somewhat narrow bill with a wide pale bluish or**

**yellowish base.** BODY.—**Very long tarsi.** WINGS.—**When perched, primaries extend a short distance beyond the tertials (short primary projection). In flight, the broad wings have parallel front and trailing edges and very blunt, rounded wingtips.** *Note:* Although wings of juveniles are narrower than adults', they are still quite broad.

**ADULT TRAITS:** HEAD.—**Black bill has a wide, pale yellowish basal area on both mandibles. The cere is orangish yellow or yellow. Pale yellowish or whitish lores. The forehead is black.** BODY.—Grayish black. Leg feathers are sometimes narrowly tipped with white. **Long yellow tarsi. In flight, the long tarsi project the feet beyond the black undertail coverts.** WINGS.—**Wingtips are somewhat shorter than tail tip when perched. In flight, wings are very broad. Whitish mottled patch on the base of the outer two primaries (p9 and 10). Undersides of secondaries are very pale rufous or grayish with narrow black barring and mottling with a wide black band on the trailing edge.** Upperwing coverts are more brownish than the body; secondaries are marbled with gray. TAIL.—**A single broad white band on the mid-section on the dorsal and ventral sides and a narrow white terminal band. When birds are perched dorsal side and is seen, the very broad secondaries typically conceal the white dorsal tail band. In flight, the feet extend into the mid-section of the broad white band on the undertail.**

**Adult male:** HEAD.—Dark grayish black or black. Area below the eyes is always black. BODY.—Grayish black or black with each feather edged with grayish. Plumage appears darker, more blackish than on females.

**Adult female:** HEAD.—**Some have a variable-sized white patch between the eyes and gape that connects to the whitish lores. The white patch may be extensive and extend behind the gape and eyes.** BODY.—Has a more brownish cast than on adult males. WINGS.—Upperwing coverts are typically more brownish than on adult males. TAIL.—Some birds have a narrow white basal band that is visible only on the underside of a widely fanned tail. This band is often difficult to see because this portion of the tail is shadowed by the broad wings and appears gray.

**BASIC (SUBADULT) TRAITS:** Like respective sex of adult on the head, body, and tail. May exhibit a small amount of juvenile-like tawny streaking or mottling on sides of the neck and back. All of the tail molts into adult feathering. WINGS.—Old, worn juvenile remiges retained on the outer one to three primaries (p8–10), possibly one or two secondaries (s4 and sometimes s8 and/or 9), and a few tawny underwing coverts. **A very distinct white patch is on the basal area of the old, worn juvenile outer primaries.** *Note:* Data based on a specimen collected in mid-Jan. in Guatemala (Denver Museum of Nature and Science, Denver, Colo.), an individual photographed in late Feb. in Nayarit, Mexico; and photograph in Wheeler and Clark (1995).

**JUVENILE TRAITS:** HEAD.—**Black bill has a wide, pale bluish or greenish basal region.** Very pale greenish or yellowish cere. Medium brown irises. **Short, thick dark eyeline, a short tawny or white supercilium, and a broad, dark malar stripe that extends onto the side of the neck.** BODY.—Tawny underparts are narrowly streaked with dark brown; underparts fade to white by late winter. **The flanks are uniformly dark brown. Whitish, finely barred leg feathers.** Dark brown upperparts and uppertail coverts are edged and speckled with tawny in fresh plumage but wear to a nearly uniform brown plumage by late winter. WINGS.—**In flight, the wings are broad. Upperwing has a white or tawny panel on the basal region of the inner primaries. Underwing has a large white or tawny panel and/or window on the basal area of all primaries and darker, rufous-colored secondaries. All remiges are finely barred.** TAIL.—**Moderately long. White with numerous irregularly shaped black bands and a moderately wide black subterminal band.** *Note:* Tawny areas of the head and underparts fade to whitish by winter.

**ABNORMAL PLUMAGES:** None documented in North America.

**HABITAT: Summer.**—Arid, hot regions with permanently flowing, clear, shallow streams and rivers. All regions have semi-open or dense groves of gallery forests of mature cottonwood or sycamore trees. Sometimes found in areas with isolated, tall mature trees. Breeding areas can be in rugged, deep canyon terrain or open, gentle riparian terrain. In Arizona and New Mexico, elevation ranges from 3,000 to 5,000 ft. (900–1,500 m), but in parts of Texas lower elevations are occupied. Breeds up to 5,300 ft. (1,600 m) in Sonora, Mexico.

**Winter.**—Found in freshwater wooded riparian areas but without regard to tree size. In Mexico, also inhabits mangroves swamps in brackish and salt water. Climate can be arid, semiarid, or humid.

**HABITS:** Temperment varies from moderately wary to tame. Generally a very reclusive species. Perches on exposed or concealed branches, rocks, and on the ground. In the U.S., rarely perches on utility poles and does not perch on poles or trees along highways. In Mexico, regularly perches on utility poles and wires along highways.

**FEEDING:** Strictly a perch hunter. Primarily forages in shallow, clear water. When foraging, perches for extended periods on near-water branches of standing trees and waterside fallen branches and rocks, or slowly walks along the shore. Clear water provides optimum hunting success. Hunting success may be temporarily impeded during periods of silty water caused by spring runoff and heavy summer rains. Small fish, amphibians, and reptiles are primary food items. Small mammals and birds and large insects are occasionally eaten. Along deep lakes, rivers, and streams in Mexico, Common Black-Hawks capture small fish swimming near the surface by flying out from a perch and snatching them with their feet. Will also plunge feet-first and submerge most of their underparts when capturing fish that are farther below the surface. May also pirate fish caught by Neotropical Cormorants.

**FLIGHT:** Although appearing awkward because of immensely wide wings, Common Black-Hawks are quite agile. Powered flight is with moderately slow, deep wingbeats. Wings are held on a flat plane when soaring and gliding. Although this species is often seen flying very low, leisure and courtship flights may occur at high altitudes. Common Black-Hawks hover briefly when hunting over deep water and may also kite briefly in wind deflections along cliff or woodland edges.

**VOICE:** Loud, high-pitched, haunting, metallic *kree, kree, kree, he-he-he*. The last *he-he-he* sequence decreases on volume. Also, may do a partial sequence of *kree, kree, kree*. Calls sound much like a laugh and similar to some calls of the Bald Eagle. Highly vocal during the breeding season and moderately vocal during other seasons.

**STATUS AND DISTRIBUTION: Summer.**—*Very uncommon* to *uncommon* and very local in the U.S. Found in very low density even in optimum habitat. Approximately 220–250 pairs in cen. Arizona, sw. New Mexico, and w. Texas. Population is probably stable but guarded. Highly dependent on sustained, uncontaminated riparian habitat with sufficient old-growth cottonwood and sycamore trees for nest sites. Status is not federally listed in the U.S. *Arizona:* Uncommon. Proposed listing as Threatened or Endangered. Found in drainages of the Mogollon Rim and Bill Williams watershed in Yavapai, Gila, n. Pinal, Graham, and Greenlee Cos. Northward extension occurs, in very low numbers, along the Virgin River. Eighty to ninety percent of U.S. pairs reside in this state. *New Mexico:* Uncommon. Listed as Endangered. Found along the San Francisco, Gila, and Mimbres Rivers in sw. Catron, Grant, and n. Hidalgo Cos. Has bred in n. Bernalillo Co. Recent breeding in Lincoln Co. *Texas:* Very uncommon. Listed as Threatened. Primarily found in the Davis Mts., Jeff Davis Co., but has also nested irregularly in Brewster, Tom Green, and Val Verde Cos. Has bred in Lubbock Co. *Utah:* Very uncommon to rare. Breeds irregularly along the Virgin River in Washington Co. *Mexico*: Fairly common. More common along the Pacific Slope.

**Extralimital summer.**—Records are widespread. *California*: An adult at Thousand Palms, Riverside Co., in mid-Apr. 1985. An adult for the second state record from late Mar. to early May 1997 at the north end of the Salton Sea in Riverside Co. *Nevada*: At least seven records in Apr., May, and Sep.: (1) Corn Creek near Las Vegas, Clark Co., in late May 1978; (2) Davis Dam near Laughlin, Clark Co., in late Sep. 1979; (3) Meadow Valley Wash near Elgin, Lincoln Co., in 1979 (no month given); (4) Las Vegas, Clark Co., in mid-Sep. 1980; (5) Fort Mohave in early Apr. 1980; (6) Davis Dam, Clark Co., in early Apr. 1984; (7) n. Clark

Co. in mid-May 2001. *Texas:* An adult was in Lubbock and Amarillo in Apr. 1999. *Colorado:* (1) Douglas Co. in Jun. 1980; (2) Baca Co. in Jun. 1991; (3) Durango, La Plata Co., in early Apr. 2001 (adult); (4) Montrose, Montrose Co., in Jun. 2001 (adult); and (5) Las Animas Co. in mid-May 2001 (subadult). *Minnesota*: One record of a bird molting into adult at Bemidji in Sep. 1978. Origin of this bird is questionable as several species of Mexican raptors were regularly sold in U.S. through the 1970s.

**Winter.**—U.S. birds probably winter in Mexico. One nestling banded in Arizona was found in Durango, Mexico, the following winter. In w. Mexico, northern winter range reaches the southern portion of Rio Yaqui in s. Sonora. Occasional along lower Rio Grande in s. Texas.

**Movements.**—Migrations of the U.S. population are short to moderate in distance. *Fall migration*: Oct. for all ages. *Spring migration*: Mar. to early Apr. for adults. Juveniles probably return to areas near natal locales in May and Jun.

**NESTING:** Mar. to mid-Sep.

**Courtship (flight).**—*High-circling, sky-dancing, leg-dangling*, and *exaggerated-flapping* (*see* chapter 5). Often vocalizes in courtship displays except in sky-dancing.

Nest trees are near streams or rivers and situated in groves of dominant, old-growth sycamores or cottonwoods that line the water's edge; occasionally in isolated tall trees. Nests are placed in primary or secondary crotches 30–90 ft. (9–27 m) high but average 45–75 ft. (14–22 m). An old nest may be reused or a new one built, often in same tree. Regularly builds on top of old Zone-tailed Hawk or Cooper's Hawk nests. Twigs and small branches for building nests are taken from the nest tree. All twigs and branches are broken off with bill. Nests are decorated with greenery. One or 2 eggs are laid in mid- to late Apr. and are incubated mainly by female for about 37 days. Fratricide occurs on occasion. Youngsters branch in 33–47 days and fledge in 41–52 days (Jul. to early Aug.). They are dependent on parents for another 50 days, until about 110 days old (from late Aug. to early Sep.). Fledglings remain in natal territory until the fall migration. Nests are often placed in close proximity to Zone-tailed Hawk nests without territorial interaction. Breeding probably does not occur until 3 years old. *Note:* Except during egg-laying and incubation, Common Black-Hawks can withstand a moderate amount of human disturbance during the nesting cycle, but they abandon nests if disturbed too often.

**CONSERVATION:** Major concern is the preservation of fragile riparian breeding habitat in the U.S. In Arizona and Texas, numerous pairs are in areas that have become popular recreational locales and are susceptible to considerable human pressure during the nesting season.

**Mortality.**—Electrocution by power lines. Illegal shooting occurs on small scale in U.S. and Mexico. In Sonora, Mexico, riparian habitat destruction of large trees adjacent to agricultural conversion areas exists.

**SIMILAR SPECIES:** COMPARED TO ADULT.—(1) **Black Vulture.**—FLIGHT.—Similar shape and size. Large white panel on upper- and underside of outer primaries. The secondaries are uniformly black. Rapid, snappy wingbeats. Solid black tail. **(2) Zone-tailed Hawk, adult.**—May share identical riparian and canyon habitat; regularly nests and seen in close proximity. PERCHED.—All-black bill, but may have very narrow pale bluish basal region. White forehead and lores. Wingtips extend to, or slightly beyond, the tail tip. Short tarsi. Uppertail bands are gray. The single white band on the underside of closed tail is identical. FLIGHT.—Moderately narrow wings. Undersides of remiges uniformly gray and completely barred. Wings held in dihedral; unstable flight, rocks back and forth. Feet extend into mid-section of black undertail coverts. Tail as in Perched. Tail separable when fanned, with two or more fairly broad white bands. Use caution on dorsal side: widely fanned tail exposes white inner bands, making bands appear white. Voice: *keeyah.* **(3) Rough-legged Hawk, adult.**—FLIGHT.—Narrow wings. Underside of remiges white with some barring. Underside of tail similar with single broad white band. Feet extend into black undertail coverts. Range overlap in Oct. and Mar. **(4) Golden Eagle, juvenile.**—Compared to dark underwing type, underwing is uniformly dark. Tail pattern, with one broad white band, is similar. Feet extend into undertail coverts. COMPARED TO JUVENILE.—**(5) Gray Hawk, juvenile.**—PERCHED.—Smaller, but size is deceptive, especially of a large female. Similar head pattern with dark eyeline and malar mark. Similar ventral markings, but flanks are streaked. Bill is all dark. Tail is brown with dark bands. Wingtips much shorter than the tail tip. **(6) Red-tailed Hawk, juvenile intermediate morph.**—PERCHED.—Lacks dark eyeline. Dark malar stripe only on jaw region. Pale irises. FLIGHT.—Upperwing surface has very large, pale-colored panel on most of primaries. Narrow, neat dark tail bands.

**OTHER NAMES:** Black-Hawk. *Spanish:* Aguilillа-Negra Menor. *French:* Unknown.

---

REFERENCES: Alcorn 1988; Arizona Breeding Bird Atlas 1993–99; del Hoyo et a1. 1994; Howell and Webb 1995; Johnsgard 1990; Maxwell and Husak 1999; Palmer 1988; Rottenborn and Morlan 2000; Russell and Monson 1998; Schnell 1994, 1998; Small 1994; Wheeler and Clark 1995.

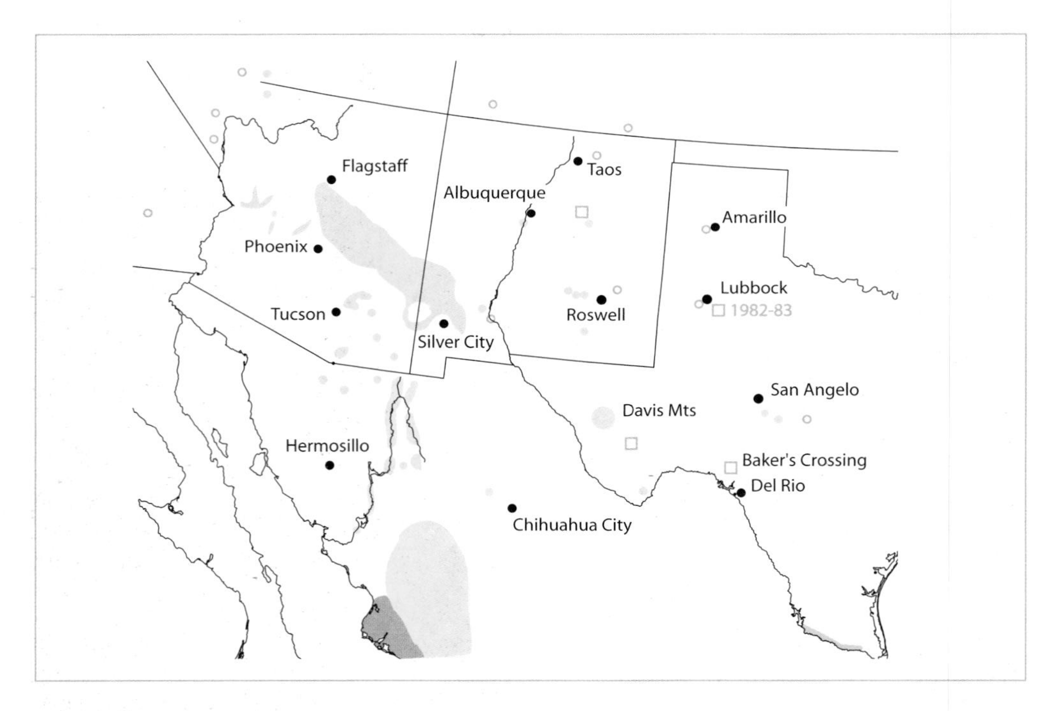


**COMMON BLACK-HAWK,** *Buteogallus anthracinus:* Very uncommon to uncommon. Very local. 220–250 pairs in U.S.

**Plate 205. Common Black-Hawk, adult male** [**Feb.**] ▪ Bill black with pale yellow base. Yellow cere and pale yellow lores. Black head, including under eyes. ▪ Body black with grayish feather edgings. Long yellow tarsi. White tips on leg feathers. ▪ Brownish black wing coverts. Secondaries marbled with gray. ▪ Broad white tail band barely shows behind broad secondaries. ▪ *Note:* Nayarit, Mexico. Males typically more blackish than females.

**Plate 206. Common Black-Hawk, adult female** [**Feb.**] ▪ Bill black with pale yellow base. Yellow cere and pale yellow lores. Whitish area below eyes. ▪ Grayish black body with brownish tone. Long yellow tarsi. ▪ Brownish wing coverts. Secondaries marbled with gray. ▪ Broad white tail band barely shows on top and bottom of tail. ▪ *Note:* Nayarit, Mexico.

**Plate 207. Common Black-Hawk, adult male** [**Feb.**] ▪ Black body. ▪ Broad wings have pale rufous-brown or grayish remiges that are narrowly barred; wide black band on rear edge. ▪ Broad white band on mid-tail; narrow white terminal band. Narrow white tips on uppertail coverts. ▪ *Note:* Nayarit, Mexico.

**Plate 208. Common Black-Hawk, adult female** [**Apr.**] ▪ Grayish black body. ▪ Broad wings with pale rufous remiges that are narrowly barred; wide black band on rear edge. Whitish patch on outer 2 or 3 primaries. Six "fingers" on outer primaries. ▪ Black tail has broad white mid-tail band, narrow white terminal band. ▪ *Note:* Graham Co., Ariz.

**Plate 209. Common Black-Hawk, juvenile [Feb.]** ▪ Short, thick, dark eyeline; dark malar stripe extends onto sides of neck. Pale blue base of bill. Yellow cere. ▪ White streaking on back. Dark brown scapulars. ▪ Dark brown wing coverts with some pale mottling. Finely barred secondaries. ▪ White tail with black bands. ▪ *Note:* Nayarit, Mexico.

**Plate 210. Common Black-Hawk, juvenile [Jan.]** ▪ Short, thick, dark eyeline; dark malar stripe extends onto sides of neck; pale, streaked neck and hindneck. Pale blue base of bill. Yellow cere. ▪ Dark brown upperparts. Streaked white breast and belly, solid brown flanks, barred leg feathers. Long tarsi. ▪ *Note:* Veracruz, Mexico.

**Plate 211. Common Black-Hawk, juvenile [Feb.]** ▪ Dark malar stripe extends onto sides of neck. ▪ Streaked underparts with dark flanks, barred leg feathers. ▪ Broad wings have 6 "fingers" on outer primaries. Pale window on inner primaries; secondaries tawny-rufous, narrowly barred. ▪ White tail has wide black terminal band with irregular, narrow inner black bands. ▪ *Note:* Nayarit, Mexico.

**Plate 212. Common Black-Hawk, juvenile [Feb.]** ▪ Dark brown upperparts; black rump. ▪ Finely barred remiges; pale tawny or white panel on dorsal side of primaries (not shown at this angle). ▪ White tail has wide black terminal band with irregular, narrow inner black bands. Black uppertail coverts. ▪ *Note:* Nayarit, Mexico. Same bird as on plates 209 and 211.

## HARRIS'S HAWK

### *(Parabuteo unicinctus)*

**AGES:** Adult and juvenile. Sexes are similar in adults. In juveniles, there is considerable variation on the ventral markings.

**MOLT:** Acciptridae wing and tail molt pattern (*see* chapter 4). First prebasic molt begins on innermost primary (p1) and continues outward. Secondary molt begins after several primaries have molted (Hamerstrom and Hamerstrom 1978 [photos]). After primaries have started molting, head, breast, back, forward scapulars, and upperwing coverts begin molting. Molt then begins on the belly, flanks, and leg feathers. All regions then molt in unison. Molt may be apparent in any month, depending on when hatched. Subsequent prebasic molts may also be active in any month but are mainly noted spring through fall. Food availability and variable nesting cycles may affect molt timing in adults.

**SUBSPECIES:** Polytypic with three races. Two races, *P. u. harrisi* and *P. u. superior*, are currently recognized in North America. The validity of splitting the two North American races, which is based only on subtle differences in size, is doubtful (Bednarz 1988, 1995). *P. u. superior*, of w. Mexico, California, and Arizona, is not consistently larger than *harrisi*. Following Bednarz (1988, 1995) they are treated here as a single race (*harrisi*). Nominate *P. u. unicinctus* is patchily distributed throughout South America to cen. Argentina.

**COLOR MORPHS:** None.

**SIZE:** A large raptor. Within geographic populations, sexually dimorphic with little overlap. Males are generally much smaller than females and visibly dimorphic under field conditions. Listed sizes are an average for both sexes: sexual distinction for the various major populations is available only for wing chord and tail length (*see* Bednarz 1995). Length: 18–23 in. (46–58 cm); wingspan: 40–47 in. (102–119 cm).

**SPECIES TRAITS:** HEAD.—**Bill is short and deep. Distal third (mainly the hook portion) is blackish, the basal two-thirds pale bluish. Yellow cere blends with the yellow lores. Prominent yellowish green skin on supraorbital ridge and orbital ring.** Iris color is orangish brown, medium brown, or dark brown. BODY.—**White uppertail and undertail coverts** (*see* Tail). Long yellow tarsi. WINGS.—**Rufous "shoulders" on all upperwing coverts. Rufous median and lesser underwing coverts and axillaries. In flight, wings are moderately long and broad. The trailing edge of the secondaries is bowed and the wingtips are very rounded. Six "fingers" are visible on the outer primaries in flight.** TAIL.—**Long. Black on dorsal surface with a white basal area on the outer rectrices and blending as a "white" unit with upper- and undertail covert. When perched, wingtips extend half way down the tail.**

**ADULT TRAITS:** HEAD.—Cere, lores, and orbital skin are bright yellow; however, lores are sometimes whitish yellow. Uniformly dark brown head. BODY.—Uniformly dark brown back, scapulars, breast, belly, and flanks. **Solid rufous leg feathers.** Bright yellow tarsi. WINGS.—**Uniformly dark gray remiges with black "fingers" on the outer primaries.** Uniformly dark brown dorsal surface of secondaries and greater coverts. TAIL.—**Black on both dorsal and ventral surfaces with a very broad white terminal band.**

**JUVENILE TRAITS:** HEAD.—Cere and lores are medium or bright yellow, and the orbital skin and supraorbital ridge skin are greenish yellow. Head is dark brown with some faint, pale tawny streaking on supercilium, auricular region, and sides of the nape. BODY.—Considerable variation on ventral markings. Legs are light grayish, pale yellow, or bright yellow and probably brighten with age. WINGS.—Greater upperwing coverts and secondaries are faintly or moderately barred. Greater primary underwing coverts are white with gray barring. **Inner four primaries and basal region of the outer six primaries are white and form a white "panel," or "window," that contrasts with the darker, pale or medium gray, finely barred secondaries.** Wings, though broad, are somewhat narrower than those of adults. TAIL.—**Ventral surface is pale or medium gray, crossed with numerous thin dark bands, and has either a narrow or broad white terminal band.**

**JUVENILE (LIGHTLY MARKED TYPE):** BODY

(ventral).—Belly, lower belly, and forward portion of flanks are white and narrowly streaked with dark brown. The white area of the feather is 50% or greater than brown area. Lower belly is often paler because of fewer dark markings, which instead of streaking may be spotted, arrowhead-shaped, or barred. Dark brown markings on distal portion of flanks are arrowhead-shaped or bars. *Streak-bellied type:* Breast varies from being nearly solid dark brown or slightly streaked with white and forming a very distinct dark bib against the pale belly and flanks. *Streaked type:* Breast is uniformly streaked along with the rest of underparts and lacks a bib demarcation between breast and belly. On both variations, the leg feathers are white with narrow rufous barring.
**JUVENILE (HEAVILY MARKED TYPE):** BODY (ventral).—Dark brown belly, lower belly, and flanks are narrowly streaked or mottled with white. The white area of each feather is less than 50% of the brown area and contrasts a moderate or minimum amount with the darker bib of the breast. Leg feathers are solid rufous or have narrow, white barring.
**ABNORMAL PLUMAGES:** None documented.
**HABITAT: Permanent resident.**—*Arizona:* Seasonally hot, arid deserts with semi-open paloverde-cacti thornscrub, particularly with paloverde-saguaro. Also thornscrub areas with scattered tall cottonwood trees along riparian tracts. In the southeast, semi-open mesquite-hackberry thornscrub savannahs. *New Mexico:* Arid and seasonally hot mesquite-cacti and mesquite-oak thornscrub savannahs. *Texas:* Arid and seasonally hot interior regions with semi-open or fairly dense tracts of mesquite-cacti thornscrub, and mesquite savannahs in some northwestern regions. Coastal areas are humid and seasonally hot, with lush, semi-open savannahs or fairly dense tracts of thornscrub with mesquite-cacti and scattered tracts of live oaks and other tall trees. *Mexico:* Similar to many of the above regions in the U.S.

In all geographic areas, access to water is a necessity. Territories encompass a permanent water source such as a ground seepage, stream, river, lake, or livestock water tank.
**HABITS:** Tame, except in New Mexico, where fairly wary. A very social species. Breeding occurs in pairs or in groups of up to seven birds of related or unrelated individuals of both sexes. Nonbreeding groups may comprise up to 11 birds. In groups, a hierarchy system is practiced: a dominant pair (alpha male/female) and subordinate "helpers." Helpers are assigned to three categories: (1) *Related helpers* (auxiliary/gamma helpers) are offspring, most often males, of the alpha pair that may stay with their parents for up to 3 years. This occurs in all geographic populations. (2) *Unrelated helpers* (beta helpers) are either juveniles or adults from other regional populations. This mainly occurs in Arizona, occasionally in New Mexico. (3) *Alpha-2 females* are subordinate to alpha females but are dominant over alpha males and other individuals in a group. This category is seen only in some Arizona groups.

"Backstanding," with one hawk perched on top of another, is common in Arizona, uncommon in New Mexico, and rare or absent in Texas. Harris's Hawks mainly perch on exposed branches, utility poles, and posts. During hot periods, they seek densely foliaged shaded areas. They drink and bathe daily.
**FEEDING:** Mainly a perch hunter but engages in low-altitude aerial hunting. Groups hunt cooperatively to flush, intercept, and corner prey. Perch hunting involves long stints of sitting and waiting or short perching stints with short flights between perching. Aerial hunting is generally used while in random flights between short stints of perching; hunting flights rarely begin from high altitudes. Prior to a hunt, members of a group may perch together, often close to each other. Three hunting methods are used: (1) *Surprise pounce.*—The prey species is in an exposed situation and is targeted by several members of the group converging from multiple directions. (2) *Flush-and-ambush.*—Prey is hidden under dense foliage and one or more Harris's walks into the hiding spot and flushes the prey into an exposed area, and other members are perched and waiting for the prey to escape. (3) *Relay attack.*—Long chase involving rabbits in which the lead attackers are alternated within the group.

Harris's Hawks mainly hunt medium-sized prey, particularly cottontail rabbits (Desert and Eastern), and larger prey up to the size of Black-tailed Jack Rabbits. Small rodents such as wood rats, ground squirrels, and gophers;

medium- and small-sized birds such as quail and songbirds; and lizards form the varied diet. Snakes are rarely eaten.

**FLIGHT:** Wings held in an "arch" when gliding and soaring. The inner portion of the wings is held upwards and outer portion is bowed downwards. Wings are occasionally held on a flat plane when soaring. Powered flight consists of a sequence of several flaps interspersed with long glides. Flight is often at very low altitudes, and birds negotiate highly maneuverable antics with their short, broad wings and long tail. They may soar to high altitudes. Hover at very low altitude for short periods when waiting on prey hidden in dense cover. Harris's Hawks do not kite.

**VOICE:** Typically vocalizes when alarmed or disturbed. Most common alarm call is a distressed, grating, drawn-out *irrrr;* the intensity wanes toward the end of the vocalization. Also when alarmed, a drawn-out, raspy, guttural *caaa.* When copulating, females utter a series of soft chirps. Males emit a repetitive croaking note before copulating. Other calls of adults are soft and rarely heard at field distances. Nestlings and fledglings give a high-pitched, screeching *kree, kree, kree* when food-begging.

**STATUS AND DISTRIBUTION:** *Common* permanent resident in all regions. Population is stable but undergoing pressure from ongoing habitat loss. *Texas:* Considerable range reduction has occurred in southern and western portions of the state. Habitat reduction began in the 1920s with agricultural conversion and loss of mesquite habitat, particularly in lower Rio Grande valley counties of Hidalgo, Willacy, and Cameron. A considerable amount of mesquite eradication is still occurring in s. Texas and habitat is becoming fragmented. Population density is still high throughout most of s. Texas. In w. Texas, range and status are irregular and patchy, and local status varies from uncommon to very uncommon. *New Mexico:* Two main population hubs: (1) Lea and Eddy Cos., and (2) small numbers in Hidalgo and Luna Cos. A very small population was recently discovered in Otero Co. Up to three individuals have been seen near Alamogordo since 1998, including a nesting pair in the spring of 2001. A nesting pair was also found near Tularosa in the spring of 2001. *Arizona:* Range expansion has been slowly occurring possibly because of birds being displaced from previous habitat by increased urban sprawl. *Mexico:* Fairly common to common in mapped area. *California:* A small population is resident near Borrego Springs in sw. San Diego Co. Only one breeding pair is in this group, and it successfully produced young for the first time in the summer of 2000. Reintroduced resident along the lower Colorado River in Imperial Co. (*see* Conservation).

**Movements.**—Sedentary. However, juveniles and some adults are prone to engage in northward dispersal, particularly in fall and winter. Short movements typically occur in all regions, but long-distance journeys are irregular north of standard range. *California:* First documented as a breeding species in 1916, but occurred in the state in the 1850s. Breeding occurred along the lower Colorado River until the mid-1960s when alteration of the river ecosystem with dams eradicated plant communities needed to support Harris's Hawks. Reintroduction (*see* Conservation) has again made the species a resident in this region. A natural incursion of individuals from Baja California population and probably Arizona has been occurring in s. California (and n. Baja California, Mexico) since the 1990s. Over 50 sightings have occurred in Imperial, San Diego, Riverside, San Bernardino, and Los Angeles Cos. Sightings have occurred Dec.–Jul. *Nevada:* At least six records, mainly from Clark Co. in the spring. *Colorado:* Seven accepted records, from Dec. (6) and Oct. (1) in 1994 and 1995. Records occur in Jefferson (5), Larimer, and Otero Cos. *Nebraska:* Two records, the most recent in Jan. 1994 in Scotts Bluff Co. *Oklahoma:* At least seven records from 1987 to 1995, spanning Oct.–Jan. Sightings are from Muskogee, Cleveland, Jackson, Garvin, and Osage Cos. *Kansas:* Five records from 1918 to 1973 spanning Dec.–Jan., except for one breeding record in late May. Sightings have been in Douglas, Sedgwick, Mitchell, Barton, and Meade Cos. Most recent record is for Nov. and Dec. 2000 for Arkansas City, Cowley Co. Only documented breeding for the state (and most northern record in the U.S.) occurred in 1963 in Meade S.P., Meade Co. (the nest, with one nestling, was destroyed by a storm). *Missouri:* One record, Feb. 1995 in Boone Co. *Eastern*

*U.S.:* Reported from several states east of the Mississippi River; however, many records are questionable since they may be falconry escapees. *Note:* Harris's Hawk is a popular falconry species because of its docile nature. However, birds often escape and may be encountered in the wild. Most escaped individuals retain leather straps (jesses) on the legs or perhaps telemetry antennae on the tail.

**NESTING:** Breeding occurs year-round in all regions. Pairs remain together as long as mates survive. If there is ample prey, double and sometimes triple clutches are laid. Breeding may not occur if there is not adequate prey. Two peak periods for nesting: (1) winter and early spring and (2) late summer. Nest-building or nest-repairing occurs mainly in Jan. and Aug. in Arizona, Mar.–Apr. and Aug.–Sep. in New Mexico.

Some populations of Harris's Hawks have a complex breeding strategy, based on a social hierarchy. *Arizona:* Primarily polyandry, but polygyny is also reported. *New Mexico* and *Texas:* Pairs mainly practice monogamy but may still have "helpers." Polyandry may occur on a limited basis. Alpha pair always mates; attempted mating occurs with beta males. Alpha female incubates, but alpha males may assist for short periods. Auxiliary and gamma adult helpers may also incubate for short stints. Alpha-2 females and beta males may also brood for short periods. Alpha males and helpers generally hunt and ward off predators. Harris's Hawks are more likely to have multiple broods if a breeding (alpha) pair is in a social group.

**Courtship (flight).**—*Sky-dancing* (*see* chapter 5). Multiple-group display noted with three pairs involved in soaring, tail-chasing, and diving antics.

Nests are constructed of large sticks and built by alpha pairs. As many as four nests may be built in a territory; the alternate nests are often used as feeding platforms. Average nests are 18 in. (46 cm) diameter, 9 in. (23 cm) deep, and lined with fresh sprigs, grass, weeds, or feathers. Nests are placed 5 ft. (2 m) to 25 ft. (8 m) high. Trees, Saguaro Cacti, and utility poles are typical nest locations, humanmade structures are also used, as are cliffs. Nests are often built on mistletoe, which possibly aids in concealment. One to 5 eggs but mainly 3–4. The eggs are incubated for 34–35 days. Young males fledge in about 45 days, females in about 48 days. Fledglings remain in the natal territory for an additional 2–3 months, but up to 3 years if they become helpers. In groups, fledglings may join cooperative hunts 2 months after fledging.

**CONSERVATION:** No measures are implemented in Texas, New Mexico, or Arizona. *Texas:* Habitat is greatly reduced and vital mesquite habitat is being constantly bulldozed, burned, and converted to agricultural or grazing lands, highway corridors, and urban areas. Currently, there is a considerable amount of habitat being lost between San Antonio and Laredo. Historical loss of mesquite first occurred in the 1920s when the agricultural boom took hold with vegetable crops and citrus groves in the south and cotton fields in the west. Harris's Hawks are still found in isolated pockets of habitat in the lower Rio Grande valley, mainly in parks, refuges, and small, fragmented thornbush tracts. The important, ever-expanding tracts of the Lower Rio Grande NWR may help regain some previously lost habitat. An extensive amount of mesquite habitat exists north of the valley and east of Interstate 35. *Arizona:* Urban sprawl between Tucson and Phoenix is possibly affecting Harris's Hawks. However, many are adapting to suburban life as seen in Phoenix and Tucson. Suburban nesters still need ample-sized, nearby habitat tracts for foraging. *New Mexico:* Population may be affected by loss of rangeland habitat, but little else is known or substantiated. *California:* About 200 Harris's Hawks were reintroduced along the lower Colorado River in Imperial Co. from 1979 to 1989. Nestlings and fledglings were either hacked or cross-fostered with Red-tailed Hawks at Cibola NWR, Imperial NWR, and Mittry Lake. Since 1985, a few pairs have bred from the released stock. *Note:* Breeding also occurs on the Arizona side of the Colorado River. This population, however, is questionably viable and may suffer in the future because ample habitat has not been restored along the Colorado River ecosystem.

**Mortality.**—Illegal shooting, disturbance of autumn nesting birds during fall hunting season, electrocution from power lines, drowning in stock tanks, and falconry. Natural mortality occurs with predation by Great Horned Owls,

Common and Chihuahuan Ravens, Bobcats, and Coyotes.

**SIMILAR SPECIES: (1) Dark buteos.**—PERCHED.—None have rufous upperwing coverts. FLIGHT.—Only adult dark Swainson's Hawks have similar rufous underwing coverts, but they have very pointed wingtips. **(2) Hook-billed Kite (Texas only).**—FLIGHT.—Shares similar wing shape and flight mannerisms, but smaller in size. Base of tail and uppertail coverts are dark. **(3) Red-shouldered Hawk, juveniles of *B. l. alleni/extimus* (mainly Texas and California).**—PERCHED.—Similar to lightly marked type juvenile with rufous upperwing coverts, streaking on ventral area, and long tarsi. Dorsal surface of tail is distinctly banded and lacks white uppertail coverts. FLIGHT.—Similar "arched" wing position. Exhibits pale, crescent-shaped panels/windows on base of outer primaries and has distinct, narrow tail banding.

**OTHER NAMES:** Bay-winged Hawk. *Spanish:* Aguililla Conejera, Aguililla Rojinegra, Aguililla de Harris. *French:* Buse de Harris.

REFERENCES: Alcorn 1988; Arizona Breeding Bird Atlas 1993–99; Baumgartner and Baumgartner 1992; Bednarz 1983, 1987, 1988, 1995; Burt and Grossenheider 1976; Dawson and Mannan 1995; Griffin 1976; Howell and Webb 1995; McCaskie and Garrett 2000; Patten and Erickson 2000; Russell and Monson 1998; Stewart 1979; Thompson and Ely 1989; Walton et al. 1988.

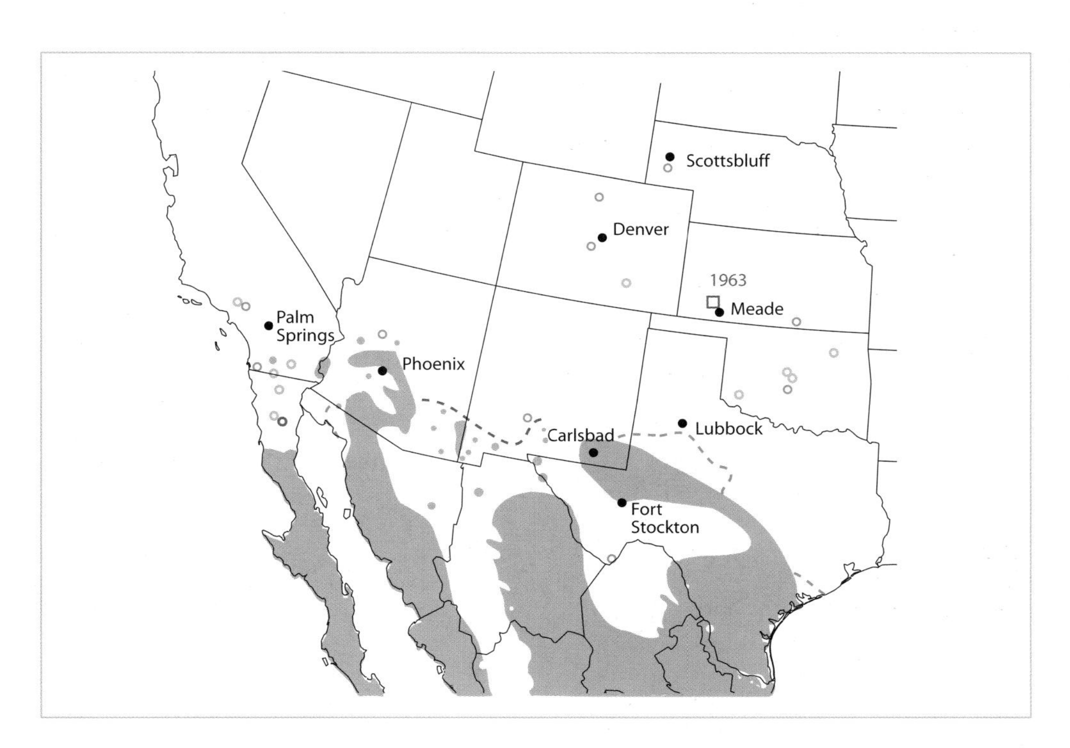


**HARRIS'S HAWK,** *Parabuteo unicinctus* (includes *P. u. harrisi* and *P. u. superior*): Common, but losing range in s. TX due to land use change. Recent incursion in s. CA. Winter dispersal within blue dashed lines.

**Plate 213. Harris's Hawk, adult** [**Feb.**] ▪ Bluish bill with dark tip; yellow cere, lores, supraorbital ridge, and eye ring. Dark brown head. ▪ Dark brown body. Rufous leg feathers. ▪ Large rufous shoulder patch. ▪ Long black tail with broad white terminal band.

**Plate 214. Harris's Hawk, adult** [**Dec.**] ▪ Dark brown body. Rufous leg feathers. ▪ Rufous underwing coverts; dark gray remiges with black primary tips. ▪ White undertail coverts blend with white base of long black tail; broad white terminal band.

**Plate 215. Harris's Hawk, adult** [**Feb.**] ▪ Dark brown body. ▪ Large rufous shoulder patch. ▪ White uppertail coverts blend with white base of long black tail; broad white terminal band.

**Plate 216. Harris's Hawk, juvenile** (**heavily marked type**) [**Feb.**] ▪ Bluish bill with dark tip; yellow cere, lores, supraorbital ridge, and eye ring. Dark brown head. ▪ Dark brown body with small amount of white mottling on belly and flanks. Rufous leg feathers. ▪ Large rufous shoulder patch. ▪ Long tail with narrow white terminal band.

**Plate 217. Harris's Hawk, juvenile (lightly marked type)** [**Feb.**] ▪ Bluish bill with dark tip; yellow cere, lores, supraorbital ridge, and eye ring. Tawny auriculars. ▪ Streaked type underparts uniformly streaked, including breast. White leg feathers barred with rufous. White undertail coverts. ▪ Long gray tail with narrow black bands and narrow or wide white terminal band.

**Plate 218. Harris's Hawk, family group (adult female, juvenile male, and adult male)** [**Feb.**] ▪ Distinct size difference between sexes. ▪ Juvenile males often stay with parents for a few years as helpers. ▪ Juvenile is 1 year old and beginning to molt; white terminal tail band has worn off.

**Plate 219. Harris's Hawk, juvenile (heavily marked type)** [**Feb.**] ▪ Dark breast contrasts with streaked belly and flanks. Rufous leg feathers. ▪ Rufous wing coverts. Whitish primaries form pale panel. ▪ White undertail coverts blend with white base of finely banded tail; narrow white terminal band. ▪ *Note:* Beginning first prebasic molt with new (black) adult innermost primary (p1).

**Plate 220. Harris's Hawk, juvenile (lightly marked type)** [**Feb.**] ▪ Streaked type with dark breast contrasting against lightly streaked belly and flanks. White leg feathers barred rufous. ▪ Rufous wing coverts. Whitish primaries form pale panel. ▪ White undertail coverts blend with white base of finely banded tail; narrow white terminal band.

# GRAY HAWK
## *(Asturina nitida)*

**AGES:** Adult and juvenile. Adult plumage is attained when 1 year old. Juvenile plumage is held for nearly 1 year. Juveniles have considerably longer tails than adults, but the width of the wing is the same for both ages.
**MOLT:** Accipitridae wing and tail molt pattern (*see* chapter 4). First prebasic molt is a complete molt. Molt begins on the head, forward scapulars, and breast. Remiges and rectrices begin molt later. First prebasic molt from juvenile to adult plumage may begin in Jan. on some individuals in s. Texas and along the Atlantic slope in Mexico. Juveniles seen on the Pacific slope from Nayarit to Colima, Mexico, which may be resident birds or wintering Arizona birds, do not exhibit molt in late Feb. (BKW pers. obs.). In Arizona birds, head molt is often not visible, and remix and rectrix molt does not begin until mid- to late May (B. Small pers. comm.; photos). Remix and rectrix molt probably begins in mid- to late May. Birds have particularly ratty plumage in summer when in the midst of molt. Subsequent adult prebasic molts begin later, in Jun. and/or Jul., and are completed by late fall.
**SUBSPECIES:** *A. n. plagiatus* is the only race in the U.S. Primary range is Mexico, from Sonora south along the Pacific slope and from Tamaulipas and Nuevo León south along the Gulf slope; Mexican ranges join in Oaxaca. From Mexico, range extends south to nw. Costa Rica. Three additional races (following del Hoyo et al. 1994): *A. n. costaricensis* is in sw. Costa Rica to n. Colombia and w. Ecuador. Nominate *A. n. nitidus* is found in e. Colombia, e. Ecuador, e. Venezuela, the Guianas, and south to the Amazon River in Brazil. *A. n. pallidus* inhabits s.-cen. Brazil to e. Bolivia, south to Paraguay and n.-cen. Argentina. *Note:* Gray Hawk was reassigned by the AOU (1997) from the genus *Buteo* to *Asturina.* It is also closely related to Red-shouldered and Roadside hawks, both currently assigned to *Buteo.* A Gray Hawk has mated with a Red-shouldered Hawk in Texas.
**COLOR MORPHS:** None.
**SIZE:** A medium-sized raptor and a small buteo. Males are smaller than females, with little or no overlap. Measurements, however, are available only as an average of both sexes. Length: 14–18 in. (36–46 cm), with juveniles appearing longer because of longer rectrices; wingspan: 32–38 in. (81–97 cm).
**SPECIES TRAITS:** HEAD.—**Large brown irises. Large yellow cere.** BODY.—**White uppertail coverts form a narrow U-shaped mark which is visible mainly when seen from above. Long, thick tarsi and moderately long, thick toes.** WINGS.—**In flight, wings are moderately short with very rounded wingtips. Outermost primaries have five defined "fingers."**
**ADULT TRAITS:** HEAD.—White lores. Bright yellowish orange cere. **Dark brown irises.** BODY.—Breast, belly, flanks, and leg feathers are finely barred gray and white. Upperparts are gray or brownish gray; females are more brownish than males. WINGS.—**Underwing is uniformly whitish with fine dark barring on all but the base of the primaries.** TAIL.—**Fairly long. Black with a white terminal band and one moderately wide white band on the mid-section and one narrower inner white band. Some birds have a second narrow white band that is visible when tail is widely fanned when soaring.**
**JUVENILE TRAITS:** HEAD.—**Medium brown irises. Distinct wide dark eyeline, white supercilium and auricular region, and narrow dark malar stripe.** BODY.—Breast, belly, and flanks are broadly streaked with dark brown. The central area of the breast is unmarked and the sides of the breast are streaked. **Leg feathers are very finely barred with dark brown.** Dark brown back and scapulars. **White U-shaped uppertail coverts have a dark streak on the center of each feather.** WINGS.—**Secondaries and greater coverts are medium brown with distinct, narrow dark bars. There is also a small tawny strip on the base of the outer primaries that is visible as a panel on the dorsal surface. On the ventral surface, the rear edge has an indistinct narrow gray subterminal band (much less distinct than in any similar raptor).** The narrow tawny panel strip on the dorsal surface shows as a narrow pale window on the underside when viewed in translucent light. Pale wingtips are narrowly

barred. When perched, wingtips are much shorter than tail tip and only extend halfway down the tail. Underwing is uniformly white with a narrow gray trailing edge. Underside of all remiges is finely barred, including the five "fingers" on the outer primaries. TAIL.—**Very long. Medium brown on the dorsal side with a wide dark subterminal band and numerous dark, chevron-shaped bands that get progressively narrower toward the base of the tail. Dark undertail bands align neatly on all rectrices.**

**ABNORMAL PLUMAGES:** None documented.

**HABITAT: Summer.**—*Arizona:* Breeds in arid, hot regions with extensive tracts of mature thornbush of mainly mesquite and hackberry, which are adjacent to riparian gallery forests or tall, solitary trees. Prime areas are along stretches of permanent flowing water lined with tall, old-growth cottonwood trees. Gray Hawks have recently expanded their breeding habitat to dry wash areas in thornbush tracts that have tall, mature Emory Oaks and other larger tree species. *Note:* Seasonal or permanent water sources are generally needed only to supply nourishment to support tall trees required for nesting. Gray Hawks have successfully nested in close proximity to rural settlements, although not in urban areas. Elevation ranges from 2,000 to 4,300 ft. (600 to 3,100 m). *Texas:* Similar to Arizona habitat but at a lower elevation (near sea level). Big Bend N. P. area and the lower Rio Grande valley region are hot and humid. Riparian tracts with large trees, particularly cottonwoods in Big Bend N. P. and oaks in the Rio Grande valley, are favored for nesting.

**Winter.**—In Mexico, found in a wide variety of habitats, including areas that lack thornbush vegetation. Common in rural agricultural regions as well as more natural habitats. The peripheries of villages and towns are also inhabited.

**HABITS:** Moderately tame to tame. Exposed and concealed branches, utility poles, and wires are used for perches. Perches at any height but generally forages at low heights.

**FEEDING:** Perch hunter, sometimes an aerial hunter. Lizards, especially spiny lizard species, are primary prey. Also feeds on small birds and rodents and large insects. Prey is captured while it is on the ground or in low branches, in both perch and aerial hunting modes. Hunting takes place beneath canopies of large, mature thornbush tracts.

**FLIGHT:** Powered by regular cadence of several rapid, stiff-winged, shallow flaps interspersed with gliding sequences. Wings are held on a flat plane when gliding and soaring. When soaring, the front edge of the wings is rigidly perpendicular to the body. With short, rounded wings and a fairly long tail, Gray Hawks are highly maneuverable and weave around trunks and branches in the lower canopy of thornbush tracts with incredible agility. All aerial hunting is at low altitudes. Gray Hawks do not hover or kite.

**VOICE:** Extremely vocal in the breeding season. Two main calls, but there is considerable variation. All calls are clear, high-pitched, loud whistled notes. The very melodic three-syllable *kah-lee-oh* denotes territorial advertisement. It is regularly shortened to *kah-lee* and may be repeated numerous times. Also, a very drawn-out *keeoooh.*

**STATUS AND DISTRIBUTION: Summer.**—*Arizona:* Uncommon to common but very local summer resident. Approximately 80 pairs. Population is stable and appears to be slowly growing and expanding range. Preferred habitat, however, is very limited. Many pairs are expanding into isolated, albeit sometimes marginal, habitat north and west of historical areas. Major breeding strongholds are along the Santa Cruz River from Nogales north to the Santa Cruz Co. line, Sonoita Creek, and the San Pedro River south of the Gila River (mainly south of Interstate 10). The Santa Cruz River area harbors about one-third of the population. About six pairs currently nest at Buenos Aires NWR in Pima Co. Additional pairs are scattered around Nogales, Arivaca, Patagonia, and other locations with isolated pockets of cottonwoods and Emory Oaks in the southern part of the state. *Texas:* Listed as Threatened. Very uncommon to rare with only a few pairs locally breeding in the state. Big Bend N. P. normally hosts one or two pairs. Five spring and/or summer records exist for the Davis Mts., Jeff Davis Co. (including spring of 2001). In 2000, a nesting record was documented for the first time in the Davis Mts. A few pairs are permanent residents along the lower Rio Grande from Falcon Dam to

Brownsville. *Mexico* (mapped area):—Common permanent resident in appropriate habitat. **Extralimital: Summer.**—Rarely reported out of typical range. *New Mexico:* Six records from southern counties just across the Arizona border. Gray Hawks may irregularly nest in the state. All records are from the breeding season: (1) an adult and possible nest along the Gila River near Cliff, Grant Co., in Apr. 1992; (2) a juvenile in the same area in Jul. 1998; (3) a juvenile in first prebasic molt at Clanton Cienega in the Animas Valley, Hidalgo Co., in late Jul. 1994; (4 and 5) an adult and juvenile in first prebasic molt along Animas Creek in the Animas Valley, Hidalgo Co., from early Jul. to early Aug. 1996; and (6) an adult foraging on the New Mexico side of Guadalupe Canyon, Hidalgo Co. *Kansas:* An adult in mid-Apr. 1990 in Clay Co. is the state's only record (and the most northern record for the U.S.).
**Winter.**—Several Arizona-banded birds were recovered in Sinaloa, Mexico. A few winter records occur for s. Arizona. The s. Texas population may be resident, or a few may winter farther south into cen. and s. Mexico. Those breeding in Big Bend N. P. and the Davis Mts. head south for the winter. In w. Mexico, Gray Hawks winter as far north as Tonichi in the Rio Yaqui valley of se. Sonora.
**Movements.**—Arizona and n. Sonora individuals are short-distance migrants. Some Texas individuals may also be migratory, especially those of Big Bend N. P. and Davis Mts.

*Fall migration*: Sep.–Oct. Both ages migrate simultaneously.

*Spring migration*: Adults return in Mar. Returning 1-year-olds are first seen in early May. One-year-old juveniles probably return to natal areas. In Arizona, they have been seen in or near territories of adult pairs in early May, and even seen perched near adults on the same branch. The returning juveniles, however, are eventually expelled from adult territories.
**NESTING:** Late Mar.–Jul. Most birds probably do not breed until 2 years old when in full adult plumage. A male has nested with a female Red-shouldered Hawk (probable *alleni* race) at Big Bend N. P., Tex.

**Courtship (flight).**—*High-circling* and *sky-dancing* (*see* chapter 5). Extremely vocal during courtship. Moderate-sized nests are constructed of freshly broken off green twigs and lined with greenery. Nests are placed in the upper portion of large trees and well concealed by foliage. Gray Hawks typically nest in cottonwoods and Emory Oaks in Arizona and oaks or other large trees in Texas. Nests are 40–60 ft. (12–18 m) high but can be up to 100 ft. (30 m) high. The 2–3 eggs are incubated 33 days by female. Youngsters fledge in about 42 days and stay with their parents in natal territory for an undetermined period (possibly until migration). Some pairs nest near human-occupied areas. *Note:* Prolonged disturbance to nesting pairs will cause the adults to abandon nest sites.
**CONSERVATION:** Many areas are protected for this species. *Arizona*: The Nature Conservancy's Patagonia-Sonoita Creek Preserve and adjacent riparian areas owned by private landowners have given protection to several pairs nesting near Patagonia since 1975. The San Pedro Riparian National Conservation Area, established by the BLM in 1987, protects 36 miles (58 km) of cottonwood riparian habitat and numerous pairs along the San Pedro River between Benson and the Mexican border. Along the Santa Cruz River, old cottonwood trees are dying and there are no replacement younger trees for future Gray Hawk generations. *Texas:* Most pairs along the Rio Grande occur within segments of the important and ever-growing Lower Rio Grande NWR system.

Primary habitat concern, particularly in Arizona, is the preservation of an extremely fragile ecosystem.
**Mortality.**—Illegal shooting undoubtedly occurs.
**SIMILAR SPECIES:** COMPARED TO ADULT.—**(1) Hook-billed Kite, adult males.**—Texas only. PERCHED.—Greenish cere, lores, orbital regions. White irises. Underparts and upperparts of the two species are similar. Very short tarsi. FLIGHT.—Dark underwing. Tail pattern similar in the two species. COMPARED TO JUVENILE.—**(2) Cooper's Hawk, juveniles.**—PERCHED.—Pale gray or yellow irises. Nondescript tawny head. Leg feathers sometimes covered with moderately wide bars. Uppertail has three or four equal-width dark bands. FLIGHT.—Underside of all remiges including rear edge of wing, prominently marked with thick barring. Flight pattern deceptively similar with snappy wingbeats and short glides. Body shape, wing attitude similar. Front edge of

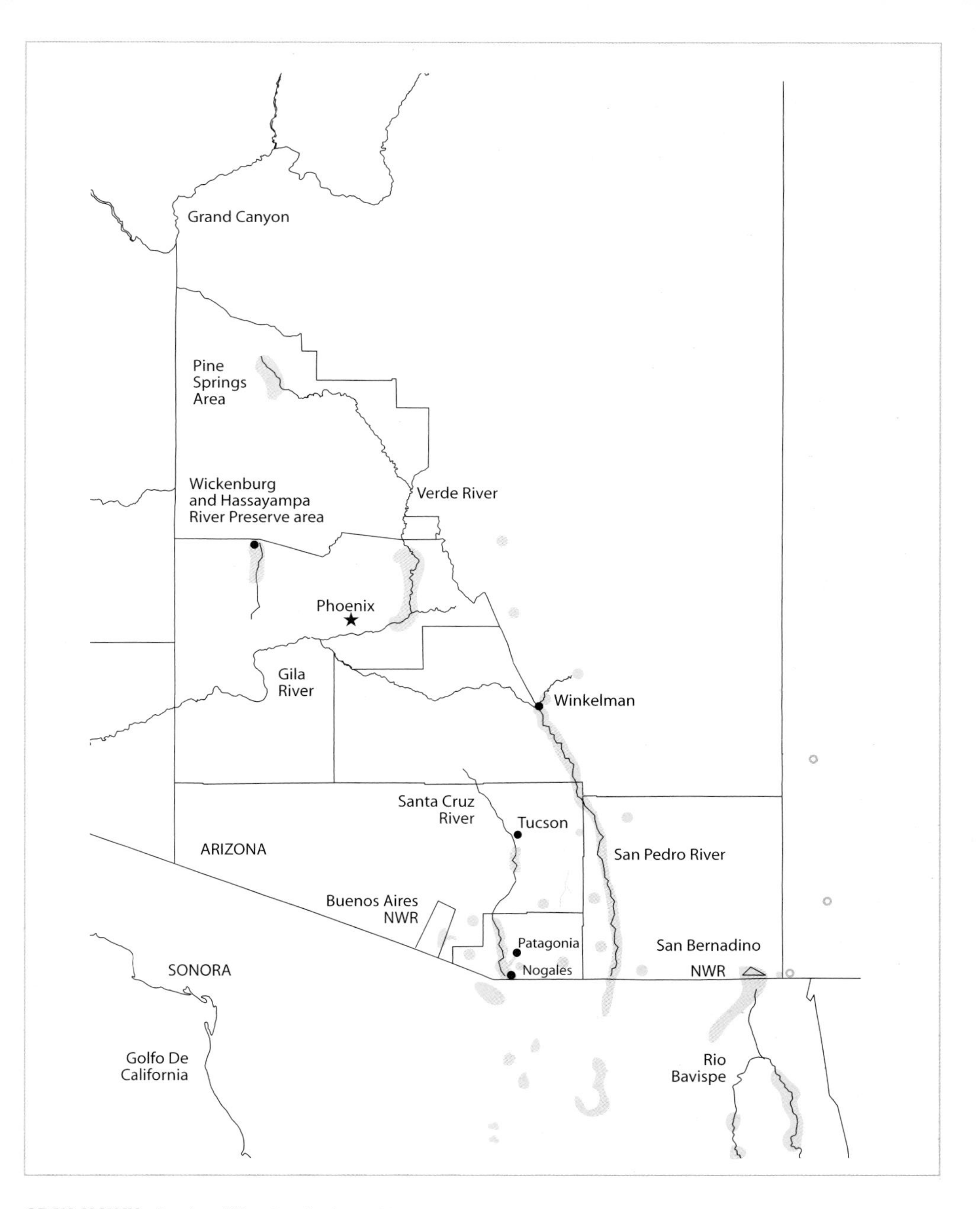


**GRAY HAWK,** *Asturina nitida:* 80 pairs in 2000.

wings also rigidly perpendicular to body angle when soaring and very similar: use caution. **(3) Red-shouldered Hawk, juveniles.**—*B. a. alleni* race in Texas; rare overlap in Arizona (*B. a. elegans*). PERCHED.—May have fairly defined dark eyeline, dark malar stripe; overall head pattern not as sharply delineated. Underparts, long tarsi, finely barred leg feathers are similar in the two species: use caution. Pale barring on upper surface of secondaries. Numerous narrow pale bands on uppertail; often rufous on basal bands. FLIGHT.—Large tawny panel/window on base of outer primaries on dorsal and ventral surfaces. Distinct, fairly wide dark gray band on rear edge of wings. **(4) Broad-winged hawk, juveniles.**—Texas range overlap primarily Aug.–Oct. and Apr.–May, but possible all winter; range overlap is rare in s. Arizona. PERCHED.—Lacks a dark eyeline; shares a dark malar stripe, dark throat stripe, and often whitish supercilium and auricular region, especially in spring. Dorsal surface of secondaries and greater coverts is uniformly dark brown and lacks dark barring. Wingtips extend three-fourths of the tail length. Leg feathers are occasionally marked with wide dark barring. Short, thick tarsi. Neatly formed dark tail bands. FLIGHT.—Minimal barring on underside of remiges but has a dark, distinct band on rear edge of wings. Outer primaries are often similarly finely barred. Pointed wingtips. Dark uppertail coverts have white borders and are not distinctly white and U-shaped.

**OTHER NAMES:** Mexican Goshawk. *Spanish:* Aguililla Gris. *French:* Unknown.

---

REFERENCES: AOU 1997; Arizona Breeding Bird Atlas 1993–1999; Clark and Wheeler 2001; del Hoyo et al. 1994; Glinski 1998a; Howell and Webb 1995; Oberholser 1974; Palmer 1988; Russell and Monson 1998; Snyder and Snyder 1991.

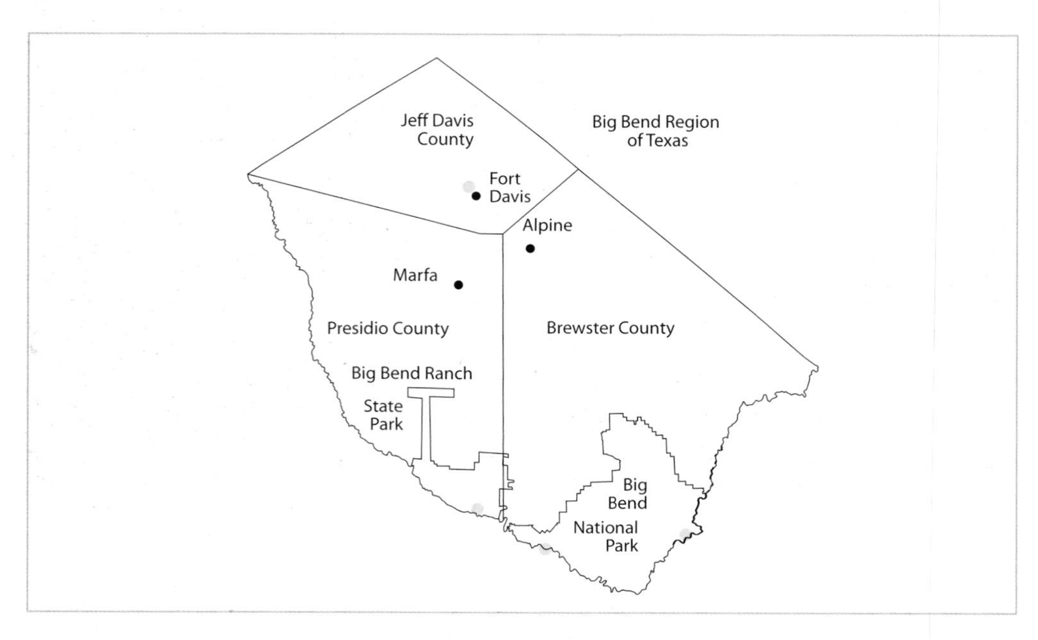


**GRAY HAWK,** *Asturina nitida:* Very uncommon to rare. Only a few pairs.

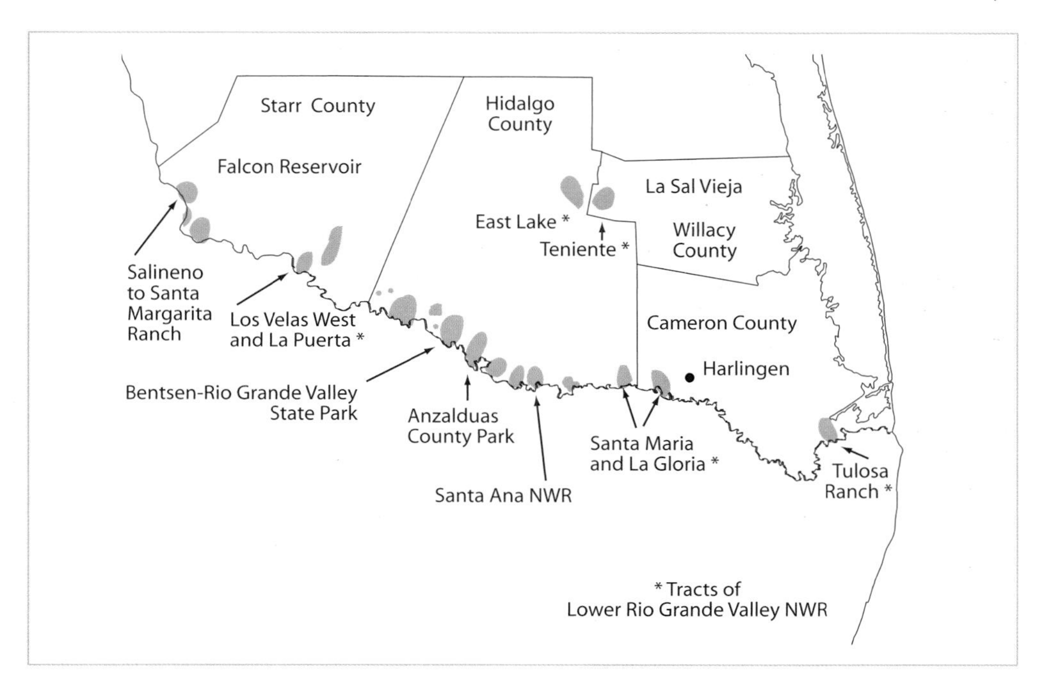


**GRAY HAWK,** *Asturina nitida.*

**Plate 221. Gray Hawk, adult [Feb.]** ▪ Black bill with bright yellow cere. Dark brown iris; pale gray head. ▪ Gray and white, finely barred underparts. Medium gray upperparts. ▪ Broad white distal tail band with black subterminal band (narrow white basal band covered by undertail coverts). ▪ *Note:* Nayarit, Mexico.

**Plate 222. Gray Hawk, adult [Feb.]** ▪ Black bill with bright yellow cere. Dark brown iris; pale gray head. ▪ Medium gray upperparts, often with brownish cast on back and scapulars, especially on females. ▪ Black tail has 2 white dorsal bands: broad distal band, narrow basal band. ▪ *Note:* Nayarit, Mexico.

**Plate 223. Gray Hawk, adult [Feb.]** ▪ White uppertail coverts. Black tail has 2 white dorsal bands: broad distal band, narrow basal band. ▪ *Note:* Nayarit, Mexico.

**Plate 224. Gray Hawk, adult [Apr.]** ▪ Gray and white, finely barred underparts. ▪ Short, rounded wings. Mainly whitish underwing with narrow black rear edge. ▪ Black tail shows 2 or 3 white bands, depending on how widely it is fanned. Single broad white band always shows. ▪ *Note:* Santa Cruz Co., Ariz.

**Plate 225. Gray Hawk, juvenile [Feb.]** ▪ Black bill with bright yellow cere. White supercilium, cheeks, and auriculars; dark eyeline, malar mark. ▪ Underparts streaked with white. Finely barred white leg feathers. ▪ Secondaries and greater coverts narrowly barred with black. Wingtips extend halfway down long tail. ▪ *Note:* Nayarit, Mexico.

**Plate 226. Gray Hawk, juvenile [Feb.]** ▪ Black bill with bright yellow cere. White supercilium, cheeks, and auriculars; dark eyeline, malar mark. ▪ Long tarsi. ▪ Secondaries and greater coverts narrowly barred with black. Wingtips extend halfway down long tail. ▪ Dark tail bands get progressively narrower towards base of tail. ▪ *Note:* Nayarit, Mexico.

**Plate 227. Gray Hawk, juvenile [Feb.]** ▪ Black bill with a bright yellow cere. White supercilium, cheeks, and auriculars; dark eyeline, malar mark. ▪ Secondaries and greater coverts narrowly barred with black. Thin panel strip on base of dark outer primaries. ▪ White uppertail coverts with small dark mark on each feather. ▪ Long tail; dark bands get progressively narrower towards base of tail. ▪ *Note:* Nayarit, Mexico.

**Plate 228. Gray Hawk, juvenile [May]** ▪ Streaked underparts with dark patch on sides of neck; barred leg feathers. ▪ Short, rounded wings. Finely barred remiges, including primaries. Thin grayish band on rear edge of wings. Thin, pale panel near base of primaries. ▪ Long tail; dark bands get progessively narrower towards base of tail. ▪ *Note:* Photographed in Santa Cruz Co., Ariz., by Brian Small.

# RED-SHOULDERED HAWK
## *(Buteo lineatus)*

**AGES:** Adult and juvenile. Sexes are similar in both age classes, but in adults females are often more distinctly marked. Adult plumage attained during the second year. Juvenile plumage is held for much of the first year. Juveniles have longer tails than adults.

**MOLT:** Accipitridae wing and tail molt pattern (*see* chapter 4). The first prebasic molt begins on the breast, then is noticeable on the back and scapulars. Molt proceeds to the remiges beginning on the innermost primary (p1). Molt on secondaries and rectrices begins somewhat later. Unlike in other similar-sized raptors, the first prebasic molt is generally a complete molt.

S. California birds may begin the first prebasic molt in mid-Nov. *B. l. alleni* may begin to molt in late winter or early spring and *B. l. lineatus* may not begin until mid- or late Apr. *B. l. alleni* end the molt in Sep. or Oct., and *lineatus* may not end until Oct. or Nov.

Subsequent adult prebasic molts begin later than the first prebasic molt for the respective latitude. Adult *alleni* are in extensive molt in early Aug. and end molt in Sep. or Oct. Adult *lineatus* are in extensive molt in mid-Sep. and end their molt by late Oct. or Nov. Molt may be well advanced on the body prior to starting on the head.

**SUBSPECIES:** Polytypic with four races in North America. Three subspecies inhabit the w. U.S. Nominate *B. l. lineatus* has a limited range in the Midwest. *B. l. alleni* is found south of *lineatus.* The demarcation between *lineatus* and *alleni* is unclear in the West. There appears to be a broad overlap zone between subspecies. *B. l. alleni* breeding range herein follows, in part, the AOU 5th check-list (1957) and extends north and west of the distribution followed by Johnsgard (1990) and Crocoll (1994). Based on specimens from the AMNH and the Oklahoma

Museum of Natural History at the University of Oklahoma in Norman, *alleni*, *lineatus*, and intergrades breed in s. Missouri and Oklahoma. Classic *alleni* breed in s. Arkansas. Birds from a historical nesting area in s.-cen. Iowa (Van Buren Co.) have *lineatus* and *alleni* traits (AMNH).

I consider *B. l. texanus* as part of *alleni*. Plumage brilliance of adults varies throughout *texanus* and *alleni* breeding ranges, including *alleni* range in the East, and juveniles of the two types are identical (data from Clark and Wheeler 2001; museum specimens; in-field photos; and photos by J. M. Economidy). According to the Texas breeding bird atlas (Benson and Arnold 2001), Red-shouldered Hawk does not breed in much of the previously published *texanus* range in s. Texas. It also does not breed in e. Mexico (Howell and Webb 1995).

*B. l. elegans* is a distinct, isolated subspecies resident mainly along the Pacific Coast. *Note:* Based on anatomical and genetic similarities with Gray Hawk (formerly *Buteo nitidus*), Palmer (1988) assigned Red-shouldered Hawk to the genus *Asturina*. Gray Hawk was assigned to the genus *Asturina*, as *Asturina nitidus* (AOU 1997), but Red-shouldered Hawk was retained in *Buteo*.

**COLOR MORPHS:** None.

**SIZE:** Medium-sized raptor. Size decreases clinally from north to south to west: *lineatus* is large, *alleni* (including *texanus* types), is midsize, and *elegans* is small (but not as small as *extimus* of s. Florida). Size differences between *lineatus* and *alleni* are difficult to detect in the field and overlap. Sexes overlap in size, with females averaging larger. Wing-chord length in millimeters for each race according to Palmer (1988): *lineatus*.—male 309–346, female 315–353; *alleni*.—male 284–330, female 281–340; *elegans*.—male 288–305, female 298–312. Total length and wingspan measurements, however, are not available for each race. Length: 15–19 in. (38–48 cm); wingspan: 37–42 in. (94–119 cm).

**SPECIES TRAITS:** HEAD.—**Distal three-fourths of the bill is black and the basal one-fourth is pale blue.** Bright yellow cere. BODY.—Long yellow tarsi. WINGS.—**Pale crescent-shaped panel on the dorsal surface and a pale crescent-shaped window on the ventral surface of the basal region of the outer primaries. Wingtips are fairly rounded, with p6–9 being nearly the same length.**

**ADULT TRAITS:** HEAD.—**Dark brown irises.** The crown and nape feathers are long and can be raised to create a large-headed appearance. **The hackles are long and spikelike when raised.** BODY.—Rufous barring on white or tawny underparts. Breast often has wider barring and may be solid rufous and form a variably distinct "bib" compared to the more narrowly barred belly and flanks. WINGS (dorsal).—**Large, dark rufous shoulder patch on the lesser upperwing coverts. White barring on the black greater upperwing coverts and secondaries. The white crescent-shaped panel on the basal region of the black outer primaries is composed of white barring.** WINGS (ventral).—**The white remiges have a wide black band on the trailing edge and are totally barred on the inner area. When backlit in translucent light, the white panel of the dorsal surface shows as a white window on the ventral surface.** The underwing coverts and axillaries are tawny with rufous streaking and barring and appear two-toned compared to the white and barred remiges. TAIL.—**Black on the dorsal surface and dark gray on the ventral surface with a white terminal band and narrow white inner bands. All races have three narrow white bands on the dorsal surface on the deck rectrices.** On the ventral surface, only one white band shows when the tail is closed; four or five show when the tail is fanned.

**ADULTS OF "EASTERN" (*B. l. lineatus*):** HEAD.—Dark brown crown and nape with a paler tawny supercilium and auricular region. Distinct dark malar mark on the lower jaw. Throat is all white, white and streaked with dark brown, or dark brown. BODY (dorsal).—Dark brown with a small amount of medium grayish and rufous on the outer edge and inner pale bar on the back and scapular feathers. BODY (ventral).—**Dark shaft streaks on the breast and dark shaft streaks or blobs on the belly and flank feathers are the classic feature of this subspecies. Females exhibit this feature more than males. Some males lack the dark markings (and appear similar to *alleni*).** Rufous barring is equal in width to the white or tawny portion on each feather. In autumn, the ventral feathers are new and fresh and are medium or dark rufous. By spring, ventral barring bleaches and fades to pale rufous or tawny (if present in the autumn, dark markings are still visible). Leg feathers are white or pale tawny and marked with thin rufous barring. The undertail coverts are white or pale tawny and are typically unmarked but may have some rufous bars. *Note:* Generally separable from other races by the dark ventral markings. If these are not present, as on many males, subspecies distinction is not reliable in the field. Many adults in Iowa and Minnesota lack dark ventral markings and look like *alleni.*

**Adult female (atypically barred type):** HEAD, BODY (dorsal), WINGS, and TAIL as on typical adults. BODY (ventral).—Breast and belly are white with broad rufous streaks and blotches with some partial rufous bars. Often has fairly wide dark shaft streaks and blotches as on typical adults. Flanks may have widely spaced rufous bars. At a distance, this type looks more streaked and spotted and not barred like a typical adult. *Note:* Uncommon to very uncommon type.

**ADULTS OF "SOUTHERN" (*B. l. alleni/texanus*):** HEAD.—The crown and nape can be dark brown and identical to *lineatus* or somewhat grayish and slightly paler. Some appear pale-headed with medium brown crown and nape and slightly or much paler tawny, grayish, or whitish supercilium and auricular region. Dark malar mark is obvious on dark-headed types, less obvious on pale-headed types. Throat as in the three variations described for *lineatus.* *Note:* Females tend to have darker heads. Heads of paler birds may nearly rival *extimus* of s. Florida (BKW pers. obs. of many birds from s. Louisiana and Brazoria Co., Tex.). BODY (dorsal).—Dark brown with a moderate amount of medium grayish and rufous on the outer edge and inner pale bar on the back and scapular feathers. BODY (ventral).—**The rufous-barred underparts lack prominent dark streaks or blobs.** At most, the breast, belly, and flanks have very thin dark shaft streaks on some feathers. The rufous barring is equal in width to the white or tawny portion on each feather. The rufous-barred underparts may be medium or dark rufous in fresh late-summer and autumn plumage but fade to pale or medium rufous or tawny by late winter or spring. Leg feathers and undertail coverts are marked as on *lineatus.* *Note:* Separable from *lineatus* that have dark ventral markings but not separable from those that lack dark markings. Pale-headed birds can be similar to *extimus.*

Darker *texanus* types: the "conspicuous dark shaft lines" on the breast as described by Palmer (1988) are visible on only one of five Texas specimens labeled as *texanus* at the AMNH. All other specimens and birds seen and photographed in the field in *texanus* range show indistinct dark breast streaking that is identical to that of *alleni* and much less obvious than the distinct dark streaking on most *lineatus.*

**ADULTS OF "CALIFORNIA" (*B. l. elegans*):** HEAD.—Tawny-rufous with dark brown streaking on the crown and nape and often appears golden colored. Malar mark is ill defined and is often grayish. BODY (dorsal).—A considerable amount of rufous on the back and scapular feathers and typically with a distinct white bar on the base of each scapular feather. BODY (ventral).—Breast is always solid dark rufous. The dark rufous barring on the belly, flanks, and leg feathers is broader than on other subspecies and creates narrow white or tawny bars between the wide rufous bars. The underparts lack thin dark shaft streaking. White or pale tawny undertail coverts are broadly barred with rufous. WINGS (dorsal).—Pronounced white barring on the black coverts and remiges. WINGS (ventral).—Dark rufous underwing coverts. TAIL.—Medium-width white bands on the black tail (wider than on eastern races).

**JUVENILE TRAITS:** HEAD.—Dark brown crown and nape with a pale tawny or whitish supercilium. Dark brown malar mark on the lower jaw. Irises are medium brown. BODY (dorsal).—Dark brown with a few irregular tawny or white blotches on the basal area of some scapular feathers. WINGS (dorsal).—**Pale crescent-shaped panel on the basal region of the outer primaries. Pale gray or white barring on the secondaries.** Rufous shoulder patch on the lesser upperwing coverts. WINGS (ventral).—**Pale panel of the dorsal surface shows as a pale window on the underwing in translucent light.**

**JUVENILES OF "EASTERN" (*B. l. lineatus*):** HEAD.—**Throat is (1) white with a broad mid-throat streak, (2) white and streaked,** or **(3) dark with a narrow white edge on each**

side. **Birds with dark throat appear hooded.** BODY (ventral).—Pale tawny or white underparts. The breast, belly, and flanks are marked with large dark brown blobs or short streaks. The lower belly has fewer dark markings or is unmarked. Leg feathers are typically unmarked, but a few birds have very small dark spots or partial bars, particularly birds inhabiting the southern edge of range (and picking up *alleni* traits). Undertail coverts are unmarked. WINGS (dorsal).—**Crescent-shaped panel is pale tawny. The secondaries typically have pale gray barring, but this may be absent on a few birds.** Rufous shoulder patch is ill defined and rarely shows on perched birds. WINGS (ventral).—White remiges have a moderately wide dusky trailing edge and narrow dark barring on most of the rest of the inner remiges. **The crescent-shaped panel shows as a pale tawny area in translucent light.** The panel may have some dark barring. The outer primaries may be uniformly dark or superimposed with darker partial barring. TAIL (dorsal).—**Dark brown with a wide dark subterminal band and five or six narrow pale bands. The distal three or four pale bands are pale gray, the basal one to three pale bands typically rufous. The distal pale band is slightly darker gray than the other pale gray bands.** TAIL (ventral).—Medium gray with narrow pale gray inner bands. The basal rufous bands show when the tail is fanned and viewed in translucent light. *Note:* Separable from *alleni* by absence of leg barring and unbarred, dark blob pattern on ventral areas.

**JUVENILES OF "SOUTHERN" (*B. l. alleni*):** HEAD.—As on *lineatus*, with the same three variations of throat markings. BODY (ventral).—Pale tawny or white underparts. Breast and belly have dark brown heart or arrowhead-shaped markings that may be barred on the inner part of the feather. These markings may elongate into streaks. The lower belly has fewer similar-shaped but smaller dark markings. The flanks have broad, widely spaced dark brown barring. The leg feathers are lightly or moderately barred with dark brown. Undertail coverts are pale tawny or white and either unmarked or partially barred. WINGS.—Dorsal and ventral surfaces are marked as on *lineatus*, but pale barring on the secondaries is consistently more pronounced and is often white. Rarely has a whitish panel on the dorsal surface of the primaries. Dark barring may be more pronounced than on *lineatus*, especially on the tawny panel and outer primaries. Nearly always has pale gray or white barring on the dorsal surface of the secondaries. TAIL.—As on *lineatus*; however, the basal rufous bands are absent on many birds and all pale bands are pale gray. *Note:* Separable from *lineatus* by leg barring and presence of flank barring.

**JUVENILES OF "CALIFORNIA" (*B. l. elegans*):** HEAD.—Dark rufous or tawny brown with paler supercilium and auricular areas. **Throat is dark brown.** BODY (dorsal).—Dark brown with irregular white blotches on the basal area of many scapular feathers. BODY (ventral).—Underparts are white and covered with dark brown markings. Breast is either (1) covered with broad, distinct heart or arrowhead-shaped markings that may expand into partial barring or (2) streaked. Flanks are broadly barred. Belly is covered with small dark blotches or arrowhead shapes. Leg feathers are thinly barred with brown or rufous-brown. Undertail coverts are white with broad rufous-brown barring. WINGS (dorsal).—**Very distinct rufous shoulder patch (more so than on juveniles of other races), but it may be mostly concealed by the scapulars when birds are perched. Very distinct white barring on the secondaries and greater upperwing coverts and white edging and spotting on the first two rows of lesser upperwing coverts. Panel on the basal outer primaries is composed of white spots.** WINGS (ventral).—Dark tawny or rufous underwing coverts contrast with the white but heavily barred remiges. TAIL (dorsal).—**Dark brown with four moderately wide whitish bands. The distal whitish band is dusky. The whitish bands often get progressively narrower towards the basal region of the tail.** TAIL (ventral).—Medium or dark gray with white bands. Only one white band shows when the tail is closed. When spread, four whitish bands show.

**ABNORMAL PLUMAGES:** Albino and partial albino birds are documented in eastern races.

**HABITAT: *B. l. lineatus*. Summer.**—Wooded lowland floodplains with mature deciduous trees and swamps and marshes adjacent to woodlands with mature deciduous or mixed deciduous-coniferous trees. Higher elevation upland, mature deciduous or mixed deciduous-coniferous woodlands with areas of water are also inhabited. This subspecies is becoming fairly common in suburban locales adjacent to or in ample-sized, moist woodlots. Climate is moderately humid to humid.

***B. l. lineatus.* Winter.**—May remain in summer habitat or occasionally move to semi-open wet or sometimes dry locations. These include forest or woodlot edges and along clear-cut shoulder areas of highways. Backyards in rural and suburban areas with birdfeeders are occasionally visited.

***B. l. lineatus.* Migration.**—Similar to summer and winter habitats, but also includes mountain ridges, large lakes, and seashores.

***B. l. alleni.* All seasons.**—Similar to habitat of *lineatus.* Common in suburban areas with proper habitat, particularly in the southern part of its range. In winter, commonly found in semi-open wet or dry areas. Areas with water-filled irrigation and drainage ditches and flooded meadows and agricultural fields, particularly rice fields, are regularly inhabited. In more arid areas, also haunts areas with livestock ponds and watering holes. Climate is humid.

***B. l. elegans.* All seasons.**—Inhabits a variety of topographic areas in California, including flatlands, hills, low mountains, and canyons. Mainly found near natural or artificial riparian areas. Found in lowland woodlands, woodlots, and groves of mature Black Oak, Coast Live Oak, eucalyptus (non-native), Western Sycamore, and various willow species. In some areas of s. California, found in agricultural plantations of Date Palms. Common in suburban and urban settings, including residential areas, campuses, golf courses, and intensively human-occupied city and county parks that have ample stands of mature, often non-native tree species. In the Coast Range Mt. of California, may nest in areas that are 3,000 ft. (900 m) in elevation. In the western foothills of the Sierra Nevada, nesting occurs at similar elevations. In the mountains of s. California, nesting regularly occurs at elevations of 4,500 ft. (1,400 m), probably to 5,500 ft. (1,700 m), and possibly higher. Higher elevation areas are in mixed forests of Black Oak, Big Cone Spruce, Canyon Live Oak, Coast Live Oak, and Ponderosa Pine. *B. l. elegans* nesting in Oregon are found in lush humid lowland riparian areas with mature trees. In Nevada, in arid lowland habitat along riparian areas with tall trees.

**HABITS:** *B. l. lineatus* is wary, *alleni* is fairly tame or tame, and *elegans* is tame. Head feathers, including the pointed hackles, are raised in cool temperatures. Any elevated perch is used, including utility-pole wires. During breeding season, high, open perches may be selected. When foraging, low, open perches are used. *B. l. alleni* and *elegans* seek sheltered, open, or concealed perches during hot periods of the day.

**FEEDING:** Mainly a perch hunter but an occasional aerial hunter. Perch hunting primarily occurs from a low or moderate-height perch in which bird drops down in a low, direct flight; also done by walking on the ground. Aerial hunting takes place when the bird is flying and a prey species happens to be in its path. Red-shouldered Hawks are generalist feeders, and diet varies with geographic region and season. Small mammals, amphibians, reptiles, and invertebrates (especially crayfish) form the bulk of the diet. Small birds, fish, and insects are occasionally preyed on. Carrion is occasionally eaten.

**FLIGHT:** Powered flight consists of a fairly regular series of moderately fast wingbeats interspersed with short glides. Soaring flight occurs regularly, particularly when courting or migrating. Wings are held on a flat plane when soaring and arched downwards when gliding. Long, impressive dives are made in courtship flights. Red-shouldered Hawks rarely hover or kite.

**VOICE:** A highly vocal raptor. Vocalizes regularly at any time of year. The most commonly heard call is a drawn-out, loud, sharp *kee-yair* or *kee-aah.* It may be repeated numerous times and is used to establish territories or give alarm. A sharp *kip* note is emitted when excited or alarmed and may be either a single or repeated note. *Keeyip* is a variation of the *kip* note. Courting birds emit a loud, drawn-out, three-syllable *kee-ann-err;* a short, squeaky *kee-aah;* or a high-pitched *scree.* This vocalization may also be repeated many times and may be altered to a *kee-yerr* or *kendrick* sound. A soft, repeated *kee* may be given by brooding or incubating females.

**STATUS AND DISTRIBUTION:** Estimated population is unknown. Population is probably stable.

*B. l. lineatus* and *alleni* have experienced population declines from historical times because of habitat loss. Wetland drainage, deforestation, and urbanization are responsible for most habitat loss. Wetlands are now more carefully managed, and forests in many regions are maturing. The altered habitat also increases competition with larger and dominant Red-tailed Hawks and Great Horned Owls. *B. l. elegans* has seen a population growth in California with the increase of available habitat of artificially planted, often non-native trees. It has experienced recent expansion into Oregon, Nevada, and Arizona.

***B. l. lineatus.*** **Summer.**—*Uncommon.* Range is sporadic in the West because of limited habitat. *Iowa:* Listed as Endangered. Breeds mainly along restricted floodplains of the Mississippi and Iowa Rivers. Isolated breeding is seen in Blackhawk and Story Cos. *Kansas:* Probable or confirmed breeding in Allen, Chautauqua, Cherokee, Labette, Linn, Montgomery, and Wilson Cos. Possible breeding in Bourbon, Jefferson, Johnson, Leavenworth, and Woodson Cos. *Minnesota:* Northern range extends to n. Becker, s. Beltrami, and s. Itasca Cos. Range then extends southward on the western edge to Mower Co., on the eastern edge to Chisago Co., and then south along the Mississippi River. Regular sightings along the Minnesota River, but breeding is not confirmed. *Missouri:* Listed as rare. Widely distributed in the southern half of the state. Isolated spots in the north-central part have confirmed breeding. Absent from the "boot" region and much of the southeastern corner of the state. *Nebraska:* Confirmed breeder since 1989 in Sarpy Co. Irregular sightings in other southeastern counties.

***B. l. lineatus.*** **Winter.**—There is withdrawal from n. Minnesota, but pairs in other states may remain fairly sedentary. A few juveniles may regularly winter in e. Texas, some to s. Texas and possibly n. Mexico.

***B. l. lineatus.*** **Movements.**—Movements are poorly defined in the West.

*Fall migration:* Juveniles move before adults. A few juveniles and adults are seen in late Oct. and early Nov. along the western edge of Lake Superior at Hawk Ridge Nature Reserve in Duluth, Minn. Small numbers are also seen at hawkwatch sites in Iowa. Moderate numbers are seen at hawkwatches in s. Texas. In Texas, movements may begin in mid-Aug. and continue through Nov. Because of small numbers, peak movements are difficult to determine.

*Spring migration:* Adults move before juveniles. There is little information.

***B. l. alleni.*** **Summer.**—*Common.* Extending breeding range westward in Oklahoma and Texas. Only a few pairs nest south of Kleberg Co., Tex. The few pairs nesting in the Rio Grande valley are in Starr and Hidalgo Cos. Nests in isolated locations as far west as Val Verde and Archer Cos, Tex.

***B. l. alleni.*** **Winter.**—Many adults remain on or near breeding grounds, and many remain paired. Juveniles may winter moderately or fairly long distances south of natal areas. In Mexico, wintering areas extend south into Veracruz, n. Puebla, and cen. Jalisco and west to Durango and e. Coahuila.

***B. l. alleni.*** **Movements.**—Few available data. Only a few birds are seen at southern hawkwatches. Many adults probably do not migrate. Juveniles probably migrate farther south than adults.

*Fall migration:* Juveniles are seen as early as mid-Aug. on the n. Texas coast at Smith Point Hawkwatch in Chambers Co. and at the Coastal Bend Hawk-Watch at Hazel Bazemore Co. Pk. near Corpus Christi. Small numbers of both ages trickle through Nov.

*Spring migration:* Few migrants are seen at spring hawkwatch sites in Texas. Two adults were migrating in early Feb. in Duval and Starr Cos, Tex.

***B. l. alleni and lineatus.*** **Extralimital movements.**—The extralimital records posted on *alleni* and *lineatus* range are not identified to subspecies by bird record committees of respective states and provinces. Records for *alleni* may include *lineatus.*

*B. l. alleni/lineatus* records: *Colorado:* 15 records in Adams, Boulder, Jefferson, Larimer, Pueblo, Yuma, and Weld Cos. Sightings span 1974 to 2001, and many counties have multiple records. There are eight records since 1990. Records exist for Jan., Feb., Apr., May, Jun., Sep., and Nov. *Kansas:* Records exist for all seasons for 17 counties: Anderson, Barton, Clay, Coffey, Cowley, Doniphan, Franklin, Harvey, Kingman, Lyon, Miami, Osage, Phillips, Riley, Sedgwick, Shawnee, and Stafford. *Texas*: Sight records exist for the following counties for all seasons: El Paso, Knox, Lubbock, Midland, Brewster (nested with a Gray Hawk at Big Bend N.P.), Potter, and Roberts.

*B. l. lineatus* records: *Manitoba:* Fairly regular in late Mar. and Apr. along the Red River south of Winnipeg and along the Assiniboine River west of Winnipeg. *North Dakota*: 12 records from 1931 to 1989 in Barnes, Fargo, Grand Forks, and Ward Cos. Five additional records from 1997 to 2000 in Fargo, Benson, McHenry, and Renville Cos. Sightings occur in Mar., Apr., May, Jun., and Jul.

***B. l. elegans.* All seasons.**—*Common.* This race has expanded rapidly north and east of typical range in California in the last three decades, especially in the last decade. *Arizona:* Early sightings occurred in 1926, 1937, and 1962. The first nesting in the state occurred in 1970 along the Colorado River in Yuma Co. Sightings became more frequent in the 1980s and 1990s and occur in most months but particularly winter. The second nesting record was in the spring of 2000 in Wickenburg, Maricopa Co. (nest failed). *California:* Primary range of *elegans* is in this state. Found on the Pacific Coast from Del Norte Co. south to San Diego Co. In the Central Valley, found from cen. Shasta Co. south to cen. Kern Co. Breeding birds are spreading eastward in the desert region and nest in suitable locations near Brawley in cen. Imperial Co., in the Owens Valley in cen. Inyo Co., near Bishop in n. Inyo Co., and along portions of the Colorado River in e. Imperial Co. *Nevada:* Breeding was confirmed for the state in early summer of 2000 at Pahranagat NWR in s.-cen. Lincoln Co. Possible breeding may occur along the Muddy River in n. Clark Co. *Oregon:* Breeding occurred in the state in the late 1800s but was not recorded again until 1971. Since then, *elegans* has been seen annually and is a confirmed breeder in various areas west of the Cascade Mts. A pair was observed in 1978 in the Rogue Valley, Jackson Co., but breeding was not confirmed and birds were not seen during summer for several years thereafter. Since the mid-1990s, breeding occurs regularly in e. Coos, e. Curry, cen. Jackson, ne. Josephine, and nw. Lane Cos. Breeding may occur in e.-cen. Crook Co. *Mexico:* Breeds in n. Baja California Norte south to Meling Ranch.

***B. l. elegans.* Movements.**—This subspecies does not migrate but is prone to disperse. Birds are seen at the Golden Gate Raptor Observatory, Marin Co., hawkwatch site on the cen. California coast during fall migration, but these are likely dispersing individuals. Adults are typically sedentary but may move moderate distances. Juveniles may disperse in any direction from natal areas. Based on banded and radio-tracked birds, juveniles typically move less than 100 miles (169 km) from natal areas. However, radio-tracked juveniles from s. California moved 220 miles (370 km) north to cen. California and east to s. Nevada. S. California radio-tracked juveniles were also plotted moving south into s. Baja California Sur, Mexico.

***B. l. elegans.* Extralimital movements.**—*Arizona:* Regularly seen in various locations from n. to s. Arizona. *British Columbia:* A record for Cassidy on the eastern side of Vancouver Island is the northernmost record for this race. *Idaho:* Six records as of 1999; most are near Boise in Ada Co. *Nevada:* A fairly regular visitor since 1969. Both ages are seen regularly in Clark and Lincoln Cos. and irregularly in Elko, Storey, and White Pine Cos. *Oregon:* Annual east of the Cascade Mts., particularly at Malheur NWR in Harney Co.; also seen in Gilliam Co. *Utah:* At least two records of juveniles and adults for w. Juab Co. (Callo and Fish Springs NWR) in the summers of 2000 and 2001; also a record for se. Iron Co. *Washington:* Extending northward in the state in fall and winter. Wintering individuals have occurred. Has been seen along the Pacific Coast in Sequim, Clallam Co., and east of the Cascade Mts. in Richland, Benton Co.

**NESTING:** *B. l. lineatus* begin nesting in Feb. in the southern latitudes and in Mar. or Apr. in northern latitudes and end in Jun. or Jul. *B. l. alleni* begin nesting in Jan. or Feb. and end in May or Jun. *B. l. elegans* begin nesting in Nov. in s. California, but it may not be until Jan. or Feb. in northern areas, and end from May to Jul.

One-year-old females adorned in juvenile plumage may pair with adult males. Most do not breed until at least 2 years old when in full adult plumage.

**Courtship (flight).**—*High-circling* and *leg-dangling* by both sexes and *sky-dancing* by males. Pairs are highly vocal in courtship flights.

Both sexes build or refurbish nests. Nests

may be used for several years. Old leafy squirrel nests or nests of other birds may be used as platforms for new nests. Heights of nests vary from 20 to 62 ft. (6–19 m). Nests are placed in a main crotch of mainly deciduous trees. If coniferous trees are used, nests are placed next to the trunk. Live trees are always used for nest sites.

Both sexes build the compact and sturdy nests, which are 18–24 in. (46–61 cm) wide and 8–12 in. (20–30 cm) deep. Greenery lines the nest. The 2–3 eggs are incubated primarily by female for about 33 days. Male performs some incubation duty but mainly provides food to female. Youngsters fledge when about 35–45 days old. Fledglings stay with parents for a few more weeks, then leave natal areas. A single brood is raised each year.

**CONSERVATION:** No direct measures are taken. Saving moist, mature woodlands and woodlots is essential to maintaining the current population. Creation of wooded wetlands creates new habitat. Maturing woodlands and woodlots in moist natural, rural, and suburban areas also create new habitat. In California, *elegans* has readily adapted to urban life with maturing of non-native tree stands.

**Mortality.**—Mainly suffers from natural causes, especially predation by Great Horned Owls. Illegal shooting occurs, and a few are hit by vehicles.

**SIMILAR SPECIES:** COMPARED TO ADULTS.—**(1) Broad-winged Hawk, adults.**—PERCHED.—Iris is orangish brown. Rufous ventral barring is typically more brownish and coarsely patterned. Dorsal surface of wing is uniformly brown. Has a single broad white or pale gray dorsal tail band or additional narrower pale band. FLIGHT.—Underside of basal region of outer primaries white and unmarked; narrow black barring on secondaries. Tail as in Perched. Voice is a high-pitched whistle, *pee-hee.* **(2) Northern Harrier, juveniles.**—FLIGHT.—At a long distance the uniformly colored tawny-rufous belly and flanks can appear similar to the rufous barring of a Red-shouldered Hawk. The dark brown median coverts and axillaries on the Northern Harrier easily separate it from a Red-shouldered Hawk. COMPARED TO JUVENILES.—**(3) Northern Goshawk, juveniles.**—PERCHED.—Pale gray iris of recently fledged birds is similar; in the fall iris is yellowish orange. Dark malar mark is absent or ill defined. Undertail coverts may have large dark marks but are never barred as on some *B. l. alleni.* Tail has three or four equal-width dark bands. FLIGHT.—Pale throat with a single dark mid-throat streak. Dark barring on the underside of the outer primaries is broad and prominent; sometimes exhibits similar pale window in translucent light on the basal areas of the outer primaries. Tail as in Perched. **(4) Broad-winged Hawk, juveniles.**—PERCHED.—Throat is white with a single dark mid-throat streak. Leg feathers are unmarked as on *B. l. lineatus*, or spotted or thickly barred and not thinly barred with brown as on *alleni* and *elegans.* Dorsal surface of secondaries is uniformly dark brown. Dorsal tail is medium brown with narrow dark bars. FLIGHT.—Throat as described in Perched. Large rectangular window on primaries and upperside is uniformly brown. When soaring, tail is pale with several narrow dark bars. **(5) Red-tailed Hawk, juveniles.**—PERCHED.—*B. j. borealis* and light morph *B. j. calurus* have pale gray or yellow irises, white unmarked breasts, dark barred dorsal surface of secondaries, and numerous narrow dark tail bands. Intermediate morph *calurus* has similar streaked breast but dark belly. Large rectangular window on underside of primaries and upperside is pale brown and contrasts with the darker secondaries and inner half of the wing. FLIGHT.—Dark patagial mark on the leading edge of the inner wing. Tail is pale brown with multiple narrow dark bands.

**OTHER NAMES:** Redshoulder, Shoulder, Texas Red (richly colored *alleni* adults). *Spanish:* Aguililla Pechirrojo, Aquililla Pecho Rojo, Gavilan Ranero. *French:* Buse à Épaulettes.

---

REFERENCES: Adamus et al. 2001; Alcorn 1988; AOU 1957; 1997; Arnold 2001; Baicich and Harrison 1997; Benson and Arnold 2001; Bloom and McCrary 1996; Busby and Zimmerman 2001; Cecil 1999; Clark and Wheeler 2001; Crocoll 1994; Gilligan et al. 1994; Glinski 1982; Grzybowski 2000; Howell and Webb 1995; Janssen 1987; Johnsgard 1990; Kent and Dinsmore 1996; Mlodinow and Tweit 2001; Oberholser 1974; Palmer 1988; Sharpe et al. 2001; Small 1994; Sutton 1967; Thompson and Ely 1989; Tyler et al. 1989.

**"EASTERN" RED-SHOULDERED HAWK,** *Buteo lineatus lineatus:* Uncommon. Very local in West. Juveniles winter into n. Mexico. Disperses spring and fall north to ND, MB and SK and west to CO.

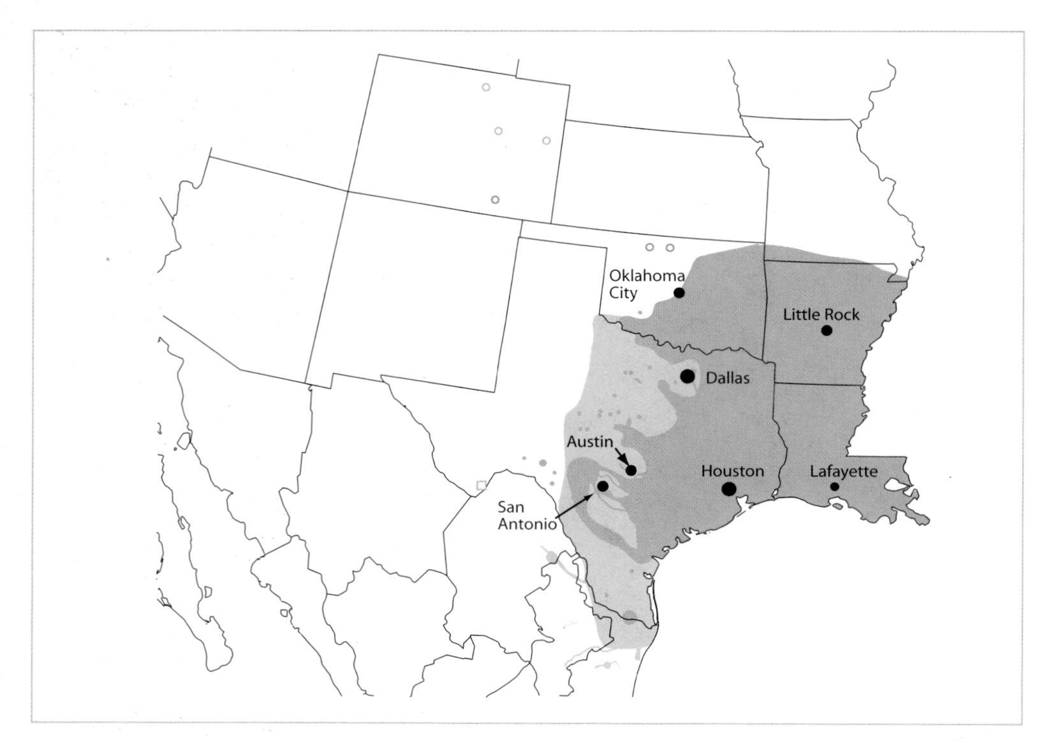


**"SOUTHERN" RED-SHOULDERED HAWK,** *Buteo lineatus alleni:* Common. "Texas" race considered with "Southern" race. Subspecies border with "Eastern" not well defined. CO records not identified by race. Multiple sightings exist for CO locations.

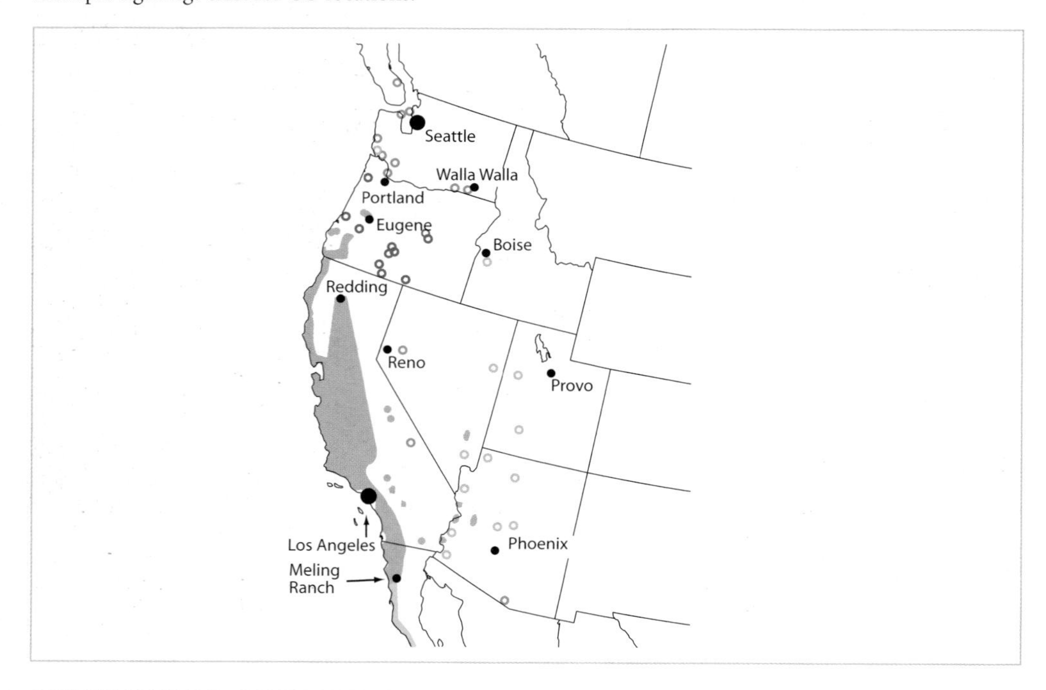


**"CALIFORNIA" RED-SHOULDERED HAWK,** *Buteo lineatus elegans:* Common. Expanding range northward and eastward from CA. Has annually visited WA since 1992.

**Plate 229. Red-shouldered Hawk, adult "Eastern" (*B. l. lineatus*) [Dec.]** ▪ Dark brown iris. Brown head; tawny supercilium and auriculars. ▪ Medium to dark rufous, thinly barred underparts; breast may be solid rufous bib. Dark shaft streaks or blobs on many ventral feathers are unique to *lineatus* but lacking on many males. Underparts fade by spring.

**Plate 230. Red-shouldered Hawk, adult "Southern" (*B. l. alleni*) [Feb.]** ▪ Dark brown iris. Tawny-brown head. ▪ Medium to dark rufous, thinly barred underparts; breast may be solid bib. Ventral areas lack dark markings; similar to some *lineatus* males. ▪ White checkering on black wings. ▪ Single white band shows on closed tail. ▪ *Note:* Female of mated pair.

**Plate 231. Red-shouldered Hawk, adult "Southern" (*B. l. alleni*) [Nov.]** ▪ Dark brown iris. Brown head with pale supercilium and auriculars. ▪ Medium to dark rufous, thinly barred underparts; breast more heavily barred. ▪ Black wings checkered with white. ▪ Single white band shows on closed tail.

**Plate 232. Red-shouldered Hawk, adult "California" (*B. l. elegans*) [Nov.]** ▪ Dark brown iris. Rufous-tawny head. ▪ Solid rufous bib; narrow white barring on rufous belly and flanks. ▪ White checkering on black wings. ▪ Single white band on closed tail.

**Plate 233. Red-shouldered Hawk, adult "California" (*B. l. elegans*) [Nov.]** ▪ Dark brown iris. Rufous-tawny head. ▪ Blackish scapulars mottled and barred with white. ▪ Rufous shoulder barely shows under scapulars. White-checkered wing coverts and remiges. ▪ Black tail has 3 broad white bands.

**Plate 234. Red-shouldered Hawk, adult "Eastern" (*B. l. lineatus*) [Jan.]** ▪ Black-and-white checkered wing with distinct, white-checkered, crescent-shaped panels near tips of primaries. Rufous shoulders. ▪ Black tail has 3 narrow white bands.

**Plate 235. Red-shouldered Hawk, adult "Eastern" (*B. l. lineatus*) [Mar.]** ▪ Brown head. ▪ Uniformly rufous-barred underparts with large black streaks or blobs unique to this race. ▪ Barred remiges with black-and-white-barred, crescent-shaped panel near tips of primaries. ▪ Single white tail band. ▪ *Note:* Photo by Jerry Liguori.

**Plate 236. Red-shouldered Hawk, adult "Southern" (*B. l. alleni*) [Feb.]** ▪ Solid dark rufous bib, rufous-barred belly, flanks, and leg feathers. ▪ Black-and-white-barred remiges; barred, white panel shows on left wing. Panels do not show if light is not at correct angle. ▪ Black tail with 1 full band, 1 partial band. ▪ *Note:* Same bird (female) as on plate 230.

**Plate 237. Red-shouldered Hawk, adult "California" (*B. l. elegans*) [Nov.]** ▪ Pale head. ▪ Solid rufous bib; narrow white barring on rufous flanks and belly. ▪ Barred remiges. White-barred window on base of outer primaries.

**Plate 238. Red-shouldered Hawk, adult "Southern" (*B. l. alleni*) [Feb.]** ▪ Wings held in bowed position when gliding. ▪ *Note:* Carrying crayfish in bill.

**Plate 239. Red-shouldered Hawk, juvenile "Southern" (*B. l. alleni*) [Jan.]** ▪ Medium brown iris. Throat often dark. ▪ Flanks barred with brown. ▪ Pale gray barring on secondaries. Rufous shoulder barely shows on juveniles. ▪ Numerous narrow pale bands; basal pale bands may be rufous. ▪ *Note: B. l. lineatus* (not shown perched) similar but streaked or spotted on flanks.

**Plate 240. Red-shouldered Hawk, juvenile "Southern" (*B. l. alleni*) [Jan.]** ▪ Brown head with paler supercilium and auriculars. Medium brown iris. Throat often dark. ▪ Streaked or partially barred breast, spotted belly, and barred flanks. Leg feathers (not shown) finely barred. ▪ *Note:* Same bird as on plate 239.

**Plate 241. Red-shouldered Hawk, juvenile "California" (*B. l. elegans*) [Nov.]** ▪ Brown head, including throat. ▪ Streaked breast, barred belly and flanks. Thinly barred leg feathers (not shown). ▪ Secondaries barred white or pale gray. Coverts mottled with white. ▪ *Note:* Starting first prebasic molt with a few adult feathers on upper flanks and breast.

**Plate 242. Red-shouldered Hawk, juvenile "California" (*B. l. elegans*) [Nov.]** ▪ Brown head, including throat. ▪ Partial white barring on brown scapulars. ▪ Secondaries barred white or pale gray. ▪ Three narrow, brownish white bands on brown tail. ▪ *Note:* Same bird as on plate 241.

**Plate 243. Red-shouldered Hawk, juvenile "Southern" (*B. l. alleni*) [Nov.]** ▪ Medium brown iris. Throat often dark. ▪ Brown-barred underparts. Finely barred leg feathers. ▪ Tawny crescent-shaped panel on outer 6 primaries. Pale gray barring on secondaries. Rufous shoulder barely shows. ▪ Pale gray bands on tail. ▪ *Note: B. l. lineatus* has identical tawny panel on wings.

**Plate 244. Red-shouldered Hawk, juvenile "Eastern" (*B. l. lineatus*) [Oct.]** ▪ Brown head and throat. ▪ Brown streaks or blobs on underparts; leg feathers nearly always unmarked. ▪ Tawny, crescent-shaped window on base of outer 6 primaries. Inner primaries and all secondaries barred. ▪ Numerous narrow, pale gray tail bands; often rufous bands on basal area.

**Plate 245. Red-shouldered Hawk, juvenile "California" (*B. l. elegans*) [Nov.]** ▪ Whitish barring on secondaries; white-barred, crescent-shaped panel on dorsal wing. Ventral wing barred with whitish window showing near base of darker primary tips. ▪ *Note:* Same bird as on plates 241 and 242.

# BROAD-WINGED HAWK
## *(Buteo platypterus)*

**AGES:** Adult, subadult, and juvenile. Many individuals retain a few juvenile feathers and are considered a basic I (subadult I) age. Juvenile plumage is retained the first year. Juveniles have distinctly longer tails than adults and subadults, but the width of their wings is similar to older ages.

**MOLT:** Accipitridae wing and tail molt pattern (*see* chapter 4). Rapid, fairly complete first prebasic molt from juvenile to subadult and/or adult occurs May–Oct. Molt is first visible early to mid-May with a gap at the primary-secondary junction caused by the dropped innermost primary (p1). Primary molt may be complete, or may be incomplete and retain the outer one to three juvenile feathers (p8–10). The deck rectrices (r1) commence molt shortly after primaries begin to molt. Tail molt is mainly complete by late Sep. and rarely retains a juvenile rectrix set. Secondaries begin molt soon after inner primaries and central rectrices have begun molt. A few juvenile secondaries may be retained (s4 and 8; occasionally s3, 4, 8, and 9). A few juvenile upperwing coverts also may be retained. Rump molt may be incomplete and retain old, faded brown juvenile feathers into the subadult age. Any retained juvenile feathering is retained until the next prebasic molt the following early summer. *Note:* It is not known if all 1-year-olds retain remnant juvenile feathering. A substantial number of individuals show retained juvenile feathering during fall and spring migration periods.

Subsequent prebasic molts from subadult to adult and within adults are also rapid and fairly complete. Molt begins in Jun. and ends Sep.–Oct. During fall migration, retained old feathers are often visible on the upperwing coverts and sometimes outer three primaries (p8–10, sometimes 7–10) and a secondary (s4, possibly more). There is massive feather replacement in Jun. and Jul. in all prebasic molts. Birds appear exceptionally ratty in the first prebasic molt.

**SUBSPECIES:** There are six races. *B. p. platypterus* is the only race in North America. Five additional subspecies that undoubtedly originated from *platypterus* are on islands in the Caribbean: (1) *B. p. cubanensis* in Cuba; (2) *B. p. brunnescens* in Puerto Rico; (3) *B. p. insulicola* in Antigua and Barbados; (4) *B. p. rivieri* in Dominica, St. Lucia, and Martinique; and (5) *B. p. antillarum* in Grenada, the Grenadines, St. Vincent, and Trinidad and Tobago.

**COLOR MORPHS:** Polymorphic with two color morphs: light and dark. There are no intermediate morphs. *B. p. platypterus* is the only race with color morphs. *Tail patterns of both color morphs are identical for the respective age classes.* Adult light morph has considerable variation on ventral markings. Juveniles exhibit variable markings on the ventral surface of the body and on the tail.

**SIZE:** A medium-sized raptor and a small buteo. Males average smaller than females, but there is considerable overlap and sexes are not separable in the field. Measurements are for all ages; juveniles are longer than adults because of their longer tail length. Length: 13–17 in. (33–43 cm); wingspan: 32–36 in. (81–91 cm).

**SPECIES TRAITS:** HEAD.—**Bill is all black except for a small pale bluish spot on the lower part of the upper mandible.** BODY.—Short tarsi. WINGS.—**In flight, wings are moderate in length, somewhat broad, with rather pointed wingtips (p7 and 8 longest and equal in length). Trailing edge is slightly bowed when soaring and straight-edged when gliding.** When perched, wingtips are moderately shorter than tail tip. *Note:* "Broad-winged" is a misnomer for this species. Wings are proportionately the same width as on several other buteo species.

**ADULT TRAITS:** HEAD.—**Pale orangish brown irises.** WINGS.—**Undersides of remiges are uniformly white and have a moderate amount of narrow dark barring on the secondaries and inner primaries.** Sometimes a pale rectangular window is exhibited on the primaries when the underwing is backlit. Typical of adult buteos, a broad black band is on the underside trailing edge of the remiges. Underside of primary tips has broad black tips. TAIL (dorsal).—**Black with one wide pale gray band on the mid-section of the dorsal side, sometimes an additional narrow white or pale gray band on the base.** TAIL (ventral).—**Black with one broad white band that is always visible on the mid-section. When fanned, an additional one or two narrow white inner bands may also show.** All the white bands may align neatly, or the outermost rectrix set may have narrow, irregular dark bands that do not align with the wide black and white bands on the ventral side (adult, not juvenile feathers).

**ADULT LIGHT MORPH (HEAVILY BARRED TYPE):** HEAD.—Rufous-brown with a discernibly darker malar stripe and white lores. BODY (ventral).—**White underparts are variably marked with wide barring or arrowhead-shaped markings on the breast and belly; flanks are always barred. Markings are rufous-brown, rarely bright rufous.** (1) All underparts may be uniformly marked with wide barring or arrowhead-shaped markings or (2) breast may be somewhat more densely marked, either solid or mottled, but darker than the barred belly and flanks and creates somewhat of a "bib." White flanks and leg feathers are barred with brownish rufous. BODY (dorsal).—Dark brown upperparts. Uppertail coverts are white-tipped (and often appear as a narrow tail band). WINGS.—**Dorsal surface is uniformly brown.** Pale underwing coverts are white or pale tawny with a small amount of brownish rufous markings; however, some are a rich tawny with fairly heavy markings. Axillaries are barred. *Note:* Common plumage type.

**ADULT LIGHT MORPH (PALE-BELLIED TYPE):** HEAD and TAIL.—As on heavily barred type. BODY.—Breast is virtually solid rufous brown and creates a bib that contrasts sharply with the sparsely marked white belly and flanks. Belly is (1) sparsely covered with rufous-brown arrowhead-shaped markings or (2) virtually lacks dark markings. Flanks are always covered with widely spaced, narrow or moderately wide rufous-brown bars. WINGS.—Pale tawny or white, lightly marked underwing coverts. *Note:* Common plumage type.

**ADULT LIGHT MORPH (LIGHTLY BARRED TYPE):** HEAD and TAIL.—As on heavily barred type. BODY.—All of the underparts are uniformly patterned with sparse markings. Breast is (1) covered with moderately wide or narrow rufous-brown arrowhead-shaped markings or (2) has sparse streaking. Belly is also lightly marked with small arrowhead-shaped marks or narrow bars. Flanks have widely spaced, narrow bars. Leg feathers are narrowly barred. WINGS.—Pale tawny or white, very lightly marked underwing coverts. *Note:* Uncommon plumage type.

**ADULT DARK MORPH:** HEAD.—Whitish lores; rest of head, including forehead, is dark brown. BODY.—Uniformly dark brown, including the undertail coverts. WINGS.—Dark brown underwing coverts.

**BASIC (SUBADULT) TRAITS:** HEAD.—Iris color as on adults. BODY and TAIL.—As on respective color morph and plumage type of adults, except rump may have retained juvenile feathers. WINGS.— Variable amount of juvenile feathers retained on remiges and upperwing coverts until the next molt the following summer (*see* Molt above). *Note:* Separable from adults only at close range.

**JUVENILE TRAITS:** HEAD.—Bright yellow cere; medium brown irises. WINGS.—Undersides of remiges are uniformly white with narrow dark barring on the secondaries and inner primaries. A narrow grayish band is on the trailing edge of the underwing. Underside of the primary tips is either solid gray or has narrow dark barring. When backlit, undersides of primaries often exhibit a translucent rectangular window, especially when wings are held in a glide position. *Note:* On perched birds, the wingtips are somewhat shorter than the tail tip and cover three-fourths of the tail. TAIL.—Medium brown on the dorsal surface and whitish on the ventral surface with two major patterns of black banding. (1) *Narrow-banded type:* **Three to five neatly formed, moderately narrow or narrow bands with a wide dark subterminal band.** Rarely, the dark bands on deck rectrices may be somewhat chevron-shaped. Patterns with three dark bands have moderately wide bands, patterns with four or five dark bands have narrow bands. *Note:* Common pattern. (2) *Broad-banded type:* **Two wide dark inner bands that are nearly as thick as the wide dark subterminal band—and may appear as three equal-width dark bands. On the underside, dark bands on the outermost rectrix set are much narrower and do not align with the wider dark bands.** *Note:* Fairly common pattern; can appear quite adultlike.

**JUVENILE LIGHT MORPH (LIGHTLY STREAKED TYPE):** HEAD.—Pale tawny or whitish supercilium and slightly darker tawny auricular area; dark malar stripe. Lacks a dark eyeline, and throat has a thin dark center stripe. BODY.—Breast, belly, flanks, and leg feathers are pale tawny or white and generally unmarked. Some birds are sparsely streaked on the sides of the breast and flanks. Unmarked white undertail coverts. WINGS (dorsal).—**Secondaries and greater coverts are uniformly dark brown with little or no dark barring. Primaries are the same color as the secondaries or show a slightly paler panel.** WINGS (ventral).—Unmarked whitish underwing coverts and axillaries, or axillaries may be very lightly streaked. *Note:* Fairly common plumage type. This looks to be a juvenile light morph Short-tailed Hawk, particularly in flight.

**JUVENILE LIGHT MORPH (MODERATELY STREAKED TYPE):** HEAD.—As on lightly streaked type. BODY.—Pale tawny or white underparts are moderately streaked on breast and belly and partially barred on rear portion of flanks. Belly is generally more heavily streaked than breast and shows a belly band. Leg feathers vary considerably in pattern of dark brown markings: (1) heart shaped, (2) V shaped, or (3) spotted. Unmarked white or tawny undertail coverts. Uppertail coverts have narrow white tips on each feather. WINGS.—**Upper surface as in lightly streaked type.** Underwing coverts are lightly or moderately marked, especially on the patagial area. Axillaries have dark brown, moderate-sized diamond-shaped markings or are partially barred. *Note:* Common plumage type.

**JUVENILE LIGHT MORPH (HEAVILY STREAKED TYPE):** HEAD.—As on lightly streaked type. BODY.—Breast and belly are densely marked with streaking or broad arrowhead-shaped markings (at times adultlike), and flanks are broadly barred. Leg feathers as in moderately streaked type or thickly and neatly barred. Unmarked white undertail coverts. WINGS.—**Upper surface as in lightly streaked type.** Underwing coverts are moderately or densely covered with dark markings, particularly on the patagial region. Axillaries are covered with large diamond-shaped brown markings or have partial barring. *Note:* Fairly common plumage type.

**JUVENILE LIGHT MORPH (WORN PLUMAGE, ALL TYPES):** By spring, head color may fade with a pronounced white supercilium and equally white auricular region; dark malar stripe is still apparent. *Note:* Can look like a juvenile Gray Hawk, including the dark throat streak; however, lacks the dark eyeline.

**JUVENILE DARK MORPH (ALL-DARK TYPE):** HEAD.—Whitish lores; rest of head, including forehead, is dark brown. BODY.—All of body, including upperparts, underparts, and undertail coverts, is uniformly dark brown. At close range, narrow tawny edgings on underparts may be visible on breast of some individuals. Undertail coverts have a pale tawny tip on each feather. WINGS.—Underwing coverts are either solid dark brown or covered with small, pale speckling. *Note:* Common plumage type.

**JUVENILE DARK MORPH (STREAKED TYPE):** HEAD.—All dark as on all-dark type or may have fairly distinct pale tawny superciliary lines and auriculars. BODY.—Upperparts as on all-dark type. Underparts have distinct pale tawny-rufous streaking on breast, belly, and flanks. WINGS.—Underwing coverts are tawny-rufous with a dark streak on central area of each feather and may appear more rufous than dark brown. *Note:* Uncommon to common plumage type.

**ABNORMAL PLUMAGES:** Incomplete and/or total albino adult: sight record of white adult (Clark and Wheeler 2001); sight record of adult at Duluth, Minn. (F. Nicoletti pers. comm.). Partial albino juvenile: sight record of juvenile with scattered white feathers (Clark and Wheeler 1987). *Note:* Plumages types are rare and not depicted.

**HABITAT: Summer.**—In the U.S., found in large tracts of mid-aged and mature deciduous and/or mixed woodlands that are interspersed with small openings and water sources. Primarily inhabits remote regions but may occupy areas with moderate human settlement in large wooded tracts of at least 100 acres (40 ha). Breeding habitat is at lowland elevation; however, found in moderately high areas of the Black Hills region of South Dakota.

In Canada, breeding occurs only in lowland mixed spruce-aspen forests and large woodlots. In s. Alberta and Saskatchewan, breeding occurs in isolated upland elevations that have forested habitat. Most densely populated breeding area for this species is the southern boreal forest zone that extends from the Rocky Mts. and eastward. In w. Canada, habitat varies from being primarily aspen (Dawson Creek, B.C.) to primarily spruce with some aspen (Prince George, B.C., and Fort Liard, NWT). Isolated breeding areas in the Rocky Mts. are in lowland valleys.

**Winter.**—In the U.S., found in semi-open, variable-aged semi-tropical deciduous tracts, often near human-occupied areas. In major wintering areas south of the U.S., tropical dry and wet woodlands and semi-open regions and rain forests from sea level up to montane elevations are inhabited. Wooded areas vary from small tracts to forests with openings. Regularly found in human altered areas.

**Migration.**—Broad-winged Hawks roost at night in wooded tracts. On the virtually treeless Great Plains, roosting (and foraging) birds seek wooded riparian areas, often thin wooded strips around lakes and along streams and irrigation ditches. They also roost and forage in dry shelterbelts, along edges of towns and settlements with trees, and in isolated tree stands at occupied and abandoned homesteads.

**HABITS:** A tame raptor. Solitary except during migration. Gregarious when migrating and forms variably sized flocks (kettles). Exposed or concealed branches, including telephone poles and wires, are used for perches. Very secretive in the breeding season.

**FEEDING:** A perch hunter; rarely an aerial hunter. Feeds on small rodents, amphibians, and reptiles; also large insects. Prey is captured mainly on the ground but may also be captured on outer branches. Broad-winged Hawks do not feed on carrion. Perch hunting is done from low to moderate-height perches. In live trees, the hawks perch below the canopy level. Being opportunistic hunters, migrating birds may feed at roost areas in mornings, evenings, and during periods of inclement weather when migration is temporarily suspended. Some may also hunt in one locale for a few days, even during periods of prime migrating weather. Broad-winged Hawks may also fast for extended periods when migrating.

**FLIGHT:** Powered by regular cadence of several moderately fast wingbeats interspersed with short to long-distance glides. Glides and soars with wings held on a flat plane. When migrating, Broadwings soar to extremely high altitudes and then glide for several miles. Migration fights may be at altitudes exceeding human sight, and often rise into cloud bases when thermal conditions are optimal. They begin in mid-morning and extend to late afternoon. During optimum weather, flights may occur until dusk. Hawks seeking night roosts may dive from considerable heights into selected wooded locales. Broad-winged Hawks do not hover or kite.

**VOICE:** Primary call is a loud, high-pitched, ear-piercing *pee-heeeeee.* The last syllable is very drawn out. Females and fledglings emit a high-pitched whine when food-begging. Highly vocal in the breeding season, especially if intruders trespass into nesting territories. Fairly vocal in the nonbreeding season.

**STATUS AND DISTRIBUTION:** Overall one of the most *common* breeding raptors in the U.S. and Canada. Population is probably stable. The entire population can be monitored with reasonable accuracy as virtually all Broad-winged Hawks migrate in a narrow corridor and a compressed time span through Veracruz, Mexico, in the fall. Estimated North American population has varied from 1.5 million birds in 2000 to 2.4 million in 2002.

**Summer, light morph.**—*Common.* Found throughout the entire North American breeding range.

**Canada.** Common, with a high breeding density in the southern spruce-aspen zone of the parkland and boreal forest from w. Alberta to e. Manitoba (and east to s. Québec). *British Columbia:* Very uncommon to uncommon and local. Range expansion has apparently been occurring in the last three decades. Found mainly east of the Rocky Mts. Irregular sight records date back to the mid-1960s. Regular sightings have occurred around Fort St. John since the early 1980s. Now breeding and regularly seen around Fort St. John, west to Chetwynd, south to Groundbirch, and around Dawson Creek. Breeding also occurs in and around the isolated lowlands of Prince George, north to the town of Mackenzie (this large valley region is in the Rocky Mts.). Very local in ne. British Columbia north of Fort St. John in appropriate lowland habitat. A pair was seen on the Liard Highway in late May 2001 between Fort Nelson, B.C., and Fort Liard, NWT (in British Columbia); a bird seen in the same vicinity in early Jul. 2002. New in the summer of 2002, and illustrating the expansion of this species, a pair was discovered in August near Golden, B.C. This pair represents the southernmost nesting pair in British Columbia (and is in the Rocky Mts.). *Northwest Territories:* Considered to be local and very uncommon to uncommon. The first territorial record was in late May 1973 near Fort Simpson, which is the most northern record in North America. An adult dark morph seen in late May 1998 near Fort Liard. Near Fort Liard, a nest was found in late Jun. 2000, a pair was seen in Jun. 2001 16 miles (25 km) north of the town, one adult seen near the Muskeg River, and at least two territories were found in the summer of 2002. This is the northernmost summer range extension for this species in North America. *Yukon Territory:* Probably *rare.* An adult was seen in mid-Jun. 1997 along the La Biche River in the extreme southeastern part of the territory, marking a first territorial record. *Note:* Summering/breeding areas conform to lowland aspen forests in the previously listed areas of w. Canada. There is a large extension of lowland habitat that extends from Alberta into e. British Columbia from Dawson Creek to Groundbirch and Chetwynd and around Fort St. John. There are also isolated lowland areas surrounding Prince George. There is a narrow strip of habitat extending northwest of Fort St. John and a nearly disjunct lowland region around Fort Nelson, east to at least Kotcho Lake, and north to extreme se. Yukon and sw. Northwest Territories.

**United States.** High breeding density in n. and e. Minnesota. There is a fairly high breeding density in s. Montana, Arkansas, and n. Louisiana. There is moderate breeding density in e. Oklahoma and e. Texas. Locally distributed with low nesting density in the prairie states. *North Dakota:* Common in the Turtle Mts., Rolette Co.; uncommon in the Pembina Hills, Pembina Co.; very local and rare in Stutsman, Ransom, Richland, and LaMoure Cos. *South Dakota:* Rare and local in the Black Hills, Lawrence Co.; rare in Marshall and Minnehaha Cos. *Iowa:* Scattered breeding in very low density throughout isolated wooded portions of the state. *Nebraska:* Has nested in Sarpy Co. *Kansas:* Rare. Nesting has occurred in Wyandotte Co.; possibly nests in Leavenworth and Douglas Cos.

**Extralimital summer.** *Colorado:* Nested in Larimer Co. in 1978; single seen late Jun. 1984 in Denver. *Wyoming:* Single bird seen in early Jul. in Sheridan, Sheridan Co. In both cases, late migrant juveniles.

**Summer, dark morph.—Canada.** See above data for British Columbia and the Northwest Territories in Light Morph. Rare summer resident and breeder in w. Canada. Estimated population is unknown. Even in their core breeding range, dark morphs are probably outnumbered by light morphs. Dark morph population is less than 1% of light morph's. Documented breeding is known only within the southern spruce-aspen boreal forest band across cen. Alberta. However, regular summer sightings and probable breeding have occurred for the last few years near Prince George and Fort St. John, B.C., and near Fort Liard, NWT (first record for the territory in 1998 was a dark morph adult). Isolated summer sightings and possible breeding may occur in Turtle Mountain Prov. Pk. and Brandon Hills WMA near Brandon, Man. There are no specimens or summer records for Saskatchewan.

**U.S.** Dark morphs are annual in small numbers in the fall and very small numbers in the spring in Duluth, Minn; fewer than 20 are seen in the fall, fewer than 10 in the spring. Spring sightings are occasionally recorded for w. Traverse Co., Minn., where two dark morphs were seen in late Apr. 2000. South of Minnesota, only Corpus Christi, Tex., detects fairly large numbers, with 50 or more seen each fall. The lower Rio Grande valley region of s. Texas, from Falcon S.P. to Santa Ana NWR, tallies a few during migration. There are specimens from e.-cen. Missouri and Oklahoma from the fall migration period. There are a few irregular sightings of juveniles from the Goshute Mts. in ne. Nevada. All ages are annual in small numbers in the spring along the foothills of the Rocky Mts., especially at Dinosaur Ridge Hawkwatch near Morrison, Colo. There is one photograph record of a juvenile (*see* plate 260) in late Sep. in Prowers Co., Colo. Probably annual in Prowers Co. in the fall. A juvenile was seen among migrating Swainson's Hawks in Dalhart, Tex., in late Sep. Dark morphs are annual in very small numbers at the Golden Gate Raptor Observatory in Marin Co., Calif., in the fall.

**Winter. Both color morphs.**—Primarily from Colima to Chiapas, Mexico, south through Central America and South America, in n. and e. Peru, Colombia, Venezuela, Bolivia, and s. Brazil. Broadwings arrive on the winter grounds in s. Central America and South America from late Oct. to late Nov.; those wintering in Mexico may arrive somewhat earlier.

**Extralimital winter.** *California:* Casual in coastal valleys and mountain ranges from Sonoma Co. south to San Diego Co. but has occurred north to Humbolt Co. In inland regions, mainly found in the Central Valley, Sierra Nevada, and southern desert valleys. The most northern inland record is for s. Siskiyou Co. A juvenile was on San Clemente Island from late fall to early winter in 2001. *Louisiana:* Fairly regular in winter in the southeast coastal region. *Texas:* Casual in the lower Rio Grande valley. *Colorado:* Northernmost interior U.S. record of a juvenile near Boulder in late Dec. 2001. All wintering individuals in the U.S. have been light morphs.

**Movements.**—Migrants form small to very large flocks, but for portions of the long journey sometimes migrate singly. Flocks often comprise thousands of individuals in southern latitudes. Migration lasts 4–8 weeks. Individuals crossing the Great Plains are mainly single birds or in very small flocks, and often associate with flocks of Swainson's Hawks. Dark morphs, in particular, are seen in very small numbers on the Great Plains, along w. Lake Superior in the fall and spring, and rarely west of the Plains.

*Fall migration*: Very punctual movements following strict calendar timing, with all ages moving simultaneously. Migration occurs from mid-Aug. to mid-Oct. but in s. Texas spans into early Nov. Juveniles have been seen in Hidalgo Co. in s. Texas in early Aug. A distinct peak occurs across the Canadian-U.S. border and w. Great Lakes in mid-Sep., the third week of Sep. in central latitudes, and from late Sep. to early Oct. in s. Texas. Peak numbers occur in Veracruz, Mexico, in early Oct. Juveniles trickle down before main event but also comprise most of the late stragglers. Dark morphs tend to be more prevalent during the latter part of the season and are seen from late Sep. to early Oct. in northern and mid-latitudes and from early to mid-Oct. in s. Texas.

A southwesterly course is taken in the eastern part of the West but a southerly or southeasterly course in western areas of the West. It is probable that most w. Canadian birds first arch eastward in the more wooded regions north of the Great Plains, then move southward mainly east of the Plains in the U.S. Migrating in these regions allows this woodland species to stay in partially wooded habitat for ample roosting areas rather than crossing the virtually treeless Great Plains. However, Broad-winged Hawks of all ages are seen annually in very small numbers on the Great Plains.

Fair numbers of mainly juveniles annually stray off course to the Pacific Coast. At the Golden Gate Raptor Observatory, Marin Co., in w.-cen. coastal California, 25–235 were tallied annually each fall between 1992 and 2000. Of these, 80% were juveniles, and dark morphs averaged 3% of annual totals. A few, primarily light morph juveniles, are seen at hawkwatch locations in the interior West in the fall. The Goshute Mts. in e. Nevada see 6–160 each fall; Lipan and Yaki Points at the Grand Canyon in Arizona see up to 35, with the most being seen in the latter location; and at Chelan Ridge near Chelan, Wash., observers see 2–5 each fall. Broad-winged Hawks are rarely seen at areas other than hawkwatch locations.

Migration path leads the bulk of migrants southward to the Coastal Bend area of Texas. At Corpus Christi, daily Broadwing flights often reach 300,000 birds, and annual tallies almost reach 1 million. The daily record for Corpus Christi is 446,200 on Sep., 26, 2001. From s. Texas, the migratory path continues southward into e. coastal Mexico (east of the Sierra Oriental), through Central America, and finally into South America.

*Spring migration*: Adults precede juveniles by 2–4 weeks, although a few early juveniles trickle through with the main adult flight. Adults are punctual and adhere to strict calendar timing. In s. Texas, migration spans from mid-Mar. to mid-Apr. and peaks in late Mar. to early Apr. In the cen. and n. U.S., adults move through from mid-Apr. to mid-May. Peak adult numbers on the w. Great Plains in e. Colorado are in late Apr., in early May in the Great Lakes region and along the Canadian border. Adults do not reach their northernmost breeding areas in British Columbia and the Northwest Territories until late May and early Jun. (based on limited data). Juveniles move in a protracted period from late Apr. to mid-Jun. but trickle northward throughout late Jun. and into early Jul. There are definitive peaks of movements: late April in s. Texas, mid-May on the cen. Great Plains, and late May to early Jun. in the w. Great Lakes. Juveniles may not reach northernmost natal regions until well into Jun. or even Jul. They are in obvious first prebasic molt and appear quite ratty.

Migrants head northward out of South America, into Central America, and enter the U.S. from e. coastal Mexico. In the U.S., they head north along the Coastal Bend of Texas. Most probably remain east of the Great Plains and fan out thereafter in a broad-fronted manner to breeding and natal areas. Some, however, cross the Great Plains, and a few follow the eastern edge of the Rocky Mts. A few migrants are seen each spring at hawkwatch sites in the Sandia Mts. near Albuquerque, N. Mex., and at Dinosaur Ridge near Morrison, Colo. Very few migrate within the Rocky Mts. Jordanelle Reservoir near Park City, Utah, and Mt. Lorette west of Calgary, Alb., each see fewer than five Broadwings per season. Broad-winged Hawks are rarely seen in areas west of the Great Plains in locations that are not hawkwatch sites.

**NESTING:** Begins in late Apr. in southern latitudes and early May to early Jun. in northern latitudes. Pair formation occurs after arrival on the breeding grounds. One-year-old females, still in juvenile plumage, occasionally mate with adult males. Nesting is completed from mid-Jul. in southern latitudes to early to mid-Aug., and possibly late Aug., in northern latitudes of n. Alberta and n. British Columbia.

**Courtship (flight).**—*High-circling* and *sky-dancing* (*see* chapter 5). Often vocalizes during high-circling.

Nest trees are in tracts of similar-sized trees, usually near forest openings and generally in deciduous trees, rarely in conifers. Nests are placed in the first major crotch of deciduous trees or next to the trunk of conifers 20–40 ft. (7–12 m) high. Nests is poorly constructed mass of sticks 15–17 in. (38–43 cm) in diameter, 5–12 in. (13–30 cm) deep, and lined with greenery. A new nest is built each year or an old nest may be reused. Nests are sometimes built on top of old nests of American Crows, other raptor species, or (leafy) squirrel nests. Two or 3 eggs are incubated for 28–31 days, mainly by female, but male takes over when female feeds. Nestlings branch in 29–31 days, fledge in 35–42 days, and become independent in about 70 days.

**CONSERVATION:** No measures are taken. Status is stable.

**Mortality.**—Illegal shooting still occurs in North America in the summer and especially south of the U.S. during migration and winter. Habitat loss in the West is minimal at this time but is uncertain on the winter grounds. Because of a general diet, exposure to pesticide contamination has been minimal.

**SIMILAR SPECIES:** COMPARED TO ADULT LIGHT MORPH.—(1) **Red-shouldered Hawk, adults.**—*B. l. lineatus and B. l. alleni* ranges overlap spring to fall; *B. l. elegans* overlaps in Sep., possibly winter. PERCHED.—Dark brown irises. Generally narrower, finer, brighter rufous-barred underparts. Long tarsi. Black and white checkered wings. White bands on dorsal side of tail. FLIGHT.—Black and white barred outer primaries. White panel or window on base of outer primaries. *Note:* Some have bibs as on pale-bellied type Broad-winged Hawk. COMPARED TO ADULT DARK MORPH.—(2) **Zone-tailed Hawk, adults.**—Range overlap Sep.–Oct.; Mar. and Apr. from Arizona to s. Texas. FLIGHT.—Similar tail pattern on closed tail. Undersides of remiges gray, very barred. (3) **Swainson's Hawk, dark morph adults.**—PERCHED.—White outer lores. Dark brown irises. Wingtips reach or exceed tail tip. White or tawny undertail coverts. FLIGHT.—Underside of remiges dark gray; wings held in dihedral. White or tawny undertail coverts. (4) **Rough-legged Hawk, dark morph adults.**—Range overlap mid-Sep.–Oct., late Mar.–Apr. PERCHED.—White "mask" on forehead, outer lores. Dark brown irises. Feathered tarsi. Wingtips reach or exceed tail tip. FLIGHT.—Undertail pattern identical on closed tail with one white band. Very long wings with rounded wingtips. Wings held slightly above horizontal. Hovers, kites. COMPARED TO LIGHT MORPH JUVENILE.—(5) **Cooper's Hawk, juveniles.**—PERCHED.—Pale yellowish or gray irises. Uniformly colored head without a dark malar stripe. Wingtips much shorter than the tail tip. Use caution with broad-banded type tail: similar, with three equal-width dark bands. FLIGHT.—Remiges heavily barred on undersides. (6) **Northern Goshawk, juveniles.**—PERCHED.—Yellowish orange irises; beware of recently fledged birds with gray irises. Malar stripe, if present, is very narrow and not well defined. Shares prominent pale supercilium. Pale bar on greater coverts. Tail can be similar to broad-banded type tail with thin white borders along dark bands on dorsal side. FLIGHT.—Undersides of remiges are heavily barred. (7) **Gray Hawk, juveniles.**—Range overlap in s. Texas Aug.–Oct., Apr.–May, but can be all winter. PERCHED.—Dark eyeline; distinct, dark mid-throat stripe. Use caution in spring, Broad-winged Hawk's supercilium and auriculars fade to white, but lacks Gray Hawk's dark eyeline. Wingtips are much shorter than the tail tip and extend halfway down the tail. Thin barring on leg feathers. Secondaries and greater coverts are pale brown with narrow dark bars. Chevron-shaped tail bands on dorsal side. FLIGHT.—Very rounded wingtips. Undersides of all remiges have thin barring: Broad-winged Hawks with narrow barring on outer primaries can be identical.

(8) **Red-shouldered Hawk, juveniles.**—PERCHED.—Heads of *B. l. lineatus* and *alleni* virtually identical; some have a defined dark eyeline. *B. l. lineatus* often has a dark throat; may have single dark streak on mid-throat like Broadwing. *B. l. alleni* and *B. l. elegans* have narrow barring on leg feathers and heart-shaped or barred underparts. Long tarsi. Pale barring on dorsal surface of secondaries. Pale tail bands narrower than dark ones. FLIGHT.—Panels/windows on base of outer primaries are white or tawny. Wingtips very rounded. (9) **Red-tailed Hawk (*B. j. calurus*) dark morph juveniles.**—Pale yellowish or grayish irises. Greenish cere; rarely yellow. FLIGHT.—Rounded wingtips. On upperside of wing, primaries and primary coverts paler brown than rest of wing. Hovers and kites.

(10) **Short-tailed Hawk, light morph juveniles.**—PERCHED.—Pale throat is unmarked. Thick dark eyeline. Wingtips equal to tail tip. FLIGHT.—Pale gray secondaries. Kites and parachutes. Wingtips bend upwards when kiting and soaring.

**BROAD-WINGED HAWK,** *Buteo platypterus:* Common. Virtually entire population migrates along Coastal Bend of TX in fall and spring to and from winter grounds in s. Mexico and Central and South America. Dark morphs breed in AB, BC, and probably NWT; have been seen in summer in sw. MB.

**OTHER NAMES:** Broadwing, Wing, Broadie. *Spanish:* Aguililla Alas Anchas, Aguililla Aluda. *French:* Petit Buse.

REFERENCES: Andrews and Righter 1992; Baumgartner and Baumgartner 1992; Busby and Zimmerman 2001; Bylan 1998, 1999; Campbell et al. 1990; Center for Conservation, Research, and Technology 2000; Clark and Wheeler 1987, 2001; Corman 1998; del Hoyo et al. 1994; Dodge 1988–1997; Dorn and Dorn 1990; Ducey 1988; Fish and Hull 1996; Goodrich et al. 1996; Howell and Webb 1995; Jacobs and Wilson 1997; Janssen 1987; Johnsgard 1990; Kellogg 2000; Kent and Dinsmore 1996; Kerlinger and Gauthreaux 1985; Machtans 2000; Oberholser 1974; Palmer 1988; Peterson 1995; Pulich 1988; Robbins and Easterla 1992; Salter et al. 1974; Semenchuk 1992; Sirois and McRae 1996; Small 1994; Smith 1996; Stewart 1975; Sutton 1967; Thompson and Ely 1989.

**Plate 246. Broad-winged Hawk, adult (heavily barred type) [Sep.]** ▪ Brown head. Pale orangish brown iris. ▪ Underparts marked with dark brown. Densely barred breast forms partial bib; belly and flanks thickly barred; leg feathers narrowly barred. Uniformly brown upperparts. ▪ Single broad white band shows when tail is closed.

**Plate 247. Broad-winged Hawk, adult (pale-bellied type) [Oct.]** ▪ Rufous-brown head. Pale orangish brown iris. ▪ Underparts marked with rufous-brown. Nearly solid dark breast forms bib; belly and flanks sparsely covered with arrowhead-shaped marks and thick bars. ▪ Single broad white band shows when tail is closed.

**Plate 248. Broad-winged Hawk, adult/subadult (lightly barred type)** [**Nov**] ▪ Brown head. Pale orangish brown iris. ▪ Breast marked with broad arrowhead-shaped marks, belly lightly marked, flanks sparsely barred. ▪ Single broad white band shows when tail is closed. ▪ *Note:* Palest type of ventral markings for adults and subadults. Adults and subadults are not separable when perched.

**Plate 249. Broad-winged Hawk, adult** [**Sep.**] ▪ Nearly uniformly barred underparts. ▪ Underside of remiges thinly barred; wide black band on rear edge; coverts can be tawny and moderately marked. Fairly pointed wingtips when soaring or banking ▪ Many birds show 2 broad white bands when tail is fanned.

**Plate 250. Broad-winged Hawk, adult (heavily barred type)** [**Sep.**] ▪ Uniformly barred, including breast. ▪ Underside of remiges thinly barred; wide black band on rear edge; coverts can be tawny and moderately marked. Pointed wingtips when gliding. ▪ Single broad white band shows when tail is closed.

**Plate 251. Broad-winged Hawk, subadult (lightly barred type)** [**May**] ▪ Breast sparsely streaked, belly covered with arrowhead-shaped marks, flanks sparsely barred. ▪ White underwing coverts unmarked. Outer 3 primaries are old, faded brown juvenile feathers. ▪ Single broad white band shows when tail is closed. ▪ *Note:* Adults similar. Photo by D. Edmondson.

**Plate 252. Broad-winged Hawk, subadult (heavily barred type) [May]** ▪ Underparts uniformly barred with rufous. ▪ Retains juvenile remiges; pale areas on wingtip and rear edge (p10 and s4, 8, 9). ▪ Some birds show up to 3 broad white bands; bands on outer rectrix set often do not align with inner bands. ▪ *Note:* Separable from adults at close range.

**Plate 253. Broad-winged Hawk, adult [Sep.]** ▪ Uniformly brown upperparts, including remiges and coverts. ▪ Black tail has 2–3 white or gray bands; mid-tail band is widest. White tips on uppertail coverts (may appear as white band).

**Plate 254. Broad-winged Hawk, adult dark morph [Sep.]** ▪ All of head and body uniformly dark brown. ▪ Uniformly whitish remiges thinly barred; wide black band on rear edge; coverts dark brown. ▪ Single broad white band shows when tail is closed. ▪ *Note:* Photographed at Duluth, Minn.

**Plate 255. Broad-winged Hawk, juvenile (lightly streaked type) [Nov.]** ▪ Brown head with tawny supercilium and auricular areas, black malar mark. ▪ White underparts narrowly streaked on sides of neck and flanks. Unmarked leg feathers. ▪ Wingtips shorter than tail tip.

**Plate 256. Broad-winged Hawk, juvenile (moderately streaked type) [Sep.]** ▪ White supercilium, tawny auricular areas, black malar mark. ▪ Underparts streaked on sides of neck, spotted or streaked belly, barred flanks. ▪ Secondaries and greater coverts uniformly brown. ▪ Narrow-banded type tail with wide dark subterminal band, narrow dark inner bands.

**Plate 257. Broad-winged Hawk, juvenile (heavily streaked type) [Nov.]** ▪ Brown head with white supercilium, tawny auricular areas, black malar mark, medium brown iris. ▪ Underparts densely covered with wide brown streaking on breast and belly, barring on flanks. Short tarsi.

**Plate 258. Broad-winged Hawk, juvenile [Sep.]** ▪ Brown head with white supercilium, tawny auricular areas, black malar mark, medium brown iris. ▪ Uniformly brown upperparts. ▪ Solid brown secondaries and greater coverts. Wingtips distinctly shorter than tail tip. ▪ Narrow-banded type tail has wide black subterminal band, 3 or 4 narrow inner bands. ▪ *Note:* Same bird as on plate 256.

**Plate 259. Broad-winged Hawk, juvenile [Nov.]** ▪ Brown head with white supercilium, tawny auricular areas, black malar mark, medium brown iris. ▪ Uniformly brown upperparts. ▪ Solid brown secondaries and greater coverts. Wingtips distinctly shorter than tail tip. ▪ Broad-banded type tail has 3 nearly equal-width dark bands; subterminal band often a bit wider.

**Plate 260. Broad-winged Hawk, juvenile dark morph (streaked type) [Sep.]** ▪ Dark brown head, including throat and forehead; pale lores. ▪ Dark brown underparts streaked with tawny. ▪ Narrow-banded type tail has dark subterminal tail band, narrow inner bands. ▪ *Note:* Also has all-dark type plumage with uniformly dark brown underparts. Photographed in Prowers Co., Colo.

**Plate 261. Broad-winged Hawk, juvenile (moderately streaked type) [Sep.]** ▪ Thin center stripe on throat. ▪ Underparts moderately streaked flanks barred; breast mainly unmarked. ▪ Rectangular window may show on primaries. Pointed wingtips when gliding. ▪ Narrow-banded type tail has wide subterminal band, several inner bands.

**Plate 262. Broad-winged Hawk, juvenile (lightly streaked type) [Sep.]** ▪ Thin center stripe on throat. ▪ Underparts thinly streaked on sides of neck and flanks. ▪ Remiges thinly barred; moderately wide dark rear edge. Unmarked coverts. Wingtips often barred. ▪ Broad-banded type tail has 3 wide dark bands (2 visible on closed tail); outer rectrix set has non-aligning narrow bands.

**Plate 263. Broad-winged Hawk, juvenile [Sep.]** ▪ Brown upperparts. ▪ Solid brown secondaries. ▪ Narrow-banded type tail with wide dark subterminal band, 3 or 4 narrow dark inner bands.

**Plate 264. Broad-winged Hawk, juvenile** [Sep.] ▪ Brown upperparts. ▪ Faint, dark barring on secondaries (typically solid brown); slight pale brown panel on primaries. ▪ Broad-banded type tail has 3 equal-width dark bands, 2 or 3 pale bands (somewhat adultlike). White markings on uppertail coverts.

**Plate 265. Broad-winged Hawk, migrant flock (kettle)** [**Oct.**] ▪ Migrants may gather in flocks composed of a few hawks or hundreds or thousands.

# SHORT-TAILED HAWK
## *(Buteo brachyurus)*

**AGES:** Adult and juvenile. Adult plumage is acquired when 1 year old. Juvenile plumage is worn the first year. Some juveniles have slightly longer tails than adults.

**MOLT:** Accipitridae wing and tail molt pattern (*see* chapter 4). First prebasic molt from juvenile to adult plumage probably begins in Apr. Molt is first apparent with loss of innermost primary (p1) and central rectrices. An individual in late May in Hidalgo Co., Tex., exhibited extensive molt on the inner remiges and rectrices. The first prebasic molt appears to be a complete molt. Based on Florida data, molt appears to be completed by mid-Nov. No data on subsequent prebasic adult molts except they appear to be completed by mid-Nov.

**SUBSPECIES:** Polytypic with two subspecies. *B. b. fuliginosus* inhabits Florida and is also found from Mexico south to Panama. It is a regular visitor from Mexico to s. Texas and, recently, a regular summer resident and probable breeder in s. Arizona. One additional race, nominate *B. b. brachyurus*, occurs in Colombia, w. Ecuador, east to the Guianas, Brazil, and south through e. Peru, e. Bolivia to Paraguay, and n. Argentina.

White-throated Hawk (*B. albigula*) of the Andean highlands in South America is considered to be a subspecies of *brachyurus* by some authors; however, hybrids do not occur in overlap zones, and it is probably a separate species (del Hoyo et al. 1994).

**COLOR MORPHS:** Light and dark morphs with no intermediate plumages. Minimal plumage variation occurs in adults of both morphs. There are minor plumage variations in juvenile light morphs and considerable variation in juvenile dark morphs.

**SIZE:** A medium-sized raptor and a small buteo. Males average slightly smaller than females but sexes are not separable in the field. Length: 15–17 in. (38–43 cm); wingspan: 32–41 in. (81–104 cm).

**SPECIES TRAITS:** BODY.—**Dark brown uppertail coverts.** WINGS.—When perched, wingtips of the long primaries extend to the tail tip. **In flight, long wings are fairly broad**

**and the trailing edge of the secondaries bows outward from the body and then tapers to the moderately pointed wingtips.** Wingtips become very pointed when gliding. *Note:* "Short-tailed" is a misnomer: tail length is not any shorter in proportion to overall size than it is in many buteo species.

**ADULT TRAITS:** HEAD.—**Most have a narrow white mask around forehead and outer lores next to cere; inner lores are dark. Dark brown irises are visible only at close range.** WINGS.—**Undersides of remiges are medium gray with a small white area on the basal region on the outer four primaries.** Typical of adult buteos, there is wide black band on the trailing edge of the wing. TAIL.—**Two patterns, which are most visible on the ventral surface:** (1) *Banded type:* **Moderately wide dark subterminal band and three or four complete, narrow dark inner bands. Medium brown uppertail also shows dark narrow bands.** *Note:* In Mexico, seen on most light morphs in w. Mexico, on many light morphs in e. Mexico, and on virtually all dark morphs. (2) *Partially banded type:* **Moderately wide dark subterminal band and two to four incomplete dark inner bands, or dark inner bands absent and only the dusky subterminal band is visible. Medium brown uppertail exhibits the dark subterminal band and either a hint of partial dark inner bands, or the inner tail is unmarked.** *Note:* In Mexico, seen on a few light morphs and rarely on dark morphs.

**ADULT LIGHT MORPH:** HEAD.—**Uniform dark brown, including the malar region, and appears hooded.** White throat. BODY.—Immaculate white underparts. At close range, a small rufous patch on sides of the neck may be visible. Uniformly dark brown back, scapulars, and rump, including the uppertail coverts. WINGS (dorsal).—Dark brown upperwing coverts contrast with darker remiges and create somewhat of a two-toned effect. An additional two-toned effect is created with the dark primaries and primary coverts contrasting with the somewhat paler secondaries and inner half of the wing. WINGS (ventral).—White underwing coverts contrast sharply with medium gray remiges.

**ADULT DARK MORPH:** HEAD.—Uniformly blackish brown except for narrow white mask. BODY and WINGS.—Uniformly blackish brown, including the undertail coverts and underwing coverts.

**JUVENILE TRAITS:** HEAD.—Medium brown irises are visible only at close range. Forehead and outer lores may be white and exhibit a mask; however, many birds have all-white lores. WINGS.—**Undersides of the secondaries and outer half of the primaries are pale gray and inner half of the primaries are white; forms a large white panel on the primaries (panel is much larger than adult's).** Narrow gray band on the trailing edge of remiges is typical of most juvenile buteos. TAIL.—Base color of the undertail surface is whitish, medium brown on the uppertail surface. Two main patterns of dark banding, but with considerable variation. (1) *Banded type:* **Undertail has four to seven complete, narrow dark inner bands with the subterminal band being equal in width to the inner bands, somewhat wider, or much wider than the inner bands.** Those with equal-width banding tend to have fewer dark bands on the ventral surface (4–6). Uppertail pattern of dark banding is distinct, with four or five narrow dark bands inside the wider subterminal band. *Note:* Common pattern on both color morphs in Mexico and Central America; common on Florida dark morphs and rare or absent in Florida light morphs. (2) *Partially-banded type:* **Undertail either has a moderately wide dusky subterminal band, or the subterminal band may be diffused and nearly absent. Inner portion of the undertail may have partial dark inner bands on mainly the outer rectrix sets or lack markings and be uniformly whitish (particularly if it lacks a subterminal band). Uppertail is mainly unmarked except for the darker, dusky subterminal band; however, many have small dark spots on the deck rectrices or faint, incomplete dark bands.** *Note:* Pattern is on some light and dark morphs in Mexico and Central America, on many Florida dark morphs, and on virtually all Florida light morphs. Patterns are often similar to those of adults; however, adults do not have equal-width banding or a dusky, diffused subterminal band as seen on many juveniles.

*B. b. fuliginosus* of Mexico (seen on the Atlantic and Pacific slopes) and Central America (museum specimens) typically have more distinctly banded tail patterns than population in

Florida. These boldly marked tail patterns are the most likely ones to be seen on individuals observed in the w. U.S.

**JUVENILE LIGHT MORPH:** HEAD.—**Bold face pattern: narrow, pale supercilium; short, broad, dark eyeline that extends under and behind the eye; broad dark malar mark; a narrow pale stripe on the cheek between the dark eyeline and malar. Head pattern is visible at moderate distances.** On some, head is darker with less pronounced pale supercilium and pale cheeks, but the broad dark line behind the eyes is still apparent. *Note:* **Throat is unmarked.** BODY.—(1) *Unmarked type.*—Cream-colored underparts with sparse dark streaking on sides of the neck, but the rest of the underparts are unmarked. *Note:* Common type. (2) *Streaked type.*—Sides of the neck, flanks, and mid-belly are narrowly streaked with dark brown. *Note:* Common type. Underparts of both types fade and become whitish by spring. Back, scapulars, rump, and uppertail coverts are uniformly brown. WINGS (dorsal).—Upperwing coverts are the same color brown as the back and scapulars, but greater coverts and all remiges are slightly darker and form a two-toned effect with upperwing surface. An additional two-toned effect is often created by the even darker primaries and primary coverts that contrast with the slightly paler inner half of the wing. WINGS (ventral).—Unmarked, cream-colored underwing coverts contrast somewhat with the slightly darker, pale gray secondaries.

**JUVENILE DARK MORPH (MOTTLED TYPE):** HEAD.—Blackish brown, often with slightly paler auriculars. BODY.—Uniformly blackish brown breast forms a bib which contrasts with paler, white-mottled and/or-spotted belly, flanks, and undertail coverts. WINGS.—Underwing coverts are blackish brown and mottled with white; they are much darker than the secondaries. *Note:* Common plumage type.

**JUVENILE DARK MORPH (STREAKED TYPE):** HEAD.—Blackish brown, typically with a large light patch on the auriculars and throat and often with pale, partial supercilium. BODY.—Blackish brown breast has some white streaking but still forms a distinct bib with the sharply contrasting white underparts which are narrowly streaked with dark brown. WINGS.—White underwing coverts are narrowly streaked with dark brown. Underwing coverts and secondaries appear uniformly pale grayish at a distance. *Note:* Fairly common plumage type.

**JUVENILE DARK MORPH (ALL-DARK TYPE):** HEAD.—Blackish brown. BODY.—Uniformly blackish brown. WINGS.—Uniformly blackish brown underwing coverts are much darker than the remiges. At close range, faint, light speckling may be visible on axillaries and some underwing coverts. *Note:* Rare or very uncommon plumage type.

**ABNORMAL PLUMAGES:** None known.

**HABITAT: Summer.**—*Mexico:* Semi-open and wooded areas from sea level to moderate-elevation dry tropical deciduous forests, and up to high-elevation pine-oak woodlands of 6,500 ft. (2,000 m) in Sonora. In e. Mexico, similar elevation and habitat, but locally up to 9,800 ft. (3,000 m). In s. Mexico in dry or wet tropical semi-open and wooded regions. *Arizona:* Moderately high-elevation pine-oak woodlands in "sky island" mountains to at least 6,500 ft. (2,000 m). *Texas:* Low elevation semi-open tropical deciduous forests.

**Winter.**—*Mexico:* Similar to breeding habitat but generally at lower elevations in foothills and coastal regions. Foraging regularly occurs in villages and towns and in moderately agricultural and grazing areas. Absent from intensively agricultural zones.

**Migration.**—Similar to summer and winter habitats.

**HABITS:** Tame. Most exhibit little fear of humans during the nonbreeding season when in low-altitude flight or perched. A highly aerial species. Virtually all daylight hours are spent airborne. Perching occurs only for night roosting, during inclement weather, when feeding, and occasionally for short periods after missed capture attempts. Roosting birds perch in the tree canopy and are rarely seen. *Short-tailed Hawks do not perch in trees along highways or on utility poles, wires, or posts.*

**FEEDING:** Exclusively an aerial hunter. Preys almost entirely on small birds perched on or near the ground and on outer branches of bushes and trees. Small rodents and reptiles comprise a small fraction of diet. A bird was seen carrying a lizard in Arizona. Warbler-sized birds are decapitated and the body swallowed whole. Captured prey is (1) immediately taken to a branch in the canopy of a nearby tree, partially plucked, and then eaten or (2) may be decaptitated and swallowed whole while the hawk is soaring or gliding. Unlike in most raptors, a relaxation period of perching does not occur after feeding. If feeding is done on a perch, aerial activities are immediately resumed once the prey is eaten.

Hawks capturing prey on the outer branches of trees may plunge from heights of a few hundred meters, hit the prey, and cling onto the branches, often hanging sideways or upside down until the prey is secured, then fly off with the prey. *Note:* In the state of Nayarit, Mexico, an adult was observed making a 500-ft. (150-m) vertical dive to capture a small passerine on the outer branches of a large tree. The hawk crashed into the branches, clung sideways and upside down for several seconds—flapping its wings for balance—then flew off with the prey. The hawk proceeded to decapitate and partially pluck the prey, then eat it in one gulp.

**FLIGHT:** An incredibly aerial raptor. Along with Swallow-tailed Kite, one of the most aerial raptors and the most aerial buteo. Primary flight modes are soaring, gliding, parachuting, and kiting. In all modes, wings are held on a flat plane with the primary tips flexing somewhat upwards. Powered flight is used sparingly to gain altitude quickly after prey capture, attempted capture, or in inclement weather when aerial lift conditions are poor. Typically, Short-tailed Hawks soar to high altitudes and then glide into or tack the wind when hunting. If wind velocity is sufficient, kiting is used for considerable lengths of time at moderate and high altitudes in areas with abrupt wall-like woodland edges, along cliffs, and on hillsides. Gliding and kiting occur at low altitudes in high winds, often just above treetop or rooftop level. When prey is detected, the hawks engage in (1) a long, impressive angled or vertical dive; (2) a series of short vertical or angled dives that are periodically interrupted with stints of parachuting; or (3) slowly parachute vertically downward and then make a short dive at the end. This flight process is repeated constantly throughout the day. Short-tailed Hawks do not hover.

**VOICE:** Vocalizes primarily near nest sites when disturbed by intruders and during courtship and food transfers. Call is a high-pitched, drawn-out *keeee* with a slight decrescendo slur at the end (making a somewhat two-syllable sound). On rare occasions, vocalization occurs during the nonbreeding season when two individuals of any age or sex are near each other in flight; call is a high-pitched *kree, kree, kree.* Food-begging of fledglings may be extended series of the *kree, kree, kree* call.

**STATUS AND DISTRIBUTION:** *Rare* but regular visitor to Sonora, Texas, and Arizona. *Mexico:* First recorded in Sonora in the early 1980s, but breeding has not been confirmed. Fairly common along the Atlantic and Pacific slopes in e. and w. Mexico and in s. Mexico. Also fairly common in Central America. Light morphs outnumber dark morphs in e. Mexico (no percentage available). Dark morphs comprise 69% of population in Sonora but only 40% in Nayarit and Jalisco (based on 52 birds seen in Feb. 2002).

*Texas:* Twelve accepted state records as of the summer of 2001: (1) Starr Co. for 7 days in late Jul. 1989; (2) Bentsen-Rio Grande Valley S.P., Hidalgo Co., in early Mar. 1994; (3) Santa Ana NWR, Hidalgo Co., in early Oct. 1994; (4) Lost Maples State Natural Area, Bandera Co., for 32 days from late May to mid-Jun. 1995; (5) Santa Ana NWR, Hidalgo Co., in May 1995; (6) near Dripping Springs, Hays Co. (most northern sighting), in early Jun. 1995; (7) Santa Ana NWR, Hidalgo Co., for 56 days from early Jun. to late Jul. 1996; (8) Santa Ana NWR, Hidalgo Co., in mid-Feb. 1997; (9) Bentsen-Rio Grande Valley S.P., Hidalgo Co., in late Apr. 1998; (10) Santa Ana NWR, Hidalgo Co., for 40 days from late May to late Jun. 1998; (11) Lost Maples State Natural Area, Bandera Co., for 15 days from early to late Apr. 1999; (12) Corpus Christi, Nueces Co., in mid-Oct. 1999.

*Arizona:* First photographed documentation and accepted state record were of two adults, a light and dark morph, from late Jul. through Sep. 1999 in the Huachuca Mts., Cochise Co. In the summer of 2000, one or two adults were again seen in the Huachuca Mts. In 2001, sightings became regular in early Aug., first in the Chiricahua Mts., then in the Huachuca Mts. There currently are no verified breeding records for the state. However, one or two light morph juveniles, in fresh plumage, were well documented in mid- and late Aug. 2001 in both the Huachuca and Chiricahua Mts. These sightings are indicative of breeding in the area. Probable and/or possible sight records for Arizona: Chiricahua Mts., Cochise Co.—(1) adult light morph in early Aug. 1985, (2) adult light morph in early Mar. 1990, (3) adult light morph in mid-Jun. 1997, and (4 and 5) two sightings of adult light morphs in late Mar. 1999. Huachuca Mts.—(1) adult light morph in late Jul. 1988.

All Texas and Arizona sightings are from Mexico populations (also *fuliginosus*) wandering north of typical haunts.

**Summer.**—*See above* for status.

**Winter.**—*See above* for status. Found in Mexico north to cen. Tamaulipas on the Atlantic slope and to cen. Sonora on the Pacific slope. Possibly more common on the Pacific slope. There is one winter record for s. Texas. Short-tailed Hawks vacate Arizona in the winter.

**Movements.**—Not well known in the West. *Fall migration:* Probably Sep. and Oct. *Spring migration:* Possibly occurs in Mar. for adults and through May for 1-year-old juveniles. Considerable dispersal activity occurs throughout the summer and during winter months.

**NESTING:** Nesting may have occurred in the Huachuca Mts., Cochise Co., Ariz., in the summer of 2001. A light morph juvenile in a recently fledged plumage type was discovered in mid-Aug. in the Huachuca Mts. Nesting is unlikely to occur in Texas because of marginal habitat.

**Courtship (flight).**—*High-circling* and *sky-dancing* (*see* chapter 5). Prey or nest materials are sometimes carried in courtship flights.

Based on Florida data; males bring material but nests are constructed mainly by females. Nests are bulky stick masses lined with greenery and are about 2 ft. (61 cm) in diameter and 1 ft. (30 cm) deep. They are placed in the upper portion of tall trees on side branches near the main trunk, often on bromeliads. Nest heights range from 15 to 95 ft. (5–29 m). The 2 eggs, rarely 3, are laid from mid-Mar. through Apr. and are incubated about 34 days. The fledging period is currently unknown but is probably like in other similar-sized raptors (35–45 days).

**CONSERVATION:** No measures taken. Susceptible to the effects of logging and burning of woodlands in Mexico, which are possible reasons for recent influx in the U.S.

**Mortality.**—Habitat alteration may be displacing Short-tailed Hawks in portions of Mexico. Illegal shooting in Mexico. Being an avian feeder, possibly contaminated by pesticides since DDT is still used to some extent in Mexico and is widely used in Central America.

**SIMILAR SPECIES: (1) Broad-winged Hawk.**—Range overlap Sep.–Oct. and Mar.–May in s. Texas; overlap in Arizona would be very rare. PERCHED.—Regularly perches along highways on branches, poles, and wires. Wingtips are distinctly shorter than the tail tip. FLIGHT.—Undersides of remiges uniformly white. Identical body length. Similar pointed wing shape but shorter winged; wings held on a flat plane, trailing edge is straighter: use caution. Soars and glides with wings on a flat plane. Does not kite or parachute and rarely dives. Flaps wings a moderate amount. **(1A) Adult dark morphs.**—Uniformly whitish remiges; wide white tail band on black tail. Use Flight data. **(1B) Juveniles.**—Similar grayish trailing edge of underwing; underwings have large translucent rectangular window on primaries. Uniform dark upperwing. Tail pattern can be identical: use caution. **(1C) Juvenile light morphs.**—Lightly streaked type with lack of underpart markings is a virtual look-alike. Has a dark mid-throat stripe and lacks a dark eyeline; wingtips are shorter than the tail tip. Uppertail coverts appear pale when seen from above. Use Flight data for other aerial separation. **(1D) Juvenile dark morphs.**—Uniformly whitish remiges. Tail pattern can be identical: use extreme caution Use Flight data. **(2) Swainson's Hawk.**—Range overlap spring through fall in s. Texas and Arizona; occasional overlap in winter in Texas. PERCHED.—Has similar wingtip-to-tail-tip ratio. Quite tame. Perches on very open elevated objects and ground. FLIGHT.—Wing shape similar, trailing edge more straight edged, and wingtips usually more pointed. Wings held in dihedral. Hovers but rarely parachutes. Kites and dives but flaps wings regularly. Uppertail coverts on all but dark morph adults have pale U shape when seen from above. Undertail coverts are white or tawny on all ages and color morphs. **(2A) Adult (all morphs).**—Dark gray undersides of remiges. **(2B) Adult light morphs.**—Can be very similar to adult light morph Short-tailed Hawks. Similar head with white

**SHORT-TAILED HAWK,** *Buteo brachyurus:* Rare but regular visitor to TX, AZ, and Sonora. Multiple records exist for areas noted in AZ and s. TX. Possible recent nesting in AZ. Common south of mapped area in Mexico.

mask and throat: use caution. Generally has bib on breast; a few have "split bib" or lack bib, identical to adult light morph Short-tailed Hawks. Undersides of dark gray remiges more uniform, contrast more sharply with pale coverts. Upperwing color pattern identical to Short-tailed Hawk's. Use uppertail covert data in Flight. **(2C) Adult dark morphs.**—Rufous underwing coverts. Pale undertail coverts. Head may be identical, with white mask. **(2D) Basic I (subadult I) (all morphs).**—Pronounced white spot on undersides of wing on basal region of outer two to four primaries (fall to early winter). Wide dark band on trailing edge of remiges. Use uppertail covert data in Flight. **(2E) Juvenile (all morphs).**—Undersides of remiges are pale gray, dark gray trailing edge; small white spot on base of outer primaries; dark "comma" mark on primary covert tips on underwing. Numerous narrow, equal-width dark tail bands. Use uppertail covert data in Flight. **(2F) Juvenile light morphs.**—Similar when perched. Has dark mid-throat stripe. Pale edging on dorsal feathers. Use uppertail covert data in Flight. **(2G) Juvenile dark morphs.**—Underparts heavily streaked. Undertail coverts are pale. Use uppertail covert data in Flight. **(3) Zone-tailed Hawk.**—PERCHED.—All lore area is white. Barred underside of visible underwing. FLIGHT.—Tilting, unstable flight; does note kite or parachute; wings held in dihedral; rounded wingtips. **(4) Red-tailed Hawk.**—PERCHED.—Perches along roadways on trees, poles, posts, and wires. FLIGHT.—Wingtips flex upwards a bit when soaring. Kites, parachutes, and dives; regularly flaps wings and hovers. **(4A)** ***B. j. harlani* adults.**—Range overlap Nov.–Mar. Rare in Short-tailed Hawk range. Dark morphs are similar to adult dark morph Short-tailed Hawks. Tail patterns can be identical at field distances. Mottling may not be apparent, especially on dark gray tail types. Wingtips usually shorter than tail when perched. Use Flight data. **(4B) Juvenile (all morphs).**—Greenish cere. Pale eyes. Wingtips much shorter than the tail tip. Pale brown panel on uppersides of primaries and primary greater coverts; white translucent window on undersides. **(4C) Juvenile (*B. j. calurus/fuertesi*) light morphs.**—Range overlap all year. Lacks dark eyeline. White patch on scapulars. Dark patagial mark on underwing. White U on uppertail coverts when seen from above. **(4D) Juvenile (*B. j. calurus/harlani*) dark morphs.**—Range overlap late Oct. through Mar. Use Flight data.

**OTHER NAMES:** None regularly used. *Spanish:* Aguililla Cola Corta. *French:* Unknown.

---

REFERENCES: del Hoyo et al. 1994; Howell and Webb 1995; Johnsgard 1990; Palmer 1988; Russell and Monson 1998; Snyder and Snyder 1991.

**Plate 266. Short-tailed Hawk, adult light morph [Aug.]** ▪ Dark head and white throat. ▪ White body. ▪ Medium gray remiges have white area on base of outer 4–5 primaries. Remiges contrast with white wing coverts. Wide black band on rear edge of remiges. ▪ Partially banded type tail with wide black subterminal band, partial dark inner bands (banded type is common in West).

**Plate 267. Short-tailed Hawk, adult dark morph [Feb.]** ▪ Dark brown or black head and body. ▪ Medium gray remiges have white area on base of outer 4–5 primaries. Wide black band on rear edge of remiges. ▪ Banded type tail with wide black subterminal band, several narrow dark inner bands.

**Plate 268. Short-tailed Hawk, adult dark morph [Jan.]** ▪ Dark brown or black head and body. Typically has dark inner lores (white on this bird). ▪ Medium gray remiges have white area on base of outer 4–5 primaries. Wide black band on rear edge of remiges. ▪ Partially banded type tail with wide black subterminal band, partial dark inner bands.

**Plate 269. Short-tailed Hawk, juvenile light morph (unmarked type) [Nov.]** ▪ White mask and outer lores; dark inner lores. Narrow white supercilium; short, thick, dark eyeline; white auriculars; wide dark malar mark; unmarked throat. ▪ Cream-colored underparts unmarked or streaked on neck, flanks, and mid-belly. ▪ Partially banded type tail (rare in West). ▪ *Note:* Big Pine Key, Fla.

**Plate 270. Short-tailed Hawk, juvenile dark morph (mottled type) [Nov.]** ▪ White mask and outer lores; dark inner lores. Tawny streaking on auriculars. ▪ Dark breast forms bib; dark belly and flanks mottled with white. ▪ Banded type tail with equal-width pattern of banding. ▪ *Note:* Just ate a Palm Warbler, Big Pine Key, Fla.

**Plate 271. Short-tailed Hawk, juvenile light morph (unmarked type) [Nov.]** ▪ White auriculars; wide, dark malar mark; unmarked throat. ▪ Cream-colored underparts. ▪ Pale gray secondaries and inner primaries contrast with large white region of primaries. ▪ Partially banded type tail with dusky subterminal band (rare in West). ▪ *Note:* Big Pine Key, Fla.

**Plate 272. Short-tailed Hawk, juvenile light morph (streaked type) [Oct.]** ▪ Dark head, unmarked throat. ▪ Cream-colored underparts streaked on neck, flanks, and mid-belly. ▪ Pale gray secondaries and inner primaries contrasts with large white region of primaries. Gray secondaries contrast with coverts. ▪ Banded type tail with narrow dark bands (common in West). ▪ *Note:* Veracruz, Mex.

**Plate 273. Short-tailed Hawk, juvenile dark morph (mottled type) [Nov.]** ▪ Dark brown head. ▪ Dark brown breast forms bib; white mottled belly and flanks. ▪ Pale gray secondaries and distal half of inner primaries contrasts with large white basal region of primaries. Coverts mottled with white. ▪ Partially banded type tail with dusky subterminal band.

**Plate 274. Short-tailed Hawk, juvenile dark morph (streaked type) [Nov.]** ▪ Pale auriculars. ▪ Dark, streaked breast forms bib; streaked, white belly. ▪ Pale gray secondaries and distal half of inner primaries contrasts with large white basal region of primaries. ▪ Partially banded type tail with dusky subterminal band.

**Plate 275. Short-tailed Hawk, juvenile dark morph (all-dark type) [Nov.]** ▪ Head, body, and wing coverts dark brown; small amount of white spotting only on axillaries. ▪ Pale gray secondaries and distal half of inner primaries contrast with large white basal region of primaries. ▪ Banded type tail with equal-width dark banding. ▪ *Note:* Rare plumage type.

# SWAINSON'S HAWK
## *(Buteo swainsoni)*

**AGES:** Adult, basic I (subadult I), and juvenile. Adult plumage is acquired when 2 years old in the third year of life. Subadult plumage is acquired when 1 year old and retained the second year. This plumage is either quite similar to that of juveniles or is a combination of juvenile and adult traits. In either case, the tail and underwing patterns are adultlike. Tail length and wing width are also adultlike. Sexual differences are not apparent. Juvenile plumage is held for much of the first year. Juveniles have somewhat shorter secondaries and longer rectrices, which create a narrower wing and slightly longer tail than on older ages. There are no sexual variations in this age class.

**MOLT:** Accipitridae wing and tail molt pattern (*see* chapter 4). The first prebasic molt begins when juveniles are 10.5 to 11 months old. Based on museum specimens and live birds, molt begins in late Apr. when birds are still migrating north. By early May, molt is obvious on the inner primaries and central and sometimes outer rectrices. In early May, p1–3 may be replaced, new back and scapular feathers are sprouting among old faded feathers, and there are new leg feathers. On the tail, the deck rectrices set (r1) and, a bit later, the outermost set (r6), begin growing new feathers. The tertials begin to molt at this time, but the rest of the secondaries do not molt until later in spring. Molt on the head, belly, and flanks begins in late May and Jun. Head molt may not be apparent even in early Jul. At this time, some birds have very pale and worn head feathers that have not yet been replaced with new subadult feathers. Body molt, including head, is completed in Sep. or Oct. Tail molt occurs during fall migration and is completed in Oct. or Nov. Remix molt is still occurring in Sep. and Oct. during migration. At this time, the outer two to four primaries (p7–10) are retained worn and frayed juvenile feathers, whereas the newly acquired feathers are dark with pale tips. However, remix molt may cease for a short period while on the winter grounds. A few (s4 and 8) or several juvenile secondaries are retained in Sep. and Oct., but fewer as the fall season advances.

Molt on secondaries may be continuous throughout the fall and winter until all subadult feathers are fully replaced. By early spring, subadults exhibit full adultlike secondaries by time they arrive in the U.S. Replacement of the outer two to four primaries also occurs during the winter or early spring. In spring, some subadults have fully molted into new outer primaries, but others in late May and early Jun. are often just finishing their molt on the outermost primary (p10), which is the last primary to molt (spring molt based on BKW pers. obs.; B. Sullivan and J. Liguori data on in-hand birds). *Note:* Pattern of completing all remix molt during winter and early spring is atypical for a large buteo. Most species cease their molt during the winter months and do not resume until spring.

The second prebasic molt into adult plumage appears to be a continuation of the subadult remix molt that extends from winter while on the wintering grounds or begins in early spring. Body molt begins in late spring (late May or Jun.). All areas molt during the summer and fall, and adult plumage is gained by late summer. Some individuals in the first adult plumage may still have partial head features that resemble younger birds (e.g., pale supercilium and auricular areas).

Subsequent adult prebasic molts begin during or after the nesting season. As in most species, males do not start their molt until nesting is nearly over or is over. Remix molt may be serially descendant, with two molt waves growing at the same time. Outer primaries that were not molted the previous year may start to replace feathers at the point where they left off last fall, and inner primaries may begin molting again. Also, it appears that birds may replace remiges in an irregular sequence and have two or three molt waves in a serially descendent pattern. Body molt is well under way during the Sep. and Oct. migration period in the U.S. Rectrices may also still be molting at this time. As is typical in larger raptors, body, remix, and possibly rectrix molt is not complete each year.

**SUBSPECIES:** None.

**COLOR MORPHS:** Polymorphic. For a species that lacks subspecies, Swainson's Hawk exhibits the most diverse and easily visible array of plumages of any North American raptor (rivaled only by Red-tailed Hawk; however, this species is polytypic). There is a continuum of plumage variations between light and dark morphs. There are three major color morphs: light, intermediate (rufous), and dark. To fully illustrate the clinal trend between the three major morphs, intermediate morphs are shown: light intermediate morph forms a clinal link between light and intermediate morphs, and dark intermediate morph forms a clinal link between intermediate and dark morphs. Because Swainson's Hawks have a continuum of plumage characters and numerous subtitles, it is often difficult to affix a categorical morph designation to a particular bird. This problem is compounded by the fact that subtle color variations look different when viewed at various angles and distances. Also, some plumage traits vary with sex; this is most apparent from light to intermediate morph birds. *See* Color morph distribution in Status and Distribution for discussion of geographic range of morphs.

**SIZE:** A large raptor. Males average smaller than females but there is considerable overlap. Length: 17–21 in. (43–53 cm); wingspan: 47–54 in. (119–137 cm). Juveniles appear slightly longer than subadults and adults because they have longer rectrices.

**SPECIES TRAITS:** Bill is all black except for small bluish area on the lower basal region of the upper mandible and basal area of the lower mandible. WINGS.—In fight, the front and trailing edges of the long wings are parallel from the body to the wrist area, then taper to pointed wingtips (p9 or p8 and 9 are the longest primaries). Ventral surface of the remiges is gray.

**ADULT TRAITS:** HEAD.—Dark brown irises. Yellow cere. WINGS.—**Ventral surface of the remiges is dark gray with a wide black band on the trailing edge. When perched, the wingtips extend just beyond the tail tip.** TAIL.—**Medium gray or brown, more brownish when worn. Wide black subterminal band and crossed with numerous narrow inner dark bands.** Pale gray on the ventral surface with the banded pattern highly visible.

**ADULT LIGHT MORPH TRAITS:** Sexes can be separated 95% of the time by head and breast color (*see below* Male and Female light morph). HEAD.—**White forehead and outer lores form a white mask. Inner lores are dark.** The forehead is occasionally dark. Some birds of either sex may have a partial, pale supercilium and auricular regions. BODY (ventral).—Neck and breast are variable brownish and form a distinct bib. Either sex can have a "split bib" with the central portion being white and forming an incomplete bib. These may be younger adults or individual variations on older adults. All the underparts are either white or pale tawny. The amount of markings on the underparts varies: (1) On the palest individuals, the underparts are unmarked. *Note:* Common variation in males, uncommon in females. (2) Typically, the upper belly has a few small brown or rufous specks and the flanks have partial, thin barring. *Note:* Common variation for both sexes. (3) On the heaviest marked individuals, the upper belly may be lightly marked with spots, arrowheads, or diamonds. The flanks are narrowly barred with rufous or brown. *Note:* Common variation for both sexes. Leg feathers and undertail coverts are always unmarked. BODY (dorsal).—Dark brown with a variable amount of pale tawny or rufous-tawny edges on the scapular feathers. The rump is dark brown. WINGS (ventral).—**As with the belly, the underwing coverts and axillaries are either white or uniformly tawny and contrast sharply with the dark gray remiges.** Coverts may be unmarked or have a small amount of brown or rufous spotting, particularly on the axillaries. WINGS (dorsal).—Marked as the rest of the upperparts. TAIL.—Uppertail coverts have a moderately wide white area on each feather and can appear as a white U-shaped band.

**Adult male light morph:** HEAD.—Crown and auriculars are medium gray or, rarely, brown. Nape and hindneck are rufous. BODY (ventral).—The bib is the rufous type (may appear orangish rufous or orangish brown) or rufous-brown/medium brown type. Any markings on the belly and flanks are rufous. Body (dorsal).—Pale feather edgings are often broad and distinct.

**Adult female light morph:** HEAD.—Except for the white mask, the head, nape, and hindneck are uniformly dark brown. Auriculars occasionally have a slight grayish cast. BODY (ventral).—Bib is either rufous-brown/medium brown or dark brown type. BODY (dorsal).—Moderately distinct or ill-defined pale edges on scapulars and upperparts often appear darker and more uniformly colored than on males.

**ADULT LIGHT INTERMEDIATE MORPH TRAITS:** Sexes can be separated 95% of the time by head and breast color (*see below* Male and Female light intermediate morph). HEAD.—White mask is similar to light morph's. Essentially as on light morph, but ventral areas become more extensively marked. BODY (ventral).—*Barred type:* White or tawny underparts are fully covered with moderately wide rufous or dark brown barring on the belly, lower belly, and flanks. Barring may be wide and dense on the most heavily marked birds. The flanks may have a large arrowhead-shaped mark on each feather and are more heavily marked than the belly. Leg feathers are finely barred. *Note:* Common plumage type. *Unbarred type:* Underparts may be (1) uniformly pale or medium rufous or tawny and often darker on the flanks or (2) dark tawny on the belly and flanks, with a wide dark brown streak on each feather. The lower belly, leg feathers, and undertail coverts are pale or medium tawny and usually unmarked. *Note:* Uncommon to very uncommon plumage type. BODY (dorsal).—As on light morph but with minimal amount of pale edging on the scapular feathers. WINGS (ventral).—Pale underwing coverts are white or tawny. On the barred type, axillaries are generally barred, with some barring possible on the rest of the coverts. Sometimes coverts are heavily barred. On the unbarred type, axillaries may either be pale or darker brownish. **The dark gray remiges contrast sharply with the paler coverts.** WINGS (dorsal).—As on rest of the upperparts. TAIL.—Uppertail coverts are a white U shape as on light morph.

**Adult male light intermediate morph:** HEAD.—As on light morph male with gray crown and auriculars. Nape and hindneck, however, vary but are the same color as bib (*see below*). The gray crown and auriculars may be more difficult to detect on birds with darker brown nape and hindneck. BODY (ventral).—Bib can be rufous, rufous-brown, or dark brown type. On the barred type of underparts, individuals with rufous type bibs typically have similarly colored barring on the belly, flanks, and leg feathers. Those with rufous-brown and dark brown types of bibs can either have rufous or dark brown barring.

**Adult female light intermediate morph:** HEAD.—As on light morph female with uniformly dark brown crown, auriculars, nape, and hindneck. BODY (ventral).—Bib colors are the same as described for light morph female. On the barred type of underparts, color of the barring is typically dark brown but can be rufous-brown.

**ADULT INTERMEDIATE (RUFOUS) MORPH TRAITS:** Sexes can be separated 95% of the time by head and breast color (*see below* Male and Female intermediate morph). HEAD.—As on respective sex of the previous two morphs. **White mask and throat are still well defined.** BODY (ventral).—Underparts are somewhat sexually dimorphic (*see below*). Breast is variably colored but belly, flanks, and leg feathers are always rufous. The rufous color is the result of having solid rufous feathers, but may have dark brown barring or vertical streaking superimposed over the rufous feather, or have a broad rufous barred pattern. *Note:* The dark brown barring or central feather streaking can be exceptionally broad on some birds. **Undertail coverts are contrastingly white and occasionally have small dark spots on each feather.** BODY (dorsal).—Dark brown with a minimal amount of pale edging on the feathers. WINGS (ventral).—White, tawny, or rufous underwing coverts. Axillaries are generally a bit more heavily marked and deeper colored than the coverts, and are usually the same color as the belly. **The dark gray remiges contrast sharply with the paler coverts.** WINGS (dorsal).—As on the rest of the upperparts with a minimal amount of pale edging. TAIL.—Uppertail coverts have a narrow white U shape to them, but this is somewhat less defined than on the two previous morphs.

**Adult male intermediate morph (all-rufous type):** HEAD.—Typically has gray crown and auriculars and rufous nape and hindneck as on light morph. BODY (ventral).—Front of the neck, breast, belly, flanks, and leg feathers is uniformly rufous. This type does not have a dark brown bar superimposed on the rufous feather. There may be some rufous barring on the lower belly. *Note:* Common plumage type for males.

**Adult male intermediate morph (bib type):** HEAD.—Crown and auriculars are grayish, but nape and hindneck are brown. BODY (ventral).—Some individuals have a rufous-brown or dark brown type bib on the neck and breast that contrasts with the rufous belly, flanks, and leg feathers. *Note:* Fairly common plumage type for males.

**Adult female intermediate morph:** HEAD.—Dark brown throughout. BODY (ventral).—Virtually all have rufous-brown or dark brown bib that contrasts at least somewhat with the paler rufous belly, flanks, and leg feathers. Individuals with rufous-brown bib, however, may have marginal contrast between bib and breast.

**ADULT DARK INTERMEDIATE (DARK RUFOUS) MORPH TRAITS:** Sexes can be separated some of the time by head color (*see below* Male and Female dark intermediate morph). HEAD.—White mask on the forehead may be similar to the previous morphs or reduced in size. Throat is typically dark with white streaking; rarely, it can be quite white. BODY (ventral).—Three main variations in this morph: (1) *Dark type:* Front of the neck, breast, belly, and flanks is dark brown and has no bib delineation. Lower belly and leg feathers are deep tawny or, more typically, rufous and may be barred with dark brown. (2) *Dark-bellied type:* Rufous-brown breast and subtly darker or much darker brown belly and flanks form a belly band. Leg feathers are rufous or rufous-brown as the breast. *Note:* Seen on either sex. (3) *Dark rufous type:* All of the underparts are dark brown with rufous edgings on most feathers, creating a dark rufous appearance. **On all types, the undertail coverts are either white or tawny and may be unmarked or partially barred but contrast with the darker underparts.** BODY (dorsal).—All-dark upperparts. Uppertail coverts are either all dark or have narrow tawny edging. WINGS (ventral).—Wing coverts are typically rufous or deep tawny. Axillaries are solid dark brown or heavily barred with rufous. **The dark gray remiges contrast sharply with the paler underwing coverts.** WINGS (dorsal).—Uniformly dark brown like rest of the upperparts.

**Adult male dark intermediate morph:** HEAD.—Grayish crown and auriculars may be visible only at close range. *Note:* Not readily separable from adult female dark intermediate morph.

**Adult female dark intermediate morph:** HEAD.—Uniformly dark brown except for the pale mask and whitish throat.

**ADULT DARK MORPH TRAITS:** Sexes are difficult to separated by head color. HEAD.—Generally dark brown for both sexes, but males may have a grayish tinge on the auriculars. There generally is no white mask in this morph. **The forehead is dark brown, the outer lores white, and the inner lores dark. The throat is dark brown.** Rarely may have all-white lores. Also, rarely has whitish on the forehead and throat. (Types with remnant white on the forehead and throat are probably dark intermediate morphs.) BODY (ventral and dorsal).—Uniformly dark brown. Some birds have a rufous tinge on the leg feathers and lower belly. **Undertail coverts are either white or tawny and vary from unmarked to heavily barred but contrast sharply with the dark underparts.** WINGS (ventral).—Rufous with dark brown axillaries and a dark brown diagonal line on the first row of lesser coverts. **There is still some contrast with the rufous underwing coverts and the dark gray remiges.** Rarely, underwing coverts are dark brown with some rufous mottling. Very rarely, underwing coverts are uniformly dark brown. TAIL.—Uppertail coverts are dark brown.

**BASIC I (SUBADULT I) TRAITS:** HEAD.—Yellow cere. Medium brown irises. **All morphs have a medium brown or dark brown crown, pale supercilium, narrow dark eyeline, pale auriculars, and dark malar mark that extends onto the sides of the neck and breast. In darker morphs, the pale areas are tawnier than on lighter birds.** *Note:* **All have a dark stripe down the middle of the throat.** BODY.—Ventral and dorsal areas resemble those of the respective juvenile-plumage morphs. However, all subadult morphs have a variation that has more adultlike barring on ventral areas and less pale edging on the dorsum and resembles adults. WINGS.—**Wide black subterminal band on the trailing edge. Dark gray remiges are also similar to those of adults. The outer two to four primaries are retained juvenile feathers in autumn. They are bleached white on the basal region and contrast sharply**

**against the new and darker subdult remiges. The secondaries may retain several or a few juvenile feathers (at least s4 and 8, often many more). The retained juvenile secondaries are shorter than the new subadult feathers and have a pale, worn dusky trailing edge that contrasts against the wide black band of the new subadult feathers. When perched, wingtips are either equal to the tail tip or barely longer.** TAIL.—**As adult's, with a wide black subterminal band.** *Note:* In Sep. and Oct., subadults are in extensive remix and rectrix molt and have a mix of retained juvenile and new adultlike feathers. By spring, all such feathers have molted into adultlike character.

**BASIC I (SUBADULT I) LIGHT MORPH TRAITS:** BODY (ventral).—**Distinct dark malar and dark patch on the side of the neck and breast, which may encircle the front of the breast as a mottled, partial bib.** White or tawny underparts possibly with a few dark spots, arrowhead shapes, or diamond shapes on the flanks and sometimes a few dark markings on the mid-belly area. Leg feathers and undertail coverts are unmarked. BODY (dorsal).—*Juvenile-like.*—Dark brown with broad pale edging like on a juvenile and pale, mottled scapular patch. *Adultlike.*—All dorsal areas may be nearly uniformly dark or with narrow, pale feather edging. WINGS (ventral).—White or tawny and unmarked; axillaries may also be unmarked or have a few dark spots. WINGS (dorsal).—*Juvenile-like.*—Distinct pale edging and spotting and a large pale patch on the mid-scapulars. *Adultlike.*—Coverts are practically uniformly dark with faint pale edging. TAIL.—Uppertail coverts are pale tawny or white and barred.

**BASIC I (SUBADULT I) LIGHT INTERMEDIATE MORPH TRAITS:** BODY (ventral).—Dark patch on malar, side of neck, and breast as on light morph. *Juvenile-like.*—As on light morph, but all of breast and belly are lightly spotted or streaked. Flanks have larger arrowhead- or diamond-shaped markings, often with thick barring on the inner portion of the feathers. *Adultlike.*—Flanks are thinly barred. Leg feathers may have some spots or partial barring. Undertail coverts are unmarked. BODY (dorsal).—As on light morph with the same juvenile-like and adultlike patterns. WINGS (ventral).—Underwing coverts are white or tawny and lightly covered with dark spotting, including the axillaries.

**BASIC I (SUBADULT I) INTERMEDIATE MORPH TRAITS:** BODY.—As on light morph. *Juvenile-like.*—Base color of underparts is tawny but fades to white by spring. Breast and belly are covered with a moderate amount of dark brown streaking. Markings on the flanks often have larger, broad arrowhead-shaped markings and thick barring on the inner portion of each feather. *Adultlike.*—Flanks and belly are distinctly barred with dark brown and have a dark arrowhead- or diamond-shaped mark on the center of each feather. Leg feathers are moderately barred. Undertail coverts are white or tawny and have some dark arrowhead markings. BODY (dorsal).—As on light morph for both juvenile-like and adultlike patterns. WINGS (ventral).—Underwing coverts are tawny and moderately covered with dark spotting; axillaries are barred. TAIL.—Uppertail coverts are marked but still appear quite pale tawny.

**BASIC I (SUBADULT I) DARK INTERMEDIATE MORPH TRAITS:** BODY (ventral).—Base color is usually rich tawny. There are two types of underpart patterns: (1) *Streaked type: juvenile-like.*—Moderately heavily marked with uniform streaking on the breast and belly. Flanks have large dark arrowhead- or diamond-shaped markings with thick inner-feather barring. Leg feathers are distinctly barred. *Adultlike.*—Flanks and belly are heavily barred and have a large dark streak or diamond-shaped mark on each feather. (2) *Dark-bellied type (juvenile-like only)*: Tawny breast is moderately streaked with dark brown; belly, flanks, and leg feathers are heavily streaked or very heavily and present a belly-band appearance. Lower belly is less heavily marked than flanks and belly and further accentuates the belly-band look. Undertail coverts are pale tawny or tawny and barred. WINGS (ventral).—Rich tawny base color and either moderately or fairly heavily marked, especially with a barred pattern. TAIL.—Uppertail coverts are slightly pale.

**BASIC I (SUBADULT I) DARK MORPH TRAITS:** HEAD.—As on paler morphs but supercilium, auriculars, and throat are rich tawny. BODY (ventral).—Three main variations in this morph: (1) *Streaked type (juvenile-like only)*: Identical to juvenile dark morph, with thick, dark brown uniform streaking on the breast and belly and large dark arrowhead- or diamond-shaped markings on the flanks. Flanks are nearly solid dark. Leg feathers are dark with narrow pale edging or heavily barred. (2) *Dark-bellied type (juvenile-like only)*: Similar to dark-bellied type of dark intermediate morph except breast is darker and somewhat streaked or mottled with narrow tawny edges on some belly and flank feathers. Leg feathers are virtually solid dark brown but may have narrow tawny edging. (3) *All-dark type:* Nearly uniformly dark brown with only a few pale tawny specks scattered on the underparts. Leg feathers are dark brown. On all types, undertail coverts are tawny and somewhat barred. BODY (dorsal).—Streaked type has some pale tawny edgings, but the other two types are all dark. WINGS (ventral).—Streaked type is a rich tawny with heavily marked pattern of spotting and barring; the other two types are dark with some tawny mottling. *Note:* **In the fall, the pale, bleached outer two to four retained juvenile primaries are very obvious as they contrast against the dark underside of the wing.** TAIL.—Streaked type may have some pale tawny tips on the coverts; the other two types have all-dark coverts.

**JUVENILE TRAITS (WORN PLUMAGE):** In spring and early summer, at least until mid-Jul. when about 1 year old, juveniles that were fairly pale headed in fresh plumage may exhibit extreme wearing and fading on the head feathers and become bleach-headed. This is most noticeable on the crown, which may be virtually white instead of brown, but the dark eyeline and neck patch are still apparent. May occur in any morph but is less common in dark morphs.

**JUVENILE TRAITS (FRESH PLUMAGE):** HEAD.—**Typically has yellow cere but sometimes pale greenish. Medium brown irises. Medium brown or dark brown crown, pale supercilium that connects to the pale forehead, narrow dark eyeline, pale auriculars, and a dark malar mark that extends onto the sides of the neck and breast.** *Note:* **All have a pale throat with a dark center streak. In fresh late-summer and autumn plumage, pale areas on the head are tawny but gradually become white with wearing and fading.** Any morph can have a paler or darker crown; those with pale crowns tend to become very bleached and pale by spring. WINGS (ventral).—**Underside of the remiges is nearly uniformly pale or medium gray (paler than on older ages) and covered with narrow dark barring. Basal region of the outer one to three primaries is paler and whitish. Trailing edge of the underwing has a narrow dusky band. When perched, wingtips are barely shorter than the tail tip or equal to it.** WINGS (dorsal).—Remiges are uniformly dark brownish black with somewhat paler coverts. TAIL.—**Uppertail is medium grayish or brownish and crossed by numerous narrow bands. The subterminal band is the same width as the inner bands or marginally wider.**

**JUVENILE LIGHT MORPH TRAITS:** BODY (ventral).—**Distinct dark patch extends from the malar to the side of the neck and breast. The dark patch sometimes extends onto the mid-breast as a partial bib.** Underparts are either unmarked or the flanks may have small or moderate-sized arrowhead- or diamond-shaped markings and the belly may be sprinkled with short streaks or spots. Leg feathers are unmarked. Undertail coverts are unmarked. BODY (dorsal).—Dark brown with distinct tawny edge on all feathers. A large white or tawny patch on the middle of each scapular tract. WINGS (ventral).—Wing coverts are unmarked. WINGS (dorsal).—Pale edging on all coverts. TAIL.—Uppertail coverts are pale.

**JUVENILE LIGHT INTERMEDIATE MORPH TRAITS:** BODY (ventral).—**Dark patch on the malar; side of the neck, and breast is distinct. The dark patch often extends as a partial bib across the front of the breast.** All of the belly and forward flanks are lightly spotted or streaked with dark brown; rear of the flanks may have larger arrowhead- or diamond-shaped dark markings. Leg feathers may be lightly spotted or barred. Undertail coverts are unmarked or have small arrowhead-shaped marks. BODY (dorsal).—As on light morph. WINGS (ventral).—Wing coverts are often lightly marked with dark brown, including the axillaries. WINGS (dorsal).—As on light morph. TAIL.—As on light morph.

**JUVENILE INTERMEDIATE MORPH TRAITS:** BODY (ventral).—**Dark patch on the malar, side of the neck, and breast is distinct and regularly wraps around the front of the breast**

**as a partial bib.** All of the belly and forward portion of the flanks are moderately streaked with dark brown; rear of flanks is covered with large arrowhead-shaped markings and wide barring on the inner portion of the longest feathers. Leg feathers are moderately barred with brown. Undertail coverts are partially barred. BODY (dorsal).—As on light morph but often with fewer pale areas; however, still shows pale scapular patches. WINGS (ventral).—Underwing coverts are moderately marked with dark spots, and the axillaries are barred or streaked. WINGS (dorsal).—As on light morph. TAIL.—Uppertail coverts are marked but still pale.

**JUVENILE DARK INTERMEDIATE MORPH TRAITS:** BODY (ventral).—**Dark patch on the malar, side of the neck, and breast is large and often wraps around the front of the breast as a mottled bib.** Two main types of markings: (1) *Streaked type:* Belly and forward flanks are heavily streaked and rear flanks are densely covered with large dark arrowhead- or diamond-shaped markings and wide barring on the inner portion of the longer rear feathers. (2) *Dark-bellied type:* Breast is moderately streaked, but belly and flanks are densely streaked and mottled and appear as a dark belly band. Lower belly is not as heavily streaked as belly and promotes an even more obvious belly-band look. Leg feathers of both types are heavily barred. Undertail coverts are barred. BODY (dorsal).—Similar to markings on light morph, but with narrower pale edging. WINGS (ventral).—Wing coverts are rather heavily marked, and axillaries are barred with dark brown. TAIL.—Uppertail coverts are somewhat pale.

**JUVENILE DARK MORPH TRAITS:** BODY (ventral).—Uniformly streaked on breast and belly or breast is somewhat paler with narrower streaking (but not as obvious a difference as in dark-bellied type of dark intermediate morph). Flanks are covered with broad, dark arrowhead- and diamond-shaped markings and appear virtually solid dark. Leg feathers are heavily barred. Undertail coverts are barred but distinctly paler than the rest of the underparts. BODY (dorsal).—Moderate amount of tawny pale edging on the scapulars. WINGS (ventral).—Coverts are heavily marked with dark brown, and axillaries are barred. WINGS (dorsal).—Moderate amount of pale edging on the coverts. TAIL.—Uppertail coverts may have narrow tawny tips but do not stand out as a large pale area.

**ABNORMAL PLUMAGES:** Partial albinism has been documented. Light morph adults, even very gray-headed males, occasionally have white supercilium and auricular areas and only partial bibs (BKW pers. obs.).

**HABITAT: Summer.**—Open and semi-open areas with low to moderate-height vegetation. Very arid regions and high-montane elevations are not inhabited. Swainson's Hawks have adapted, in part, to the ongoing human alteration of the western landscape. In many regions, they fare better in moderate agricultural areas than in native habitat. A requirement is that moderate amount of low-height vegetated tracts in non-agricultural areas or light agricultural lands, especially grass hay or alfalfa, be available in or adjacent to their territory, even if only in fragmented, small parcels. In certain arid regions of the West, such as Nevada, Swainson's Hawks are found mainly in irrigated light agricultural locations, particularly in areas with alfalfa and grass hay. Largest numbers still inhabit the vast, but often still fragmented, grazed pastures and prairies east of the Rocky Mts. In the Rocky Mts. and mountains of the Great Basin, found in appropriate semi-open low scrub tracts on moderately high mountains. Nesting in montane regions occurs locally up to 8,000 ft. (2,400 m) in Nevada and Utah; to 8,200 ft. (2,500 m) in nw. Wyoming in Yellowstone N.P.; to 9,000 ft. (2,700 m) in s. Wyoming; and to 9,500 ft. (2,900 m) in Colorado. Climate is arid in most of the range but is humid in Iowa, Minnesota, Missouri, Montana, and e. Texas.

Swainson's Hawk is common in rural areas throughout much of its range and fairly common in certain suburban settings. Locally isolated pairs are found in a few urban locales. They species is also being subjected to intense nesting disruption in suburban and urban areas when foraging areas succumb to continued urban sprawl.

**Winter.**—Semi-open and open regions on relatively flat terrain. Light agricultural zones, pastures, grasslands, and prairies are favored haunts in the typical wintering areas in Argentina. In other areas, also found mainly in agricultural zones, pastures, and grasslands. **Migration.**—Found in all previously listed habitats; however, foraging birds primarily feed in light agricultural locations. In Colorado, for instance, the largest numbers of foraging birds are in harvested alfalfa and fallow and stubble wheat fields. Wheat fields with stubble that has been left standing for a season are particularly attractive. In Texas, huge numbers are seen in pastures and agricultural fields infested with insects, particularly fields being harvested, plowed, or burned.

**HABITS:** Generally a tame raptor and becomes well acclimated to humans in certain areas, but disdains human intrusion in remote locations. Nesting pairs are solitary. Nesting pairs and their siblings remain as a family clan until fall migration. When breeding duties are completed, adults also become gregarious and join in multi-aged flocks for migration and winter. Nonbreeding birds generally are in flocks year-round; however, in the summer it is also common to see single birds.

Swainson's Hawks perch on the ground or any elevated object, including utility wires. Migrants prefer elevated perches, if available, for roosting. Several hawks often perch in a single tree. (More than 20 have been seen in mesquite trees in s. Texas). If elevated perches are not available, hawks readily roost on the ground, sometimes with hundreds scattered across a field. Ground-roosting birds seek open, fallow (dirt) fields or fields with very short vegetation for optimal visibility of potential predators. Even ground-perching and -roosting birds prefer to sit on top of slightly elevated clumps of earth. Migrants are also fond of water and will sit around shorelines of ponds and in the shallow water of ponds and irrigation runoff pools. During inclement weather, shelter is often sought by perching on the ground on the lee side of clumps of dirt or dense vegetation.

Western, Cassin's, and Eastern kingbirds relentlessly harass nesting Swainson's Hawks.

**FEEDING:** Perch and aerial hunter. When perch hunting, may drop down or fly from an elevated perch to capture prey or may run on the ground like a chicken adeptly pursuing large insects. Aerial hunting is a commonly used foraging method. Aerial crepuscular feeding on bats occurs in certain areas in New Mexico.

Ground squirrels and other rodents, small hares and rabbits, small birds, reptiles, and insects form the summer diet. Vertebrates comprise the bulk of the diet of breeding birds in order to supply adequate protein to nestlings and fledglings. Nonbreeding summering birds feed extensively on grasshoppers and, in some areas, ground squirrels. During migration and in winter (in South America) all ages of Swainson's Hawks feed almost exclusively on large insects, especially grasshoppers.

Swainson's Hawks capture grasshoppers and other large insects with their feet or bill. They use their feet to snatch flying insects, often in impressive acrobatic aerial maneuvers. Aerial-captured insects are eaten while the hawks are soaring or gliding. Vertebrate prey are captured with the hawk's feet and eaten on the ground at the point of capture, transported to another ground or elevated location, or taken to the nest site to be devoured. Swainson's Hawks regularly hunt mice, voles, and insects in fields that are being or have been harvested, plowed, or burned.

Feeding occurs regularly during migration. In the fall in e. Colorado, hawks may feed on grasshoppers for several days in one location. Feeding also occurs in other areas, particularly where grasshoppers and caterpillars infest fields and where fields are being burned or plowed. Aerial feeding has been observed in e. Mexico in the autumn. Based on telemetry data, California birds may stop for a week or more along the southward migration path in s. California, w. Mexico, and Central America. Northward-bound migrants also stop and feed.

In the fall, feeding may begin early in the morning prior to the day's migratory flight. Birds roosting in trees may depart their roosts at daybreak and forage in nearby fields until mid-morning. Birds that roost in fields scatter more widely about the field at daybreak to forage. Migrant flocks may begin to settle into fields for roosting and feeding in late afternoon and feed for up to 4 hours before sunset. Migrants may also fast for a few days or a week.

**FLIGHT:** Wings are held in a high or low dihedral when soaring and a modified or low dihedral when gliding. Powered flight is regularly used. Wingbeats are moderately slow but snappy. Very high altitudes are attained when migrating. Exceptionally long glides, often extending for several miles, are used when migrating in order to be energy efficient. Hovering and kiting occur frequently when hunting. Rather impressive acrobatic dives and twists accompany pursuits of insects and chasing of intruders. As noted above, migratory flights may end in late afternoon to feed or, particularly in s. Texas and farther south, may extend until dusk on optimal flight days.

**VOICE:** Vocalizes primarily when disturbed at nest sites and rarely heard elsewhere. Adults emit a loud, somewhat raspy *keeyaah, keeyaah* or *keeair, keeair*. Also may have short *kee, kee, kee* notes. Nestlings and fledglings have the typical raptor begging call of a high-pitched whining *skree, skree, skree.*

**STATUS AND DISTRIBUTION:** *Common* in the West. Estimated population of up to 1 million individuals, based on fall counts of migrants at Veracruz, Mexico (*see* Movements). Fall tallies at Veracruz have varied over the years from 273,000 to 541,000. An amazing 1,062,500 Swainson's Hawkswere counted in the fall of 2001 (much of the count verified with digital photos). This substantial increase does not reflect a population growth but is tied to count location and viewer capabilities. Numbers passing Veracruz are still thought to be undercounted, even with the 1 million birds seen in 2001. Thousands of Swainson's Hawks are often seen in late afternoon migrating in the foothills of the mountains far inland of the two official hawkwatch sites and are not tallied. *Note:* One million were again tallied in Veracruz in fall 2002.

Overall population is stable but deserving of attention. This species has undergone incredible declines in certain areas since historical times, particularly in California. However, on the Canadian prairies and probably other areas of the Great Plains, the current breeding population is probably higher than in historical times because of increased tree nesting sites with settlement of the West. In most regions, the population is currently stable or possibly decreasing. Severe population declines occurred in the 1990s, and possibly earlier, because of pesticide contamination on the wintering grounds; the contamination, however, was guickly stopped (*see* Conservation). Declining reproductive success was documented on portions of the Canadian prairies, beginning in the late 1980s in Saskatchewan and the early 1990s in Alberta. The decline was due to reduced numbers of primary prey (Richardson's Ground Squirrels) and land-use changes, including increased use of pesticides and fertilizers. The breeding population and reproduction success have now stabilized; however, they remain substantially lower than before the initial decline.

**Color morph distribution.**—Intermediate to dark morphs primarily breed west of a line (the black dashed line on the map on p. 283) from cen. Saskatchewan (just east of Saskatoon) and southward to e. Montana, e. Wyoming, and e. Colorado (rarely more than 30 miles [48 km] east of the Rocky Mts.) and through extreme e. New Mexico. A dark intermediate morph adult has been seen in Kidder Co., N.D., in summer. South Dakota logged only one dark-colored bird, an intermediate morph adult, during the state's breeding bird atlas period. Nonbreeding birds are occasionally seen in e. Colorado in summer. East of the color morph demarcation line on the range map, light and light intermediate morphs comprise 99.9% of the breeding population.

California has the highest percentage of darker birds. It is often published that 35% of the birds are "dark." However, "dark" in many cases (as seen in photographs) often includes intermediate through dark morph and not just pure dark birds. Light morphs in California comprise only 10–15% of the summering birds. Saskatchewan has 2% intermediate (rufous) morphs and 4% dark morphs. In Colorado, intermediate to dark morphs are fairly common in the mountains (west of the dashed line on the map) but are rare breeders on even the very western edge of the Great Plains. The various morphs mate freely with each other.

Autumn percent of color morphs of migrant birds have been recorded by the author for several years on the e. Plains of Colorado and in Texas. In this area, migrants may come from more western, northern, or eastern regions. Light morphs are the most common and account for 40% of the birds. Light intermediate morphs comprise roughly 23%, intermediate morphs 15%, dark intermediate morphs 12%, and dark morphs 10%. On a daily basis, however, some flocks may have a high percent of darker birds.

**Summer.**—Highest numbers and most contiguous populations are found on the Great Plains from Alberta east to w. Manitoba and southward, east of the Rocky Mts., to n. Chihuahua, n. Coahuila, n. Nuevo León, and s. Texas. Smaller populations exist in the Rocky Mts. and west of the Rockies in the more arid landscape of the Great Basin. However, local densities may be fairly high in isolated locales of the Rocky Mts. and westward in areas with good habitat. Breeding is very localized in Arizona, California, w. Colorado, s. Idaho, n. Nevada, e. Oregon, Utah, and e. Washington. In British Columbia, there is also localized breeding in the south-central and southeastern areas, including isolated breeding in the Bulkley Basin in the west-central part of the province. There are also isolated summer sightings and possible breeding around Grande Prairie (two immatures seen in Jul. 1997) and Peace River, Alberta.

Swainson's Hawks are rarely found in Alaska, the Yukon Territory, and the Northwest Territories. Verifiable sightings have infrequently occurred, but breeding is not confirmed in these areas. Two darker colored 1-year-olds spent part of May 1998 at Juneau, Alaska. Two adult dark morphs were observed at a hawkwatch near Eureka in e. Alaska in late Apr. 2001, but none were seen there in spring of 2002. Birds are occasionally seen in the Yukon Territory, as far north as Dawson.

Breeding populations east of the Great Plains are in restricted areas of n. Iowa, w. and s. Minnesota, sw. Montana, and e. Texas. *Iowa:* Primarily found in Black Hawk, Cerro Gordo, Dickinson, Hancock, Kossuth, Mitchell, Osceola, and Wright Cos. Breeding is in low density. *Minnesota:* Between 1981 and 1997, breeding has been documented in Big Stone, Dakota, Fillmore, Lac Qui Parle, Lincoln, Lyon, Kittson, Mower, Murray, Pipestone, Red Lake, Rice, Stevens, Waseca, Washington, and Winona Cos. Breeding areas in many of these counties, particularly the last two, are in isolated locations. Primary range in the state is in the southwestern counties. *Missouri:* Fairly recent breeding or summering has occurred in isolated locations in Barton, sw. Dade, w. Greene, se. Lawrence, n. Jasper, ne. McDonald, and s. Vernon Cos. Possible breeding may occur from Vernon Co. north to Jackson Co. Listed as Endangered in the state. *Kansas:* A common breeder in the western part of the state, but isolated breeding was documented in breeding bird atlas surveys in two areas in the eastern part of the state. Confirmed breeding occurs in s.-cen. Anderson, nw. Jackson, and e. Pottawatomie Cos. Summering birds, with possible breeding, have been seen in Allen, Johnson, Linn, Miami, Neosho, Shawnee, and Wilson Cos. *Texas:* Annual summering or breeding occurs in isolated small areas south and east of Houston.

South of Missouri, summering and especially breeding birds are rare. *Arkansas:* Irregular sightings confirmed in summer for the last three decades, primarily in the northwestern part of the state in Benton and Washington Cos. Nesting occurred near Fayetteville, Washington Co., in 1986 (nest failed), 1994, 1995, and 1997. *Louisiana:* Regular summer sightings of adults and younger birds near Lake Charles, Calcasieu Parish, for several years. Nesting is not confirmed for Louisiana, although adults have been seen carrying sticks. *Mexico:* Breeding occurs in a few places in the northern half of Sonora, much of the northern half of Chihuahua and Coahuila, in two isolated locations in w.-cen and s.-cen. Coahuila, and the northern third of Nuevo León.

Estimates of breeding populations in Canada and the U.S. are vague for most areas and based on surveys in the 1970s to early 1990s. For many states and provinces, the estimates appear conservative considering the consistently large numbers of fall migrants counted in Veracruz, Mexico. Estimated number of pairs for states and provinces where data are available: California 550 (550–600 in 1997; 90% reduction since historical times); Canada 20,000–50,000 (latter figure seems most appropriate; Alberta had 6,000–8,000 based on 1987 data); Colorado 400–500 (very low estimate considering that the much less common Ferruginous Hawk has 300–400 pairs in this state; unquestionably closer to several thousand pairs); New Mexico 3,000 or greater; North Dakota 5,200; Montana 1,000–3,000; Nevada 150; Oklahoma less than 300; Wyoming 1,000 (probably much higher).

One-year-olds and subadults molting into their first adult plumage spend the summer moving about singly or in small or large flocks in a nomadic fashion wherever prey is most abundant. Most do not return to natal areas.

**Winter.**—Virtually all Swainson's Hawks winter in n. and cen. Argentina, with small numbers in Uruguay and s. Paraguay. Hawks arrive on the breeding grounds from early Nov. to early Dec. Once in Argentina, they may shift around to different areas for optimal feeding. The hawks are gregarious and feed and roost in large flocks, which often number in the thousands. Telemetry data, however, has recently illustrated that many Swainson's Hawks, including adults from the same breeding regions, may winter at various locations along the migratory route to Argentina. This includes cen. and s. California, along coastal Sonora, southward along the Pacific slope of Mexico, Central America, and Colombia. In the w. U.S., a few juveniles and subadults irregularly winter along the Coastal Bend region of Texas, occasionally near Houston, and in sporadic areas in s. Louisiana. A unique but regular contingent of Swainson's Hawks spend the winter in the Sacramento-San Joaquin River Delta near Stockton, Calif. The majority are adults. This wintering area was first documented with isolated sightings in the mid-1970s and again in the early 1980s. In 1989, nearly 30 birds were seen. Since then, up to 100 have been seen each winter. Winter sightings exist for ne. Los Angeles Co. near Lancaster.

In the East, wintering juveniles and subadults are regular each year in s. Florida.

**Movements.**—A highly migratory raptor. Canadian birds may travel 7,000 miles (11,500 km) each spring and fall to wintering areas in Argentina. Among raptors, Swainson's Hawks are surpassed in migratory distance only by "Arctic" Peregrine Falcons (*Falco peregrinus tundrius*). Migration timing is very punctual. Since the mid- to late 1990s, considerable information regarding migration routes and timing has resulted from telemetry.

*Fall migration:* Southward movements may begin in late Aug. However, as far north as Alberta, it is common to see family groups on their breeding territory until mid-Sep. Based on telemetry data, some adults may not leave breeding areas until late Sep., including birds from sw. Minnesota and s. Alberta.

On the e. plains of Colorado, in Washington Co., the first migrant flocks are seen each year in early Sep. Most of these are subadults and adults (probably nonbreeders or failed breeders), along with a few juveniles. Juveniles become more prevalent after mid-Sep. in e. Colorado. By mid-Sep., flocks contain a mixture of all three ages. Peak migration period on the plains of e. Colorado occurs in late Sep. By early Oct., only small numbers are left. Stragglers, which include a large percent of adults, occur until mid-Oct. Huge numbers congregate in the Texas panhandle in late Sep. and early Oct. Peak movement in s. Texas is in the early part of mid-Oct. (Migration peak is about 1 week later than for Broad-winged Hawks.) Most pass through s. Texas on a more inland route than that taken by Broad-winged Hawks.

Massive numbers have been seen in s. Texas in cen. Webb, s. Zapata, and w. Starr Cos. In Oct. 1984, 50,000 were tallied in 1.5 hours late on Oct. 10 in Webb Co.; 8,000–10,000 roosted at Falcon S.P., Zapata Co., that night; untold thousands filled the sky in a stream of birds 0.5 mile (0.8 km) wide that stretched from horizon to horizon all day on Oct. 11; the flight finished by mid-morning on Oct. 12. An estimated 200,000–300,000 birds may have passed by during this period, most on Oct. 11.

The plains of e. Colorado, w. Kansas, and the Texas panhandle are a melting point for birds coming from more northern and western areas (based on the large number of darker birds). Individuals from more eastern regions of the West may also head towards the Great Plains; a telemetry-tracked adult from sw. Minnesota veered sharply southwest into e. Colorado before heading south into Texas. However, most Minnesota birds that were tracked with telemetry took a more conventional, nearly due southerly course into n.-cen. Texas. Two telemetry-tracked Alberta adults stayed east of the Rocky Mts. and angled southeast to s. Texas.

Migration period in Veracruz, Mexico, spans from late Sep. to early Nov. The majority of birds are clustered between early and mid-Oct. Peak is in mid-Oct (over 360,000 birds have been seen on a peak day). Stragglers occur through Nov.

As previously pointed out, a great many birds congregate on the cen. and s. Great Plains of e. Colorado, w. Kansas, and n. Texas and continue south from that point into cen. and then s. Texas, with most probably staying east of Laredo but substantially west of the Coastal Bend area. Migrants continually gather into larger and larger flocks by the time they reach s. Texas and especially s. Mexico. In Mexico, most of the population stays east of the Sierra Madre Oriental and remains along the Gulf slope in Tamaulipas and Veracruz. In s. Veracruz, most keep angling southeast into Chiapas and s. Guatemala and continue through Central America. They enter South America at Colombia, cross the Cordillera Mts. (northern part of the Andes), and stay along the eastern foothills of the Andes in cen. Colombia, e. Peru, s. Bolivia, and thence into Argentina. Or from Colombia, the hawks may angle farther east and cross w. Brazil and cen. Bolivia before dropping into Argentina. Typical migrants may stop and feed for short periods as described in Feeding. Based on birds tracked by telemetry from Alberta to Argentina, the long migration may be completed in just over 50 days.

Based on telemetry-tracked breeding birds from cen. California, this regional population often takes entirely different routes until they enter Guatemala. Migration may begin in Oct., but many stay and feed in the s. Central Valley until Nov., then leisurely head south. Migrants often stop for extended periods. Some may remain in various areas for several days at a time along the Pacific slope of Mexico, in southern interior regions of Mexico, and along Pacific coastal areas of Central America. Three main routes are taken from s. California: (1) From nw. Sonora, some continue to migrate along the Pacific slope of Mexico and join the bulk of North American migrants in s. Chiapas, Mexico, and continue south into Central America. (2) From nw. Sonora, birds may angle sharply southeast and cross the Sierra Madre Occidental, Central Plateau, and Sierra Madre Oriental in n. Mexico and end up on the Gulf slope in n. Tamaulipas where they join the bulk of North American migrants and continue south into Central America. (3) A few may not migrate at all and stay in cen. California for the winter, or head south short distances and then return north to the typical cen. California winter area. *Note:* Adults tracked with telemetry that nested in cen. California may winter from the cen. Central Valley of California south to Argentina, a winter range that spans 6,000 miles (9,700 km).

Migrant Swainson's are often mixed with large numbers of migrant Turkey Vultures and a few Broad-winged Hawks by the time they reach s. Texas. In e. Mexico and farther south, thousands of Turkey Vultures and Broad-winged Hawks may intermix with Swainson's Hawks. (The migration period of Turkey Vulture is very similar to that of Swainson's Hawk, with similar peak periods; however, the vulture migration extends a bit later.)

*Spring migration:* Swainson's Hawks wintering in Argentina may begin leaving winter grounds in early to mid-Feb., retracing the path taken in the fall to return to breeding areas in North America. Adults leave first and primarily migrate separately from subadults and juveniles. Some adults, as seen with telemetry-tracked birds from cen. California, may not leave until late Feb. or even mid-Mar. Adults peak in s. Texas from mid- to late Mar. Very early migrant adults are seen in e. Colorado in late Mar., but typical early adults are not seen until early Apr. Adults peak in e. Colorado in mid-Apr., but large numbers are seen until late Apr. and into early May. Adults begin arriving in Saskatchewan in mid-Apr. In California, the first adults arrive in early Mar. Spring adult migration is more expedient than fall migration and flock sizes are much smaller. By the time adults have reached the cen. Great Plains, flock sizes are very small.

Subadults and juveniles leave the winter grounds in Argentina during Mar. Peak migration periods are unknown. Juveniles are in s. Texas in mid-Apr. but seem to trickle north as singles and small flocks into Jun. The first 1-year-olds are seen in Colorado in early to mid-May. Younger birds also retrace the same routes taken the previous autumn; however, unlike most raptors, they may not return to natal regions.

**NESTING:** Begins in Apr. or May and ends from Jul. to mid-Sep., depending on latitude and elevation. Swainson's Hawks typically do not breed until 3 years old.; however two females that were known 2-year-olds bred in Alberta.

**Courtship (flight).**—*Sky-dancing* by males and *high-circling* by both members (*see* chapter 5).

Nesting pairs are susceptible to nest desertion if disturbed during incubation, especially in areas where they are not acclimated to humans. However, a pair built three new nests in 1 year, each one higher in the same tree to escape disturbance.

Males pick nest locations, but both sexes build new nests or refurbish old nests. Nests are typically in the top portion of trees or bushes. In less disturbed areas, nests are often in the lower and mid-section of tall trees, particularly in windy regions. Swainson's Hawks regularly place nests 3–6 ft. (1–2 m) high in bushes. However, nests may also be placed much higher, even in remote areas, depending on available trees. Around human-occupied areas, nests are always placed in tall trees and may be up to 60 ft. (18 m) high.

Live or dead bushes and trees are used as nest sites. Nest trees may be either deciduous or coniferous. Single, isolated trees in the midst of vast open stretches are common. Nests sites also include shelterbelt groves, narrow wooded riparian areas, trees on abandoned and occupied homesteads, and occasionally tall trees in suburban and urban settings. Swainson's Hawks may nest on wooden utility poles (rarely in Saskatchewan), particularly types with two parallel cross arms that supply a broad base for nests. Nests are occasionally placed on artificial platforms erected for Ferruginous Hawks.

Nests average 24 in. (60 cm) in diameter and 13 in. (32 cm) in depth. However, new nests or nests that are compressed by nestlings or are the product of poor workmanship may become flattened. Some nests, especially ones reused for several years, may become quite large. The often rather shabbily made structures are built with thin or medium-thick sticks or thistle and lined with finer material and usually greenery. Prey remains are typically on the outer area of the nest, particularly when nestlings are younger.

One to 4 eggs, but 2–3 is the typical clutch size. A 5-egg clutch was documented in Idaho, and in Saskatchewan, a first-ever brood of five successfully fledged. Most of the 28-day incubation is performed by females; however, males incubate when females are off the nest feeding. Youngsters may branch in 27–33 days and fledge in 38–46 days. Family groups generally remain intact until fall migration.

House Sparrows regularly builds nests at the base of Swainson's Hawk nests or in the same tree, without harm to the sparrows or bother to the hawks.

**CONSERVATION:** No measures taken other than preventing harmful pesticide use (*see below*). Breeding pairs thrive in moderate agricultural regions providing ample foraging habitat is available.

**Mortality.**—Pesticide contamination can be a major problem. Severe mortality was discovered in the winter of 1994–1995 in n. Argentina when 4,000 Swainson's Hawks were found dead from pesticide poisoning in one small area on their wintering grounds. In the winter of 1995–1996, biologists found 20,000 Swainson's Hawks killed by pesticides Fewer dead hawks were found in the next 2 years. The culprit was the highly toxic organophosphate insecticide monocrotophos, which was used to control grasshoppers. It had probably been used in Argentina since the late 1980s in major wintering areas of the hawk. Argentina banned all use of the insecticide in areas where Swainson's Hawks typically winter in 1996. As of the 1998–1999 winter, no dead hawks have been found. In late March 2000, monocrotophos was totally banned in Argentina. The ban was a tribute to the concerted efforts of biologists, the Argentine government, and the manufacturer of the pesticide. Since virtually the entire population of Swainson's Hawks winters in a restricted area in Argentina, continued exposure to the deadly chemical could have devastated the population of this species. Monocrotophos was not used in the U.S. or Canada.

Organophosphate insecticides are used in North America, but it is not known if they have affected Swainson's Hawks. The hawk also faces possible contamination from the well-known organochlorine pesticide DDT south of the U.S. The deadly pesticide is still used in parts of Central America. Mexico ceased most use in 2002.

As in most large raptors, illegal shooting is still a problem. The fall Swainson's Hawk migration through Colorado in Sep. overlaps with the Mourning Dove hunting season. Errant hunters take their toll on the tame raptor at this time. In Colorado, late Oct. and Nov. Swainson's Hawks are often victims of gunshot wounds and not capable of migrating. Natural predators include Great Horned Owl, Golden Eagle, Coyote, and foxes.

**SIMILAR SPECIES:** COMPARED TO ADULT LIGHT MORPH.—**(1) Short-tailed Hawk, light morph adults.**—PERCHED.—Similar head pattern with the mask. Has small rufous patch along the sides of the breast. Tail pattern often indistinct. FLIGHT.—Underside of remiges paler gray with larger whitish area on the outer four primaries. Tail pattern indistinct or only a few dark narrow inner bands. Does not hover or kite. **(2) Rough-legged Hawk, light morph lightly marked type adults.**—PERCHED.—Head pattern similar but has distinct dark malar mark and white throat is not sharply defined. Bib on breast can be similar to Swainson's: use caution. Flanks are spotted or thickly barred. Tarsi are feathered (not visible if belly feathers are fluffed over them). Dark tail banding is wide on inner tail. Wingtip-to-tail-tip ratio is similar. FLIGHT.—Underside of remiges white. Tail pattern as noted above. COMPARED TO JUVENILE LIGHT MORPH.—**(3) Red-tailed Hawk, "Western" (*B. j. calurus*) light morph, "Eastern" (*B. j. borealis*), and "Fuertes" (*B. j. fuertesi*) juveniles.**—PERCHED.—Iris color is pale yellow or gray. Cere typically more greenish. Lacks dark eyeline and single dark streak down the middle of the throat. Pale scapular patches are similar to Swainson's. Wingtips are distinctly shorter than the tail tip. Dark tail banding is fairly wide. FLIGHT.—Dark patagial mark on leading edge of underwing. Has pale brown upperside of the primaries. Underside of remiges white. Tail pattern as noted above. **(3A) Red-tailed Hawk, "Krider's" (*B. j. borealis*) juveniles.**—Very similar to pale-headed 1-year-old Swainson's: use caution. PERCHED.—Rarely has a dark eyeline. White scapular patches are similar. Tail is whitish and crossed by narrow dark banding on distal two-thirds. Wingtips much shorter than the tail tip. COMPARED TO ADULT INTERMEDIATE/DARK INTERMEDIATE MORPH.—**(3B) Red-tailed Hawk, intermediate morph "Western" (*B. j. calurus*) adults.**—PERCHED.—Forehead is dark and all of lores are white; dark throat. Rufous breast and dark belly similar to dark-bellied type adult dark intermediate morph Swainson's. Rufous tail. FLIGHT.—Dark patagial mark on underside of leading edge of the wings. White or pale gray remiges. Dark undertail coverts. Rufous tail. **(4) Ferruginous Hawk, intermediate/dark intermediate morph adults.**—PERCHED.—Dark forehead and throat and all-white lores. Large yellow gape. Underparts similar, but undertail coverts are dark. Feathered tarsi. Tail lacks banding. Wingtips are shorter than tail tip. FLIGHT.—White underside of remiges. Dark undertail coverts. Tail as noted above. COMPARED TO ADULT DARK MORPH.—**(5) Broad-winged Hawk, dark morph adults/juveniles.**—PERCHED.—All pale lores. Wingtips are shorter than the tail tip. Wide pale band across mid-tail (adult) or moderately wide dark bands (juvenile). FLIGHT.—White underside of the remiges. Dark undertail coverts. Tail patterns as noted above. Does not hover or kite. **(6) Red-tailed Hawk, dark morph "Western" (*B. j. calurus*) adults.**—PERCHED.—All of lores are pale. Rufous tail. FLIGHT.—Pale gray underside of remiges. Rufous tail. **(6A) Red-tailed Hawk, dark morph "Western" (*B. j. calurus*) juveniles.**—PERCHED.—Pale lores. Pale yellow or gray iris. Wingtips are shorter than the tail tip. Prominent fairly wide dark tail banding. FLIGHT.—White underside of the remiges. Pale brown panel on dorsal side of primaries. Dark undertail coverts. COMPARED TO JUVENILE AND SUBADULT DARK INTERMEDIATE/DARK MORPH.—**(7) Northern Harrier, adult females.**—PERCHED.—Older birds have pale irises; younger similar brownish. Lacks dark eyeline. Wingtips much shorter than the tail tip. FLIGHT.—Dorsal view similar but white on uppertail coverts more extensive and forms a larger white patch. Dihedral wing attitude is similar. Pale underside of the remiges. Tail banding is very wide. **(8) Red-tailed Hawk, intermediate/dark morph "Western" (*B. j. calurus*) and "Harlan's" (*B. j. harlani*) juveniles.**—PERCHED.—Pale yellow or gray iris. Dark throat on *calurus*, pale on *harlani* but lacks dark center streak. Lacks dark eyeline. Wingtips are distinctly shorter than the tail tip. Dark tail banding is fairly wide. Intermediate morph of the two species very similar with the dark streaking on the ventral areas: use caution. FLIGHT.—White underside of the remiges. Pale brown panel on upper surface of the primaries. Undertail coverts are the same darker color as rest of the underparts. Tail banding is fairly wide.

**OTHER NAMES:** None regularly used. *Spanish:* Aguililla de Swainson, Aquililla Cuaresmera. *French:* Buse de Swainson.

---

REFERENCES: Adamus et al. 2001; Andrews and Righter 1992; Arizona Breeding Bird Atlas 1993–1999; Baicich and Harrison 1997; Baumgartner and Baumgartner 1992; Bloom 1980; Busby and Zimmerman 2001; Bylan 1998, 1999; Contreras-Balderas and Montiel-de la Garza 1999; Dodge 1988–1997; Dorn and Dorn 1990; Ducey 1988; England et al. 1997; Faanes and Lingle 1995; Fuller et al. 1998; Fung 1999; Holt 2000; Houston 1974, 1995, 1998; Houston and Fung 1999; Houston and Schmutz 1995; Howell and Webb 1995; Jacobs and Wilson 1997; James 1994; James and Neal 1986; Janssen 1987; Kellogg 2000; Kent and Dinsmore 1996; Kingery 1998; Line 1996; Martell et al. 1998; McKinley and Mattox 2001; Montana Bird Distribution Committee 1996; Munro and Reid 1982; Oberholser 1974; Peterson 1995; Raptor Center 1999b; Robbins and Esterla 1992; Russell and Monson 1998; Schmutz 1996; Schmutz et al. 2001; Semenchuk 1992; Sharpe 1902; Small 1994; Smith 1996; Smith et al. 1997; Stewart 1975; Sutton 1967; Tucker 1999; Woodbridge 1997.

**SWAINSON'S HAWK,** *Buteo swainsoni:* Common. 1 million. Locally dense in some areas, sparse in others. Rufous and dark morphs breed mainly west of dotted line. Small numbers winter in CA, s. TX, s. FL, and Mexico. Most winter in Argentina. Very rare in summer in AK, YK, and NT.

**Plate 276. Swainson's Hawk, adult male light morph** [Sep.] ▪ Gray head with rufous nape; white forehead, outer lores, and throat. ▪ Rufous bib; white belly, flanks, and leg feathers. Upperparts often edged with tawny.

**Plate 277. Swainson's Hawk, adult female light morph** [Apr.] ▪ Brown head; white forehead, outer lores, and throat. ▪ Rufous-brown bib. White belly, flanks, and leg feathers unmarked, or partial barring on flanks and leg feathers. ▪ *Note:* Palest brown bib color for females. Both sexes may have faint supercilium and pale auricular areas with dark malar.

**Plate 278. Swainson's Hawk, adult female light morph** [Jul.] ▪ Brown head; white forehead, outer lores, and throat. ▪ Dark brown bib. White belly, flanks, and leg feathers; flanks usually lightly barred, leg feathers sometimes lightly barred. ▪ Wingtips extend past tail tip. ▪ *Note:* This is heaviest amount of ventral barring for light morph.

**Plate 279. Swainson's Hawk, adult male light intermediate morph** [Jul.] ▪ Gray head; white throat, forehead, and outer lores. ▪ Rufous bib. White underparts vary from lightly to heavily barred with rufous on belly, flanks, and leg feathers. White undertail coverts. ▪ *Note:* This is least amount of ventral barring for this morph.

**Plate 280. Swainson's Hawk, adult male light intermediate morph [Jun.]** ▪ Gray head; white outer lores and throat (forehead typically a white). ▪ Dark brown bib. White underparts vary from lightly to heavily barred with dark brown or rufous on belly, flanks, and leg feathers. Flank feathers may have rufous wash. ▪ *Note:* Palest morph for a male to have dark brown bib.

**Plate 281. Swainson's Hawk, adult female light intermediate morph [Aug.]** ▪ Brown head; white outer lores and throat (forehead typically white). ▪ Dark brown bib. White or tawny underparts vary from lightly to heavily barred on flanks, belly, and leg feathers. Unmarked tawny undertail coverts. ▪ *Note:* Heaviest amount of markings for this morph.

**Plate 282. Swainson's Hawk, adult male intermediate (rufous) morph (all-rufous type) [Sep.]** ▪ Gray head; white forehead, outer lores, and throat. ▪ Uniformly rufous underparts; somewhat barred on leg feathers and lower belly. White undertail coverts. ▪ *Note:* Only males have all-rufous underparts.

**Plate 283. Swainson's Hawk, adult female intermediate (rufous) morph [Sep.]** ▪ Brown head; white forehead, outer lores, and throat. ▪ Dark brown bib contrasts with rufous belly, flanks, and leg feathers. Brown barring often superimposed on rufous feathers. ▪ *Note:* Either sex can have contrasting rufous-brown or dark brown bib.

**Plate 284. Swainson's Hawk, ▪ adult male dark intermediate (dark rufous) morph** [**Sep.**] ▪ Gray head; white forehead, outer lores, and throat. ▪ Dark brown underparts, including leg feathers; rufous tinge on lower belly; white undertail coverts. ▪ *Note:* For a fairly dark bird, this individual is quite pale headed.

**Plate 285. Swainson's Hawk, adult female dark intermediate (dark rufous) morph** [**Jul.**] ▪ Dark brown head with white outer lores and throat; forehead dark. ▪ Breast, belly, and flanks dark brown; parts of belly and all lower belly and leg feathers rufous. White undertail coverts. ▪ *Note:* Similar to dark morph except for pale throat and rufous on portions of ventral areas.

**Plate 286. Swainson's Hawk, adult (sex unknown) dark morph** [**Sep.**] ▪ Dark brown head with white lores (typically white only on outer lores). ▪ Uniformly dark brown body; leg feathers often have slight rufous tinge. Undertail coverts pale tawny, often barred. ▪ Wingtips extend to tail tip.

**Plate 287. Swainson's Hawk, adult male light morph** [**Sep.**] ▪ Gray head, white throat. ▪ Rufous bib; white belly, flanks, and leg feathers. ▪ Dark gray remiges contrast with white coverts. Long pointed wings. ▪ Wide black subterminal tail band, narrow inner bands.

**Plate 288. Swainson's Hawk, adult female light morph** [**Aug.**] ▪ White throat. ▪ Dark brown bib. Type with tawny underparts. ▪ Dark gray remiges contrast with tawny coverts. Long pointed wings. ▪ Wide black subterminal tail band, narrow inner bands. ▪ *Note:* Tawny underparts fairly common on light morphs.

**Plate 289. Swainson's Hawk, adult female light morph** [**Jul.**] ▪ White throat. ▪ Split bib type with incomplete brown bib; white underparts. ▪ Dark gray remiges contrast with tawny coverts. Long pointed wings. ▪ Wide black subterminal tail band, narrow inner bands. ▪ *Note:* Uncommon for either sex to have split bib.

**Plate 290. Swainson's Hawk, adult male light intermediate morph** [**Sep.**] ▪ Gray head, white throat. ▪ Breast, upper flanks, and belly rufous. Lower flanks and much of belly white, barred with rufous. White undertail coverts. ▪ Dark gray remiges contrast with white coverts. Axillaries barred with rufous. ▪ Wide dark subterminal tail band, narrow inner bands.

**Plate 291. Swainson's Hawk, adult female light intermediate morph** [**Sep.**] ▪ Brown head; white forehead, outer lores, and throat. ▪ Dark brown bib. Tawny belly, flanks, and leg feathers barred with dark brown. ▪ Dark gray remiges contrast with tawny coverts. Axillaries barred. Long pointed wings. ▪ *Note:* Male may be similar, including dark brown bib, but typically has gray head.

**Plate 292. Swainson's Hawk, adult male intermediate (rufous) morph (all-rufous type) [Sep.]** ▪ Gray head, white throat. ▪ Uniformly rufous underparts except white undertail coverts. ▪ Dark gray remiges contrast with tawny-rufous coverts. Axillaries rufous. ▪ Wide black subterminal tail band, narrow inner bands.

**Plate 293. Swainson's Hawk, adult female intermediate (rufous) morph [Sep.]** ▪ Dark brown bib; rufous belly, flanks, and leg feathers (often with darker brown barring). Pale tawny undertail coverts. ▪ Dark gray remiges contrast with pale tawny coverts. Axillaries barred. ▪ Wide black subterminal tail band, narrow inner bands.

**Plate 294. Swainson's Hawk, adult (sex unknown) dark intermediate (dark rufous) morph (dark-bellied type) [Sep.]** ▪ Dark head with some whitish streaking on throat. ▪ Rufous breast, dark brown belly and flanks (creates belly band), rufous leg feathers and lower belly. White undertail coverts. ▪ Dark gray remiges contrast with tawny-rufous coverts. Brown axillaries.

**Plate 295. Swainson's Hawk, adult female dark intermediate morph (dark type) [Sep.]** ▪ Brown head with whitish throat. ▪ Dark brown breast, belly, and flanks. Lower belly and leg feathers rufous. White undertail coverts. ▪ Dark gray remiges contrast with rufous coverts. Axillaries barred. ▪ Wide black subterminal tail band, narrow inner bands.

**Plate 296. Swainson's Hawk, adult (unknown sex) dark morph** [Sep.] ▪ Dark brown head. ▪ Dark brown underparts, including leg feathers. White undertail coverts partially barred. ▪ Dark gray remiges contrast with tawny-rufous coverts, which have dark diagonal bar. Brown axillaries. ▪ *Note:* A few have all-dark underwing coverts.

**Plate 297. Swainson's Hawk, subadult (2-year-old) light morph** [May] ▪ Lightly spotted underparts. ▪ Remiges full adultlike dark gray, but just replacing outermost primary (p10) for first time (1/3 grown). Dark gray remiges contrast with pale coverts. ▪ Tail adultlike with wide black subterminal band. ▪ *Note:* Bird is completing molt that began previous spring.

**Plate 298. Swainson's Hawk, subadult light morph (juvenile-like)** [Aug] ▪ Yellow cere. Medium brown iris. Pale head with thin dark eyeline, dark malar mark. ▪ Underparts have dark patch on sides of neck that merges with dark malar mark; rest of underparts lightly streaked. ▪ A few retained juvenile feathers on wings. Wingtips slightly shorter than tail tip.

**Plate 299. Swainson's Hawk, subadult light morph (adultlike)** [Sep.] ▪ Yellow cere, nearly all-black bill. Medium brown iris. Dark head with pale supercilium, thin dark eyeline. ▪ Retains a few juvenile secondaries (paler, frayed feathers). Wingtips nearly reach tail tip. ▪ Wide black subterminal tail band, narrow inner bands.

**Plate 300. Swainson's Hawk, subadult dark morph (dark-bellied type) [Sep.]** ▪ Medium or dark brown iris. Yellow cere. Thin dark eyeline; dark malar. ▪ Tawny streaked or mottled breast; solid dark brown belly and flanks; heavily barred leg feathers. White undertail coverts. ▪ Wingtips nearly reach tail tip. ▪ Wide black subterminal tail band.

**Plate 301. Swainson's Hawk, subadult dark morph (all-dark type) [Sep.]** ▪ Yellow cere. Dark brown iris. Thin dark eyeline; dark malar. ▪ Dark underparts have a few tawny spots. White undertail coverts lightly barred. ▪ Wide black subterminal tail band, narrow inner bands.

**Plate 302. Swainson's Hawk, subadult light morph (juvenile-like) [Sep.]** ▪ Pale head. ▪ Dark patch on sides of neck. ▪ Dark gray remiges contrast with white outer 3 primaries (retained juvenile feathers) and coverts. A few retained juvenile secondaries (shorter, lack wide black band). ▪ New rectrices have wide black subterminal band.

**Plate 303. Swainson's Hawk, subadult light intermediate morph (juvenile-like) [Sep.]** ▪ Underparts moderately marked, including barred leg feathers. ▪ Dark gray remiges contrast with white outer 3 primaries (retained juvenile feathers) and coverts. Several retained juvenile secondaries (shorter, lack wide black band). ▪ Wide black subterminal tail band.

**Plate 304. Swainson's Hawk, subadult intermediate morph (adultlike) [Sep.]** ▪ Rufous-tawny underparts have dark patch on neck and barring on belly, flanks, and leg feathers. ▪ Dark gray remiges contrast with outer 3 primaries (retained juvenile feathers) and coverts. A few retained juvenile secondaries (shorter, lack wide black band). ▪ r2 and 5 rectrix sets partially grown.

**Plate 305. Swainson's Hawk, subadult dark morph (streaked type) [Sep.]** ▪ White throat with center stripe. ▪ Heavily streaked underparts. White undertail coverts. ▪ Dark gray remiges contrast with white outer 4 primaries (old juvenile feathers). Several retained juvenile secondaries (shorter, lack wide black band). ▪ Wide black subterminal tail band on new rectrices.

**Plate 306. Swainson's Hawk, juvenile light intermediate morph (1-year-old) [Jul.]** ▪ White head. Thin dark eyeline; dark inner lores. Dark malar mark merges with dark neck patch. Pale brown iris; yellow cere. ▪ White underparts have some dark markings; barred leg feathers. ▪ Wingtips reach tail tip. ▪ *Note:* Bleached, worn plumage.

**Plate 307. Swainson's Hawk, juvenile light morph (1-year-old) [May]** ▪ White head. Thin dark eyeline; dark malar mark merges with dark neck patch. ▪ Lightly marked white underparts. ▪ Pale gray remiges contrast with white coverts. Molt on inner 2 primaries (p1 and 2 are dropped). ▪ Narrow, equal-width tail bands. ▪ *Note:* First stage of molt takes a year for many birds to complete.

**Plate 308. Swainson's Hawk, juvenile light morph** [**Sep.**] ▪ Thin dark eyeline, dark malar mark. All-black bill. Yellow cere. Pale head. ▪ Upperparts broadly fringed. Dark patch on side of breast merges with malar mark. Lightly spotted flanks; rest of underparts unmarked, including leg feathers. ▪ Wingtips reach tail tip. ▪ *Note:* Palest type of light morph.

**Plate 309. Swainson's Hawk, juvenile light intermediate morph** [**Sep.**] ▪ Thin dark eyeline, dark malar mark. Medium brown iris. All-black bill, yellow cere. ▪ Upperparts broadly fringed with tawny. Large dark patch on sides of neck and breast merges with dark malar mark. Moderately streaked underparts; lightly barred leg feathers. Unmarked undertail coverts. ▪ Wingtips reach tail tip.

**Plate 310. Swainson's Hawk, juvenile intermediate (rufous) morph** [**Sep.**] ▪ Thin dark eyeline; dark inner lores. Dark malar. Medium brown iris. All-black bill. Yellow cere. Wide dark center-throat streak. ▪ Dark neck patch merges with malar mark. Moderately streaked breast and belly, moderately barred flanks and leg feathers. ▪ *Note:* Same bird as on plates 314 and 317.

**Plate 311. Swainson's Hawk, juvenile dark intermediate (dark rufous) morph** [**Sep.**] ▪ Thin dark eyeline; dark inner lores. Dark malar. Medium brown iris. All-black bill. Yellow cere. ▪ Dark neck patch merges with dark malar mark. Heavily streaked breast and belly, heavily barred flanks and leg feathers. Narrowly edged upperparts.

**Plate 312. Swainson's Hawk, juvenile dark intermediate (dark rufous) morph (dark-bellied type) [Sep.]** ▪ Thin dark eyeline; dark inner lores. Pale forehead, supercilium, and auriculars. Yellow cere. ▪ Moderately streaked breast and lower belly paler than heavily streaked belly and barred flanks and form dark band across belly and flanks.

**Plate 313. Swainson's Hawk, juvenile dark morph [Sep.]** ▪ Thin dark eyeline; dark inner lores. Yellow cere. Dark malar mark merges with dark sides of breast. ▪ Some tawny mottling on center of breast and upper belly. Heavily barred leg feathers. Barred, white undertail coverts. Upperparts have little pale edging. ▪ Wingtips somewhat shorter than tail tip.

**Plate 314. Swainson's Hawk, juvenile intermediate (rufous) morph [Sep.]** ▪ Thin dark eyeline; dark inner lores. Dark malar. Medium brown iris. All-black bill. Yellow cere. Pale forehead, supercilium, and auriculars. ▪ Upperparts broadly edged with tawny; pale patch on mid-scapulars. ▪ Wingtips nearly reach tail tip. ▪ Narrow, equal-width dark tail bands.

**Plate 315. Swainson's Hawk, juvenile light morph [Sep.]** ▪ Underparts lightly spotted or unmarked. Large dark patch on sides of breast merges with dark malar mark. ▪ Pointed wingtips. Medium gray remiges contrast with unmarked underwing coverts; unmarked axillaries. Whitish area on outer 3 primaries. ▪ Numerous equal-width dark tail bands.

**Plate 316. Swainson's Hawk, juvenile light intermediate morph [Sep.]** ▪ Rich tawny underparts lightly spotted or streaked, including leg feathers. Large dark patch on sides of breast merges with dark malar mark. ▪ Pointed wingtips. Medium gray remiges contrast with lightly marked underwing coverts; lightly marked axillaries. Whitish area on outer 3 or 4 primaries. ▪ Numerous equal-width dark tail bands.

**Plate 317. Swainson's Hawk, juvenile intermediate (rufous) morph [Sep.]** ▪ Dark malar merges with dark neck patch. Tawny underparts moderately streaked; leg feathers barred. ▪ Pointed wingtips. Medium gray remiges contrast with tawny coverts; barred axillaries. Whitish area on outer 3 or 4 primaries. ▪ Subterminal tail band slightly wider than inner bands. ▪ *Note:* Same bird as on plates 310 and 314.

**Plate 318. Swainson's Hawk, juvenile dark morph [Sep.]** ▪ Pale throat. ▪ Heavily streaked underparts with less streaking on center of breast. White undertail coverts. ▪ Pointed wingtips. Medium gray remiges do not contrast with heavily marked underwing coverts; heavily barred or dark axillaries.

**Plate 319. Swainson's Hawk, migrant flock (kettle) [Sep.]** ▪ Migrants gather in flocks comprising of a few hawks or hundreds or thousands.

## WHITE-TAILED HAWK

### *(Buteo albicaudatus)*

**AGES:** Adult, basic II (subadult II), basic I (subadult I), and juvenile. Adult plumage is first attained when probably 3 years old, but it may take somewhat longer to achieve the full definitive status. "Young adults" are probably birds in the first stage of having attained adult plumage.

Subadult II plumage is gained when about 2 years old and is retained for most of the third year. This age was formerly known as "younger adults" (*see* Wheeler and Clark 1995). Plumage progression into adult age class is poorly understood. This age class has two main plumage variations.

Subadult I is achieved when 1 year old and is held for much of the second year. However, some birds appear to begin the transition from subadult I into subadult II during mid-winter. Other individuals are still in old, worn subadult I feathering in late Apr. This age class has three main plumage variations.

Juvenile plumage is held most of the first year. This age class has three main plumage variations. Juveniles have much narrower wings and longer tails than older ages.

White-tailed Hawks have incredibly variable plumages in the juvenile and subadult age classes, and this complicates plumage issues. There are no banded, known-aged individuals for definitive data. All data here are based on knowledge of sequential molt, particularly of the remiges and rectrices, and plumage features of in-field individuals and museum specimens.

**MOLT:** Accipitridae wing and tail molt pattern (*see* chapter 4). First prebasic molt begins in Mar., with the innermost primaries. Several primaries molt prior to the start of molt on the secondaries. Rectrices begin molting shortly after the primaries with replacement of the deck set (r1). By mid- to late May, half of the primaries but only a few secondaries have molted (s1, s5, and tertials). Molt is usually completed from Oct. to early Dec., at which time the subadult I plumage is attained. The outer two primaries (p9 and 10) and two secondaries (s4 and 8) may not molt the first year.

The second prebasic molt into subdult II may begin with replacement of the rectrices in Dec. or Jan. Molting birds in Apr. and May have either full tail characters of this molt and age stage or only partial rectrix molt into this class. Remix molt begins for the second time. Many birds do not complete the molt on the outer three or so primaries, which are now retained subadult I remiges. Molt appears to be completed by early Dec. Further study is needed for this molt and plumage stage.

The third prebasic molt signifies change into full adult character. It takes place from late spring through fall. All wintertime birds have completed their molt.

**SUBSPECIES:** Three races are described, but plumage differences are poorly defined and White-tailed Hawk is considered monotypic by some taxonomists (*see* Palmer 1988 and Color Morphs, below). In the U.S., *B. a. hypospodius* is found in s. Texas, where it is at the northern periphery of its range. *Hypospodius* also inhabits disjunct locations in Mexico and Central America. The U.S. and Mexican ranges connect at n. Tamaulipas and extreme e. Nuevo León and extend south to n. Veracruz. There are isolated population pockets in Mexico in s. Veracruz, n. Tabasco, on the Yucatán Peninsula; also found in Belize. Farther south on the Gulf slope, it is found in n. Honduras and n. Nicaragua. On the Pacific slope of Mexico, it is found in sw. Sonora, Sinaloa, from s. Guadalajara south through w. Michoacán, and from n. and e. Oaxaca and s. Chiapas south through s. Guatemala and w. El Salvador. This race also occupies areas of w. Nicaragua, nw. Costa Rica, and Pacific coastal areas of Panama.

*B. a. colonus* is a small race that ranges from e. Colombia to Suriname and south to n. Brazil north of the Amazon Basin. It also inhabits a few islands in the s. Caribbean: Aruba, Bonaire, Curaçao, Marajo, Margarita, St. Vincent, and Trinidad. There is one sight record for Tobago.

Nominate *B. a. albicaudatus* is found in South America in se. Peru, e. Bolivia, se. Brazil, Paraguay, and the northern two-thirds of Argentina. Similar in size to *hypospodius.* Adults may lack rufous on the scapulars, and flanks have more distinct barring than in *hypospodius.* White throat and gray dorsal areas of adults are similar to *hypospodius.* Basic II age also has rufous on the scapulars and is similar to same-aged *hypospodius.*

**COLOR MORPHS:** None found in *B. a. hypospodius.* A dark morph regularly occurs in *colonus.* All of the body is dark gray except the leg feathers, which are barred with rufous, and rufous on the upperwing coverts and scapulars. Dark morphs are rare in *albicaudatus.* Body and leg feathers are completely dark brownish black and the upperwing coverts and scapulars lack a rufous patch. Color morphs in these two races separate them from *hypospodius* and probably warrant its current subspecies recognition. (Descriptions of the dark morphs are based on Ferguson-Lees and Christie 2001).

**SIZE:** A large raptor. Juveniles are longer than older ages because they have longer rectrices. Tails of juveniles are 1–1.5 in. (25–37 mm) longer than in older ages. Males average smaller than females but overlap. Length: 18–22 in. (46–56 cm); wingspan: 49–53 in. (124–135 cm).

**SPECIES TRAITS:** HEAD.—Bluish green cere. Dark brown irises. BODY.—Long yellow tarsi. WINGS.—Long and pointed. In flight, wings are narrow and "pinched in" at the junction of the body, bow outward on the trailing edge of the broad secondaries, then taper to pointed wingtips. The longest primaries are p7 and 8. On perched birds, the wingtips extend well beyond the tail tip.

**ADULT TRAITS (OLDER):** HEAD.—Cere is sometimes yellow on males. Throat, lores, and forehead are white. Rest of the head is medium gray. BODY (dorsal).—Medium gray back and scapulars with rufous on the rear two-thirds of the scapulars. Females tend to be slightly darker on the gray areas. Lower back and rump are white but visible only when seen from above. BODY (ventral).—White. On the palest birds, mainly males, only the flanks have pencil-line-thin gray or rufous barring on the flanks. On most birds, pencil-line-thin barring also covers the belly and leg feathers. WINGS (dorsal).—Dark gray with a rufous patch on the patagial area from the wrists to the shoulder. Secondaries have narrow gray barring. WINGS (ventral).—White underwing coverts are finely barred with gray or rufous. Axillaries are more heavily barred than the underwing coverts. Secondaries are pale gray, and the primaries are dark gray except the white basal region of the outer two or three primaries. All remiges are narrowly barred with black. The trailing edge of the remiges has a wide black bar. TAIL.—White uppertail coverts. White with a broad black subterminal band. Moderately spaced pencil-line-thin black barring is visible at close range.

**ADULT TRAITS (YOUNGER):** HEAD.—Throat is medium gray but chin is white. BODY, WINGS, and TAIL.— As on older adults.

**BASIC II (SUBADULT II) TRAITS:** All of the head, including the throat, is black. Chin is often white. Lores are white. BODY (dorsal).—Black back and scapulars. Rear two-thirds of scapulars are rufous or partially rufous and similar to those of adults. Lower back and rump are white and mottled with blackish. BODY (ventral).—Somewhat variable with two main types (*see below*). WINGS (dorsal).—Black coverts have a distinct adultlike rufous patch on the patagial region. WINGS (ventral).—White underwing coverts are variably marked with two main types (*see below*). Remiges as on adults. TAIL.—Uppertail coverts are white. Rectrices are mainly white, particularly the deck set, and have pencil-line-thin barring. Some inner and outer sets are mottled with dingy grayish and have partial barring or dark mottling. Rectrices have a wide or moderately wide subterminal black band; it can be the same width as on adults but is often a bit narrower and not as neatly formed.

**Basic II (subadult II) (lightly marked type):** BODY.—Breast is white and unmarked. Flanks are lightly barred or may have some blackish or rufous blotches. Belly and leg feathers are white and lightly barred or blotched with black or rufous. WINGS (ventral).—All underwing coverts are lightly barred with black or rufous. The axillaries are moderately barred with black or rufous. *Note:* Common plumage type.

**Basic II (subadult II) (moderately marked type):** BODY (ventral).—Breast is white and unmarked or may have black streaks or blotches on the sides of the breast. Flanks are heavily mottled or barred with black and rufous but may be nearly solid black with a small amount of white mottling. Belly and leg feathers are white and moderately or heavily barred or blotched with black and rufous. WINGS (ventral).—All underwing coverts are moderately or heavily barred with black or rufous. The lesser underwing coverts on the forward portion of the wing may be rufous. Axillaries are heavily barred with black or rufous, are darker than the rest of the underwing, and can appear as dark "armpits." *Note:* A common plumage type.

**BASIC I (SUBADULT I) TRAITS:** HEAD.—Uniformly brownish black head, including the forehead and throat. Some birds have small tawny markings on the supercilium and auriculars and are similar to some juveniles. BODY (dorsal).—Brownish black back and scapulars but lacks the rufous color on the scapulars of older ages. Lower back and rump are brownish black with some white mottling. BODY (ventral).—Highly variable with three main types (*see below*). Most birds gain a variable amount of rufous on the plumage. WINGS (dorsal).—Brownish black upperwing coverts have a distinct rufous patch on the patagial region. WINGS (ventral).—Three main types, but all have some rufous coloration on the lesser underwing coverts (*see below*). Most have a wide black band on the trailing edge of the remiges. A few have a dusky band with darker thin barring on the trailing edge. Remiges are similar to those of older ages but do not have as distinct a difference between the dark secondaries and pale primaries: secondaries are pale gray and primaries medium gray. Many birds retain faded and worn pale brown juvenile remiges on the outer one to three primaries (p8–10). Old juvenile feathers are often retained at two locations on the secondaries (s4 and 8); old feathers are shorter than the new subadult feathers and lack the wide dark trailing edge of subadults. TAIL.—Uppertail coverts are white. On the dorsal surface, tail is grayish on new feathers but wears and fades to brownish on older feathers. Subterminal band is moderately narrow and dusky with ill-defined edges. Inner portion of the rectrices is covered with moderately spaced pencil-line-thin barring, mottling, or a combination of barring and mottling.

**Basic I (Subadult I) (Lightly Marked Type):** BODY (ventral).—The breast is white with a narrow black strip along the sides of the breast. Flanks and belly are white and lightly or moderately barred or blotched with black or rufous or a mixture of both colors. Leg feathers are lightly barred. Undertail coverts are white and generally unmarked. WINGS (ventral).—Lesser underwing coverts are moderately barred with rufous or are solid rufous. Median and greater underwing coverts are barred with rufous or black. Axillaries are white with black barring and can be fairly dark. *Note:* A common plumage type.

**Basic I (Subadult I) (Moderately Marked Type):** BODY (ventral).—The breast is white with a narrow black strip along the sides of the breast. Two variations of underparts: (1) *Black type.*—Flanks and belly are black with some white and rufous mottling and barring. Lower belly is white with black or rufous mottling or barring; leg feathers are white and barred with black or rufous. *Note:* Common plumage type. (2) *Rufous type.*—Flanks, belly, and leg feathers are nearly solid rufous with some black or white mottling; or the sides of the neck, flanks, and upper belly may be black with rufous on the lower half of the belly and leg feathers. Undertail coverts are white but may have narrow dark bars or blotches. *Note:* Uncommon plumage type. WINGS (ventral).—Lesser underwing coverts are solid rufous. Median and greater underwing coverts are extensively barred or mottled black or rufous. Axillaries are black with narrow white barring or can be solid black.

**Basic I (Subadult I) (Heavily Marked Type):** BODY (ventral).—Brownish black breast with moderate-sized or narrow white center streak or patch. Flanks, belly, and leg feathers are uniformly brownish black. Lower belly sometimes has pale mottling. Undertail coverts are white and heavily barred or blotched with black. WINGS (ventral).—Rufous lesser underwing coverts. Median and greater underwing coverts are uniformly brownish black but may have pale mottling.

Axillaries are black. *Note:* An uncommon plumage type.

**JUVENILE TRAITS:** HEAD.—Brownish black with three main variations: (1) *Lightly marked type.*—Large tawny or white patch extends from the rear supercilium onto the auriculars and sides of the head below the auriculars. There is an isolated dark patch on the center of the throat with a white strip on each side of the dark patch along the lower jaw. (2) *Moderately marked type.*—Three separate tawny or white patches: on the rear supercilium, auriculars, and sides of the head. There is an isolated dark patch on the center throat with a white strip on each side of the dark patch along the lower jaw. (3) *Heavily marked type.*—Small partial or faint tawny or white patches on the supercilium and auriculars; head may also be nearly all dark. Throat is uniformly dark. BODY (dorsal and ventral).—Brownish black dorsal regions with a variable amount of tawny edging and blotches. Highly variable ventral markings with four main patterns (*see* respective dorsal and ventral types). Lower back and rump are blackish with some white speckling. WINGS (dorsal).—Blackish brown with a small area on the patagial region that has tawny edging. This tawny patch area can be moderately or very distinct. Pale tawny edgings on coverts often wear off by mid-winter or spring. Upper surface of the remiges is uniformly colored. WINGS (ventral).—Highly variable markings on the underwing coverts (*see* types). Underside of the remiges is narrowly barred on all but the basal region of the outermost primaries and has a narrow dusky trailing edge. Basal region of the outer three or four primaries is white. There are two variations of patterns on the remiges: (1) uniformly pale gray except the basal region of the outermost primaries and (2) similar to older ages in having paler secondaries and darker primaries: pale gray secondaries and medium gray primaries. TAIL.—Uppertail coverts are white and may have a small dark pattern on some feathers. Upper surface of tail is pale gray in fresh plumage but fades and wears to medium brown. Tail is covered with numerous narrow dark bands. Because this species often perches and feeds on the ground, the tips of the rectrices often become frayed and broken by late winter or spring.

**Juvenile (Lightly Marked Type):** HEAD.—Nearly always the lightly marked type (*see above*) of pale markings. The dark throat is conspicuous. BODY (dorsal).—Brownish black with tawny-edged scalloping and large tawny spots on the basal region of the scapulars. Most pale edgings wear off by spring. BODY (ventral).—All of the breast is white and unmarked except for a moderately wide brownish black area on the sides of the breast. Flanks are moderately spotted or blotched with brownish black, and the belly is lightly spotted. Leg feathers are lightly spotted. WINGS (ventral).—Axillaries are heavily marked with black and always darker than the rest of the underwing and appear as dark "armpits." Lesser underwing coverts are lightly or moderately spotted. Middle underwing coverts are white with a dark spot on the tip of each feather and form a broad white bar on the underwing. *Note:* An uncommon to fairly common plumage type.

**Juvenile (Moderately Marked Type):** HEAD.—Pale markings are lightly marked, moderately marked, or heavily marked types (*see above*). The dark throat is conspicuous. BODY (dorsal).—Brownish black with narrow tawny edging on the scapulars. Tawny spots on the basal region of the scapular feathers are small or moderate in size. BODY (ventral).—Breast is white and either unmarked or has sparse dark streaks. The sides of the breast are uniformly dark. Flanks and belly are brownish black with some white mottling. Leg feathers are white with a moderate amount of dark spotting or blotching. Undertail coverts are white and lightly or moderately marked with barring or blotching. WINGS (ventral).—Two main patterns: (1) mainly dark with scattered pale mottling on all coverts or (2) brownish black or rufous on all lesser coverts with a broad white band on the median wing coverts. Axillaries are solid dark or dark with a small amount of pale mottling; they are always darker than the rest of the underwing and appear as dark "armpits." *Note:* A common plumage type.

**Juvenile (Heavily Marked Type):** HEAD.—Most have the heavily marked type (*see* *above*) pale markings. BODY (dorsal).—Even in recently fledged plumage, the scapulars have only narrow tawny edges. The pale basal spots on the scapulars are small and rarely show. BODY (ventral).—Brownish black breast, flanks, and belly. There is a short, narrow white stripe or patch on the breast. The flanks and belly have a small tawny tip on most feathers. Undertail coverts are white with large dark markings. WINGS (ventral).—Brownish black with some scattered white mottling on all coverts. *Note:* Common plumage type.

**Juvenile (All-Dark Type):** HEAD.—All dark. BODY (dorsal).—Uniformly brownish black; however, recently fledged birds have narrow tawny edging on the scapulars. BODY (ventral).—Uniformly brownish black and lacks a white mark on the breast. The white breast patch may also be reduced and may not be visible at long distances. In fresh plumage, the flanks and belly may have small tawny tips on most feathers. White undertail coverts are extensively marked and appear mostly dark. WINGS (ventral).—Underwing coverts are mainly dark. *Note:* A rare plumage type. At certain angles and long distances, a heavily marked type may appear as an all-dark type.

**ABNORMAL PLUMAGES:** None documented.

**HABITAT:** Breeding pairs inhabit open and semi-open coastal and inland grasslands, savannahs, and thornscrub areas. Bushes and small trees such as mesquite, yucca, and live oak grow either singly or in small groups in these habitats. Much of the habitat is on moderately grazed lands. During all seasons, nonbreeding birds are found in these habitats. Younger birds commonly use moderate agricultural areas. Extensive agricultural areas such as e. Nueces Co., Tex., are not inhabited. For short periods, all ages regularly use microhabitats at naturally occurring and controlled fires and plowing and harvesting activities. Controlled fires, plowing, and harvesting typically occur from fall to early spring. In much of the range, the climate is seasonally hot and humid. In Brooks, Duval, Hidalgo, Jim Hogg, Jim Wells, and Starr Co., Tex., the climate is also seasonally hot but arid.

**HABITS:** A tame raptor. Breeding pairs, however, are skittish at their nest sites. Nonbreeding birds are highly gregarious. Breeding pairs are solitary but may become gregarious in the nonbreeding season. All types of exposed natural and artificial structures, including posts, utility poles, and wires, are used for perches.

**FEEDING:** Perch and aerial hunter. Opportunistically feeds on a wide variety of prey: large insects, reptiles, amphibians, rodents, rabbits, and small and moderately large birds. Insects may be captured and eaten while in flight. Most other prey are taken to a perch or nest site or eaten on the ground. White-tailed Hawks regularly attend fields that are being burned, plowed, or harvested. It is common to have up to 20 and uncommon to have up to 60 White-taileds accompany such activities. The largest congregations occur in areas with abundant prey. White-tailed Hawks prey on dislodged prey and on passerines that are drawn to the area, especially Great-tailed Grackles. Hawks may remain at burned, plowed, or harvested areas for a few hours or a few days. The largest numbers are seen the first day; fewer remain or are drawn to the area in successive days. At burn areas, White-tailed Hawks feed above the burning area on flying insects and in front of and behind the fire for rodents and passerines. At plowed areas, they follow behind plowing equipment. (Burning and plowing also attract large numbers of Crested Caracaras, Harris's Hawks, Turkey Vultures, and Red-tailed and Swainson's hawks.) White-tailed Hawks regularly feed on carrion. Crested Caracaras often pirate live prey and carrion from White-tailed Hawks.

**FLIGHT:** Powered flight is an irregular series of moderately slow wingbeats interspersed with gliding sequences. Wings held in a high dihedral when soaring or gliding. White-tailed Hawks often hover and kite when foraging.

**VOICE:** Rarely heard except when disturbed around nest sites, when agitated by close human presence away from a nest site, or when being harassed by pirating Crested Caracaras. Primary mildly agitated call is either a soft, short, somewhat nasal and high-pitched *kair* or *kee-aah*. Both calls are from nonbreeding individuals.

**STATUS AND DISTRIBUTION:** In the U.S., only found in coastal Texas. Fairly common it its limited range but classified as *Threatened* in Texas. Population is thought to be stable but guarded. Population estimates vary from 200 to 400 pairs.

White-tailed Hawks suffered considerable range reduction in the lower Rio Grande valley because of agricultural conversion and urbanization beginning in the 1920s. Also, from the late 1800s to the early 1900s, mesquite thornbush spread throughout formerly open areas and reduced habitat in areas that were not converted to farmlands. Partially because of the eradication of mesquite in many areas, White-tailed Hawks have regained many breeding areas since the 1960s. The population decline that occurred from late 1940s to early 1970s may be attributed to organochlorine pesticides, particularly DDT. Eggshell thickness, which can by affected by such pesticides, decreased during this time.

The species is currently affected in the northern portion of its range by intense urban sprawl, particularly in Fort Bend, Harris, and Waller Cos. Because of intense agricultural land use, adequate habitat is lacking in large portions of Calhoun and Nueces Cos. Agriculture and woodlands fragment habitat in other counties. However, an extensive amount of habitat is on large, protected private ranches in Kenedy and Kleberg Cos. In Mexico, fairly common but local in Tamaulipas and in most other areas listed above under Subspecies. Rare and does not breed in Sonora. The population in Sonora has declined in the last few decades. Breeding range in Nuevo León and n. Tamaulipas is affected by a massive amount of agriculture.

**Summer.**—Inhabits the following Texas counties: s. Austin, nw. and s. Brazoria, extreme se. Bee, Brooks, portions of Calhoun, e. Cameron, ne. Chambers, s. Colorado, extreme se. Dewitt, s. Duval, w. Fort Bend, e. Goliad, se. Lavaca, w. Harris (may have succumbed to intense urbanization), an isolated spot in se. Harris, n. Hidalgo, Jackson, e. Jim Hogg, Jim Wells, Kenedy, Matagorda, w. Nueces, Refugio, San Patricio, ne. Starr, Victoria, and Willacy.

**Winter.**—Range in Texas extends somewhat west and south of breeding counties. White-tailed Hawks are regular in small numbers in the following counties: s. and n. Bee, Duval, cen. Goliad, s. McMullen, Starr, Webb, and Zapata. Larger numbers winter in the Borderplex region of Cameron and Hidalgo Cos. A few juveniles and subadults winter on South Padre Island.

**Movements.**—Breeding pairs are mainly sedentary. Moderate numbers of adults, however, attend burned and plowed areas substantial distances from breeding habitat. Immatures tend to disperse or become nomadic from fall through spring; a few may actually migrate unknown distances southward in the fall. A few southbound birds are seen in the fall at Smith Point in Chambers Co. and at Hazel Bazemore Co. Pk. west of Corpus Christi, Tex. These birds may be part of the dispersal or of nomadic excursions and not true migrants. Migrants are not seen at hawkwatches in Veracruz, Mexico.

**Extralimital movements.**—*Texas:* Irregularly disperses to w. and n. Texas in fall and winter. Records exist for the following counties: Brazos, Brewster (Big Bend N. P.), Bexar, Callahan, Dallas, Llano, Travis, and Val Verde. *Louisiana:* (1) Immature collected in sw. Calcasieu Parish in mid-Nov. 1888. (2) Adult in Jefferson Davis Parish from mid-Dec. 1995 to mid-Jan. 1996. Possibly the same bird at the same locale in Nov. and Dec. 1996, from late Nov. 1999 to mid-Jan. 2000, and again in the winter of 2001. (3) Subadult (video of the bird) in Jefferson Davis Parish in Feb. 2001. *Arizona:* Undocumented nesting in 1889, and one specimen collected that year near Phoenix. Recent-era sightings, but none are documented with photographs: (1) A single bird in Maricopa Co. in the winter of 1954–1955, (2) a single bird in Pima Co. from Dec. 1964 through Jan. 1965, (3) a single bird in Santa Cruz Co. from early Jan. to early Feb. 1971, and (4) a single bird in Cochise Co. in mid-May 1974.

**NESTING:** Adult pairs may begin occupying eventual breeding territories in Dec. Nest-building does not begin until Jan. or Feb.

**Courtship (flight).**—*High-circling* is the only display (*see* chapter 5). (Most large raptors perform sky-dancing.)

**Courtship (perched).**—*Grass-pulling.* Both sexes of a pair fly low to the ground and then land. Both perch erect; however, the male leans over and pulls grass and stems with his bill but does not uproot or break them off. The female is near but does not look directly at the male. This display occurs during nest-building but ends before egg-laying. No other buteo performs this behavior.

Both sexes build nests. Nests are placed on the top portion of bushes or small trees. They average 9 ft. (3 m) high but can be as low as 1.6 ft. (0.5 m) and as high as 41 ft. (12 m). Nests may be somewhat oblong in shape and range from about 20 to 36 in. (52–91 cm) long but only slightly less in width. Old nests are often reused and become larger with each use. Nests are constructed of sticks and twigs and lined with small, green leafy twigs and grasses. Two eggs are typically laid, occasionally 3 and rarely 1 or 4. Incubation, lasting 29–32 days, is mainly done by females, but males assist for short periods. Nestlings fledge in 49–53 days. It is suggested by some authors that youngsters remain with their parents until the next breeding season. However, it has been documented that juveniles may become independent by early Aug., and all juveniles seen in autumn are fully independent.

**CONSERVATION:** No measures are taken. A few pairs nest on national wildlife refuges along the Coastal Bend region of Texas. Many pairs nest on fairly protected ranches, but most are on fragmented parcels with minimal habitat protection. White-tailed Hawks do not nest on artificial nest platforms.

**Mortality.**—Illegal shooting is a problem with any raptor and undoubtedly occurs with White-tailed Hawks. An adult was photographed that had one leg neatly severed, perhaps the result of a leg-hold trap.

**SIMILAR SPECIES:** COMPARED TO ADULT.—**(1) Swainson's Hawk, adults.**— FLIGHT.—Similar gray underside of the remiges, but gray color is uniform throughout all of the remiges. Similar dihedral when soaring and gliding. Narrow white band on the uppertail coverts. Black subterminal tail band is only moderately wide with discernable inner narrow dark bands. COMPARED TO SUBADULT II.—**(2) "Harlan's" Red-tailed Hawk (*B. j. harlani*), light intermediate and intermediate morph white-tailed type adults.**—PERCHED.—Very skittish. Throat is usually pale. White breast is normally streaked or mottled throughout but can be very similar to that of a White-tailed that has streaking on the sides of the breast. Wingtips are shorter than tail tip or, at most, equal to tail tip. FLIGHT.—Throat is usually pale. Lesser upperwing coverts are dark. Dark or rufous uppertail coverts. White tail has a moderately wide, irregular black subterminal band (mottling may not be visible or may be lacking). COMPARED TO SUBADULT I.—**(3) Swainson's Hawk, dark morph adults.**—Similar to heavily marked type White-tailed Hawks. PERCHED.—Dark inner half of the lores. Wingtips barely longer than tail tip. Lesser upperwing coverts are dark. FLIGHT.—Wing attitude, gray underside of remiges, and rufous underwing coverts are similar. Uppertail coverts are dark. **(4) "Harlan's" Red-tailed Hawk (*B. j. harlani*), light, intermediate, and dark intermediate morph gray-tailed type adults.**—PERCHED.— Same as #2. FLIGHT.—Same as #2 except tail, which is similar in being grayish with narrow barring or mottling, but the subterminal band is generally more irregular than on White-tailed Hawks. COMPARED TO JUVENILE.—**(5) Swainson's Hawk, intermediate, dark intermediate, and dark morph juveniles and subadults.**—Similar to moderately marked, heavily marked, and all-dark types of White-tailed Hawks. PERCHED.—Pale throat with a narrow, dark center streak. Wingtips are somewhat shorter than or equal to the tail tip. FLIGHT.—Wing attitude and gray underside of the remiges are similar. Pale throat. Uppertail coverts are pale but heavily marked. Dark inner tail bands are quite distinct. Subadults have a wide black band on the trailing edge of the remiges and a fairly wide black subterminal tail band. **(6) Red-tailed Hawk, light morph (includes *B. j. fuertesi*, *B. j. calurus*, and *B. j. borealis*) juveniles.**—Very similar to lightly marked and moderately marked type White-tailed Hawks. PERCHED.—Pale throat on *fuertesi* and *borealis;* dark throat on *calurus* is similar but is not edged with a white stripe on each side of jaw. Irises are pale yellow or gray. Greenish cere color is similar. Wingtips are distinctly shorter than tail tip. FLIGHT.—Throat data as in Perched. Axillaries are marked but are pale and do not contrast with the rest of the underwing. Pale panel on dorsal surface of primaries; pale panel or window on underside of inner primaries in translucent light. Dark patagial mark on leading edge of the underwing: use caution as some White-tailed Hawks also have a hint of being darker on the patagial region. Narrow, dark tail banding is fairly obvious. **(7) Red-tailed Hawk (includes *B. j. calurus* and *B. j. harlani*) intermediate, dark intermediate, and dark morph juveniles.**—PERCHED.—Iris and cere colors as in #6. Throat dark, but is not edged with a narrow white stripe. Wingtips as in #6. FLIGHT.—Pale panel on dorsal surface of primaries; often pale window or panel in translucent light on the ventral surface. Mainly dark uppertail coverts. **(8) Ferruginous Hawk, rufous and dark morph adults and juveniles.**—PERCHED.—Large yellow cere and gape. Irises often pale, except on some adults, which are dark. Feathered tarsi. Wingtips are shorter than tail tip. FLIGHT.—Pale panel on dorsal surface of the primaries; often shows as a pale window on the underside in translucent light. Distinct black (adults) or gray (juveniles) tips on "fingers" on the ventral surface of the outermost primaries. Rufous or dark uppertail coverts.

**OTHER NAMES:** None used. Formerly called Sennett's White-tailed Hawk. *Spanish*: Aguililla Cola Blanca, Aguililla Coliblanca, Gavilan Coliblanco. *French:* Buse à que Blanche.

REFERENCES: Baicich and Harrison 1997; Bent 1961; Clark and Wheeler 2001; Farquhar 1992; Ferguson-Lees and Christie 2001; Howell and Webb 1995; Johnsgard 1990; Kellogg 2000; Monson 1998; Palmer 1988; Russell and Monson 1998; Sexton 2001; Wheeler and Clark 1995, 1999.

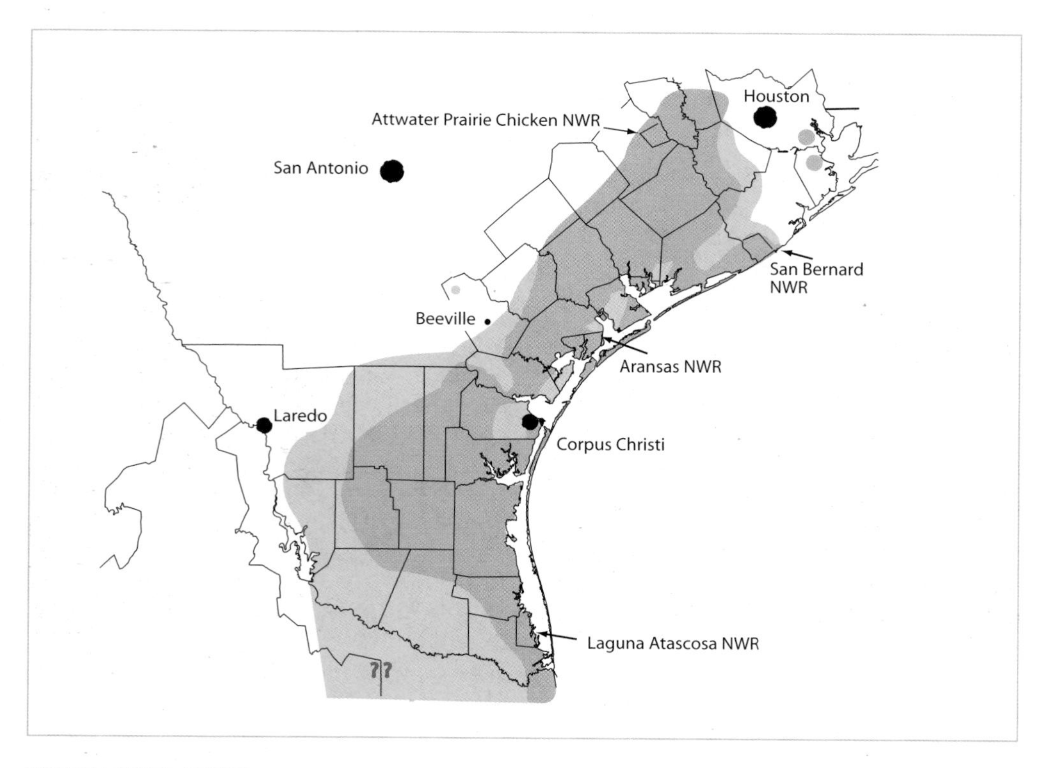


**WHITE-TAILED HAWK,** *Buteo albicaudatus:* Texas classifies as Threatened. Disperses short distances north, west, and south of winter range. Farming and urbanization limit breeding range.

**Plate 320. White-tailed Hawk, adult [Feb.]** ▪ Medium gray head, white throat. Pale greenish yellow cere. ▪ White underparts with thin barring on flanks. Long tarsi. ▪ Gray wings with rufous shoulder patch. Wingtips extend far past tail tip. ▪ Short white tail has wide black subterminal band. ▪ *Note:* Probable male by pale color and reduced ventral markings.

**Plate 321. White-tailed Hawk, adult [Jan.]** ▪ Medium gray head, white throat. Pale bluish green cere. ▪ White underparts. Gray back and forward scapulars; rufous on rear scapulars. ▪ Black wings with rufous shoulder patch. Wingtips extend far past tail tip. ▪ Short white tail has wide black subterminal band. ▪ *Note:* Probable female by dark wing color.

**Plate 322. White-tailed Hawk, adult [Feb.]** ▪ Gray back and forward scapulars; rufous rear scapulars. White lower back. ▪ Gray wing with rufous shoulder patch. White lower back. ▪ White uppertail coverts and tail. Wide black subterminal tail band. ▪ *Note:* Probable male by pale upperwings.

**Plate 323. White-tailed Hawk, adult [Feb.]** ▪ Gray head, white throat. ▪ White underparts. ▪ Long, broad, pointed wings bow outward on rear edge, taper next to body. Inner primaries dark gray, secondaries pale gray; wide black band on rear edge. Coverts white, finely barred. ▪ Short white tail has wide black subterminal band. ▪ *Note:* Probable male by lightly marked body.

**Plate 324. White-tailed Hawk, young adult** [**Feb.**] ▪ Gray head and throat. ▪ White underparts finely barred on belly, flanks, and leg feathers. ▪ Long, broad, pointed wings bow outward on rear edge, taper next to body. Dark gray primaries, pale gray secondaries; wide black band on rear edge. Coverts white, finely barred. ▪ Short white tail has wide black subterminal band.

**Plate 325. White-tailed Hawk, basic II** (**subadult II**) (**lightly marked type**) [**Nov.**] ▪ Black head and throat. ▪ White underparts moderately barred on belly, flanks, and leg feathers. ▪ Long, broad, pointed wings bow outward on rear edge, taper next to body. Dark gray primaries, pale gray secondaries. ▪ Short white tail has moderately wide black subterminal band.

**Plate 326. White-tailed Hawk, basic II** (**subadult II**) (**moderately marked type**) [**Feb.**] ▪ Black head and throat. ▪ White breast, black flanks, barred or blotched belly, barred leg feathers. ▪ Dark gray primaries, pale gray secondaries. Coverts often barred with rufous ▪ Short white tail has moderately wide black subterminal band.

**Plate 327. White-tailed Hawk, basic I** (**subadult I**) (**moderately marked type**) [**Feb.**] ▪ Black head and throat; may have small pale areas. Pale green cere. ▪ White breast; dark flanks; barred or blotched belly. Black back and scapulars. ▪ Black wings with rufous shoulder patch. ▪ Wingtips extend far past tail tip.

**Plate 328. White-tailed Hawk, basic I (subadult I) (heavily marked type)** [**Feb.**] ▪ Black head and throat; may have small pale areas. Pale green cere. ▪ Black underparts except for small white breast patch; white undertail coverts. Long tarsi. ▪ Rufous shoulder patch. Wingtips extend far past tail tip.

**Plate 329. White-tailed Hawk, basic I (subadult I) (heavily marked type)** [**Feb.**] ▪ Black head. ▪ Black body (white breast patch not visible at this angle). ▪ Rufous shoulder patch. ▪ White uppertail coverts; gray tail with dusky subterminal band.

**Plate 330. White-tailed Hawk, basic I (subadult I) (lightly marked type)** [**Jan.**] ▪ Black head and throat. ▪ White underparts lightly barred or mottled on belly, flanks, and leg feathers. ▪ Long, broad, pointed wings bow outward on rear edge, taper next to body. Wide black band on rear edge. Coverts marked with rufous. ▪ Gray tail has moderately wide dusky subterminal band.

**Plate 331. White-tailed Hawk, basic I (subadult I) (moderately marked type)** [**Feb.**] ▪ Black head and throat. ▪ Rufous type underparts have white breast, black and rufous belly and flanks, rufous leg feathers; white undertail coverts. ▪ Wide black band on rear edge of wing. Rufous lesser coverts. ▪ Gray tail has moderately wide dusky subterminal band.

**Plate 332. White-tailed Hawk, basic I (subadult I) (moderately marked type)** [**Feb.**] ▪ Black head and throat. ▪ Black type underparts have white breast; black-marked belly and flanks; barred, white leg feathers; white undertail coverts. ▪ Wide black band on rear edge of wing. Rufous lesser coverts. ▪ Gray tail has moderately wide dusky subterminal band.

**Plate 333. White-tailed Hawk, basic I (subadult I) (heavily marked type)** [**Feb.**] ▪ Black head and throat. ▪ Black underparts except for small white patch on breast; barred, white undertail coverts. ▪ Wide black band on rear edge of wing. Partially rufous lesser coverts. ▪ Gray tail has moderately wide dusky subterminal band.

**Plate 334. White-tailed Hawk, juvenile (moderately marked type)** [**Feb.**] ▪ Lightly marked type head with large white patch on sides of head. Dark throat. Wide pale blue base of bill. ▪ Large white patch on center of breast; black belly and flanks; white, partially barred leg feathers; white undertail coverts. ▪ Wingtips extend past tail tip.

**Plate 335. White-tailed Hawk, juvenile (moderately marked type)** [**Feb.**] ▪ Moderately marked type head with 3 white patches: supercilium, auricular area, and side of neck. Pale greenish cere. White lores; dark throat. Wide pale blue base of bill. ▪ Tawny-edged shoulder patch visible on some birds. ▪ Wingtips extend past tail tip.

**Plate 336. White-tailed Hawk, juvenile (heavily marked type) [Feb.]** ▪ Heavily marked type head with 2 or 3 small white patches or all black. Pale greenish cere. White lores; dark throat. Wide pale blue base of bill. ▪ Body black except for small white patch on center of breast (not visible at this angle). ▪ Wingtips extend past tail tip.

**Plate 337. White-tailed Hawk, juvenile (lightly or moderately marked type) [Feb.]** ▪ White head patch. ▪ Dark brown or black upperparts, often with white mottling on lower back and rump. ▪ Pointed wingtips. Uniformly dark upperwing. ▪ White uppertail coverts. Pale brown tail with numerous narrow dark bands.

**Plate 338. White-tailed Hawk juvenile (lightly marked type) [Feb.]** ▪ Black throat with white edge stripe. Large white breast patch; black mottling on belly and flanks; white leg feathers barred. ▪ Long, narrow, pointed wings bow outward on rear edge, then taper next to body. Black axillaries; broad white bar on median coverts. ▪ Numerous narrow tail bands.

**Plate 339. White-tailed Hawk, juvenile (moderately marked type) [Feb.]** ▪ Black throat with white edge stripe. ▪ Moderate-sized white breast patch. White-mottled black underparts. White undertail coverts. ▪ Long, narrow, pointed wings bow outward on rear edge, taper next to body. White-mottled black axillaries and coverts. ▪ Numerous narrow tail bands.

**Plate 340. White-tailed Hawk, juvenile (moderately marked type)** [**Feb.**] ▪ Black throat with white edge stripe. Moderate-sized white breast patch. ▪ Black underparts; spotted leg feathers. White undertail coverts. ▪ Black axillaries; broad white band on median coverts. ▪ Numerous narrow tail bands.

**Plate 341. White-tailed Hawk, juvenile (heavily marked type)** [**Feb.**] ▪ Black head. ▪ Small white patch on black breast; white-mottled belly, flanks, and leg feathers; white undertail coverts barred. ▪ Mainly black axillaries and coverts. ▪ Numerous narrow tail bands.

---

# ZONE-TAILED HAWK
## *(Buteo albonotatus)*

**AGES:** Adult and juvenile. Adult plumage is acquired when 1 year old. Juvenile plumage is worn the first year. Juveniles have somewhat longer tails than adults, but wing width is identical for both ages.

**MOLT:** Accipitridae wing and tail molt pattern (*see* chapter 4). First prebasic molt from juvenile to adult occurs from Apr. to late fall. By mid-May, the inner three or four primaries and some head and scapular feathers are replaced, but the juvenile secondaries and rectrices are still retained. Secondaries and rectrices molt somewhat later. Subsequent annual adult prebasic molts appear to be partially complete. They begin in Jun. on females and Jul. on males and are completed by late fall. Adult prebasic molts may retain a few adult remiges and wing coverts from the previous year.

**SUBSPECIES:** Monotypic.

**COLOR MORPHS:** None.

**SIZE:** A large raptor. Sexes are dimorphic with no overlap for wing chord and weight (Palmer 1988). Wingspan and total length probably do not overlap. Measurements are not available for the sexual differences; males are on smaller spectrum, females on larger. Sexual dimorphism is not readily apparent in the field unless a pair of birds are together. Length: 19–21 in. (48–53 cm); wingspan: 48–55 in. (123–140 cm).

**SPECIES TRAITS:** HEAD.—Black bill with small, pale bluish area on the basal part of both mandibles. Bright yellow cere. **White lores and forehead.** Dark brown irises. BODY.—Moderate-length tarsi. In flight, black undertail coverts extend beyond feet.

WINGS.—When perched, wingtips extend far beyond tertials (long primary projection). **In flight, wings are moderately wide with parallel front and rear edges; rounded wingtips have five "fingers." Undersides of the remiges are completely barred.** *Note:* This is the only North American buteo that has only a dark-colored plumage for all ages.

**ADULT TRAITS:** HEAD.—Grayish black. BODY.—Grayish black. WINGS.—Upper- and underwing coverts are grayish black. **Undersides of the remiges are medium gray and completely barred with narrow black bars and a wide black band on the trailing edge.** When perched, the long primaries extend past the tail tip. When bird is perched and viewed from the front, the undersides of the outer primaries are often visible and exhibit narrow gray and black barring. TAIL (dorsal).—**Black with two pale gray bands: a wide band on the mid-section and a narrow inner band.** *Female only:* **Black with a wide pale gray band on the mid-section and two progressively narrower pale gray inner bands.** *Note:* When the tail is widely fanned, the white inner webs of the rectrices are often partially exposed and the typically gray dorsal bands may appear white. This is especially the case on summertime molting birds which are often missing rectrices and show an even more accentuated white band. TAIL (ventral).—Pattern is somewhat sexually dimorphic on the ventral surface when viewed on a widely fanned tail, but there is some overlap. Both sexes exhibit a single broad, white mid-tail band. The narrower, inner white bands on the basal region are covered by the long black undertail coverts. *Male only:* **Two white bands. Wide black subterminal band with a broad white mid-tail band and one narrow inner white band.** *Female only:* **Three to five white bands. Broad white mid-tail band and three or four additional, progressively narrower inner white bands.** *Male or Female:* **Three white bands. Broad white mid-tail band and two additional, progressively narrower inner white bands.** *Note:* This pattern was formerly thought to be a female-only trait; however, accurately sexed specimens verify a shared three-white-banded pattern for both sexes (AMNH). The innermost white band, however, does not extend onto the outer web of the outermost rectrix set (r6) on males, but this is difficult to see in the field.

**JUVENILE TRAITS:** HEAD.—Iris is dark brown. Head is black or has a variable amount of white mottling on the nape and hindneck. BODY.—Plumage is black (darker, more blackish than on adults). Underparts may be uniformly black or have small white speckling throughout. White speckling is often difficult to see at field distances. WINGS.—**Upperwing coverts are brownish and not as black as the head and underparts. Whitish undersides of the remiges are completely covered with narrow black barring; trailing edge has a narrow gray band.** When perched, the long primaries extend to the tail tip. TAIL.—**Uppertail is brown with a moderately wide black subterminal band and five or six chevron-shaped dark bands that get progressively wider on the basal region. Black bands on the dorsal surface of the central rectrices do not align and form an irregular, chevron-shaped pattern.** On a few birds, the dorsal black-banded pattern is indistinct. Whitish undertail has a moderately wide dark subterminal band with up to seven narrow, equal-width, neatly formed black bands.

**JUVENILE (WORN PLUMAGE):** HEAD and BODY.—From spring to early summer when 1 year old and beginning the first prebasic molt into adult plumage, body feathers have faded from black to blackish brown. If fresh blackish plumage had any white speckling, it is still retained. *Note:* Not depicted.

**ABNORMAL PLUMAGES:** None documented.

**HABITAT: Summer.**—Arid, hot-climate, semi-open remote regions. Topography varies from flat to rugged terrain. Breeding pairs typically are found below 7,000 ft. (2100 m) in montane elevations of Arizona and New Mexico. Foraging may occur up to 8,500 ft. (2,600 m). Canyons, dry washes, rivers, and creeks that have ample seasonal moisture to support tall mature cottonwoods, sycamores, pines (mainly Ponderosa), Emory Oak, Velvet Ash, and palo verde species comprise primary breeding areas. Nesting areas may be in regions with isolated trees, small clusters of trees, gallery forests along riparian corridors, or densely forested regions. Foraging habitat may be identical to nesting areas, but this species commonly hunts over sparsely vegetated, often open, treeless expanses. Less frequently, may forage in rural locales and occasionally edges of suburban areas. In Mexico, found in pine-oak woodlands up to 6,900 ft. (2,100 m).

**Winter.**—Semi-open areas with flat or hilly terrain may be inhabited in California and Texas. In Mexico, where most of the U.S. population winters, found in natural areas of semi-open and wooded tropical dry and tropical wet low montane areas, foothills, and coastal areas. Villages, towns, and grazing and moderate agricultural lands are also regularly inhabited. Climate is warm or hot and varies from arid to humid in all wintering areas.

**Migration.**—Similar to winter habitat.

**HABITS:** Moderately tame to tame raptor. Generally has minimal contact with humans in the nesting season, but wintering birds may have regular contact with humans. Nesting birds are typically very aggressive towards intruders entering the nesting territories. A highly aerial species and rarely perches except at nest sites and night roosts. Exposed branches are used for perches. Rarely perches along highways on utility poles or trees. Mainly a solitary species. In the U.S. and Mexico, however, Zone-tailed Hawks coincidentally accompany soaring Common Black, Gray, Red-tailed, and Short-tailed hawks and Turkey and Black vultures. In some areas, they may roost with Turkey and Black vultures.

**FEEDING:** Aerial hunter. Considerable distances are often traveled from nesting and roosting areas to forage. Hunting is done up to altitudes of 300 ft. (100 m). Large lizards, birds up to the size of quail, and small rodents are major prey items. Zone-tailed Hawks may specialize in particular prey types in various regions. Amphibians and large terrestrial insects are occasionally preyed on. Most prey are captured on the ground, but avian prey may be taken while perched on outer branches of bushes and trees. Prey are captured by high-speed, angled dives. Foraging may be done in a small area for several minutes, often making several dives in pursuit of prey. *Note:* Based on observations of 37 Zone-tailed Hawks (36 adults, 1 juvenile) in Feb. 2002 in Jalisco and Nayarit, Mexico, and one actively hunting bird in Apr. 1998 in s. New Mexico, Zone-tailed Hawks appear to be mainly solitary hunters (BKW pers. obs.).

**FLIGHT:** A highly aerial species rivaled only by Short-tailed Hawk and Swallow-tailed Kite. Much of the day is spent airborne. Since foraging techniques are very aerial, all flight mannerisms involve energy-saving, aerodynamically efficient gliding and soaring modes. Energy-expending powered flight is rarely used. Hawks may glide and soar for more than an hour without resorting to powered, flapping flight. Powered flight is used sparingly to gain altitude quickly, for low-altitude stabilization, during interactions with other raptors, or when chasing nest-site intruders. Wings are held in a high dihedral when soaring and in a modified dihedral when gliding (flight modes most suited for low-altitude flight). The unstable side-to-side rocking mannerism is caused by a bird being buffeted by unstable air currents. Maintaining sufficient atmospheric lift at low altitudes is best accomplished by dihedral wing attitude. Flight is steady in calm winds. Leisurely soaring flights and migration movements may occur at high altitudes.

Short- and moderate-length angled dives are made when pursuing prey. In moderate-altitude hunting, repeated, looping circles, either gliding or soaring, may be initiated in a general area to survey for prey. If prey is spotted, a stoop is made. Several diving attempts may be made in a small area over a period of several minutes. Or when prey is detected, a wide gliding loop may be made prior to stooping in order to gain aerial stability of diving into the wind. Low-altitude hunting flights consists of gliding low over undulating terrain in order to obtain ample lift without having to expend energy flapping. To gain altitude, Zone-tailed Hawks soar. They do not hover or kite and never "dip" their wings as vultures do.

**Hypothesis of Turkey Vulture mimicry:** It is uncertain if Zone-tailed Hawks mimic harmless Turkey Vultures in color, shape, and flight to deceive prey. Although Zone-tailed Hawks superficially resemble Turkey

Vultures in appearance and flight, other raptors can also look very Turkey Vulture-like, holding their wings in a high dihedral, rocking from side to side in flight, and having dark bodies. Willis (1963) proposed that Zone-tailed Hawks are aggressive mimics of Turkey Vultures and use their color, shape, and flight to gain a hunting advantage. Mueller (1972) suggested that the flight mannerism is due to using energy-efficient aerodynamics rather than mimicry. Snyder and Snyder (1991) and H. Snyder (*in* Palmer 1988) suggest that mimicry occurs at times in s. Arizona. However, if mimicry is practiced, it is an uncommon and localized phenomenon.

*Color.* Adult and juvenile plumages of Zone-tailed Hawks are superficially alike with a similar blackish body, but wing and tail patterns are much different. Zone-tailed Hawk has similar plumages for adults and juveniles. Willis (1963) suggested that this similarity parallels the similar plumages of adult and juvenile Turkey Vultures in order to deceive prey. Juvenile Zone-tailed Hawks, however, have considerably paler undersides on the remiges and rectrices than do adult Zone-tailed Hawks or Turkey Vultures. Three western raptors with multiple color morphs—Swainson's, Red-tailed, and Ferruginous hawks—also have quite similar (and dark colored) adult and juvenile plumages. Dark morph Swainson's Hawks, in particular, can appear quite vulturelike. Harris's Hawk, also of the arid southwest, has similarly colored adult and juvenile plumages, but it is not a mimic. Golden Eagle, which may have virtually identical adult and immature plumages, is also extremely Turkey Vulture-like in color, size, shape, and flight; however, it is not considered a mimic.

*Size and shape.* The size and shape of the two species are somewhat similar yet also markedly different. Turkey Vultures average 6 in. (15 cm) longer in length and 16 in. (41 cm) longer in wingspan than Zone-tailed Hawks. Wing shape is also different: Zone-tailed Hawks have five "fingers" on the outer primaries and Turkey Vultures have six. This difference creates a more rounded wingtip on vultures. Tail shape is also different: square-shaped on Zone-tailed Hawk and wedge-shaped on Turkey Vulture.

*Flight and behavior.* Zone-tailed Hawks and Turkey Vultures are both extremely aerial species and depend on energy-conserving flight modes to endure extended periods of flight (*see* Flight). Zone-tailed Hawks are even more aerial than Turkey Vultures. Zone-tailed, however, do not fly like Turkey Vultures only when hunting; the hawks use the similar flight mannerism at all times. It is a flight mode of aerodynamic efficiency. Not all flight mannerisms of Zone-tailed Hawks are identical to Turkey Vultures. Vultures regularly "dip" their wings for aerial stability; Zone-tailed Hawks never do this (BKW pers. obs.).

Zone-tailed Hawks are regularly seen with Turkey Vultures in the U.S. and Mexico, mainly near roosting areas and when gaining altitude in thermal soaring flight. Soaring raptors regularly join other raptor species, and other bird species for that matter, in rising thermals of warm air in order to use the energy-saving lift of air masses. Especially in Mexico, Zone-tailed Hawks may also join Black Vultures and other birds of prey when soaring in thermals. Other raptors and birds regularly accompany Turkey Vultures in high-altitude flights. Migrating Swainson's and Broad-winged hawks commonly mix with migrating Turkey Vultures.

*Feeding.* Some observers have described a few cases in which Zone-tailed Hawks were seemingly using harmless, carrion-eating Turkey Vultures as "decoys" in order to approach potential prey at a closer range (Palmer 1988; Snyder and Synder 1991). Yet none of the 37 Zone-tailed Hawks observed in Jalisco and Nayarit, Mexico, were seen foraging with vultures (BKW pers. obs.). Any association with either Black or Turkey vultures appeared to be coincidental. If a Zone-tailed Hawk was with a group of soaring vultures or raptors, it always broke away from the group and hunted in a solo manner once it gained enough altitude to glide away. The hawks also often hunted at altitudes that would not appear to be advantageous for mimicry.

*Other.* A pair of American Avocets were observed in 2001 in Albany Co., Wyo., giving the identical alarm call and posture to a leisurely passing Turkey Vulture as they did to a similar-distance, ground-squirrel-hunting Swainson's

Hawk. The hawk showed no interest in the avocets and was interested only in its typical abundant prey, Richardson's Ground Squirrels (BKW pers. obs.). Swainson's Hawks do not prey on adult American Avocets (C. S. Houston pers. comm.). This incident illustrates, at least in one case, that an avian species reacted in the same way to a harmless Swainson's Hawk and Turkey Vulture as it normally would to a true predator.

Most smaller creatures react to Turkey Vultures as they do to any large, raptorlike bird: they show territorial aggression or fear. In nature, size is dominant and small creatures typically harass or are afraid of larger creatures. Flycatchers and mockingbirds harass Turkey Vultures as they would any bird of prey (BKW pers. obs.; O. Carmona pers. comm.). Even small birds such as hummingbirds and swallows, which would never be prey for Zone-tailed Hawks, harass the hawks. Mississippi Kites, which probably have never suffered depredation by Turkey Vultures, were observed on numerous occasions harassing vultures flying high above kite breeding areas in Prowers Co., Colo. (BKW pers. obs.).

Passerines in some regions of Arizona acclimate to Turkey Vultures' presence and neither harass nor show alarm (T. Corman pers. comm.).

In s. Texas, O. Carmona (pers. comm.) saw Scissor-tailed Flycatchers and Cave Swallows single out a Zone-tailed Hawk among a group of Turkey Vultures and harass the hawk until it left the area. He also saw a pair of Zone-tailed Hawks evoke pandemonium among passerines. At San Blas in Nayarit, Mexico, a pair of nesting Gray Hawks singled out an adult Zone-tailed Hawk among a group of Black and Turkey vultures leisurely flying along the updrafts of a cliff face and forced the Zonetail out of their territory (BKW pers. obs.).

Zone-tailed Hawk is absent as a breeding species from much of Central America, an area where Turkey Vultures are very common year-round. Vultures are also common throughout all of temperate North America, and expanding rapidly in a variety of habitats. Recent range expansion of Zone-tailed Hawks is minimal, and all expansion occurs within typical rugged habitat, and only slightly north of normal range areas. Zone-tailed Hawks are restricted to areas with a hot climate, arid breeding habitat, and abundant prey.

**VOICE:** Vocalizes only when agitated in nest territories. A drawn-out, high-pitched, somewhat nasal *keeyah* is the most typical call. Females emit a whining note when courting.

**STATUS AND DISTRIBUTION: Summer. U.S.—** Overall, *very uncommon* and locally distributed. Northward range expansion is occurring in California and Arizona. The estimated population is 200–300 nesting pairs. Previous status estimates appear to be low.

*Arizona*: Very uncommon to uncommon. Bulk of the U.S. population resides in this state. The Arizona Breeding Bird Atlas data had 76 atlas blocks with probable or confirmed breeding and 96 blocks with possible breeding. Birds seen in the possible blocks could be breeding adults, nonbreeding adults, or nonbreeding 1-year-olds. Some probable and confirmed blocks had at least two nesting pairs.

The atlas survey covered one-sixth of an atlas block (a block was 10 sq. miles [25.9 sq. km]). Thus, there may have been numerous pairs not accounted for in potential habitat in unsurveyed regions, particularly in rugged, often difficult to access terrain that this species regularly inhabits. Conservative estimates, based on a minimum of 76 confirmed or probable pairs and on only half of the 96 "possible" birds (assuming many were single nonbreeders), indicate there may be a minimum of 125 pairs nesting in the state. More realistically, there may be 150–200 pairs when taking into account the multiple pairs in surveyed block regions and numerous atlas blocks and regions that were not surveyed. *Note:* Common Black-Hawk has an estimated U.S. population of 220–250 pairs, most of which are in Arizona. Since Zone-tailed Hawk has considerably broader habitat and prey preferences, it seems likely that it would be as abundant as, or more abundant than, Common Black-Hawk in Arizona.

*Texas:* Very uncommon. State listed as Threatened. Perhaps 50 pairs nest in the Trans-Pecos and Edwards Plateau regions. Breeding occurs in the following counties: w. Bandera, Brewster (Big Bend N. P.), Edwards, Jeff Davis (Davis Mts.), w. Kerr, n. Kinney, nw. Medina, s.

Pecos, Presidio (Sierra Vieja and Chianti Mts. and eastward), Real, s. Sutton, Terrell, nw. Uvalde, and Val Verde. Lampasas Co. may have nesting pairs.

*New Mexico:* Very uncommon. Possibly 25–50 pairs. Most reside in the southern half of the state in montane regions. Breeding occurs in the following counties: Catron (Apache and Cibola N.F.), Eddy (Guadalupe Mts.), Grant (along the Gila River, Gila N.F., and Burro Mts.), Hidalgo (Coronado N.F., Animas Mts., and Alamo Hueco Mts.), Lincoln (Capitan Mts.), Los Alamos (Jemez Mts.), Luna (Cedar Mts.), Harding (one pair along the Canadian River), Otero (Sacramento and Guadalupe Mts.), w. Sierra (Cibola N.F.), and w. Socorro (Cibola N.F.).

*California:* Rare. Resident in the mountains of San Diego Co. with at least one nesting pair and possibly more. First discovered in s. California in 1962 (specimen from San Diego Co.). Irregular sightings in the early to mid-1900s. A fairly regular species since the 1970s in San Diego Co. First documented nesting was from 1979 to 1981 in the Santa Rosa Mts. in San Diego Co. Numerous sightings have occurred in all months in San Diego and Orange Cos. since 1990 (purple dashed line on range map) and a few sightings in Los Angeles Co.

**Summer. South of U.S.:** Sparsely distributed and uncommon to locally fairly common breeder in Mexico: Baja California, cen. Sonora to Nuevo León, and south to San Luis Potosí and Durango. Absent as a breeding species in much of Central America. Uncommon in South America from Colombia and Venezuela south to Brazil.

**Winter.**—Zone-tailed Hawks remain solitary and are distributed in very low density. Virtually all the U.S. population winters south of breeding range. *Arizona:* Rare; regular near Tucson. *Texas:* A few irregularly winter in counties east of the breeding range: Bastrop (Bastrop S.P. first documented in 1979), Bell, Colorado (Attwater Prairie Chicken NWR), Tom Green (rarely), and Victoria. Virtually all sightings are along the Colorado River. Regular winter visitor in Starr and Hidalgo Cos. along the lower Rio Grande. An adult observed in mid-Feb. and late Oct. 2002 near Huntsville, Walker Co., is the most eastern record for Texas. *California:* Mainly found in San Diego Co. but also fairly regular in Orange Co. Single birds are regular north of their breeding range along coastal areas, especially Goleta and Santa Barbara in Santa Barbara Co. An adult was first documented in Goleta in Dec. 1993. As of the winter of 2001–2002 an adult had wintered there for nine consecutive years. As of 2002, an adult has spent every consecutive winter since 1999 in Ojai, Ventura Co. *Arizona:* Zone-tailed Hawks rarely winter in the state. *Mexico:* Primary wintering area of U.S. breeding birds is in s. Baja, s. Sonora, cen. Tamaulipas, and southward, rarely as far north as w.-cen and n.-cen. Sonora. A nestling banded in Pinal Co., Ariz., was found dead 2 years later in Michoacán, Mexico. in Mar.

**Movements.**—Short- to moderate-distance migrant. Only a few individuals are seen during migration in the U.S. A few hawkwatches see a handful of birds. Migrants are solitary birds; however, in Veracruz, Mexico, Zone-tailed Hawks are sometimes seen among flocks of Turkey Vultures and Swainson's and Broad-winged hawks.

*Fall migration:* Late Aug. to early Nov. Both ages probably move simultaneously. *Arizona*: Seen in late Sep. at two hawkwatches at Grand Canyon N. P. *Texas:* A few appear to move southeast from breeding areas in the Edwards Plateau to coastal areas, then southward. Adults have been seen at Hazel Bazemore Co. Pk. near Corpus Christi in late Aug. but are mainly seen from early Oct. to early Nov. *California:* There appears to be a northward movement in late summer and fall for some birds breeding in San Diego Co. and possibly from Baja California, Mexico. Individuals wintering near Goleta and Ojai, Santa Barbara and Ventura Cos., respectively, have appeared as early as late Aug. and early Sep. (and remained until spring). There is a coastal Orange Co. record in mid-Aug. *New Mexico:* A few individuals pass the Manzano Mts. hawkwatch south of Albuquerque during mid- to late Sep. *Mexico:* Migrants pass the hawkwatch sites near Veracruz from early Sep. to mid-Nov. and peak from late Sep. to early Oct. In the fall of 2002, 138 were tallied, which has been an average number for the last few years.

*Spring migration:* Adults migrate from late Feb. to early Apr. Most Arizona and probably most s. New Mexico birds arrive on breeding territories by mid-Mar. Migration may extend as late as early May in n. New Mexico; a few adult females have been seen at the hawkwatch in late May in the Sandia Mts., Albuquerque. It is unknown if these late-arriving females are breeders or nonbreeders. Adult males precede females in all areas. Returning juveniles arrive in the U.S. in May and Jun.

**Extralimital movements. U.S.**—*California:* Sightings beyond typically inhabited areas in southern counties: California City, Kern Co., in late May 1994; Weldon, Kern Co., in early Oct. 1994; Furnace Creek in Death Valley N. P., Inyo Co., in late May 1997; San Luis Obispo, San Luis Obispo Co., in mid-Oct. 1997; a juvenile in El Centro, Imperial Co., from early Dec. to late Feb. 1997; adult(s) in San Bernardino N. F., San Bernardino Co., in summer 1999. Single adults (possibly the same bird) were reported in 2000 in Bolinas, Marin Co., on May 7 and in Andrew Molera S.P., Monterey Co., on May 10. The Marin Co. sighting is the most northern sighting for the state. (The California Birds Records Committee has yet to review these last two sightings.) *Nevada:* Occurs irregularly in spring and summer in the southern part of the state. In 2000, one spent from spring through much of the summer at Pahranagat NWR, Lincoln Co. *Utah:* Three summer records as of 1999 in southern counties. *Colorado:* (1) Mid-Jul. 1999 in Mesa Co., (2) late Apr. in Pueblo, and (3) adult female in early Aug. 2002 in Boulder Co. (well documented). *Nova Scotia:* One bizarre (photographed) record from late Sep. to early Oct. 1976. *Louisiana:* One record from St. Bernard Parish in late Dec. 1984. The bird was photographed but later captured and found to have embedded shotgun pellets and died a few weeks later. *Florida:* One (also bizarre) record of an adult on Boca Grande Key in mid-Dec. 2000 (documented on video among a group of Turkey Vultures). (Most raptors on the Keys associate with the thousands of migrant and wintering Turkey Vultures.) If accepted, this would be Florida's first state record.

**NESTING:** Late Mar. through Aug.

**Courtship (flight).**—*High-circling* and *sky-dancing* (*see* chapter 5).

Except for palo verde species, nest trees are generally very tall. Trees are solitary growths, in groves of riparian gallery forests, or in forested montane areas. Cliff ledges are often used in Texas. The same territories and nests may be used for many years, and there are often several old nests in a territory. Size of the nest varies with use but averages 24 in. (61 cm) in diameter. New nests can be shallow and old nests quite deep. Nests are often rather flimsy structures composed of thin sticks and decorated with greenery. (Nests are similar to Common Black-Hawk's but have smaller sticks than most Red-tailed Hawk and Chihuahuan Raven nests.) Nest heights range from 40 to 100 ft. (12–30 m), but most are 50–85 ft. (15–26 m). Regardless of height, nests are placed in protected, shaded canopies. Most are in upper primary or secondary forks, but they are sometimes in the lower mid-section in a major fork. In palo verde trees, nests are often placed atop mistletoe. Two eggs are laid and incubated mainly by female for 35 days. Youngsters fledge in 42–49 days. Fledglings depend on parents for a few weeks after fledging and probably remain in natal territories until fall migration. Zone-tailed Hawks often nests in close proximity to Common Black-Hawks without territorial disputes.

**CONSERVATION:** No measures taken. Preservation of nesting areas in Arizonia, New Mexico, and Texas is necessary. Many nests are located in popular recreational canyons.

**Mortality.**—Illegal shooting occurs in the U.S. and south of the U.S. Human disturbance and habitat alteration in recreational canyon areas causes some nest failures in the U.S.

**SIMILAR SPECIES:** COMPARED TO ADULT.—**(1) Turkey Vulture.**—FLIGHT.—Similar, but regularly "dips" wings. Small head: red on adults and gray on juveniles. Unmarked, uniformly gray undersides of wings and tail. Six "fingers" on outer primaries. Wedge-shaped tail. **(2) Common Black-Hawk, adults.**—PERCHED.—Black forehead. Pale yellowish base on bill. Wingtips are somewhat shorter than tail tip; short primary projection beyond the tertials. Long tarsi. FLIGHT.—Broad wings. Undersides of primaries are solid black; white area on base of outer two primaries (p9 and 10). Long tarsi and feet extend past the black undertail coverts into the broad, white tail band. Voice is high-pitched, metallic, laughing-like call. (3) **Swainson's Hawk, dark morph adults.**—PERCHED.—Perches along highways. Voice is similar. FLIGHT.—Similar size and flight. Often rocks back and forth when soaring and gliding. Pale undertail coverts; pale undertail with numerous narrow dark bands. (4) **Red-tailed Hawk, dark morph *B. j. harlani* adults.**—Range overlap Oct. to early Apr. Dark gray-tailed type similar. PERCHED.—Sometimes has white forehead: use caution. Wingtips usually shorter than tail tip (but never beyond). FLIGHT.—Undersides of remiges are barred; no barring on basal area of the outer four primaries. Flaps wings regularly and hovers and kites. Uppertail is dark gray with black terminal band (mottling is not always visible). Perches along highways. Very wary. (5) **Rough-legged Hawk, dark morph adults.**—Range overlap Oct.–Mar. PERCHED.—Distinct white mask on the forehead and outer lores; inner lores are dark. Feathered tarsi. Shares similar wingtip-to-tail-tip ratio; wing tips equal to tail tip or longer. May share identical tail patterns on dorsal and ventral sides. Perches along highways. FLIGHT.—Remiges are whitish and lack barring on inner primaries. Similar wing shape; wings held in slight dihedral; steady flight. Flaps wings regularly and hovers and kites. Tail pattern can be identical, especially on the undersides: use caution. COMPARED TO JUVENILE.—(6) **Red-tailed Hawk, dark morph *B. j. calurus* juveniles.**—Range overlap all year. PERCHED.—May have similar whitish forehead. Large pale bluish area on base of bill. Greenish cere, rarely yellow. Pale yellow to pale brown irises. Dark brown plumage similar to worn-plumage Zone-tailed Hawks. Wingtips are distinctly shorter than tail tip. Tail pattern is similar but has more pronounced, neatly formed equal-width dark bands. Regularly perches along highways. FLIGHT.—Pale window or panel on the primaries, most visible on the dorsal surface. Moderate barring on secondaries; primaries usually solid dark. (7) **Red-tailed Hawk, intermediate and dark morph *B. j. harlani* juveniles.**—Range overlap Oct. to early Apr. PERCHED/FLIGHT.—Similar brownish or blackish body with possible white speckling: use caution. Underwing is barred on the outer primaries but unbarred on the basal region of primaries. Tail pattern similar and can be identical. All other data like #6.

**OTHER NAMES:** Zonetail. *Spanish*: Aguililla Aura. *French*: Unknown.

---

REFERENCES: Alcorn 1988; Arizona Breeding Bird Atlas 1993–1999; Bent 1961; Call 1978; del Hoyo et al. 1994; Howell and Webb 1995; Johnsgard 1990; Kellogg 2000; Ligon 1961; Mueller 1972; Oberholser 1974; Palmer 1988; Rottenborn and Morlan 2000; Russell and Monson 1998; Snyder 1998b; Snyder and Snyder 1991; Tufts 1986; Willis 1963.

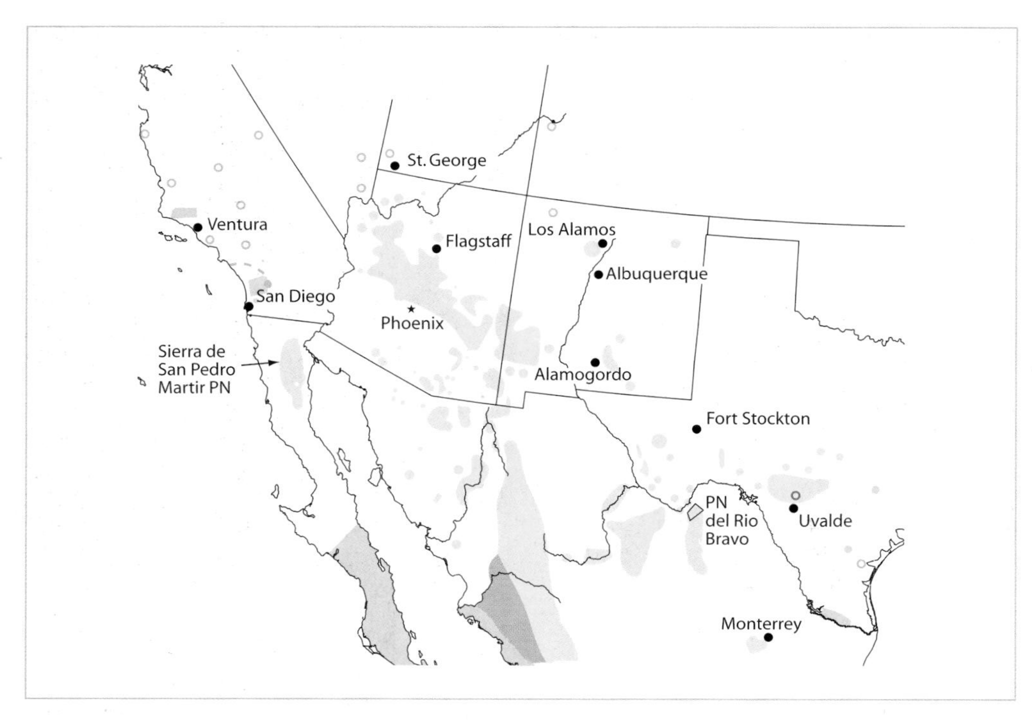


**ZONE-TAILED HAWK,** *Buteo albonotatus:* Very uncommon. Very local. 200–300 pairs in U.S.

**Plate 342. Zone-tailed Hawk, adult male** [Apr.] ▪ Black bill with small blue area on base of mandibles; yellow cere. White forehead and lores. ▪ Grayish black plumage. ▪ Long wings extend past tail tip. White barring on underside of primaries. ▪ Single broad white band on closed tail.

**Plate 343. Zone-tailed Hawk, adult** [Feb.] ▪ Grayish black plumage. ▪ Long, moderately wide wings with parallel front and rear edges. ▪ Two gray inner tail bands: 1 wide, 1 narrow. Bands white on inner feather web. Both sexes may have 2 gray dorsal bands. ▪ *Note:* Nayarit, Mexico.

**Plate 344. Zone-tailed Hawk, adult male** [Apr.] ▪ Grayish black plumage. ▪ Gray remiges heavily barred, including all primaries; wide black band on rear edge of wing. ▪ Single broad white tail band on closed tail (on both sexes). A second white, narrow inner band may partially show. Five "fingers" on outer primaries.

**Plate 345. Zone-tailed Hawk, adult female** [Apr.] ▪ Grayish black plumage. ▪ Gray remiges heavily barred, including primaries. Parallel front and rear edges of wings. ▪ Females have 3–5 white bands on underside of tail. White bands get progressively narrower towards inner tail. ▪ *Note:* Males show 2–3 white bands when tail is fanned.

**Plate 346. Zone-tailed Hawk, juvenile** [Jul.] ▪ Bill black except for small blue spot on upper mandible; yellow cere. Dark brown iris. ▪ Some have small white specks on underparts and hind-neck. ▪ Wings equal tail tip. ▪ Wide black subterminal band with narrow inner bands. *Note:* Photographed by Ned Harris in Santa Cruz Co., Ariz.

**Plate 347. Zone-tailed Hawk, juvenile**[Jul.] ▪ Black body; may have some small white spots. ▪ White remiges finely barred throughout; narrow dark rear edge. Parallel front and rear edges of wings. ▪ Wide dark subterminal band with narrow dark inner bands. ▪ *Note:* Plumage fades to blackish brown by spring.

# RED-TAILED HAWK
## *(Buteo jamaicensis)*

**AGES:** Adult, basic I (subadult I), and juvenile. Sexes are similar for all ages, but males tend to be less heavily marked. Adult plumage is attained when 2 years old. Iris color gradually darkens with age. Birds in full adult plumage when 2 years old may still have rather pale irises (*see* Mindell 1985). Subadult plumage is attained when 1 year old and is similar to adult plumage, but birds have pale irises, retain some juvenile remiges, and regularly retain a few juvenile rectrices. Juvenile plumage is held for much of the first year. Juveniles have shorter remiges, which create narrower wings and longer tail than in older ages.

**MOLT:** Accipitridae wing and tail molt pattern (*see* chapter 4). The first prebasic molt begins when about 11 months old. Southern races may begin molting in Jan., northern races in Apr. or May. Molt first appears on the innermost primary (p1) and shortly thereafter on the central two rectrices (r1). It then extends to the back, scapulars, secondaries, underparts. The molt is incomplete, and all birds retain some juvenile remiges, upperwing coverts, and regularly a few rectrices. The outer three primaries (p8–10) and middle secondaries (s4 and 8, or also s7) are typically retained juvenile feathers. On slow-molting individuals, the outer five primaries (p6–10), several secondaries (s2–4 and s6–8), and two or three sets of rectrices (r3–5) are retained juvenile feathers. Juvenile feathering is retained until the second prebasic molt which begins the next spring or summer.

Subsequent annual prebasic molts begin later than the first prebasic molt. Females begin before males. Southern races begin molting in spring, northern races in late spring and early summer. Body and wing molt is incomplete each year, with adults having a mixture of worn and fresh feathering. Wing molt, particularly on the primaries, is serially descendant with at least two molt waves occurring at the same time. The upperwing coverts, in particular, retain a large number of the previous year's feathers. Tail molt is often but not always complete each year.

**SUBSPECIES:** Polytypic, with 13 accepted races and 1 proposed race in North and Central America. Five of these races inhabit various regions of the w. U.S. and Canada: (1) "Eastern" *B. j. borealis*, which includes the pale morph, "Krider's," which is not considered a subspecies (Clark and Wheeler 2001; Wheeler and Clark 1995); (2) "Fuertes" *B. j. fuertesi*; (3) "Western" *B. j. calurus*; (4) "Alaskan" *B. j. alascensis*, which is marginally separable from *calurus*; and (5) "Harlan's" *B. j. harlani*. A sixth subspecies inhabiting the e. U.S., *B. j. umbrinus*, is in Florida and on the larger islands of the Bahamas (Grand Bahama, Abaco, Andros).

The population of the ne. U.S. and e. and cen. Canada was proposed to be a separate subspecies, *B. j. abieticola*, by Todd (1950) and supported by Dickerman and Parkes (1987). Proposed *abieticola* individuals are darker and more heavily marked than more southern *B. j. borealis* and have some or several *B. j. calurus*-like traits. Heavily marked individuals, as described by Todd, breed in the boreal forest from Québec to the Northwest Territories. However, birds that fit this description occur infrequently and do not support a recognizable separate population (R. Dickerman pers. comm.). An adult photographed in the northeastern part of Red-tailed Hawk range in Goose Bay, Labrador, had features similar to most southern *borealis* (photo by B. Mactavish). A few adults seen in New Brunswick, Nova Scotia, and Churchill, Man., also resemble southern *borealis* (J. Harrison pers. comm.). Also, adults with traits seen mainly in northern *borealis* or intergrades (e.g., dark throat) occasionally occur in the heart of typical *borealis* range in New Jersey (Liguori 2001 and pers. comm.).

I find many of these northern birds to be similar to many *borealis* x *calurus* intergrades that breed along the eastern edge of the Rocky Mts. from New Mexico to Alberta. Similar-looking birds can regularly be found in all of *calurus* winter range. The thousands of migrant Redtails I have observed and photographed at Duluth, Minn., over the past decade seemed to be pure *borealis* or *calurus* or had a variable amount of intergrade characters. Until further evidence is provided, I do not recognize *abieticola*. Some individuals of the moist boreal forest, per Gloger's Rule, are heavily marked/northern variations of *borealis* or are *borealis* x *calurus* intergrades.

*Note on North American subspecies:* There is a considerable individual variation within each subspecies. Adjacent races intergrade, often over broad areas. *B. j. borealis*-looking birds occasionally occur within *calurus* breeding range (Liguori 2001). Intergrade features are more apparent in adults and subadults than in juveniles.

*Additional subspecies:* Following del Hoyo et al. (1994), Ferguson-Lees and Christi (2001), Palmer (1998), and Preston and Beane (1993), there are an additional seven accepted subspecies. South of the U.S., Red-tailed Hawks are resident on the Caribbean islands, in Central America, and in Mexico. Subspecies are based on minor or moderate differences in plumage, size, and geography. (1) *B. j. socorroensis* comprises of a small population on isolated Socorro Island, Mexico, in the Revillagigedo group of islands south of Baja California. Similar to *calurus* but has heavier, more powerful feet and legs. (2) *B. j. fumosus* is on the isolated Islas Tres Marias off the coast of Nayarit, Mexico. Similar to *calurus* but more rusty brown on ventral areas. (3) *B. j. hadropus*, of the highlands from Jalisco south to Oaxaca, Mexico, is similar to *fuertesi* but smaller and has rufous barring on belly, flanks, and legs. (4) *B. j. kemsiesi* inhabits the isolated highlands from Chiapas, Mexico, and s. Belize to n. Nicaragua. Similar to *hadropus* but has rufous barring only on the leg feathers. (5) *B. j. costaricensis* is in the isolated highlands of Costa Rica and n. Panama. A richly colored race. (6) *B. j. solitudinis* is found on Cuba and Isla de la Juventud. Similar to *umbrinus* but smaller. (7) Nominate *B. j. jamaicensis* is found on Hispaniola, Jamaica, Puerto Rico, and smaller Caribbean island east to St. Kitts. Similar to *umbrinus* but has a more rufous wash on the ventral areas, darker and less mottled dorsal areas, and leg feathers always barred with rufous. Smaller than *solitudinis*.

Additionally, there is one proposed race, *B. j. suttoni* (Dickerman 1994), which is similar to *calurus* but smaller and more rusty brown on the underparts.

**COLOR MORPHS:** *B. j. calurus* and *B. j. harlani* are polymorphic with a continuum of plumage trends from light to dark morph: light, light intermediate, intermediate (rufous colored in *calurus*), dark intermediate, and dark. *B. j. borealis* is polymorphic with a pale morph called "Krider's" Hawk. *B. j. hadropus* of cen. Mexico is also polymorpic with a dark morph (one specimen; Palmer 1988); however, it is not known if there is a plumage continuum.

**SIZE:** A large raptor. Males average smaller than females but overlap. Subspecies subtly vary in size, particularly in wing chord and wingspan. Wing-chord measurements listed here are averages given by Preston and Beane (1993).—*B. j. borealis* 370 mm for males and 390 mm for females; "Krider's" morph of *borealis* 379 mm for males and 412 mm for females; *B. j. calurus* 387 mm for males and 412 mm for females; *B. j. fuertesi* 393 mm for males and 431 mm for females; *B. j. harlani* 381 mm for males and 408 mm for females; *B. j. alascensis* 347 mm for males and 361 mm for females; *B. j. suttoni* 376 mm for males and 403 mm for females. Wingspan varies subtly according to wing chord: the larger the wing chord, the longer the wingspan. Basic wingspan and total length measurements mainly include *borealis* and *calurus*. Lengths for *borealis* are on the smaller end of the spectrum, for "Krider's" and *harlani* in the middle, and for *calurus* on the larger end. *Fuertesi* and *alascensis* may be somewhat larger or smaller, respectively, than the listed sizes. Length: 17–22 in. (43–56 cm); wingspan: 43–56 in. (109–142 cm).

**SPECIES TRAITS:** HEAD.—Narrow white or pale gray swath under the eyes. Distal three-fourths of bill is blackish, basal one-fourth is pale blue. Lores are white or pale gray on all races and color morphs. WINGS.—In flight, rounded wingtips with p6–8 being nearly the same length and the secondaries bowing outward.

**ADULT TRAITS:** HEAD.—Greenish or yellow cere. Irises vary from pale orangish or yellowish brown to dark brown with age. WINGS.—Trailing edge of the remiges has a wide black band.

**ADULT LIGHT MORPH TRAITS:** BODY.—White, pale tawny, or rufous underparts. WINGS.—**Black, dark brown, or rufous mark on the patagial region on the leading edge of the underside of the wings.**

**ADULTS OF "EASTERN" (*B. j. borealis*):** HEAD.—Medium or dark brown crown and nape, pale tawny or white supercilium, pale or medium-tawny or tawny-rufous auriculars, and dark brown or black malar mark. Throat is white and typically unmarked; rarely streaked or with a dark mark on the central area. The white throat typically merges with the front of the neck and breast but may have a narrow brown collar that separates it from the neck. BODY (dorsal).—Dark brown back. Forward and rear scapulars are dark brown, and each scapular has a large pale white, tawny, or pale gray patch on the mid-section. On perched birds, the patches form a pale V when viewed from the rear; on flying birds, they show as two pale patches when viewed from above. BODY (ventral).—Often pale tawny in fresh autumn plumage but may fade to white by winter. There is a rufous-brown patch on the sides of the neck and breast. Breast is unmarked. Flanks are always narrowly barred with rufous or dark brown and often have a small or large dark central-feather streak or diamond-shaped mark on each feather. The amount of markings on the mid-belly is highly variable (birds breeding on the Great Plains tend to be less heavily marked, and males of all areas are typically less heavily marked than females): (1) *Lightly marked type.*—Belly lacks dark markings, but flanks are always narrowly barred with brown or rufous-brown. *Note:* Fairly common variation throughout all of *borealis* range, including all of the East; mainly on males. Beware; this type can be as lightly marked on the ventral areas as a classic *fuertesi.* (2) *Moderately marked type.*—Belly has a moderate amount of dark streaking and forms a fairly distinct belly band. *Note:* Common variation. (3) *Heavily marked type.*—Belly has a large amount of dark streaking and forms a distinct belly band that can appear as a nearly solid black band. *Note:* Fairly common to common variation, particularly on females. The lower belly is pale tawny or white and unmarked. Leg feathers on all types are pale tawny or white and unmarked. Undertail coverts are immaculate pale tawny or white. WINGS (dorsal).—Dark brown upperwing coverts with pale tawny spotting on median and first row of lesser upperwing coverts. WINGS (ventral).—**Pale tawny or white underwing coverts with a dark brown patagial mark.** Axillaries are unmarked or have small dark spots on the tip of each feather. Underside of the secondaries and inner four or five primaries are narrowly barred; outer primaries are barred or solid black. *Note:* On perched birds, the wingtips fall somewhat short of the tail (on or just above the black subterminal band). TAIL.—**Dorsal tail is medium or dark rufous but fades to pale or medium rufous or rufous-tawny by spring. There is a narrow or moderately wide black subterminal band and a white terminal band.** Occasionally, the black subterminal band is absent. The uppertail coverts are white. Undertail is pinkish or pale gray. In translucent lighting, it shows as rufous.

**ADULTS OF "KRIDER'S" (pale morph of *B. j. borealis*):** HEAD.—In the purest form, the head is all white except for some dark streaking on the nape. Many birds have a small brown patch on the crown and a partial dark malar mark (these may be *borealis/calurus* intergrades). Throat is always white and unmarked and never has a dark collar. BODY (dorsal).—Dark brown back and forward and rear scapulars are edged with tawny-rufous. A large white mottled area covers the middle two-thirds of the scapulars. When perched, the white scapular areas form a large white V. When seen from above, two large white patches show on the scapulars. BODY (ventral).—Sides of the neck and breast have a pale rufous patch. Breast, flanks, belly, lower belly, leg feathers, and undertail coverts are pure white and unmarked. *Note:* Pure "Krider's" lack dark markings on the belly or flanks. WINGS (dorsal).—Dark brown upperwing coverts are extensively mottled with white on the median and distal lesser upperwing coverts. Basal lesser upperwing coverts are often edged with tawny-rufous and may appear to from a rufous shoulder patch. Adults occasionally exhibit a large pale brown panel on the inner primaries that is similar to the panel of juveniles. WINGS (ventral).—**White underwing coverts are unmarked except for a small or moderate-sized brown or rufous patagial mark.** The secondaries and inner primaries are narrowly barred; the "fingers" on the outer primaries are broadly barred. *Note:* On perched birds, the wingtips fall somewhat short of the tail tip (on or just above the thin black subterminal band). TAIL.—**Distal third is pale rufous or pale tawny-rufous, inner**

**two-thirds white. A narrow brown subterminal band is always present.** Uppertail coverts are white.

**ADULTS OF "FUERTES" (*B. j. fuertesi*):** HEAD.—Dark brown crown and nape, medium-tawny or tawny-rufous supercilium, medium or dark tawny-rufous auriculars, and dark brown or black malar mark. Throat is typically white but may be white with dark streaking or rarely all dark. Rarely has a dark collar. *Note:* Head is generally dark (but with a white throat) as on *calurus* but is sometimes paler as on *borealis*. BODY (dorsal).—Dark brown back. Scapulars are dark brown with a moderate-sized or small white or tawny patch on the central region. On perched birds, the pale patch may form a V and, when seen from above, shows as two pale patches. BODY (ventral).—White underparts. Sides of the neck and breast have a rufous-brown patch. Flanks usually have narrow streaking and rufous barring or are unmarked. Belly is typically lightly marked with short streaks or spots; occasionally unmarked. *Note:* Belly and flanks are identically marked as many lightly marked adult *borealis*, including birds (mainly males) from all parts of the East. Leg feathers and undertail coverts are immaculate white. WINGS (dorsal).—Upperwing coverts are uniformly dark brown and lack the pale markings on the median and lesser coverts seen on most light morph Redtails. WINGS (ventral).—**White or pale tawny with a large dark brown patagial mark.** The secondaries and inner four or five primaries are narrowly barred. The outer primaries are solid black or partially barred. *Note:* On perched birds, the wingtips reach or are barely shorter than the tail tip. TAIL.—**Dorsal surface is medium-rufous in fresh plumage but fades to pale rufous or pale rufous-tawny (often a paler rufous than in other races). The black subterminal band is narrow or absent.** Lacks inner tail bands. Undertail is pinkish or pale gray; in translucent light it becomes rufous. Uppertail coverts are pale rufous.

**ADULTS OF LIGHT MORPH "WESTERN" (*B. j. calurus*):** HEAD.—Dark brown crown and nape, medium-tawny or tawny-rufous supercilium, medium-tawny or dark tawny-rufous auriculars, and a dark brown or black malar mark. Throat is typically dark brown or black. However, it may be white and streaked with brown or black and have a dark collar or a large dark area on the central part of the throat with a white strip on the outer edges. Throat is sometimes white and unmarked except for a dark collar. BODY (dorsal).—Dark brown back. Scapulars are dark brown on the forward and rear areas and have a moderate-sized or small tawny or grayish patch on the central area. On perched birds, the two pale patches may form a V when viewed from the rear; when seen from above, two pale patches show. BODY (ventral).—(1) Rufous type, which has been considered type specimen for this subspecies, has a uniform pale rufous wash on the underparts, including the leg feathers and undertail coverts. (2) White type has whitish underparts, including leg feathers and undertail coverts. The sides of the neck and breast have a rufous brown patch. Breast is sometimes narrowly streaked with brown or rufous. Flanks are always barred with dark rufous or brown. Belly is variably marked. Variations of belly and flank markings: (1) *Lightly marked type.*—Small amount of dark brown streaks or diamond-shaped marks that form an indistinct belly band. *Note:* A fairly common variation. (2) *Moderately marked type.*—Moderate amount of dark brown streaks or diamond-shaped marks, often with narrow rufous barring that forms a fairly distinct belly band **(no other subspecies exhibits barring on the belly).** *Note:* A common variation. (3) *Heavily marked type.*—Extensive dark markings that may form a virtually solid black band across the belly and flanks. *Note:* A common variation. The lower belly is pale rufous and often barred with dark rufous. Leg feathers on all *calurus* have moderate or extensive rufous or dark brown barring or are solid rufous and darker than the rest of the underparts. The undertail coverts are often barred with rufous or dark brown. WINGS (dorsal).—Dark brown upperwing coverts have a small amount of tawny spotting on the median and first row of lesser upperwing coverts. WINGS (ventral).—**Pale rufous or white, depending on color type, with large dark brown patagial mark.** Axillaries are barred with dark brown. The secondaries and inner four or five primaries are covered with narrow or moderately wide barring. The outer primaries are solid black or barred. *Note:* On perched birds, the wingtips reach or are barely shorter than the tail tip. TAIL.—Dorsal tail is medium or dark rufous in fresh plumage but fades to pale or medium rufous or rufous-tawny by spring. Tail variations: (1) *Unbanded type.*—Lacks a black subterminal band. *Note:* A very uncommon variation. (2) *Single-banded type.*—Has only a black subterminal band that varies from being narrow to wide (up to 1 in.[25 mm]). *Note:* A fairly common type. (3) *Partially banded type.*—Moderate-width or wide black subterminal band and a few incomplete narrow black inner bands. Partial banding is typically on the central and outer rectrices. *Note:* A common type. (4) *Banded type.*—Moderate-width or wide black subterminal band with narrow, complete black bands on all of the inner tail. *Note:* A common type. Variations 3 and 4 account for 50% of tail variations. The uppertail coverts are typically rufous, occasionally partially white, and often barred with black. The undertail is pinkish or pale gray but rufous when seen in translucent light. Tail banding does not show well on the undertail unless seen in translucent light.

**ADULTS OF "SUTTON'S" (*B. J. suttoni*):** Based on data from Dickerman 1994. HEAD.—Dark brown with a minimal amount of tawny-rufous on the supercilium and auricular regions (darker than *calurus* and *fuertesi*). Throat is white with dark streaking or occasionally is dark brown. BODY (dorsal).—Dark brown with minimal pale mottling on the scapulars. BODY (ventral).—Rufous washed and has fine barring on the flanks and a minimum or moderate amount of dark spotting or streaking on the belly. Leg feathers are finely barred. WINGS (dorsal and ventral).—Similar to *calurus*, but possibly less pale mottling on upperwing coverts and similar to *fuertesi.* TAIL.—Rufous dorsal surface with only a narrow black subterminal band. Uppertail coverts are rufous and unbarred. *Note:* Not depicted.

**ADULTS OF "ALASKAN" (*B. j. alascensis*):** Data based on adult breeding-season specimens ($n$ = 5) from Graham Island of the Queen Charlotte Islands, B.C. (AMNH and RBCM). HEAD.—Blackish brown crown and nape, dark rufous supercilium and auriculars, and blackish malar mark. Throat is always dark brown. BODY (dorsal).—Blackish brown back and scapulars with a small mottled gray area on the mid-scapulars. BODY (ventral).—Each tawny breast feather is marked with a large arrowhead or heart-shaped dark rufous pattern. Overall appearance of the breast is somewhat mottled or nearly solid dark rufous. The flanks are covered with large black streaks, blobs, and thick bars. The pale rufous or tawny belly is heavily barred with dark rufous and may have dark streaks and blobs (belly less heavily marked on males). The lower belly is pale tawny and heavily barred with dark rufous. Pale tawny leg feathers are broadly barred with dark rufous. The undertail coverts are pale tawny and may be unmarked or rufous barred. Overall appearance of the underparts is a dark rufous breast with some pale mottling, moderately heavy or heavy blackish belly band, and paler lower part of the body. WINGS (dorsal).—Uniformly blackish brown upperwing coverts. WINGS (ventral).—**Richly tawny and rufous with a large, distinct black patagial mark.** *Note:* The wingtip-to-tail-tip ratio of perched birds is not known; presumably similar to *calurus.* TAIL.—**Dark rufous on the dorsal surface with a wide black subterminal band and may either be partially banded or fully banded with narrow black inner bands.** *Note:* Some may appear similar to intermediate (rufous) morph *calurus*; however, the rufous areas on *calurus* are typically more uniform. Light intermediate morph *calurus* may be very similar because some have slightly darker rufous breasts, but dorsal areas are not be as uniformly dark or ventral areas as barred as on *alascensis. Additional Note:* Not depicted.

**ADULTS OF LIGHT MORPH "HARLAN'S" (*B. j. harlani*):** HEAD.—Brownish black crown and nape, stark white supercilium and auriculars, and black malar mark. Throat is sometimes white and streaked with a dark collar. BODY (dorsal).—Brownish black (not a warm brown as on most races). Small or moderate-sized white patches on the mid-scapulars. Dark markings in the white scapular areas are typically more streaked and not as barred or diamond-shaped as on other races. BODY (ventral).—Stark white underparts. Flanks always marked with blackish streaks or blobs but sometimes narrowly barred. Belly is moderately or heavily streaked or blobbed; very rarely is the belly unmarked (only one [specimen] has been seen that lacked belly markings). Leg feathers are immaculate white but may be partially barred with brown on more heavily marked birds. Undertail coverts are immaculate white. WINGS (dorsal).—Uniformly blackish brown upperwing coverts. WINGS (ventral).—**White underwing coverts have a distinct black patagial mark.** Patterns of the underside of the white or pale gray remiges are highly variable and apply to all color morphs of *harlani*: (1) *Barred type.*—All remiges are neatly or irregularly barred; outer primary "fingers" are broadly barred or solid black. (2) *Mottled type.*—All remiges are mottled and speckled with black and gray; outer primary "fingers" are usually solid black but may have some barring. (3) *Unmarked type.*—Remiges lack dark markings; outer primary "fingers" are solid black. (4) *Barred and mottled type.*—Remiges are a mixture of barring and mottling; outer primary "fingers" are solid black or broadly barred. (5) *Partially marked type.*—Remiges have a mixture of barred and/or mottled and unmarked feathers; outer primary "fingers" are generally solid black. Pattern may vary in having a few or several feathers marked. *Note:* Contrary to previously published data, the "fingers" on the outer five primaries of adult *harlani* are just as likely to be solid black as barred. *Additional Note:* On perched birds, the wingtips are typically somewhat shorter than the tail tip; however, the wingtips nearly reach the tail tip on some birds. TAIL.—There is an incredible array of dorsal tail patterns for *harlani.* When seen at close range, most *harlani* have some tinge of rufous on the dorsal surface of the distal portion of the rectrices. It is also common for very *harlani*-looking adults to have rufous on the distal half of the rectrices. If more than half of the tail is rufous, a bird is likely an intergrade with *calurus* or *alascensis.* Variations are less apparent on the ventral surface: most are white or pale gray. When seen in translucent light, the dorsal color shows on the ventral surface. Dorsal tail variations (for all color morphs): (1) *White type.*—(1a) White and lack dark markings or (1b) have some dark marbling and mottling. The black subterminal or terminal band is moderately wide or wide and irregularly marbled. *Note:* A common type (possibly slightly more common than other types). (2) *White-rufous type.*—The basal half is white and the distal half is rufous. Tail may be marbled or mottled with dark markings or unmarked. *Note:* A fairly common type. (3) *Pale gray type.*—(3a) Nearly plain pale gray or (3b) pale gray with moderate or extensive dark *marbling* and mottling. The black subterminal or terminal band is moderately wide or wide and irregularly marbled or mottled. *Note:* A common type. (4) *Pale gray-rufous type.*—The basal half is pale gray and the distal half is rufous. The tail may be marbled and mottled with black or unmarked. *Note:* A fairly common type. (5) *Medium/dark gray type.*—(5a) Nearly plain medium or dark gray or (5b) medium or dark gray and covered with black marbling and mottling. The black subterminal or terminal band is moderately wide or wide. *Note:* A common type. (6) *Medium/dark gray-rufous type.*—The distal half of the tail is rufous; however, the rufous is often mixed with the darker gray and a brownish cast is created. The tail may be marbled or mottled with black or unmarked. *Note:* A common type. (7) *White banded type.*—Rectrices are white with a fairly neat, moderately wide black subterminal band and several narrow, complete black bands on the inner tail. There is often a tinge of rufous along the edges of some black bands. *Note:* A fairly common type. (8) *Pale gray banded type.*—As in #7 but with a pale gray base color. *Note:* An uncommon type. (9) *Medium/dark gray banded type.*—As in #7 and 8 but with a medium or dark gray base color. *Note:* An uncommon type. (10) *Mottled-barred type.*—A mixture of dark marbling and barring on white and gray base, often with a rufous wash on the distal part of the tail. Dark patterns may be irregularly distributed on a few rectrices: some being marbled or mottled, some barred, or some fairly regularly patterned on most or all feathers. In many cases, the barred pattern on a feather becomes diffused and is an irregular mottled or spotted design. Uppertail coverts are white, gray, or rufous.

**ADULT LIGHT MORPH INTERGRADE TRAITS:** Below are some basic plumage traits that appear to be shared features between adjacent races. The degree of intergrading, however, is infinite. **Heavily marked/northern type *borealis:*** Head is dark as on most *calurus;* throat is white, white and streaked with a dark collar, or all dark. Breast and underparts are white, or the breast is often tawny-rufous or quite rufous and streaked. (The darker tawny or rufous breast on some individuals separates them from *calurus*, which typically have uniformly colored breasts and underparts.) Belly band, consisting of the flanks and belly, is moderately or heavily marked with black. Belly is not barred with rufous as on many *calurus*. Leg feathers vary from unmarked to heavily barred with rufous or brown (similar to *calurus*). Rufous tail often has a moderately wide or wide black subterminal band and may have partial or complete narrow black inner bands.

**Intergrade of *borealis* x *calurus*:** Faint or partial leg barring signifies acquisition of *calurus* traits. Many have a *calurus*-like streaked or dark throat. The ventral regions may remain whitish and lightly marked as on many *borealis* (even those with some leg barring); some may obtain *calurus*-like faint rufous barring on the belly. Tail often shows some partial dark inner banding, particularly on the deck and outer rectrices.

**Intergrade of "Krider's" x typical *borealis*:** Intergrade characters generally affect the head or tail. Some may have moderately dark heads as on typical *borealis* but have a whitish tail like typical "Krider's," and vice versa. As a rule, the dark crown and malar mark are present on darker headed birds. The supercilium may be a more pronounced white on birds with darker heads.

**Intergrade of "Krider's" x *calurus*:** Variations as described for *borealis* but often with partial dark bands on those with a whitish tail.

**Intergrade of *calurus* x *fuertesi*:** Similar darkish heads but often with streaked or dark throat as on *calurus*. Pale and lightly marked underparts may have a light rufous wash. Leg feathers are typically partially barred (a *calurus* trait). Tail may be partially banded (a *calurus* trait).

**Intergrade of *calurus* x *harlani*:** Head is typically similar to *harlani*: stark white and dark brown. Sides of the neck and breast are also *harlani*-like in that they lack rufous tones. Scapular are often barred and have diamond-shaped markings as on *calurus* and not streaked as on *harlani*. Tail is more than 50% rufous with white or gray mottling on the basal region next to the uppertail coverts and rufous on the distal region. In flight, the underside of the remiges may show *harlani* features of mottling or lack of markings. Dark patagial mark may have *calurus*-like rufous tones (*see* Mindell 1985).

**Intergrade of *calurus* x *alascensis*:** Pale rufous or whitish ventral body with partial arrowhead or diamond-shaped marks on the breast. Throat may be streaked white and brown. Pale scapular patches are moderate in size. Seen on breeding-season specimens from Vancouver Island, B.C.

**ADULT LIGHT INTERMEDIATE MORPH TRAITS:** Pertains to "Western" *B. j. calurus* and "Harlan's" *B. j. harlani*. These morphs are more heavily marked than their respective light morphs in each race and form links in the continuum from light to dark morph.

**ADULTS OF LIGHT INTERMEDIATE (LIGHT RUFOUS) MORPH "WESTERN" (*B. j. calurus*):** HEAD.—Dark headed as on most light morph *calurus*, including the throat. BODY (dorsal).—As on light morph *calurus* with small or moderate-sized pale tawny or gray scapular patches. BODY (ventral).—All of the underparts are washed with medium rufous (darker than light morph but paler than intermediate morph). The underparts may be uniformly medium rufous or the breast may be somewhat darker than the leg feathers, lower belly, and undertail coverts. Breast is sometimes partially streaked with dark brown. Variations of belly and flank markings: (1) *Lightly marked type.*—Small amount of dark brown streaks or diamond-shaped marks form an indistinct belly band and underparts appear virtually solid medium rufous. *Note:* A common variation. (2) *Moderately marked type.*—Moderate amount of dark brown streaking or diamond-shaped marks, often with narrow dark rufous or dark brown barring that forms a fairly distinct belly band. *Note:* A common variation. (3) *Heavily marked type.*—Extensive markings form a virtually solid blackish band. *Note:* A common variation. Leg feathers are always barred with dark rufous or brown. WINGS (dorsal).—As on light morph *calurus* with little if any pale markings on the medium upperwing coverts. WINGS (ventral).—**Underwing coverts are medium rufous with prominent, large black patagial marks.** Remiges are barred except on the outer four or five primaries; primary tips are solid black or broadly barred. *Note:* On perched birds, the wingtips reach the tip of the tail. TAIL.—Same variations as described for Adults of Light Morph "Western" except the unbanded type is rare. Uppertail coverts are medium rufous and often barred.

**ADULTS OF LIGHT INTERMEDIATE MORPH "HARLAN'S" (*B. j. harlani*):** HEAD.—As on light morph *harlani* with a broad white supercilium, white auricular region, dark malar, and white throat. Supercilium and auricular regions typically form a large white area and, except for the dark crown, make all of the head appear white. BODY (dorsal).—Brownish black with a moderate amount of white on the two mid-scapular patches. BODY (ventral).—The white breast is narrowly streaked with brownish black. The dark streaking covers less than 50% of the breast area. Flanks and belly are heavily mottled and blotched with brownish black and form a distinct belly band. The leg feathers are thickly barred with black and white. The undertail coverts are white with some dark barring and accentuate the dark belly band of the darker flanks and belly. WINGS (dorsal).—Brownish black without pale markings on the upperwing coverts. WINGS (ventral).—Remiges are marked in any of the five variations described for Adults of Light Morph "Harlan's." Underwing coverts are heavily mottled with black and white and exhibit a fairly distinct dark patagial mark. Axillaries are black with white barring. *Note:* On perched birds, the wingtips are somewhat shorter than the tail tip. TAIL.—Any of the variations described for Adults of Light Morph "Harlan's." The uppertail coverts are white, gray, or rufous.

**ADULT LIGHT INTERMEDIATE INTERGRADE TRAITS:** No data.

**ADULT INTERMEDIATE MORPH TRAITS:** Pertains to "Western" *B. j. calurus* and "Harlan's" *B. j. harlani.* These morphs races are more heavily marked than their respective light intermediate morphs in each race and form links in the continuum of plumages from light to dark morph.

**ADULTS OF INTERMEDIATE (RUFOUS) MORPH "WESTERN" (*B. j. calurus*):** HEAD.—Dark brown crown, nape, and throat. The supercilium and auricular regions are dark rufous-brown. Overall, the head appears dark rufous-brown. BODY (dorsal).—Dark brown with at most only a small tawny mottled patch on each mid-scapular region. BODY (ventral).—Breast is dark rufous but may have some dark streaking. Belly and flanks are variably marked: (1) *Lightly marked type.*—Small amount of dark brown streaks or diamond-shaped marks form an indistinct belly band, and underparts appear virtually solid dark rufous. *Note:* An uncommon variation. (2) *Moderately marked type.*—Moderate amount of dark brown streaks or diamond-shaped marks, often with narrow dark rufous or dark brown barring that forms a fairly distinct belly band. *Note:* A common variation. (3) *Heavily marked type.*—Extensive dark markings form a virtually solid blackish band. Leg feathers are always barred with dark rufous or dark brown. *Note:* A common variation. The lower belly is dark rufous and may have some darker rufous or dark brown barring. The undertail coverts are dark rufous and barred with darker rufous or dark brown. WINGS (dorsal).—Dark brown upperwing coverts lack pale markings on the median and lesser coverts. WINGS (ventral).—Dark rufous underwing coverts with a large, distinct black patagial mark. Axillaries are often barred with white. Remiges are barred except on the outer four or five primaries; primary tips are solid black or broadly barred. *Note:* On perched birds, the wingtips reach the tail tip. TAIL.—Same variations as described for Adults of Light Morph "Western" except the unbanded type is very rare. Uppertail coverts are dark rufous and often barred.

**ADULTS OF INTERMEDIATE MORPH "HARLAN'S" (*B. j. harlani*):** HEAD.—Uniformly brownish black, or the supercilium and auriculars may have white streaking or mottling on them. The throat is white or white with black streaking. The white swath under the eyes is often very apparent. Auriculars may also have a variable amount of white on them. Forehead is often white. BODY (dorsal).—Brownish black with somewhat pale brownish oblong streaks on the outer edges of the lower scapulars. BODY (ventral).—*See* types below. The undertail coverts are brownish black with white markings. WINGS (dorsal).—Mainly uniformly brownish black on all coverts. WINGS (ventral).—Brownish black coverts are speckled with white. The axillaries may be solid black or barred or speckled with white. *Note:* There is no dark patagial distinction as seen on paler *harlani* morphs. Remiges may be any of the five variations described for Adults of Light

Morph "Harlan's." *Note:* On perched birds, the wingtips typically are somewhat shorter than the tail tip. TAIL.—Same patterns as described for Adults of Light Morph "Harlan's." Uppertail coverts may be white, gray, or rufous.

**ADULT INTERMEDIATE MORPH "HARLAN'S" (STREAK-BREASTED TYPE):** BODY (ventral).—The breast has an equal amount of white and black streaking or mottling. The belly, lower belly, and flanks are abruptly darker brownish black and are either solid dark or speckled with white. Leg feathers are either solid brownish black or black or have white speckling or narrow barring. *Note:* A common type.

**ADULT INTERMEDIATE MORPH "HARLAN'S" (STREAKED TYPE):** BODY (ventral).—The breast, belly, and flanks are uniformly marked with nearly an equal amount of white and brownish black streaking (*see* Wood 1932). *Note:* An uncommon to very uncommon type. Not depicted.

**ADULT INTERMEDIATE MORPH INTERGRADE TRAITS:** Shared traits between *calurus* and *harlani.* Intergrade of *calurus* x *harlani:* (1) Dark brown head and throat as on *calurus* but with small white supercilium. Breast is dark rufous with a moderate amount of white *harlani* mottling or streaking. Rest of body and wings as on *calurus.* Tail may be any of the previously described *harlani* or *calurus* types or a mixture of both subspecies. (2) Body as on intermediate (rufous) morph *calurus,* but tail is any of the *harlani* variations. *Note:* If there is a mix of *harlani* traits on the tail, it occurs on the basal half or two-thirds, with *calurus* traits on the distal half or third.

**ADULT DARK INTERMEDIATE MORPH TRAITS:** Pertains to "Western" *B. j. calurus* and "Harlan's" *B. j. harlani.* These morphs races are more heavily marked than their respective intermediate morphs and form links in the continuum of plumages from light to dark morph.

**ADULTS OF DARK INTERMEDIATE (DARK RUFOUS) MORPH "WESTERN" (*B. j. calurus*):** HEAD.—Dark brown with a hint of dark rufous on supercilium and auricular regions. Throat is always dark brown. The forehead may have some white on it. BODY (dorsal).—Uniformly dark brown. BODY (ventral).—Dark brown with dark rufous outer edges on most feathers and slightly paler and more rufous than the belly and flanks. Belly and flanks are uniformly dark brown. Leg feathers are dark brown with narrow rufous edges on most feathers. The undertail coverts are dark brown with rufous tips or barring on most feathers. WINGS (dorsal).—Uniformly dark brown coverts. WINGS (ventral).—Underwing coverts are mainly dark brown but may have a rufous tinge; however, there is not a dark patagial distinction. Axillaries may be barred with white or gray. Remiges are barred except on the outer four or five primaries; primary tips are solid black or broadly barred. *Note:* On perched birds, the wingtips reach the tail tip. TAIL.—Same patterns as described for Adults of Light Morph "Western" except the unbanded type does not occur in this morph. Uppertail coverts are dark or may have some rufous tinge.

**ADULTS OF DARK INTERMEDIATE MORPH "HARLAN'S" (*B. j. harlani*):** HEAD.—Uniformly brownish black but may have white patch on the supercilium. The throat is white or white with black streaking. Forehead may have some white on it. BODY (dorsal).—Uniformly brownish black or with slightly paler oblong grayish areas on the outer edges of the lower scapulars. BODY (ventral).—*See* types below. WINGS (dorsal).—Mainly solid brownish black. WINGS (ventral).—Underwing coverts are solid brownish black or have a small amount of white speckling. Axillaries are solid dark or barred with white. Pattern on the remiges as described for Adults of Light Morph "Harlan's." *Note:* On perched birds, the wingtips typically are somewhat shorter than the tail tip. TAIL.—Patterns as described for Adults of Light Morph "Harlan's." Uppertail coverts are dark but may have a rufous tinge.

**ADULT DARK INTERMEDIATE MORPH "HARLAN'S" (STREAK-BREASTED TYPE):** BODY (ventral).—Brownish black with a small amount of white streaking or mottling on the breast. The white markings cover less than 50% of the breast surface. Rest of the underparts uniformly brownish black or may have a small amount of white speckling. *Note:* A common type.

**ADULT DARK INTERMEDIATE MORPH "HARLAN'S" (SPOT-BELLIED TYPE):** BODY (ventral).—Breast is solid brownish black and belly and flanks are moderately spotted or blotched with white. The leg feathers may have a small amount of white barring. WINGS (ventral).—

Underwing coverts tend to have some white speckling and white barring on the axillaries. *Note:* An uncommon type.

**ADULT DARK INTERMEDIATE MORPH INTERGRADE TRAITS:** Shared traits between *calurus* and *harlani.* Intergrade of *calurus* x *harlani:* Dark brown head and throat as on *calurus* but may have a small white supercilium. Breast is dark rufous or dark rufous with a small amount of white mottling or streaking. Rest of body and wings as on *calurus.* Tail may be any of the previously described *harlani* or *calurus* types or is mixture of both subspecies. If there is a mix of *harlani* traits, it occurs on the basal half or two-thirds, with *calurus* traits on the distal half or third.

**ADULT DARK MORPH TRAITS:** Pertains to "Western" *B. j. calurus* and "Harlan's" *B. j. harlani.* These morphs are more heavily marked than their respective dark intermediate morphs and form the final links in the continuum of plumages from light to dark morph.

**ADULTS OF DARK MORPH "WESTERN" (*B. j. calurus*):** HEAD and BODY.—Typically a uniformly warm dark brown but occasionally more blackish brown (perhaps having *harlani* genes). The undertail coverts have rufous tips on most feathers. WINGS (dorsal).—Uniformly warm dark brown. WINGS (ventral).—Dark brown underwing coverts; axillaries sometimes faintly barred with paler brown or gray. Remiges are barred except on the outer four or five primaries; primary tips are solid black or broadly barred. *Note:* On perched birds, the wingtips reach the tail tip. TAIL.—Same patterns as described for Adults of Light Morph "Western" except the unbanded type does not occur in this morph. Uppertail coverts are dark brown.

**ADULTS OF DARK MORPH "HARLAN'S" (*B. j. harlani*):** HEAD.—Uniformly brownish black, including the throat. The forehead is dark. BODY.—Uniformly brownish black or sometimes a warm dark brown. WINGS (dorsal).—Uniformly brownish black. WINGS (ventral).—The underwing coverts are solid brownish black. Axillaries are solid dark or have faint grayish barring. Remiges have the same patterns described for Adults of Light Morph "Harlan's" *Note:* On perched birds, the wingtips are typically somewhat shorter than the tail tip. TAIL.—Same patterns as described for Adults of Light Morph "Harlan's." The banded types appear to be more common in this morph than any other. Medium/dark gray type is the most common non-banded variation.

**ADULT DARK MORPH INTERGRADE TRAITS:** Shared traits between *calurus* and *harlani.* Intergrade of *calurus* x *harlani:* Dark brown or brownish black head, body, and underwing coverts. Tail may be any of the previously described *harlani* or *calurus* types or is mixture of both subspecies. If there is a mix of *harlani* traits, it occurs on the basal half or two-thirds, with *calurus* traits on the distal part.

**BASIC (SUBADULT) TRAITS (ALL SUBSPECIES AND COLOR MORPHS):** Superficially similar to adults of respective race and morph. At close and moderate distances, subadults can easily be separated from adults by head and wing features. HEAD.—Iris is pale: yellowish brown, orangish brown, or brown. However, younger adults may also have pale irises. WINGS.—Retain some faded brown juvenile upperwing coverts, but these are not separable from old, faded adult feathers that are from partial prebasic molts. In flight, retained juvenile remiges are quite visible on the dorsal and ventral sides. On the dorsal side, retained juvenile remiges are paler brown than most adult remiges. The outer three to five primaries (typically p8–10, rarely p6–10) are worn, frayed, and pale juvenile feathers. If these outer primaries are barred, the juvenile barring is narrower than on most adjacent adult primaries. On the secondaries, the retained feathers are typically s4 and 8, but also s7 and sometimes more. The juvenile feathers have a narrow grayish subterminal band versus the wide black band of adults and are shorter than adjacent adult feathers. TAIL.—Some birds retain scattered juvenile rectrices.

**JUVENILE TRAITS:** HEAD.—Iris is pale: yellow, gray, or brown; rarely medium brown. **Cere is pale bluish green or pale green.** WINGS (dorsal).—**In flight, the primaries and respective greater coverts are paler than the secondaries and their respective coverts and create a large pale panel on the outer half of the wing. All subspecies and morphs have distinct black barring on the secondaries: narrow on light morphs, fairly wide on darker morphs.** WINGS (ventral).—In flight in translucent light, a pale rectangular window is created by the pale panel of the dorsal surface of the primaries. The subterminal band on the remiges is narrow and medium gray. *Note:* On all races, the wingtips of perched birds are moderately shorter or much shorter than the tail tip. TAIL.—Two main patterns: **(1) Numerous narrow to moderately wide equal-width black bands, which are being narrower than the pale bands,** or **(2) numerous narrow or moderately wide equal-width black bands with a wider black subterminal band.** Base color of the dorsal pale bands is variable (*see* description for each subspecies/morph). Ventral color on all types is pale grayish.

**JUVENILE LIGHT MORPH TRAITS:** BODY (dorsal).—White markings on the mid-scapular region. BODY (ventral).—All have some sort of dark belly and flank markings. WINGS (ventral).—**All have a lightly to heavily marked dark area on the patagial region.** TAIL.—Uppertail coverts are white with dark marks on the central region.

**JUVENILES OF "EASTERN" (*B. j. borealis*):** HEAD.—Dark brown crown, nape, and malar mark with a broad white supercilium and brownish or tawny auricular region. White throat. On more heavily marked birds, the white throat is narrowly streaked with dark brown or has a narrow dark brown collar that separates it from the breast. BODY (dorsal).—Warm dark brown back and scapulars with a large white patch on the middle area of each scapular. The dark mark on the central part of each scapular is diamond-shaped. BODY (ventral).—White underparts are variably marked on the belly, flanks, and leg feathers. The breast is white and unmarked on older juveniles. Recently fledged birds have a tawny wash that quickly fades after fledging (this tawny color may be evident on late-fledged birds until Dec.). Belly and flanks are marked with dark brown markings: (1) *Lightly marked type.*—Dark arrowhead or diamond-shaped markings with a broad dark inner-feather bar on each flank feather and small dark streaks or diamond-shaped marks on the belly that create a poorly formed belly band. *Note:* A common variation. (2) *Moderately marked type.*—Flanks as described for #1 but belly has moderately large streaks or diamond-shaped marks that form a fairly distinct dark belly band. *Note:* A common variation. (3) *Heavily marked type.*—Flanks are more broadly marked than #1 and the belly has large dark streaks or diamond-shaped marks that form a distinct, nearly solid belly band. *Note:* A fairly common variation (typical on many individuals from the northern part of their range). Lower belly is white and unmarked. White leg feathers are unmarked, lightly spotted or barred, or fairly heavily barred. *Note*: Amount of leg markings corresponds to amount of belly markings. Undertail coverts are white and unmarked. WINGS (dorsal).—**Pale panel on the primaries and respective greater upperwing coverts are pale brown.** Prominent white markings on the median and on the first one or two rows of lesser upperwing coverts. WINGS (ventral).—**Moderately large dark brown patagial marks.** There are a few scattered dark marks on other underwing coverts; axillaries have a few dark marks on them. The "fingers" on the outer five primaries are either solid dark gray or barred. TAIL.—Both banded patterns as described under Juvenile Traits are common except dark bands are always narrow. Dorsal tail colors: (1) *Medium brown type.—Note:* A common variation. (2) *Rufous-brown type.*—A small or moderate amount of rufous or orangish rufous mixed with the brown. *Note:* A fairly common variation. (3) *Rufous type.*—Tail is totally orangish rufous or rufous and nearly rivals the color of adult tail. *Note:* Uncommon variation, but more common in *borealis* than in other subspecies.

**JUVENILES OF "KRIDER'S" (pale morph of *B. j. borealis*):** HEAD.—**In the purest form, the head is immaculate white and lacks any dark markings except for a small amount of dark streaking on the nape.** Birds that still appear quite pure may also have a white head but with a small dark area on the crown, a partial dark malar mark, and the streaked nape. BODY (dorsal).—Medium warm brown back and scapulars with a large white mottled area on the mid- and lower region of each scapular. The dark mark on the central part of each scapular is diamond-shaped. The uppertail coverts are white with a small amount of dark spotting or barring. BODY (ventral).—White underparts with small dark spots on the flanks and belly. *Note:* Juveniles always have some dark markings on the belly and flanks. Leg feathers and undertail coverts are white and unmarked. WINGS (dorsal).—**Pale panel on the primaries and respective greater upperwing coverts is pale brown or white.** Medium warm brown with large white markings on the median and the first two or more rows of lesser upperwing coverts. The greater upperwing coverts may have some white on them. WINGS (ventral).—The medium brown secondaries are distinctly barred with dark brown. The white underwing coverts are virtually white and unmarked except for a small dark strip on the tips of the primary greater coverts and the ill-defined dark streaking on the patagium. The remiges are narrowly barred, including the outer five "fingers" of the primaries. TAIL.—The distal half to two-thirds is pale brown, the basal half to third white. The brownish region is marked with numerous narrow dark brown bands; the white region either is unbanded or partially banded.

**JUVENILES OF "FUERTES" (*B. j. fuertesi*):** HEAD and BODY.—Similar to *borealis* in nearly all aspects. The belly is similar to lightly marked type *borealis*, but the small dark markings are often V-shaped rather than streaked or diamond-shaped as on *borealis*. The flanks may have less pronounced inner-feather barring than many *borealis* and are often spotted or streaked. Leg feathers may be unmarked or narrowly barred. WINGS.—Since the wings are longer than in *borealis*, the wingtips reach closer to the tail tip than they do on *borealis*. TAIL.—Dorsal surface is mainly medium brown and rarely has a rufous tinge.

**JUVENILES OF LIGHT MORPH "WESTERN" (*B. j. calurus*):** HEAD.—Dark brown crown, nape, and malar mark with a tawny supercilium and dark tawny auricular region. Throat is all dark, dark with a pale edge on each side, or white and streaked with dark; occasionally it is white. Head appears to be overall dark and darker than on juvenile *borealis*. BODY (dorsal).—Warm dark brown with small or moderate-sized white or grayish patches on the mid-scapular region. The dark marks on the central part of each scapular feather are diamond-shaped. BODY (ventral).—Breast is white on older juveniles but tawny on recently fledged birds; it is mainly unmarked except for streaking on each side. The belly and flanks are either the moderately marked or heavily marked type as described for Juveniles of Eastern. Lower belly feathers are lightly or moderately spotted or partially barred with dark brown. Leg feathers are moderately or heavily barred with dark brown. Undertail coverts are white and lightly or moderately barred. WINGS (dorsal).—**The pale panel on the primaries and respective coverts is pale brown.** Warm dark brown with a small amount of pale grayish or tawny marks on the median and first row of lesser upperwing coverts. WINGS (ventral).—**Large dark brown patagial mark.** The underwing coverts are moderately marked and the axillaries are barred. The "fingers" on the outer five primaries may be solid dark gray or barred. TAIL.—The two banded patterns described for Juvenile Traits: pattern is narrow to moderately wide. Base dorsal color as described for Juveniles of Eastern, but rufous type is very uncommon in *calurus*. Uppertail coverts are dark with some pale markings.

**JUVENILES OF "SUTTON'S" (*B. j. suttoni*):** Similar to *calurus* but may be more lightly marked on the ventral areas (Dickerman 1994). *Note:* Not depicted.

**JUVENILES OF "ALASKAN" (*B. j. alascensis*):** Data based on specimens ($n = 6$; AMNH), from Graham Island of the Queen Charlotte Islands, B.C. HEAD.—Brownish black crown, nape, and malar marks. Small pale whitish or tawny supercilium and dark tawny auriculars. Throat white or white with some dark streaks (not dark as on most *calurus*). Overall, head is quite dark. BODY (dorsal).—Brownish black (similar to *harlani* and darker than other races) with small white patches on the mid-scapular region. BODY (ventral).—The white breast is mainly unmarked except for streaking on the sides of the breast. The flanks are marked with large dark arrowhead or diamond shapes with a broad bar on the inner part of each feather. The belly is the heavily marked type with broad streaks, heart shapes, or diamond shapes. The lower belly has a moderate amount of dark spotting and partial barring. The white leg feathers are broadly barred with dark brown. The undertail coverts are white and moderately or heavily barred. WINGS (dorsal).—**The pale panel on the primaries and respective coverts is pale brown.** Little if any pale markings on the dark median and lesser upperwing coverts. WINGS (ventral).—**Large dark brown patagial mark.** Underwing coverts are heavily marked and the axillaries are broadly barred. TAIL.—Medium brown dorsal color with banded pattern as described for Juvenile Traits; dark bands are moderately wide to fairly wide. *Note:* Tail bands are wider than on most light morph *calurus.* Uppertail coverts are dark with narrow white tips and a central-feather white bar. *Note:* Not depicted.

**JUVENILES OF LIGHT MORPH "HARLAN'S" (*B. j. harlani*):** HEAD.—Crown and nape are streaked with brownish black and the dark malar mark is pronounced. The supercilium and auricular region blend as a large white unit. The throat is white and unmarked. Overall, the head appears quite white. BODY (dorsal).—Brownish black back and scapulars. The scapulars have moderate-sized white patches, and the dark mark on the central part of each feather is either a diamond shape or a streak. BODY (ventral).—White with moderately marked or heavily marked type of brownish black streaking on the belly and flanks (markings are arrowhead, diamond-shaped, or barred on most other races). Leg feathers are white and unmarked. WINGS (dorsal).—**The pale panel on the primaries and respective greater coverts is pale brown or white.** Brownish black with white markings on the median and some lesser coverts. WINGS (ventral).—Large dark patagial mark on the white underwing coverts. The "fingers" on the outer five primaries are always barred. TAIL.—Three main colors of pale banding: (1) medium brown, (2) pale brown, or (3) white on part of the central retrices and sometimes other rectrices. Some rectrices may have a rufous wash on them. *See* Mindell 1985 for additional variations. As in other races, the dark banding may be equal width, or the subterminal band wider. Dark tail bands vary from narrow to fairly wide but are usually narrower than the pale bands, and are often a wavy or chevron-shaped pattern. A dark "spike" often shows along the distal feather shaft, but light morphs are less likely to have black rectrix tips as seen on darker morphs of juvenile *harlani.* Uppertail coverts are white with dark markings. Undertail coverts are white and unmarked.

**JUVENILE LIGHT MORPH INTERGRADE TRAITS:** Intergrade features between subspecies and morphs of most juveniles are difficult to see. Intergrade characters of other adjacent races are too difficult to differentiate.

**Intergrade of typical *borealis* x "Krider's":** Subspecies features are mainly seen on the head and tail. (1) Birds may have a darker head as on typical *borealis* but a white tail as on "Krider's"; (2) the head may be quite white but with a dark crown and partial or full malar marks, and the tail brown as on typical *borealis*; or (3) crown may be white but with dark malar and white or brown tails.

**Intergrade of *calurus* x "Krider's":** As on any of the three variations described for *borealis* x "Krider's," but the leg feathers may be quite barred and tail more widely banded.

**Intergrade of *borealis* x *calurus*:** Difficult to separate from the respective races because there is so much variation. Streaked throat, heavy belly band, distinctly barred leg feathers, and partial barring on the undertail coverts.

**JUVENILE LIGHT INTERMEDIATE MORPH TRAITS:** Pertains to "Western" *B. j. calurus.* Field-observed and museum specimens of *B. j. harlani* have not been seen for this morph; however, it undoubtedly exists. This morph of *B. j. calurus* is more heavily marked than the light morph and forms a link in the continuum of plumages.

**JUVENILES OF LIGHT INTERMEDIATE (LIGHT RUFOUS) MORPH "WESTERN" (*B. j. calurus*):** HEAD.—As in juvenile light morph, but throat almost always dark. BODY (dorsal).—As in juvenile light morph. BODY (ventral).—The white or pale tawny breast is narrowly streaked with dark brown with the streaking covering less than 50% of the surface. The belly and flanks are heavily marked and form a distinct dark mottled band. White or pale tawny leg feathers are heavily barred. The undertail coverts are white or pale tawny and partially barred. WINGS (dorsal).—**The pale panel on the primaries and respective coverts is pale brown.** There is a small amount of pale spotting on the median and first rows of lesser upperwing coverts. WINGS (ventral).—**Large dark brown patagial mark, but it is somewhat masked by the heavily marked underwing coverts.** Axillaries are barred. The "fingers" on the outer five primaries are dark gray. TAIL.—As described for Juvenile Traits; black banded pattern is narrow to moderately wide. Rarely has rufous type tail. Broad white terminal tail band. Uppertail coverts are dark with narrow pale tips.

**JUVENILES OF LIGHT INTERMEDIATE MORPH "HARLAN'S" (*B. j. harlani*):** Probably similar to respective color morph of *calurus* but more brownish black and with a white throat.

**JUVENILE INTERMEDIATE MORPH TRAITS:** Pertains to "Western" *B. j. calurus* and "Harlan's" *B. j. harlani.* These morphs are more heavily marked than their respective light intermediate morphs and form links in the continuum of plumages from light to dark morph.

**JUVENILES OF INTERMEDIATE (RUFOUS) MORPH "WESTERN" (*B. j. calurus*):** Dark crown, nape, and malar marks with a medium tawny-rufous supercilium and auricular region. Throat is always dark brown. The forehead is dark. BODY (dorsal).—Nearly uniformly dark warm brown back and scapulars, but scapulars may have a small amount of pale tawny markings. BODY (ventral).—Breast is pale tawny or tawny with dark brown streaking covering 50% of the surface. Belly and flanks are nearly solid dark brown with some tawny speckling. Leg feathers are heavily barred. Tawny undertail coverts are heavily barred. WINGS (dorsal).—Dark coverts with pale spots only on the median coverts. WINGS (ventral).—Underwing coverts are uniformly and heavily marked with dark brown and tawny. There is no distinction of a dark patagial mark because the coverts are so heavily marked. The "fingers" on the outer five primaries are dark gray. TAIL.—As described for Juvenle Traits; dorsal bands are narrow to moderately wide; rarely has rufous type color. Moderately wide or wide white terminal tail band. Uppertail coverts are dark with a small amount of tawny edging.

**JUVENILES OF INTERMEDIATE MORPH "HARLAN'S" (*B. j. harlani*):** Crown, nape, and malar marks are brownish black; supercilium and auricular region are white. The white supercilium and auriculars may blend as a large white area on the side of the head. The throat is always white or white with narrow dark streaking. The forehead is often white. BODY (dorsal).—**Brownish black with white oval-shaped markings on the outer edges on the lower scapulars: the center of each scapular has a dark streak (diamond-shaped on other races).** BODY (ventral).—*See* types below. Leg feathers are brownish black and streaked with white. WINGS (dorsal).—**The pale panel on the primaries and respective greater coverts is pale brown or often quite white.** White oval-shaped markings on the median and first one or two rows of lesser coverts with a dark streak down the center of each feather. WINGS (ventral).—Brownish black with white speckling on the underwing coverts and white barring on the axillaries. There is no distinction of a dark patagial mark because the underwing coverts are dark and speckled. The "fingers" on the outer five primaries are distinctly barred. TAIL.—As described for Juveniles of Light Morph "Harlan's." However, the black bands are often fairly wide and not equal in width. On the inner half of the tail, the pale bands may be narrower than the dark bands. A black "spike" runs along the feather shaft from the black subterminal band to the tip of most or all rectrices. On many birds, the dark spike expands to cover all of the tip of each rectrix (tail tip is white on other races). *Note:* The spike pattern is more prevalent on darker morphs than on light morphs.

**JUVENILE INTERMEDIATE MORPH "HARLAN'S" (STREAK-BREASTED TYPE):** BODY (ventral).—Breast is equally streaked brownish black and white. The belly and flanks are brownish black and speckled with white and are abruptly darker than the breast. *Note:* A common type.

**JUVENILE INTERMEDIATE MORPH "HARLAN'S" (STREAKED TYPE):** BODY (ventral).—The breast, belly, and flanks are nearly equally and uniformly streaked with brownish black and white. *Note:* An uncommon type.

**JUVENILE INTERMEDIATE MORPH INTERGRADE TRAITS:** Refers mainly to individuals with subtle variations between *harlani* and *calurus* that are difficult to assign to a particular race. Involves birds with dark throats as on *calurus* and barred outer primaries as on *harlani,* and different patterns on the tail tip. Plumages are generally a warm dark brown as in *calurus.*

**JUVENILE DARK INTERMEDIATE MORPH TRAITS:** Pertains to *B. j. calurus* and *B. j. harlani.* These morphs are more heavily marked than their respective intermediate morphs and form links in the continuum of plumages from light to dark morph.

**JUVENILES OF DARK INTERMEDIATE (DARK RUFOUS) MORPH "WESTERN" (*B. j. calurus*):**

HEAD.—Dark brown with somewhat paler tawny-rufous supercilium and auricular areas. Throat is dark brown. Forehead is mainly dark but can be white. BODY (dorsal).—Uniformly warm dark brown but at close distance shows faint dark tawny-rufous mottling on the scapulars. BODY (ventral).—The breast feathers are narrowly edged with tawny or tawny-rufous and may appear quite rufous at moderate and long distances. The belly, flanks, and leg feathers are solid warm dark brown or may have a small amount of tawny-rufous edging on many feathers. The breast always appears somewhat paler than the rest of the underparts. WINGS (dorsal).—Nearly uniformly warm dark brown with a small amount of pale spotting on the median upperwing coverts. WINGS (ventral).—The underwing coverts are dark brown with a small amount of pale edging or mottling. TAIL.—As described for Juvenile Traits; black banded pattern is moderately wide to fairly wide. Moderate-width white terminal tail band.

**JUVENILES OF DARK INTERMEDIATE MORPH "HARLAN'S" (*B. j. harlani*):** HEAD.—Brownish black with a white throat and sometimes a white supercilium patch. The forehead is mainly black. BODY (dorsal).—Gray oval shapes on the scapulars with a black center streak on each feather. BODY (ventral).—*See* types below. WINGS (dorsal).—White oval shapes on median and some lesser upperwing coverts. WINGS (ventral).—Brownish black underwing coverts are lightly speckled with white; axillaries are barred with white. TAIL.—As described for Juvenile Intermediate Morph "Harlan's"; the black "spike" regularly extends onto the tip of most or all rectrices.

**JUVENILE DARK INTERMEDIATE MORPH "HARLAN'S" (STREAK-BREASTED TYPE):** BODY (ventral).—Brownish black breast has a small amount of white streaking or blotching. Rest of underparts is uniformly brownish black or has a small amount of white speckling.

**JUVENILE DARK INTERMEDIATE MORPH "HARLAN'S" (SPOT-BELLIED TYPE):** BODY.—The breast is solid brownish black and the belly and flanks are spotted or blotched with white. The leg feathers are narrowly streaked with white.

**JUVENILE DARK INTERMEDIATE MORPH INTERGRADE TRAITS:** Refers mainly to individuals with subtle variations between *harlani* and *calurus* that are difficult to assign to a particular race. Involves birds with dark throats as on *calurus* and barred outer primaries as on *harlani* and vice versa, and different patterns on the tail tip. Plumages are generally a warm dark brown as in *calurus*.

**JUVENILE DARK MORPH TRAITS:** Pertains to *B. j. calurus* and *B. j. harlani*. These morphs are more heavily marked than their respective dark intermediate morphs and form the final links in the continuum of plumages from light to dark morph.

**JUVENILES OF DARK MORPH "WESTERN" (*B. j. calurus*):** HEAD.—Uniformly warm dark brown. BODY (dorsal).—Uniformly warm dark brown with slightly paler tawny-rufous blotches on the scapulars. WINGS (dorsal).—Warm dark brown with tawny spotting on the median coverts. WINGS (ventral).—Uniformly dark brown but may have faint tawny-rufous speckling on some feathers. TAIL.—As described for Juvenile Traits; black tail bands are moderately wide or fairly wide. Narrow white terminal band.

**JUVENILES OF DARK MORPH "HARLAN'S" (*B. j. harlani*):** HEAD.—Uniformly brownish black, including the throat. The forehead is black. BODY (dorsal).—Brownish black with grayish oval shapes on the outer edges and a back center streak on most scapular feathers. BODY (ventral).—Uniformly brownish black and lacks any white speckling. WINGS (dorsal).—Brownish black and may have white spots on the median coverts. WINGS (ventral).—Uniformly black, including the axillaries. TAIL.—As described for Juvenile Intermediate Morph "Harlan's"; the black "spikes" regularly extend to tips of rectices.

**JUVENILE DARK MORPH INTERGRADE TRAITS:** There is a considerable amount of intergrading, and the differences are very subtle.

**ABNORMAL PLUMAGES:** Albinism of various types is more prevalent in Red-tailed Hawks than in any other raptor. H. Kendall (pers. comm.) has obtained data on over 550 cases of albinism in Red-tailed Hawks in recent years. Albinism is most commonly seen in birds from California, New York, Pennsylvania, and Wisconsin (H. Kendall pers. comm.). All but one record of albinism has been on a light morph. *Note:* Albinism types are based on descriptions in Pettingill (1970); *see* chapter 2. *Total albinism:*

All-white plumage with pink irises and fleshy parts. Extremely rare type of albinism and known in Red-tails from only one adult. *Incomplete albinism:* All white or nearly all white with a few dark feathers. Some fleshy areas are paler than on normal birds. On these types of Red-tails, the irises are a normal dark brown or may be bluish, and bill, cere, and feet are also normal color. The talons, however, are nearly always pink. Uncommon type of albinism. At a distance looks like total albinism. Seen on adults. *Partial albinism:* Plumage can be nearly all white with a few dark feathers or with any amount of white feathering among normal feathering. If partially white, then scattered white feathers are often on the head, remiges, and upperparts. Ventral areas are often normal. The white areas of the plumage may be the total feather, or a portion of various feathers may lack pigmentation and be white. Most common type of albinism. Seen on adults. Seen on one migrant dark morph *calurus* adult at Duluth, Minn., that had a white collar and scattered white feathers on its dark body and remiges (Nicoletti et al. 1998). *Imperfect albinism* (dilute plumage): Brown and rufous colored areas of normal birds are tan colored in this type of albinism. Seen on adults and juveniles. Uncommon type of albinism. *Note*: A hybrid adult dark morph *harlani* x dark morph Rough-legged Hawk was photographed in Prowers Co., Colo., in Dec. 1999. This bird had head and bill features and proportions, including wing shape, like a Red-tailed Hawk but had fully feathered tarsi and tail pattern as on a dark morph Rough-legged Hawk (BKW pers. obs./photos).

**HABITAT:** Each subspecies has adapted to a particular geographic region during the breeding season, but habitat is often quite variable. The different subspecies often share similar winter habitats.

***B. j. borealis.* Summer.**—East and south of the Great Plains, found in semi-open, dry upland, flat and hilly regions with small woodlots of tall trees, and amid large openings surrounded by mature woodlands. Climate is warm and moderately humid to humid. On the hot and dry Great Plains, inhabits open regions with tree growths along narrow riparian zones of lake shores, streams, rivers, and coulees. Also found amid rural farmlands with small artificially planted groves of trees around homesteads and with shelterbelts. Also inhabits semi-open areas of the cool and moist boreal forest.

***B. j. borealis.* Winter.**—Similar to summer habitat but found at more southern latitudes. On the s. Great Plains, regularly found in locales that have few if any trees. It trees are lacking, then ample elevated perches such as utility poles and fence posts are present for roosting and foraging. Tends to winter in moist lowland areas of swamps and fresh- or saltwater marshes. Light or moderate agricultural areas are favored. The narrow open grassy or weedy corridors along highways that extend through densely wooded regions of the southeastern part of the West are prime habitat. Regular in rural and suburban centers.

***B. j. borealis.* Migration.**—Similar to summer and winter habitat but often found on very open plains that lack tall grasses and trees.

**"Krider's" (*B. j. borealis*). Summer.**—Open, arid short grass or moderately humid tall grass prairies that are dotted with scattered stands or single growths of mainly tall deciduous trees growing along narrow riparian corridors of rivers, streams, ponds, or lakes. Also found in regions with rolling, open short grass or tall grass that have sheltered coulees with small stands of tall deciduous or coniferous trees.

**"Krider's" (*B. j. borealis*). Winter.**—Tends to favor less open and more moist regions during winter. Some inhabit semi-open flat or rolling grasslands and light or moderate agricultural areas from s. Iowa through n.-cen. Texas. However, the humid and lush semi-open areas in s. Louisiana and e. Texas are favored wintering locations. The semi-open areas in this region may be meadows, pastures, or light to intensely agricultural fields, particularly rice fields.

**"Krider's" (*B. j. borealis*). Migration.**—Similar habitat as in summer and winter. For short periods, found in treeless expanses on the Great Plains.

***B. j. calurus.* Summer.**—Highly variable depending on geographic region. Inhabits semi-open moist, moderate, or high montane regions with coniferous and, to a lesser extent, deciduous trees to nearly timberline elevation; semi-open moist or dry low montane or hilly regions with coniferous or deciduous trees or scrub; arid short grass plains or mountain valleys with single or small stands of coniferous or deciduous trees; or in arid rugged canyon areas with cliffs that are among semi-open coniferous forests or barren lands. Birds in s. Arizona and s. New Mexico tend to breed at moderate to high elevations. Also inhabits rural, suburban, and sometimes urban centers in various geographic areas and climates.

***B. j. calurus.*** **Winter.**—Similar to summer areas and often at fairly high elevations. Open regions lacking trees are regularly inhabited as long as suitable elevated perches such as utility poles, posts, or cacti are available. Light or moderate agricultural areas are favored. A few winter in semi-open and open areas with colonies of Black-tailed Prairie Dogs. This subspecies inhabits regions with deciduous trees more so than during the breeding season.

***B. j. calurus.*** **Migration.**—Found mainly in summer and winter areas. Birds crossing the vast expanse of the Great Plains are likely to be found in arid regions that lack trees.

***B. j. fuertesi.*** **All seasons.**—Along the Coastal Bend of Texas, moist, humid, lowland, lush coastal scrub thornscrub and semi-open brush and deciduous wooded areas are inhabited. From interior s. Texas and westward, habitat is seasonally hot and arid and semi-open to open. Vegetation consists of low to moderate-height thornscrub and larger deciduous trees along narrow riparian corridors and, at higher elevations, coniferous bushes and trees. In the desert regions, cacti comprise much of the tall vegetation. Birds in s. Arizona and s. New Mexico tend to breed at low elevation deserts.

***B. j. harlani.*** **Summer.**—Found mainly in semi-open and forested regions of White and Black spruce. Spruce forests consist of widely separated, moderate-height White Spruce in drier locations and very widely separated, short, scraggly Black Spruce in wet locations (farther south, spruce forests consist of dense stands of tall to very tall trees). Paper Birch and Quaking Aspen are often mixed with spruce. Areas near lakes, rivers, and streams are often preferred because White Spruce, which is main nesting tree, grows larger along riparian areas. Terrain varies from flat lowlands to low montane elevations.

***B. j. harlani.*** **Winter.**—Mainly found in semi-open wooded areas of mainly deciduous trees. Regularly found in open areas with single deciduous trees or small groves of deciduous or coniferous trees or utility poles or fence posts. Regions with light to moderate upland agriculture, grasslands, and pastures are typically inhabited. Smaller numbers are found in dry thornscrub or wet swamp and marsh habitat. Terrain varies from flat lowlands to flat highlands to hilly areas in lower elevation regions. Rural locations are regular winter haunts, but rarely found in suburban centers. Found in arid and humid climates.

***B. j. harlani.*** **Migration.**—Wooded or open low elevation to montane regions consisting of any tree type. Individuals crossing the Great Plains are often seen in areas void of trees.

***B. j. alascensis.*** **All seasons.**—Semi-open areas among dense coastal and island rainforests. Terrain is often steep and rugged. Primary trees in or bordering meadows and forest openings are Red Alder, Douglas Fir, Western Hemlock, and Sitka Spruce. A few birds may winter in drier, less humid areas south of typical breeding areas.

**HABITS:** *B. j. borealis* varies from tame to wary (usually tamer in the West than in the East). "Krider's" morph of *borealis* is wary on the breeding grounds and fairly tame to wary on the winter grounds. *B. j. calurus* varies from tame to wary. *B. j. fuertesi* is tame to fairly tame. *B. j. harlani* varies from fairly tame to wary on the breeding grounds but is almost always wary on the winter grounds (this is the wariest subspecies); juveniles vary from tame to wary on the winter grounds. *B. j. alascensis* temperament is not known.

Red-tailed Hawks perch on exposed, elevated natural and human-made perches, including utility wires. Hawks that are eating often sit on the ground. When perched in trees, they may sit on the top-most branches or on mid- and lower outer branches. This is a solitary species except during winter in locations with high prey density. In these areas, large numbers, especially juveniles, may loosely assemble. In other areas, territories are established and defended, particularly by adults. Hackles and nape feathers are erected when agitated or in cool temperatures.

**FEEDING:** Perch and aerial hunter. All subspecies have a generalist diet and hunt all types of small and medium-sized amphibians, birds, mammals, and reptiles. Large insects may also be eaten. Prey is captured on the ground or on outer branches of bushes and trees. Avian prey is occasionally captured in low-altitude flight. Red-tailed Hawks regularly feed on carrion, particularly during lean periods in winter.

**FLIGHT:** Wings are held at a low dihedral when soaring and on a flat plane when gliding. Long-winged subspecies, such as *calurus* and *fuertesi*, have a more distinct dihedral because the outer primaries flex upwards more than in shorter winged subspecies. Powered flight is used regularly and is interspersed with gliding sequences. At times, there is a fairly regular cadence of flapping and gliding. Wingbeats are moderately slow. Red-tailed Hawks regularly hover and kite. Aerial forging occurs when soaring, hovering, or kiting. Angled dives, sometimes from long distances, are made to capture prey. Legs are often extended downwards—with wings nearly closed—for stability in the mid- and last part of a dive.

**VOICE:** All subspecies have similar calls. A fairly vocal raptor in all seasons. Mainly heard when agitated. Adults and subadults have discernibly different tones than juveniles. *Adults and subadults:* Drawn-out, hoarse, and raspy whistled *skee-ah*, *skeerh* or *squeer*. *Juveniles:* Similar vocalization as older birds but lacks the hoarse, raspy tones and is a clear, high-pitched whistled call. Adults also emit a raspy squeal, *chee-aack* or *chee-aah*, at any time of year; mated pairs utter a low-pitched, nasal *gank* call to each other. Nestlings and fledglings repeatedly utter clear, high-pitched *klee, klee* or *kree, kree* notes when food-begging.

**STATUS AND DISTRIBUTION:** Overall population of Red-tailed Hawks is stable. It is one of the most common raptors with an estimated population of 1 million birds. The status of the different subspecies and color morphs varies.

***B. j. borealis.*** **Summer.**—Overall *common.* Regularly distributed throughout much of breeding range. Breeds south of the boreal forest from Alberta to cen. Manitoba and east of the Rocky Mts. from Alberta to ne. New Mexico. Irregularly distributed and fairly common on the Great Plains and uncommon in n. Manitoba. An extensive amount of intergrading occurs with *calurus* on the northern and western range borders and with *fuertesi* on the southern range border. Breeding range on the Plains has expanded greatly since colonization of the West in the late 1800s. Dammed rivers and reservoirs and suppression of range fires have allowed substantial growth and encroachment of trees into formerly treeless regions. Tree expansion on the Plains created additional breeding locations.

***B. j. borealis.*** **Winter.**—Withdraws from the northern third of breeding range. A few hardy birds winter in isolated locations in Alberta, Minnesota, Montana, and North Dakota. Winter range extends to the Rocky Mts. in Colorado and south through much of Texas and into e. Mexico. Wintering birds do not occur in New Mexico; however, a few are seen in the state during spring migration.

***B. j. borealis.*** **Movements.**—Most birds from the northern half of the breeding range probably migrate because of severe weather and lack of prey.

*Late summer northward dispersal:* After fledging, some juveniles embark on a northward journey in Aug. This interesting facet is not as well documented in the West as it is in the East. Banding data showed a juvenile that was fledged in Oklahoma had journeyed to Manitoba by fall. Also, a juvenile banded in Aug. near Rochester, N.Y., was found a few days later in Manitoba. Actual southward migration begins after the northward dispersal, in Sep. or Oct.

*Fall migration:* Juveniles migrate prior to subadults and adults in the eastern part of the West, but in the western part of range both ages move simultaneously. *Eastern region* (east of the Great Plains).—A few juveniles are tallied when hawkwatches begin in mid-Aug. Juveniles steadily increase throughout Sep. and peak in early to mid-Oct. Adults move later and, in Minnesota, peak in late Oct. with numbers decreasing substantially after early Nov. South of Minnesota, juveniles may peak in mid- to late Oct. and adults in early to mid-Nov. *Western region* (Great Plains and west).—Juveniles begin moving in mid-Aug. and adults by late Aug. Both ages spread across the Great Plains during Sep. and Oct. In late Sep. and early Oct., the n. Great Plains are filled with incredible numbers of juveniles. Peak movements on the cen. Great Plains occur from mid- to late Oct., but large numbers exist into mid-Nov.

*Spring migration:* Adults move north before

juveniles. Adults begin moving in early Feb. in southern latitudes and by late Feb. in mid-latitudes. They peak in southern and middle latitudes in mid- to late Mar. Some adults, possibly subadults, migrate through mid- to late Apr. in southern and middle latitudes. Adults peak along w. Lake Superior at Duluth, Minn., in early to mid-Apr. Juveniles migrate from Mar. to Jun. and peak in mid- to late Apr. Straggling juveniles may continue moving into early or mid-Jun. at northern latitudes.

**"Krider's" (*B. j. borealis*). Summer.**—*Very uncommon* to *uncommon.* Thinly and sporadically distributed throughout breeding range and are far outnumbered by typical plumaged *borealis.* Possibly formerly more common in the eastern part of range. Previously nested east to Minneapolis, Minn., and was possibly more widespread in Iowa. Intergrades with *calurus* along the western and northern parts of range.

**"Krider's" (*B. j. borealis*). Winter.**—Thinly and sporadically distributed, but population becomes more condensed than during summer. Most winter south of s. Kansas and s. Missouri. Largest numbers winter in s. and e. Texas and s. Louisiana. Southern extent of winter range is unknown; a few may extend into ne. Mexico. *Note:* Winter records are not documented for Colorado and New Mexico. ("Krider's" are mistaken for similar-looking light morph *B. j. harlani.*)

**"Krider's" (*B. j. borealis*). Movements.**—Migratory. Only a few are seen at hawkwatches. *Fall migration:* Similar to previously described Eastern Region movements of *borealis.* A juvenile was seen in Prowers Co., Colo., in mid-Nov. *Spring migration:* Similar to spring movements of *borealis.*

***B. j. calurus.* Summer.**—*Common.* Regularly distributed throughout breeding range. Not found above timberline (blank areas on range map). Breeding range may extend southward at high elevations in n. Sonora and n. Chihuahua (intermediate and dark morphs are common at higher elevations in s. Arizona). There are broad overlapping intergrade zones with all adjacent subspecies. Many birds from e. Alaska are intergrades with *harlani.* Also intergrades extensively with *harlani* in the Yukon Territory and n. British Columbia. At an e. Alaska spring hawkwatch, only 8% of Red-tailed Hawks were *calurus*; all others were *harlani.* Although somewhat isolated by the formidable Coast Range Mts., *calurus* intergrades with *alascensis* along portions of w. British Columbia and on Vancouver Island and adjacent islands of British Columbia. An extensive amount of ntergrading occurs with *borealis* along the eastern part of range and with "Krider's" morph of *borealis* from Manitoba to Wyoming. *B. j. calurus* intergrades with *fuertesi* in s. Arizona and s. New Mexico (*fuertesi* is generally found at lower elevations). A dark morph *calurus* has mated with a *fuertesi* in s. New Mexico.

***B. j. calurus.* Winter.**—Withdraws from Alaska and most Canadian areas. Winters in all of the w. U.S. breeding range and south into cen. Mexico. Large numbers winter south and east of breeding range and some extend into the eastern U.S. A few hardy birds winter in isolated locations on the n. Great Plains.

***B. j. calurus.* Movements.**—Migratory in much of range, particularly the northern half.

*Fall migration: B. j. calurus* has a strong tendency to migrate southeast of breeding and/or natal areas. Saskatchewan-banded birds all moved southeast of the province, including to Wisconsin and Georgia. Juveniles begin moving in mid-Aug. in all areas of the West. *Eastern region* (east of the Great Plains).—Juvenile numbers build throughout Sep. and peak in early to mid-Oct. In Minnesota, adult numbers build in early Oct., peak in late Oct., and continue in smaller numbers through early Nov. South of Minnesota, juveniles peak in mid- to late Oct., adults in early to mid-Nov. *Western region* (Great Plains and west).—Juvenile peak movements occur in late Sep. and early Oct. Adults begin moving out of montane elevations down to valleys and onto the w. Great Plains in late Aug. Adult movement continues through Sep. and birds spread east and south throughout the Plains (often moving with Swainson's Hawks). Peak adult movement on the Plains is in mid- to late Oct. but can extend until early to mid-Nov. for individuals migrating from more northern latitudes. Large numbers of dark morphs and those with *harlani* traits are often seen in Nov. In the intermountain region of the U.S., migration is early: juveniles peak in late Sep., adults in early Oct. Juveniles banded as nestlings in s. California were found to migrate north and spend one or possibly two winters in Idaho, ne. Nev., Ore., Utah, or Wash.

*Spring migration:* In southern latitudes, adult migration may begin in early Feb. In mid-latitudes adults may begin moving by mid- to late Feb. Peak adult period at mid-latitudes is mid- to late Mar. On Cape Flattery in nw. Washington, the first adults are seen in early to mid-Mar., and they peak in early Apr. Subadults may migrate a bit later than adults. Juveniles begin moving in Mar. and peak in mid- to late Apr. in most areas. At northern latitudes, small numbers of juveniles continue migrating through May and into Jun.

***B. j. fuertesi.* All seasons.**—*Common.* Sedentary. There are very few pure *fuertesi* in the depicted Arizona range. Most birds in this region are intergrades with *calurus.* They are primarily found at lower elevations, particularly where their range overlaps with *calurus.*

***B. j. harlani.* Summer.**—*Uncommon.* Sparsely distributed in Alaska, the Yukon Territory, and n. British Columbia. Inhabits wooded regions south of the Brooks Range, east of the Seward Peninsula, and east of the barrens of Bristol Bay and Norton Sound. Found mainly along rivers and lakes, especially along the Yukon and Kuskokwim Rivers. On the Yukon River, range extends to the town of Galena. Ninety-two percent of spring Red-tailed Hawk migrants counted at a hawkwatch near Eureka in e. Alaska in the spring of 2002 were *harlani.*

There is no formidable geographic barrier to isolate *harlani* from *calurus.* Because of this, intergrading with *calurus* occurs commonly in n. British Columbia, the Yukon Territory, and e. Alaska to at least Fairbanks. Pure *harlani,* however, can occur in all of the depicted range. Some intergrading probably occurs with *alascensis* in very northern part of the Alaskan panhandle. *Note:* Even though there are superficial similarities between light morph *harlani* and "Krider's" *borealis,* they are quite different and there is no genetic flow between them. The nearest *harlani* of the boreal forest of n. British Columbia are separated by at least 800 miles (1,300 km) from the nearest possible "Krider's" on the Great Plains of Alberta.

***B. j. harlani.* Winter.**—Widely separated from the breeding range. Historical and current core winter range encompasses sw. Iowa south through e. Kansas, w. Missouri, Oklahoma, and w. Arkansas. Smaller numbers winter in e. Texas and Louisiana. Adaptable as Red-tailed Hawks are, they have readily acclimated to new wintering areas created by light and moderate types of agriculture and an increase in trees on the cen. Great Plains and mountain valleys. Large and dense populations now occur in e. Colorado, along the Front Range, the South Platte River, and particularly the Arkansas River (especially Prowers Co.). Fairly large numbers occur in e. Nebraska and along the Platte River in Nebraska. Small numbers regularly winter in isolated locations in all other western states and in very southern British Columbia.

***B. j. harlani.* Movements.**—Highly migratory.

*Fall migration:* All migrate in a southeasterly direction to wintering grounds. As described above in Winter, migrants can be widely scattered throughout the West; however, the majority still angle sharply southeast to major wintering areas east of the Great Plains. Migration may begin in mid-Aug. for juveniles and by late Aug. for subadults and adults. Most birds depart breeding areas by late Sep. Although juveniles usually migrate a bit earlier than older birds, subadults have been the earliest migrants seen in e. Colorado. In late Sep., migrant *harlani* are commonly seen in North Dakota. By late Sep., the earliest birds have reached typical winter grounds (by Sep. 25 in Prowers Co., Colo.). Some adults are on the winter grounds by early Oct., many have arrived by late Oct. or early Nov., and most have arrived by mid- to late Nov. Movements may continue into Dec. in the southern part of winter range. Peak movements across the cen. Great Plains occurs from late Oct. to mid-Nov.

*Spring migration:* Earliest adult migrants leave the winter grounds in late Feb., and most leave by mid- to late Mar. A few adults linger on wintering areas until mid-Apr. Adults are seen in e. Alaska by early Apr. Peak adult flights in e. Alaska are in late Apr. (data from J. Bouton, P. and C. Fritz). Juveniles begin moving after adults. Most leave the winter grounds by mid-Apr.

***B. j alascensis.* Summer.**—*Uncommon.* Sparsely distributed in its restricted range. Found from Yakutat, Alaska, south to Vancouver Island, B.C. However, those on Vancouver Island and adjacent islands and coastal mainland of British Columbia appear to have considerable intergrade traits with *calurus* (based on museum specimens and photographs). This subspecies is isolated to a certain extent from *calurus* by the formidable Coast Range Mts. of w. British Columbia and numerous large glaciers. There may be a small amount of intergrading with *harlani* in the very northern part of range. Classic-looking individuals inhabit the isolated Queen Charlotte Islands of British Columbia.

***B. j. alascensis.* Winter.**—Based on Christmas Bird Counts and knowledgeable persons, virtually all *alascensis* withdraw from Alaskan range in winter. They are rarely seen in winter in core breeding areas of se. Alaska. Red-tailed Hawks are seen on the Queen Charlotte Islands during winter and on all Christmas Bird Counts. The population on the Queen Charlotte Islands is probably sedentary and may have an influx of wintering birds from nearby se. Alaskan islands.

***B. j. alascensis.* Movements.**—The Alaskan population migrates but there is no data on timing. Presumably, the migration period is similar to these of other similar-latitude subspecies. Some birds may migrate south of British Columbia. However, photographs of fall migrants captured on the Olympic Peninsula, Wash., do not indicate such a movement.

**Color morph status for *B. j. calurus* and *B. j. harlani*.**—All morphs breed throughout each respective subspecies' range. Morph ratios locally and regionally may be higher or lower than listed.

*B. j. calurus*: The darker the color morph, the less common it is. Light morph.—Common. Far outnumbers darker morphs and may comprise at least 70% of the population. Light intermediate morph.—Uncommon. Intermediate morph.—Uncommon. In winter, more common west of the Rocky Mts. and outnumbered by darker morphs east of the Rockies. Dark intermediate morph.—Uncommon. In winter, found in equal numbers in all areas. Dark morph.—Uncommon. In winter, more common east of the Rocky Mts. Comprises 5–10% of the population.

*B. j. harlani*: The light and dark extremes are the least common. Light morph.—Uncommon and accounts for about 4% of *harlani* population based on 2002 spring hawkwatch data from e. Alaska; however up to 15% of local winter populations (e.g., in Prowers Co., Colo.) may be light morph. Light intermediate morph.—Uncommon and only slightly more numerous than light morph. Intermediate morph.—Common. The most common color morph and may comprise 50% of the population. Dark intermediate morph.—Fairly common. Somewhat less common than intermediate morph. Dark morph.—Uncommon and may comprise about 5% of the population.

**NESTING:** Some pairs of all races remain united all year, northern birds often pair only for nesting season. *B. j. borealis.*—Begins Jan. to Mar. as far north as Colorado and Iowa, Apr. in n. U.S. and Canada (includes "Krider's"); ends Jun. to Aug. *B. j. calurus.*—Begins Jan. to Mar. at low elevation southern and central areas, high elevation and northern pairs Mar. to May; ends June to Aug. *B. j. fuertesi.*—Begins Jan. to Mar.; ends May to June. *B. j. harlani.*—Begins late Apr. and May; ends Aug. to mid-Sept. *B. j. alascensis.*—Unknown.

**Courtship (flight).**—*High-circling* by one or both sexes, *leg-dangling* by one or both sexes, *talon-grappling* by both sexes, and *sky-dancing* by males.

**Courtship (perched).**—*High-perching.* See chapter 5 for courtship descriptions.

Large stick nests are built by both sexes. A new nest may be built or a may reuse a previous year's nest. Nests may be built in a week or less, constructed mainly in the morning. Old nests may be used for several years and become quite deep. Nests are 28–30 in. (71–76 cm) in diameter. New nests may be 4–6 in. (10–15 cm) deep; old nests may be much deeper. Finer material and greenery are added to the nest throughout the nesting season.

Nests are located where there is an expansive view of the landscape and are typically placed 35–90 ft. (11–27 m) high, but *B. j. calurus* and *B. j. fuertesi* nests can regularly be as low as 15 ft. (5 m). On the Saskatchewan prairie, a nest of *B. j. borealis/calurus* was only 1 ft. (0.3 m) high in a small bush. Most nests of *borealis* (including "Krider's") are placed in trees; all nests of *fuertesi*, *harlani*, and *alascensis* are placed in trees. Tree nests are built in the dominant species of the area and are placed in the upper part of live or dead trees. In deciduous trees, nests are placed at the first major fork of the trunk or in a fairly major fork of side branches. In conifers, nests are located near the trunk for optimum support. Nests are placed in isolated single trees, in small woodlots, or on the edge of a large woodland. Cliff ledges as well as trees are readily used as nest sites by *calurus*. Tall cacti are often used as nesting areas for *fuertesi* and some *calurus*. All races except *harlani* and *alascensis* commonly place their nests on cross bars of utility poles. Other artificial structures, such as building ledges, metal poles, and windmills, are used by *borealis*, *calurus*, and *fuertesi*. Tall White Spruces are the typical nesting location for *harlani*.

The 2–3 eggs, rarely 1–5, are incubated by both sexes for 28–35 days. Youngsters branch before fledging in 42–46 days. The smaller males fledge before the larger females. Fledglings depend on parents another 30–70 days.

**CONSERVATION:** Since this is a highly adaptable species, no measures are needed.

**Mortality.**—Illegal shooting, electrocution from utility lines, and collisions with vehicles are major factors. There are some local problems with organophosphate poisoning. Great Horned Owls prey on all ages of Red-tailed Hawks. Coyotes and foxes prey on fledglings perched on the ground. Golden Eagles occasionally kill Red-tailed Hawks that harass them.

**SIMILAR SPECIES:** COMPARED TO JUVENILE LIGHT MORPH (including *harlani*).—**(1) Red-shouldered Hawk, juveniles.**—Irises darker brownish. Bright yellow cere. Ventral areas are uniformly streaked or barred. Pale tail bands are narrower than dark bands. FIGHT.—Pale tawny panels and windows on base of primaries. Lacks dark patagial marks on the underwing. **(2) Broad-wing Hawk, juveniles.**—PERCHED.—Irises are darker brownish. Bright yellow cere. Often have similarly marked belly band. Dorsal surface of secondaries is dark brown and unbarred. Scapulars are dark brown and lack white patches. FLIGHT.—Similar pale windows on the ventral wing; pale only on primaries on the upperwing. Does not hover or kite. **(3) Swainson's Hawk, light morph juveniles.**—PERCHED.—Irises usually darker medium brown. Cere is usually bright yellow but can be a similar greenish. Inner half of lores dark. Similar white mottling on mid-scapulars. Large dark brown patch on the sides of the neck and breast. Wingtips equal or nearly equal to tail tip. FLIGHT.—Medium gray remiges on the ventral wing. Lacks dark patagial marks. **(4) Rough-legged Hawk, light morphs.**—PERCHED.—Yellow cere. Inner half of lores dark. Belly markings often quite similar, especially mottled pattern of many adults. Tarsi are feathered to the toes. Wingtips equal to or extend beyond tail tip. FLIGHT.—Pale window on underside of primaries similar, but when seen from above it is only on primaries and not primary greater coverts. White uppertail coverts similar to those of Red-tail but usually more extensive white on the basal region. COMPARED TO ADULT/JUVENILE "KRIDER'S".—**(5) Swainson's Hawk, one-year-olds.**—PERCHED.—Thin dark eyeline. Medium brown irises. Dark patch on the sides of the neck and breast. Wingtips nearly equal to or equal tail tip. FLIGHT.—Medium gray remiges. Lacks dark patagial mark on the underwing. Tail is uniformly medium brown on the dorsal surface. **(6) Ferruginous Hawk, light morph juveniles.**—PERCHED.—Yellow cere. Thick dark eyeline may extend under eyes. Feathered tarsi. Dorsal surface of secondaries is plain dark brown and unbanded. Dark dorsal tail bands are fairly wide, and rarely more than three or four dark bands. Scapulars are mainly dark brown with little if any white mottling. FLIGHT.—Small dark tips on the outer primaries (barred wingtips on all juvenile "Krider's"). COMPARED TO JUVENILE INTERMEDIATE MORPH (including *harlani*).—**(7) Rough-legged Hawk, light/intermediate morphs.**—Breast is similarly streaked. PERCHED.—Yellow cere. Dark inner half of lores. Thin dark eyeline. Dorsal surface of secondaries is plain brown and unbarred. Wingtips extend to tail tip. FLIGHT.—Large dark carpal patch on the wrist of the underwing. COMPARED TO JUVENILE DARK MORPH (including *harlani*).—**(8) Broad-winged Hawk, dark morph juveniles.**—PERCHED.—Similar head and body features. Dorsal surface of the secondaries lacks black barring. FLIGHT.—Underside of remiges lightly barred, not heavily barred as on *calurus* or *harlani*. Dorsal surface of wing may be pale only on the primaries, not on the primary greater coverts. **(9) Swainson's Hawk, dark morph adults.**—PERCHED.—Yellow cere. White spot on outer half of lores. Wingtips equal to tail tip. FLIGHT.—Rufous underwing coverts. Dark gray underside of the remiges. Undertail coverts are distinctly paler than the body. **(10) Rough-legged Hawk, all ages.**—PERCHED.—As described for #4. COMPARED TO ADULT INTERMEDIATE TO DARK MORPH "HARLAN'S".—**(11) Rough-legged Hawk, adult female intermediate and dark morph.**—PERCHED.—Dark inner half of lores. Thin dark eyeline. Tarsi are feathered to the toes. Wingtips equal to tail tip. Breast pattern deceptively similar: both *harlani* and Rough-leg can have whitish mottling (intermediate morph) or lack mottling on the breast (dark morph). Tail pattern of *harlani* with medium/dark gray type dorsal tail pattern can be identical to that of Rough-legs (dark mottling is not always present on *harlani* tail). Rough-legs typically have a neat, broad black subterminal tail band; however, this is also present on a few *harlani* with gray tails. FLIGHT.—Underwing coverts and axillaries may be white spotted or barred and identical to some *harlani*: use caution. Most darker morphs of Rough-legs have some rufous on the underwing coverts with a black carpal patch on the wrist area; lacking on *harlani*.

**OTHER NAMES:** Redtail, Tail, Black Warrior (*B. j. harlani*). *Spanish:* Aguililla Colirroja, Aguililla Cola Roja, Aguililla Parda. *French:* Buse à Queue Rousse.

---

REFERENCES: Adamus et al. 2001; Baicich and Harrison 1997; Busby and Zimmerman 2001; Bylan 1998, 1999; Campbell et al. 1990; Cecil 1999; Clark and Wheeler 2001; del Hoyo et al. 1994; Dickerman 1994; Dickerman and Parkes 1987; Dodge 1988–1997; Dorn and Dorn 1990; Dotson and Mindell 1979; Ferguson-Lees and Christi 2001; Gibson and Kessel 1997; Godfrey 1986; Houston 1967, 1983; Houston and Bechard 1983; Howell and Webb 1995; Jacobs and Wilson 1997; James and Neal 1996; Janssen 1987; Kellogg 2000; Kent and Dinsmore 1996; Kingery 1998; Liguori 2001; Meehan et al. 1997; Mindell 1985; Montana Bird Distribution Committee 1996; Nicoletti et al. 1998; Oberholser 1974; Palmer 1988; Peterson 1995; Pettingill 1970; Pittaway 1993; Preston and Beane 1993; Russell and Monson 1998; Semenchuk 1992; Smith 1996; Steenhof et al. 1984; Stewart 1975; Sutton 1967; Todd 1950, 1963; Wheeler and Clark 1995; Wood 1932.

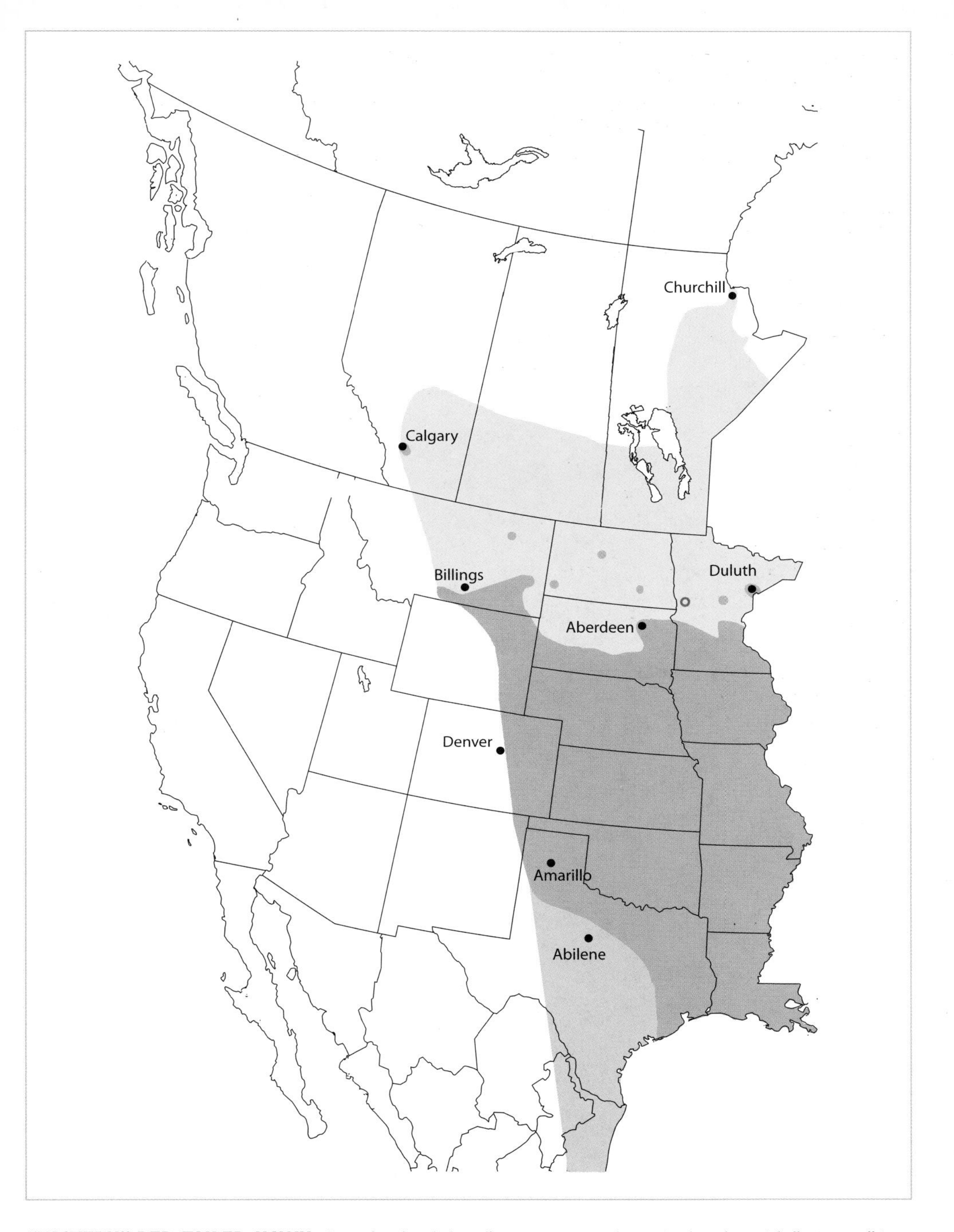


**"EASTERN" RED-TAILED HAWK,** ***Buteo jamaicensis borealis:*** Common. Subspecies border with "Western" is unclear in boreal forest region of Manitoba. Intergrades with "Western" and "Fuertes" at respective range borders.

**"KRIDER'S" RED-TAILED HAWK,** *Buteo jamaicensis borealis* (Pale Morph): Very uncommon to uncommon. Sparsely distributed among typical "Eastern" in summer and intergrades with "Eastern." Intergrades with "Western" on w. and n. edges of range.

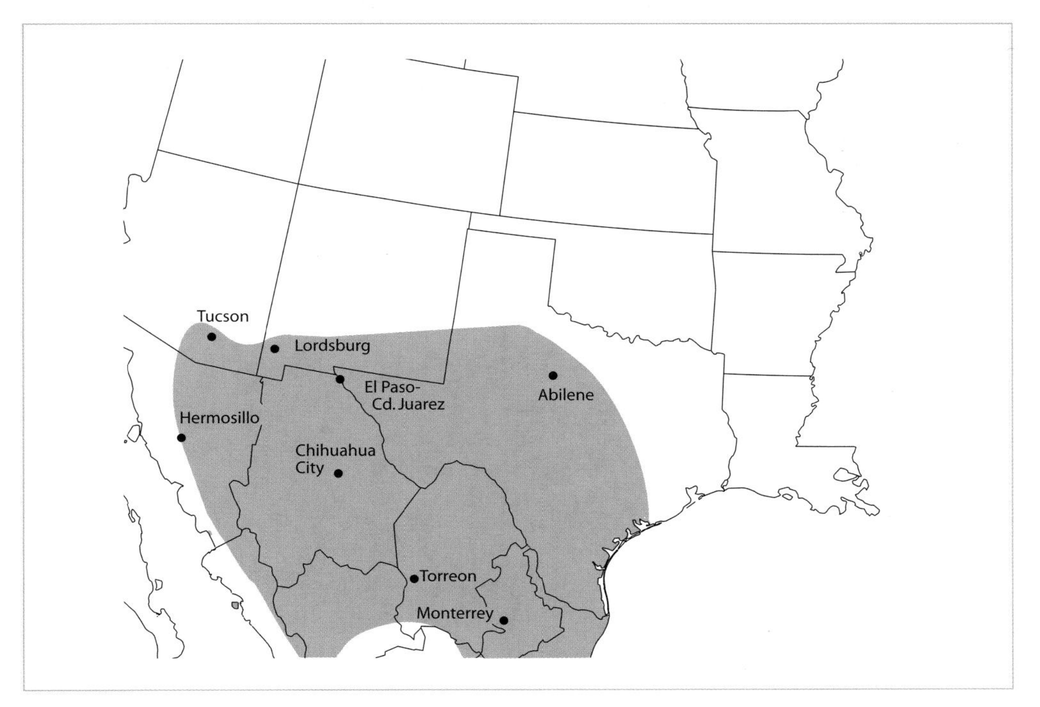


**"FUERTES" RED-TAILED HAWK,** *Buteo jamaicensis fuertesi:* Common. Intergrades with "Eastern" in TX border zone and "Western" in all of NM and AZ zone.

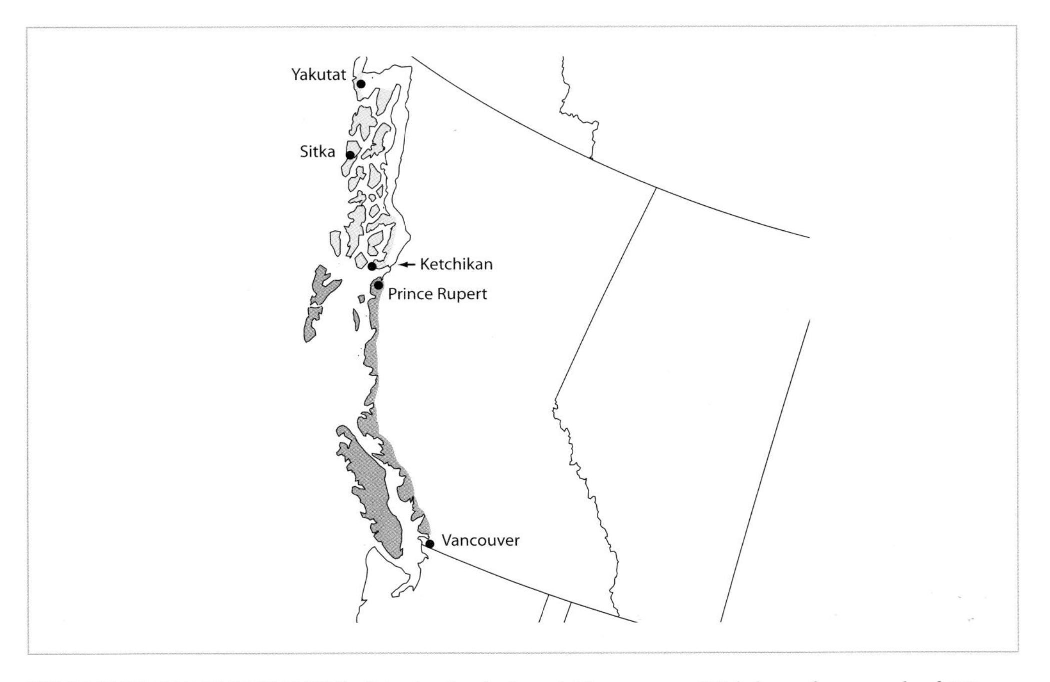


**"ALASKAN" RED-TAILED HAWK,** *Buteo jamaicensis alascensis:* Uncommon. Withdraws from much of AK range in winter. Some may winter farther south than shown. Probable minor intergrading with "Harlan's" in n. part of range and considerable intergrading with "Western" in s. part of range.

**"WESTERN" RED-TAILED HAWK,** *Buteo jamaicensis calurus:* Common. Intergrades with "Harlan's" in n. BC, YK, e. AK, and possibly sparingly in the rest of "Harlan's" range. Common in winter range. Proposed subspecies in s. Baja California for smaller "Sutton's" race, *B. j. suttoni.*

**"HARLAN'S" HAWK,** *Buteo jamaicensis harlani:* Uncommon. Sparsely distributed in summer. Moderately dense in core winter range in e. CO, e. NE, e. KS, OK, w. AR, and w. MO. Uncommon to rare elsewhere in winter.

**Plate 348. Red-tailed Hawk, adult "Eastern" (*B. j. borealis*) [Dec.]** ▪ White throat. Young adults have pale irises. ▪ Lightly marked type has white underparts with barred flanks, unmarked belly. Unmarked leg feathers. ▪ Wingtips shorter than tail tip. ▪ Rufous tail with black subterminal band. ▪ *Note:* Pale bellied types found throughout breeding range.

**Plate 349. Red-tailed Hawk, adult "Eastern" (*B. j. borealis*) [Dec.]** ▪ Tawny-brown head with dark crown and malar. Streaked throat with dark collar. Old adults have dark irises. ▪ Large blotchy white patch on scapulars. Unmarked leg feathers. ▪ Pale mottling on brown wing coverts. Wingtips shorter than tail tip. ▪ Rufous tail with black subterminal band.

**Plate 350. Red-tailed Hawk, adult "Eastern" (*B. j. borealis*) [Feb.]** ▪ Tawny-brown head with dark crown and malar. White throat with dark collar. Old adults have dark irises. ▪ White underparts are moderately marked type with black streaking on belly, black barring on flanks. Lower belly, leg feathers, and undertail coverts unmarked.

**Plate 351. Red-tailed Hawk, adult "Krider's" (pale morph of *B. j. borealis*) [Feb.]** ▪ White head with dark malar. Old adults have dark irises. ▪ Large blotchy white patch on scapulars. White blotches on coverts. Wingtips shorter than tail tip. ▪ Tail pinkish on distal 1/3, white on basal 2/3; thin subterminal band. ▪ *Note:* Purest types have all-white head.

**Plate 352. Red-tailed Hawk, adult "Fuertes" (*B. j. fuertesi*) [Nov.]** ▪ Dark tawny-brown head. White throat. Old adults have dark irises. ▪ White underparts may have some dark barring on flanks and streaks on belly, but can be unmarked. Scapulars (not shown) have small to moderate white or gray patches. Uniformly dark wing coverts. ▪ Wingtips reach tail tip.

**Plate 353. Red-tailed Hawk, adult "Western" (*B. j. calurus*) light morph [Nov.]** ▪ Dark tawny-brown head. Dark throat. Young adults have pale irises. ▪ Small to moderate-sized blotchy tawny patch on scapulars. ▪ Minimal pale mottling on wing coverts. Wingtips reach near tail tip. ▪ Rufous tail is banded type with full, narrow bands; wide black subterminal band.

**Plate 354. Red-tailed Hawk, adult "Western" (*B. j. calurus*) light morph [Feb.]** ▪ Dark tawny-brown head. Dark throat. Old adults have dark irises. ▪ Small to moderate-sized blotchy tawny patch on scapulars. ▪ Wingtips reach tail tip. ▪ Rufous tail is partially banded type with narrow bands on portion of rectrices; narrow black subterminal band. Orangish rufous uppertail coverts on most *calurus*.

**Plate 355. Red-tailed Hawk, adult "Western" (*B. j. calurus*) light morph [Nov.]** ▪ Dark tawny-brown head. Dark throat. Young adults have pale irises. ▪ Pale tawny underparts are heavily marked type. Barring on belly and lower belly is classic *calurus* trait. Tawny leg feathers barred. ▪ *Note:* Heaviest marked type of light morph.

**Plate 356. Red-tailed Hawk, adult "Western" (*B. j. calurus*) light intermediate morph** [Nov.] ▪ Dark tawny-brown head. Dark throat. Old adults have dark irises. ▪ Pale rufous underparts are typically heavily marked type. Barred, pale rufous lower belly and leg feathers.

**Plate 357. Red-tailed Hawk, adult "Western" (*B. j. calurus*) intermediate (rufous) morph** [Oct.] ▪ Dark tawny-brown head. Dark throat. Old adults have dark irises. ▪ Rufous breast (rufous leg feathers, lower belly, and undertail coverts hidden by perch). Belly and flanks vary from mottled with dark brown to solid dark brown.

**Plate 358. Red-tailed Hawk, adult "Western" (*B. j. calurus*) dark intermediate (dark rufous) morph** [Nov.] ▪ Dark brown head and throat. Old adults have dark irises. ▪ Rufous tinge on breast; belly, flanks, and leg feathers dark brown. Undertail coverts may be rufous. ▪ Upperparts dark brown.

**Plate 359. Red-tailed Hawk, adult "Western" dark morph** [Feb.] ▪ Uniformly dark brown head. Old adults have dark irises. ▪ Uniformly dark brown body. ▪ Rufous tail is banded type but can be partially banded or unbanded.

**Plate 360. Red-tailed Hawk, adult "Harlan's" (*B. j. harlani*) light morph** [Mar.] ▪ Blackish head with white supercilium and spectacles. Throat appears dark (from shadow) but is white with dark streaking. ▪ Stark white underparts lack rufous tinge on sides of neck on light morphs of other races. Large black blobs on belly, partial bars on flanks.

**Plate 361. Red-tailed Hawk, adult "Harlan's" (*B. j. harlani*) light morph** [Feb.] ▪ Brownish black head with white supercilium and auricular forms white spectacle look. White throat. ▪ White blotches on scapulars form a V. ▪ Wingtips shorter than tail tip. ▪ Tail mainly gray *harlani*; possible *calurus* influence with partly rufous tail and neat subterminal band.

**Plate 362. Red-tailed Hawk, adult "Harlan's" (*B. j. harlani*) light intermediate morph** [Feb.] ▪ Mainly white head, including throat; dark crown and malar. Old adults have dark irises. ▪ White breast sparsely streaked. Belly and flanks blackish with white speckling. Undertail coverts white.

**Plate 363. Red-tailed Hawk, adult "Harlan's" (*B. j. harlani*) intermediate morph** [Nov.] ▪ Dark head with white forehead and partial supercilium; white throat. Old adults have dark irises. ▪ White breast broadly streaked. Belly and flanks blackish with white speckling. Undertail coverts black with white barring.

**Plate 364. Red-tailed Hawk, adult "Harlan's" (*B. j. harlani*) intermediate morph** [Nov.] ▪ Dark head with white forehead and throat. Old adults have dark irises. ▪ White streaking on hindneck. Back and scapulars dark. ▪ Wingtips shorter than tail tip. ▪ White type tail. Irregular black terminal band.

**Plate 365. Red-tailed Hawk, adult "Harlan's" (*B. j. harlani*) dark intermediate morph** [Nov.] ▪ Black head with white streaking on throat. Old adults have dark irises. ▪ Black breast sparsely streaked/mottled with white. Rest of body black. ▪ Wingtips shorter than tail tip. ▪ Medium/dark gray type tail.

**Plate 366. Red-tailed Hawk, adult "Harlan's" (*B. j. harlani*) dark morph** [Dec.] ▪ Old adults have dark irises. ▪ Black body. ▪ Wingtips shorter than tail tip. ▪ Medium/dark gray banded type tail. Gray with several dark bands and neat, wide, black subterminal band. ▪ *Note:* Uncommon *harlani* tail pattern.

**Plate 367. Red-tailed Hawk, adult "Eastern" x "Western" (*B. j. borealis* x *B. j. calurus*) light morph** [Nov.] ▪ Tawny-brown head and white throat as on *borealis.* ▪ Large white blotchy scapulars form V as on *borealis.* ▪ Extensive pale mottling on wing coverts as on *borealis.* ▪ Rufous tail is partially banded type as on *calurus.*

**Plate 368. Red-tailed Hawk, adult "Harlan's" x "Western" (*B. j. harlani* x *B. j. calurus*) [Dec.]** ▪ Dark rufous-brown head and throat as on *calurus* with white *harlani* supercilium. ▪ Rufous breast as on intermediate morph *calurus* with small amount of white mottling as on dark intermediate morph *harlani*. Dark brown belly and flanks. ▪ Dorsal tail (not shown) pale gray as on *harlani*.

**Plate 369. Red-tailed Hawk, adult "Eastern" (*B. j. borealis*) [Oct.]** ▪ White throat with dark collar. ▪ Pale tawny underparts with moderately marked type belly and flanks. Unmarked leg feathers. ▪ Black patagial mark on wing. ▪ Rufous tail with a dark subterminal band.

**Plate 370. Red-tailed Hawk, adult "Eastern" (*B. j. borealis*) [Oct.]** ▪ Dark throat. ▪ Pale tawny underparts with heavily marked type belly and flanks. Unmarked leg feathers. ▪ Black patagial mark on wing. ▪ Rufous tail with dark subterminal band. ▪ *Note:* Adults occasionally have dark throats.

**Plate 371. Red-tailed Hawk, adult "Eastern" (*B. j. borealis*) [Oct.]** ▪ Large white patch on each scapular. ▪ White uppertail coverts; rufous tail with dark subterminal band.

**Plate 372. Red-tailed Hawk, adult "Krider's" (pale morph of *B. j. borealis*) [Nov.]** ▪ White throat. Partial dark collar. ▪ Immaculate white underparts. ▪ Rufous-brown patagial mark. Barred remiges, including "fingers" on outer primaries; wide black band on rear edge. ▪ Mainly white tail with neat, thin, dark subterminal band.

**Plate 373. Red-tailed Hawk, adult "Fuertes" (*B. j. fuertesi*) [Feb.]** ▪ White throat may be streaked; sometimes all-dark throat. ▪ White underparts typically faintly barred on flanks but may be unmarked. Unmarked leg feathers. ▪ Black patagial mark. ▪ Rufous tail often lacks thin dark subterminal band. ▪ *Note: B. j. borealis* that lack belly markings are similar.

**Plate 374. Red-tailed Hawk, adult "Western" (*B. j. calurus*) light morph [Dec.]** ▪ White streaking on dark throat. ▪ Lightly marked type underparts with minimal belly and flank markings. Narrow rufous barring on all areas but breast, including leg feathers. ▪ Black patagial marks; wide black band on trailing edge of wings. ▪ Partially banded type rufous tail is faintly banded.

**Plate 375. Red-tailed Hawk, adult "Western" (*B. j. calurus*) light morph [Dec.]** ▪ Dark throat. ▪ Heavily marked type underparts with heavy belly and flank markings. Narrow rufous barring on belly, lower belly, leg feathers, and undertail coverts. ▪ Black patagial marks; wide black band on trailing edge of wings. ▪ Banded type rufous tail.

**Plate 376. Red-tailed Hawk, adult "Western" (*B. j. calurus*) light morph [Jan.]** ▪ Dark throat. ▪ Heavily marked type underparts with heavy belly and flank markings. Narrow rufous barring on belly, lower belly, and undertail coverts; leg feathers distinctly rufous-barred. ▪ Black patagial marks; wide black band on trailing edge of wings. ▪ Tail grayish or pinkish when closed.

**Plate 377. Red-tailed Hawk, adult "Western" (*B. j. calurus*) light morph [Dec.]** ▪ Dark throat. ▪ Heavily marked white type underparts white with heavy belly and flank markings. Faint rufous barring on leg feathers. ▪ Black patagial marks; wide black band on trailing edge of wings. ▪ Tail grayish or pinkish at certain angles.

**Plate 378. Red-tailed Hawk, adult "Western" (*B. j. calurus*) intermediate (rufous) morph [Nov.]** ▪ Dark throat. ▪ Rufous breast and leg feathers; mottled black belly and flanks. ▪ Rufous underwing coverts with black patagial mark. ▪ Partially banded type rufous tail.

**Plate 379. Red-tailed Hawk, adult "Western" (*B. j. calurus*) dark intermediate (dark rufous) morph [Mar.]** ▪ Dark rufous-brown head, including throat. ▪ Dark rufous-brown breast, black belly, flanks, leg feathers. ▪ Blackish underwing coverts. Wide black rear edge of remiges. ▪ Banded type rufous tail.

**Plate 380. Red-tailed Hawk, adult "Western" (*B. j. calurus*) dark morph** [Nov.] ▪ Head and all of body dark brown, occasionally black. ▪ Wide black band on rear edge of remiges. Outer remiges may be barred. ▪ Banded type rufous tail.

**Plate 381. Red-tailed Hawk, adult "Harlan's" (*B. j. harlani*) light morph** [Dec.] ▪ Pale spectacles, white throat. ▪ Stark white underparts with lightly marked type black blobs on belly and flanks. ▪ Black patagial mark. Barred type remiges, including outer primaries. ▪ Pale grayish tail with irregular black subterminal band.

**Plate 382. Red-tailed Hawk, adult "Harlan's" (*B. j. harlani*) light morph** [Jan.] ▪ Dark head, white throat. ▪ Stark white underparts with heavily marked type black blobs on belly and barred flanks. Barred leg feathers. ▪ Black patagial mark. Barred type remiges, including outer primaries. ▪ White tail with irregular black subterminal band.

**Plate 383. Red-tailed Hawk, adult "Harlan's" (*B. j. harlani*) light morph** [Mar.] ▪ White spectacles. ▪ White underparts with moderately marked type black barring on flanks and blobs on belly. ▪ Black patagial mark. Unmarked type remiges lack barring or mottling; black outer primaries. Wide black band on rear edge of remiges. ▪ Pale gray type tail; irregular black subterminal band.

**Plate 384. Red-tailed Hawk, adult "Harlan's" (*B. j. harlani*) light morph** [Mar.] ▪ White spectacles. ▪ Classic types lack dark barring on white part of scapulars. ▪ Pale gray-rufous type tail with gray mottling on inner half, rufous on outer half; irregular black subterminal band. ▪ *Note:* Even though rufous in tail, still classic *harlani.*

**Plate 385. Red-tailed Hawk, adult "Harlan's" (*B. j. harlani*) light intermediate morph** [Jan.] ▪ White head with dark neck and malar. ▪ Sparsely streaked white breast. Black belly and flanks speckled with white. White leg feathers heavily barred with black. White undertail coverts. ▪ Barred type remiges, including outer primaries. Black patagial marks. ▪ White type tail.

**Plate 386. Red-tailed Hawk, adult "Harlan's" (*B. j. harlani*) intermediate morph** [Jan.] ▪ Whitish throat. ▪ Breast mottled white and black. Rest of underparts, including undertail coverts, black with some white speckling. ▪ Barred and mottled type remiges; black outer primaries. ▪ Medium/dark gray type tail.

**Plate 387. Red-tailed Hawk, adult "Harlan's" (*B. j. harlani*) intermediate morph** [Dec.] ▪ Whitish throat. ▪ Breast mottled white and black. Rest of underparts black with white spotting. ▪ Mottled type remiges have fine gray mottling visible only at close range; black outer primaries. ▪ Mottled-barred type tail with rufous tinge on some rectrices.

**Plate 388. Red-tailed Hawk, adult "Harlan's" (*B. j. harlani*) dark intermediate morph** [Dec.] ▪ Black head with white speckling. ▪ Black underparts, including undertail coverts; small amount of white mottling on breast. ▪ Barred type remiges; black outer remiges. ▪ Dark gray type tail. ▪ *Note:* Same bird as on plate 365.

**Plate 389. Red-tailed Hawk, subadult "Harlan's" (*B. j. harlani*) dark intermediate morph** [Nov.] ▪ Black head. ▪ Black body with some white mottling on breast. ▪ Barred type remiges. Outer 3 primaries and 2 or 3 secondaries are juvenile. Juvenile primaries brown and narrowly barred; secondaries lack wide black band. ▪ Pale gray-rufous type adult tail.

**Plate 390. Red-tailed Hawk, adult "Harlan's" (*B. j. harlani*) dark intermediate morph** [Nov.] ▪ White throat. ▪ Spot-bellied type has black breast with white spotting on black belly, flanks, leg feathers, and undertail coverts. ▪ Barred type remiges; barred outer remiges. ▪ Pale gray banded type tail appears white in translucent light; neat black subterminal band. ▪ *Note:* Same bird as on plate 394.

**Plate 391. Red-tailed Hawk, adult "Harlan's" (*B. j. harlani*) dark morph** [Mar.] ▪ Head and all of body black or brownish black. ▪ Barred type remiges; black outer primaries. Black wing coverts. ▪ Mottled-barred type medium/dark gray tail (most dark morphs have darker gray tails). Irregular wide black terminal band (some rectrices broken).

**Plate 392. Red-tailed Hawk, adult "Harlan's" (*B. j. harlani*) intermediate morph** [Dec.] ▪ Dark upperparts. Whitish uppertail coverts. ▪ White or pale gray type tail with irregular black areas on several rectrices. Irregular wide black terminal band. ▪ *Note:* Same bird as on plate 363.

**Plate 393. Red-tailed Hawk, adult "Harlan's" (*B. j. harlani*) dark intermediate morph** [Jan.] ▪ Dark head with some white speckling; whitish throat. ▪ Mainly dark upperparts. ▪ Medium/dark gray type tail with irregular black terminal band.

**Plate 394. Red-tailed Hawk, adult "Harlan's" (*B. j. harlani*) dark intermediate morph** [Nov.] ▪ Pale gray banded type tail with fairly neat wide black subterminal band. ▪ *Note:* Same bird as on plate 390.

**Plate 395. Red-tailed Hawk, adult "Eastern" x "Western" (*B. j. borealis* x *B. j. calurus*)** [Nov.] ▪ White throat as on *borealis.* ▪ Unmarked leg feathers as on *borealis.* ▪ Black patagial mark. ▪ Banded type rufous tail as on *calurus.* ▪ *Note:* Intergrades may also have dark throats and partially barred leg feathers.

**Plate 396. Red-tailed Hawk, adult "Western" x "Harlan's" (*B. j. calurus* x *B. j. harlani*)** [Feb.] ▪ Pale throat. ▪ Rufous breast of intermediate morph *calurus* with white *harlani* streaking. ▪ Black patagial mark. Barred remiges, including outer primaries. ▪ Mottled-barred type *harlani* tail is white on underside, brownish on upperside; neat subterminal band.

**Plate 397. Red-tailed Hawk, adult "Western" light morph (partial albino)** [Nov.] ▪ Body normal for this morph. ▪ Partially white or all-white feathers on some remiges. ▪ Tail normal. ▪ *Note:* Partial albinism affects remiges and dorsal areas more than other areas.

**Plate 398. Red-tailed Hawk, adult "Eastern" (*B. j. borealis*) (partial albino)** [Nov.] ▪ Head and body white with a few scattered dark feathers. ▪ Remiges white with a few partially marked normal feathers. ▪ Tail a mixture of rufous and white.

**Plate 399. Red-tailed Hawk, juvenile "Eastern" (*B. j. borealis*)** [Nov.] ▪ Broad white supercilium, white throat, pale iris. Greenish yellow cere. ▪ Large white patch on each scapular may form V. ▪ Extensive white mottling on wing coverts. Wingtips much shorter than tail tip. ▪ Pale brown tail has numerous thin, equal-width dark bands.

**Plate 400. Red-tailed Hawk, juvenile "Krider's" (pale morph of *B. j. borealis*)** [**Feb.**] ▪ White head with brown streaking on nape and hindneck, pale iris, greenish yellow cere. ▪ Belly streaked, flanks barred. Unmarked leg feathers. ▪ Distinct dark barring on secondaries. Wingtips much shorter than tail tip. ▪ *Note:* Palest type.

**Plate 401. Red-tailed Hawk, juvenile "Krider's" (pale morph of *B. j. borealis*)** [**Nov.**] ▪ Head white with brownish sides. ▪ White underparts with typical streaked belly, barred flanks. ▪ Distinct, thin dark barring on secondaries. Extensive white mottling on wing coverts. ▪ White tail has thin, equal-width bands.

**Plate 402. Red-tailed Hawk, juvenile "Western" (*B. j. calurus*) light morph.** [**Sep.**] ▪ Narrow white supercilium, dark tawny auricular region, dark throat, pale iris. Greenish yellow cere. ▪ Small white patch on each scapular may form partial V. ▪ Minimal white mottling on wing coverts. Wingtips shorter than tail tip. ▪ Pale brown tail has numerous moderately thin, equal-width dark bands.

**Plate 403. Red-tailed Hawk, juvenile "Western" (*B. j. calurus*) light intermediate (light rufous) morph** [**Nov.**] ▪ Partial tawny supercilium, dark tawny auricular region, dark throat, pale iris. Greenish yellow cere. ▪ White breast with narrow brown streaking; less on center region. Nearly solid brown band on belly and flanks. Barred leg feathers and undertail coverts.

**Plate 404. Red-tailed Hawk, juvenile "Western" (*B. j. calurus*) intermediate (rufous) morph** [**Nov.**] ▪ Dark tawny supercilium and auricular regions, dark throat, pale iris. Greenish cere. ▪ Tawny breast with wide brown streaking. Nearly solid brown band on belly and flanks. Heavily barred leg feathers and undertail coverts. ▪ Moderately thin dark tail bands.

**Plate 405. Red-tailed Hawk, juvenile "Western" (*B. j. calurus*) intermediate (rufous) morph** [**Dec.**] ▪ Dark tawny supercilium and auricular regions, dark throat, pale iris. Greenish cere. ▪ Small white blotches on scapulars. ▪ Wingtips shorter than tail tip. ▪ Numerous thin, equal-width dark bands on pale brown tail (white terminal band worn off).

**Plate 406. Red-tailed Hawk, juvenile "Western" (*B. j. calurus*) dark intermediate (dark rufous) morph** [Dec.] ▪ Dark tawny supercilium and auricular regions, dark throat, pale iris. Greenish cere. ▪ Dark brown body with tawny-rufous edges on breast feathers. ▪ Wingtips shorter than tail tip. ▪ Moderately thin tail bands, but subterminal band sometimes a bit wider (on all morphs).

**Plate 407. Red-tailed Hawk, juvenile "Western" (*B. j. calurus*) dark morph** [Nov.] ▪ Dark brown head and throat. Pale iris. Greenish yellow cere. ▪ Uniformly dark brown body. ▪ Moderately wide equal-width tail bands; narrow white terminal band.

**Plate 408. Red-tailed Hawk, juvenile "Western" (*B. j. calurus*) dark morph** [**Nov.**] ▪ Dark brown head and throat. Pale iris. Greenish yellow cere. ▪ Uniformly dark brown body. ▪ Moderately wide equal-width tail bands; white terminal band absent.

**Plate 409. Red-tailed Hawk, juvenile "Harlan's" (*B. j. harlani*) light morph** [**Dec.**] ▪ White head, including throat, with brownish crown and malar. Pale iris. Greenish yellow cere. ▪ Brownish black upperparts. ▪ Pale brown tail has wide or narrow often irregular dark bands; may have dark "spike" along shafts.

**Plate 410. Red-tailed Hawk, juvenile "Harlan's" (*B. j. harlani*) intermediate morph (streak-breasted type)** [**Jan.**] ▪ Black head with white throat, supercilium, and forehead. Pale iris. ▪ White breast streaked with black. Rest of underparts black and speckled with white. ▪ Dark "spike" along feather shaft below dark subterminal band may extend onto tail tip.

**Plate 411. Red-tailed Hawk, juvenile "Harlan's" (*B. j. harlani*) intermediate morph (streaked type)** [**Dec.**] ▪ Black head with white speckling and throat streaking. Pale iris, greenish yellow cere. ▪ Underparts, including leg feathers, streaked with black and white. ▪ Dark "spike" visible on outermost rectrix and extends onto tail tip. Equal-width dark tail bands, or subterminal band wider.

**Plate 412. Red-tailed Hawk, juvenile "Harlan's" (*B. j. harlani*) intermediate morph (any type) [Jan.]** ▪ Black head (dark throat is possible intergrade with *calurus*). Pale iris. Greenish cere. ▪ Scapular markings are classic white ovals with dark T shape. ▪ Dark shaft streaks on feather tips on wing coverts and secondaries. ▪ Wavy or V-shaped equal-width tail bands.

**Plate 413. Red-tailed Hawk, juvenile "Harlan's" (*B. j. harlani*) dark intermediate morph (streak-breasted type) [Dec.]** ▪ Black head with short supercilium. Pale iris, greenish yellow cere. ▪ Black body with small amount of white streaking on breast. ▪ Equal-width dark tail bands.

**Plate 414. Red-tailed Hawk, juvenile "Harlan's" (*B. j. harlani*) dark intermediate morph (spot-bellied type)** [Dec.] ▪ Black head. Pale iris, greenish yellow cere. ▪ Black breast; white spotting on black belly, flanks, and leg feathers.

**Plate 415. Red-tailed Hawk, juvenile "Harlan's" (*B. j. harlani*) dark morph** [Jan.] ▪ Black head. Pale iris, greenish yellow cere. ▪ Black body. ▪ Wingtips much shorter than tail tip. ▪ Dark "spikes" below dark subterminal band extends onto tail tip. ▪ *Note:* Not readily separable from dark morph *calurus* except by tail tip.

**Plate 416. Red-tailed Hawk, juvenile "Eastern" (*B. j. borealis*) [Sep.]** ▪ Large pale panel on upperwing with pale primaries and primary coverts contrasting with darker secondaries and secondary upperwing coverts. ▪ *Note:* All juveniles have pale panel on upperwing.

**Plate 417. Red-tailed Hawk, juvenile "Eastern" (*B. j. borealis*) [Oct.]** ▪ White throat. ▪ Breast white but is tawny when recently fledged. Moderately marked type belly and flanks. Leg feathers may be spotted, barred, or unmarked. ▪ Dark patagial mark. Lightly barred axillaries. White, rectangular panel. ▪ Thin tail bands; subterminal band wider.

**Plate 418. Red-tailed Hawk, juvenile "Eastern" (*B. j. borealis*) [Oct.]** ▪ Many juveniles have rufous tail with numerous narrow, equal-width dark bands; subterminal band sometimes wider. White uppertail coverts barred. ▪ *Note:* Rufous-colored tails common in *borealis*, uncommon in light morph *calurus* and darker morphs of *calurus*.

**Plate 419. Red-tailed Hawk, juvenile "Krider's" (pale morph of *B. j. borealis*) [Jan.]** ▪ White head. ▪ White underparts have small amount of spotting or streaking on belly and arrowhead markings or barring on flanks. Unmarked leg feathers. ▪ Faint dark patagial mark. Outer primaries finely barred. ▪ Thin, equal-width dark bands on distal half of white tail.

**Plate 420. Red-tailed Hawk, juvenile "Krider's" (pale morph of *B. j. borealis*)** [**Jan.**] ▪ White head. ▪ White underparts have few markings. Unmarked leg feathers. ▪ White underwing has faint dark patagial mark. Outer primaries finely barred. White panel on dorsal surface of inner primaries. ▪ Thin, equal-width dark bands on distal half of white tail. ▪ *Note:* Same bird as on plate 419.

**Plate 421. Red-tailed Hawk, juvenile "Fuertes" (*B. j. fuertesi*)** [**Feb.**] ▪ White throat. ▪ White underparts lightly marked with spots or arrowhead shapes, including flanks; sometimes with V-shaped markings. Lightly spotted leg feathers. ▪ Faint dark patagial mark on underwing. Axillaries nearly unmarked. ▪ Thin, equal-width dark tail bands. ▪ *Note:* Similar to lightly marked *borealis.*

**Plate 422. Red-tailed Hawk, juvenile "Western" (*B. j. calurus*) light morph** [**Nov.**] ▪ Streaked or dark throat. ▪ White underparts are heavily marked type on belly and flanks. Heavily barred leg feathers. Lightly barred undertail coverts. ▪ Large, dark patagial mark; heavily marked wing coverts, including barred axillaries.

**Plate 423. Red-tailed Hawk, juvenile "Western" (*B. j. calurus*) light intermediate (light rufous) morph** [**Nov.**] ▪ Heavily marked type on belly and flanks. Breast partially streaked. Heavily barred leg feathers. Moderately barred undertail coverts. ▪ Dark patagial mark partially obscured by heavily marked wing coverts; heavily barred axillaries.

**Plate 424. Red-tailed Hawk, juvenile "Western" (*B. j. calurus*) intermediate (rufous) morph [Nov.]** ▪ Heavily marked type on belly and flanks. Breast streaked. Heavily barred leg feathers and undertail coverts. ▪ Dark patagial mark nearly obscured by heavily marked wing coverts; heavily barred axillaries. Thin tail bands with wider subterminal band.

**Plate 425. Red-tailed Hawk, juvenile "Harlan's" (*B. j. harlani*) intermediate morph** [**Dec.**] ▪ Large whitish panel on primaries and primary coverts cuts wing in half. ▪ Moderately wide dark tail bands; subterminal band wider. ▪ *Note:* Dorsal wing panel of similar looking darker morphs of *calurus* is pale brown.

**Plate 426. Red-tailed Hawk, juvenile "Harlan's" (*B. j. harlani*) intermediate morph** (**streak-breasted type**) [**Dec.**] ▪ White throat. ▪ Black and white streaked breast; rest of underparts black with small amount of white speckling, including on undertail coverts. ▪ Barred outer primaries. White speckling on black wing coverts. ▪ Moderately wide dark tail bands; subterminal band wider.

**Plate 427. Red-tailed Hawk, juvenile "Harlan's" (*B. j. harlani*) dark intermediate morph** (**spot-bellied type**) [**Dec.**] ▪ White throat with center streak. ▪ Black underparts spotted with white on belly, flanks, and leg feathers. ▪ Barred outer primaries. Dark "spikes" on tips of secondaries. Equal-width dark tail bands; dark "spikes" on tips of rectrices.

**Plate 428. Red-tailed Hawk, juvenile "Harlan's" (*B. j. harlani*) dark intermediate morph (streak-breasted type)** [**Nov.**] ▪ White throat. ▪ Black underparts lightly streaked with white on breast. ▪ Barred outer primaries. Dark "spikes" on tips of secondaries. Equal-width dark tail bands; dark "spikes" on tips of rectrices.

**Plate 429. Red-tailed Hawk, juvenile "Western" (*B. j. calurus*) dark intermediate (dark rufous) morph** [**Nov.**] ▪ Pale to medium brown rectangular-shaped panel on inner primaries and primary coverts. Outer primaries dark. ▪ Brown tail has narrow, equal-width bands; uppertail coverts dark. ▪ *Note:* Dark morph similar but dark on underparts.

**Plate 430. Red-tailed Hawk, juvenile "Harlan's" (*B. j. harlani*) dark morph** [**Nov.**] ▪ Head and body uniformly dark brown or black. ▪ Barred outer primaries. ▪ Equal-width dark tail bands. ▪ *Note:* Dark morph *calurus* similar but has solid dark outer primaries. Much intergrading between *harlani* and *calurus.*

# FERRUGINOUS HAWK
## *(Buteo regalis)*

**AGES:** Adult and juvenile. Unlike other large buteos, Ferruginous Hawk does not have a subadult plumage with retained juvenile feathering. Juvenile plumage is held for most of the first year. Juveniles have somewhat longer tails than adults.
**MOLT:** Acciptridae wing and tail molt pattern (*see* chapter 4). The first prebasic molt begins on the innermost primary (p1) as early as mid-Apr. on some birds but may not occur until late Apr. or early May in very northern populations. By mid-May, many have replaced p1 and 2 and have dropped old p3 and 4. Body molt on the back, scapulars, underparts, and leg feathers begins in Jun. but is not really apparent until Jul. Molt is rapid in Jul. and Aug and is completed, with total replacement of all juvenile feathering, in Oct. or Nov.

Subsequent prebasic molts begin on breeding birds in May or Jun. for females and in Jun. or Jul. for males. Females may replace at least half their primaries by the time nesting duties are completed, but males rarely begin molting until nesting is over. Adults are in the final stage of molt and replacing the outer one or two primaries (p9 and 10) in late Oct. Molt is generally completed in Nov. Annual molt is fairly complete but may retain a few of the previous year's feathers on the body and wing coverts. Remiges appear to completely molt each year, which is unusual for such a large raptor.
**SUBSPECIES:** Monotypic.
**COLOR MORPHS:** Adults have four color morphs: light, intermediate (rufous), dark intermediate (dark rufous), and dark. Light morphs have major plumage variations. Intermediate and dark intermediate morphs have minor plumage variations. Dark morphs have virtually none. There is no clinal variation between light and intermediate morphs. There is a clinal variation between intermediate and dark intermediate morphs and between dark intermediate and dark morphs.

Juveniles have three color morphs: light, intermediate (rufous), and dark. The dark morph is only subtly different than the intermediate, and they are not separable at moderate and long distances. *Note:* Previous literature considered the intermediate (rufous) morph as a dark morph. The true dark morph plumage was not recognized. However, the plumage differences between intermediate and true dark morphs are as obvious for Ferruginous Hawk as for Swainson's and Red-tailed hawks.
**SIZE:** A large raptor. Length: 20–26 in. (51–66 cm); wingspan: 53–60 in. (135–168 cm). Juveniles may appear longer because of their longer tail. Some males can appear deceptively small.
**SPECIES TRAITS:** HEAD.—**Black bill with a small pale blue area on the distal lower region of the upper mandible and on the base of the lower mandible. Large yellow cere and gape. The long gape extends to the halfway point under the eyes.** BODY.—Feathered tarsi. Thick toes. WINGS.—**When perched, the pale gray outer webs on the primaries form a distinct pale gray border on the outer edge on the folded wings. In flight, the underside of the outer five primaries (the "fingers") has small black tips. When perched, the wingtips are somewhat shorter than the tail tip. In flight, wings are moderately wide with parallel front and rear edges that taper to fairly pointed wingtips.**
**ADULT TRAITS:** HEAD.—Iris color varies from pale orangish brown to dark brown and may darken with age. WINGS (ventral).—**The outer five primaries have moderately small, sharply defined black tips. Trailing edge of the inner primaries and secondaries has a moderately wide gray band, and often a narrow black band superimposed on the gray band (most adult buteos have a wide black band).** The primaries do not have narrow barring on the inner portion of the feathers, and the secondaries may have only minimal narrow dark barring. WINGS (dorsal).—**Medium gray secondaries are crossed with narrow black bars. Primaries have a pale gray outer web and a white inner web and in flight create a whitish panel on the dorsal surface and a pale window on the ventral surface.** TAIL (ventral).—**Uniformly white.** A few birds have a partial, irregular dark subterminal band.
**ADULT LIGHT MORPH TRAITS:** HEAD.—**Short, broad dark eyeline may connect to rear of the**

**gape. The throat is always white. There are three main variations of head color:** (1) *Pale gray type.*—**Pale gray with narrow dark streaking.** *Note:* Mainly on males (J. Watson pers. comm.; BKW pers. obs.). (2) *Pale brown type.*—**Pale brown with narrow or moderate-width dark streaking.** (3) *Medium brown type.*—**Moderate brown with moderate-width or wide dark streaking. Dark eyeline still apparent. None of the variations exhibit a dark malar mark.** BODY (ventral).—White, but may have a tawny wash on the breast during any time of the year and appear similar to most recently fledged juveniles. Breast, belly, and flanks can be: (1) *Lightly marked type.*—Unmarked or have a tawny wash on the breast. (2) *Moderately marked type.*—**Breast may be white or tawny and either unmarked or has some narrow rufous or brown streaking. Belly is moderately spotted or barred with rufous or brown. Flanks are white and barred with rufous or brown, or rear half of flanks are rufous with dark brown barring.** (3) *Heavily marked type.*—**White or tawny breast is moderately or heavily streaked with rufous or brown; rarely is unmarked. Belly is heavily barred with rufous or brown. Flanks are rufous and barred with dark brown barring.** Leg feathers, including tarsi, are: (1) *All-white type.*—White and unmarked. *Note:* Rare type. (2) *White type.*—**White with sparse brown spotting or barring.** *Note:* Fairly common type. (3) *Rufous type.*—**Rufous with dark brown barring.** *Note:* Common type. Undertail coverts are white and unmarked on all birds. BODY (dorsal).—Back is dark brown. **Scapulars are mainly rufous but may fade to tawny-rufous by spring.** There is a moderately wide or wide dark brown streak on the central part of each scapular. WINGS (ventral).—The white underwing coverts are: (1) *Unmarked type.*—White and lacking dark markings or have only a small amount of rufous streaking on the lesser coverts on the patagial area. (2) *Moderately marked type.*—**Lesser coverts on the patagial region have a moderate amount of rufous streaking and spotting. The median coverts also have a moderate-sized rufous tip and possibly some barring on each feather. The wrist area and primary lesser coverts are unmarked. Axillaries are mainly white with a minimal amount of rufous markings.** (3) *Heavily marked type.*—**The lesser coverts on the patagial region are solid rufous. The median coverts are rufous on the distal half of each feather. The underwing appears two-toned with nearly solidly rufous coverts and pale remiges. The wrist area is unmarked and the primary lesser coverts are moderately marked. Axillaries are white with a moderate amount of rufous on the tip of each feather.** The tips of the primary greater coverts on all types are white on the basal region and dark gray on the tips. WINGS (dorsal).—**Lesser and median coverts are rufous with narrow dark streaking and appear as a large rufous shoulder region.** TAIL (dorsal).—Four main color variations on the outer webs of the rectrices: (1) *White type.*—**White and unmarked.** (2) *Gray type.*—**Gray with dark brown or black mottling on the distal half or two-thirds.** (3) *Rufous type.*—**Rufous with dark brown or black mottling on the distal half or two-thirds.** (4) *Mixed type.*—**Variable mix of gray or rufous and dark brown or black mottling.** Sometimes appears brownish. All types have white uppertail coverts with a small amount of rufous or brown spotting or barring or may be rufous with dark brown markings.

**ADULT LIGHT MORPH (LIGHTLY MARKED TYPE):** HEAD.—Pale gray or pale brown types. BODY (ventral).—Lightly marked type. Leg feathers are white type or rarely all-white type. BODY (dorsal).—Dark markings on the scapulars are moderately narrow or moderate in width. Scapulars often appear very rufous. WINGS (ventral).—Unmarked type. TAIL (dorsal).—Mainly white or gray type patterns. Rufous pigmentation is typically reduced. Uppertail coverts are mainly white and have only minimal dark markings. *Note:* Fairly common plumage type; it has a reduced amount of rufous pigmentation.

**ADULT LIGHT MORPH (MODERATELY MARKED TYPE):** HEAD.—Pale gray or pale brown types. BODY (ventral).—Lightly or moderately marked types. BODY (dorsal).—Dark markings on the scapulars are moderate in width and appear quite rufous. WINGS (ventral).—Moderately marked type. TAIL (dorsal).—Any of the four types, but rarely white type. Uppertail coverts may either be white or rufous. *Note:* Common plumage type; it has a moderate amount of rufous pigmentation.

**ADULT LIGHT MORPH (HEAVILY MARKED TYPE):** HEAD.—Medium brown type. BODY (ventral).—Heavily marked type. BODY (dorsal).—Scapulars are rufous but the dark brown central feather streak is often wide. WINGS (ventral).—Heavily marked type. TAIL (dorsal).—Mainly rufous or mixed types. Uppertail coverts are typically quite rufous. *Note:* Uncommon plumage type; it is saturated with rufous pigmentation.

**ADULT INTERMEDIATE (RUFOUS) MORPH TRAITS:** HEAD.—Rufous-brown, dark brown, or dark gray. Gray-headed birds may be males (J. Watson pers. comm.; BKW pers. obs.). All of the lore region is white and the forehead is dark. **The large yellow gape is prominent.** BODY (ventral).—Variably marked on the breast, but the belly and flanks are always rufous: (1) Breast is rufous and makes all of the underparts rufous. (2) Breast may be dark brown and contrast with the rufous belly and flanks. The breast may be like a bib or create a hooded appearance along with the dark head. Both types often have dark brown or black streaking or partial barring on the belly and flanks. Both types may also have a variable amount of white streaking or mottling on the breast. BODY (dorsal).—Back is dark brown. Scapulars are dark brown and each feather is narrowly edged with rufous. WINGS (ventral).—Uniformly rufous underwing coverts. The dark band on the trailing edge of the remiges may be more distinct than on many light morphs. The white basal region and dark tips of the primary greater coverts form a white "comma" (comparison only to dark Rough-legged Hawks). WINGS (dorsal).—The dark brown lesser upperwing coverts are edged with rufous and form a distinct rufous shoulder patch. Secondaries and primaries are marked as in Adult Traits. White and unmarked. TAIL (dorsal).—**Uniformly medium gray, but sometimes has a small amount of rufous or brown intermixed. At close range, may exhibit a partial mottled pattern.** Uppertail coverts are rufous and very obvious as they contrast against the dark dorsal body and gray tail.

**ADULT DARK INTERMEDIATE (DARK RUFOUS) MORPH TRAITS:** This morph is a link between intermediate and dark morphs and shares traits of both. HEAD.—Same three colors as on intermediate morphs. **Yellow gape is prominent.** BODY (ventral).—Two variations: (1) *Rufous breasted type.*—Rufous breast contrasts sharply with dark head and blackish brown belly, flanks, and leg feathers. (2) *All-dark type.*—All of the underparts are uniformly dark brown with a small amount of rufous speckling. Breast may have a small amount of white streaking or mottling. BODY (dorsal).—Dark brown with narrow rufous edging on all scapular. WINGS (ventral).—All have blackish brown underwing coverts with a small amount of rufous speckling. White "comma" on the base of the primary greater coverts (comparison only to dark morph Rough-legged Hawks). WINGS (dorsal).—Narrow rufous edging on the lesser upperwing coverts; forms a moderately distinct rufous shoulder patch. TAIL (dorsal).—**Uniformly medium gray but may have a small amount of rufous or brown wash and sometimes darker mottling.** Uppertail coverts are dark brown with possibly a small amount of rufous edging or mottling. *Note:* This morph has not been previously described or depicted.

**ADULT DARK MORPH TRAITS:** HEAD.—Blackish brown with dark forehead and white lores. **Yellow gape is prominent.** BODY (ventral).—Uniformly blackish brown and does not exhibit white mottling on the breast. BODY (dorsal).—Uniformly blackish brown and does not have rufous edging. WINGS (Ventral).—Underwing coverts are uniformly blackish brown. The white "comma" is fairly obvious on the basal area of the primary greater coverts (comparison only to dark Rough-legged Hawks). WINGS (dorsal).—Upperwing coverts are uniformly blackish brown and do not show rufous on the shoulder region. TAIL (dorsal).—**Uniformly medium gray. May exhibit some brownish but not rufous.**

**JUVENILE TRAITS:** HEAD.—Iris color varies: pale gray, pale brown, or pale yellow. WINGS (ventral).—The five outer primaries are pale gray with a small dark gray tip. The underside of the remiges is white and has a grayish border on the trailing edge with little if any narrow dark barring. WINGS (dorsal).—Outer web of each primary feather is pale gray and the inner web is white; as in adults, forms a distinct whitish panel on the dorsal surface and a pale window on the ventral surface. The greater primary coverts are darker than the outer webs of the primaries but paler than the secondaries. The dorsal surface of the secondaries is medium brown with three or four rows of narrow dark bars.

**JUVENILE LIGHT MORPH TRAITS:** HEAD.—Dark brown crown, forehead, and nape. The supercilium is tawny or white. The cheeks and auriculars are tawny or white. **Short, thick dark eyeline often connects to rear of the gape. There is no dark malar mark.** BODY (ventral).—White undersides. Sides of the neck and breast may be sparsely streaked. The belly can be (1) unmarked, (2) moderately spotted and form a slight belly band, or (3) heavily spotted and form a distinct belly band. The flanks of all birds are moderately or heavily marked with dark spots or arrowhead shapes. **In flight, the large leg muscles condense the flank feathers and force the markings into an elongated dark patch. Tibia feathers on upper leg are either lightly or moderately spotted. Tarsi feathers are either white and spotted with brown or all brown.** The undertail coverts are white and unmarked. BODY (dorsal).—Dark brown with narrow tawny edges on the scapular. Small white spots may occasionally show on the basal area of some scapular. WINGS (ventral).—White underwing coverts and axillaries are variably marked with dark brown: (1) unmarked, (2) lightly spotted, or (3) moderately spotted. WINGS (dorsal).—Upperwing coverts are dark brown and have tawny or tawny-rufous edges on most feathers. TAIL (ventral).—Basal third to fourth is white and unmarked; distal region is pale gray with a moderately wide dusky subterminal band and two to four narrow, often partial, dark inner bands. TAIL (dorsal).—Basal third to fourth is white, distal region is gray in fresh plumage and medium brown in fairly worn plumage. The distal gray or brown region has a moderately wide dark subterminal band and two to four, often partial, dark inner bands. The white uppertail coverts have a dark spot on each feather.

**JUVENILE LIGHT MORPH TRAITS (WORN PLUMAGE):** By spring and early summer, 1-year-olds may have rather worn and faded plumage. HEAD.—**The crown, forehead, and nape may wear and fade to white and the entire head may be white except for the wide dark brown eyeline.** BODY (dorsal).—All pale feather edgings wear off and the upperparts become uniformly brown. TAIL.—Tip can become quite frayed and broken. Juveniles, in particular, brace themselves with tail while standing on the ground while feeding. This is noticeable by mid-winter and becomes more prevalent by spring.

**JUVENILE LIGHT MORPH TRAITS (RECENTLY FLEDGED):** HEAD.—Supercilium and auricular regions are dark tawny. The head can appear quite brown and the dark eyeline is not always prevalent until more fading occurs. BODY (ventral).—The front of the neck and breast is either pale tawny or tawny and forms a bib. This wears off to a great extent by Sep., but some individuals still exhibit traces of it in Oct. BODY (dorsal).—Upperparts are dark brown and edged with tawny-rufous. WINGS and TAIL.—As in Juvenile Traits.

**JUVENILE INTERMEDIATE (RUFOUS) MORPH TRAITS:** HEAD.—Uniformly tawny-brown, including the forehead. Lores are white. BODY (ventral).—Neck and breast are tawny-brown and along with the tawny-brown head give a "hooded" appearance. The head and breast contrast sharply against the dark brown belly, flanks, and leg feathers. At close range, the belly, flanks, and leg feathers may show narrow tawny edging on many feathers. BODY (dorsal).—Mainly uniformly dark brown. WINGS (ventral).—Dark brown underwing coverts are often mottled with tawny or white. A white "comma" is visible on most birds (comparison only to dark morph Rough-legged Hawks). TAIL.—Two patterns that are most visible on the dorsal surface: (1) *Unbanded type.*—Medium gray or brownish gray on the entire dorsal surface. There is a moderately wide darker, dusky subterminal band. The ventral surface is white with a dusky subterminal band. (2) *Partially banded/banded types.*—Medium gray or grayish brown dorsal surface. There is a moderately wide darker, dusky subterminal band and three or four partial or complete narrow or fairly narrow dark inner bands. On the banded type, the dark inner tail banding is on all rectrices but becomes less distinct on the outer sets. On the partially banded type, only the deck rectrices (r1) have somewhat distinct dark banding. The ventral surface is white with a dusky moderately wide subterminal band and partial or fairly complete narrow dusky inner bands. By late winter and spring, the dorsal grayish sheen wears off and becomes quite brownish. The uppertail coverts are dark brown.

**JUVENILE DARK MORPH TRAITS:** HEAD.—Uniformly dark brown, including the forehead. Lores are white. BODY (ventral).—Uniformly dark brown. The breast feathers may have a hint of tawny edging but not enough to create the pale bib as on intermediate morphs. BODY (dorsal).—Uniformly dark brown. WINGS (ventral).—Uniformly dark brown underwing coverts. White "comma" shows on most birds (comparison only to dark morph Rough-legged Hawks). WINGS (dorsal).—Upperwing coverts are uniformly dark brown. TAIL.—As on intermediate morphs. Uppertail coverts are dark brown. *Note:* Not separable from intermediate morphs unless seen at close range and in good light.

**ABNORMAL PLUMAGES:** None documented.

**HABITAT: Summer.**—Remote, open, arid, short-grass prairies, cattle pastures, and sagebrush scrub. Terrain varies from flatlands to gentle rolling hills, large hills, and badlands. If the habitat is void of trees, it has buttes, large boulders, and rock spires for nest sites. If trees are present, either coniferous or deciduous types may be used for nest sites. Nest sites may be single trees, in small natural tracts, in artificial shelterbelt groves, or occasionally on the perimeter of a pinon-juniper woodlands. Nesting also occurs on the perimeters of Ponderosa Pine, Limber Pine, or Quaking Aspen stands. On the central and northern prairies, ponds and lakes dot the landscape. In some regions, a small amount of agriculture may be interspersed among natural areas. Partially human-altered regions generally contain artificial structures that are also used for nest sites. Ferruginous Hawks are typically intolerant of humans in their breeding territories. However, if undisturbed, some pairs nest within 0.5 mile (0.8 km) of active homesteads.

In the s. and cen. Rocky Mts., breeding habitat extends in a few areas to 7,500 ft. (2,300 m). The highest nesting elevations in the s. Rockies are on the Plains of San Agustin (7,100 ft. [2,200 m] and near Datril (7,500 ft. [2,300 m], both in Catron Co., New Mexico. Nesting pairs in Albany, Carbon, and Sweetwater Cos., Wyo., often breed above 7,000 ft. (2,100 m) and approach the maximum elevation. In Montana, nesting occurs in isolated areas to 7,200 ft. (2,200 m). Identical habitat and prey are found at slightly higher elevations (8,100 ft. [2,500 m]) in portions of the cen. Rocky Mts. of Colorado, but nesting does not occur in these areas.

**Winter.**—Inhabits similar, remote, open areas as during the summer, but generally at more southern latitudes and lower elevations. Semi-open, human-altered locations, however, are regularly inhabited in this season. In e. Colorado, wintering birds regularly occur in suitable habitat at 5,300 ft. (1,600 m) and fairly regularly in limited areas of high plains up to 6,000 ft. (1,800 m). In ne. New Mexico, regularly found on the high plains at 6,600 ft. (2,000 m) in Colfax Co. and a few in Catron Co. Winter climate is generally arid but is moderately humid for populations wintering in the eastern winter range and humid in the southeastern winter range of coastal Texas.

Ferruginous Hawks readily acclimate to rural and suburban areas during the winter. They are common in these locations in e. Colorado, w. Kansas, and n. Texas where prey is plentiful, particularly at Black-tailed Prairie Dog towns. Although avoided during the breeding season, extensive agricultural areas become important winter habitat in areas where Black-tailed Prairie Dogs are absent. Wheat and irrigated alfalfa and grass hay fields often support a large prey base of small rodents.

**Migration.**—Found in habitats and regions described for summer and winter. During this transient period, appropriate habitat up to 8,100 ft. (2,500 m) may be regularly used. Ferruginous Hawks are common at this elevation in the autumn in Jackson Co. in n.-cen. Colorado.

**HABITS:** Wary during the breeding season and tame during other seasons. Ferruginous Hawks readily perch on the ground and on most elevated natural and artificial structures. *They do not perch on utility wires.* When it is very windy, juveniles especially will perch or lie on the ground on the lee side of isolated Soapweed Yucca, sage brush, or prairie dog holes. On hot summer days, adults will perch on the ground in the shadow of fence posts or utility poles.

Nesting pairs are solitary, but in the nonbreeding season Ferruginous Hawks are somewhat gregarious and several will gather at prime feeding areas. Over 30 may gather at large Black-tailed Prairie Dog towns. Winter night roosts in trees may host several Ferruginous Hawks plus a few Red-tailed Hawks, Rough-legged Hawks, and Bald and Golden eagles. In areas without trees, Ferruginous Hawks will roost on the ground.

Most raptors fly away at the sound of gunfire. Ferruginous Hawks, however, may fly towards the source of gunfire at prairie dog towns to prey on prairie dogs killed by shooters.

Ferruginous Hawks will abandon nests if disturbed prior to hatching of eggs.

The large gapes allow Ferruginous Hawks to dissipate heat, especially in nestlings which often sit baking in exposed, hot nest locations.

**FEEDING:** Perch and aerial hunter. Perch hunts are a direct method of foraging. They are launched from elevated objects or from the ground at prairie dog and gopher colonies. At Black-tailed Prairie Dog colonies, hawks may stand by a burrow and grab emerging occupants or grab tunneling gophers that are pushing up dirt. Aerial hunts are an indirect method of foraging (*see* Flight).

Ferruginous Hawks are opportunistic and feed on prey of all sizes and types. In the breeding season, they often rely on the dominant mammal of the region. Breeding success and population fluctuations are often linked to the abundance of the dominant prey. The hawks feed on a greater variety of prey in the breeding season in spring and summer than in the fall and winter. Primary prey species include ground squirrels, prairie dogs, jackrabbits, rabbits, pocket gophers, and kangaroo rats. Secondary prey species, in descending order of importance, include other small rodents, birds, reptiles (except in Washington State, rarely feeds on snakes), and large insects such as Lubber Grasshoppers.

Ground squirrels inhabit the hawk's breeding range. Most species are unavailable as prey during portions of the year because they estivate during hot seasons and hibernate in cold seasons. Richardson's and Thirteen-lined ground squirrels are dominant prey in the provinces and states that have the larger breeding populations of Ferruginous Hawks. Richardson's are the dominant prey in Alberta, Montana, N. Dakota, Saskatchewan, and Wyoming. Thirteen-lined are an important prey in e. Colorado. Belding and Townsend ground squirrels are dominant prey in the Great Basin. Other species are locally important in the breeding season. The endangered Southern Idaho Ground Squirrel is an important prey species in s. Idaho.

Four species of prairie dogs inhabit the main summer and winter range of Ferruginous Hawk. A fifth species, the Mexican Prairie Dog, inhabits a small area in cen. Mexico south of the Ferruginous Hawk's main winter range. All species are colonial. Human persecution has dramatically reduced—and continues to reduce—prairie dog populations. Three species estivate during part of the summer and hibernate during winter and afford only seasonal prey availability for the hawks. Black-tailed Prairie Dogs do not estivate or hibernate and afford a staple diet in all seasons. During periods of inclement winter weather, Black-taileds may remain submerged in burrows for a few days and temporarily affect the hawk's local food supply.

Black-tailed Prairie Dogs are the most widespread species. They are colonial and establish towns and/or colonies in closely knit units; a collection of towns and/or colonies form a complex. However, with considerable fragmentation, very few complexes remain (the largest is near Janos in n. Chihuahua, Mexico). Black-taileds are found east of the Rocky Mts. from n. Montana to e. N. Dakota and south through w. Texas and n. Chihuahua. Black-taileds inhabit half of the Ferruginous Hawk's winter range and were their staple winter prey until being nearly decimated. Due to intense persecution, Black-tailed range and population have been reduced 99% from historical times. In Feb. 2000, the USFWS warranted listing the Black-tailed Prairie Dog as Threatened under the Endangered Species Act of 1973. However, due to the priority of other species, the Service placed it on a candidate list to be monitored annually. Despite listing, persecution continues! The deadly sylvatic plague, which was introduced to prairie dogs by fleas and rats that came over on Asian ships in the early 1900s, has spread throughout much of the prairie dog's range wiping out entire colonies.

White-tailed Prairie Dogs inhabit more arid

and higher elevations than the Black-tailed and are found in n. Colorado, cen. and w. Wyoming, and ne. Utah. Ferruginous Hawks prey heavily on this species in the breeding season. Gunnison's Prairie Dogs, also of arid regions, are found in a small part of the hawk's range in w. Colorado, ne. New Mexico, n. Arizona, and se. Utah. Utah Prairie Dog is a Threatened Species. It is a minor prey species in the arid lands of s.-cen. Utah in the breeding season.

Two jackrabbit species inhabit most parts of the Ferruginous Hawk's range. The White-tailed Jackrabbit is in most of the hawk's breeding range and a small part of its winter range. White-taileds are major prey in many areas, especially west of the Rocky Mts. The Black-tailed Jackrabbit is in the southern half of the hawk's breeding range and all of its winter range. It is a major prey species, particularly in winter. Desert, Eastern, and Mountain cottontail rabbits live in portions of the Ferruginous Hawk's summer and winter ranges. Cottontails form a minor to moderate percentage of the summer prey and are fairly important winter prey. Pocket gophers and kangaroo rats may be minor or major breeding-season food sources, depending on locality. They are minor prey species in winter.

Except in the breeding season, Ferruginous Hawks typically eat prey at the site of capture. At large prairie dog towns, several Red-tailed Hawks and Bald and Golden eagles may also vie for prey. If eagles are present, they pirate prey from the smaller Ferruginous Hawks. If eagles are not present, Ferruginous Hawks vigorously defend their prey from conspecifics and Red-tailed Hawks. The most aggressive raptor mantles the prey and feeds while others perch on the ground nearby and wait for leftovers. Eighteen Ferruginous Hawks and four Red-tailed Hawks have been seen vying for prey at a large Black-tailed Prairie Dog town in e. Colorado. (A large Red-tailed Hawk may dominate over a Ferruginous Hawk when feeding.)

**FLIGHT:** Wings are held in a high dihedral when soaring and in a high or modified dihedral when gliding. Powered flight consists of moderately slow wingbeats interspersed with irregular gliding sequences. The upstroke has a quick snapping movement. Hunting flights: (1) *High-altitude flight.*—Soaring or gliding at high altitudes and then making a long dive for the capture. (2) *Low-altitude flight.*—Moderate-speed, low-level powered and gliding flight and then making a short dive for the capture. (3) *Surprise-and-flush flight.*—High-speed, low-level powered and gliding flight, often undulating by nearly skimming the ground and then swooping upwards, then making a fast low-angle dive for capture.

**VOICE:** Quite vocal in the breeding season when disturbed at nest sites. Rarely heard in other seasons unless agitated. Vocalizes when prey is being pirated by eagles and in feeding confrontations with conspecifics and Red-tailed Hawks. Has a rather soft, clear, high-pitched, and drawn-out *kreeaah* or *keeeoh.* The call is sometimes louder and more forceful.

**STATUS AND DISTRIBUTION:** Overall, an *uncommon* raptor. Populations trends have concerned wildlife officials in several states and provinces. In 1980, the Committee on the Status of Endangered Wildlife in Canada listed the species as Threatened in Canada. In 1992, the USFWS was petitioned to list it as Threatened under the Endangered Species Act but determined there was not sufficient evidence to warrant the request. In 1995, Canada downlisted the species to Vulnerable because of an increasing and stabilizing population. A few states still consider the Ferruginous Hawk as Threatened or of uncertain status. Oregon designates it as Endangered and the BLM as Sensitive. Arizona and Colorado categorizes it as a Species of Special Concern.

Ferruginous Hawks, along with several other threatened or endangered species of birds and mammals, are unique adaptations to the prairie ecosystem, especially the Great Plains. Since the late 1800s, the Plains, in particular, have succumbed to agriculture, urbanization, and oil, gas, and mineral exploration and recovery. These activities have greatly reduced breeding areas and winter prey for the Ferruginous Hawk. In Canada, the prevention of grassland fires has allowed aspen parkland to encroach southward onto the prairies and reduced Ferruginous Hawk breeding habitat. The Ferruginous Hawk's future remains a concern at the start of the 21st century.

**Summer.**—Regionally and locally, populations

vary from rare to common. In 1992, the USFWS estimated 5,220–6,000 pairs in the U.S. and Canada. Number of pairs estimated in each state and province (updated status in parenthesis): Alberta 1,772, Arizona 25, British Columbia 2, California 1, Colorado 300–400, Idaho 100, Kansas 50–100, Manitoba 50, Montana 190–450, Nebraska 35, Nevada 240, New Mexico 35 (80–100), North Dakota 200, Oklahoma 20–33, Oregon 250, Saskatchewan: 750, South Dakota 400, Texas 0 (4), Utah 190–300, Washington 62, and Wyoming 800.

*Alberta:* Supports the largest and densest population in the species' range. Found east of the Rocky Mts. and south of Camrose in the southeastern part of the province. Highest concentrations are around Hanna. Population has increased since the 1980s. *Arizona:* The small population is found in three areas (1) southern Navajo (near Winslow) and Apache Cos., (2) w. Coconino Co. south of the Grand Canyon (has the most pairs), and (3) ne. Mohave Co. north of the Grand Canyon. Much of the northeastern part of the state is arid and overgrazed and has few raptors of any species. *British Columbia:* One pair in 1992 in the s.-cen. part of the province; possibly bred through the 1990s. *California:* One pair in Lassen Co. Habitat is adequate to support several pairs in the county. *Colorado:* Inhabits a large portion of the e. Plains. Large areas of the Plains are farmed and uninhabited. Absent from the heavily populated and extensively farmed Front Range area (Denver to Fort Collins). One or two pairs breed in isolated areas of Park and Alamosa Cos. A few breed in w. Garfield and Mesa Cos. Several pairs breed in cen. Moffat Co. May breed in sw. Montezuma Co. *Idaho:* Nests in the very southern part of the state south of the mountains. Absent from much of the extensively agricultural areas east of a line from Rexburg to Burley. Habitat loss and reduction of prey in the Snake River Birds of Prey National Conservation Area may affect breeding population in this area. The endangered Southern Idaho Ground Squirrel now occupies less than half of its historical range due to habitat loss. *Kansas:* Breeds in sw. Gove, s. Grant, s. Logan, se. Meade, sw. Morton, sw. Trego, and se. Wallace Cos. Possibly breeds, but more likely nonbreeding birds may be found in s. Clark, se. Greeley, cen. Hamilton, se. Kearny, sw. Meade, sw. Ness, ne. Scott, Stevens, and se. Trego Cos. Isolated sightings of probable nonbreeding birds occur in the south-central part of the state in se. Commanche and cen. Barber Cos. *Manitoba:* Breeds in the southwestern corner of the province. Extralimital sightings occur east to Winnipeg. A large percent of pairs nest on artificial platforms. *Montana:* Pairs are widely spread east of the Continental Divide. Absent from much of the northeastern part of the state due to extensive agriculture. Also absent from the many upland "sky island" wooded areas that dot the Great Plains in the eastern part of the state. *Nebraska:* The few pairs that nest in the state probably occur in Banner, Box Butte, Cherry, Dawes, Garden, Hooker, Kimball, Lincoln, and Sioux Cos. Possibly nests in Sheridan Co. *Nevada:* Most of the breeding population is found north of U.S. Highway 50. Nevada's habitat consists of inhabited, fragmented, low-elevation sagebrush valleys that are interspersed between high-elevation uninhabited "sky island" mountains. Population declines occurred in the 1990s. *New Mexico:* Nesting has occurred in 21 counties. High number of pairs are in the Estancia Valley in Torrance Co. and on the prairies in the northeastern part of the state north of I-40 and east of I-25. Breeding areas are fragmented in the western part of the state. The Plains of San Agustin in Catron Co. is an isolated area that historically had a large population when prairie dogs were more numerous. In the 1980s, breeding occurred south of typical areas in Eddy, Lea, and Otero Cos. Absent from the heavily agricultural east-central part of the state, montane regions, and portions of the very arid western part of the state. Range is considered to be similar to that of historical times. *North Dakota:* Found throughout much of the state except in heavily agricultural regions in the east. Mainly found west of a line from Cavalier Co. south to Nelson Co., then south to sw. Dickey Co. Recent breeding occurred in Grand Forks Co. Possibly breeds in s. Barnes Co. In Kidder Co., declines have occurred in the last few decades. *Oregon:* Breeding or possible breeding in Baker, Gillman, n. and s. Harney, n. and s. Lake, Malheur, Morrow, Sherman, Umatilla, Wallowa, and se. Wasco Cos. May breed in portions of Klamath Co. *Oklahoma:* Inhabits portions of Beaver,

Cimmaron, Ellis, and Texas Cos. *Saskatchewan:* Inhabits the southwestern corner of the province and a small part of the southeastern corner. Before the southward encroachment of aspen parkland, range extended farther north. The population is stable but experiencing low reproductive success. In recent years, the low population of Richardson's Ground Squirrels, the dominant prey, has food-stressed pairs. *South Dakota:* Breeding occurs mainly west of a line from sw. Brown Co. south to sw. Tripp Co. Due to extensive agriculture, absent as a breeder in much of the eastern third of the state. Isolated breeding occurs in ne. Brown, sw. Day, and se. Roberts Cos. Isolated pairs inhabit se. Spink Co. *Texas:* A few breeding pairs in Dallam, Hansford, and Moore Cos. Regularly seen but breeding not confirmed in Hartley and Deaf Smith Cos. Oldham Co. has occasional sightings, but much of the county is on inaccessible private ranches. Sherman Co. has had summer sightings. *Utah:* Most pairs are in the western part of the state. Absent from very arid, forested, and montane regions. Isolated pairs inhabit the eastern part of the state. Population declines occurred in the 1990s in n. Utah. Breeding range is poorly known. *Washington:* The small population is in the south-central and southeastern parts of the state. Breeding occurs or possibly occurs in Adams, Benton, n. Columbia, e.-cen. Douglas, Franklin, s. and e.-cen. Grant, s.-cen. Klickitat, sw. Lincoln, w. Walla Walla, and e. Yakima Cos. *Wyoming:* Much of range is on minimally disturbed and stable grazing lands and mirrors historical distribution. Population is stable. Numbers may have declined in se. Wyoming in the 1990s.

**Winter.**—Locally common where prey density is high but otherwise uncommon to rare. Areas with Black-tailed Prairie Dogs support the highest numbers of hawks. The hawks acclimate to humans and human-altered habitats. Adults predominate in the northern winter range, and juveniles are more common in eastern and southern winter range.

Most birds that breed east of the Rocky Mts. remain east of the mountains in winter. Juveniles banded in Saskatchewan winter as far south as n. Mexico. Banded individuals from Alberta winter mainly in Texas, but some winter in Arizona, Colorado, Kansas, and New Mexico. Some Alberta birds winter as far south as s. Durango and e. Nuevo León, Mexico. Banded juveniles from ne. Colorado were found in Texas, n. Mexico, and one in California. High winter densities are in the following areas: Front Range area of e. Colorado, Prowers Co. in se. Colorado (which has large Black-tailed Prairie Dog towns), w. Kansas, Texas panhandle, Oklahoma panhandle, Cochise Co. in se. Arizona, n. Los Angeles Co. in California, and near Janos in n. Chihuahua, Mexico (which supports the largest Black-tailed Prairie Dog complex in North America).

Other wintering areas on the range map have low winter densities and few birds. Small numbers regularly winter in valleys of nw. Montana (which also have numerous Red-tailed and Rough-legged hawks). A few birds regularly winter in isolated areas of s. Idaho. Irregular wintering occurs in parts of e. Montana, e. Kansas, and several areas of Wyoming. Very irregular wintering occurs east of the Cascade Mts. in Washington and Oregon. Highly irregular west of the Cascades in nw. Oregon.

*Winter range in Mexico.*—Baja California Norte, n. and coastal Sonora, on the Mexican Plateau between the Sierra Madre Occidental and Sierra Madre Oriental south to the state of Guanajunto and the Central Volcanic Belt Mts. A juvenile seen near Texcoco in the State of México represents the southernmost record in Mexico.

**Extralimital winter.**—Casual in winter in Arkansas, Iowa, and Louisiana; rare in Missouri. *Arkansas:* (1) Jan. 1965 in Prairie Co. and (2) mid-Feb. 1996 in Pope Co. *Iowa:* (1) mid-Feb. 1997 in Cerro Gordo Co. (juvenile light morph), (2) early Dec. 1999 in Pottawattamie Co., and (3) mid-Dec. in Pottawattamie Co. *Louisiana,* all of juveniles: (1) mid-Dec. 1989 in Acadia Parish, (2) mid-Jan. to early Feb. 1989 in Calcasieu Parish, and (3) early to mid-Jan. 1996 in Cameron Parish. *Missouri:* (1) early Dec. 1993 in Holt Co. (juvenile light morph); since 1995, three additional records: (2 and 3) two birds in Dec. at Prairie S.P. in Barton Co., (4) one at Squaw Creek NWR in Holt Co., and (5) one near Montrose in Henry Co.

**Movements.**—Very few Ferruginous Hawks are seen at western hawkwatch sites. Most migration data were based on banding records. Rather astounding data are surfacing with recent telemetry studies. Migration data are also based on 15 years of observations on the Great Plains.

*Post-breeding dispersal:* Based on telemetry data, adults in Washington State disperse northeasterly and cross the Continental Divide after nesting. Juveniles from all regions may disperse in any direction after fledging. Most dispersing birds head to the northern Great Plains of Montana and Canada where there is a late-summer abundance of juvenile Richardson's Ground Squirrels (adult ground squirrels estivate in Jul.).

After completing nesting, four telemetry-tracked adults from s.-cen. Washington headed northeast in early to mid-Jul. to the Great Plains of n.-cen. Montana, se. Alberta, and sw. Saskatchewan. They stayed on the n. Plains until early Sep. to early Oct. when they began their southward migration.

*Fall migration:* The first migrants arrive from late Aug. to early Sep. along Colorado's Front Range. Migrants are seen in Sep. and Oct. on the e. Plains of Colorado. Peak movement period on the Plains appears to be from early to mid-Oct. All birds have left Alberta by late Oct. A continual influx of birds is seen in Nov. and Dec. in se. Colorado. Nov. and Dec. movements may be due to food supply.

The four telemetry-tracked Washington adults took interesting routes from dispersal areas to winter areas. Two birds headed southwest and arrived in late Oct. in the Central Valley of California. One bird also headed southwest to n. California, then went back east and in late Oct. arrived in w.-cen. Nevada. The fourth bird headed southeast and arrived in late Oct. in the e. Oklahoma panhandle. Alberta, Colorado, and Saskatchewan banded birds stayed east of the Rocky Mts. and headed south and east of breeding and natal areas.

*Spring migration:* Adults migrate from mid-Feb. through Mar. Some juveniles migrate in Mar., but many continue into Apr. and May. "Northward" movements are no doubt highly variable, as seen with telemetry-tracked and banded individuals making diagonal movements in the fall.

**Extralimital movements.**—Regular in w. Minnesota in spring and fall and probably an annual migrant in some southwestern counties. Irregular movements occur through Iowa with five spring and fall records: (1) late Oct. 1973 in Page Co. (light morph adult), (2) late Oct. 1977 in Pocahontas Co. (juvenile light morph banded in North Dakota), (3) late Oct. 1983 in Page Co. (adult light morph), (4) late Sep. 1984 in Marshall Co. (juvenile light morph wing-tagged from North Dakota), and (5) late Mar. 1995 in Louisa Co. (dark morph adult).

Very irregular movements occur through Arkansas, Louisiana, and Missouri. *Arkansas:* (1) mid-Apr. 1998 in Sebastion Co. and (2) late Oct. 1999 in Pope Co. *Louisiana,* all of juveniles: (1) late Oct. 1988 in Cameron Parish, (2) late Oct. 1989 in Acadia Parish, and (3) early to mid-Nov. 2001 in Jefferson Davis Parish. *Minnesota:* (1) Hawk Ridge Nature Reserve, Duluth, in late Aug. 1984 and (2) Duluth in mid-Apr. 2002 (adult light morph). *Missouri:* one spring record and six fall records. All are light morph adults except one: (1) late Apr. 1949 in Platte Co., (2) early Nov. 1953 near Kansas City (dark morph), (3) late Nov. 1960 near St. Louis, (4) early Oct. 1970 in Andrew Co., (5) early Oct. 1972 in Nodaway Co., (6) mid-Oct. 1979 in Cedar Co., and (7) mid-Oct. 1981 in Vernon Co.

**Color morph status.**—Intermediate, dark intermediate, and dark morphs have not been categorized separately in previous literature or personal correspondence. Most previous data concerning "dark" morphs invariably included intermediate and dark intermediate morphs as well as pure dark morphs. The highest percentage of breeding "dark" morphs are in Alberta and Saskatchewan. "Dark" morphs are substantially less common in the breeding season in the U.S. Known percentages of dark birds from the breeding season: Alberta 9, Saskatchewan 7 (based on data by C. S. Houston: 296 adults from 1999, 2000, and 2001), Idaho 4, and Colorado 3. Other states have fewer nesting darker birds.

Based on data from 15 winter seasons (author's data), intermediate and dark intermediate morphs comprise only 5% of the birds on the Great Plains. True dark morphs comprise only 1% of the wintering population.

During the breeding season, light-light pairs are most common; light-dark pairs are common; and dark-dark pairs are very rare, even in Canada.

**NESTING:** Begins in late Feb. to late Mar. and ends from mid-Jun. to late Aug., depending on latitude and elevation. Adults arrive on breeding grounds in late Feb. and early Mar. in e. Colorado and by late Mar. in s. Saskatchewan. Based on known-aged birds from Alberta, Ferruginous Hawks do not breed until 3 years old.

**Courtship (flight).**—*High-soaring* (mutual) and *sky-dancing* by males. Both sexes perform *leg-dangling* and *cartwheeling* (with males approaching females, then locking talons and tumbling).

**Courtship (perched).**—*High-perching* by males. *See* chapter 5 for description of displays.

Ferruginous Hawks build their own nests or may refurbish nests of other large raptors, particularly nests of Swainson's and Red-tailed hawks. New nests may also be constructed on top of old Black-billed Magpie nests. Both sexes build nests, but refurbished structures often have only a minimal amount of work done to them. Nests vary in size. New nests may be small and old nests can be large. Old nests, which can be used for decades, can rival the size of eagle nests. Average nest diameter is 24–42 in. (61–107 cm). Nest depths vary considerably: new nests or old nests of other raptors can be rather flat; old nests are tall columnar or conical shapes. Pairs often have alternate nests in their territories. Pairs may rotate between nests each breeding season or use a favorite nest for several seasons. A tree nest in w.-cen. Saskatchewan was used for 32 consecutive years by a few different pairs, and successfully fledged young for the entire period. Another tree nest in Saskatchewan was used for 22 consecutive years by different pairs. A tree nest in se. Wyoming was used for at least 19 years.

Nests are placed up to 55 ft. (17 m) high, on the tops of the following elevated structures: live or dead bushes, including sagebrush; wooden or metal utility poles; non-functioning windmills; haystacks; abandoned buildings; large rocks; rock pinnacles; and artificial platforms. Nests may be placed on the mid-height areas of metal transmission towers. Nests are also placed on the ground on gently or steeply angled hillsides, on knolls, on the top edge of a low embankment or low cliff, and in some regions on barren flat areas. Most nest sites are in exposed areas and offer little or no protection for the nestlings from the sun and elements. In trees, nests can be located in upper or lower area sections. Even in large trees, nests can be placed high or low. Nests that are placed above the ground may have as many sticks littering the ground below the nest as in the nest itself. All nest locations offer panoramic views of the surrounding open landscape. Tree nests are in isolated trees, on the periphery of a natural-growing cluster of trees, or in artificially planted shelterbelt groves. Nest trees may also protrude out of shallow water of a pond or lake edge.

Nests built on flat ground are an uncommon or rare phenomenon in most regions. This occurs more in North Dakota than in any other state or province. Nearly half the nests in North Dakota are located on the ground.

Nests are built of sticks. In treeless regions, sagebrush sticks are most commonly used. In areas with trees, tree sticks are used as well. Bones are often added to the nest structure. Horse and cow dung (formerly Bison dung), grasses, and bark comprise the inner lining of nests. Greenery of small twigs and branches may also be added to the lining.

Two to 4 eggs comprise the typical clutch, but 5 eggs are occasionally laid. Highly unusual clutches of 8 eggs have occurred. Eggs are incubated by both sexes for 32 or 33 days. Youngsters fledge in 38–50 days, with females taking longest. Most branch long before they can fly. Tree-nesting branchers perch on outer branches away from the nest or, more typically, fall, jump, or glide to the ground and stay there, often considerable distances from their nest. Ground-nesting branchers simply walk away from the nest and may also remain a considerable distance from it. Fratricide among young nestlings may occur when food is scarce.

**CONSERVATION:** Since the mid-1970s, provinces and some states have been erecting nest platforms to boost breeding potential. Suitable elevated natural nesting structures are becoming scarce in some prairie areas, and platforms generate more and safer nest sites. Anti-predator shields on nest platforms deter terrestrial predators. In Canada, trees are also planted to ensure future nest sites. Existing nest trees are protected from damage from cattle.

The U.S. Government subsidizes farmers to let acreage return to its natural state or to plant grasses. Utility poles are being modified in areas with large raptor concentrations to prevent electrocution. Lands purchased by state and local governments and nonprofit wildlife organizations give migrating and wintering Ferruginous Hawks a place to feed.

**Mortality.**—Ferruginous Hawks suffer from illegal shooting, particularly during upland gamebird seasons in the fall. Vehicles may hit hawks that feed on road-killed prey. Electrocution from utility poles and wires is prevalent. Since Ferruginous Hawks readily feed on carrion, secondary poisoning meant for varmints and organophosphate poisons used to control insect infestations in cattle indirectly kill scavengers feeding on animals that have died. Cattle break down dead nest trees by rubbing against them and may topple nests and nestlings. Golden Eagles and Great Horned Owls kill adults and youngsters. Coyotes and foxes may kill branchers on the ground.

**SIMILAR SPECIES:** COMPARED TO ADULT LIGHT MORPHS (mainly lightly marked type.—**(1) Red-tailed Hawk (*B. j. borealis*), "Krider's" adults.**—PERCHED.—Head lacks a dark eyeline. Upperparts are very similar, but "Krider's" has large white patches on the mid-scapulars. Tarsi are bare; leg feathers are unmarked. Dark subterminal tail band is thin and neat. Otherwise tail is very similar. FLIGHT.—Rufous patagial mark on the underwing may be similar to the rufous patagial on Ferruginous Hawks: use caution. Underside of the outer primaries is fully barred. Tail pattern as in Perched. COMPARED TO JUVENILE LIGHT MORPHS.—**(2) Red-tailed Hawk (*B. j. borealis*), "Krider's" juveniles.**—PERCHED.—Lacks dark eyeline. Cere is greenish. Large white patch on the mid-scapulars. Belly band markings are similar. Tarsi are bare. Tail is similar but dark banding is very narrow. Wingtip-to-tail-tip ratio is similar. FLIGHT.—Underside of the outer primaries is fully barred. Lacks a dark mark on the sides of the flanks. Tail pattern as in Perched. **(3) Red-tailed Hawk (*B. j. calurus* light morph *and B. j. fuertesi*), juveniles.**—PERCHED.—All have dark malar marks and lack a dark eyeline. Greenish ceres. Throats are often streaked or dark. White patches on mid-scapulars. Tails are brown and fully banded with narrow dark bands. FLIGHT.—All have distinct, dark brown patagial marks on the underwing. Tail pattern as in Perched. COMPARED TO INTERMEDIATE AND DARK INTERMEDIATE MORPH ADULTS.—**(4) Swainson's Hawk, intermediate and dark intermediate morph adults.**—PERCHED.—Forehead and outer lores are white. Thin yellow gape. Underparts are very similar. Bare tarsi. White undertail coverts. Wingtips reach or extend beyond tail tip. Tail is dark and fully banded. FLIGHT.—Dark gray underside of remiges. Undertail coverts and tail pattern as in Perched. **(5) Red-tailed Hawk (*B. j. calurus*), intermediate morph adults.**—PERCHED.—Cere is often greenish. Bare tarsi. Rufous dorsal surface of tail with a neat black subterminal band; often multiple inner dark bands. FLIGHT.—Dark brown patagial mark on the underside of the wings. All-dark or barred outer primaries. Tail as in Perched. COMPARED TO INTERMEDIATE MORPH JUVENILES AND DARK MORPH ADULTS.—**(6) Swainson's Hawk, dark morph adults.**—PERCHED.—White spot on outer lores. Dark body is similar. Bare tarsi. White or tawny undertail coverts, and may be barred. Wingtips extend to or just beyond tail tip. Tail is fully banded. FLIGHT.—Dark gray underside of the remiges. Undertail coverts and tail as in Perched. **(7) Red-tailed Hawk (*B. j. calurus*), dark morph adult.**—PERCHED.—Bare tarsi. Rufous dorsal tail surface with neat black subterminal band; often multiple inner dark bands. FLIGHT.—All-dark or fully barred outer primaries. Tail as in Perched. **(8) Rough-legged Hawk, dark morph juveniles.**—PERCHED.—Yellow gape can be similar. White forehead and outer lores. Feathered tarsi are similar. Wingtips equal or nearly equal to tail tip. Dorsal and ventral tail patterns can be identical. FLIGHT.—Underside of outer primaries extensively dark. All of primary greater underwing coverts are typically dark; some have whitish comma as on Ferruginous. Pale wing panels/windows are similar. Tail pattern as in Perched.

**OTHER NAMES:** Ferrug. *Spanish:* Aguililla Real. *French:* Buse Rouileuse.

---

REFERENCES: Adamus et al. 2001; Andrews and Righter 1992; Arizona Breeding Bird Atlas 1993–2000; Bechard 1997; Bechard and Schumtz 1995; Burt and Grossenheider 1976; Busby and Zimmerman 2000; Cartron 2001; Cecil 1997; Dodge 1988–1997; Dold 1998; Dorn and Dorn 1990; Ducey 1988; Hawks Aloft 2000; Herron et al. 1985; Houston

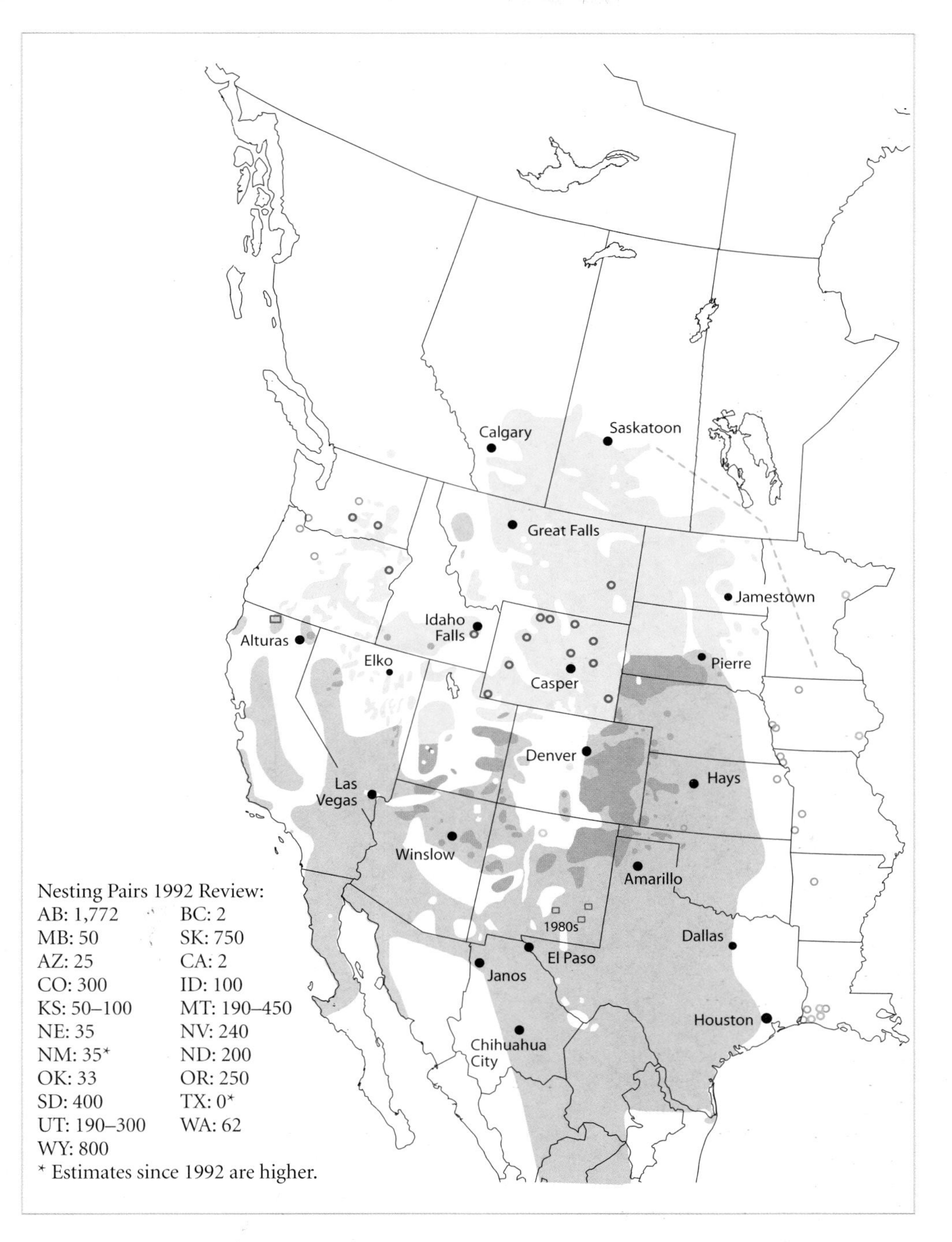


**FERRUGINOUS HAWK,** *Buteo regalis:* Uncommon. Breeding density is fairly high in certain locales. In breeding season, largest percent of rufous and dark morphs are found in AB and SK. Regular in w. MN in spring and fall. Winters south to México state.

1982, 1985, 1995a, 1995b; Houston and Bechard 1983, 1984; Houston et al. 1998; Howell and Webb 1995; Idaho Conservation Data Center 1999; James and Neal 1986; Kent 1997; Kent and Dinsmore 1996; Kingery 1998; Long 1998; Mlodinow and Tweit 2001; Montana Bird Distribution Committee 1996; New Mexico Natural Heritage Database 1996; Oberholser 1974; Olendorff 1993; Palmer 1988; Peterson 1995; Pierce 1988; Robbins and Esterla 1992; Schumtz and Fyfe 1987; Semenchuk 1992; Seyffert 2001; Sharpe et al. 2001; Small 1994; Smith 1996; Smith et al. 1997; Stewart 1975; USFWS 1992, 1999a, 2000a, 2001b; Watson 1999; Wheeler and Clark 1995.

**Plate 431. Ferruginous Hawk, adult female light morph (moderately marked type)** [**Mar.**] ▪ Black bill, yellow cere, and large yellow gape that extends to mid-eye. Brown head with thick dark eyeline and white throat; dark malar mark absent. ▪ Rufous upperparts. ▪ Wingtips shorter than tail tip. ▪ Dorsal side of tail may be gray, rufous, white, or a mixture.

**Plate 432. Ferruginous Hawk, adult female light morph (lightly marked type)** [**Dec.**] ▪ Black bill, yellow cere, and large yellow gape that extends to mid-eye. Pale brown head with thick dark eyeline and white throat; dark malar mark absent. ▪ Leg feathers white with dark brown barring; rufous tarsi feathered to toes. Unmarked or lightly marked underparts.

**Plate 433. Ferruginous Hawk, adult male light morph (moderately marked type) [Jan.]** ▪ Black bill, yellow cere, and large yellow gape that extends to mid-eye. Pale gray head with thick dark eyeline and white throat. ▪ Rufous upperparts. Rufous leg feathers; rufous tarsi feathered to toes. White underparts lightly speckled on belly and flanks; lacks dark malar mark.

**Plate 434. Ferruginous Hawk, adult female light morph (moderately marked type) [Jun.]** ▪ Black bill, yellow cere, and large yellow gape that extends to mid-eye. Brown head with thick dark eyeline. ▪ Rufous upperparts. Rufous leg feathers; rufous tarsi feathered to toes. White underparts moderately marked on belly and flanks.

**Plate 435. Ferruginous Hawk, adult female light morph (heavily marked type) [Jan.]** ▪ Black bill, yellow cere, and large yellow gape. A brown head with thick dark eyeline and white throat. ▪ Pale tawny breast heavily streaked; belly and flanks rufous with dark barring; white lower belly lightly barred. Leg feathers hidden by lower belly. ▪ *Note:* Most heavily marked type.

**Plate 436. Ferruginous Hawk, adult light morph [Dec.]** ▪ Rufous back and scapulars. ▪ Rufous upperwing coverts. Gray remiges with white rectangular panel on primaries formed by pale gray outer webs, white inner webs. ▪ *Note:* Typical dorsal markings on all light morph adults.

**Plate 437. Ferruginous Hawk, adult light morph (lightly marked type) [Jul.]** ▪ White leg feathers lightly barred. Tarsi rufous and contrast only slightly with white underparts. ▪ White remiges have small black tips, gray rear edge. ▪ White coverts moderately marked with rufous; may be lightly marked or unmarked. ▪ Tail white on ventral side.

**Plate 438. Ferruginous Hawk, adult light morph (moderately marked type) [Jun.]** ▪ Rufous leg feathers and tarsi form V that contrasts with white underparts; belly and flanks moderately barred. ▪ White remiges have small black tips, gray rear edge. Wing coverts heavily marked with rufous. ▪ Tail white on ventral side.

**Plate 439. Ferruginous Hawk, adult light morph (heavily marked type) [Oct.]** ▪ Rufous leg feathers and tarsi form V that contrasts with underparts; belly and flanks heavily barred and mottled; breast washed with tawny and has narrow streaking. ▪ White remiges have small black tips. Wing coverts heavily marked with rufous.

**Plate 440. Ferruginous Hawk, adult male intermediate (rufous) morph (Mar.]** ▪ Grayish head, including forehead and throat. Black bill and yellow cere. ▪ Small rufous markings on brown back and scapulars. ▪ Rufous upperwing covert edges. Pale gray outer edge of primaries. Wingtips shorter than tail tip. ▪ Tail mainly gray on dorsal side.

**Plate 441. Ferruginous Hawk, adult female intermediate (rufous) morph [Jan.]** ▪ Brown head, including forehead and throat. Black bill, yellow cere, large yellow gape. ▪ Rufous underparts. Leg feathers covered by lower belly.

**Plate 442. Ferruginous Hawk, adult female intermediate (rufous) morph [Feb.]** ▪ Brown head, including forehead and throat. Black bill, yellow cere, large yellow gape. ▪ Rufous underparts with white mottling on breast. Tarsi feathered to toes. ▪ Pale gray outer edges on primaries. Rufous edges on wing coverts. ▪ White ventral side of tail.

**Plate 443. Ferruginous Hawk, adult female intermediate (rufous) morph [Feb.]** ▪ Rufous underparts with white mottling on breast. ▪ Rufous underwing coverts. White remiges have small black tips, gray rear edge. ▪ White ventral side of tail. ▪ *Note:* Same birds as on plate 442.

**Plate 444. Ferruginous Hawk, adult dark intermediate (dark rufous) morph [Jan.]** ▪ Brown head, including throat. ▪ Rufous-breasted type with rufous breast and black belly, flanks, leg feathers, and tarsi. ▪ White remiges have small black tips. Black underwing coverts. ▪ White ventral side of tail.

**Plate 445. Ferruginous Hawk, adult dark morph [Feb.]** ▪ Uniformly dark brown head and body, including upperwing coverts and uppertail coverts. Large yellow gape. ▪ White remiges have small black tips. Dark brown wing coverts. ▪ *Note:* Rare morph constituting 1% of population.

**Plate 446. Ferruginous Hawk, juvenile light morph [Dec.]** ▪ Thick dark eyeline. Black bill, yellow cere, large yellow gape ▪ Brown upperparts. ▪ Pale gray outer edges on primaries. Wingtips shorter than tail tip. ▪ White base on brown tail; 3 or 4 irregular dark bands with subterminal band widest.

**Plate 447. Ferruginous Hawk, juvenile (1-year-old) light morph [Jul.]** ▪ White head with thick dark eyeline. Black bill, yellow cere, large yellow gape. ▪ White underparts, including leg feathers and tarsi. New barred adult feather on left leg. ▪ *Note:* Full adult plumage attained by mid-fall. Similar species lack thick dark eyeline.

**Plate 448. Ferruginous Hawk, juvenile (old nestling) light morph [Jul.]** ▪ Tawny-brown head with thick, dark brown eyeline. Black bill, yellow cere, large yellow gape. ▪ Tawny bib on breast fades to white by late summer. ▪ Dark brown upperparts edged with tawny. ▪ *Note:* This bird fledged a few days later.

**Plate 449. Ferruginous Hawk, juvenile light morph** [Dec.] ▪ Dark patch on flanks. ▪ White rectangular panel on primaries; primary coverts dark. ▪ Basal 1/3 of tail white and blends with white uppertail coverts which have dark spot on each feather. Distal 2/3 of tail brown with dark subterminal band and 2–3 narrower inner dark bands.

**Plate 450. Ferruginous Hawk, juvenile light morph** [Mar.] ▪ White underparts moderately spotted on belly, flanks, and leg feathers; tarsi dark. ▪ White remiges have small dark tips. White coverts moderately spotted. White rectangular window on primaries. ▪ Distal 2/3 of tail gray. Many have broken tail tips by late winter.

**Plate 451. Ferruginous Hawk, juvenile light morph** [Jan.] ▪ White underparts lightly spotted on flanks and leg feathers; belly unmarked. ▪ White remiges have small dark tips. Minimal dark barring on secondaries. White coverts lightly spotted. White rectangular window on primaries. ▪ Distal 2/3 of tail gray with dark subterminal band and 2–3 narrower inner dark bands.

**Plate 452. Ferruginous Hawk, juvenile light morph** [Nov.] ▪ White underparts heavily spotted on the belly, flanks, and leg feathers; tarsi brown. ▪ White remiges have small dark tips. Minimal dark barring on secondaries. White coverts heavily spotted, especially on patagial region. White rectangular window on primaries. ▪ Distal 2/3 of tail gray with dark subterminal band.

**Plate 453. Ferruginous Hawk, juvenile intermediate/dark morph** [**Jan.**] ▪ Tawny-brown head, including forehead and throat. Black bill, yellow cere, large yellow gape. ▪ Uniformly dark brown upperparts. ▪ Pale gray outer edge on primaries (separates it from similar dark buteos). ▪ Dorsal side of tail brown with irregular dark bands.

**Plate 454. Ferruginous Hawk, juvenile intermediate/dark morph** [**Dec.**] ▪ Dark brown upperparts. ▪ White rectangular panel on primaries; primary coverts often fairly pale. ▪ Gray tail has moderately wide dark subterminal band; inner tail may be partially banded or unbanded.

**Plate 455. Ferruginous Hawk, juvenile intermediate morph** [**Feb.**] ▪ Tawny-rufous head. ▪ Tawny-rufous breast; rest of underparts dark brown. ▪ White remiges with small, dark wingtips. ▪ White ventral side of tail has gray subterminal band, often a few thin, gray inner bands.

# ROUGH-LEGGED HAWK
## *(Buteo lagopus)*

**AGES:** Adult, basic I (subadult I), and juvenile. There is a vast array of individual and sexual variation within each age class (*see* Color Morphs). Adults and subadults are somewhat sexually dimorphic; however, there are considerable overlap of sexual characters within each age class. Adult characters are first attained when 2 years old. Subadult features are gained when 1 year old and retained for 1 year. Subadults have a few of their own age distinctions (mainly iris color) but may exhibit plumage features of their own, of adults, or of juveniles. Most subadults retain a few juvenile remiges (*see* Molt). Juvenile age class is held the first year. Juveniles have shorter secondaries, which create a narrower wing and make the tail appear longer than in older ages.

**MOLT:** Accipitridae wing and tail molt pattern (*see* chapter 4). The first prebasic molt begins on juveniles from late Apr. through May and ends by late Nov. Molt begins with the loss of the innermost primary (p1). It then begins on the tail, then on the secondaries and body. The first prebasic molt is mostly but not totally complete on the body. Molt on the wings, including remiges and coverts, is incomplete. Primaries retain up to three old, worn, and faded juvenile feathers (p8–10). Secondaries often retain juvenile feathers at s4 (sometimes also s3), s8, and often s9. The respective greater upperwing coverts for the retained juvenile primaries and secondaries are also generally retained juvenile feathering. A few median and lesser upperwing coverts may also be retained worn and faded juvenile feathers. The tail occasionally retains one or two juvenile rectrices. *Note:* Retained juvenile feathering is held throughout the fall, winter, and spring of the second year during the subadult age period.

The second prebasic molt is fairly complete but also retains old feathering. Molt begins in Jun. or Jul., depending on sex, breeding status, and breeding success, and ends in late Oct. or Nov. This molt replaces most or all previously retained juvenile feathering but may not replace all subadult feathers. Molt begins with sequential replacement of the outer juvenile primaries, from inner to outer feathers, then many of the rest of the inner primaries are replaced, also in an inner to outer sequence. Old juvenile secondaries are also replaced prior to molt on other secondaries. Molt on the tail appears to be mainly complete.

Additional prebasic molts are also fairly complete but retain a variable amount of the previous year's feathers on the body and wings. Remiges often have serially descendent molt and replace the primaries in an irregular order. Tail molt appears to be mainly complete. Timing is the same as the second prebasic molt.

**SUBSPECIES:** Polytypic with three races, but only one, *B. l. sanctijohannis*, inhabits North America.

Nominate *B. l. lagopus* breeds in tundra and sometimes in the northern boreal forest from Norway east through n. Russia to the Ural Mts. Winters along eastern coastal areas of the United Kingdom and from n. France southeast to n. Italy and east to Turkey, then east to the Caspian Sea. Similar to the larger *sanctijohannis* but does not have a dark morph.

*B. l. kamtschatkensis* breeds in the tundra region east of the Ural Mts. of n. Siberia and east to the Bering Sea to Kamtchatka and the Kuril Islands. Winters south to the e. Caspian Sea to Turkmenistan then east through n. China, Korea, and Japan. It is the largest race and is also fairly similar to *sanctijohannis* and only marginally different than *lagopus*. A dark morph also does not exist for this subspecies.

**COLOR MORPHS:** Polymorphic. Light and dark morphs occur for all ages; however, adult females, subadult females, and juveniles also have an intermediate morph that forms a clinal link between light and dark morphs. For unknown reasons, adult and subadult males do not have an intermediate morph. All ages and sexes exhibit extraordinary variation within each morph.

Adult light morphs are sexually diagnostic 95% of the time based on plumage. Recent research shows that with adult dark morphs, only some males can be accurately separated from females. Adult dark morphs have extensive overlapping plumage traits, particularly on tail patterns (BKW research at Bell Museum of Natural History, Minneapolis, Minn.; C. Olson unpubl. data from Montana; J. Bouton and D. Tetlow unpubl. data from e. New York). In the field, juvenile light and dark morphs usually cannot be identified to sex.

Plumage variation and regional status of the morphs adhere to Gloger's Rule (*see* Status and Distribution: Color morph status and distribution). As noted in Subspecies, dark morphs are absent in the two Eurasian subspecies.

**SIZE:** A large raptor. Males average smaller than females but sexes overlap. Length: 18–23 in. (46–58 cm); wingspan: 48–56 in. (122–142 cm).

**SPECIES TRAITS:** HEAD.—**Small bill. Cere and narrow, long gape are bright yellow. Inner half of lores dark and outer half of lores and forehead pale and form a mask.** Gape is narrow but prominent and extends under the eyes. Large white area on the nape with a center black spot. BODY.—**Feathered tarsi. Small feet.** WINGS.—**In flight, long and moderately wide with parallel front and trailing edges. The five "fingers" form a blunt wingtip. The long wings are either equal to or extend somewhat beyond the tip of the tail when perched.**

**ADULT TRAITS:** HEAD.—Dark brown irises. WINGS.—Moderately wide or wide black subterminal band on the trailing edge of the ventral surface. The dorsal surface of the inner and outer webs of the primaries is brownish. Dorsal surface of the secondaries is gray and barred with black. TAIL.—**Ventral surface has a neat, wide black subterminal band.**

**ADULT LIGHT MORPH TRAITS:** HEAD.—Thin dark eyeline. Pale supercilium and auricular regions. BODY.—Variable amount of dark markings on the neck and breast; markings are often heavier on the sides of the neck and breast. Flanks are spotted, barred, or solid blackish brown. Leg feathers are spotted or barred. Uppertail coverts have a moderate-sized or large dark diamond-shaped mark on each feather with a pale tawny or white edge and appear largely as a dark area. WINGS.—Primary greater underwing coverts are dark gray or black. TAIL.—Variable-width white basal region with a broad grayish or brownish distal band and a darker subterminal band on the dorsal surface and the neat black subterminal band on the ventral surface.

**Adult male light morph (lightly marked type):** HEAD.—Varies from being quite pale tawny or whitish with a minimal or moderate amount of dark streaking on the crown and malar with minimal distinction of the white mask to being moderately heavily marked with a distinct mask. BODY (ventral).—Base color of the underparts varies from white to pale tawny. Neck and upper breast can be (1) lightly streaked and lacking a defined bib, (2) **moderately streaked and mottled and forming a fairly distinct bib,** (3) **sometimes heavily streaked and mottled,** or (4) **uniformly dark with a distinct bib.** Base color of the ventral areas is white or pale tawny. Lower breast area is unmarked and lacks a defined necklace. Belly is unmarked or lightly spotted. Flanks are white or gray and lightly spotted, moderately spotted, or barred. Moderately barred leg feathers form a V in flight. Undertail coverts are white and unmarked. BODY (dorsal).—Upperparts appear pale and speckled with dark brown. The dark mark on the central portion of each feather is small and surrounded by an extensive amount of pale gray, and often a slight tawny-rufous wash. WINGS (ventral).—The underwing coverts and axillaries are white or tawny and lightly spotted, but the patagial region is often more densely marked. The long feathers of the first row of lesser underwing coverts may exhibit a somewhat distinct pale band because of the lack of markings. The carpal region may (1) have a small dark area on the primary greater coverts with the rest of carpal area being pale and sparingly marked with dark spots or (2) **be uniformly dark and form a large dark square patch.** WINGS (dorsal).—Upperwing coverts are pale grayish and mottled like the rest of the upperparts. The lesser coverts may be edged with a considerable amount of rufous. TAIL.—**Likely to have only the single wide black subterminal band or may have one or two additional narrow black bands; rarely more bands.** *Note:* As a general rule, males with pale bellies are also paler and less heavily marked on other body areas. Also called white-bellied type. Common plumage type in the West.

**Adult male light morph (moderately marked type):** HEAD.—Moderately heavily marked to heavily marked and appearing moderately dark headed or dark headed. Distinct white mask. BODY (ventral).—Base color of the underparts varies from white to a rich tawny. Neck and breast are moderately or heavily marked and form a distinct bib as described above in #1 and 2 types of lightly marked type. **Lower breast has a narrow necklace consisting of an unmarked or lightly marked narrow band separating the dark breast from the moderately marked belly and flanks. Belly is moderately barred and blotched with black. Flanks are white or dark gray and distinctly barred with black.** Leg feathers are moderately or heavily barred. Undertail coverts are unmarked or some feathers may have a small dark mark. BODY (dorsal).—Upperparts are pale gray, often with a pale tawny-rufous wash, with a small dark mark on the central portion of each feather or a larger dark mark on the central area of each feather and not as grayish looking. WINGS (ventral).—The underwing coverts are white or tawny and moderately or heavily spotted, particularly the patagial area. A distinct broad, unmarked pale bar exists on the first row of lesser underwing coverts. The carpal region may (1) have a small dark area on the primary greater coverts with a heavily mottled carpal area that blends with the rest of the underwing coverts and obscures the carpal patch or (2) be entirely dark, forming a distinct black square shape. WINGS (dorsal).—The upperwing coverts are lightly or moderately marked like the rest of the upperparts. TAIL.—May have at least one or two dark inner bands but likely to have the fully banded pattern of four or five dark bands. The deck set has the four or five dark bands, and there may be six or more dark bands on the outer rectrix sets. Inner dark bands may all be equal in width or get progressively narrower towards the basal region of the tail. On the fully banded pattern, the white basal region of the tail is less pronounced because of the additional banding. *Note:* Common plumage type in the West.

**Adult male light morph (heavily marked type):** HEAD.—As on moderately marked type. BODY (ventral).—Base color of the underparts is either a rich tawny, as in many adult females, or whitish. Front of neck and breast are extensively marked and appear as a dark bib. An unmarked or lightly marked narrow band creates the necklace that separates the dark breast from the dark belly and flanks. Belly is heavily barred or blotched but not uniformly blackish brown. **Flanks are dark and may have faint gray barring but can also be uniformly blackish brown like on adult females.** Leg feathers are heavily barred. Undertail coverts are likely to have a dark bar on each feather. BODY (dorsal).—Upperparts have a large dark mark on the central part of each feather with grayish and often tawny-rufous wash on the outer edges and do not appear as grayish as in most lightly marked type adult males. WINGS (ventral).—The underwing coverts are extensively marked and quite dark, but there is still a pale bar on the first row of lesser underwing coverts because of a lack of markings. Carpal region is (1) heavily mottled and blends with the rest of the underwing coverts, or (2) solid black and somewhat distinct. WINGS (dorsal).—Marked like the rest of the upperparts. TAIL.—As in moderately marked type, but more likely to be fully banded pattern. *Note:* Fairly common plumage type in the West. *Also note:* Not depicted.

**Adult female light morph (lightly marked type):** HEAD.—Quite pale with little if any dark malar mark and only a light amount of streaking on the crown and hindneck. Pale mask is ill defined because the head is so pale. BODY (ventral).—Base color of the underparts is pale tawny but may fade to nearly white by spring. Lightly streaked neck and upper breast do not have a necklace demarcation. Belly is unmarked or may be sparsely marked with small dark spots and blotches. **Flanks are uniformly blackish brown or may have light tips on most feathers but are not spotted or barred as on similarly marked males.** Leg feathers are lightly covered with dark spots, bars, or diamond-shaped markings. Undertail coverts are unmarked. BODY (dorsal).—A dark brown streak is on the central part of each feather and the outer portion is edged with tawny or tawny-rufous. WINGS (ventral).—**Pale tawny underwing coverts and axillaries are lightly marked and have a distinct square black carpal patch.** Carpal area is sometimes lightly mottled with tawny. WINGS (dorsal).—Marked as on rest of the upperparts. TAIL.—Typically has a single neat black subterminal band. There is considerable white on the base. *Note:* Common plumage type in the West.

**Adult female light morph (moderately marked type):** HEAD.—Moderately streaked on the crown and hindneck and has a fairly distinct pale mask. Dark malar is moderately defined and often most distinct on the very lower portion of the lower jaw. BODY (ventral).—Base color of the underparts is a pale or rich tawny and often fades to whitish or pale tawny by spring. Neck and upper breast are lightly or moderately streaked. Birds with lightly streaked breasts do not have a bib or necklace; those with moderately marked bibs have a moderate bib and necklace demarcation. Flanks are always either solid brownish black or each feather is tipped with tawny. Two variations of belly markings: (1) *Dark bellied type.*—Uniformly blackish brown but may have a small tawny tip on each feather or (2) *Split-bellied type.*—Mid-section of the dark belly is split by a narrow unmarked tawny or mottled strip. Leg feathers are moderately barred. Undertail coverts are unmarked. BODY (dorsal).—As in lightly marked type, but often darker, with narrower pale edgings. WINGS (ventral).—As in "lightly marked type," but often with a fairly rich tawny base color. Axillaries are tawny with a few dark marks. WINGS (dorsal).—Similarly marked as rest of the upperparts. TAIL.—Often has up to three additional narrow partial or complete dark basal tail bands but not fully banded as on many males. White area on the basal region of the dorsal surface of the tail can be very reduced in width, especially on the two deck rectrices. *Note:* Common plumage type in the West.

**Adult female light morph (heavily marked type):** HEAD.—Heavily streaked with dark brown on the crown and hindneck with a distinct pale mask. Dark malar mark is well defined on all of the lower jaw. Pale supercilium and auricular areas are obvious. BODY (ventral).—Base color of the underparts is a rich tawny but fades to pale tawny or even whitish by spring. Bib is heavily marked with streaks or blotches and can be virtually solid dark. Necklace is well defined. **Flanks and belly are uniformly blackish brown but may have a small tawny tip on each feather.** Leg feathers are heavily barred. Undertail coverts may have a dark mark or bar on most feathers. BODY (dorsal).—As in previously described types. WINGS (ventral).—**Underwing coverts are a rich tawny and fairly heavily marked, but the dark carpal patch is still obvious.** Axillaries are tawny and marked with dark brown but still pale. WINGS (dorsal).—Marked similarly as rest of the upperparts. TAIL.—Regularly has up to three narrow partial or complete inner dark bands but not fully banded as on many males. White basal area of the tail is often very reduced, and white may be lacking on the deck rectrix set. *Note:* Fairly common plumage type in the West.

**Adult intermediate morph traits (female only):** HEAD.—Fairly pale and similar to moderately and heavily marked types of light morph with a broad pale supercilium and auricular areas. White mask is fairly distinct. BODY (ventral).—All breast feathers have tawny or tawny-rufous edging and may appear streaked or mottled, with little or no necklace demarcation. Breast is paler than belly and flanks. Uniformly blackish brown belly and flanks. Leg feathers are dark with some tawny markings or pale and heavily barred. Undertail coverts can be unmarked or have a moderate amount of dark markings. BODY (dorsal).—Dark brown with a minimal amount of pale edging on the scapulars. WINGS (ventral).—**Axillaries are dark brown (tawny colored on light morphs). The underwing coverts are tawny-rufous and more rufous than on light morphs but not as rufous as on many dark morphs. Distinct large square black carpal patch.** WINGS (dorsal).—Dark coverts with some pale edging. TAIL.—Ventral surface is marked as on many light morphs, but the white basal region is very narrow. Dorsal surface is medium or dark brown with a narrow white band at the very base of the tail. May have multiple dark bands. Uppertail coverts are tawny or dark. *Note:* Rare plumage type in the West. This is the interim female plumage between a heavily marked type light morph and brown type dark morph, in particular pale-headed variations of brown type dark morph (*see below*).

**Adult dark morph traits:** HEAD.—**Very defined white mask.** Yellow gape is narrow but obvious as it contrasts against the dark head. BODY, WINGS, and TAIL.—Shared sexual traits (*see below*).

**Adult brown type (male/female):** HEAD.—Variable amount of pale tawny on the supercilium and auricular regions: (1) all dark without pale supercilium and auricular areas (both sexes), (2) small, pale tawny supercilium and auricular areas with a narrow dark eyeline (both sexes), and (3) large pale tawny supercilium and auricular areas with a distinct narrow dark eyeline (mainly female). White mask is obvious on all but #3. BODY.—Neck and breast have a variable amount of tawny or rufous-tawny streaking and mottling; rarely have whitish mottling. As a rule, the paler the head, the more streaked and paler the breast. There is often a rather sharp demarcation line between the more tawny-brown neck and breast and the darker, uniformly blackish brown belly and flanks. On some individuals, there is some tawny or white speckling at the junction of the breast and belly (which is the necklace area on light morphs). Leg feathers are dark brown. Undertail coverts are dark with pale tips. Dorsal region is solid blackish brown or may have considerable pale gray mottling on the scapulars (more typical of males). Uppertail coverts are dark brown but may have a pale tip on each feather. WINGS (ventral).—Underwing coverts vary in color and pattern of markings: (1) uniformly dark brown, including the axillaries, with little or no demarcation of the carpal region; (2) **slightly or moderately rufous with a distinct black square carpal patch and dark brown axillaries;** (3) **very rufous with dark brown axillaries and an obvious black carpal patch (predominately females);** or (4) variably speckled with tawny or white but the carpal area is unmarked and ill defined and the axillaries are often speckled or barred with white. *Note:* In general, birds with paler heads tend to have more rufous on the underwing coverts and those with darker heads have little or no rufous on the coverts. WINGS (dorsal).—Mainly uniformly dark on the coverts. TAIL.—Four patterns. One type is only on females and only visible on the dorsal surface. The other three are shared by both sexes and are visible on both the ventral and dorsal surfaces. All have the neat, wide black subterminal band. (1) *Unbanded type* (females only).—**Ventral surface is pale gray and unmarked inside the neat, black subterminal band;** dorsal surface is uniformly medium brown. (2) *Partially banded type* (many females, some males).—Ventral surface is pale gray with partial, narrow gray bands; dorsal surface is medium brown with two or three faintly darker bands, mainly on the inner rectrix sets. (3) *Banded type* (many females, many males).—Ventral surface is pale gray with distinct, narrow dark gray bands on all the inner tail; dorsal surface is dark brown with three fairly distinct narrow pale gray bands (pattern is most pronounced on the deck and inner rectrix sets). (4) *Distinctly banded type* (many males, a few females).—**Ventral surface is gray or black with distinct, narrow white bands; dorsal surface is dark brown or black with three distinct, narrow, pale gray or white bands.** *Note:* Uncommon plumage type in the West. In-hand photographs of a large female (wing chord at the top end of the female category) banded at Point Peninsula area in Jefferson Co., N.Y., provided the first absolute documentation of an adult female with distinctly banded type tail pattern. Similar types also in Bell Museum of Natural History (also females at large size spectrum for their sex), unpublished photographs of a bird found shot by C. Olson in w. Montana, and several in-field photographed individuals that appeared very large and otherwise female-like.

**Adult male black type:** HEAD.—**Pronounced white mask.** Very rarely, the white forehead may be reduced or lacking on the blackest types. Head is all black or may have a slight brownish cast. Yellow gape is quite obvious. BODY (ventral).—Black. Rarely, there is sparse white spotting at the junction of the breast and belly at the necklace region of paler birds. Leg feathers may have white spotting. Often have white spotting or barring on each undertail covert feather. BODY (dorsal).—Two main types with a clinal trend between them: (1) mainly all-black back and scapulars with minimal gray outer feather edging or mottling; (2) extensively mottled with pale gray and can nearly rival the gray dorsum of some light morphs. Uppertail coverts are either black with a white tip on each feather or may have white inner bars on each feather. WINGS (ventral).—Underwing coverts are uniformly black and do not show distinction of the carpal area. Some individuals have white spotting on the coverts and barring on the axillaries. TAIL.—**Black tail is distinctly banded type.** On the most melanin-saturated birds, the three dorsal gray or white bands may be reduced and only appear as partial bands. *Note:* Uncommon plumage type in the West.

**BASIC I (SUBADULT I) TRAITS:** HEAD.—Iris is medium brown. BODY.—May possess any of the adult variations previously described. Males tend to acquire female-like traits; however, males may also molt into any of the five adult male plumage variations described above. WINGS (ventral).—Retains up to three juvenile feathers on the outermost primaries (p8–10) and two or three secondaries (s4, 8, and 9). *See* Molt. Retained feathers are frayed and pale brown. On the secondaries, the old juvenile feathers are shorter than the newly acquired adultlike feathers and lack the more distinct dark subterminal band. A few or many faded and worn juvenile feathers are retained on the upperwing coverts. WINGS (dorsal).—The inner web of the primaries is whitish or white and forms a distinct panel on the dorsal surface, and is similar to juveniles. Also, since many juveniles exhibit a white area on the outer web on the base of some of the outer primaries, and subadults typically retain these feathers, subadults may show a white patch on the base on the dorsal surface of the outer primaries. TAIL.—Generally molts into adult characters; however, a few retain one or two juvenile rectrices. Such feathers are longer than the newly acquired feathers and lack the black subterminal band.

**BASIC (SUBADULT) LIGHT MORPH TRAITS:** *See* male and female data below.

**Basic (subadult) male light morph (all types):** Males exhibit any of the three adult male light morph plumage and tail variations described above, including lightly marked type (BKW pers. obs./photographs showing distinct molt pattern and retained juvenile feathering; unpubl. data and in-hand photos from C. Olson). The lightly marked type is very uncommon; the other two are fairly common or common. Many males also have female-like characters, including a uniformly blackish brown belly and flanks. Many males cannot be separated in the field from subadult or adult females unless seen at close range to see molt pattern or iris color. Many subadults have bib markings that extend to the front of the breast and belly and lack the unmarked necklace that adults have.

**Basic (subadult) female light morph (all types): Identical to the plumage variations of adult light morph females, including the blackish brown belly band color.** Separated from adult females at close range by iris color and retained juvenile feathering.

**BASIC (SUBADULT) INTERMEDIATE MORPH TRAITS (FEMALE ONLY):** No data. Use iris color and retained juvenile feathering to age properly.

**BASIC (SUBADULT) DARK MORPH TRAITS:** Females adhere to the same variations described for adult female dark morphs. For males, several specimens and photographs substantiate brown and black types.

**JUVENILE TRAITS:** Sexes are similar. HEAD.—Pale yellow or brown iris. WINGS (ventral).—Gray subterminal band on the trailing edge of the wing. The "finger" primaries are generally all black or have fairly large black area on the tip. In translucent light, a pale window may show on the primaries. WINGS (dorsal).—The inner web of each primary is white and shows as a whitish panel. On many individuals, p7, 8, and 9 may be white on the basal area of the outer web next to the greater primary coverts. TAIL.—The ventral surface has a diffused gray subterminal band. A darker smudge may occasionally be superimposed on this gray band but is never as defined as the neat black band on subadults and adults. Either sex may have partial or fairly complete bands on the dorsal surface.

**JUVENILE LIGHT MORPH TRAITS:** HEAD.—Thin dark eyeline. Tawny head is lightly or moderately streaked on the crown and hindneck. There is little if any demarcation of a dark malar mark on individuals with pale heads; those with darker heads may have a faint dark malar mark. BODY (ventral).—Base color of the underparts is pale or medium tawny. Neck and breast vary from sparsely to heavily streaked. Streaking extends to the top of the dark belly and flanks and is most dense on the sides of the neck and breast. **Flanks and belly are mainly uniformly warm dark brown and form a belly band (belly band is not the blackish brown of subadult and adult females).** However, many birds have a narrow tawny strip or mottling down the center of the belly. Leg feathers are tawny and may be unmarked, lightly spotted, or heavily spotted. BODY (dorsal).—Tawny or slightly grayish outer edges with a wide dark streak on the mid-section of each dorsal feather. Some males are perhaps more grayish on the dorsal regions than are females. WINGS (ventral).—Underwing coverts and axillaries are tawny and lightly marked with dark spots. A large dark brown square patch adorns the carpal area of the underwing. WINGS (dorsal).—Pattern is similar to the dorsal areas. Middle and first row of lesser coverts have a broad pale tawny or grayish edge. TAIL (ventral).—Grayish on the distal half or two-thirds and often a darker gray on the subterminal band region. The basal half or third of the tail is white. TAIL (dorsal).—Uppertail coverts are generally white with small dark central spot or streak and much paler than on most adults. However, more heavily marked birds have tawny or white uppertail coverts with a large dark spot or diamond-shaped mark on most feathers. Dorsal surface of the tail is medium brown on the distal half or two-thirds and white on the basal half or third. On heavily marked birds, the white basal area may be limited to the basal fourth or less. Dark tail banding is often present but not well defined. As in adults, either sex can have a variable number of partial or complete narrow dark inner bands.

**JUVENILE INTERMEDIATE MORPH TRAITS:** HEAD.—Pale as on juvenile light morph with a thin dark eyeline. The white mask is often nondescript since the head is rather pale. BODY.—Neck and breast are extensively streaked with brown. Uniformly brown belly and flanks. Leg feathers are typically heavily marked but may be lightly marked. Undertail coverts may be unmarked or have moderate-sized dark markings. WINGS.—**Underwing coverts are rufous-tawny and more rufous than on light morphs. Axillaries are brown (tawny in light morphs). Square-shaped black carpal patch is obvious.** TAIL (ventral).—Like light morph's. TAIL (dorsal).—Uppertail coverts are tawny and may have dark markings on each feather. Basal area of the tail has a narrow white band.

**JUVENILE DARK MORPH TRAITS:** Two main plumage variations in either sex; *see below.* TAIL.—Three main variations. Patterns can be for either sex and are most noticeable on the dorsal surface: (1) *Unbanded type.*—Dorsal surface is medium brown with a slightly darker smudge on the subterminal area. Ventral surface is pale gray with a medium gray subterminal smudge. (2) *Partially banded type.*—Medium brown dorsal surface with a darker subterminal smudge. There are two to four narrow or moderately wide partial dark bands on the inner portion of the tail. The ventral surface is pale gray with the darker subterminal smudge and with narrow, partial gray inner bands. (3) *Banded type.*—Dorsal surface is dark brown with three delineated pale gray bands. The pale gray bands may be narrower or wider than the dark bands.

**Juvenile dark morph brown type:** HEAD.—Varies from very pale and nearly rivaling those of light morphs to moderately dark with reduced pale supercilium and auricular areas and a fairly defined white mask. BODY (ventral).—Neck and breast feathers are variably edged and mottled with tawny or tawny-rufous. Belly and flanks are uniformly warm dark brown and contrast sharply with the paler tawny-rufous neck and breast. Leg feathers are brown. Undertail coverts are brown with pale tips. BODY (dorsal).—Uniformly brown. WINGS (ventral).—**Dark brown axillaries, rufous coverts, and a sharply defined large black carpal patch.** WINGS (dorsal).—Dark brown with a pale panel on the primaries. TAIL.—Uppertail coverts are brown. *Note:* Common plumage type in the West.

**Juvenile dark morph black type:** HEAD.—Dark brown with a distinct white mask. Rarely has a pale throat. BODY.—Uniformly dark brown throughout, including leg feathers and undertail coverts. TAIL.—Uppertail coverts are dark brown. *Note:* Rare plumage type for males and females in the West. Probable female type based on only one specimen from Bell Museum of Natural History (wing chord measured at the small end of female range and labeled as a female).

**ABNORMAL PLUMAGES:** A hybrid adult dark morph Rough-legged Hawk x dark morph "Harlan's" Red-tailed Hawk (*Buteo jamaicensis harlani*) was photographed in Prowers Co., Colo., in Dec. 1999. This bird had head and bill features and proportions, including wing shape, like a Red-tailed Hawk but fully feathered tarsi and tail pattern like a dark morph Rough-legged Hawk (BKW pers. obs./photos). Imperfect albino (dilute plumage) was photographed in Ohio in the winter of 2000. *Note:* In summer, the plumage of 1-year-olds, in particular, can be very worn and bleached and appear very pale.

**HABITAT: Summer.**—Breeds mainly on the tundra but also in semi-open spruce wood-

lands. On the tundra, areas with cliffs, embankments, or rocky outcrops serve as nest sites.
**Winter.**—Semi-open and open prairies; meadows; idle fields; harvested agricultural areas, especially grass hay and alfalfa fields; pastures; freshwater, brackish, and saltwater marshes; and swamps. Absent from high montane areas. Found up to 7,500 ft. (2,300 m) on the high plains of s. Wyoming and possibly a few montane valleys of the Rocky Mts. Regular on the plains at 6,600 ft. (2,000 m) in ne. New Mexico.
**Migration.**—All areas described for Winter. Wooded locales may also be inhabited for short periods.

**HABITS:** Varies from wary to tame. Gregarious in spring, fall, and winter. Communal roosts are often formed and may roost with Red-tailed Hawks, Ferruginous Hawks, and Bald and Golden eagles. Perches on any type of elevated perch, including utility wires. Readily perches on the ground. In strong winds, may seek sheltered areas on the ground: vegetation clumps, fence posts, or clumps of dirt.

**FEEDING:** Perch and aerial hunter. Feeds extensively on lemmings during the summer. Also readily preys on other small rodents, young hares, and small and medium-small birds up to ptarmigan size. In the nonbreeding season, occasionally feeds on fish, reptiles, amphibians, and insects. Readily feeds on carrion in all seasons and commonly seen along roads at roadkills.

**FLIGHT:** Powered by slow wingbeats interspersed with irregular glide sequences. Wings are held in a low dihedral when soaring and in a modified dihedral when gliding. Often hovers and kites when hunting. Legs are often lowered when hovering. Large expanses of open water are regularly crossed during migration.

**VOICE:** Rarely heard on the winter grounds. May vocalize if agitated or forced to move from a perch. Primary call is a plaintive *keeaah.* On the breeding grounds, nesting pairs are highly vocal towards intruders at their nests sites.

**STATUS AND DISTRIBUTION:** *Fairly common.* Population is stable but fluctuates locally and regionally at irregular intervals because of cyclic prey abundance and other factors.
**Color morph status and distribution.**—Following Gloger's Rule, dark morphs, including the darkest types (e.g., black type adult males), are not as common in the West as they are in the East. The arctic breeding regions in the West are generally more arid than in the East. Likewise, dark morphs are less common in the winter in w. Canada and the w. U.S. than in e. Canada and the e. U.S. The north-central Canadian tundra in Nunavut and the Northwest Territories, in particular, is more arid than similar habitat in e. Canada and the percent of dark morphs decreases substantially in this region. Dark morphs are also virtually absent in the high-arctic latitudes of Nunavut and the Northwest Territories. In portions of Alaska, however, breeding populations of dark morphs become more numerous and in certain locales perhaps rival or exceed ratios seen in the East. Wintering areas in s.-cen. Canada and the w. U.S. also exhibit a similar, virtually clinal reduction of dark morphs westward across the Great Plains and Rocky Mts. There is a probable increase of dark morphs in the Pacific Coast states.

In the eastern part of the West, F. Nicoletti found dark morphs comprise 20% of the migrants seen at Hawk Ridge Nature Reserve, Duluth, Minn. Because of the small number of dark morphs in portions of the central Arctic, the percent of dark morphs drops substantially and only averages 7% of the wintering population on the w. Great Plains (BKW pers. obs.). C. Olson found wintering dark morphs in nw. Montana average only 5% of the population. Based on data from P. and C. Fritz,154 migrants seen at a hawkwatch near Eureka in e. Alaska in the spring of 2002, 71% were light morph and 29% were dark morph (similar percentages seen in e. U.S. in winter). Intermediate morphs are very rarely seen in the West.
**Summer.**—Breeds from the tree line and northward in Nunavut, Northwest Territories, and the Yukon Territory. Nests in various open areas below and south of tree line elevation in Alaska. Rare along Hudson Bay in ne. Manitoba. Juveniles are fairly regular until early and mid-Jun. in the northern border states and s. Canada.
**Winter.**—Locally dense in many regions in mapped winter range. Absent from high montane elevations. Females of all ages winter at more northern latitudes than males. Females are more common in winter at a latitude from n. California to s. Wyoming and northward. Most birds in Montana, the Dakotas, and

Wyoming are females. Males outnumber females in s. Colorado, s. Kansas, n. New Mexico, Oklahoma, and n. Texas. Deep snow, which conceals prey, may force Rough-legged Hawks to vacate certain areas in winter. Subadults and adults have strong site fidelity and often return to the same winter territory year after year. In some areas, juveniles may outnumber adults, even in northern wintering areas. On the w. Great Plains, adults always outnumber juveniles. Based on banding data, Rough-legged Hawks wintering in California come from breeding and natal areas in Alaska.

Isolated but regular wintering occurs on Kodiak Island and the Seward Peninsula, Alaska. Also, isolated wintering in e.-cen. British Columbia and w.-cen. Alberta. May be absent from many areas of northern half of Minnesota, especially during winters of deep snow (map shows maximum winter range).

Irregular or very irregular in winter on the Aleutian Islands, in sw. Yukon Territory, and in n. British Columbia. Juveniles are very irregular south of typical winter range along the Coastal Bend of s. Texas. Adult males winter as far south as s. Arizona and juveniles occasionally show up in n. Sonora, Mexico.

**Migration.**—Rough-legged Hawks engage in short or moderate-length migrations. Males of all ages migrate farther south than females. Juveniles, which are probably males, occasionally engage in fairly long movements south of typical winter areas. Moderate-length over-water crossing are readily undertaken.

*Fall migration:* Autumn migration begins in late Aug. or Sep. from high-latitude arctic breeding areas. Hawkwatch sites in the Rocky Mts. in Montana and along the western edge of Lake Superior at Duluth, Minn., are the only locations that tally large numbers of migrants. The first migrants may be seen in the n. U.S. by late Sep., but more typically by early Oct. Juveniles are often the first to appear, but sometimes subadult and adult females appear first. For adults, this is possibly due to poor breeding success and a low prey base. Adult females precede adult males. Peak flights in Montana and Minnesota are typically in late Oct., but may extend until early Nov. and occasionally into late Nov. and early Dec. Numbers taper off by mid-Nov., and with barely a trickle of movement thereafter.

The first migrants on the cen. Great Plains in Colorado are virtually always subadult and adult females. These females may appear as early as early Oct., but generally not until mid-Oct. Subadult and adult males appear later, mainly in Nov. and Dec. There is a continual southward movement of all ages throughout Nov. and Dec. on the cen. and s. Great Plains. There is also a predictable late-season southward push for all ages in late Dec. and early Jan.

*Spring migration:* Adults begin some northward movement in late Feb. on the Great Plains. Noticeable movement is apparent in early and mid-Mar. for adults. Most adults have departed the plains of e. Colorado by late Mar. Juveniles continue moving northward through Apr. with stragglers seen on the cen. Plains into early May. In the northern border states and s. Canada, juveniles may linger well into Jun. and a few may spend the summer.

**NESTING:** Begins in May or early Jun. depending on latitude and typically ends in Jul. or Aug. but may extend into early Sep.

**Courtship (flight).**—*High-circling* by both sexes and *sky-dancing* by males (*see* chapter 5).

Nests are fairly large, constructed of sticks, chunks of wood, twigs, and bones, and are up to 30 in. (76 cm) in diameter and 15 in. (34 cm) deep. The inner nest is lined with soft vegetation, fur, and feathers from prey. Nests are often reused, and pairs may have alternate nests in a territory. New nests may be small but become larger with reuse. Males bring material and females build the nest. Nest sites are on cliffs (do not require an overhang above the nest), embankments, rock outcrops, knolls, and even flat ground. Trees are occasionally used. Tree nests may be up to 30 ft. (9 m) high. Incubation of the 2–6 eggs, rarely 7, is mainly by females. Males supply food to females and nestlings. Eggs hatch in 28–31 days and youngsters fledge in 34–45 days.

**CONSERVATION:** Since a stable species, no measures are taken. Breeding areas are not usually affected by human pressure. Wintering birds are often partial to moderately human-altered areas and regularly inhabit agricultural areas, particularly alfalfa fields.

**Mortality.**—Illegal shooting is prevalent in the West. Electrocution occurs frequently from utility wires. Many birds are hit by vehicles when feeding on roadkills. Oil contamination from bathing in open oil pits occurs.

**SIMILAR SPECIES:** COMPARED TO LIGHT MORPH.—(1) **Northern Harrier, all ages.**—PERCHED.—Wingtips are shorter than tail tip. FLIGHT.—Share similar white uppertail coverts with many juvenile Roughlegs but does not have white on the base of the tail. Dorsal surface of tail has wide dark bands. Flight mannerism similar, especially with larger-sized female harriers since wings are held in dihedral when soaring and a modified dihedral when gliding. **(2) Swainson's Hawk, light morph adults and subadults.**—Regular range overlap during migration periods in the West, mainly in late Mar./Apr. and late Sep./Oct. PERCHED.—(2A) Adults are similar to adult male Roughlegs, with a defined bib on the breast, including the white mask on the forehead and outer lores and wingtip-to-tail-tip ratio. Dorsal color of Swainson's is more brownish; tarsi are bare; tail banding is narrow and uniform in width (but with similar wide dark subterminal band). (2B) Subadult Swainson's are similar to adult male Roughlegs, with a partial bib consisting of streaking on the sides of the neck and breast. Use features listed above, including tail pattern, to separate. **(3) Ferruginous Hawk, juveniles.**—PERCHED.—Dark eyeline is much thicker and extends to the gape. Belly and flanks are spotted like adult male lightly marked type Roughlegs. Feathered legs similar. Tail pattern similar to juvenile: use caution. Wingtips distinctly shorter than tail tip. FLIGHT.—All flight mannerisms similar; wingbeats have more snappy upwards beat on Ferruginous. Wing panels/windows and tail pattern similar only to juvenile Roughlegs, but Ferruginous has a white belly, not a dark band. COMPARED TO DARK MORPH.—**(4) Swainson's Hawk, dark morphs of all ages.**—Range overlap as for #2. PERCHED.—(4A) Adults are similar in having pale outer lores but rarely have pale forehead (creating a mask). Bare tarsi. White or tawny undertail coverts. Dark banding on the tail is very narrow. (4B) Immature ages are very similar to paler headed Roughlegs, especially in head pattern (mask and thin dark eyeline), but have bare tarsi and pale undertail coverts. FLIGHT.—All ages have medium or dark gray undersides of the remiges and pale undertail coverts. **(5) "Harlan's" Red-tailed Hawk (*B. j. harlani*), dark morph or dark intermediate morph gray-type-tail adults.**—Range overlap in Alaska in summer and in all of w. U.S. in migration and winter. PERCHED.—Lores are completely white. May have similar white nape area with the black nape spot. May have pale breast mottling like some dark Roughlegs. Bare tarsi. Wingtips are somewhat shorter than tail tip. Tail pattern nearly identical to juvenile and adult female Roughlegs with unbanded type dorsal tail pattern. FLIGHT.—Secondaries are more broadly bowed. Subterminal/terminal black tail band is typically irregular and not as neatly formed as on Roughleg. **(6) Western Red-tailed Hawk (*B. j. calurus*) and Harlan's Red-tailed Hawk (*B. j. harlani*), dark morph juveniles.**—Range overlap as in #5. PERCHED.—Bare tarsi. Wingtips are much shorter than the tail tip. FLIGHT.—Dark "fingers" on all of the underside of the outer primaries on *calurus* are similar to Roughlegs; *harlani* has barred outer primaries. Both Red-tailed races have more than four dark bands on the tail than Roughlegs. **(7) "Western" Red-tailed Hawk (*B. j. calurus*) light morph and "Eastern" Red-tailed Hawk (*B. j. borealis*), juveniles.**—PERCHED.—Many have a solid dark band on the belly and flanks and appear similar to Roughlegs. Wingtips are shorter than tail tip. Cere is greenish. **(8) Ferruginous Hawk, rufous and dark morph adults and juveniles.**—PERCHED.—Forehead always dark; lores completely white. Large gape but may not appear any more obvious or larger than on Roughlegs. Both species have feathered tarsi. Outer edge of folded primaries is pale gray. Wingtips shorter than tail tip. FLIGHT.—(8A) Adults have small, distinct black outer primary tips; tail is uniformly white on ventral surface. (8B) Juveniles have small gray outer primary tips (all black on juvenile Roughlegs); tail can be identical for both species.

**OTHER NAMES:** Roughleg, Roughie. *Spanish:* Aguililla Ártica. *French:* Buse Pattue.

---

REFERENCES: Baicich and Harrison 1997; Clark 1999; Dodge 1988–1997; Dotson and Mindell 1979; Forsman 1999; Garrison and Bloom 1993; Godfrey 1986; Howell and Webb 1995; Kellogg 2000; Oberholser 1974; Olson and Arsenault 2000; Palmer 1988; Russell and Monson 1998; Wheeler and Clark 1995.

**ROUGH-LEGGED HAWK,** *Buteo lagopus sanctijohannis:* Fairly common. Numbers vary with fluctuations in summer prey abundance. Rare south of regular winter range. Females winter farther north than males.

**Plate 456. Rough-legged Hawk, adult male light morph (lightly marked type)** [**Dec.**] ▪ White forehead and outer lores, thin dark eyeline, dark brown iris. ▪ Underparts streaked on breast, unmarked on upper belly, unmarked or lightly spotted or barred on mid-belly, flanks, and leg feathers. Feathered tarsi. ▪ *Note:* Palest type.

**Plate 457. Rough-legged Hawk, adult male light morph (lightly marked type)** [**Jan.**] ▪ White forehead and outer lores, thin dark eyeline, dark malar mark, dark brown iris. ▪ Dark mottled bib; white belly; lightly spotted or barred flanks. ▪ Wingtips extend past tail tip. ▪ Wide black subterminal tail band; narrow inner dark band.

**Plate 458. Rough-legged Hawk, adult male light morph (moderately marked type)** [**Mar.**] ▪ White forehead and outer lores, thin dark eyeline, dark brown iris. ▪ Dark mottled bib; white necklace on upper belly; mid-belly, flanks, and leg feathers are barred. Feathered tarsi. ▪ Wide black subterminal tail band; narrow inner dark band.

**Plate 459. Rough-legged Hawk, adult male light morph (moderately marked type)** [**Nov.**] ▪ Dark head, whitish throat with white forehead and outer lores, thin dark eyeline, dark brown iris. ▪ Black upperparts mottled with gray. ▪ Tail has wide black subterminal band, 3 progressively narrower inner dark bands; base of tail white.

**Plate 460. Rough-legged Hawk, adult male light morph (moderately marked type) [Feb.]** ▪ Black upperparts mottled with gray. ▪ Gray remiges barred with black, including wide black band on rear edge. Coverts mottled with gray. ▪ Tail has wide black subterminal band, 3 progressively narrower inner dark bands; base of tail white.

**Plate 461. Rough-legged Hawk, adult male light morph (lightly marked type) [Dec.]** ▪ Sides of breast streaked. Flanks lightly barred, leg feathers barred. Tarsi feathered. ▪ Lacks black carpal patch; black only on primary greater coverts. ▪ Wide black subterminal tail band; narrow dark inner band. ▪ *Note:* Same bird as on plate 456.

**Plate 462. Rough-legged Hawk, adult male light morph (lightly marked type) [Dec.]** ▪ Dark mottled bib; white belly unmarked, flanks lightly spotted and barred. ▪ Large black carpal patch on wrists. White remiges lightly barred. ▪ White underside of tail has wide black subterminal band, narrow partial inner black bands.

**Plate 463. Rough-legged Hawk, adult male light morph (moderately marked type) [Apr.]** ▪ Dark mottled bib; whitish necklace; belly, flanks, and leg feathers barred gray and black. ▪ Large black carpal patch. ▪ Tail white with wide black subterminal band, several progressively narrower inner dark bands.

**Plate 464. Rough-legged Hawk, adult female light morph (lightly marked type) [Jan.]** ▪ Pale head with thin dark eyeline; dark malar mark; white forehead and outer lores; dark brown iris. ▪ Lightly mottled bib, solid dark flanks, lightly mottled belly. Brown upperparts. ▪ Wingtips equal to tail tip. ▪ *Note:* Similarly marked males have spotted or barred flanks, gray upperparts.

**Plate 465. Rough-legged Hawk, adult female light morph (moderately marked type) [Mar.]** ▪ White forehead and outer lores blend with pale crown; dark brown iris. ▪ Mottled bib; broad white necklace on upper belly. Split-bellied type has split down center of blackish belly; solid brownish black flanks. Barred leg feathers; tarsi feathered to toes. ▪ Wingtips extend past tail tip.

**Plate 466. Rough-legged Hawk, adult female light morph (moderately marked type) [Dec.]** ▪ Thin dark eyeline; dark malar mark; white forehead that blends with pale crown; dark brown iris. ▪ Streaked/mottled bib has minimal necklace demarcation. Dark-bellied type with brownish black band on belly and flanks. Lightly barred leg feathers. ▪ Single, wide black subterminal tail band.

**Plate 467. Rough-legged Hawk, adult female light morph (heavily marked type) [Dec.]** ▪ White outer lores and forehead; dark brown iris. ▪ Nearly solid dark bib; narrow, mottled tawny necklace; dark-bellied type brownish black belly and flanks. Barred leg feathers. White undertail coverts. Brown upperparts. ▪ *Note:* Heaviest marked type.

**Plate 468. Rough-legged Hawk, adult female light morph (moderately marked type) [Mar.]** ▪ Pale head with thin dark eyeline; dark brown iris. ▪ Dark brown upperparts have some pale mottling on scapulars. ▪ Wingtips equal to tail tip. ▪ White base on tail. Some may have narrow white base on tail or lack white base on dorsal surface. ▪ *Note:* Same bird as on plate 465.

**Plate 469. Rough-legged Hawk, adult female light morph (all types) [Dec.]** ▪ Brown upperparts. ▪ Uniformly brown remiges but may have slightly paler panel on primaries. ▪ Basal 1/3 of tail white. ▪ *Note:* Some moderately and heavily marked types have 3 narrow inner dark tail bands.

**Plate 470. Rough-legged Hawk, adult female light morph (moderately marked type) [Nov.]** ▪ Lightly streaked bib; broad necklace; split-bellied type with tawny center strip on belly, solid brownish black flanks. Barred leg feathers. ▪ Large black carpal patch. ▪ Wide black subterminal tail band with narrow partial inner band.

**Plate 471. Rough-legged Hawk, adult female light morph (moderately marked type) [Nov.]** ▪ Dark, mottled bib; thin tawny necklace; dark-bellied type solid belly and flanks. ▪ Large black carpal patch. ▪ Wide black subterminal tail band.

**Plate 472. Rough-legged Hawk, adult male/female dark morph (brown type) [Nov.]** ▪ White outer lores and forehead. Brown head. Dark brown iris. ▪ Brown breast often mottled with tawny-rufous; belly and flanks often darker brownish black. May have white or tawny speckling on breast, especially at breast-belly junction. Feathered tarsi. ▪ Wingtips extend past tail tip.

**Plate 473. Rough-legged Hawk, adult male/female dark morph (brown type) [Mar.]** ▪ Pale-headed type with large tawny/white supercilium and auricular areas. White outer lores and forehead. Thin dark eyeline. Dark brown iris. ▪ Brown breast mottled with tawny; belly and flanks brownish black.

**Plate 474 Rough-legged Hawk, adult male dark morph (black type) [Nov.]** ▪ White outer lores and forehead. Black head. Dark brown iris. ▪ Black underparts, including undertail coverts. ▪ Distinctly banded type tail with narrow black and white bands, wide black subterminal band. ▪ Wingtips extend past tail tip.

**Plate 475. Rough-legged Hawk, adult male/female dark morph (brown type) [Nov.]** ▪ White outer lores and forehead. Dark brown iris. ▪ Scapulars mottled with gray (females are uniformly brown). ▪ Wingtips extend to tail tip. ▪ Distinctly banded type tail has 3 or 4 narrow white or pale gray bands. ▪ *Note:* Some adult females have similarly banded tail.

**Plate 476. Rough-legged Hawk, adult male/female dark morph (any type) [Jan.]** ▪ Distinctly banded type tail can be found on either sex: 3 or 4 white or pale gray bands on black tail. Common trait on males, uncommon on females.

**Plate 477. Rough-legged Hawk, adult male/female dark morph (brown type) [Nov.]** ▪ Partially banded type tail found on either sex: 3 or 4 indistinct, narrow pale gray on gray tail; wide black subterminal band. Only females can have unbanded type plain gray dorsal tail surface with wide black subterminal band.

**Plate 478. Rough-legged Hawk, adult male/female dark morph (brown type) [Mar.]** ▪ Dark brown underparts. ▪ Large black carpal patches somewhat darker than brownish wingcoverts. Dark brown axillaries often barred. ▪ Distinctly banded type tail with wide black subterminal band and up to 5 narrow black and white bands. ▪ *Note:* Sexed as probable female by large size.

**Plate 479. Rough-legged Hawk, adult male/female dark morph (brown type) [Nov.]** ▪ Pale supercilium and auriculars. ▪ Dark brown underparts. ▪ Large black carpal patches somewhat darker than brownish wing coverts. Axillaries often barred. ▪ Partially banded type tail with wide black subterminal band, thin dark inner bands. ▪ *Note:* Common tail pattern for both sexes.

**Plate 480. Rough-legged Hawk, adult male/female dark morph (brown type)** [**Nov.**] ▪ Tawny-rufous streaking on breast; brownish black belly, flanks, and leg feathers. Large black carpal patches contrast with rufous wingcoverts. ▪ Single wide black subterminal tail band. ▪ *Note:* Palest type of dark morph.

**Plate 481. Rough-legged Hawk, adult male dark morph (black type)** [**Feb.**] ▪ Uniformly black head, body, and underwing coverts. ▪ Distinctly banded type tail with 3 or 4 narrow black and white bands, wide black subterminal band. Remiges may be more heavily barred than on brown type dark morphs.

**Plate 482. Rough-legged Hawk, subadult male light morph** [**Dec.**] ▪ Medium brown iris. White outer lores and forehead. ▪ Breast streaking may extend to top of dark belly and flanks (as on juveniles). Flanks may be solid brownish black or barred. ▪ Single wide black subterminal tail band. ▪ *Note:* Males often acquire female-like traits.

**Plate 483. Rough-legged Hawk, subadult male/female light morph** [**Jan.**] ▪ Similar to adult but retains several faded, worn, brown juvenile upperwing coverts and remiges. Retained juvenile secondaries are shorter than new adult feathers and lack wide black rear band. (Retained juvenile outer primaries not visible at this angle.) ▪ White band on inner half of tail.

**Plate 484. Rough-legged Hawk, subadult male/female light morph** [Mar.] ▪ Dark-bellied type band on belly and flanks. ▪ Large black carpal patches. Retains juvenile remiges; primaries faded brownish, secondaries shorter, lack black band. Left wing has 2 retained juvenile primaries, right wing has 3 retained primaries. Secondaries s3 and 4, s7 and 8 are retained juvenile feathers.

**Plate 485. Rough-legged Hawk, subadult male/female dark morph (brown type)** [**Nov.**] ▪ Pale head. ▪ Tawny-rufous streaking on breast; brownish black belly, flanks, and leg feathers. ▪ Large black carpal patches; axillaries dark brown. Outer 3 primaries faded brown juvenile; secondaries s4 and 9 are juvenile feathers. ▪ Partially banded type tail.

**Plate 486. Rough-legged Hawk, juvenile light morph** [**Feb.**] ▪ Pale brown iris; thin dark eyeline. ▪ Light to heavy breast streaking extends to belly and flanks. Belly and flanks brown and lack blackish tone of adults and subadults. ▪ Secondaries dark brown (barred on older ages). Wingtips extend to tail tip. ▪ Gray undertail.

**Plate 487. Rough-legged Hawk, juvenile light morph** [**Feb.**] ▪ Pale brown iris; thin dark eyeline; partial dark malar mark. ▪ Brown upperparts with pale edges on scapular. Dark flanks. Lightly streaked leg feathers; tarsi feathered to toes. ▪ Long wings extend to tail tip. Dark secondaries; pale edges on coverts. ▪ White base of tail.

**Plate 488. Rough-legged Hawk, juvenile light morph [Jan.]** ▪ Dark brown band on belly and flanks; spotted leg feathers. ▪ Large black carpal patches, often mottled. Nearly unmarked coverts, including axillaries. Rectangular window on primaries. Gray rear edge of wings. ▪ Gray band on distal half of tail, white on inner half.

**Plate 489. Rough-legged Hawk, juvenile light morph [Dec.]** ▪ White rectangular panel on primaries. Dark secondaries. ▪ Basal 1/3 of tail white, distal 2/3 brown. Subterminal region sometimes a bit darker.

**Plate 490. Rough-legged Hawk, juvenile intermediate morph [Oct.]** ▪ Heavy breast streaking extends to solid dark belly and flank band. Tawny leg feathers spotted. White undertail coverts. ▪ Large black carpal patches; rufous-tawny coverts; dark brown axillaries. ▪ Basal 1/3 of tail is white. ▪ *Note:* Similar to heavier marked light morph but axillaries dark, coverts more rufous.

**Plate 491. Rough-legged Hawk, juvenile dark morph (brown type) [Jan.]** ▪ Pale, tawny-streaked head with thin dark eyeline; white outer lores and forehead form white mask. Yellow cere. Pale brown iris. ▪ Tawny-rufous streaking on breast contrasts with solid brown belly and flanks. Tawny tarsi.

**Plate 492. Rough-legged Hawk, juvenile dark morph (brown type) [Feb.]** ▪ Pale, tawny-streaked head with thin dark eyeline; white outer lores and forehead form white mask. Yellow cere. Pale brown iris. ▪ Brown upperparts with dark secondaries. ▪ Brown dorsal surface of tail may be unmarked, partially banded, or have 3 distinct pale gray bands.

**Plate 493. Rough-legged Hawk, juvenile dark morph (brown type) [Feb.]** ▪ Tawny head. ▪ Tawny speckling on breast; dark brown belly, flanks, and leg feathers. White tips on dark under-tail coverts. ▪ Large black carpal patches contrast with rufous coverts; axillaries dark brown. All wingtips are dark.

**Plate 494. Rough-legged Hawk, juvenile dark morph (black type) [Mar.]** ▪ Uniformly brownish black head, body, and wing coverts. ▪ Banded type gray tail has 3 narrow gray bands; subterminal band slightly darker. All wingtips are dark. ▪ *Note:* Common plumage on males, very uncommon on females. Missing a primary on left wing.

## GOLDEN EAGLE
### *(Aquila chrysaetos)*

**AGES:** Adult, four interim basic/subadult ages (basic/subadult I–IV) that correspond to 1- to 4-year-old birds that are in their second to fifth years of life, and juvenile birds in their first year of life. Individual plumage variation is minimal.

Adult plumage is usually attained as a 5-year-old when in its sixth year of life. Sexes are similar, but tail patterns are sometimes marginally different.

Subadults slowly alter from juvenile to adult plumage characters through partial annual molts. Sexes are similar; but based on sex, tail patterns of older subadults are sometimes marginally different. Subadult age characters are most prevalent on the remiges and rectrices, and are most visible in flight and when seen from a dorsal angle when perched. Basic/subadult III and IV are not readily separable from adults, particularly when perched. Basic/subadult II is fairly separable from other subadults and adults when in flight. Basic/subadult I is readily separable from all other ages when flying or perching.

Juvenile plumage is retained for the first year. Sexes are similar. Remiges and rectrices are longer and more pointed at tips than in older ages. Because of this, their wings are broader and tails are longer than older ages.

**MOLT:** Accipitridae wing and tail molt pattern. Wing and tail molt is complex and variable, and it is important to understand the various molt centers and molt waves (*see* chapter 4). Molt timing varies with latitude, diet, and other factors. Molt may be suspended when nesting or migrating. Only a portion of body, remix, and rectrix feathers are replaced each year. Young subadults have one or two molt waves in progress at the same time. Older subadults and adults may have two or three molt waves in progress, with two or three ages of remiges and rectrices.

Molt data are based on Edelstam (1984); Forsman (1999); W. S. Clark unpubl. data, pers. comm.; and BKW pers. obs. of wild birds and museum specimens.

**First prebasic molt.**—Molt takes place during the second year of life when a basic/subadult I. Molt generally begins in Jun. and ends in Oct. or Nov. This first molt can be very slow, and many individuals replace only a small fraction of body, remix, and rectrix feathers. Most of the head and neck feathers are replaced. On most birds, a large amount of breast, belly, and flank feathers are replaced; however, most leg feathers and undertail coverts are not molted. Unusually slow-molting birds may still replace most head and neck feathers, but few, if any, feathers on the ventral regions. Many back and scapular feathers are molted. Dorsal and ventral areas appear blotchy, with new dark feathers among old and faded juvenile feathers.

Remix molt begins on the innermost primary (p1) and advances outwardly. Slow-molting birds molt only the inner three primaries (p1–3), but many molt p1–5. Slow-molting birds may molt some tertials, usually at least the innermost one (s17), but sometimes all three (s17–15, in that order). Many also molt outermost, middle, and inner secondaries (s1 or sometimes s1 and 2, s5, or sometimes s5 and 6, and s14 or sometimes s14 and 13). The respective greater upperwing coverts molt with the molting secondaries; however, lesser upperwing coverts often molt ahead of respective remiges and greater upperwing coverts. Unmolted remiges are worn and faded juvenile feathers.

At most, only a few underwing coverts are replaced. New feathers can be more adultlike or retain similar juvenile characters (e.g., white areas). Only a few upperwing coverts are replaced, and the dorsal wing appears largely worn and bleached with a few new, fresh dark feathers.

Rectrix molt is often very slow. Slow-molting birds replace only the deck set (r1). The outermost set (r6) may also be replaced. Some birds replace the r2 set, which is somewhat atypical of Accipitridae sequence. All other rectrices are retained, worn juvenile feathers. New rectrices have a mix of adult and juvenile characters.

**Second prebasic molt.**—Molt takes place during the third year of life when a basic/subadult II. Molt begins mainly in spring but can begin in mid-winter. A few are molting in Feb. and

Mar. and most are molting by Apr. All body feathers are fully replaced, but not all of the subadult/basic I feathers are replaced, so the body appears blotchy.

Remix molt begins where it left off in the previous molt. Molt may become irregular and not follow prescribed sequence. On the primaries, it begins with s4 or 6, then replaces p5–8 or p7–9 for the first time. P9 and 10 or only p10 are retained, very old worn and faded brown juvenile feathers. P1–3 are replaced for the second time and are newer and darker than the other primaries. *Note:* There are now three ages of feathers on the primaries. On the secondaries, s8 and 9, or s9 (which is the last secondary to molt), are often retained juvenile, but may also retain s10 and 11. All other secondaries were replaced either in the first prebasic molt or in this molt. S4 is one of the newest and darkest secondaries as it was one of the last to molt and thus not subject to wearing and fading. The three tertials (s15–17) may be replaced for the second time. *Note:* There are now three ages of secondaries. Newly molted feathers are adultlike.

Most underwing and upperwing coverts are replaced. The mix of old and new upperwing coverts gives the dorsal surface of the wing a blotchy appearance.

Rectrix molt also begins where it left off in the previous molt, and r1 set may molt for the second time. Most or all rectrices are replaced by the end of the molt, with r2–5 or r3–5 being replaced for the first time. During part of the year, there are three ages of rectrices: juvenile, subadult I, and subadult II. Newly molted feathers are fairly adultlike.

**Third prebasic molt.**—Molt takes place during the fourth year of life when a basic/subadult III. Molt occurs from Feb. or Mar. through Nov., but can occur in winter. Some body feathers are partially replaced for a second or third time.

Remix molt continues from where it left off in the previous molt. P9 and 10 or p10 are replaced for the first time, and bird finally replaces all juvenile primaries. Inner primary molt may continue, starting at p2–4, and p1 may be replaced for the third time. All secondaries, including s9, are fully replaced. The sequence of molt waves may begin again with s1 and 5 being replaced for the second time. However, irregular sequences of secondary molt occur, and it becomes difficult to label an exact molt sequence. Newly molted feathers are very adultlike.

Rectrix molt continues from where it left off or becomes irregular in sequence. Only a few feathers are molted, and there is a mixture of new, dark, and adultlike rectrices and old, faded, and juvenile-like rectrices. R3 and 4 sets are often old subadult/basic I or II and have more juvenile-like characters.

**Fourth prebasic molt.**—Molt takes place during the fifth year of life, when a basic/subadult IV. Molt takes place from Feb. or Mar. through Nov., but can occur all winter. The body keeps replacing some feathers and still retains a blotchy appearance due to a mixture of feather ages.

Remix and rectrix molt often becomes highly irregular with multiple molt waves in progress. All newly molted feathers are nearly adultlike, but may still retain some juvenile-like traits on r2–5.

**Subsequent prebasic molts.**—Molt takes place annually from the sixth year of life; occurs from Feb. or Mar. through Nov., but can be all winter. As with previous molts, only a portion of body, wing, and tail feathers are replaced each year. With multi-ages of feathers, this plumage nearly always appears blotchy.

**SUBSPECIES:** Polytypic, but only one race, *A. c. canadensis*, inhabits North America. This race is also found from cen. Siberia to ne. Russia.

Four additional subspecies inhabit the Palearctic region: (1) Nominate *A. c. chrysaetos* is in Europe and east to cen. Russia. Medium-sized and palest of the races. (2) *A. c. homeyeri* is in n. Africa east to Crete, the Middle East, Saudi Arabia, and Iran. Smaller and darker than nominate and has a darker nape. (3) *A. c. daphanea* is in e. Iran to cen. China. Averages largest of the races and intermediate in color between the two previous subspecies. (4) *A. c. japonica* inhabits Japan and Korea. This is the smallest and darkest race. Its nape feathers are quite rufous and have white on the inner webs of the rectrices.

**COLOR MORPHS:** None.

**SIZE:** A large raptor. Males average smaller but overlap with females. Juveniles are somewhat longer than older ages because of their longer tails. In flight, they may present the illusion of being larger because of their broader secondaries. Length: 27–33 in. (69–84 cm); wingspan: 72–87 in. (183–221 cm).

**SPECIES TRAITS:** HEAD.—**The distal half of the bill is blackish and the basal half is pale blue. Prominent yellow cere and gape. Gape extends to the rear of the eyes. Nape and hindneck are pale tawny or golden and contrast with the dark crown, auriculars, and front of the neck.** BODY.—Dark brown, including the leg feathers. **Feathered tarsi are white or buff-colored.** The undertail coverts are either dark brown or tawny rufous. The long feathers of the undertail coverts often conceal tail patterns on perched and gliding birds. WINGS (dorsal).—**Six "fingers" on the outer primaries.** Median and one or two rows of lesser upperwing coverts are paler than the rest of the coverts and remiges. WINGS (ventral).—Dark brown underwing coverts and axillaries contrast with paler grayish remiges and create a variable two-toned appearance. **There is a tawny patch on the wrist area and a tawny front edge of the inner half of the wing.**

**ADULT TRAITS:** HEAD.—Iris color may be yellow, yellowish brown, orangish brown, or medium brown. Iris is rarely dark brown (W. S. Clark unpubl. data). **Tawny nape has a mixture of new dark and faded old feathering but always appears quite pale.** BODY.—Since molt is not complete on all body feathers each year, dorsal and ventral areas appear blotched with a mixture of new and old, dark, and faded feathers. WINGS (dorsal).—There is a distinct pale bar on the median and first one or two rows of lesser upperwing coverts caused by excessively worn and faded feathers. The dark tertials and inner greater upperwing coverts are often marbled with pale gray designs. WINGS (ventral).—The inner four primaries and all secondaries have a wide black band on the trailing edge. These same remiges typically have a pale gray marbled or barred pattern. The underwing appears distinctly two-toned with pale remiges. *Note:* Molting birds, particularly in summer and early fall, often have white blotches on the underwing coverts caused by missing feathers. TAIL.—Three basic patterns that are somewhat sexually dimorphic and visible on dorsal and ventral surfaces: **(1) *Wide-banded type* (mainly females).—Wide black or brown terminal band with two or three gray bands that get progressively narrower towards the basal part of the tail. Irregularly formed moderately wide or narrow black or brown bars separate the gray bands. (2) *Narrow-banded type* (males and females).—Black or brown tail has three narrow gray bands with a wide black or brown terminal band. Gray bands are often irregularly formed. (3) *Partial-banded type* (mainly males).—Black or brown tail has narrow and incompletely formed pale gray bands. Sometimes only a single pale band shows. The pale bands may be very diffuse or lacking, making the tail appear solid dark.**

**BASIC IV (SUBADULT IV) TRAITS:** HEAD—As on adults, including any of the various iris colors. BODY.—As on adults, with blotchy array of faded old feathers mixed with new dark feathers. WINGS.—Mainly as on adults. Individuals that had extensive white on the remiges as juveniles and younger subadults may still have a very small amount of white on the basal area of the inner eight primaries. The white feather area may be a narrow strip on the inside edge or a small spot. Newly molted remiges are mainly adultlike and have little if any white. Upperwing coverts are adultlike with the faded and worn tawny bar. *Note:* Due to missing feathers, molting birds may show a variable amount of white blotching on the underwing coverts. TAIL.—As on adults except may have a very small white areas on some rectrices, particularly r2–5 or r3–5. White areas may be visible only when the tail is fanned.

**BASIC III (SUBADULT III) TRAITS:** HEAD.—As on adults, including iris color; however, irises are rarely yellow. BODY.—As on adults and subadult IV, with faded old feathers mixed with new dark ones. WINGS.—Primarily adultlike since all juvenile remiges have been replaced. Similar to subadult IV, but may have larger white areas on the underside of the inner eight primaries and possibly on some secondaries. All juvenile remiges are replaced in this age class. Outermost primaries (p9 and 10 or just p10) are new and dark. New subadult remiges may have white areas on the feathers. Upperwing coverts are adultlike with the faded and worn tawny bar. *Note:* Molting birds may show a variable amount of white blotching on the underwing coverts caused by missing feathers. TAIL.—Rectrices are a mix of two or three ages: subadult I, II, and III or subadult II and III. Feather pattern is a mix of adultlike gray barring and juvenile-like white areas, especially retained subadult I feathers (one or two feathers in r2–5 or r3–5 sets).

**BASIC II (SUBADULT II) TRAITS:** HEAD.—As on adults, but iris color is similar to subadult III. BODY.—As on adults and older subadults, with faded old feathers mixed with new dark ones. WINGS.—Somewhat adultlike but has a mix of very old, faded, and worn retained juvenile remiges. Outermost one or two primaries (p9 or 10, or both) are frayed and worn, pale brown juvenile; this condition is a distinctive trait of this age class. All other primaries are newer subadult feathers; the inner four have a wide black band on the trailing edge as on older subadults and adults. Most secondaries are newer subadult and appear fairly adultlike, with the broad black band on the trailing edge and inner gray marbling or barring. New subadult remiges may also have a considerable amount of white on them. Retained juvenile secondaries (p9, sometimes p4, and maybe p10 and 11) are a paler brownish and lack the black terminal band. Retained juvenile feathers are most visible when in flight. Upperwing coverts are like older subadults and adults with the worn and faded tawny bar and a mix of old and new feathers. The underwing coverts are fully molted for the first time. *Note:* Molting birds may have a variable amount of white blotches on the underwing coverts caused by missing feathers. TAIL.—Rectrices are a mix of three ages: juvenile, subadult I, and subadult II. The deck set (r1) is often quite adultlike or may have a mix of juvenile-like whitish with mottling and marbling patterns on the basal region. The outermost set (r6) is also quite adultlike. There is considerable rectrix molt in the early and mid-stages of this age class. The r3 set is replaced for the first time with adultlike feathers, and the r2 set may have already been replaced or will replaced shortly after r3. The r4 and 5 sets are retained juvenile for a while but usually molt by the end of the age class and may molt simultaneously with r3 and 2 sets. Overall appearance of the tail is a dark adultlike center and outer areas with white juvenile-like feathering isolated in two sections between the dark areas.

**BASIC I (SUBADULT I) TRAITS:** HEAD.—As on adults, but iris color is likely to be brownish. BODY.—As on older subadults and adults, with faded old feathers mixed with new dark feathers. The leg feathers are mainly retained juvenile and are paler than the rest of the body, which has attained many new and darker feathers. WINGS.—Pattern of old and new feathering makes this age class easy to categorize. The new inner primaries, p1–3 or p1–5, are easy to see when in flight. The secondaries may be nearly all faded brown retained juvenile feathers. Juveniles that had extensive amount of white as juveniles still show it in this age class. **Perched birds are also easy to age because much of the upperwing coverts is uniformly faded and worn juvenile feathering, with only a scattering of new dark feathers. The pale tawny bar is not readily apparent because of uniform extreme wearing and fading. On some, the newly molted tertials and s14, and respective greater upperwing coverts, are darker and contrast with the paler retained juvenile secondaries.** The outer half of the first row of lesser upperwing coverts may molt ahead of the respective greater upperwing coverts and appear as a dark strip. Large white areas may appear on the basal region of the inner greater upperwing coverts and some first-row lesser upperwing coverts due to molting, missing feathers. The underwing coverts are largely juvenile. *Note:* The underwing coverts may show a variable amount of white blotches due to missing feathers. TAIL.—The deck set (r1) always molts into subadult character and is often fairly adultlike. On birds with more adultlike r1 sets, the tail is split by the two dark rectrices contrasting with the white retained juvenile outer sets. On some, the r6 set also has adultlike darker color and markings and may outline the tail. Most new subadult feathers have partial dark barring, marbling, or mottling at the junction of the basal white area and dark distal area. The dark terminal band on the retained juvenile rectrices is faded brown, and much of each pointed feather tip is worn, so tail appears more rounded than in fresh juvenile plumage.

**JUVENILE TRAITS:** HEAD.—By fall, exhibits the pale tawny nape and hindneck as on older ages. Because of minimal fading and wearing, the nape and hindneck are dark tawny in mid- and late summer on nestlings and recently fledged birds. Iris color is medium or dark brown. BODY.—Since all feathers are the same age with uniform wearing and fading, the body is uniformly dark brown. WINGS.—The greater upperwing coverts, first two rows of upperwing coverts, and tertials are slightly paler than the rest of the dorsal wing surface. (This is the same region on which older ages have the pale tawny bar.) The remiges have three patterns: (1) *Extensive white type.*—A large white area on ventral surface of the inner four to eight primaries and a white strip on the basal region of most secondaries. On the dorsal surface, there is a large square white patch on the inner five or so primaries. *Note:* Uncommon type. (2) *Moderate white type.*—A medium-sized white patch on the ventral surface of the inner three to five primaries and sometimes a small white patch on the inner one to three secondaries. On the dorsal surface, there is a small white patch on the inner one to three primaries. *Note:* Common type. (3) *All-dark type.*—Ventral surface of the remiges is uniformly dark. At most, there may be a sliver of white on the innermost primary. All types may have faint, narrow barring on the inner three or four primaries. TAIL.—Somewhat wedge-shaped. Broad dark terminal band that is neatly separated from a broad inner white band. At close range, dark mottling shows on the distal white area next to the dark band. On some, grayish marbling adorns the dark band on the deck set. Two variations: (1) *Moderate white type.*—The dark and white bands are equal width. When the tail is closed, the long undertail coverts reach or extend slightly into the dark terminal band. *Note:* Common type. (2) *Extensive white type.*—The dark band is fairly narrow. When the tail is closed, a white area is always visible between the long undertail coverts and the dark band. *Note:* Uncommon type.

**BARTHELEMYI VARIANT:** Unusual plumage found on all ages and sexes (Palmer 1988).—As in typical birds but there is a small to large white epaulete on the front of each scapular. *Note:* Rare plumage type. A subadult (basic) I was photographed by the author in Oct. 2002 in Duluth, Minn.

**ABNORMAL PLUMAGES:** Partial albinism is rare (Clark and Wheeler 2001).

**HABITAT: Summer.**—Mainly found in remote open and semi-open hilly and montane areas, including above timberline. May inhabit locales with light agricultural use, but rarely found in rural areas. Nesting birds favor locations with embankments or cliffs for nest sites. Areas with flat or moderate terrain but having scattered large trees are also inhabited. Nonbreeding birds may be found in open or semi-open areas that have elevated perches. Climate varies from hot to cool and from arid to moderately humid; less commonly found in humid regions. **Winter.**—Habitat is similar to summer but below timberline. Moderate agricultural areas are also inhabited, including rural locations. **Migration.**—Similar habitats as for summer and winter.

**HABITS:** Golden Eagles vary from being tame to wary in the West. They may perch or fly for long periods. Eagles often sit on level ground but prefer elevated natural and artificial structures, particularly large metal towers and wooden utility poles. In treeless hilly areas, eagles perch on the ground near or at the top of hills. When windy, eagles sit on the lee side and below hilltops.

A solitary species; however, pairs remain together year-round. They typically perch near each other throughout the year, often on the same branch or utility pole cross bar.

In winter, a few Golden Eagles are found in areas where Bald Eagles congregate. Golden Eagles sometimes roost with Bald Eagles, Ferruginous Hawks, Red-tailed Hawks, and Rough-legged Hawks.

**FEEDING:** Perch and aerial hunter. Golden Eagles often kill prey much larger and heavier than themselves. Pairs may cooperatively hunt but younger eagles hunt singly. Hares, ground squirrels, marmots, prairie dogs, and rabbits form the bulk of their mammalian diet in summer. Non-hibernating species such as Black-tailed and White-tailed jack rabbits, Black-tailed Prairie Dogs, and cottontail rabbits are a mainstay in winter. Herons, upland gamebirds, and waterfowl are regionally and seasonally important. Young Bighorn and Dall sheep, Elk, Mule and White-tailed deer, and Pronghorn are regularly preyed upon. Old Pronghorn are also preyed upon, but mainly in winter. Arctic and Red foxes and Coyotes are also hunted. Fish, reptiles (mainly non-poisonous snakes and turtles), rodents, and small birds form a small part of the eagle's diet.

Very large birds such as Whooping Cranes are rarely taken. Eagles also kill smaller raptors.

Birds are captured while airborne or on the ground.

Carrion is readily eaten, particularly in winter.

**FLIGHT:** Wings are held in a low dihedral when soaring and gliding. At times, wings may be raised to a high dihedral when soaring. Kiting regularly occurs in strong winds. Wings are held on a flat plane when kiting. Powered flight is used frequently. The upstroke and downstroke are an equidistant up-and-down motion. Flapping sequences may be irregularly interspersed with periods of gliding, or birds may flap for considerable lengths of time.

Perch hunting involves a direct attack on prey by diving from a high perch or launching into flight and coursing low over the ground. Aerial hunting may be initiated from high altitudes while soaring or kiting and from low altitudes while kiting or in random surprise-and-flush flights.

**VOICE:** Virtually silent. Rarely heard in the wild, even when approached at nest sites. Captive birds emit a moderately high-pitched yapping *yeh, yeh, yeh* when excited or agitated.

**STATUS AND DISTRIBUTION:** *U.S. and Canada.*—Overall *uncommon*. Population is stable. Estimated population in North America is 100,000 birds. Local declines are occurring in certain areas, especially in the Snake River Birds of Prey National Conservation Area near Boise, Idaho. Declines have occurred in the National Conservation Area since 1971 because of habitat changes and loss of prey.

*Mexico.*—Designated as an Endangered Species. In the northern states as shown on the range map. Southern states: *San Luis Potosí*: Northern and southern parts of the state. *Zacatecas*: Southeastern, southern, central, and western parts of the state. *Guanajuato*: Near Sierra de Santa Rosa and the city of Pozos, in the Sierra Gordo Mts. near San Jose Iturbide, and the northern part of the state. *Aguascalientes*: Near the border with Jalisco and Zacatecas. *Hidalgo*: El Chico N.P. near Pachuca. *Estado de Mexíco*: On the inactive Nevado de Toluca Volcano. *Nuevo León* (below mapped area): Near Doctor Arroyo. *Tamaulipas* (below mapped area): Between Juamave and Bustamante. *Oaxaca*: Near Tlaxiaco (probably southern limit of Golden Eagle range in Mexico).

**Summer.**—Widely distributed in Alaska, w. Canada, and west of the Great Plains. Highest breeding density is in the Diablo Mts. in w.-cen. California. Lesser numbers are found on the Great Plains and Trans-Pecos region of w. Texas. Breeding has not occurred in Manitoba since the 1800s, but are irregularly found along the Hudson Bay. Very few inhabit the humid coastal region of British Columbia and southeastern Alaska.

Sparsely distributed in the highlands of Mexico. Breeds from ne. Sonora and nw. Coahuila south to w. Oaxaca.

**Winter.**—Departs much of Alaska and Canada. Regular on Kodiak Island, Alaska, and sw. Yukon and British Columbia. Irregular in Denali N.P., Alaska, and other areas of s. Yukon Territory. Regular on the cen. and s. Great Plains, and fairly regular in portions of Arkansas, Iowa, Minnesota, and Missouri. In mild winters, a few birds are found within the blue dashed linear pattern on the range map in w. Minnesota and the Dakotas. Rare in Louisiana and e. Texas. In Mexico, mainly stays in the high elevations of the Sierra Madre Occidental and Oriental Mountains and the Mexican Plateau. Wintering birds occur throughout Baja California.

**Movements.**—Alaskan and Canadian birds are highly migratory. More southern birds are moderately migratory. Virtually all juveniles migrate but many southern adults remain on or near breeding areas. Incredible fall and spring migration concentrations occur in the Rocky Mts. west and south of Calgary, Alberta, and north of Bozeman, Mont. Nearly 600 Golden Eagles may be seen on peak days in the fall and over 400 in the spring at both areas.

*Fall migration*: Juveniles begin moving in mid- to late Aug. A few juveniles are seen on the cen. Great Plains in mid- to late Sep.

In the Canadian Rocky Mts., juveniles peak in early to mid-Oct. then taper off by late Oct. and early Nov. Older birds also comprise the Oct. peaks, but they continue moving in small numbers through Nov. and into early or mid-Dec.

In the U.S. Rocky Mts. and Minnesota, peak numbers of mainly juveniles and some older birds occur from mid- to late Oct. Movement of mainly older birds continues into Nov. in

Minnesota but extends through Dec. on the Great Plains and west of the Plains.

*Spring migration*: Adults move before juveniles and subadults. Adults begin migrating in early Feb. in the U.S. and arrive in s. Alberta by mid-Feb. The first adults are seen near Eureka in e. Alaska in mid-Mar. In the s. Rocky Mts., adults peak in early Mar., and in s. Canada, they peak in mid- to late Mar. Near Eureka, Alaska, peak numbers of mainly older birds are seen from late Mar. to early Apr. In the contiguous U.S. and s. Canada, younger ages move from Mar. through mid-May. They peak in early to mid-Apr. in the contiguous U.S. and from mid- to late Apr. in s. Canada. The first significant movement of younger ages does not occur until mid-Apr. in e. Alaska.

**NESTING:** Highly variable depending on latitude and elevation. Texas birds begin nest-building in Jan. Those nesting at high montane elevations and in Alaska and Canada may not begin until late May or early Jun.

**Courtship (flight).**—*Sky-dancing* by males, which may turn into *talon-grappling* by both sexes. Both sexes engage in *high-circling*.

Nests are built on low embankments, high cliffs, or trees 10–100 ft. (3–30 m) above ground. Both sexes build the structure. Nesting materials are mainly thick branches, but may also be thin branches, twigs, and weed stalks. Greenery is often added. (An unusual nest in the Texas Panhandle was constructed of tumbleweed on a utility pole.) Material is added onto the nest during the course of use. Nests are typically reused for many seasons and often for many decades, and may become very large. Nesting territories have several alternate nests.

New nests on embankments and cliffs may be mere shallow layer of sticks; those in trees are much deeper. Average new nests are 2.5–3 ft. (0.75–1 m) and old nests are 5–6 ft. (1.5–2 m) in diameter. Old nests are typically around 6 ft. (2 m) and occasionally up to 10 ft. (3 m) deep. Greenery is added during the nesting season.

Two eggs are generally laid, but may be only 1. Eggs are laid at 3- to 4-day intervals. Females perform most of the 43- to 45-day incubation task beginning with the laying of the first egg. Fratricide (siblicide) typically occurs. If there are two nestlings, usually only the older survives. Fledging takes place when they are 63–70 days old. Fledglings stay with their parents for several additional weeks, but are independent by the fall migration.

**CONSERVATION: Restoration programs.**—Strong efforts have been made to enforce illegal poisoning, shooting, and trapping. With a stable population in most regions, there have been few programs assisting population growth. Golden Eagles were not greatly affected during the organochlorine pesticide era of the mid-1940s to early 1970s. Urban sprawl is affecting nesting and feeding areas in many parts of the West.

A Golden Eagle reintroduction program created in cen. Kansas ran from 1986 to 1999, with 60 birds being released. However, no birds have returned to breed. Eaglets were supplied mainly by zoos and a few from nests from healthy populations in Colorado and Wyoming. From 1986 to 1996, reintroduction occurred at Wilson Reservoir in Russell Co.; in 1997 it was moved to Kanopolis S.P. in Ellsworth Co. because of a larger Black-tailed Prairie Dog population.

**Protective laws.**—Golden Eagles were first protected from wanton killing in 1962, when the Bald Eagle Act of 1940 was amended to include them. In 1986, the U.S. Supreme Court held that Native Americans must abide by the slated protective eagle laws. Since the 1970s, the USFWS has collected eagle parts and feathers from zoos, rehabilitators, and state agencies; the collection is housed at its National Eagle Repository in Commerce City, Colo. The USFWS distributes the parts and feathers to Native American tribes for ceremonial uses. The repository annually receives about 900 Golden and Bald eagle carcasses for distribution.

**Mortality.**—Poisoning meant for varmints was a major cause of mortality. Organophosphate poisoning still causes mortality, particularly topical chemicals legally used on cattle to control insect infestations, but can poison scavengers if cattle die. Shooting was formerly a major threat but is now a minor threat due to passage of the amended Bald and Golden Eagle Act in 1962 to cover Golden Eagles. Shooting regularly occurred from airplanes. Leg-hold traps meant for Coyotes and other furbearers cause mortality. Electrocution from utility poles is an ongoing problem. Collisions with wind turbines in w.-cen. California kills nearly

50 Golden Eagles each year. A few birds struck by vehicles when feeding on carrion along highways. Rock climbers in a few locations may disturb nesting pairs and cause nest abandonment. Being at the top of the food chain, eagles have few natural predators.

**SIMILAR SPECIES:** (1) **Turkey Vulture, juveniles.**—FLIGHT.—Head is small. Six "fingers" on the outer primaries. Rocks back and forth when soaring and gliding and does not kite. Tail uniformly dark gray. (2) **Bald Eagle, juveniles.**—PERCHED.—Bill and cere are black. Nape is usually dark brown, but sometimes has tawny tips on the feathers and appears similar. Throat is whitish. In fall through spring, belly is tawny. Dorsal body is paler than the head and neck. Bare yellow tarsi not always separable from whitish feathered tarsi of Golden Eagles. FLIGHT.—Extensive white on the underwing coverts and axillaries; however, be cautious of molting Golden Eagles as they may have some white mottling in similar areas. On white-tailed types, outer edge of tail (r6 rectrix set) is black and may be similar to some subadult Golden Eagles. There are six "fingers" on the outer primaries. High upstroke motion of wings when in powered flight. When soaring, wings are held on a flat plane. (3) **Bald Eagle, subadult II and III (heavily marked types).**—PERCHED.—Bills are often grayish with yellow ceres. Nape and hindneck may appear golden colored, but usually have distinctly darker auricular patches. FLIGHT.—White areas on underwing coverts and axillaries may be limited and similar to molting Golden Eagles, with white mottling caused by missing feathers. Otherwise, use flight data as described for juveniles. (4) **Red-tailed Hawk, dark-morph juveniles.**—PERCHED.—Greenish cere and nondescript greenish or yellowish gape. Tail has numerous dark bars. FLIGHT.—White ventral surface of the remiges. Tail as in Perched. (5) **Rough-legged Hawk, brown type adults and juveniles.**—PERCHED.—Large females are deceptively similar in size to small male Golden Eagles. Head often appears tawny or golden, but has a thin dark eyeline. White mask on forehead. Bill and feet are small. Wingtips equal to the tail tip. Diameter of a Rough-legged Hawk is equal or less than the diameter of most utility poles; on eagles, it is larger than most poles. FLIGHT.—Five "fingers" on the outer primaries. Underside of remiges are white or pale gray. Ventral tail is similar; dorsal tail is dark or has narrow gray or white bands. (6) **Ferruginous Hawk, rufous and dark morphs.**—PERCHED.—Large females are deceptively similar in size to small Golden Eagles. Bright yellow cere and large yellow gape are similar. Bill is all black. Rufous-morph juveniles can especially exhibit a tawny head color similar. Feathered tarsi are similar to Golden Eagle. Pale gray outer edge of primaries is distinct. Use "diameter" as described in #5. FLIGHT.—Underside of remiges is white; upperside has square whitish panel on the primaries.

**OTHER NAMES:** Golden. *Spanish*: Aquila Real. *French*: Aigle Royal.

---

REFERENCES: Adamus et al. 2001; Baicich and Harrison 1997; Burt and Grossenheider 1976; Busby and Zimmerman 2001; Bylan 1998; Campbell et al. 1990; Cecil 1999; Clark and Wheeler 2001; Dodge 1988–1997; Edelstam 1984; Faanes and Lingle 1995; Ferguson-Lees and Christi 2001; Forsman 1999; Fuller et al. 1995; Godfrey 1986; Howell and Webb 1995; Janssen 1987; Kellogg 2000; Kent and Dinsmore 1996; Kingery 1998; Kochert et al. 1999; Millar 2002; Montana Bird Distribution Committee 1996; New Mexico Natural Heritage Database 1996; Oberholser 1974; Palmer 1988; Peterson 1995; Pierce 1998; Rideout and Swepston 1984; Rodriquez-Estrella et al. 1991; Semenchuk 1992; Seyffert 2001; Sharpe et al. 2001; Sherrington 1993; Smith 1996; Smith et al. 2002; Steenhof et al. 1984; Stewart 1975; Swepston and Rideout 1984; Santa Cruz Predatory Bird Research Group 1995, 1999b; USFWS 1999a, 2000a, 2001b.

**GOLDEN EAGLE,** *Aquila chrysaetos canadensis:* Uncommon. Sparsely distributed in all seasons. In winter, rarely found in unmapped areas south to Gulf of Mexico. In mild winters, extends to dashed line in MN and ND. Last nest in MB was in the 19th Century.

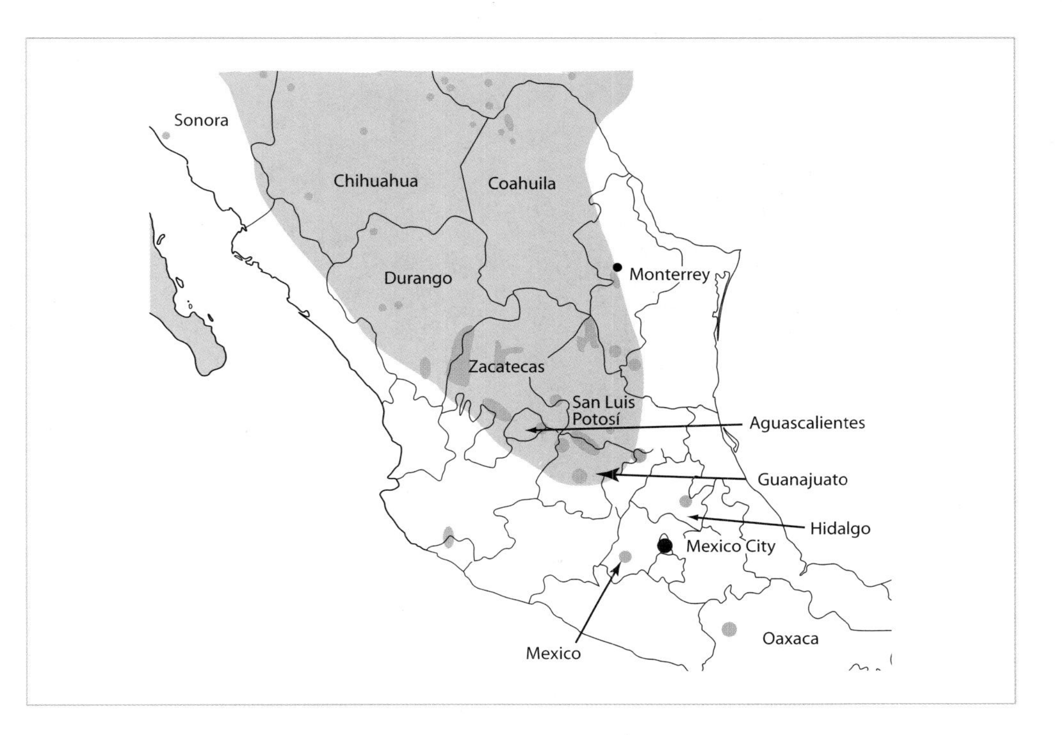


**GOLDEN EAGLE,** *Aquila chrysaetos canadensis:* Mexico classifies the Golden Eagle, its national symbol, as endangered.

**Plate 495. Golden Eagle, adult** [Dec.] ▪ Bill is black on outer half, pale blue on inner half; yellow cere, large yellow gape. Tawny nape and hindneck (nape hidden at this head angle). ▪ Dark brown body. ▪ Pale bar of bleached, worn feathers on upperwing coverts. ▪ Dark tail has 2 or 3 gray bands.

**Plate 496. Golden Eagle, adult** [Nov.] ▪ Tawny nape and hindneck. ▪ Dark brown upperparts. ▪ Pale bar of bleached, worn feathers on upper wing coverts.

**Plate 497. Golden Eagle, adult** [Oct.] ▪ Dark brown body (white mottling on breast is disarranged feathers from a distended crop). ▪ Gray barring on dark gray remiges; wide dark band on rear edge of wings. ▪ Partial-banded type tail with indistinct, partial, pale gray bands (mainly males).

**Plate 498. Golden Eagle, adult** [Mar.] ▪ Tawny hindneck is visible at most angles. ▪ Dark brown body; tawny undertail coverts. ▪ Gray barring on dark gray remiges; wide dark band on rear edge of wings. ▪ Wide-banded type tail with wide gray bands and a wide black terminal band (both sexes, but mainly females).

**Plate 499. Golden Eagle, adult** [June] ▪ Tawny hindneck. ▪ Dark brown body. ▪ Gray barring on dark gray remiges; wide dark band on rear edge of wings. Adults and subadults are in extensive molt in summer. Remiges are ratty; coverts may have white blotches due to molting feathers. ▪ Narrow-banded type tail with 3 distinct narrow gray bands.

**Plate 500. Golden Eagle, subadult III/IV** [Dec.] ▪ Bill is black on outer half, pale blue on inner half; yellow cere, large yellow gape. Tawny nape and hindneck. ▪ Dark brown upperparts. ▪ Brownish tail is mainly adultlike with gray banding, but has white areas on very basal part of some rectrices. ▪ *Note:* Older subadults are difficult to age when perched.

**Plate 501. Golden Eagle, subadult I** [**Dec.**] ▪ Bill is black on outer half, pale blue on inner half; yellow cere, large yellow gape. Tawny nape and hindneck. ▪ Wings are 90% retained juvenile feathering. A few new (dark) coverts; inner 3 secondaries and greater coverts are new (dark) feathers. White area on inner coverts. ▪ New, marbled mid-rectrix set; rest of tail is juvenile.

**Plate 502. Golden Eagle, subadult IV** [**Mar.**] ▪ Tawny hindneck (partly visible). ▪ Dark brown body has patches of faded old feathers. ▪ Adultlike remiges with a sliver of white on some remiges. ▪ Tail is mainly narrow-banded type with narrow gray bands. ▪ *Note:* Age based on fact that all remiges have been replaced at least once.

**Plate 503. Golden Eagle, subadult IV** [**Nov.**] ▪ Tawny nape and hindneck are partly visible. ▪ Dark brown body. ▪ Adultlike remiges with a sliver of white on some remiges. ▪ Tail is mainly narrow-banded type with narrow gray bands. ▪ *Note:* Age based on fact that all remiges have been replaced at least once.

**Plate 504. Golden Eagle, subadult II** [**Jan.**] ▪ Tawny nape and hindneck. ▪ Dark brown body with patches of faded old feathers. ▪ Outer 2 primaries and 2 middle secondaries are retained, worn, brown, long juvenile feathers. Small white areas on remiges. ▪ White on basal part of rectrices. ▪ *Note:* Aged by retained juvenile remiges.

**Plate 505. Golden Eagle, subadult I [Oct.]** ▪ Tawny nape, hindneck. ▪ Dark brown body has a few faded old juvenile feathers. ▪ Secondaries are retained juvenile feathers. Inner 3 primaries, with some white on them, are new subadult; all others are retained juvenile feathers. ▪ All visible rectrices are retained juvenile feathers, with a white band on inner half of the tail; deck set is new.

**Plate 506. Golden Eagle, subadult I [Oct.]** ▪ Tawny nape and hindneck. ▪ Wing coverts are mainly faded, worn juvenile feathers with a few new (dark) subadult feathers; tawny bar on inner coverts. Secondaries are all retained juvenile feathers; 3 inner primaries are new subadult feathers, all other primaries are juvenile feathers. ▪ New gray and black subadult deck rectrix set.

**Plate 507. Golden Eagle, juvenile [Dec.]** ▪ Black tip and pale blue base of bill; yellow cere, large yellow gape. Tawny nape and hindneck. ▪ Wing has a slightly paler brown bar on inner greater coverts and median coverts and first row of lesser coverts. ▪ Tail is white on basal half and black on distal half.

**Plate 508. Golden Eagle, juvenile [Dec.]** ▪ Tawny hindneck is visible. ▪ Uniformly dark brown body. ▪ Dark gray remiges have extensive white type white patch on the inner primaries and basal area of most secondaries. ▪ Extensive white type tail with inner white band extending behind long undertail coverts. Black band on distal third of tail.

**Plate 509. Golden Eagle, juvenile** [Apr.] ▪ Tawny hindneck is visible. ▪ Uniformly dark brown body. ▪ Dark gray remiges have moderate white type white patch on inner primaries. Most juveniles have faint gray barring on tips of inner few primaries. ▪ Extensive white type tail with the inner white band extending behind long undertail coverts. Black band on distal third of tail.

**Plate 510. Golden Eagle, juvenile** [Nov.] ▪ Tawny auriculars and hindneck are visible. ▪ Uniformly dark brown body. ▪ Dark gray remiges are the all-dark type that lack white markings. Most juveniles have faint gray barring on tips of inner few primaries. ▪ Moderate white type tail, with inner white band partially covered by long undertail coverts. Black band on distal half of tail.

**Plate 511. Golden Eagle, juvenile** [Oct.] ▪ Tawny nape and hindneck. ▪ A pale brown bar extending from inner greater coverts onto median coverts and first row of lesser coverts. Eagles with white underwing patches also have white on dorsal surface of inner primaries. ▪ Tail is white on basal half, black on distal half. (Grayish area is an atypical pattern.)

# CRESTED CARACARA
## *(Caracara cheriway)*

**AGES:** Adult, basic I (subadult), and juvenile. Adult plumage is acquired when 2 years old. Basic I (subadult) is a separate plumage that does not retain juvenile feathering and is acquired when 1 year old. Juvenile plumage is worn the first year.

**MOLT:** Falconidae wing and tail molt pattern (*see* chapter 4). First prebasic molt from juvenile to subadult is complete. As in most falcons, the first prebasic molt begins on the head. Head and body molt is virtually complete prior to the start of wing and tail molt. Molt may be active any time of the year except Dec. (Morrison 1996).

Subsequent prebasic molts from subadult to adult and within adults are probably fairly complete and occur mainly Apr.–Oct.

**SUBSPECIES:** Monotypic. At least four subspecies formerly recognized. Recent data (Dove and Banks 1999, AOU 2000) places former subspecies at full species level.

Former subspecies, *Caracara plancus audubonii*, of the U.S., Mexico, Central America (to w. Panama), Trinidad, and Isla de la Juventud; *C. p. cheriway* of e. Panama to n. South America (north of the Amazon Basin); and *C. p. pallidus* of Islas Tres Marias, Mexico are now considered as one species: Crested Caracara. Former *C. p. plancus* of South America (south of the Amazon Basin) to Tierra del Fuego and the Falkland Islands is now called the Southern Caracara (*C. plancus*). Former *C. p. lutosus*, of Guadalupe Island and s. Baja, Mexico, is now called the Guadalupe Caracara (*Polyborus lutosus*).

**COLOR MORPHS:** None.

**SIZE:** A large raptor. Males average somewhat smaller than females but are not separable in field. Length: 21–24 in. (53–61 cm); wingspan: 46–52 in. (117–132 cm).

**SPECIES TRAITS:** HEAD.—**Thick, pale blue bill. Bare fleshy cere, lores, orbital area, and chin.** Color intensity of fleshy parts varies with mood level: brightest during confrontations and palest during fright; color change occurs rapidly. **Dark crown and bushy crest on nape. Long, pale neck.** BODY.—Black and/or dark brown. **White tail coverts. Very long tarsi.** WINGS.—**In flight, long, moderately narrow wings have parallel front and trailing edges and rounded wingtips. White panel on both dorsal and ventral surfaces of the outer six primaries (p5–10).** Wingtips equal to tail tip when perched. Undersides of the secondaries are covered with pale, very narrow barring (visible only at close range). TAIL.—**Long, with a broad black and/or brown terminal band and white on the distal area, with numerous, very narrow dark bands.**

**ADULT TRAITS:** HEAD.—**Fleshy cere, lores, orbital area, and chin vary from orange to yellow.** Pale orangish or pale brown irises. **Black crown and crest. Cheeks, auriculars, and upper neck are white.** Lower neck is buff colored and finely barred with black. BODY.—**Buff-colored breast and back are finely barred with black.** Belly, flanks, leg feathers, and scapulars are black. Bright yellow tarsi. WINGS.—**Black with white panels on the primaries.** TAIL.—Wide black terminal band.

**BASIC I (SUBADULT I) TRAITS:** HEAD.—**Fleshy cere, lores, orbital area, and chin vary from dark pink to dull yellow.** Pale brown irises. **Dark brown crown and crest (rufous streaking is not visible in the field). Cheeks and auriculars are whitish. Neck is a rich tawny-buff.** BODY.—**Breast and back are a rich tawny-buff with short bar-type markings.** Belly, flanks, leg feathers, and scapulars are dark brown. Tarsi are medium yellow or bright yellow. WINGS.—**Dark brown with white panels on the primaries.** TAIL.—Wide dark brown terminal band.

**JUVENILE TRAITS:** HEAD.—**Fleshy cere, lores, orbital area, and chin vary from dark pink to pale pink or pale purple.** Pale brown irises. **Dark brown crown and crest. Cheeks, auriculars, and all of neck are a rich tawny-buff.** BODY.—**Breast and back are a rich tawny-buff and streaked with brown.** Belly, flanks, leg feathers, and scapulars are dark brown. Pale grayish tarsi. WINGS.—**Dark brown with white panels on primaries. Median upperwing coverts and one or two rows of lesser upperwing coverts have broad white tips.** TAIL.—Wide dark brown terminal band.

**ABNORMAL PLUMAGES:** None documented in North America.

**HABITAT:** *Texas.*—Semi-open mesquite-thorn-scrub regions. Also, semi-open moderate agricultural areas and lush savannahs interspersed with bushes and tall trees. Climate varies seasonally from hot to temperate. Humid in coastal areas and arid in inland areas. *Arizona.*—Semi-open paloverde-saguaro thorn-scrub. Climate is arid and seasonally hot or temperate.

**HABITS:** Moderately tame to tame raptor. A gregarious species, especially juveniles and subadults. Adults have strong mate fidelity and remain paired year-round. Strong, long-lived family ties. Mated pairs and offspring often perch side by side. An extremely aggressive species and dominates over both Black and Turkey vultures and other raptors in communal feeding situations. Caracaras also pirate from other caracaras and large raptors such as White-tailed and Ferruginous hawks.

Exposed branches, particularly treetops, poles, and posts are used for perches, as are concealed branches, especially to obtain shade during hot periods of the day. A very terrestrial species.

*Head-throwback display,* accompanied by vocalization, is an animated behavioral trait. Allopreening occurs between individuals of a pair and between parents and offspring. Allopreening also occurs infrequently between Black Vultures and juvenile and subadult Crested Caracaras. Caracaras initiate the preening behavior by moving closer to a nearby vulture, then bowing their heads to have their napes preened and straightening their heads to have their throats preened.

**FEEDING:** An aerial and terrestrial scavenger and hunter. An opportunistic raptor. Caracaras often fly along highways looking for carrion. This behavior is most common in the early morning. Often feeds at open-pit garbage dumps where food is plentiful and easy to get. Small, live prey items include rodents, amphibians, reptiles, birds, fish, and large insects. Prey is detected while in low-altitude flight or while walking on the ground. In Texas, Caracara feed on dislodged prey at range fires and in freshly plowed fields. Caracaras use their incredibly dexterous feet to turn over chunks of dirt and cow dung in order to locate prey. Regularly pirates food from other caracaras and other raptors, even larger species (*see* Habits). Behavioral actions, such as *head-throwback display,* are common when feeding in groups.

**FLIGHT:** Powered by long series of steady, shallow, and moderately slow wingbeats interspersed with short glides. Usually holds a straight course when flying but will sometimes bank erratically. Wings are bowed somewhat downward when gliding. Wings are held on a flat plane or bowed slightly downward when soaring. Typically, caracaras are rather unstable when soaring, and powered flight may be used for a short periods in each revolution. Flight, however, can be quite stable, and bird will make numerous soaring revolutions without flapping. Leisurely soaring flights rise to high altitudes. Crested Caracaras do not hover or kite.

**VOICE:** Generally silent except in agonistic confrontations. *Rattle* call is a guttural, staccato, and raspy vocalization accompanying the *head-throwback display. Cackle* is similar, but without an animated display. *Cre-ak* is emitted by adults when young are food-begging or when males court females. *Wuck* is a soft call emitted when an adult approaches another adult or when delivering food to nestlings.

**STATUS AND DISTRIBUTION: Permanent resident.**—*Texas: Common.* Estimated population is unknown but appears stable, healthy, and possibly increasing. Fragmented small populations exist west, north, and east of their core range. Suffered massive habitat destruction in the lower Rio Grande valley with agricultural conversion that began in the early 1920s. Caracaras are highly adaptable and acclimate to moderately altered habitat; however, extensive urban sprawl and habitat alteration are possibly affecting them near some metropolitan areas.

Core range consists of virtually all counties in s. Texas. The northern periphery of their core range includes the following counties: sw. Kinney, s. Maverick, s. Zavala, s. Frio, Atascosa, e. Bexar, s. Comal, s. Hays, s. Travis, Bastrop, s. Lee, s. Burleson, Washington, Waller, w. Harris, nw. Fort Bend, w. Wharton, Matagorda, and s. Brazoria.

Isolated breeding or seasonal dispersal has been documented in the following counties: s. Anderson, Bell, s. Bandera, e. Blanco, Bosque, Chambers, Comal, Delta, sw. Edwards, se. Ellis, e. Falls, Fannin, se. Gillespie, Hopkins, Hunt, e. Kaufman, ne. and s. Leon, se. Llano, w. Limestone, n. Hays, Jefferson, se. McLennan, cen. Medina, sw. Navarro, s. Rains, s. Williamson, s. Milam, and Van Zandt.

*Arizona: Uncommon* to *common* but very local. The total breeding population of 20–25 pairs is all within the Tohono O'odam Nation Indian Reservation in Pima Co. The Tohono O'odam Nation covers 2.8 million acres (1 million ha), although not all of it is caracara habitat. Numbers, though small, are seemingly stable, with minimal habitat alteration and persecution.

*Louisiana:* Recent breeding expansion into Cameron Parish. Nesting also suspected for Calcasieu Parish.

**Winter.**—Pairs remain intact and primarily on territory. Immatures not in family groups, particularly subadults, are nomadic within typical range and congregate where food is most plentiful.

**Movements.**—Dispersal mainly by juveniles and subadults. Mated pairs are generally sedentary. *Texas:* Dispersal extends north to Grayson Co. and west to Lampasas and Uvalde Cos. *Arizona:* Immatures may wander into adjacent counties north and east of Pima Co. Individuals have also visited Casa Grande and Red Rock in Pinal Co. and east to Cochise Co.

**Extralimital sightings.**—Occurs north to cen. New Mexico and n.-cen. Oklahoma (Alfalfa Co.). *Accidental:* the only state record is from Trinidad, Las Animas Co., Colo. in late Sep. 1997; Yellowstone N.P. in mid-Sep. 1984.

*Mexico* (mapped area): *Common.*

**NESTING:** *Texas.*—Begins in a span of Jan.–Mar. and ends from May to Jul. *Arizona.*—From late Mar. through August.

**Courtship (flight).**—Not known to occur.

**Courtship (perched).**—*Allopreening* (*see* chapter 5).

Nests are built by both adults and average 28 in. (71 cm) in diameter. Nests are placed in the top sections of tall bushes or trees; in Arizona, often in Saguaro Cacti. Nests are 8–50 ft. (3–15 m) high. Nests are not lined, but accumulate prey debris. Nests are reused in consecutive years and can become quite large. One to 4 eggs, but normally 2 or 3. The eggs are incubated by both adults for 32–33 days. Youngsters fledge in about 56 days and are fed by their parents until about 116 days old. Sometimes produces double broods.

Juveniles stay with parents an additional 4–7 months, often until the beginning of the next breeding season. Intact family groups are common sight and are often seen perched side by side.

**CONSERVATION:** No measures are implemented. Population, as a whole, is stable. Arizona population is on non-threatened Indian lands.

**Mortality.**—Illegal shooting and varmint poisoning undoubtedly occur. Some are hit by vehicles when feeding along highways. Large numbers were formerly killed in vulture traps in the early to mid-1900s.

**SIMILAR SPECIES:** **(1) Black Vulture.**—Similar white panels on outer primaries; black body, tail. **(2) Bald Eagle, immatures.**—Primarily older subadults. PERCHED.—Pale heads with similar-looking darker crowns, but with dark eyelines; dark bodies and white tails with dark terminal band are similar to caracaras. Large grayish or yellowish bills. Short yellow tarsi. FLIGHT.—Dark primaries with white only on inner two or three feathers. Very slow wingbeats with a high upstroke and a shallow downstroke.

**OTHER NAMES:** Caracara, Mexican Eagle. *Spanish*: Caracara Quebrantahuesos, Quelele, and Totache. *French*: Caracara Commun.

---

REFERENCES: AOU 2000; Arizona Breeding Bird Atlas 1993–1999; Baumgartner and Baumgartner 1992; Clark and Wheeler 2001; Dove and Banks 1999; Howell and Webb 1995; Johnsgard 1990; Morrison 1996; Oberholser 1974; Palmer 1988; Pulich 1988; Russell and Monson 1998; Wood and Schnell 1984.

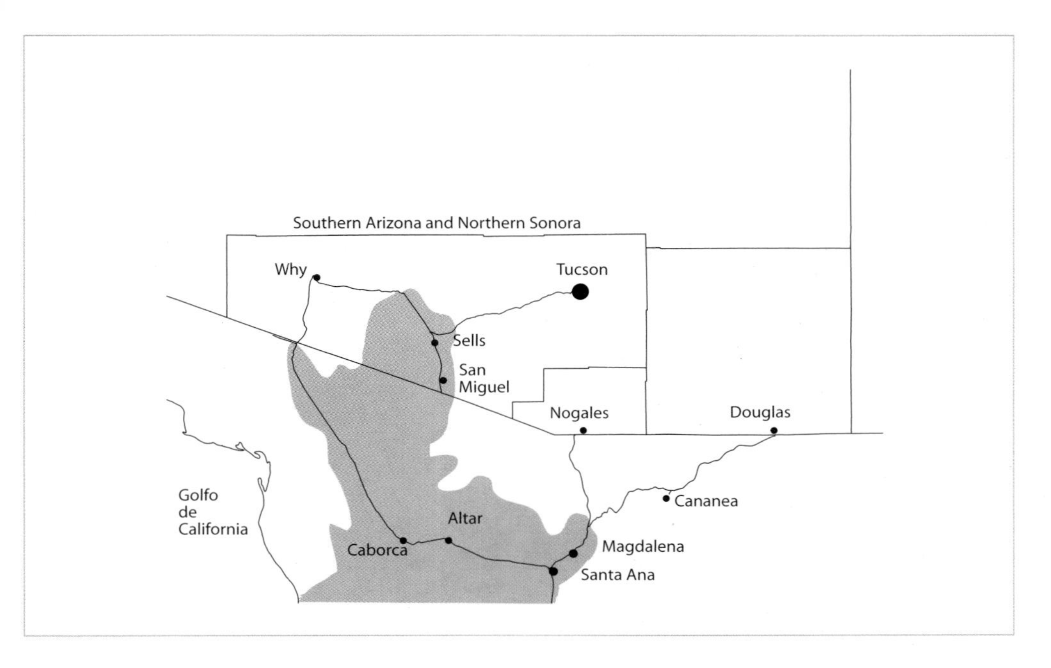


**CRESTED CARACARA,** *Caracara cheriway:* Uncommon to common. In winter, disperses west to Colorado River, north to Phoenix and rarely Flagstaff, and east to the New Mexico border.

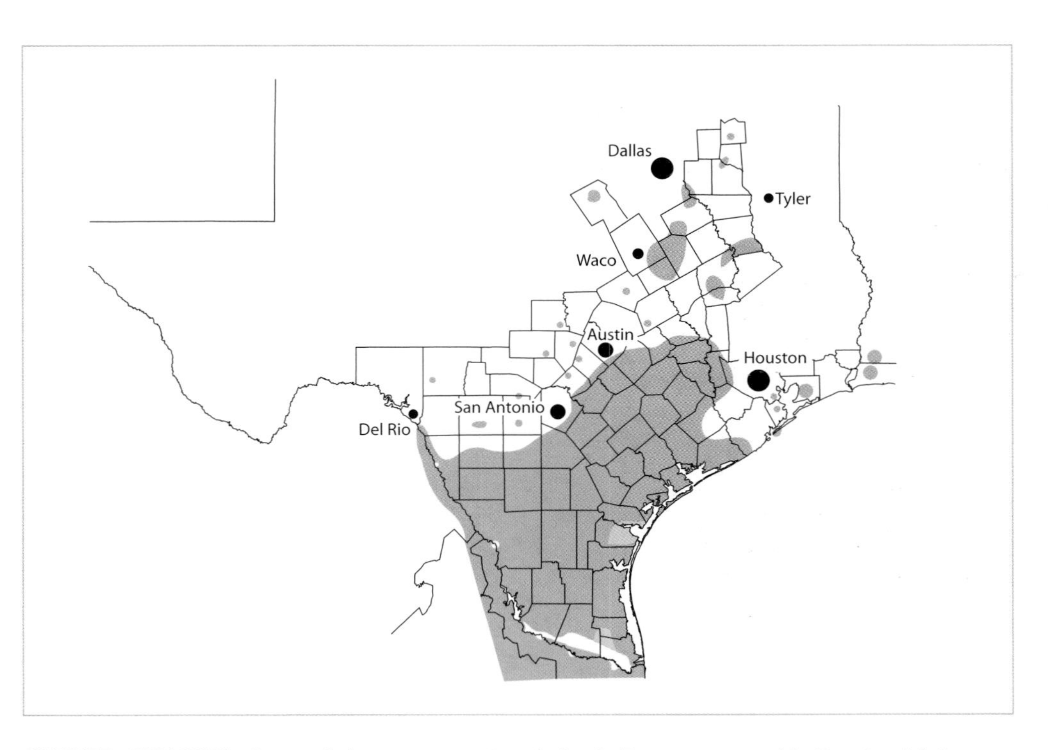


**CRESTED CARACARA,** *Caracara cheriway:* Common. Population in Texas appears stable. Regular sightings north to Fannin Co.

**Plate 512. Crested Caracara, adult** [May] ▪ Pale blue bill, orange facial skin. Black crown with a bushy nape. ▪ White upper neck. White or tawny lower neck and breast and white back finely barred with black. Black belly, flanks, and scapulars. Long yellow tarsi. ▪ Black wings. Wingtips extend to tail tip.

**Plate 513. Crested Caracara, adult (pair) [Oct.]** ▪ Sexes alike. Some birds have dark tawny lower necks and breasts that are finely barred with black. ▪ Pairs remained together year-round and often perch side by side. ▪ *Note:* Yellowish sacks on their breasts are the skin of distended crops, visible after extensive feeding.

**Plate 514. Crested Caracara, adult** [**Feb.**] ▪ Pale blue bill, orange facial skin, and black crown. ▪ Long white neck. White lower neck and breast finely barred with black. Black belly and flanks, white undertail coverts. Long yellow tarsi. ▪ Large white window on outer 6 primaries. ▪ Finely banded long white tail has a wide black terminal band.

**Plate 515. Crested Caracara, adult** [May] ▪ Black crown, long white neck. ▪ Finely barred white back. Black scapulars and rump. ▪ Black wings except for large white panel on outer 6 primaries. ▪ Finely banded long white tail has a wide black terminal band.

**Plate 516. Crested Caracara, adult (1st prebasic molt) [Aug.]** ▪ Pale blue bill, orange facial skin. Black crown with bushy nape. ▪ Scapulars are a mix of brown subadult and new black adult feathers. ▪ Wing coverts are mainly brown subadult with some new black adult feathers. ▪ Middle rectrices are new adult. ▪ *Note:* Head and breast as on adult.

**Plate 517. Crested Caracara, subadult/basic I [Feb.]** ▪ Pale blue bill, pink facial skin. Dark brown crown and bushy nape. ▪ Buff-colored upper neck. Buff-colored lower neck (and breast) marked with partial, fine, dark brown barring. Dark brown upperparts. Long yellow tarsi. ▪ White tail has thin brown inner bands. ▪ *Note:* Some have pale yellow facial skin.

**Plate 518. Crested Caracara, juvenile [Aug.]** ▪ Pale blue bill, pink facial skin. Dark brown crown and bushy nape. ▪ Unmarked buff-colored upper neck; buff-colored lower neck, breast, and back streaked with brown. Dark brown scapulars, belly, flanks, and leg feathers. Pale gray tarsi. ▪ Median coverts and some lesser coverts have white tips. ▪ Long white tail is finely banded with brown.

**Plate 519. Crested Caracara, juvenile [Aug.]** ▪ Dark brown crown, long buff-colored neck. ▪ Dark brown upperparts. ▪ Dark brown wings have a large white panel on outer 6 primaries. ▪ Long white tail finely banded, with a wide dark brown terminal band.

## AMERICAN KESTREL
### *(Falco sparverius)*

**AGES:** Adult and juvenile. Sexes of both ages are dimorphic in plumage.
**MOLT:** Falconidae wing and tail molt pattern (*see* chapter 4). Juveniles go through a first prebasic body molt during Aug.–Nov. Some late-hatched individuals may still be molting after this period. Molt is visibly apparent at field distances only on males; plumage markings are not altered in the molt on females. Juvenile males usually attain an adult plumage by end of the molt; however, many retain small amount of retained juvenile feathering on the breast and sometimes other areas. As with most falcons, molt begins on the head, is completed on the head, and then continues down the body. Wings and tail do not molt until kestrel is 1 year old. Females of both ages have virtually identical plumages. Adult females can be separated from juvenile females Jun.–Nov. by presence of wing and tail molt in adults.

Subsequent prebasic molts begin in late spring and/or summer and completed Oct.–Nov. Females begin molt earlier than males. Latitude and double brooding affect may molt timing.

The following molt sequence of remiges and rectrices is based on Palmer (1988). Sequence of primary molt (with some variation): p4, 5, 6, 7, 2, 8, 9, 10, and 1. Secondary molt sequence: s5, 6, 7, 8, 4, 9, 3, 10, 1, and 2. Sequence of rectrix molt: r1, 2, 3, 6, 4, and 5.
**SUBSPECIES:** Polytypic. Nominate race, *F. s. sparverius*, breeds throughout most of North America, except in the se. U.S. In Mexico, *sparverius* breeds from e. Sonora and w. Coahuila south to n. Michoacán.

The Southeastern American Kestrel, *F. s. paulus*, barely extends its range into the West. *Paulus* inhabits extreme e. Texas and much of Louisiana. In the East, *paulus* is found in the southern portions of Mississippi, Alabama, Georgia, e. South Carolina, and Florida. *F. s. paulus* intergrades with *sparverius* at the northern periphery of its range.

The San Lucas Kestrel, *F. s. peninsularis*, inhabits the southern two-thirds of Baja California, w. Sonora, and Sinaloa, Mexico. There are a few probable records for s. California and s.-cen. Arizona that are based on wing-chord measurements (Rea 1983, P. Unitt pers. comm.). Plumage based on description by E. A. Mearns, who originally described this race in Bent (1961) and Friedmann (1950). Range and habitat also based on Bent (1961).

There are 14 additional races: (1) *F. s. tropicalis* is found from s. Mexico to n. Honduras; (2) *F. s. nicaraguensis* is in the lowland pine savannas of Honduras and Nicaragua; (3) *F. s. sparverioides* is on Cuba, Isle of Pines, and the Bahamas; (4) *F. s. dominicensis* is on Hispaniola; (5) *F. s. caribaearum* is on the Caribbean islands from Puerto Rico to Grenada; (6) *F. s. brevipennis* is on the islands of Aruba, Curacao, and Bonaire; (7) *F. s. isabellinus* is found from Venezuela to n. Brazil; (8) *F. s. ochraceus* is in the mountains of e. Colombia and nw. Venezuela; (9) *F. s. caucae* is in the mountains of w. Colombia; (10) *F. s. aequatorialis* is in subtropical regions of w. Ecuador; (11) *F. s. peruvianus* is in subtropical areas of sw. Ecuador, Peru, and n. Chile; (12) *F. s. fernandensis* is on Robinson Crusoe Island of the Juan Fernandez Islands, Chile; (13) *F. s. cinnamominus* is in se. Peru, Chile, south to Tierra del Fuego, Argentina; and (14) *F. s. cearae* from ne. Brazil west to e. Bolivia.
**COLOR MORPHS:** Polymorphic but not in any of the three North American races. *F. s. sparverioides* of Cuba and adjacent islands has a dark morph that is mainly rufous.
**SIZE:** A small raptor. Sexually dimorphic, but with some overlap. Males average smaller than females. Measurements are for *F. s. sparverius. F. s. paulus* and *F. s. peninsularis* average smaller than *sparverius*; however, there is some overlap with small *sparverius*. Subtle size differences, however, are not readily discernible in the field. MALE.—Length: 8–10 in. (20–25 cm); wingspan: 20–22 in. (51–56 cm). FEMALE.—Length: 9–11 in. (23–28 cm); wingspan: 21–24 in. (53–61 cm).
**SPECIES TRAITS:** HEAD.—**Two narrow black vertical facial markings on the white sides of head: one below the eyes and one on the rear portion of auriculars. Two black spots (ocelli) on each side of the white or orangish nape.** Bluish gray top of head with

a rufous center crown patch; rufous patch is sometimes lacking. Crown patch may have dark streaking. WINGS.—**In flight, wings are moderately long and fairly narrow and taper to pointed wingtips.** In flight, underside of wings is uniformly pale and barred or spotted with black. Wingtips on perched birds are distinctly shorter than the tail tip.

**ADULT TRAITS:** Orange nape surrounds the black ocelli spots.

**Adult Male (*F. s. sparverius*):** BODY.—**Lower half of the rufous back and scapulars have black cross bars. Underparts are orangish on the breast and orangish or whitish on the flanks, and white on belly and lower belly.** Flanks and belly have a moderate amount of black spotting, with the largest spots on the flanks. *Note:* Some northern individuals have reduced black spotting on the belly and flanks and nearly rival the reduced amount of markings as on *F. s. paulus.* WINGS.—**Upper surfaces of the secondaries and coverts are bluish gray with small black spots. Whitish undersides of wings are finely marked with black barring and spotting. In flight and when backlit, trailing edge of primaries shows a row of white spots.** TAIL.—**Rufous, except the outer rectrix set, which is white with one or two black spots or bars and a wide black subterminal band.** *Aberrant tail patterns:* (1) Deck rectrices have blue on much of the set, but the rest of the tail is the typical rufous color. (2) Deck rectrices are the typical rufous, but rectrix sets r2–4 have rufous on the very basal area, and the distal two-thirds of r2–4 and r5 and r6 are crossed with two or three black and white bands. (3) Like example #2, but deck rectrices may have black and white bands on the distal portion. (4) Like examples #2 or 3, but deck rectrices may have blue on the distal half or on all of the set or blue is on small portions of adjacent rectrice.

**Adult Male "Southeastern" (*F. s. paulus*):** HEAD.—**As on *sparverius*, but rufous crown patch is likely to be reduced or lacking.** BODY.—**Back is plain rufous; scapulars have a small amount of black spotting or barring only on the extreme distal part, often only on the most distal feather. Underparts have little if any black spotting on the flanks and belly.** WINGS.—**Upper surface of the wings has little if any black spotting on the coverts.** TAIL.—**As on *sparverius* and probably with identical aberrations.**

**Adult Male "San Lucas" (*F. s. peninsularis*):** HEAD.—Similar head markings to *sparverius.* BODY.—Rufous color of the back and scapulars are paler than on *sparverius.* There is a reduced amount of black barring on the back and scapulars and reduced black spotting on the belly and flanks. Underparts have a yellowish cast. WINGS.—Grayish blue on the upper surface of the wings is somewhat paler than on *sparverius.* Black spotting on upperwing coverts is reduced and nearly lacking, and there is a minimal amount of black spotting and barring on the underwing. TAIL.—As on *sparverius* but paler rufous.

**Adult Female (*F. s. sparverius and E. s. paulus*):** BODY.—**Back and scapulars are rufous with variable-width dark brown barring. Barring can be either** (1) **equal-width dark brown bars or** (2) **narrow dark brown bars.** Underparts are white or buff-colored, with variable-width rufous streaking. WINGS.—**Upperwing coverts are rufous with dark barring as on the rest of the upperparts.** TAIL.—**Rufous with numerous, narrow dark brown or black bars; the dark subterminal band sometimes much wider.**

**Adult Female "San Lucas" (*F. s. peninsularis*):** HEAD.—Irises are yellow, not dark brown as on other North American races. BODY.—Underparts more yellowish than on *sparverius.* Upperparts are a paler rufous than on *sparverius.* TAIL.—Pale rufous.

**FIRST ADULT MALE PLUMAGE (ALL RACES):** The first prebasic molt may not be complete on some birds and they may retain a few juvenile feathers that were not molted during the late summer and autumn. This plumage spans winter and spring. HEAD.—Some whitish feathering may be on the orange nape surrounding the black ocelli spots. BODY.—The breast is orangish like on adults but often retains a few dark juvenile spots or streaks. The flanks and belly are typically heavily spotted. Back and scapulars are typically marked as on adult males but may have a few black bars on the back and forward parts of the scapulars.

**JUVENILE TRAITS:** Juveniles have whitish napes surrounding the black ocelli spots. Northern populations more likely to have larger rufous crown patches and more dark crown streaking than do southern latitude birds (Smallwood et al. 1999).

**Juvenile male (*F. s. sparverius*) Early Stage:** BODY (based on Smallwood et al. 1999).—**Rufous back and scapulars vary in the extent of black barring:** (1) **completely barred with black (all northern-latitude individuals) or** (2) **barred only on the lower one-half of the back and scapulars (some mid- and lower-latitude individuals). Whitish or orangish underparts are variably marked on breast, belly, and flanks:** (1) **Small spots on the breast and belly with large spots on the flanks;** (2) **short, narrow dark streaks on the breast and belly with large flanks spots; or** (3) **large spots on the breast and belly and very large spots on the flanks.** WINGS.—**As on adult male.** TAIL.—**As on adult male and with identical aberrations.**

**Juvenile male (*F. s. paulus*) early stage:** HEAD.—Rufous crown patch on blue crown is often reduced or lacking. BODY (based on Miller and Smallwood 1997).—*Upperparts:* Nearly half of the individuals studied were barred only on the distal half of the back and scapulars; the rest were either fully barred (as in most juvenile *sparverius*) or were unmarked (similar to adult male *paulus*). *Underparts:* Vary in the amount of spotting or streaking, as seen in juvenile male *sparverius*. WINGS.—more heavily marked birds will have black spotting on upperwing coverts (as in *sparverius*); others may have little if any spotting.

**JUVENILE MALE (*F. s. peninsularis*):** Similar to juvenile male *sparverius* but paler.

**Juvenile male late stage:** HEAD.—Nape mostly molts into the tawny color of adults. BODY.—Remnant black barring may be on the upper scapulars and back. There are a few dark streaks or spots on the breast from non-molted juvenile feathers.

**Juvenile female (races are similar):** Plumage identical to adult female's. Juvenile told from adult by lack of wing and tail molt during summer and fall.

**ABNORMAL PLUMAGES: Aberrant-barred juveniles (both sexes).**—Dark brown back and scapulars (and upperwing coverts on females) with narrow pale tawny or rufous tips on each feather. All other anatomical areas are normal. Several records. Rare plumage type. **Partial albino.**—Scattered white feathers on body (Palmer 1988). Several records. Rare. **Incomplete/total albino.**—All-white plumage (Palmer 1988). Very rare; few records. Very rare. **Melanistic.**—Extremely rare plumage. One record for the U.S. from Freer, Texas, in mid-Mar. 1988 (Palmer 1989; P. Palmer pers. comm.). **Gynandromorph.**—At least two records of females with partial or total malelike plumage characters (Palmer 1988). *Note:* Plumages are not depicted of any aberrant types. *Additional note:* There are records of oil-stained individuals that appear to be melanistic!

**HABITAT: *F. s. sparverius.* Summer.**—Breeds in variety of undisturbed or moderately agricultural, semi-open, and open upland habitats. All areas have elevated natural and human-made structures with cavities. Holes and crevices on cliffs, rock formations, trees, buildings, nest boxes, and woodpecker holes in utility poles are used for nest sites. Also, kestrels readily adapt to urban and suburban areas that have some open habitat for foraging and cavities for nesting. Inhabits both humid and arid regions.

***F. s. sparverius.* Winter.**—Can be identical to breeding areas, but without a need for cavities, only elevated perches. Regularly found in open prairie regions with elevated perches.

***F. s. sparverius.* Migration.**—Similar to summer and winter, but are also found in moist lowland habitat.

"Southeastern" race, *F. s. paulus*, is a **permanent resident.** This race is found in low-elevation dry sandhill areas with semi-open, old-aged pine woodlands. Most are Longleaf Pine woodlands that have open understories and are adjacent to meadows. Dead trees with large holes and crevices are used for nest sites. Also found in semi-open rural areas with buildings, nest boxes, and utility poles to provide nest sites. Small numbers inhabit suburban and urban areas. Found only in very humid regions.

"San Lucas" race, *F. s. peninsularis*, is a **permanent resident.** This subspecies occupies arid, lowland regions dotted with cacti.

**HABITS:** Generally fairly wary, though in some areas, becomes very acclimated to humans. *Tail is pumped down and up repeatedly.* Pumps tail downward then back up to the angle of the body, primarily after landing, but may occur any time while perched. Any type of exposed elevated perch is used, notably, treetops and utility wires, especially along highways. *Falcons habitually elevate throat feathers.* Kestrels are very gregarious after breeding season and migrants and wintering birds often gather in loosely formed groups. Kestrels engage in highly antagonistic behavior towards other raptors, especially larger species.

**FEEDING:** Perch and aerial hunter. Insects, small rodents and songbirds, small amphibians and reptiles, and occasionally small bats form their varied diet. Avian prey may be captured while they are on the ground and occasionally while airborne at low altitude. Very small prey, mainly insects, may be eaten while kestrel is airborne. Most prey is taken to elevated perches and devoured. Migrating *sparverius* feed extensively on Green Darner Dragonflies. Carrion is rarely eaten.

**FLIGHT:** Powered by erratic, flitting sequences of lithe wingbeats interspersed with gliding. Often does a short series of quick flicks of the wings between glides. Wings are held on a flat plane or bowed downward a bit, with primary tips flexing slightly upward when gliding. Wings are held on a flat plane soaring. Hovering occurs regularly when hunting. Kestrels may kite for short periods between hovering sequences. When kiting, wings may be held in a low dihedral. Low to moderate-altitudes stoops are made to capture prey.

**VOICE:** Quite vocal at all seasons. Vocalizes during antagonistic encounters and especially during courtship activities. High-pitched, rapid *klee-klee-klee-klee* (also interpreted as *killy-killy-killy-killy*) is the most common vocalization. During courtship, an equally high-pitched, drawn-out, wavering whine, *kree, kree, kree*; a short, high-pitched guttural trill; and a short *kree* by females during pre-copulation.

**STATUS AND DISTRIBUTION:** ***F. s. sparverius.*** **Summer.**—*Common.* Estimated population probably exceeds 2 million birds. The most common raptor in the West. Population is stable, thriving, and probably increasing in much of the West. Readily adapts to an ever-changing, human-altered environment, provided suitable nest sites and foraging habitats are retained or created.

***F. s. sparverius.*** **Winter.**—Northern populations of *sparverius* winter far south of breeding grounds. Some central- and southern-latitude U.S. adults may remain on breeding territories with possible year-round monogamy. Some may winter only a short distance south of breeding areas. Juveniles born at similar latitudes may also winter near natal areas. In migratory populations, males winter at more northern latitudes than do females. Kestrels are generally solitary and establish territories. Large, loosely gathered groups may form in prime feeding areas. Adults have a high degree of territorial fidelity, and many adults return to the same locale each winter. High winter densities occur in s. Louisiana and from Texas to California. *Sparverius* is *fairly common* to *common* in w. Mexico in winter.

"Southeastern" race, *F. s. paulus*, is a *very uncommon* **permanent resident.**—Found only in a small portion of the West. Breeds in most of Louisiana and perhaps a small portion of e. Texas. Habitat destruction of semi-open, mature pine woodlands appears to be the primary reason for reduced numbers in much of this subspecies' breeding range.

***F. s. paulus.*** **Winter.**—Mainly sedentary. A few winter somewhat farther south of breeding range; occasionally along coastal Texas and rarely as far south as Veracruz, Mexico.

San Lucas race, *F. s. peninsularis*, is *uncommon* to *fairly common* **permanent resident** in its Mexican range as described in Subspecies. A few individuals labeled to this race, based on wing-chord measurements, exist for all seasons in Pima and Pinal C.S. in s.-cen. Arizona. S. California also has records that are also based on small wing-chord measurements. Each of the following California locations has one documented specimen: San Diego, San Diego Co.; Seeley, Imperial Co.; and the Laguna Dam, Imperial Co.

***F. s. sparverius.*** **Movements.**—Variable length migrations. Central- and southern-latitude breeding adults may migrate only short distances or may not migrate at all. In e. Colorado, paired birds are seen on territory year-round. Northern individuals, especially juveniles, migrate considerable distances and "leap-frog" southern birds.

*Fall migration*: Begins in mid-Aug. and peaks in mid-Sep. for juveniles and late Sep. for adults. Migration ends abruptly in early Oct.

*Spring migration*: Begins in late Feb. in southern areas, peaks in mid-Apr., and tapers off by mid-May. Males generally precede females.
***F. s. paulus.* Movements.**—Mainly sedentary, although a few may travel short distances as described in Winter.
***F. s. peninsularis.* Movements.**—Sedentary. There may some northward dispersal into s. Arizona and s. California.
**NESTING:** *F. s. sparverius* begins nesting in late Feb. to late May; ends Jun.–Sep., depending on latitude and elevation. *Note:* Copulation has been observed in early Jan. in e.-cen. Colorado. Many southern and central U.S. pairs are laying eggs or incubating when northern birds are in their peak of migration at the same latitude.

**Courtship (flight).**—*High-circling, sky-dancing,* and *flutter-flight* (*see* chapter 5). Vocalization often accompanies courting flights, especially *flutter-flight.*

**Courtship (perched).**—*Food transfers.*

No nest is built. Nest sites are cavities in natural crevices in trees and rocks, woodpecker holes, and human-made crevices in building structures or nest boxes. Very adaptable in nest sites, but competes with several other species of cavity-nesting birds and small mammals. No materials are added to nests. Nest sites are normally 10–30 ft. (3–9 m) high, but can be much higher. Four or 5 eggs, but sometimes 6 eggs. Both sexes incubate the eggs for 29 or 30 days. Youngsters fledge in 30 days and are independent in about 51 days. After becoming independent, juveniles and some adults form groups of up to 20 individuals derived from local populations by mid- to late summer. Double clutches are common in early-nesting pairs.

*F. s. paulus* (based on Georgia/Florida data) usually begins nest activities in mid-Apr., but can be as early as late Mar. and as late as mid-May. First broods fledge by early Jul. All other data similar to *sparverius.*

*F. s. peninsularis* are documented to be in the midst of incubating eggs their 2–4 eggs in mid-May in Baja California Sur, Mexico. Nests are in woodpecker holes in giant cacti.
**CONSERVATION:** Human alteration of the West—with artificial tree planting; containment of rivers, which has allowed substantial tree growth, especially cavity-prone cottonwood species; buildings; and utility poles—has greatly increased nest-site availability. In some areas, especially in the Midwest, private or state-managed nest-box programs assist breeding potential. Kestrels compete with other cavity-nesting species, particularly small owls, woodpeckers, bluebirds, European Starlings, and squirrels.
**Mortality.**—Illegal shooting occurs on a limited basis in North America and south of the U.S. Undoubtedly suffers from some pesticide contamination from organophosphate insect and rodent poisoning. Some collide with vehicles along highways.
**SIMILAR SPECIES:** None of the species listed below pump its tail when perched. **(1) Sharp-shinned Hawk.**—FLIGHT.—Similar shape and size. Sharp-shinned Hawks have more rounded-tipped, broad-shaped wings, deceptively pointed when gliding. Powered flight is a punctual flap-glide sequence. Juveniles may have rufous-streaked underparts; three or four wide dark tail bands. **(2) Aplomado Falcon.**—Much larger than kestrel. Range overlap only in s. Texas, sw. New Mexico, and Chihuahua, Mexico. PERCHED.—Single vertical facial mark below eyes. Uniformly gray or brown upperparts. Dark band on belly and flanks. Numerous, very narrow pale tail bands. FLIGHT.—Dark underwings; distinct white edge on trailing edge of secondaries. Dark band on belly and flanks. Dark tail has several very narrow white bands. **(3) Merlin.**—PERCHED.—Male same size as female American Kestrel. Single dark vertical facial mark below eye or lacks facial markings (*Falco columbarius richardsonii*). Solid blue or brown upperparts. FLIGHT.—Powered flight is with steady, powerful wingbeats. Dark tail has three or four narrow white/buff-colored bands. Voice harsh and rapid *kee-kee-kee.* **(4) Peregrine.**—PERCHED.—Considerably larger. Single dark vertical facial mark below eye. Uniformly dark brown or blue upperparts. Wingtips nearly equal or equal to tail tip. FLIGHT.—Powered flight is with steady, powerful wingbeats interspersed with long, stable glides. Underwing uniformly medium- or dark colored. Rarely vocalizes except at nest sites. **(5) Prairie Falcon.**—PERCHED.—Single dark vertical facial stripe below eye. Uniformly brown upperparts. Tail bands are indistinct. FLIGHT.—Dark axillaries.
**OTHER NAMES:** Kestrel, "K" bird, Little Kestrel (*paulus*); formerly called Sparrow Hawk. *Span-*

*ish*: Cernicalo, Cernicalo Americano. *French*: Crécerelle d'Amérique.

REFERENCES: Andrews and Righter 1992; Arnold 2001; Benson and Arnold 2001; Bent 1961; Breen and Parrish 1996, 1997; Call 1978; Campbell et al. 1990; Clark 2001; del Hoyo et al. 1994; Dodge 1988–1997; Dorn and Dorn 1990; Dotson and Mindell 1979; Ducey 1988; Environment Canada 2001; Friedmann 1950; Harrison 1979; Herron et al. 1985; Hoffman and Collopy 1988; Howell and Webb 1995; Jacobs and Wilson 1997; Johnsgard 1990; Kent and Dinsmore 1996; Kingery 1998; Lane and Fischer 1997; Miller and Smallwood 1997; Montana Bird Distribution Committee 1996; Nicoletti 1997; Oberholser 1974; Palmer 1988; Palmer 1989; Peterson 1995; Pulich 1988; Rea 1983; Robbins and Easterla 1992; Russell and Monson 1998; Semenchuk 1992; Small 1994; Smallwood 1990; Smallwood et al. 1999; Smith 1996; Sutton 1967; Varland et al. 1991, 1993; Varland and Loughin 1992, 1993.

**AMERICAN KESTREL,** ***Falco sparverius:*** Common. "Southeastern" American Kestrel, *F. s. paulus*, is found in e. TX, LA, and the se. U.S. "San Lucas" Kestrel, *F. s. peninsularius*, is found in Baja, Sonora, and Sinaloa.

**Plate 520. American Kestrel, adult male** [Mar.] ▪ Two black facial stripes. Gray crown with rufous patch; tawny nape. ▪ Unmarked tawny breast; tawny belly and flanks are very lightly spotted (palest type). Sparsely barred on lower half of scapulars. ▪ *Note: F. s. paulus* is similar but has less dark spotting and barring.

**Plate 521. American Kestrel, adult male** [Dec.] ▪ Two black facial stripes; gray crown with rufous patch; tawny nape. ▪ Mainly unmarked pale tawny breast; white flanks and belly are heavily spotted. Sparsely barred on lower half of scapulars. ▪ Large black spots on bluish gray wing coverts. ▪ *Note:* First-adult with a few small dark specks of juvenile feathers on breast.

**Plate 522. American Kestrel, adult male** [Feb.] ▪ Two black facial stripes; gray crown with rufous patch; tawny nape. ▪ Tawny breast; white flanks and belly are heavily spotted. Sparsely barred on lower half of scapulars. ▪ Large black spots on bluish gray wing coverts. ▪ Rufous tail with a wide black subterminal band.

**Plate 523. American Kestrel, adult male** [Apr.] ▪ Tawny breast is unmarked, belly and flanks are spotted. ▪ Underwing is moderately spotted and barred. White band on rear edge of secondaries and a row of white spots on rear part of the primaries. ▪ Rufous tail with a wide black subterminal band.

**Plate 524. American Kestrel, adult male** [**Nov.**] ▪ Rufous back, scapulars, and rump. ▪ Bluish gray upperwing coverts; black primaries with a row of white spots on rear edge. ▪ Rufous tail with a wide black subterminal band.

**Plate 525. American Kestrel, juvenile male** (**late stage**) [**Oct.**] ▪ Two black facial stripes; gray crown with small rufous patch; tawny nape. ▪ Tawny breast is lightly spotted with remnant juvenile feathering; tawny belly and flanks are heavily spotted. Moderate barring on most of back and scapulars. ▪ Rufous tail has a wide black subterminal band.

**Plate 526. American Kestrel, juvenile male** (**early stage**) [**Dec.**] ▪ Two black facial stripes; gray crown lacks a rufous patch; white nape. ▪ White underparts are narrowly streaked on breast and belly and heavily streaked on flanks. Moderate barring on all back and scapular feathers. ▪ Bluish gray wing coverts are heavily spotted.

**Plate 527. American Kestrel, juvenile male** (**early stage**) [**Sep.**] ▪ Two black facial stripes. ▪ Streaked and spotted white underparts. ▪ Whitish underwing is moderately spotted and barred. White band on rear edge of secondaries and a row of white spots on rear part of primaries. ▪ Rufous tail has a wide black subterminal band; black bars are sometimes on a few or several rectrices.

**Plate 528. American Kestrel, adult female** [**Apr.**] ▪ Two black facial stripes; gray crown has a large rufous patch; black ocelli spot on tawny nape. ▪ Underparts are streaked with rufous. Rufous brown upperparts are barred with black. ▪ Rufous brown tail is narrowly banded with black. ▪ *Note:* Juvenile females are similar in fall, but have white napes.

**Plate 529. American Kestrel, juvenile female** [**Sep.**] ▪ Two black facial stripes: below eyes and on rear of auriculars; white cheeks; gray crown has a small rufous patch; white nape. ▪ Whitish underparts are streaked with rufous or sometimes brown. ▪ *Note:* Nape will molt into tawny adult color by mid- to late fall.

**Plate 530. American Kestrel, adult/juvenile female** [**Sep.**] ▪ Rufous streaking on underparts. ▪ Tawny spotting on the white, lightly barred remiges. Rufous markings on coverts. ▪ Rufous tail is marked with numerous thin black bands; subterminal band often wider.

**Plate 531. American Kestrel, adult female** [**Mar.**] ▪ Two black facial stripes. ▪ Rufous-streaked underparts. ▪ White remiges are lightly barred; coverts are marked with rufous. ▪ Narrowly banded tail is pinkish when closed. Black subterminal band is wider than inner bands. ▪ *Note:* Tawny nape color of an adult female is bleached in this photograph.

# MERLIN
## *(Falco columbarius)*

**AGES:** Adult and juvenile. Adult sexes are dimorphic in color. Juveniles have minor sexually dimorphic traits on the dorsal color and tail pattern. Adult plumage is gained during the second year. However, on adult females, it may take a few years to attain the "adult" color on the rump and uppertail coverts. Juvenile plumage is held for much of the first year.

Adult and juvenile females have nearly identical plumages. Adult females are separable from juvenile females at long distances in late summer and fall by molt and retention of old, faded feathers. Adult females are difficult to separate from juvenile females after molt is completed in late fall.

**MOLT:** Falconidae wing and tail molt (*see* chapter 4). The first prebasic body molt varies somewhat from larger falcon species in that molt on the upper body precedes molt on the head.

Based on males, whose molt is highly visible, the first prebasic molt begins from early to late Apr.; possibly in very late Mar. on some (BKW pers. obs. of captured birds and museum specimens). One-year-old males in late Apr. and early May show definitive signs of molt. A small amount of bluish gray adult male feathering is mixed with worn and faded brown dorsal feathering on the nape, back, and forward and mid-scapulars; a few new feathers are also on the breast. Remix molt begins later in May, starting at p4, and advancing in both directions. Beginning at s5, the secondaries also molt in both directions. Tertial molt begins on the innermost and molts outwardly. Molt on the upperwing coverts begins after remix molt begins. Rectrix molt, starting with r1 (deck set), also begins later, after the primaries begin their molt. All juvenile remiges and rectrices are replaced in the first prebasic molt. First-year males that begin the first prebasic molt before nesting will suspend the molt until nesting is over (M. Solensky pers. comm.). Males of *F. c. columbarius* and *F. c. richardsonii* may still have retained juvenile upperwing coverts in mid-Sep. Molt is rapid and is generally completed in Oct.

Subsequent prebasic molts begin later in late May or Jun. for adult females and in late Jun. or Jul. for adult males. Molt is complete each year. It ends from Sep. to early Oct. for males and from Oct. to early Nov. for females. During autumn migration, females still exhibit a fair amount of old, worn upperwing coverts and dorsal feathers in early and mid-Oct. The outermost primary (p10), which is the last primary to be replaced, is typically still growing on females in mid- to late Oct. The outermost secondary (s1), which is the last secondary to be replaced, may also still be growing in mid- to late Oct. Rectrix molt is completed before remix molt is done.

**SUBSPECIES:** Three races breed in three major geographical regions of the U.S. and Canada. All three subspecies are found in the West. The three subspecies are prime examples of Gloger's Rule: darkest in humid regions and palest in arid regions. The moderately dark nominate "Boreal" race, *F. c. columbarius*, breeds in the moderately humid boreal forest and taiga zones. ("Boreal" means "northern" and the term typifies the range of this race, which is in more northern latitudes than the other two races and primarily in the boreal forest zone.) The pale "Richardson's" race, *F. c. richardsonii*, breeds on the arid n. Great Plains and winters in arid regions. The dark "Black" race, *F. c. suckleyi*, breeds in the humid rain forests of the Pacific Northwest and winters sparingly elsewhere. Winter range of *suckleyi* is based, in part, on Haney and White (1999) for Utah and California and J. Schmitt (pers. comm.) for California.

*F. c. columbarius* may intergrade with *suckleyi* in a small area in sw. British Columbia and possibly at points along the Coast Range Mts. of British Columbia. *F. c. columbarius* intergrades with *richardsonii* at the periphery of their ranges in Montana, Alberta, Saskatchewan, Manitoba, and Minnesota.

Six additional races breed in the n. Palearctic region of Europe and Russia and winter mainly in the s. Palearctic. Subspecies data are primarily based on Ferguson-Lees and Christi 2001: (1) *F. c. subaesalon* breeds in Iceland and winters w. Iceland and in Ireland and w. England and occasionally eastward to Norway and w. Europe. (2) *F. c. aesalon* breeds on the Faeroe Islands and from n. Europe to w. Siberia and winters from the Mediterranean region eastward to Iran. Similar in color to *subaesalon* but smaller. (3) *F. c. pallidus* is found in w. Siberia and Kazakhstan and winters from e. Turkey to nw. India. Palest and largest of the Eurasian subspecies. (4) *F. c. insignis* breeds in cen. and e. Siberia and winters in n. India, Korea, and Japan. Pale colored but somewhat darker than *pallidus.* (5) *F. c. pacificus* breeds in e. Siberia and winters from n. China to Japan. Darker than the previous two Siberian races. (6) *F. c. lymani* breeds in the mountains of cen. Asia and winters in China. Similar to *insignis* but has distinctly longer wings.

**COLOR MORPHS:** None.

**SIZE:** A small raptor. Sexually dimorphic with no size overlap between sexes within each subspecies. The respective sexes of *F. c. columbarius* and *F. c. suckleyi* average the same size. *F. c. richardsonii* average slightly larger (and heavier) than the other two races. Wing-chord and tail lengths for *richardsonii* average one-half in. (12.5 mm) longer than the other two races; wingspan data are not available. *F. c. richardsonii* average at the top end of the respective measurements: MALE.—Length: 9–11 in. (23–28 cm); wingspan: 21–23 in. (53–58 cm). FEMALE.—Length: 11–12 in. (28–30 cm); wingspan: 24–27 in. (61–68 cm).

**SPECIES TRAITS:** HEAD.—Dark brown irises. Crown feathers have a black shaft on each feather. Yellow cere, orbital skin, tarsi, and feet. BODY (dorsal).—Each back, scapular, and rump feather has a black shaft. BODY (ventral).—Fairly long tarsi and toes. WINGS.—**Moderately long wings. In flight, the broad secondaries taper to the pointed wingtips. When perched, wingtips are distinctly shorter than the tail tip.**

**ADULT TRAITS:** The sexes do not have any similarly colored plumage features.

**ADULTS OF "BOREAL" (*F. c. columbarius*):** HEAD.—The long, thin supercilium connects with the white forehead and extends over the eyes and auriculars. Faint dark malar mark. Pale whitish or tawny cheeks and dark auriculars. A narrow dark brown eyeline extends over the auriculars. Unmarked white throat. Wide black shaft streaks on crown are apparent at close range. Pale mottling on the nape. BODY (ventral).—Moderately wide dark brown streaking on the breast, belly, and forward flanks; rear flanks have a broad arrowhead-shaped mark on each feather with a wide inner bar. Leg feathers have a thin dark brown streak on each feather. Undertail coverts also have a dark feather shaft streak on each feather. WINGS (dorsal).—Small pale spots on the outer web of the inner primaries. WINGS (ventral).—**The dark remiges have moderate-sized pale spots that form a barred pattern. The pale spotted pattern covers 50 percent of the dark gray surface of each feather. Underwing coverts are dark brown with white spotting and mottling. The underwings appear dark because of the moderate-sized pale spots. The underwing coverts are dark brown with white spots on the median coverts and tawny spots on the lesser coverts.** TAIL (ventral).—**Three or four distinct, moderately wide pale bands. White terminal band is moderately wide with a short dark spike extending along the shaft of the basal part of the white feather tip.**

**Adult male *columbarius:*** HEAD.—Cere and orbital skin are medium yellow. Medium bluish gray crown and auriculars. Supercilium is white and well defined. White or tawny mottling on the dark nape. BODY (dorsal).—Medium bluish gray back, scapulars, and rump. The bluish gray on the dorsal areas graduate to a paler color from the back to the rump (M. Solensky pers. comm.). BODY (ventral).—Often has a tawny or tawny-rufous wash on the breast, lower belly, and undertail coverts. Dark barring on the distal flanks may be grayish. **Leg feathers are tawny or tawny-rufous.** Tarsi and feet are orangish yellow. WINGS (dorsal).—Medium bluish gray. The primaries are blackish and the inner primaries have small bluish gray spots on the outer web. WINGS (ventral).—**The pale spots on the black remiges are white.** TAIL (dorsal).—**Black with two or three sharply defined narrow bluish gray bands. Black band on the subterminal region is wider than the inner black bands.** TAIL (ventral).—**Black with three white bands.**

**Adult female *columbarius:*** HEAD.—Cere and orbital skin are medium yellow. Medium brown crown and auriculars. Crown may be grayish. Crown has moderately wide black shaft streaking. Supercilium is pale tawny.

Tawny mottling on the dark nape. BODY (dorsal).—Medium brown back and scapulars. In fresh plumage, each feather may have a very narrow tawny-rufous edge; otherwise, plumage is uniformly brown. Rump is grayish, but is only visible in good light in flight from above, if the wings are drooped when perched, or if viewed in the hand. BODY (ventral).—Uniformly tawny or white, including the leg feathers, with brown streaking. WINGS (dorsal).—Medium brown. Dark brown primaries typically have small, pale tawny spots on the outer web on the inner primaries. WINGS (ventral).—**The pale spots on the brownish black remiges are tawny-rufous.** TAIL (dorsal).—Uppertail coverts (and rump) are grayish. Medium brown with a wide darker blackish brown subterminal band. **There are three variations of pale banding on the dorsal surface: (1) Lacks pale dorsal bands.** *Note:* Fairly common pattern. **(2) Two or three moderately wide tawny or sometimes slightly grayish bands on the deck rectrices and often on the inner two or three rectrix sets, but the outer three to five rectrix sets are unbanded.** *Note:* Fairly common pattern. **(3) Tail is fully banded by two or three sharply defined pale tawny or sometimes slightly grayish bands.** *Note:* Common pattern. TAIL (ventral).—**Dark brown with pale tawny-rufous bands on all three types.**

**Adult female (dark type) *columbarius:*** HEAD.—Cere and orbital skin are medium yellow. Dark brown crown and auriculars. Wide dark crown streaking is apparent. The crown is rarely solid dark as on *F. c. suckleyi* females. The cheeks may be dark brown like on *suckleyi* or somewhat streaked with tawny or white. Dark malar mark is fairly distinct. Thin, short, pale tawny supercilium over the eyes. Dark nape has a small amount of pale mottling. BODY (dorsal).—Dark brown back and scapulars. Rump is grayish. Tarsi and feet are medium yellow. BODY (ventral).—Tawny or whitish with wide dark brown streaking on the breast, belly, and lower belly; flanks are broadly barred with dark brown. Whitish or tawny leg feathers have a narrow or moderately wide dark streak on each feather. Undertail coverts have a moderately wide dark brown feather shaft streak, most pronounced on the distal feathers. WINGS (dorsal).—Dark brown and lacks pale spots on the outer webs of the inner primaries. WINGS (ventral).—**As on typical adult female *columbarius*, with moderate-sized pale tawny spots (which separates them from very similar *suckleyi*).** TAIL (dorsal).—**Same three patterns as described for typical adult female *columbarius*.** TAIL (ventral).—**As on typical adult female *columbarius*, but tawny-rufous bands may be narrower.** *Note:* Data based on photographs of breeding females from Michigan (K. T. Karlson, plate 535) and Wisconsin (W. S. Clark, p. 257 in Clark and Wheeler 2001) and a migrant specimen from Colorado (Denver Museum of Nature and Science).

**ADULTS OF PALE TYPE OF "BOREAL" (*F. c. columbarius*):** Like respective sex of typical *columbarius* but paler on the dorsal regions: medium pale bluish gray on males and medium pale brown on females. They are somewhat darker than the respective sex of *F. c. richardsonii*. *Note:* Many are simply paler plumage variants, others are intergrades with *richardsonii*. Only adult female/juvenile female is depicted in accompanying photograph.

**ADULTS OF "BLACK" (*F. c. suckleyi*):** HEAD.—Dark crown, cheeks, and auriculars. The forehead is dark. At most, there is a very small pale supercilium patch over the eyes. The nape is solid dark or may have a small amount of pale mottling. The dark malar merges with the dark cheeks and auriculars. White or pale tawny throat is streaked with dark brown or black. BODY (ventral).—Broadly and densely streaked on the breast, belly, and lower belly. At a distance, the underparts appear nearly solid dark. The flanks are broadly barred and have a large arrowhead-shaped mark on the tip of each distal flank feather. Leg feathers are thickly streaked. The undertail coverts typically have a large arrowhead or diamond-shaped dark mark on each feather. The basal dark mark typically broadens into a bar shape, particularly on the distal coverts. WINGS (dorsal).—Outer webs of all primaries lack pale spots. WINGS (ventral).—**On the darkest individuals, the remiges are uniformly dark and lack pale spots. Many have a few very small pale spots on the outer one to three primaries. On the darkest individuals, the secondaries are all dark. Many have very small pale spots on some or all of the secondaries. If pale**

**spotting is on the primaries and secondaries, it covers much less than 50 percent of the dark gray feather surface.** *Note:* The reduction or absence of pale underwing spotting is a trademark of this subspecies. Birds that are possible intergrades with *F. c. columbarius* have moderately small pale spotting on all the remiges. TAIL.—Pale bands are reduced in size and number or are absent on the dorsal surface. Narrow partial or complete pale bands on the ventral surface. White terminal band is very narrow with the dark "spike" extending along the feather shaft to the tip of each feather.

**Adult male *suckleyi:*** HEAD.—Dark bluish gray crown, cheeks, auriculars, and nape. The dark feather streaking is visible on the crown. The nape is sometimes mottled with tawny or rufous-tawny. BODY (dorsal).—Dark bluish gray back, scapulars and rump. The dark feather shaft is visible. *Note:* The bluish gray dorsum is slightly or moderately darker than on classic adult male *F. c. columbarius.* BODY (ventral).—Base color to the breast is rufous-tawny, belly is whitish, and the lower belly is whitish or rufous tawny. **Leg feathers are rufous.** Dark markings on the distal flanks may be dark bluish gray. Undertail coverts may also be rufous-tawny. WINGS (dorsal).—Coverts are dark bluish gray like the rest of the upperparts. The remiges are uniformly black. WINGS (ventral).—If a small amount of pale spotting is present, it is whitish. Underwing coverts are dark brown and spotted with white on the median coverts and tawny on the lesser coverts. TAIL (dorsal).—**Either uniformly black or may have a single row of bluish gray bands on the mid- or basal region.** TAIL (ventral).—**Black with one or two rows of partial white bands on a fanned tail and only a single band on a closed tail.** *Note:* Not known to be depicted in published photographs.

**Adult female *suckleyi:*** HEAD.—Crown, cheeks, auriculars, and nape are uniformly dark brown or brownish black. Paler birds may have some visible dark crown streaking. Dark shaft streaks are typically not visible because the feathers are so dark. Nape is likely to be uniformly dark, but may have some tawny mottling. BODY (dorsal).—Uniformly dark brown; some are brownish black. Based on museum specimens, the rump is grayish on some birds; however, this is difficult to see in the field. BODY (ventral).—Whitish or tawny base color of all of the underparts, with very dense and thick streaking. WINGS (dorsal). Uniformly dark brown or brownish black like the rest of the upperparts. WINGS (ventral).—**If pale spotting is present on the remiges, it is pale tawny.** Underwing coverts are dark brown and spotted with white. TAIL (dorsal).—**Typically uniformly dark brown with a wide black subterminal band. Some may have a hint of one or two partial bands.** Uppertail coverts (and rump) are grayish on some birds. TAIL (ventral).—**Brownish black with one or two partial or complete, very narrow pale tawny-rufous bands on a fanned tail and one pale band on a closed tail.**

**ADULTS OF "RICHARDSON'S" (*F. c. richardsonii*):** HEAD.—The long and moderately wide white supercilium connects with the white forehead and extends over the eyes and auriculars. Thin dark crown streaking is apparent on all birds. The dark malar mark is absent or very ill-defined. A narrow dark brown eyeline extends over the auriculars. The cheeks are white and the auriculars are either white or have a slightly darker tinge. Throat is white and unmarked. BODY (dorsal).—The black feather shafts are very obvious because the plumage is pale. BODY (ventral).—Underparts are mainly white and covered with narrow or moderate-width dark brown or rufous-brown streaking. The distal half of the flanks are barred. Leg feathers are either thinly streaked or are unmarked. The undertail coverts are either immaculate white or have a very thin dark streak on the shaft of each feather. WINGS (dorsal).—**Large pale spots on the outer web of all primaries.** WINGS (ventral).—**Large pale spots on the dark remiges. The large pale spots cover more than 50 percent of the surface area of each feather. The underwings appear pale because of the large spots.** TAIL (ventral).—**Distinct, wide white bands and a broad white terminal band. When the tail is closed, only one pale band shows on the mid-tail next to the tips of the undertail coverts.**

**Adult male *richardsonii:*** Crown is pale bluish gray with dark streaking. Nape is whitish or tawny. BODY (dorsal).—**Pale bluish gray back, scapulars and rump.** BODY (ventral).—

**Leg feathers are pale tawny-rufous.** Dark markings on the rear portion of the flanks are sometimes bluish gray. WINGS (dorsal).—Bluish gray secondaries and coverts. The primaries are blackish with large bluish gray spots on the outer web of all feathers. WINGS (ventral).—**The large pale spots on the black remiges are white.** TAIL (dorsal).—**Two patterns: (1) Wide black subterminal band with three moderately wide white, very pale gray, or pale bluish gray bands. Two or three darker bands cross between the three pale bands. The dark band above the wide subterminal band is black; the other one or two bands may be partially black or bluish gray. If the inner dark bands are bluish gray, the inner pale bands will still show if they are white or very pale bluish gray. The pale inner tail banding will not show if the pale bands are also bluish gray. (2) Tail is black with the wide black subterminal area and three moderately wide white, very pale bluish gray, or pale bluish gray inner bands.** *Note:* Both variations are common. TAIL (ventral).—When fanned, the black tail shows four or five white bands.

**Adult female *richardsonii:*** HEAD.—Crown is pale brown with dark streaking. Nape is extensively mottled with white. BODY (dorsal).—Pale brown back and scapulars. Each back and scapular feather is neatly edged with very pale brown. Some have large white or tawny spots on the basal region of many scapular feathers and appear blotched if the feathers are fluffed. According to Warkentin et al. (1992), it may take adult females at least 4 years to attain the grayish rump and uppertail coverts. The grayish cast on the rump and uppertail coverts are difficult to see in the field. BODY (ventral).—Leg feathers are white like the rest of the underparts. WINGS (dorsal).—Upperwing coverts are pale brown and edged with very pale brown. Many coverts have small very pale tawny spots on them. Greater coverts may appeared somewhat barred because of pale tawny spots. The secondaries may have pale tawny spotting, and the primaries have large pale tawny spots on the outer web of all feathers. WINGS (ventral).—**Large pale spots are pale tawny-rufous.** TAIL (dorsal).—Uppertail coverts as described in dorsal body. **Two patterns on the dorsal tail surface: (1) Medium brown with a fairly wide blackish brown subterminal band. Three moderately wide white or very pale tawny bands neatly cross the tail above the subterminal band.** *Note:* Common pattern. **(2) Medium or brown with a fairly wide blackish brown subterminal band. The area above the subterminal band may have three faint bands on a few central rectrix sets, but the pale bands are absent on most of the rest of the rectrices.** *Note:* Uncommon pattern. TAIL (ventral).—**Medium brown with a dark brown subterminal band. The medium brown area above the subterminal band is crossed by four or five very pale, neat tawny-rufous bands when the tail is fanned.**

**JUVENILE TRAITS:** HEAD.—Cere and orbital skin are medium greenish yellowish or yellow. Brown crown. Pale mottling on the dark nape. Throat is white and unmarked. BODY (dorsal).—Brown back, scapulars, and rump. BODY (ventral).—Tarsi and feet are medium yellow. WINGS (dorsal).—Brown wing coverts. WINGS (ventral).—**Pale tawny-rufous spotting on the dark remiges.**

**JUVENILES OF "BOREAL" (*F. c. columbarius*):** HEAD.—Crown feathers often have pale tawny-rufous edges with a dark center shaft streak. BODY (dorsal).—Medium brownish back, scapulars, and rump. BODY (ventral).—Medium thick dark brown streaking on the breast, belly, and lower belly. Forward region of the flanks are streaked with dark brown and the rear portion have a broad arrowhead-shaped mark on the tip of each feather and broad inner-feather dark bars. Leg feathers have thin dark brown streaking. Undertail coverts have a thin dark brown streak along the shaft of each feather. WINGS (dorsal).—Small pale tawny spots on the outer web of the inner primaries. WINGS (ventral).—**Underside of the brownish black remiges have moderate-sized pale tawny-rufous spots that form a barred pattern. The pale spots cover 50 percent of the surface area of each feather. The underwings appear dark because of the moderate-sized pale spots.** TAIL (dorsal).—**Medium brown with a wide blackish brown subterminal band. Typically have two or three fairly distinct pale bands on the dorsal surface.** TAIL (ventral).—**Three or four narrow pale tawny-rufous bands. Three bands show**

**on a fanned tail on the outer part of the tail, but only one band shows next to the tips of the undertail coverts on a closed tail.**

**Juvenile male *columbarius:*** HEAD, BODY, and WINGS (dorsal).—Medium brown back, scapulars, and rump have a grayish sheen. The grayish sheen on the feathers is visible at moderate range and in good light. Upperparts appear slightly darker than on females. TAIL (dorsal).—**Two or three narrow pale gray bands. The pale bands are sometimes diffused and tail appears partially banded.** *Note:* Separated from adult and juvenile females by their smaller size and more grayish dorsal tail bands.

**Juvenile female *columbarius:*** HEAD, and BODY, and WINGS (dorsal).—Medium brown upperparts are a warm brown and similar to adult females. Rump is the same shade of brown as the rest of the upperparts. TAIL (dorsal).—**Uppertail covers are the same shade of brown as the rump and rest of the dorsal areas. Two or three narrow pale tawny bands. The pale bands are typically neat and complete.** *Note:* Separable from juvenile males by larger size and tawny dorsal bands. Separable from adult females only in late summer and fall by lack of molt.

**JUVENILES OF PALE TYPE OF "BOREAL" (*F. c. columbarius*):** As in pale type of adult female *columbarius* in being medium pale brown on the dorsal region. *Note:* Some of these types are paler variants, others are probable intergrades between dark typical *columbarius* and pale *richardsonii.* Only adult/juvenile female is depicted in accompanying photographs.

**JUVENILES OF "BLACK" (*F. c. suckleyi*):** Identical to adult female *suckleyi* plumage in virtually all aspects. Rump is always the same brown color as the rest of the dorsal areas.

*Note:* With practice, males are separable from either age of females by their smaller size. Females are separable from adult females only in late summer and fall by lack of molt.

**JUVENILES OF "RICHARDSON'S" (*F. c. richardsonii*):** Identical to adult female in nearly all aspects. BODY (dorsal).—Rump is the same pale brown color as the back and scapulars; however, this is also the case on many younger adult females (Warkentin et al. 1992). TAIL (dorsal).—Males sometimes have a slight grayish tone on the whitish tawny dorsal tail bands. *Note:* With practice, males are separable from adult and juvenile females by their smaller size. Females are separable from adults only in late summer and fall by lack of molt.

**ABNORMAL PLUMAGES: Incomplete albino (dilute plumage).**—No records in the West, but a few records of captured-for-banding individuals and sight records from the fall in the Mid-Atlantic region (Clark and Wheeler 2001, B. Sullivan pers. comm.). One was an adult male, the others have been "brown" juveniles or adult females. What are typically medium or dark plumage features are very pale tawny-brown in this type of albino. The birds appear very pale whitish with very pale brownish dorsal areas and ventral streaking. Iris color is the typical dark brown and the cere, tarsi, and foot color is also normal yellow.

**HABITAT: *F. c. columbarius.* Summer.**—Breeds or summers in the following habitat types: (1) Low-elevation boreal forest zone with semi-open areas of medium-height and tall conifers and hardwoods adjacent to lakes and occasionally large meadows. Regularly nests on wooded islands in lakes. Found throughout much of the forested regions of n. Minnesota, Alaska, and Canada. Merlin may be found in human-inhabited or undisturbed areas. (2) Low-montane boreal forest zone with semi-open areas of conifers adjacent to lakes, ponds, streams, and large meadows. Regularly nests on wooded islands in lakes. Found in various low montane regions of Oregon, the Cascade Mts. of Washington, the Rocky Mts. of Montana; the Coast Range and Rocky Mts. in Canada, and higher elevations in Alaska. (3) Taiga zone with semi-open and open expanses of dwarf spruce and deciduous scrub adjacent to lakes and ponds in Alaska and Canada. (4) "Medium shrub" habitat of the tundra zone characterized as open areas with scattered spruce and deciduous shrubs, particularly dwarf birch and alder. Merlin are mainly found nesting on slopes adjacent to lakes and ponds. This area is north of the stunted-tree region of the taiga. (5) Suburban and urban zones with tall conifers that have nests of American Crows. Merlin are increasingly using these areas in the West.

All breeding areas are in moderately humid or humid regions.

***F. c. columbarius.* Winter.**—A variety of habitats may be used: seashores, lakeshores,

marshes, tidal flats, suburban and urban areas, semi-open woodlands. Coniferous trees are favored roost locations. The various habitats may be in temperate, subtropical, or tropical zones.
***F. c. columbarius.* Migration.**—Found in habitats similar to those of Winter. Small numbers are also on high mountain ridges.
***F. c. richardsonii.* Summer.**—Nests in the following habitats on the arid northern Great Plains: (1) In ravines with single or small groves of deciduous trees in Canada. (2) Along rivers, lakes, and ponds with single or small groves of deciduous and rarely coniferous trees in Canada, North Dakota, South Dakota, and Minnesota. (3) Occasionally on open prairie in single deciduous trees in Canada. (4) On rural prairies with deciduous shelterbelt groves, often adjacent to abandoned homesteads in Canada. (5) In Montana, w. Nebraska, North Dakota, South Dakota, and Wyoming, moderate-elevation hills and buttes with small, semi-open stands of Ponderosa Pine and in Ponderosa Pine savannas. To a lesser extent in Wyoming, at higher elevations in small stands of Limber Pine that are adjacent to open areas. Pine stands may be mixed with some aspen and other deciduous bushes and trees. Most nesting areas are located on the sides of hills or buttes. Agricultural areas may be intermixed with prairie grasslands. (6) In suburban and urban areas in Canada, North Dakota, and Minnesota with small conifer stands or single conifer trees in cemeteries, city parks, business parks, and residential locations that nave old nests of American Crows.

Coniferous trees with old, abandoned nests of American Crows are used in suburban and urban regions. In all other regions, *richardsonii* are almost totally dependent on deciduous trees with old abandoned, more sheltered nests of Black-billed Magpies.
***F. c. richardsonii.* Winter.**—Found in all habitats listed for Summer. Regularly found in towns and cities throughout their entire winter range, but without regard to tree type. Also inhabits open plains without trees. In these open areas, they are usually found in locations with fence posts and utility poles for perching and roosting. Rural farmlands are regular haunts, particularly wheat fields, which draw large numbers of Horned Larks. Occupied and abandoned rural homesteads with a few trees are favorite haunts. *F. c. richardsonii* are nearly always found in arid habitat. They are not found in montane areas. Many winter west of the Continental Divide and on the Mexican Plateau in n. and cen. Mexico.
***F. c. richardsonii.* Migration.**—Found in all habitats listed for Summer and Winter.
***F. c. suckleyi.* Summer.**—Breeds in the humid rainforests of British Columbia and se. Alaska. Forests may be mixed conifer-hardwood, but are primarily conifers like Douglas Fir, Western Hemlock, and Western Red Cedar. The forests may be contiguous or have large openings of meadows and lakes. In Courtenay and Comox on Vancouver Island, they breed in city parks.
***F. c. suckleyi.* Winter.**—Similar to breeding areas for most Merlin. Those wintering along the Pacific Coast may be in semi-open humid areas in Oregon and n. California and more arid areas in the Central Valley and desert valleys and urban areas of s. California (e.g., Palm Springs, San Diego). The few birds that winter in Utah are in semi-open arid areas. Rural and urban locations are inhabited in all wintering areas. On the Pacific Coast, some birds inhabit open tidal flats, fields, and pastures.
***F. c. suckleyi.* Migration.**—Found in all locations listed for Summer and Winter.
**HABITS:** All subspecies are fairly tame to tame. Perches mainly on elevated exposed branches, wires, and other suitable structures. They may also perch near or on the ground. If available, night-roosting birds seek coniferous trees for optimum shelter from the elements and predators. In the winter, *richardsonii* may roost during the day, often only a few feet off the ground, accipiter-like in dense thickets in isolated deciduous trees and conifers. Mainly a solitary species, but a few birds may gather at night roosts.

Merlin regularly elevate forehead and throat feathers. Nape feathers are also raised.

Merlin are very antagonistic raptors and constantly harass larger raptors and birds too large to kill. They often hunt at dawn and dusk.
**FEEDING:** Perch and aerial hunter. Perch hunting is direct attack initiated from high, exposed perches. Aerial hunting is mainly by high-speed, low- or moderate-altitude, surprise-and-flush forays. Merlin primarily feed on small avian species that are captured in flight. Urban

nesting and wintering Merlin of all races feed extensively on House Sparrows. Dragonflies, particularly the large Green Darners, are a major prey of *F. c. columbarius* during fall migration. In Canada, wintering *F. c. richardsonii* in urban areas also feed extensively on Bohemian Waxwings, which are drawn into urban centers because of berry-laden trees. Horned Larks are the primary prey of breeding and wintering *richardsonii* in much of their range. In summer in se. Montana, *richardsonii* also feed extensively on Lark Buntings, Mountain Bluebirds, and Vesper Sparrows. Red Crossbills, Western Meadowlarks, and Chestnut-collared Longspurs also make up a fair amount of their prey in Montana. Small rodents and other vertebrates form a very small portion of their diet. Being crepuscular, Merlin often feed on small bats. Merlin rarely pirate food from other raptors: A juvenile male *richardsonii* was observed pirating a vole from a female American Kestrel. Rarely feeds on carrion.

Insect prey may be eaten while gliding or soaring or taken to a perch or to the nest. All other types of prey are eaten at the point of capture on the ground, taken to a perch, or taken to the nest. Kills prey by breaking the neck with its notched bill. All prey are decapitated; avian prey are plucked and wings are torn off. Legs and leg bones may be swallowed whole or the meat is eaten and legs and bones are discarded. Nesting pairs will cache prey that is not immediately eaten.

All subspecies of Merlin typically forage in open areas that give them the advantage of high-speed attacks on prey that cannot seek shelter. *F. c. suckleyi* may hunt avian prey above the forest canopy. *F. c. richardsonii* hunt avian prey over open short vegetation expanses. *F. c. columbarius* regularly hunt avian prey over large bodies of water and either force the prey into the water or the prey willingly dives near or on the surface of the water to try to escape.

Attack methods: (1) *Tail-chase.*—High-speed flights following behind and trying to overtake avian prey. Flights occur at ground-skimming or treetop altitudes or sometimes higher. (2) *Ringing-flight.*—A variation of *tail-chase* in which the prey rises upward in a spiral with the Merlin following closely behind. (3) *Dive.*—Another variation of *tail-chase* in which the Merlin rises slightly above the prey, then makes short dives to capture it. (4) *Undulating-flight.*—A low-altitude, sneak-attack flight in which Merlin emulate the bouncing flight of songbirds.

**FLIGHT:** Powered by a long series of steady, rapid, powerful wingbeats that are interspersed with short glide sequences. Flight is very fast. Merlin are capable of quick, high-speed maneuvers. Soars infrequently. Wings are held on a flat plane when gliding and soaring. Merlin performs undulating-flight at high speeds with quick, flicking wingbeats on the downward part of the undulation, then closes its wings into the body on the upward part of the undulation. The flight is similar to the bouncing flight of most songbirds and woodpeckers.

**VOICE:** Typically heard, agitated call is a rapid, high-pitched *ki-ki-ki-ki-ki-ki*. When courting, both sexes emit a *tic* call. A *chrr* note is given by males before copulating, and sometimes emitted by females. Conspecifics that come near a breeding female also emit the *chrr* call. A food-begging, whining *kree-kree-kree* is given by adult females during food transfers with mates; also given by youngsters.

**STATUS AND DISTRIBUTION:** The three subspecies are overall stable and possibly increasing. *F. c. columbarius* and *F. c. richardsonii* are overall *uncommon* but can be locally common and *F. c. suckleyi* are *uncommon*. Since Merlin primarily feed on migrant birds, the organochlorine pesticide era produced thinned eggshells and affected reproduction during this period. This was particularly noticeable in *richardsonii* breeding in Canada, where large amounts of organochlorine pesticides were used in farming areas in the 1950s to 1960s. Canadian *richardsonii* populations began rebounding on their own beginning in the mid-1970s.

Merlin of all races readily adapt to human alteration of the environment, particularly by nesting and wintering in urban and rural areas. After American Crows began colonizing urban locations, Merlin followed because of ample food provided by the huge numbers of House Sparrows and other birds and nest sites provided by crows.

***F. c. columbarius.* Summer.**—Sparsely distributed throughout mapped range. *Alaska* and *Canada*: Primary breeding range. *Idaho:* Possible rare breeder in n. montane areas. *Minnesota:* As part of the boreal forest, this state has numerous pairs, including several in suburban and urban areas of Duluth. There are one or two birds breeding, often paired with *richardsonii*, in suburban Minneapolis-St. Paul. *Montana:* Rare breeder in montane areas. *Oregon:* Numerous, recent breeding-season sightings based on breeding-bird atlas work in coastal and interior montane locations. However, breeding has not been confirmed in nearly a century. Breeding is probable in Gilliam, Grant, Polk, and Umatilla Cos. At least 16 other counties had casual summertime sightings of mainly single birds. Most sightings are of irregular occurrence and distribution, and most are probably 1-year-olds and nonbreeders. *Washington:* Status is uncertain. Breeding is not confirmed for this subspecies. Possible rare breeder and summer visitor in the Cascade Mts. and very northeastern areas. *Wyoming:* There are no breeding records for this race.

***F. c. columbarius.* Winter.**—Sparsely distributed in their limited winter range in the U.S. and British Columbia. Regular in winter on Kodiak Island, Alaska. Largest numbers winter along coastal regions of the U.S. Very sparse wintering occurs in the interior states. Status in the interior sections of Oregon is uncertain.

***F. c. columbarius.* Migration.**—Largest numbers are seen along the coastal regions, but migrants can be encountered in very small numbers throughout the interior provinces and states. Migration poorly defined in the West since there are few concentration barriers.

*Fall migration*: Appears to mirror that of the East, with juveniles preceding adults. Juveniles peak in mid-Sep. Adults peak in mid-Oct.

*Spring migration*: Mainly seen in Apr. Adults precede 1-year-old juveniles.

***F. c. richardsonii.* Summer.**—*Canada:* Largest numbers and highest breeding densities are in Canada. Many in Winnipeg are possibly intergrades with *columbarius.* In Canada, urban nesting of *richardsonii* began in 1963 but did not occur in residential areas until 1970. Merlin became regular breeders in Saskatoon, Saskatchewan, by 1971. Thereafter, natural colonization rapidly took place, and numerous pairs now nest in many towns and cities. In Regina, Saskatchewan, urban nesting began in the 1970s with a few released birds. *Minnesota:* This subspecies began nesting in the state in 1998 near Lake Bronson in Kittson Co. Nesting occurred in Roseau, Roseau Co., in 1999. Nesting was documented in Minneapolis-St. Paul in 2000; however, an injured adult male *richardsonii* was found in the spring of 1998. In the summer of 2001, four pairs nested in the Twin Cities: 4 males and 2 females were *richardsonii*, 1 female an intergrade with *columbarius*, and 1 female was a dark *columbarius. Montana:* Sparsely distributed in the eastern part of the state east of the Continental Divide in isolated hills with Ponderosa Pine and Ponderosa Pine-prairie savannahs. *Nebraska:* Breeds in Ponderosa Pine areas in Dawes and Sioux Cos. Breeding was first confirmed in Dawes Co. in 1975. *North Dakota:* Very rare and sparse breeder in the western and northern parts of the state in mainly Ponderosa Pine savannas. In the east, has bred in Grand Forks since 1998; in 2002, five pairs nested in the city. Nesting pairs have also expanded into urban Dickenson, Minot, and Jamestown (since 2000). Pairs also nest in a rural Ramsey Co. *South Dakota:* Breeds in isolated locations in a few western counties. Several pairs breed in Harding Co. A few pairs breed in Custer (Black Hills National Forest), Fall River, Meade, and Shannon Cos. They possibly breed in Jackson Co. *Wyoming:* Thirty-five known territories, with about 70 percent occupation. Merlin breed sparsely mainly in the northeastern part of the state. Irregularly found south of the Big Horn Mts. to the Shirley Basin region in n. Carbon Co. Isolated pairs breed west of the Bighorn Mts. Nesting has occurred in the city of Cody. Breeding also occurs in the foothills near Pinedale in Sublette Co. Until the late 1980s, nesting occurred along the Green River north of Green River, Sweetwater Co.

***F. c. richardsonii.* Winter.**—A few remain in their breeding areas, particularly in urban areas of Canada. Merlin probably did not winter in Canada before House Sparrows became established in urban areas in 1900. Wintering was rare in Canada even until the mid-1950s; since then, Merlin have remained, being attracted to the irregular, but often large winter influxes of Bohemian Waxwings that are drawn by a food supply of now-maturing artificially planted, decorative berry bushes and trees.

Substantial numbers winter on the cen. and s. Great Plains, with the largest concentrations occurring in e. Colorado. Fairly large numbers winter west of the Rocky Mts.

In Mexico, *richardsonii* range mainly extends south on the Mexican Plateau to the north-central border of Guerrero and n. Puebla.

Many Merlin establish territories in areas with consistently high prey densities, particularly in natural and rural areas with huge flocks of Horned Larks and in urban areas with large numbers of House Sparrows and Bohemian Waxwings. Merlin may be nomadic and, where Horned Larks are abundant, several may accompany the roving flocks that often number in the thousands. Juveniles, in particular, may be in groups of up to four birds. Some adults appear to have formed pairs by late winter while on the winter grounds.

***F. c. richardsonii.* Migration.**—Data are primarily based on author's 14 years of sightings on the Plains of e. Colorado.

*Fall migration*: The first juveniles arrive in early Sep. Adult males may appear as early as mid-Sep., but most do not arrive until later. A large influx of all ages and sexes, but particularly adult males, occurs in mid- to late Oct. Movements become difficult to assess after Oct. because of the species' nomadic nature.

*Spring migration*: Adult males begin leaving the plains in Feb. Peak departure of adult males on the plains is in late Feb. and early Mar. Adult males are rarely seen after mid-Mar. In se. Montana, adult males arrive on breeding grounds in mid-Mar. and adult females arrive in early Apr. Though difficult to determine because of overwintering birds, the previous earliest arrival on Canadian breeding grounds was in late Feb. in s. Alberta and late Mar. in s. Saskatchewan. Most adults arrive on Canadian breeding grounds by early Apr. Juveniles begin leaving the winter grounds in Mar. and continue though Apr. and rarely into early May. In Colorado, Merlin are difficult to find after early Apr. and very difficult to find after mid-Apr.

***F. c. suckleyi.* Summer.**—The Merlin is found in low and moderate elevations on Vancouver Island, particularly the eastern part of the island, and the immediate coastal mainland of British Columbia. Its range is disjunct along the northern coast of British Columbia then resumes again in se. Alaska. It is absent from the Queen Charlotte Islands, B.C.

***F. c. suckleyi.* Winter.**—Most remain on breeding regions. A fair number of juveniles, particularly females, and adult females, winter southward along the Pacific Coast and in moderate-climate interior valleys of Oregon and California. They are fairly regular to San Diego, Calif. A few also fairly regular in cen. Utah, which is substantially east of previously recognized winter range. Most are females of both ages and a few are adult males. The region from Ogden south to Provo, Utah, has the most consistent wintering population. Other birds have been near Nephi, Juab Co., and Delta, Millard Co.

This race is rare or accidental elsewhere in the West: Abiquiu, Arriba Co., N. Mex., in mid-Dec. 1976 (specimen at University of Puget Sound, Tacoma, Wash.); juvenile female on Galveston Island, Galveston Co., Texas, in late Dec. 1994 (photograph); and a juvenile female in Bee Co., in late Jan. 2000 (photograph by J. Jackson showing the distinct, dark ventral surface of the remiges). There are records for Missoula, Mont., and possible records for Fallon, Nev.

***F. c. suckleyi.* Migration.**—Little available data.

*Fall migration*: Based on a few specimens from Washington State, several were from mid-Sep. and Oct., the typical migration period for the other two races.

*Spring migration*: Very little data. A juvenile female was captured by a falconer in Valley Center in San Diego Co., Calif., in early Apr.

**NESTING:** *F. c. columbarius.*—Early nesting pairs are on southern breeding grounds in early Apr.; a few remain on breeding grounds. In more northern regions, arrival may be throughout Apr. and into May. Courtship occurs from early Apr. to late May. *F. c. richardsonii.*—Arrival on breeding grounds and courtship begin in Mar. and Apr.; some remain on breeding grounds. *F. c. suckleyi.*—No data.

Breeding may occur when 1 year old. Pairs may be formed of adults, one adult and one yearling, or yearlings. Nesting success is highest when both pair members are adults. Pairs regularly return to the same territory but often use a different nest site.

**Courtship (flight).**—Both members of a pair perform *mutual high-circling* and *aerial food transfers*. Only males perform the following elaborate courting routines: *flutter-flight, power-diving, power-flight, rocking-glide,* and *slow-landing.*

**Courtship (perched).**—*High-perching* is used by males. *Food-begging* is performed by females. *Food transfers* and *nest displays* are performed by both sexes. Copulation often follows *food transfers. See* chapter 5 for courtship descriptions.

No nest is built. Merlin nest in abandoned nests of other species of birds. Typical clutches are 3–5 eggs, but occasionally 2–7, are laid. The 30-day incubation is performed mainly by females. Youngsters fledge in 28–30 days but may remain near the nest for another 2 weeks.

*F. c. columbarius.*—A majority of nests in the Lower 48 and Canada are in old nests of American Crows. Tree-nesting birds may occasionally use the tops of flattened leafy squirrel nests or the tops of broken tree trunks. North of the American Crow's range in Northwest Territories, Nunavut, and Alaska, tree nests may also be in the nests of raptors or in Common Ravens. In the taiga and tundra, where trees are small or nonexistent, nests are typically on the ground under dense deciduous bushes and scrub spruce. This subspecies rarely nests on cliff ledges. Tree nests are situated in the densely foliaged upper part of a conifer. Tree nest heights range from 12 ft. (4 m) to 81 ft. (25 m).

*F. c. richardsonii.*—Nests only in trees. Urban pairs in Canada and the U.S. use nests of American Crows that are in tall, dense spruce. Rural pairs in Canada primarily use nests of Black-billed Magpies that are in deciduous trees or bushes. Rural pairs in Montana and Wyoming mainly use Black-billed Magpie nests in Ponderosa Pines and sometimes Limber Pines. Crow nests in deciduous trees and bushes are regularly used in rural Canada, but are very rarely used in Montana or Wyoming. Pairs using the stick-built, dome-shaped magpie nests use the sheltered interior nesting area, but they may rarely nest on top of the dome. Rarely uses abandoned nests of Swainson's Hawks in Canada; this practice is unknown in the U.S.

Based on Saskatchewan pairs nesting in urban spruce and rural deciduous trees and bushes, nests are 7–50 ft. (2–15 m) high and average 16 ft. (5 m) high. Crow nests in urban spruce trees are generally high nests. Crow nests in rural deciduous trees and bushes can be low or moderately high. Magpie nests are in any type of tree and range from low to moderately high.

*F. c. suckleyi.*—Nests only in tall coniferous trees. Primarily uses abandoned nests of Steller's Jays and Northwestern Crows. The cup-shaped, twig-made nests of Steller's Jays are typically 10–30 ft. (3–10 m) but may be 100 ft. (30 m) high. Nest heights are not available for the stick nest of the crow.

**CONSERVATION:** Except for a few *F. c. richardsonii* hacked in Regina, Saskatchewan, in 1979, all Merlin populations rebounded on their own after DDT was banned. As the Merlin is a predator of birds and at the top of the food chain, the ban on organochlorine pesticide use, particularly DDT, assisted the recovery of its populations in the last three decades. Merlin, mainly *F. c. columbarius* and *F. c. richardsonii,* rebounded on their own after DDT was banned in Canada and the U.S. This lethal pesticide was first used in 1946 in Canada and the U.S.

Organophosphate pesticides, although not as persistent in the environment as organochlorine pesticides, can be deadly to some species that Merlin eat.

Urban nesting began after American Crows and Black-billed Magpies began nesting in urban areas.

**Mortality.**—Natural mortality of all ages due to depredation by large hawks and owls. Mammalian predators cause egg and nestling mortality. During migration, large gulls kill overwater migrants. Carbamate pesticides may have contributed to some mortality among Merlin. Newer-generation pyrethroid pesticides are harmless to birds and mammals, but affect aquatic insects. Some may collide with wires and windows. Shooting is another threat.

**SIMILAR SPECIES:** COMPARED TO ADULT MALES.—**(1) Sharp-shinned Hawk, adults.**—Similar size. PERCHED.—Orange or red irises. Underparts are rufous barred. Long, thin tarsi. Upperparts are similar bluish gray. Tail has 3 or 4 black bands on dorsal side. FLIGHT.—Pale remiges with black barring. Rufous-barred underparts. Undertail is pale with narrow black bands. COMPARED TO *COLUMBARIUS* ADULT FEMALES/JUVENILES.—**(2) American Kestrel, females.**—Similar size. PERCHED.—Two black facial stripes. Upperparts barred with rufous and dark brown. Ventral areas rufous streaked. Tail is rufous with numerous, narrow black bands. FLIGHT.—Similar, especially when in high-speed powered flight. Wings are narrow on the secondaries; wingbeats generally lithe and wispy. Wingbeats can be deceptively powerful at times; use caution! Ventral side of remiges are pale barred. Tail is pale or bright rufous with multiple, narrow dark bands. **(3) Peregrine Falcon, *F. p. anatum* and *F. p. tundrius* juveniles.**—Much larger. Sharply defined dark malar stripe. Pale bluish cere and orbital skin. Ventral and dorsal markings and color are similar, but pale dorsal feather edgings may be very defined (especially *F. p. tundrius*). Wingtips equal or nearly equal to tail tip. Tail may be similar, with pale tawny bands on both ventral and dorsal areas. **(4) Sharp-shinned Hawk, juveniles.**—Similar size. PERCHED.—Yellow irises. Ventral and dorsal areas similar. Long, thin tarsi. Tail has three or four wide, equal-width black bands. Lacks white terminal band. FLIGHT.—Pale underside of the remiges with black barring. Pale underside of tail with narrow black barring. COMPARED TO *RICHARDSONII* ADULT FEMALES/JUVENILES.—**(5) Prairie Falcon.**—PERCHED.—Thin dark malar stripe and white patch behind the eyes. Pale brown dorsal color similar; ventral streaking of juvenile is similar to Merlin but large dark flanks. Wingtip-to-tail tip ratio similar. Dorsal tail unbanded or numerous narrow pale bands. FLIGHT.—Axillaries are dark. COMPARED TO *SUCKLEYI* ADULT FEMALES/JUVENILES.—**(6) Sharp-shinned Hawk, *Accipiter striatus perobscurus* juveniles.**—PERCHED.—Similar dark brown dorsal and densely streaked ventral. Yellow irises. Three or four black wide bands on dorsal surface of brown tail. FLIGHT.—Broadly barred pale underside of the remiges. Pale ventral surface of the tail with black bands. **(7) Peregrine Falcon, *F. p. pealei* juveniles.**—Much larger size. PERCHED.—Pale bluish cere and orbital skin. Dorsal and ventral color and markings similar. Wingtips reach the tail tip. FLIGHT.—Underside of the remiges are dark with distinct pale spotting and barring.

**OTHER NAMES:** Merl, merlin (female), jack (male), blue jack (adult male), Taiga Merlin (*columbarius*), Prairie Merlin (*richardsonii*); formerly called Pigeon Hawk. *Spanish*: Esmerejon. *French*: Faucon Emerillon.

---

REFERENCES: Adamus et al. 2001; Baicich and Harrison 1997; Becker 1984; Bent 1961; Bylan 1998; Campbell et al. 1990; Cecil 1999; Clark and Wheeler 2001; Dinsmore 1990; Dodge 1988–1997; Dorn and Dorn 1990; Dotson and Mindell 1979; Environment Canada 2001; Ferguson-Lees and Christi 2001; Gibson and Kessel 1997; Godfrey 1986; Haney and White 1999; Howell and Webb 1995; Houston 1981; Idaho Conservation Data Center 1999; Janssen 1987; Kaufman 1996; Kellogg 2000; Kessel 1989; Martin 2000; Montana Bird Distribution Committee 1996; Oberholser 1974; Palmer 1988; Peterson 1995; Sailer 1987; Semenchuk 1992; Sharpe et al. 2001; Sieg and Becker 1990; Sharpe et al. 2001; Small 1994; Smith 1996; Smith et al. 1997; Smith et al. 2002; Sodhi and Oliphant 1993; Sodhi et al. 1993; Solensky 1997, 2000, 2001; Stewart 1975; Svingen 2000; Warkentin and James 1990; Warkentin et al. 1992; Wheeler and Clark 1995.

**"BOREAL" MERLIN,** *Falco columbarius columbarius:* Uncommon. Sparsely distributed in all seasons. Uncommon to rare in winter in CO, NE, MO, and IA. Mainly winters south of U.S. to n. South America. Most "Boreal" Merlin wintering in U.S. and Canada are females. Breeding is not confirmed in OR or WA. Rare in winter on coastal AK.

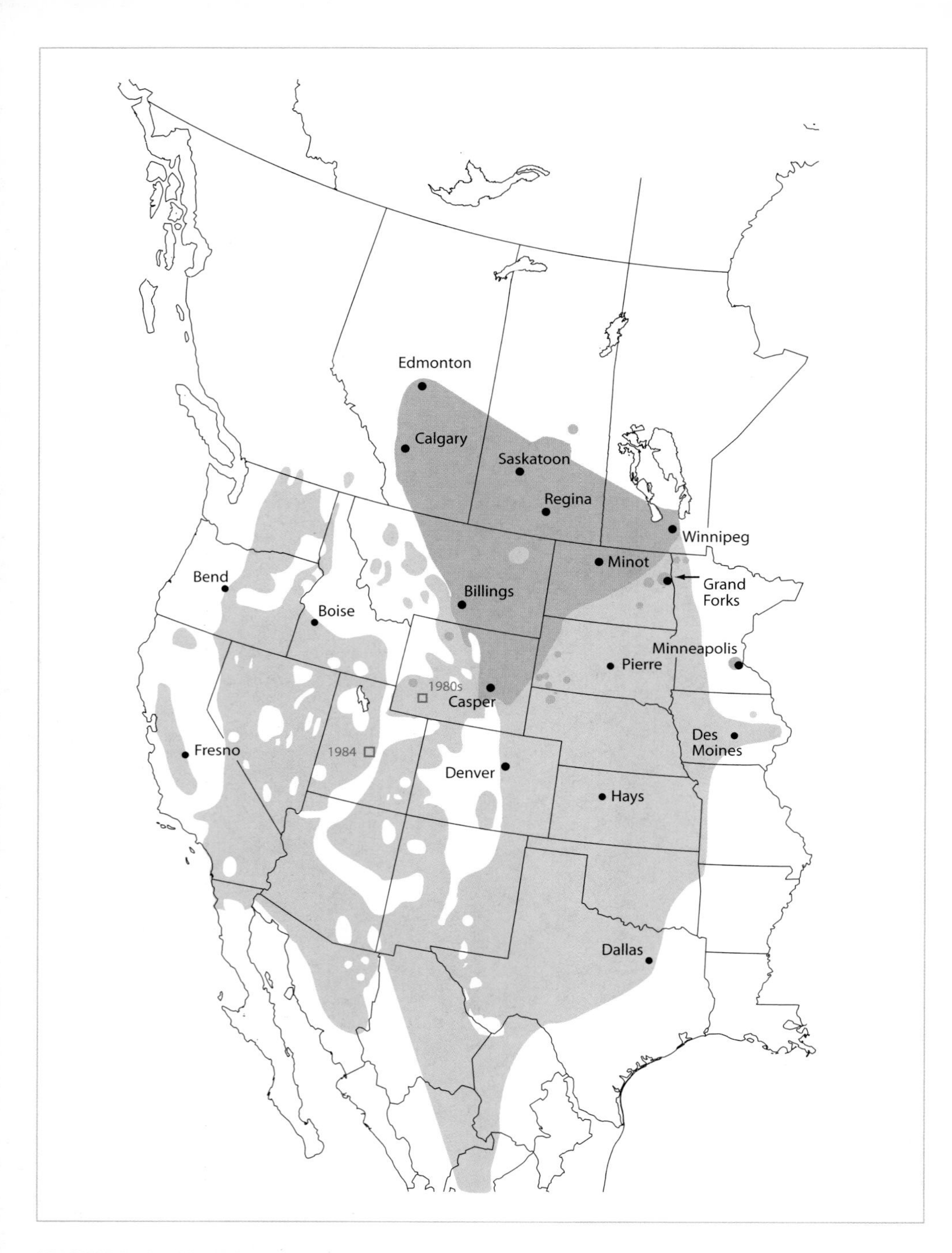


**"RICHARDSON'S" MERLIN,** *Falco columbarius richardsonii:* Uncommon. Urban raptor in many areas in summer and winter. Core winter range includes breeding areas and cen. Great Plains. Intergrades with "Boreal" Merlin on mainly northern edges of range.

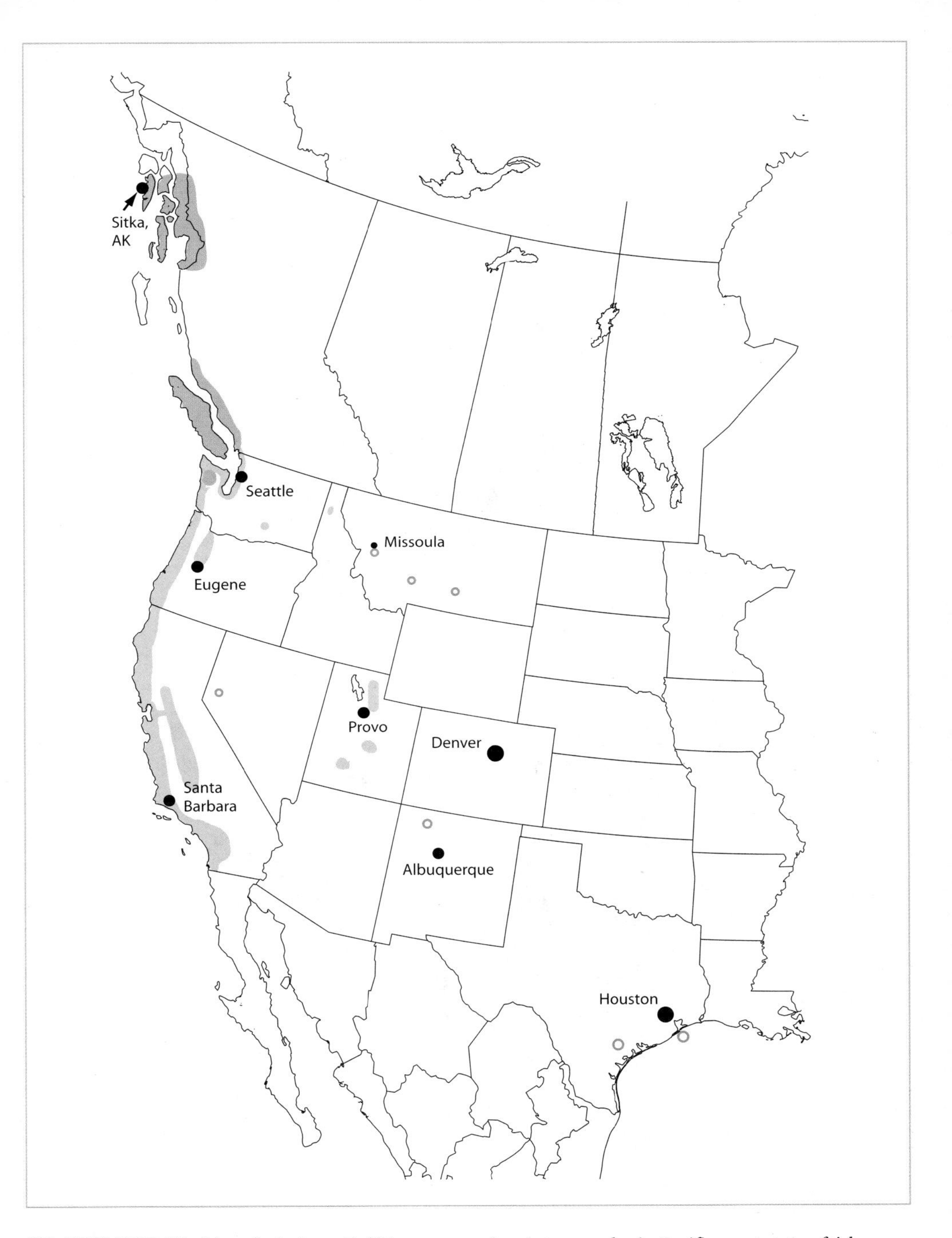


**"BLACK" MERLIN,** *Falco columbarius suckleyi:* Uncommon. In winter, regular in Pacific coast states, fairly regular to UT, very irregular to MT and TX. Most birds wintering south of WA are females.

**Plate 532. Merlin, adult male "Boreal" (*F. c. columbarius*) [Oct.]** ▪ Thin dark eyeline; faint, dark malar mark; pale supercilium and cheeks; yellow orbital ring. ▪ Medium bluish gray upperparts. Streaked on breast and belly and barred on flanks. Lightly streaked rufous tawny leg feathers. ▪ Black tail has 2 or 3 narrow bluish gray bands.

**Plate 533. Merlin, adult male "Boreal" (*F. c. columbarius*) [Oct.]** ▪ White or tawny underparts are moderately streaked with brown; rufous-tawny leg feathers. ▪ Moderately dark underwings: white spotting/barring covers 50% of the surface of each remix. ▪ Black tail has 3 or 4 narrow white bands.

**Plate 534. Merlin, adult female "Boreal" (*F. c. columbarius*) [Feb.]** ▪ Thin dark eyeline; faint dark malar mark; thin white supercilium. ▪ White underparts are moderately streaked with brown. Leg feathers and undertail coverts are white and lightly streaked. Medium brown or dark brown upperparts.

**Plate 535. Merlin, adult female "Boreal" (dark type) (*F. c. columbarius*) [May]** ▪ Dark head with partial white supercilium. ▪ Heavily marked underparts are streaked on breast and belly and barred on flanks. Dark brown upperparts. ▪ *Note:* Underside of remiges are spotted as on typical *columbarius.* Photographed in Paradise Co., Mich., by Kevin T. Karlson.

**Plate 536. Merlin, adult/juvenile female "Boreal" (pale type) (*F. c. columbarius*) [Dec.]** ▪ Faint, dark malar mark and dark auricular patch; thin dark eyeline; thin white supercilium. ▪ Moderately marked white underparts. Upperparts are paler than on typical *columbarius*. ▪ Adult females often have indistinct pale dorsal tail bands.

**Plate 537. Merlin, adult female "Boreal" (*F. c. columbarius*) [Oct.]** ▪ Upperparts are medium or dark brown and mottled with patches of old, faded brown feathers when molting from summer to mid-fall. Three narrow tawny tail bands. ▪ *Note:* Grayish rump difficult to see in the field.

**Plate 538. Merlin, juvenile male "Boreal" (*F. c. columbarius*) [Oct.]** ▪ Faint, dark malar mark and dark auricular patch; thin dark eyeline; thin tawny supercilium. Yellow orbital skin. ▪ Leg feathers and undertail coverts are lightly streaked. Upperparts of males are grayish brown. ▪ Dorsal tail bands are gray.

**Plate 539. Merlin, juvenile female "Boreal" (*F. c. columbarius*) [Oct.]** ▪ Faint, dark malar mark and dark auricular patch; thin dark eyeline; thin tawny supercilium. ▪ Moderately marked white or pale tawny underparts. Medium brown or dark brown upperparts. ▪ *Note:* Separable in summer and fall from adult females by lack of molt; from juvenile males by size and dorsal color.

**Plate 540. Merlin, juvenile female "Boreal" (*F. c. columbarius*) [Sep.]** ▪ Streaked underparts. ▪ Moderately dark underwings: tawny spotting/barring covers 50% of the surface of each remix. ▪ Dark tail has 3 or 4 thin tawny bands and a white terminal band.

**Plate 541. Merlin, juvenile male "Boreal" (*F. c. columbarius*) [Oct.]** ▪ Brown upperparts, often with a grayish sheen. ▪ Tawny spotting on primaries. ▪ Dark tail has 3 or 4 thin whitish or grayish bands (tawny on females).

**Plate 542. Merlin, adult/juvenile female "Black" (*F. c. suckleyi*) [Dec.]** ▪ Dark head. ▪ Heavily marked white underparts. Leg feathers are moderately streaked. Undertail coverts are marked with broad arrowhead shapes/bars. Blackish brown upperparts. ▪ Single, partial, pale band on closed tail. ▪ *Note:* Underside of remiges (not shown) lack pale spots or have small pale spots.

**Plate 543. Merlin, adult male "Richardson's" (*F. c. richardsonii*) [Feb.]** ▪ Thin dark eyeline; thin white supercilium, very faint dark malar mark. Pale bluish gray crown. ▪ Lightly marked white underparts. Pale tawny-rufous leg feathers. Pale bluish gray upperparts. ▪ Bluish spotting on primaries. ▪ *Note:* Captured a Horned Lark, a major prey species.

**Plate 544. Merlin, adult male "Richardson's"** (***F. c. richardsonii***) **[Mar.]** ▪ Lightly streaked white underparts. Pale tawny leg feathers. ▪ Pale underwings: white spotting/barring on dark remiges covers greater than 50% of the surface of each remix. Pointed wingtips. ▪ Black tail has 3 or 4 broad white bands.

**Plate 545. Merlin, adult/juvenile female "Richardson's"** (***F. c. richardsonii***) **[Nov.]** ▪ Thin dark eyeline; thin white supercilium; faint dark malar mark. ▪ Lightly marked white underparts. Pale brown upperparts are edged with pale tawny; often spotted with white or tawny. ▪ Secondaries and greater coverts are barred with tawny. ▪ Tail has broad white tail bands.

**Plate 546. Merlin, adult/juvenile female "Richardson's"** (***F. c. richardsonii***) **[Nov.]** ▪ Lightly streaked white underparts. White leg feathers. ▪ Pale underwings: white spotting/barring on dark remiges covers greater than 50% of the surface of each remix. Pointed wingtips. ▪ Brownish tail has 3 or 4 broad white bands and a wide black subterminal band.

**Plate 547. Merlin, adult/juvenile female "Richardson's"** (***F. c. richardsonii***) **[Feb.]** ▪ Pale brown upperparts are edged with pale tawny. ▪ Primaries, secondaries and greater coverts are spotted with tawny. ▪ Tail has 3 broad white/pale tawny bands and a broad white terminal band. Subterminal dark band is black.

# APLOMADO FALCON
## *(Falco femoralis)*

**AGES:** Adult and juvenile. Adult plumage is acquired at 1 year of age. Adults have somewhat sexually dimorphic plumage markings, mainly on the breast, but some overlap occurs. Juvenile plumage is retained for much of the first year.

**MOLT:** Falconidae wing and tail molt pattern (*see* chapter 4). First prebasic molt from juvenile to adult begins in a span from winter to early spring when less than 1 year old (possibly when 8 months old). Molt is well under way when 1 year old. As in most falcons, the first prebasic molt begins on the head, breast, and back. Adult features are attained on the head, breast, and back prior to molting the belly, flanks, and scapulars. Much of the body is molted before wings and tail begin molting. No data is available on sequence of primary molt. During spring and early summer, molting birds will be adultlike on the head, breast, and back; however, juvenile feathering is retained on the belly, flanks, wings, wing coverts, and tail. Molt into adult plumage is completed by mid- to late fall.

Subsequent annual prebasic molts begin later than does the first prebasic molt. In nesting adults in s. Texas, females may have replaced p4 and p5 by mid-May; males have not started their molt at this time. In adult prebasic molts, all anatomical areas proceed in unison once primary molt has begun.

**SUBSPECIES:** *F. c. septentrionalis* historically bred in the U.S. in se. Arizona, s. New Mexico, and s. Texas until becoming extirpated in 1952. Due to reintroduction programs, this race is once again a breeding resident in s. Texas; also through natural recolonization, a pair was discovered nesting in New Mexico the spring of 2001 (*see* Status and Distribution). In Mexico, *septentrionalis* is sparingly found in n. and cen. Chihuahua, se. San Luis Potosí, s. Coahuila, s. Tamaulipas, Veracruz to w. Campeche, and possibly on the Pacific slope in Guatemala.

Two additional races: (1) Nominate *F. c. femoralis* from Nicaragua and Belize south to Colombia, e. Bolivia, and continuing south to Tierra del Fuego; and (2) *F. c. pichinchae* from sw. Colombia, Ecuador, Peru, w. Bolivia south to n. Chile and nw. Argentina.

**COLOR MORPHS:** None; however, the intensity of underparts coloration varies due to having either fresh, newly molted feathers or worn and faded feathers. There is also some individual variation. Fledglings and recently fledged juveniles are quite richly colored (*see* Juvenile Traits).

**NOTE:** All captive-released individuals in Texas are recently fledged birds raised in captivity. Released fledglings and many nestlings of wild-breeding pairs have a USFWS band on one tarsi and a colored, numbered band on the other. Captive juveniles are released from mid-Jun. through Aug. Not all former-released wild-nesting pairs are located, and some nestlings are not banded because of the logistics of locating all nest sites and banding them within the safe-banding period.

**SIZE:** Medium-sized raptor. Somewhat sexually dimorphic but some overlap may occur. Length: 14–18 in. (36–46 cm); wingspan: 31–40 in. (79–102 cm).

**SPECIES TRAITS:** HEAD.—**Narrow white or tawny supercilium that begins over the eye and forms a "V" on the nape. A moderately wide blackish eyeline merges under the eye with a narrow black malar stripe.** BODY.—**Dark, narrow belly band and flanks form a "cummerbund."** Moderately long tarsi and toes. WINGS.—Long and moderately narrow. In flight, wings taper to fairly pointed tips on the primaries (p8 and 9 are the same length). **Underwings appear uniformly dark except for a pale tawny strip on the leading edge of the patagial and carpal regions. Underside of remiges are dark gray and spotted or barred with white. Secondaries have a distinct, broad white border on the trailing edge that is most visible in flight.** This white band is often concealed by the long greater secondary coverts when perched. TAIL.—**Long, black, and crossed with numerous, very narrow whitish bands on both dorsal and ventral sides.**

**ADULT TRAITS:** HEAD.—Bluish gray crown. Supercilium is white with a tawny wash on the rear portion. Bright yellow or orangish yellow cere and orbital areas. BODY.—White or pale tawny breast. Bluish gray upperparts. **Belly**

**band and flanks are black with very narrow white scalloped edges. Lower belly, undertail coverts, and leg feathers are a rich tawny or rufous-tawny.** Bright yellow tarsi. WINGS.—Bluish gray on the dorsal surface. TAIL.—**Five very narrow white bands on both the dorsal and ventral surfaces.**

**Adult male:** HEAD.—Orangish yellow or yellowish orange cere and orbital areas. BODY.—Unmarked white breast. Bluish gray upperparts.

**Adult female:** HEAD.—Yellow or orangish yellow cere and orbital areas. BODY.—Central area of the breast has a cluster of very narrow dark brown streaks, but a few have unmarked breasts. Leg feathers, lower belly, and undertail coverts are sometimes a fairly pale tawny. Bluish gray upper parts may have a subtle brownish cast.

**SUBADULT TRAITS (FIRST PREBASIC MOLT STAGE):** HEAD.—As in adult. BODY.—Breast, back, scapulars, rump, and uppertail coverts as on adults. Flanks and belly band are moderately dark brown, retained juvenile feathering until late in the molt stage. Leg feathers and undertail coverts are also very pale, worn, tawny colored retained juvenile feathers. WINGS.—Upperwing coverts are faded medium brown retained juvenile feathers. Remiges have mix of some adult amongst the juvenile feathers. TAIL.—Retained juvenile but with incoming adult deck rectrices in latter stage. *Note:* Degrees of this molt stage are exhibited from late winter through early summer.

**JUVENILE TRAITS:** HEAD.—Supercilium is pale tawny but may fade to nearly white. Pale yellowish cere and pale bluish orbital region. The color of the cere and orbital regions gradually increases to a brighter yellow toward end of the age cycle. BODY.—**Breast is pale tawny and variably streaked with dark brown. Breast has moderate-width or wide streaks that are clustered on the upper and central region, but the lower breast is unmarked. Breast streaking connects to the forward area of the dark flanks.** *Note:* Streaking may appear as a large dark area on the central area of the breast on heavily marked individuals. Dark brown upperparts. Upperparts are narrowly edged with tawny-rufous, which gradually wears off and becomes uniformly dark brown. **Belly band and flanks are dark brown with narrow tawny streaking. Lower belly, undertail coverts, and leg feathers vary from orange to pale tawny and are usually much duller than on most adults, but similar to some faded plumaged adult females.** Tarsi and feet vary from pale yellow to medium yellow. TAIL.—**Six or seven very narrow pale grayish bands.** Deck rectrices and adjacent rectrices are sometimes unmarked, but the ventral surface is fully banded on all rectrices (female specimen from Cameron Co., Texas, AMNH).

**JUVENILE (RECENTLY FLEDGED):** HEAD.—Cere and orbital skin are pale bluish. Supercilium and auriculars are a rich tawny. BODY.—Breast, leg feathers, lower belly, and undertail coverts are a rich tawny. Dark brown upperpart feathers are edged with tawny-rufous. *Note:* Captive-released individuals may be exceptionally richly tawny on respective anatomical regions due to minimal exposure to sunlight. Foot color as on older juveniles. TAIL.—As on older juveniles.

**ABNORMAL PLUMAGES:** None documented.

**HABITAT:** Open grasslands with a few widely scattered bushes or small trees that are used for perches and nest sites. In s. Texas, bushes and trees are mainly yucca, mesquite, and rose bushes; in s. New Mexico and n. Chihuahua, they are mainly yucca. Elevation varies from sea level on humid, lush coastal savannahs and near-coastal islands in Texas up to relatively flat, yucca-studded arid interior plains at 5,500 ft. (1,700 m) in s. New Mexico and n. Chihuahua.

**HABITS:** A tame raptor. Perches on exposed branches, poles, posts, and wires. Aplomado Falcons also use concealed perches within foliage, particularly to seek shade during hot periods of the day.

Mated pairs have a strong bond and remain together year-round. Aplomado Falcons are often antagonistic towards other raptors and other bird species.

**FEEDING:** A perch hunter. Medium- and large songbirds make up the bulk of their diet. Large flying insects, small mammals, and reptiles are also preyed upon. White-tailed Kites, which share year-round Aplomado range in s. Texas, and Northern Harriers, which share Aplomado range from fall through spring in Texas and New Mexico, are primary victims of Aplomado

Falcon pirate acts. Other birds also fall victim to their pirate ventures; e.g., has been seen stealing crayfish from a Little Blue Heron in Mexico. Many nonavian prey species are thought to be obtained from pirate acts. Aplomado Falcons have been seen following hunting Coyotes, and they pursue Coyote-flushed avian prey in Chihuahua, Mexico. Airborne attacks are launched from a perch. Aerial pursuits are high-speed, low-altitude tail-chases to capture prey while it is airborne or on the ground. Prey may be pursued on foot into dense vegetation. Pairs often hunt cooperatively. High-altitude avian prey may be pursued and forced down to a lower altitude for capture. Generally hunts in early morning and late afternoon. Aplomado Falcons do not feed on carrion.

**FLIGHT:** Soars and glides with wings held on a flat plane. Powered flight consists of moderate-speed wingbeats interspersed with irregular glide sequences. Soaring flights may rise to high altitudes. Hunting flights are fast and direct.

**VOICE:** When agitated, a high-pitched, somewhat toylike, squeaky, rapid chatter: *cack-cack-cack-cack.* Duration of the call varies. Vocalizes when agitated near nest sites, during courtship, and sometimes when flushing prey. High-pitched *chips* and low-pitched *chup* notes are emitted when courting.

**STATUS AND DISTRIBUTION:** *Texas: Currently* (1986 to present).—*Very uncommon* **permanent resident.** Aplomado Falcons have been designated a federally Endangered Species since 1986. Current population in Calhoun, Cameron, Kenedy, Kleberg, Matagorda, San Patricio, and Willacy Cos. are the result of an intense reintroduction program (*see* Conservation). Beginning in 2002, a released population is being created in s. Culbertson Co. *Recently* (1950–1985).—*Casual.* Periodic sightings occurred on an irregular basis in portions of their former range in Brewster, Presidio, Cameron, Starr, Hidalgo, and Aransas Cos. *Historically* (pre-1950).—*Uncommon.* However, possibly *locally common* in Cameron Co. Numbers steadily decreased beginning in the early 1900s. Last recorded nesting in the state, with the egg set collected, was in 1941 in s. Brooks Co. The last specimen was collected in 1949 near Riviera, Kleberg Co.

*New Mexico: Currently* (1987 to present).—Confirmed breeding of a naturally recolonized pair in sw. Luna Co. in the spring and early summer of 2001. The nest, however, suffered depredation and failed. In 2002, a wild pair successfully raised a brood of three youngsters in the same area, which marked the first successful natural nesting to occur in the U.S. since 1952. The 2001 and 2002 breeding is an optimistic sign that the breeding population in New Mexico may slowly increase over the next several years from natural recolonization of birds coming from Chihuahua, Mexico. Previously, a *rare* annual visitor since 1987. At least one individual has been seen each year; however, there are often multiple records. Sightings are primarily from historical areas from Hidalgo Co. east to Lea Co. and north along the Rio Grande region from Dona Ana Co. to Bernalillo Co. The majority of recent sightings are from Dona Ana, Grant, Hidalgo, Luna, Otero, and Sierra Cos. From early Aug. to early Oct. 2000, five individuals were seen: (1) east of Las Cruces, Dona Ana Co., in early Aug.; (2) an adult at Isaack's Lake, Dona Ana Co., from late Aug. to early Sept.; (3) a juvenile near Muzzle Lake, Dona Ana Co., in early Sep.; (4) a juvenile in Otero Co. in mid-Sep. of a bird banded in n. Chihuahua, Mexico (*see* Movements); and (5) a bird south of Hachita, Grant Co., in early Oct. *Recently* (1953–1987).—*Casual* visitor. There were no sightings recorded from 1953 to 1961. A few very irregular sightings occurred from 1962 to 1987. *Historically* (pre-1953).—An *uncommon* breeding resident. The population began to diminish in the early 1900s. The last specimen was collected in 1939 and the last documented nesting, and the last nesting in the U.S., was south of Deming near the village of Hermanas, Luna Co., in 1952.

*Arizona: Currently/recently* (1900s).—*Accidental,* with only a few sporadic sightings in Cochise Co. since 1900. *Historically* (pre-1900).—A *very rare* resident. All historical nesting records are from Cochise Co. The last nesting occurred in 1887.

*Mexico* (mapped region): *Chihuahua.*—*Very uncommon* with localized breeding clusters. Some recently discovered breeding populations are very close to the Texas and New Mexico borders. Nesting areas or locations with year-round sightings: Ahumada, Ascencion, Laguna de Santa Maria, Casas Grandes, Gomez Farias, Guzman, Janos, Mata Ortiz, and Sueco, and the counties of Coyame and Tinaja Verde Ranch. There are also numerous sightings of single birds in the northern part of the state. There are possibly at least 40 pairs in the state. *Coahuila.*—Recent nesting in the southern part of the state 46 miles (74 km) west of Saltillo. The nest, however, was destroyed by fire in 1999. *Sonora.*—Observed in the county of Bacerac in the summer of 2001. *Tamaulipas.*—Breeding occurs sparingly south of the mapped area along coastal regions south of the Rio Bravo and inland 112 miles (180 km) south of Reynosa in the Sierra de San Carlos.

**Movements.**—Irregular dispersal excursions are undertaken by some individuals. Dispersal may occur in any season. Northward dispersal was evident, with periodic sightings in Texas and New Mexico, from individuals roaming north of the Mexican border after the Aplomado Falcon disappeared as a breeding species in those states, in 1941 and 1952, respectively. Breeding pairs are typically sedentary.

In Texas released birds disperse west and north, in all seasons, from release sites to Starr, Duval, Lavaca, and Chambers Cos.

Northward dispersal of some individuals from Chihuahua, Mexico, into New Mexico has been an annual event since 1987. A fledgling banded in Chihuahua in May 1999 moved 180 miles (290 km) north into Otero Co., N.M. by mid-Sep.

**NESTING:** Feb.–Jul. Nesting pairs are often in loosely assembled colonies. In captive-released Texas birds, pairs may form when nearly 1 year old; however, breeding has not been documented with these younger individuals. Pairs may be comprised of 1-year-olds or mixed 1-year-olds and adults. Breeding normally does not occur until 2 or 3 years of age.

**Courtship (flight).**—*High-circling* and *power-diving* (*see* chapter 5). Highly vocal when courting.

No nest is built. Abandoned nests of other raptors are used. Nests of Chihuahuan Raven are extensively used. (The raven is very common and found throughout all of falcon's historical and current range.) As in former times, the reintroduced Texas population also uses abandoned nests of White-tailed Kite, White-tailed Hawk, Crested Caracara, and to a lesser extent, nests of Harris's Hawk and Red-tailed Hawk (*Buteo jamaicensis fuertesi*). Nests are primarily in yucca, mesquite, or low bushes 4–17 ft (1–5 m) high. In 1999, the first ground nesting was reported on Matagorda Island, Texas. Nest also on human-made platforms and utility poles, some of which have platform structures attached to them. These structures reduce mortality from ground predators. Nests on utility poles are up to 60 ft. (18 m) high.

In Chihuahua, nests average 9 ft. (3 m) high and are mainly in old Chihuahuan Raven nests in yucca.

Two or 3 eggs are laid. Captive birds lay 3 or 4 eggs. The incubation period is 31–32 days and is done by both adults. Youngsters fledge in about 35 days and are independent in about 65 days.

**CONSERVATION:** *South Texas.*—Some of the current Texas population is the result of the first captive-release hacking program that ran from 1985 to 1989, with 22 falcons being reintroduced. Most of the current Texas population is from the reinstated, intense reintroduction effort that began in 1993. Reintroduction is expected to continue until a viable breeding population of 60 pairs is attained.

Reintroduction is by The Peregrine Fund, Boise, Idaho, in alliance with the USFWS, Texas Parks and Wildlife Department, U.S. Coast Guard, American Electric Power Co., and private landowners. Over 1 million acres of protected habitat has been designated in s. Texas for reintroduction. Reintroduction is taking place at Laguna Atascosa NWR, Cameron Co., and private, state, and federal lands in Cameron, Kenedy, San Patricio, Matagorda, and Calhoun Cos. (including Matagorda Island).

In 1977 and 1978, eight Aplomado Falcons were obtained from various native populations of *septentrionalis* in Mexico and captive-bred at the Chihuahua Desert Research Institute in Alpine, Texas. The first captive-bred falcon was produced at this facility in 1982. In 1983, the captive Aplomado Falcons were transferred to facilities previously used by The Peregrine Fund at The University of California at Santa Cruz, Calif. In 1988 and 1989, The Peregrine Fund captured 20 more wild falcons from Mexico for captive-breeding purposes. Aplomado Falcons were captive bred at the Santa

Cruz facility until 1990. Since then, captive breeding occurs only at The Peregrine Fund's facility in Boise.

The first reintroduction-era wild breeding occurred in Cameron Co. in 1995. Five pairs nested in 1998. Nineteen pairs were located, with eight pairs nesting in 1999. Thirty pairs were located in 2000, with 18 nesting. Thirty-three pairs were found in 2001, with 22 nesting; however, at least 8 pairs failed to produce young. There were 37 pairs in s. Texas in 2002.

On Matagorda Island, 10 pairs are known, with six pairs attempting to nest in 2001. Thirty-five Aplomados were released on Matagorda Island between 1996 and 1999; however, many individuals of pairs on the island have come from mainland release sites. *Note:* Many of the "pairs" include nonbreeding-age 1-year-olds.

As of the summer of 2001, The Peregrine Fund has released 702 captive-raised fledgling Aplomado Falcons. At least 100 falcons are expected to be released each year (124 were released in the summer of 2001). The Aplomado Falcon population in Texas is increasing annually as reintroduction efforts take hold.

*W. Texas.*—The first captive-raised Aplomado Falcons were released in the Trans-Pecos region in Jul. 2002. The first release of 16 juveniles was conducted near Van horn, Culbertson Co. The USFWS and The Peregrine Fund are coordinating and financing the reintroduction in the Trans-Pecos region. Private landowners are providing access to lands for the released population. Additional releases are planned for the next few years in the high-elevation grasslands of this region, which were inhabited by Aplomados before their disappearance several decades ago.

*New Mexico.*—In mid-Jun. 2002, the New Mexico State Game Commission approved of a captive-release program to be conducted by The Peregrine Fund. Numerous wild falcons have been annually dispersing north from breeding colonies in Chihuahua, for the last several years.

*Mexico (northern).*—Breeding colonies in n. Chihuahua, from natural colonization from more southern populations in Mexico, has occurred since the late 1980s.

**Mortality.**—Currently exists from natural causes, particularly of eggs and nestlings at predator-accessible nest sites, especially at low bush and tree sites. Great Horned Owls, Chihuahuan Ravens, Raccoons, and rats are major predators. Depredation is a major obstacle in the reintroduction efforts of all ages of falcons, particularly of hacked juveniles that do not have parental protection. Matagorda Island, Texas, for instance, is used as a release site because of lesser amount of depredation by Great Horned Owls.

Historical decline was noted in Texas and New Mexico in the early 1900s. Reason for decline is subject of considerable debate.

*Habitat alteration*: S. Texas and portions of the west-central part of the state underwent massive habitat alteration with agricultural conversion beginning in the 1920s. Most habitat alteration occurred in the midst of this species' last stronghold. Areas left "natural" were not subjected to naturally occurring range fires that formerly kept dense thornscrub (mesquite) habitat to a minimum, thus restricting suitable open habitat. Serious overgrazing by cattle, which began in the late 1880s, may have caused enough grassland deterioration to be detrimental to sustaining ample prey species. Grazing, however, has been moderated since the 1970s. *Note:* Most of the Aplomado Falcon's current range in Texas, New Mexico, and Mexico is on grasslands that are moderately grazed by cattle.

*Pesticide contamination*: Organochlorine pesticides, notably DDT, first used in the U.S. in 1946, and extensively used in the 1950s, may have dealt lethal blows to remnant population. Like other bird-eating raptors, Aplomado Falcons are at the top of the deadly food chain affected by organochlorine pesticides. Aplomado Falcons may have been affected not only by locally contaminated resident avian prey, but also by seasonal migrant species.

Although the Aplomado Falcon's decline began prior to the pesticide era, those few that remained in the U.S. into the late 1940s and early 1950s may have been affected. The last known breeding in Texas was in 1941, which was prior to the use of DDT; however, pesticides may have affected remnant single birds and kept pairs from being viably reproductive, since sporadic sightings of single birds occurred in s. Texas through the DDT era. With the last breeding in 1952 in New Mexico, several years after DDT was first used, the falcon may have been affected in this region by feeding on migrant and wintering avian prey.

*Collecting* (specimens and eggs): Collectors, both professional and amateur, intensely shot birds and collected egg sets of this species from 1890 to 1949. Substantial numbers of birds and eggs were taken from local areas. Intense collecting at the very periphery of a species' range, especially one that was not very common, may have had a negative impact. Being on the extreme range periphery, replacement individuals are not readily available to fill gaps of lost birds, as would be the case in the core range. *Note:* Mexican populations were also subjected to intense collecting, and considerable habitat loss, and currently experience high levels of pesticides.

**SIMILAR SPECIES:** (1) **Mississippi Kite.**—Range overlap in migration in Apr. and May and Aug. to early Oct. PERCHED.—Lacks dark malar stripe. Wingtips extend past tail tip. FLIGHT.—Share similar white band on trailing edge of secondaries and long pointed wings. Adult/subadult underparts are gray, juveniles are streaked. (2) **American Kestrel.**—PERCHED.—Two dark vertical stripes on face. Underparts uniformly pale. Rufous back and tail. FLIGHT.—Uniformly pale underwing. Rufous tail. (3) **Merlin.**—Range overlap Sep.–Apr. PERCHED.—Nondescript dark malar stripe. Uniformly streaked underparts. Tail has three or four pale bands. FLIGHT.—Dark trailing edge of underwing. Underparts and tail as in Perched. (4) **Peregrine Falcon.**—Range overlap all year in interior and Sep.–May on coastal areas. PERCHED.—*Falco peregrinus tundrius* have a narrow dark malar stripe, especially juveniles. Pale supercilium, and often a pale "V" on nape; use caution! Barred belly and flanks (adults), uniformly streaked underparts (juveniles). Wingtips nearly equal or equal tail tip. FLIGHT.—Pale underparts (adults), uniformly dark underparts (juveniles). (5) **Prairie Falcon.**—Range overlap in w. Texas all year; coastal Texas in winter (Prairie Falcon is rare in coastal areas). PERCHED.—Share similar head pattern: single dark malar stripe and pale supercilium often extends as a well defined "V" onto the nape; use caution! Whitish breast and belly. Juveniles have dark flanks similar to the dark flank area on Aplomados; use extreme caution! Wingtips extend to near tail tip. FLIGHT.—Distinct dark axillaries and median underwing coverts can be deceptively similar to an Aplomado's, but remiges are pale; use caution!

**OTHER NAMES:** Aplomado. *Spanish:* Halcon Aplomado, Halcon Fajado.

---

REFERENCES: del Hoyo et al. 1994; Hector 1987; Howell and Webb 1995; Hunt 1983; Keddy-Hector 1998; Laack 1995; Lemieux 1995; Montoya et al. 1997; New Mexico State Game Commisson 2002; Oberholser 1974; Perez et al. 1996; Snyder and Snyder 1991; The Peregrine Fund 1999a, 2001; White et al. 1995; D. Williams 2002; S. O. Williams 1997, 2000; Young et al. 1999.

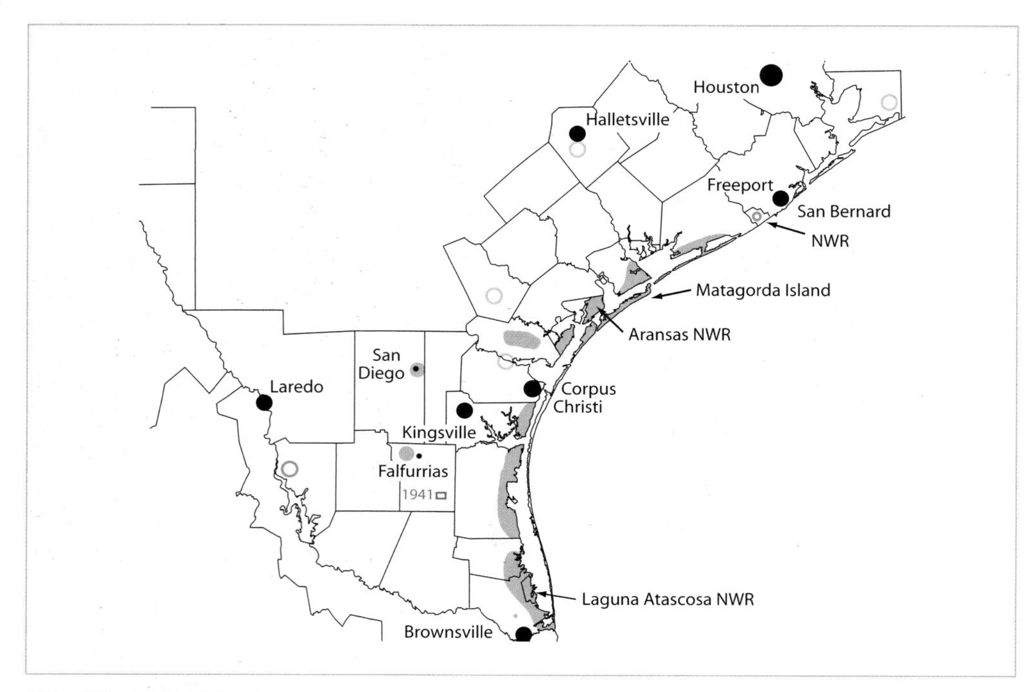


**APLOMADO FALCON,** *Falco femoralis:* Aplomado Falcons have been reintroduced into s. and coastal TX since 1993 to produce the present range. 33 territorial and 22 nesting pairs in 2001. Goal: 60 nesting pairs.

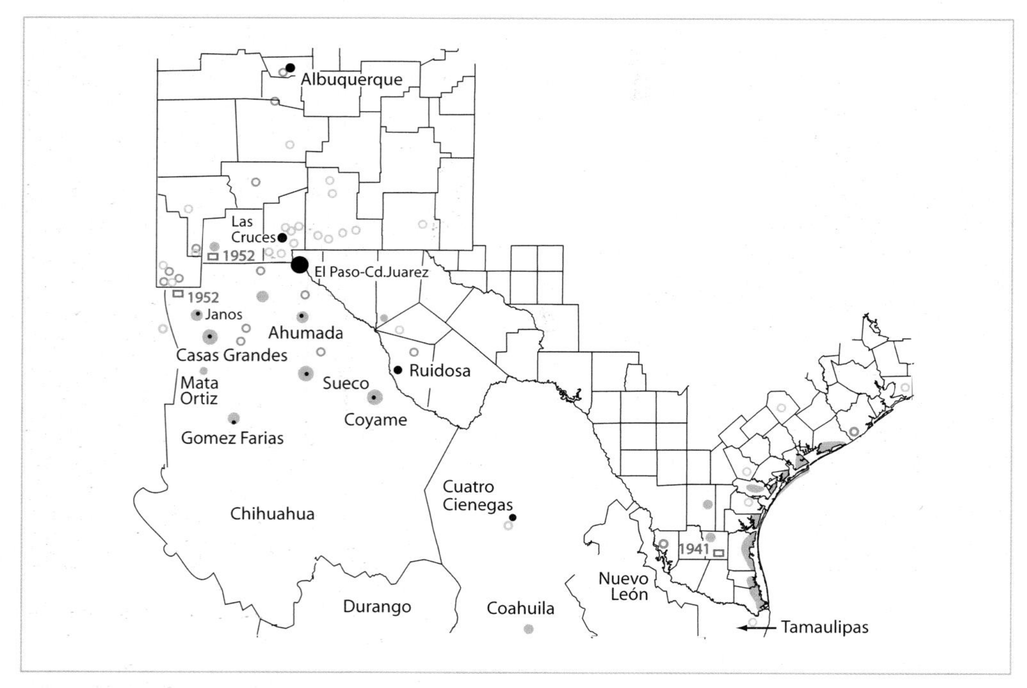


**APLOMADO FALCON,** *Falco femoralis:* Endangered in U.S. and Mexico. Remnant population of over 40 pairs exists in Chihuahua. Aplomados there disperse into NM and w. TX. Nested in NM in 2001–02, first nests in U.S. since 1952.

**Plate 548. Aplomado Falcon, adult male** [May] ▪ Narrow black malar mark, narrow black auricular stripe, white supercilium. Yellow orbital skin. ▪ White breast; black belly and flanks have narrow white feather edges; tawny-rufous leg feathers, lower belly, and undertail coverts. ▪ Long black tail has many thin white bands. ▪ *Note:* Captive-released. Cameron Co., Tex.

**Plate 549. Aplomado Falcon, adult male** [May] ▪ White breast; black belly and flanks have narrow white feather edges; tawny-rufous leg feathers, lower belly, and undertail coverts. ▪ Dark underwing with white bar on rear edge of secondaries. ▪ Long black tail has many thin white bands. ▪ *Note:* Captive-released. Cameron Co., Tex.

**Plate 550. Aplomado Falcon, adult female** [May] ▪ Dark streaking on center of white breast; black belly and flanks; tawny-rufous leg feathers, lower belly, and undertail coverts. ▪ Dark underwing with white bar on rear edge of secondaries. ▪ Long black tail has many thin white bands. ▪ *Note:* Captive-released. Cameron Co., Tex.

**Plate 551. Aplomado Falcon, subadult female** [May] ▪ Narrow black malar mark, white supercilium. Yellow orbital skin. ▪ Dark streaks on white breast; black belly and flanks; tawny leg feathers, lower belly, and undertail coverts. Dark brown upperparts. ▪ Long black tail has many thin white bands. ▪ *Note:* Captive-released. Cameron Co., Tex. Head and breast are adult plumage.

**Plate 552. Aplomado Falcon, subadult female [May]** ▪ White supercilium extends as a V onto nape. ▪ *Note:* Same bird as on plate 551. First prebasic molt on large falcons begins on head, and is completed on head and breast before continuing onto rest of body, wings, and tail. Head and breast are adult plumage, rest of body and wings are juvenile feathering.

**Plate 553. Aplomado Falcon, subadult male [Jan.]** ▪ Narrow black malar mark, narrow black auricular stripe, white supercilium. Pale yellow orbital skin. ▪ Juvenile streaking on breast; black belly and flanks; pale tawny leg feathers, lower belly, and undertail coverts. ▪ *Note:* Chiapas, Mexico. Head is adult plumage, breast is partially adult; rest is juvenile feathering.

**Plate 554. Aplomado Falcon, juvenile female [Oct.]** ▪ Narrow black malar mark, narrow black auricular stripe, pale tawny supercilium. Pale blue cere and orbital skin. ▪ Thick dark streaks on pale tawny breast; black belly and flanks; pale tawny leg feathers, lower belly, and undertail coverts. Dark brown upperparts are thinly edged with tawny. ▪ *Note:* Veracruz, Mexico.

# PEREGRINE FALCON

## (*Falco peregrinus*)

**AGES:** Adult, subadult, and juvenile. Adults are somewhat sexually dimorphic in color, but juveniles are similarly marked. Juveniles have broader wings and longer tails than do adults. Subadult is an interim molt stage that retains a portion of old juvenile and new adult feathering in a distinct pattern during the first prebasic molt.

**MOLT:** Falconidae wing and tail molt pattern (*see* chapter 4).

The first prebasic molt begins when the birds are about 10–14 months old. An incomplete molt attains 95–98 percent of the adult plumage. On migratory birds, the molt is suspended until they have reached the winter grounds. A few juvenile feathers are often retained on the upperwing coverts, belly, scapulars, and rump until the second prebasic molt. Molt varies with the latitude and elevation at which juveniles were born. Molt typically begins in Apr. or early May in the continental U.S. and Canada; however, it may begin as early as Jan. or Feb. (N. J. Schmitt pers. comm., museum specimens) or as late as mid- to late Jun. (J. Zipp photos, BKW pers. obs.). Molt appears to begin latest on mated but nonbreeding 1-year-old females. Birds of the Northwestern Pacific region often do not begin to molt until mid- to late May. Arctic birds do not begin to molt until late May or Jun. Near-adult plumage is attained by Oct. to Dec.

The first prebasic molt begins on the head, nearly completes on the head (except on the crown), then expands down the back, forward and mid-scapulars, and the breast. Molt then continues down the rear portion of the body. Once molt is nearly complete on the head and partially complete on the back, scapulars, and breast, the primaries begin to molt. Primary molt begins, as on all falcons, on p4, then molts in both directions in the following order (after Palmer 1988): 5, 3, 6, 7, 2, 8, 1, 9, and 10. Secondary molt, beginning with s5 and sequentially molting in either direction, begins shortly after primary molt begins. The outermost secondaries are the last to molt, with s1 being the last. Rectrix molt begins after primary molt is well under way. Rectrix molt is in the following sequence: r1, 2, 3, 6, 4, and 5. The greater and median upperwing coverts molt in unison with the replaced remiges. The last feather tract to molt is the lesser upperwing coverts.

By mid-summer, much of the body and remiges may be adultlike, but the upperwing coverts and tail are still mainly juvenile and are in a classic "subadult" stage.

Subsequent prebasic molts begin later than the first prebasic molt. Nonbreeding birds begin molting before breeding birds. Molt is annual and complete each year. Females begin molting when incubating and males mainly during the latter part of the nesting period or after the young have fledged. On migratory birds, molt is suspended during migration, then it is resumed and finished on the winter grounds. Molt sequence on the remiges and rectrices is the same as for the first prebasic molt; however, all areas of the body may molt in unison. Molt is completed from Oct. to early Jan., with the outermost primaries being the last anatomical area to molt. Some adults of *F. p. tundrius* and *F. p. anatum* are still growing a new outermost primary (p10) in early Jan. (based on AMNH specimens).

**SUBSPECIES:** Polytypic. **Native subspecies.**—Three subspecies are found the w. U.S. and Canada: (1) "American" Peregrine Falcon, western *F. p. anatum*, inhabits regions west of the Great Plains, east of the northwest Pacific, and south of the Arctic. (2) "Peale's" Peregrine Falcons, *F. p. pealei*, is on the Pacific coastal islands and headlands from Washington to w. Russia. (3) "Arctic" Peregrine Falcon, *F. p. tundrius*, breeds in the Arctic of Alaska, Canada, and Greenland. This race was not recognized as a separate subspecies until 1968 (White 1968).

A small breeding population of western *anatum* hung on during the DDT era of the late 1940s to early 1970s in the w. U.S. and w. Canada. The population of *pealei* was minimally affected by the organochlorine pesticide era. The Arctic breeding but very southerly wintering *tundrius* was moderately affected during this pesticide era.

**Former subspecies in the West.**—East of the Great Plains, the last eastern pair of *F. p. ana-*

*tum*, the "Rock Peregrine," nested at a historical site on a cliff along the Mississippi River in ne. Iowa in 1967 (Kent and Dinsmore 1996). However, eastern *anatum* were extirpated east of the river by 1962. Eastern *anatum* were overall darker and larger than their western counterpart.

**Reintroduced subspecies west of the Great Plains.**—*United States*: The captive-released stock reintroduced back into the w. U.S. came from original, captive-raised western *F. p. anatum* from the w. U.S. (J. Linthicum; J. Craig pers. comm.).

*Canada*: The captive-released stock reintroduced in all of Canada are from captive-raised western *F. p. anatum* (G. Holroyd pers. comm.). The original breeding stock came from nests in 1970 in s. Alberta, nw. Northwest Territories (Mackenzie River), and the s. Yukon; a few probable eastern *anatum* were taken from Labrador (Fyfe 1988).

**Introduced subspecies east of the Great Plains in the U.S.**—The current restoration-era population east of the Great Plains in the U.S. is the product of mixed lineage and is not a subspecies. Herein, it is called the "Eastern" Peregrine and does not have a subspecies designation.

*Note:* A few U.S. birds of mixed lineage have immigrated into s. Manitoba and w. and cen. Ontario and bred with Canada's released *F. p. anatum*. Also, some Canadian-released *anatum* have entered the n. U.S. and bred.

The breeding birds used to create this introduced population came from subspecies from w. and n. North America and Europe. The subspecies used for captive releases in Kansas, Iowa, Minnesota, Missouri, and Nebraska were carefully tracked and their survival and breeding success tabulated. *See* conservation for percentages of the different subspecies that were released and those surviving to breed (from Tordoff 2000).

If a true subspecies does eventually evolve, it may take hundreds of years for the conglomeration of races to sufficiently intermix to have genetic similarity among 75 percent of the population, which is required for a subspecies to be recognized. (Most subspecies are formed from one or two adjacent races.)

*See below* for data on subspecies and conservation for reintroduction details.

As described above, two foreign subspecies were used in building the current population of "Eastern" Peregrine Falcon in states listed above that are west of the Great Plains (two additional foreign races were used in the East). (1) Nominate *F. p. peregrinus* is found in most of Europe south of the Arctic and north of the Pyrenees, Balkan, and Himalayan Mts. to cen. Russia. It may winter to the Mediterranean region and Iberian Peninsula. A large Peregrine. Often similar in appearance to North American *anatum*. Head is not as black and malar mark not as broad on the cheeks and auriculars as *anatum*. The ventral areas vary from whitish to medium pinkish tawny and the breast is moderately spotted on males and heavily spotted or with large teardrop-shaped markings on females. Juveniles are similar to *anatum* but have more distinct pale edgings on the dorsum. *See* Forsman (1999) for photographs. (2) *F. p. brookei* is resident from nw. Africa and Spain, and east through the Middle East and the Caspian Sea region of Iran. Dark and heavily marked race, but much smaller than *anatum*. Head is black with solid black cheeks and auriculars; nape often has rufous patches. Breast and belly are often deep pinkish rufous with heavily spotted breast and narrowly but densely barred belly and flanks. Juveniles are similar to *anatum* but are much smaller. *See* Forsman (1999) for photographs.

Although not used in the West or Midwest, *F. p. macropus* of se. Australia and *F. p. cassini* of s. South America made up a small portion of the lineage of released stock in the e. U.S. (White et al. 1995; C. M. White pers. comm.).

**Additional subspecies in the world.**—There are 11 additional subspecies in Europe and Asia: (1) *F. p. calidus* inhabits the Arctic of Eurasia. Winters to throughout Africa, s. Asia, Indonesia, New Guinea, and the Philippines. Similar pale color of *tundrius* of North America and Greenland, but the breasts of most females have dash marks or spots. (2) *F. p. japonensis* of ne. Siberia and n. Japan is similar to *calidus* but somewhat darker, and smaller and even darker on the islands of the Sea of Okhotsk (possible separate race *F. p. pleskei*). (3) *F. p. minor* is resident south of the Sahara Desert in Africa. Small race with dark head and broad malar marks; may have rufous on the nape. Moderately marked buffy ventral region.

(4) *F. p. radama* is resident on Madagascar and the Comoro Islands. Similar to *minor* but smaller, darker head, and more heavily marked on ventral areas. (5) *F. p. peregrinator* is mainly resident from Pakistan to se. China and n. North Vietnam. Very blackish dorsal and broad malar marks. Ventral is deep pinkish rufous and lightly to fairly heavily barred on flanks. (6) *F. p. ernesti* is resident on w. and peninsular Thailand, Indonesia, Philippines, New Guinea, and the Bismarck Archipelago. Small race. Darkest race. Very black head and dorsum and very grayish and heavily barred ventral region. (7) *F. p. nesiotes* is sedentary on Fiji, New Caledonia, and Vanuatu. Small and dark race like *ernesti* but more rufous on ventral region. (8) *F. p. madens* is a rare resident on the Cape Verde Islands off nw. Africa. Large and very brownish on the dorsum with much rufous on the nape. (9) *F. p. furuitii* is endangered and may be extinct on Volcano and possibly Bonin Islands, Japan. Similar to *pealei*; juvenile not as dark and heavily marked. (10) *F. p. pelegrinoides* and (11) *F. p. babylonicus* have typically been listed as subspecies of Peregrine Falcon; however, they are considered a possible separate species, the Barbary Falcon (*F. pelegrinoides*), by some recent authors. The pale western form, *pelegrinoides*, is resident in arid regions of n. Africa, Middle East, and the Arabian Peninsula. The even paler and more rufous-headed eastern form, *babylonicus*, is found in the arid regions from Iran to w. Mongolia. Color markings and patterns are similar to but paler than those of all Peregrine races; voice is identical to Peregrine's (Clark 1999). *F. p. pelegrinoides* apparently do not intergrade with the Spanish Peregrine, *F. p. brookei*, in Morocco. However, current DNA samples align them with other Peregrines (White et al. 1995). A falconer's escaped female *pelegrinoides* bred with a male *anatum* Peregrine in cen. Mexico in 2001 and 2002 (R. Padilla-Borja pers. comm.). It is considered a subspecies of Peregrine by Forsman (1999), but Clark (1999) and Ferguson-Lees and Christi (2001) consider it a probable species. Museum specimens appear as pale, more rufous-headed Peregrines; however, partial rufous infusion on the head is also found on some other subspecies of Peregrine. Until taxonomic studies are more conclusive, *F. p. pelegrinoides* and *F. p. babylonicus* are considered here as a subspecies of Peregrine Falcon.

**COLOR MORPHS:** None in the three North American races. *F. p. cassini* of South America has a rare pale "Kleinschmidt's" morph that inhabits the Straits of Magellan in South America.

**SIZE:** Medium-sized raptor. Males are smaller than females, with no overlap. Juveniles are often longer than adults because many have longer rectrices. In flight, juveniles also appear bulkier than adults because of having broader remiges. Sizes below are for *F. p. tundrius* and *F. p. pealei*; wing chord but not total length and wingspan data are available for *F. p. anatum*. Sizes for *F. p. tundrius*: MALE.—Length: 14–16 in. (36–41 cm), some juveniles may be on the larger end of the spectrum because of having longer tails; wingspan: 37–39 in. (94–99 cm); wing chord: MALE.—292–330 mm. FEMALE.—Length: 16–18 in. (36–46 cm); wingspan: 40–46 in. (102–117 cm); wing chord: 331–368 mm. Sizes for *F. p. pealei* (D. Varland unpubl. data on migrant/winter birds in w. Washington): *F. p. pealei*: MALE ($n = 4$ adults, 4 juveniles; all measures similar).—Length: 16.3 in. (41.4 cm); wingspan: 36.2 in. (92.1 cm); wing chord: 320–345 mm. FEMALE ($n = 6$ adults, 7 juveniles).—Length: 19.1 in. (48 cm); wingspan: 43.6 in. (110.8 cm); wing chord 363–391 mm. Wing chord for western *F. p. anatum* (C. M. White pers. comm.): Smaller in size south of the boreal forest of the West; extirpated eastern birds were larger than most western birds. Male: 291–322 mm, female: 333–365 mm.

Since "Eastern" Peregrines released in the Midwest and e. U.S. are a mix of bloodlines and encompass the spectrum of sizes from the largest race, *pealei*, to the smallest races, *brookei* and *macropus*, measurements are not applicable. However, wing-chord lengths for each respective race are listed below for general size comparison. "Eastern" types involve a mixture of the following subspecies, including western *F. p. anatum* listed above: *F. p. peregrinus*.—Male: 289–334 mm, female: 339–375 mm; *F. p. brookei*.—Male: 275–312 mm, female: 306–355 mm. Two other small foreign races, *F. p. cassini* and *F. p. macropus*, were released in the e. U.S., but apparently not in the areas covered in this book.

**SPECIES TRAITS:** HEAD.—Dark malar mark (mustache) that extends below the eyes and onto the lower mandible. Bare orbital skin. BODY.—Long toes. When perched, the outer toe typically bends inward and lays sideways, and the inner toes may cross over each other. WINGS.—**In flight, wings are long and taper to pointed wingtips. When perched, wingtips extend to tail tip or just short of the tail tip.**
**ADULT TRAITS:** HEAD.—Yellow cere and orbital skin, which are generally brighter on males. **Black malar mark may be one of four types:** (1) *Very wide type.*—**Malar mark covers most or all of the auriculars and creates a black "helmet" with all of the head being black.** (2) *Wide type.*—**A wide columnar shape and covers most of the auriculars except for a small area on the lower one-third of the auriculars.** (3) *Moderately wide type.*—**Moderately wide columnar shape with the lower one-half to two-thirds of the auriculars being white.** (4) *Narrow type.*—Narrow black stripe, often with a break at the gape. On all types, the crown is dark gray or black and the nape is black. BODY (dorsal).—**Males are bluish gray and distinctly barred and females are darker, more brownish gray or brownish black, and often have ill-defined barring and may be uniformly colored. When viewed from above in flight, the lower back, rump, and uppertail coverts are paler and more bluish than the rest of the upperparts and dorsal tail surface.** Males exhibit the pale lower back, rump, and uppertail coverts more than females do. Females are typically more distinctly barred with black on these areas, and the pale bluish color is not as apparent. BODY (ventral).—Breast is unmarked or covered with black spots or dashes. Flanks are distinctly barred, and the belly is partially barred or spotted. Leg feathers are covered with black barring. Yellow feet and tarsi are generally brighter on males. WINGS (dorsal).—Dorsal surface is like the back and is more bluish on males and darker and more brownish on females. WINGS (ventral).—**The dark gray underside of the remiges is equally barred throughout with either pale tawny-rufous or white. The underwing coverts and axillaries are narrowly barred and either pale tawny-rufous or white. The underwing appears uniformly grayish at a distance.** TAIL.—**The dorsal surface of the tail is distinctly darker than the bluish back, rump, and uppertail coverts. It is also darker than much of the upperparts except the back and head. On the dorsal surface, the tails of males are often dark blue with equal-width blue and black bands or have narrower black bands; some males and most females have black tails with narrow blue or bluish brown bands. Both sexes have broad white terminal bands. The black bands may get progressively narrower toward the basal region of the tail. The ventral surface of the tail is darker than the rest of the underparts.**

*Note:* At a distance, the dorsal surface of Peregrines appears black on the head, back, and tail and medium bluish or grayish on the wings and forward body; on males and many females, the pale blue back, rump, and uppertail coverts are very obvious.
**ADULTS OF "ARCTIC" (*F. p. tundrius*):** HEAD.—Broad white forehead and white lores. Black malar mark is moderately wide type. The auriculars are white on the lower half and medium gray or dark gray on the upper half. The crown is medium gray or dark gray; rarely black. At close range, birds with medium and dark gray crowns have distinct black shaft streaking. Nape and hindneck are black and sometimes have a small or moderate amount of white or rufous mottling or patches on the nape. BODY (ventral).—White breast, pale pinkish tawny belly, white lower belly, and pale grayish flanks and leg feathers. WINGS.—On perched birds, wingtips equal to the tail tip. WINGS (ventral).—White barring on the remiges, and the underwing coverts are white and barred with black.
**Adult male (*F. p. tundrius*):** HEAD.—The upper auriculars and crown are typically medium gray. Black nape and hindneck contrast with the rest of the paler bluish gray upperparts. Nape rarely has pale mottling. BODY (dorsal).—Medium bluish gray with moderately wide black cross bars on the back and front half of the scapulars. The rear half of the scapulars may be medium or pale bluish gray and have faint, partial blackish crossbars or are nearly unmarked bluish gray. The pale bluish gray lower back and rump may be unmarked or have partial, narrow blackish cross bars. The pale bluish gray uppertail coverts are faintly barred with black. Overall dorsal appearance of

males is medium bluish gray. BODY (ventral).—Unmarked breast. Rarely, breast is finely dashed with small dark markings on the lower half and very rarely on all of the breast. Flanks are covered with narrow black bars. Average birds have only a few small dark spots or partial bars on the belly. The palest types can be nearly unmarked. The lower belly is either thinly barred or unmarked. The leg feathers are covered with very narrow dark bars. WINGS (dorsal).—Blackish on the forward upperwing coverts, but medium bluish gray with narrow black bars on all coverts. The tertials are often pale bluish gray with faint, narrow blackish barring. The tertials may be unbarred on the palest males. The tertials blend with the pale bluish back, rump, and uppertail coverts. TAIL.—**Dorsal surface is blackish and distinctly darker than the uppertail coverts, rump, tertials, and back.**

**Adult female (*F. p. tundrius*):** HEAD.—Upper half of the auriculars and crown are dark gray, but are occasionally medium gray or black. Nape and hindneck are black. Females are likely to have whitish or rufous mottling or patches on the nape. BODY (dorsal).—Back and forward half of the scapulars can either be solid black or black with narrow bluish brown cross bars and outer feather edges. The rear half of the scapulars are a paler bluish black with broad medium bluish or medium bluish brown cross bars and outer feather edges. Overall appearance is more blackish and brownish than on males. The lower back and rump are moderately paler and more bluish than the rest of the upperparts and broadly barred with black. BODY (ventral).—The breast is typically unmarked; however, heavily marked individuals have a small amount of narrow dashes or spots on the lower half of the breast and sometimes on all of the breast. Flanks are marked with narrow or moderately wide black bars. The belly is fully covered with partial black bars or spots. The lower belly is thinly barred. The leg feathers are narrowly barred with black. WINGS (dorsal).—The lesser upperwing coverts are quite black and have a minimal amount of pale bluish edgings on the feathers. Pale cross-barring becomes more evident on the median and greater secondary coverts. TAIL.—**The blackish dorsal tail surface is moderately darker than the uppertail coverts, rump, and back.**

**ADULTS OF "ARCTIC" (PALE TYPE of *F. p. tundrius*):** Information herein is based on photographs and data supplied on breeding birds from Rankin Inlet, Nunavut (G. Court pers. comm.). A rare plumage type or morph that exists on a very small percentage of birds inhabiting the cen. Canadian Arctic in Nunavut. It is estimated that fewer than 3 percent of females and a lesser percentage of males exhibit this pale plumage. Photographs of this type are of known-age females that are greater than 2 years old (and mated with typical adult males). This rare plumage was first noted by White (1968). *Note:* Migrant and wintering birds found in normal *tundrius* areas.

**Adult male pale type (*F. p. tundrius*):** As described below for females, but the upperparts are more bluish and not brown. *Note:* Male is not depicted in accompanying photographs.

**Adult female pale type (*F. p. tundrius*):** HEAD.—Forehead and central crown is white. There is a dark brown U-shaped mark that starts above the eyes and loops around the rear of the crown. Lores are white. An irregularly shaped narrow type or moderately wide type black malar mark extends under the eyes (generally narrower than on typical adults). The dark malar does not connect to the narrow dark brown eyeline that runs above the all-white auriculars and extends onto the white nape as a partial dark brown "V." Fleshy areas are the typical yellow. BODY and WINGS (dorsal).—In early summer, the medium or dark brown back, scapulars, and upperwing coverts have white or pale brown crossbars and outer edges. Incoming new feathers have a slight grayish cast to them. The lower back and rump may be bluish and barred with brown. BODY (ventral).—White. Breast is unmarked. Belly has a few dark specks, and the lower belly is unmarked. The forward flanks are lightly spotted and the rear flanks are sparsely marked with thin partial bars. The leg feathers and undertail coverts are white and unmarked. WINGS (ventral).—Similar to typical *tundrius* with barred remiges and underwing coverts. TAIL.—Uppertail coverts may be pale bluish and barred. Tail is dark brown or blackish and barred with bluish brown. Undertail is darker than rest of the ventral regions.

**ADULTS OF "AMERICAN" (*F. p. anatum*):** HEAD.—Forehead is solid black or has a narrow white band. Lores are typically all black even if the forehead is white. Black malar mark can be any of the three types: very wide, wide, or moderately wide (and similar to *tundrius*). *Note:* The moderately wide type malar did not occur on former eastern *anatum*. Malar mark averages narrower on birds from nw. Canada and Alaska. The crown, nape, and hindneck are black. Nape is always black and lacks pale mottling. Black malar mark is sometimes a moderately wide columnar shape with white on the lower one-third of the auriculars. The crown is sometimes dark gray. BODY (ventral).—Two main types: (1) *Rufous type.*—Breast, belly, and lower belly are uniformly pale or medium tawny-rufous; flanks are pale or medium gray or are rufous on the forward half. (2) *White type.*—White breast, pale grayish flanks and leg feathers, pale tawny-rufous belly, and white lower belly. Leg feathers are sometimes pale tawny-rufous. WINGS.—On perched birds, wingtips nearly equal or equal to the tail tip. WINGS (ventral).—The black-barred underwing coverts are either tawny-rufous or white, and the primaries can be barred with tawny-rufous or white. Birds with tawny-rufous on the underwing are similarly marked on their ventral areas.

**Adult male (*F. p. anatum*):** HEAD.—Black malar mark either of two types: very wide type or wide type. Moderately wide columnar-shaped malar marks are rare on males. BODY (dorsal).—Black back and forward half of scapulars. They may have faint bluish barring and pale outer edges on some forward scapulars. Rear half of the scapulars are a paler medium bluish gray with distinct black cross bars. The lower back, rump, and uppertail coverts are medium blue and barred with black. BODY (ventral).—Breast, belly, and lower belly are either uniformly tawny-rufous or white. Breast is typically unmarked. Flanks are distinctly barred with black. Belly is moderately marked with small cross bars or spots, but is occasionally very lightly marked. Leg feathers are moderately marked with narrow bars. WINGS (dorsal).—Lesser upperwing coverts are black with narrow bluish gray edges. The median and greater secondary coverts have bluish gray barring the outer edges. The tertials are usually slightly paler than the rest of the wing coverts and have fairly distinct black cross bars. TAIL.—**The dark bluish or blackish banded tail contrasts sharply with the medium bluish gray uppertail coverts, rump, and back.**

**Adult female (*F. p. anatum*):** HEAD.—**Black malar mark can be any of the three types: very wide type, wide type, or moderately wide type.** If malar mark is a wide type or moderately wide type, then the forehead is often white. The moderately wide type of black malar mark occurs throughout the range, but seems to be especially common on females inhabiting the Rocky Mts. (including from pre-captive-release era). BODY (dorsal).—Back and forward half of the scapulars are black. The rear half of the scapulars is also black and may have grayish brown cross-barring or lack pale barring and be solid blackish gray or blackish brown. Overall appearance of the dorsal region is much darker and more brownish than on males, and darker than on most *tundrius* females. **The lower back, rump, and uppertail coverts are somewhat more bluish than the back and scapulars and are barred with black.** BODY (ventral).—Those with whitish ventral areas seem to be more prevalent in the northern part of their range or where they intergrade with *pealei*. Breast is either unmarked, partially marked on the lower half, or fully marked. Breast markings consist of small dashes or small or medium-sized spots. The flanks are moderately to heavily barred. The lower belly is moderately barred. Leg feathers are moderately barred. WINGS (dorsal).—The lesser upperwing coverts are mainly black and unmarked or, at most, have narrow bluish brown edges. The median and greater secondary coverts are somewhat barred with bluish brown. The tertials are distinctly marked with bluish brown and black bars. TAIL.—**The black tail with pale bands is somewhat darker than the bluish-barred uppertail coverts and rump.**

**ADULTS OF "PEALE'S" (*F. p. pealei*):** HEAD.—**Black malar mark either is the wide type or moderately wide type. Rarely, both sexes may have a black malar that is a very wide type and covers all of the auriculars (and similar to many *anatum*; may be intergrades with *anatum*).** The crown is medium gray, dark gray, or black. The nape and hindneck are black. White auriculars are usually spotted or dashed with black. Forehead varies from being a moderately wide white patch or is solid dark. Lores typically are white if the forehead is white and black if the forehead is dark. However, on occasion the forehead has narrow band of white and the lores are black. BODY (dorsal).—As in all North American races, males are bluish and females more brownish. The lower back, rump, and uppertail coverts are more bluish and paler than the rest of the upperparts. BODY (ventral).—White breast, white or pale grayish flanks, pale yellowish belly and lower belly (pinkish tawny on other races), and white or pale gray leg feathers and undertail coverts. Variably marked with black on the breast, flanks, and belly *(see below* Types). WINGS and TAIL.—Marked as in Adult Traits. On perched birds, the wingtips are barely shorter than the tail tip. WINGS (ventral).—Underwing covers are white and heavily barred with black. Remiges are barred with white.

*Note:* Three plumage variations are described: lightly marked type, moderately marked type, and heavily marked type. Lightly marked type is uncommon and local east of the Aleutian Islands, Alaska. Some breed on the Queen Charlotte Islands, B.C., and in the Gulf of Alaska (Campbell et al. 1990, C. M. White pers. comm., BKW pers. obs.). Moderately marked type is the common type from the Gulf of Alaska and southward. Heavily marked type is uncommon east of the Aleutians, but is the only type on the Aleutians and in Russia.

**Adult male (lightly marked type *F. p. pealei*):** HEAD.—Black malar is typically moderately wide type; however, the malar is often somewhat narrower and extends nearly to the rear of the eyes and all of the auriculars may be white. The white auriculars may have little if any dark spotting. Crown is medium or dark gray. Forehead and lores are white. The black nape is unmarked. BODY (dorsal).—Black back is partially barred with bluish gray. The scapulars are barred black and bluish gray (and similar to a more heavily marked adult male *tundrius*). Lower back, rump, and uppertail coverts are paler bluish and fully barred. BODY (ventral).—Breast is unmarked or lightly dashed or spotted with black. Belly and flanks are moderately marked with black barring. Leg feathers are lightly or moderately barred with black barring. WINGS and TAIL.—As in Adult Traits.

**Adult male (moderately marked type *F. p. pealei*):** HEAD.—Black malar is typically moderately wide type. Crown is medium or dark gray. Forehead and lores are white, but lores may be black. The white auriculars are moderately spotted with black. BODY (dorsal).—Black back is partially barred with bluish gray. The scapulars are barred black and bluish gray (similar to a more heavily marked adult male *tundrius*). Lower back, rump, and uppertail coverts are paler bluish and fully barred. BODY (ventral).—Breast is unmarked or lightly dashed or spotted with black. Belly and flanks are heavily marked with fairly wide black barring. Leg feathers are barred with wide black barring. WINGS and TAIL.—As in Adult Traits.

**Adult male (heavily marked type *F. p. pealei*):** Black malar mark is the wide type, but may be the very wide type. Dark gray or black crown. Forehead is mainly dark or have a very narrow white band. The white auriculars are moderately or heavily spotted with black. BODY (dorsal).—As on lightly marked type. BODY (ventral).—Breast is heavily spotted and almost forms streaks. Belly and flanks are very heavily barred with very wide black barring. Leg feathers are very heavily marked with black barring. WINGS and TAIL.—As in Adult Traits.

**Adult female (lightly marked type *F. p. pealei*):** HEAD.—Black malar mark is the moderately wide type; but may be somewhat narrower and extend nearly to the rear of the eyes, making the entire auricular region white (classic plumage example; *see* Campbell et al. 1990 of nesting adult female at Langara Island of the Queen Charlotte Islands). The white auriculars have little if any dark spotting. Crown is medium or dark gray. Forehead and lores are white. The nape has a moderate amount of white or pale tawny mottling. BODY (dorsal).—Black back has a small amount of pale grayish barring. The scapulars are distinctly barred black and grayish or brownish, or are only partially barred and dark. The lower back, rump, and uppertail coverts are more bluish than the back, scapulars, upperwing coverts, and tail. BODY (ventral).—Breast is unmarked or lightly dashed or spotted. Belly and flanks are heavily barred with black. Leg feathers are moderately barred. WINGS and TAIL.—As in Adult Traits.

**Adult female (moderately marked type *F. p. pealei*):** HEAD.—Black malar mark is the moderately wide type. Crown is medium or dark gray. Forehead and lores are mainly white, but lores may be dark. The nape is dark. The white auriculars are moderately spotted with black. BODY (dorsal).—Black back has little if any pale grayish barring and may be virtually solid blackish gray or blackish brown. The scapulars are distinctly barred with black and grayish or brownish gray, but may be partially barred and dark. Lower back, rump, and uppertail coverts as in lightly marked type. BODY (ventral).—Breast is moderately spotted with black. Belly and flanks are heavily barred with black. Leg feathers are heavily barred. WINGS and TAIL.—As in Adult Traits.

**Adult female (heavily marked type *F. p. pealei*):** HEAD.—Black malar is typically the wide type, but may be the very wide type. Crown is dark gray or black. Auriculars are heavily spotted with black. BODY (dorsal).—As on lightly marked type, but likely to lack discernible grayish barring on the scapulars and will appear blackish gray or blackish brown. Lower back, rump, and uppertail coverts are marginally paler and more bluish than the rest of the upperparts. BODY (ventral).—Breast is heavily spotted and nearly streaked. Belly and flanks are very heavily and thickly barred with black. Leg feathers are very heavily barred. *Note:* Average female more heavily marked than males and the most heavily marked North American type. WINGS and TAIL.—As in Adult Traits.

**ADULTS OF "EASTERN" (no subspecies designation):** It is difficult to describe plumages since several subspecies were used in the captive-breeding programs, with a vast array of intergrade characters prevailing. HEAD.—Black malar mark varies from moderately wide type with a large amount of white on the auriculars as on all *tundrius* and some western *anatum*, wide type with small or moderate white auriculars as on *pealei* and *peregrinus*, or very wide type with all-black helmets and dark foreheads as on many *anatum*, *brookei*, *cassini*, and *macropus*. BODY (dorsal).—Varies from being barred with medium bluish (males) or blackish gray or brownish with a moderate or minimal amount of paler bluish or brownish cross-barring (females). Some females will be virtually solid blackish on the dorsum because of the lack of cross-barring. The lower back, rump, and uppertail coverts will be paler bluish on all birds since this is a distinct Peregrine trait, but is more noticeable on males than on females. BODY (ventral).—Variable from being white to deep tawny-rufous as on the original eastern *anatum* and on *brookei* and some *macropus* types. Breast may vary from being unmarked to heavily spotted as on *brookei*, *pealei*, *peregrinus*, and some western *anatum*. Flanks are barred on all races. Belly may be lightly to heavily spotted or barred. TAIL.—**All races have dark dorsal and ventral tail surfaces that contrast with the paler bluish uppertail coverts, rump, and lower back.**

*Note:* The female (plates 560 and 561; banded known-aged bird) is nearly identical to female *peregrinus* of Europe; however, in her fresh first-adult plumage in autumn, this bird was very rufous on the underparts and similar to *anatum*. Her juvenile plumage was very heavily marked and similar to *pealei*.

**SUBADULT TRAITS (ALL SUBSPECIES):** This plumage stage may occur from mid-spring to mid-summer. HEAD.—Fully adult with black malar and black or gray cap and nape. A few brown juvenile feathers may be scattered on the crown region. Black malar area molts into full adult feathering before other areas. Cere and orbital skin can be fairly bright yellow on males but nearly always pale yellow or greenish yellow on females. BODY (dorsal).—Nearly full or full adult bluish or grayish feathering. Lower back and rump are often largely retained brown juvenile feathering (seen only from above in flight). BODY (ventral).—Largely adultlike, but all have a small amount of brown juvenile streaking retained on the mainly adultlike spotted or barred belly and forward flanks. Leg feathers are primarily barred adult type. Legs and tarsi are medium or bright yellow and usually brighter on males. WINGS (dorsal and ventral).—Ninety-five percent of the body molts before much of the wing feathers begin molting to adult feathering. A few primaries will be dark and neat adult feathers, but most will be worn and faded brown juvenile feathers. Virtually all of the upper- and underwing coverts will be retained juvenile. On the dorsal surface, the worn brown upperwing coverts contrast with the new adult bluish or grayish scapulars and back. TAIL.—Retained brown juvenile feathers until the two central deck rectrices start growing in as new adult feathers. The white tips of the worn juvenile rectrices are mostly worn off.

**JUVENILE TRAITS:** HEAD.—Cere and orbital skin are pale blue throughout fall. Cere and orbital skin gradually change to pale yellow by mid- to late winter and some, especially males, may be bright yellow by late winter or spring. Black malar mark is variable in width: (1) *Wide type.*—Wide black columnar shape with a small amount of tawny or white on the lower one-third of the auriculars. (2) *Moderately wide type.*—Moderately wide columnar shape, and often with a break in the dark mark at the gape region. There is a dark strip on the top one-third of the auriculars and the lower two-thirds are tawny or white. (3) *Narrow type.*—Narrow dark columnar shape, and very often with a break in the dark mark at the gape region with all of the auriculars being tawny or white. BODY.—Dorsal areas are dark brown. Tawny underparts are variably streaked with dark brown. In the fall, the pale-colored legs and tarsi are variable in color: bluish, grayish, greenish, or yellowish; however, they can be medium yellow. By mid- to late winter, legs and tarsi turn to medium or bright yellow. WINGS.—On perched birds, wingtips are somewhat shorter than (about 12 mm) or equal to the tail tip.

**JUVENILES OF "ARCTIC" (*F. p. tundrius*):** HEAD.—**Black malar mark may be either moderately wide type or occasionally the narrow type.** The black malar mark merges with the broad dark brown or black eyeline that runs along the very upper part of the auriculars and behind the eyes above the auriculars, and extends onto the nape. The lower two-thirds or one-half of the auriculars is pale tawny and unmarked. Broad pale tawny forehead. There is a large dark brown patch on the crown of the head, and the forward part extends ahead of the eyes and touches the orbital skin; most feathers are narrowly edged with pale tawny. A moderately wide pale tawny supercilium is situated between the brown crown patch and the broad dark brown or black eyeline that begins above the eyes and extends onto the nape. On the darker-headed birds, the crown is virtually all dark and the supercilium is short. Much of the nape is pale tawny and mottled with dark brown. BODY (dorsal).—The dark brown back and scapulars are moderately edged with pale tawny and create a moderately scalloped look. BODY (ventral).—Pale tawny breast, belly, and lower belly are narrowly streaked with dark brown. On the forward section, the flanks are streaked and on the distal section they are covered with large arrowhead marks with a broad bar on the basal region of each feather. The tawny leg feathers are covered with narrow dark brown streaks. The tawny undertail coverts marked with very narrow shaft streaks or are narrowly barred with dark brown. WINGS (dorsal).—Dark brown with moderately wide tawny scalloped edgings on all coverts. WINGS (ventral).—Dark gray remiges are marked with moderately wide pale tawny-rufous spots or bars. The tawny coverts are streaked and barred with dark brown. Overall appearance is a uniformly marked under wing. TAIL.—Dark brown or grayish on the dorsal surface with four or five narrow tawny-rufous bands and a broad white terminal band. The ventral surface is dark gray with several narrow tawny-rufous bands. *Note:* A very common type for either sex.

**JUVENILES OF "ARCTIC" (LIGHTLY MARKED/BLONDE TYPE *F. p. tundrius*):** HEAD.—The black malar mark is the narrow type. The auriculars are pale tawny and may have a narrow rufous-brown strip on the top edge. The moderately wide or narrow dark brown eyeline above the auriculars extends onto the nape. The forehead and crown are either uniformly pale tawny or there is a narrow brown region on the rear of the crown. BODY (dorsal).—The dark brown back and scapulars are broadly edged with pale tawny, creating a scalloped appearance. BODY (ventral).—The pale tawny underparts have very narrow dark brown streaking on the breast, belly, and lower belly. The flanks are narrowly streaked on the forward part and have small brown arrowhead-shaped markings on the distal part. Leg feathers are very narrowly streaked. The undertail coverts can have a thin bar on each feather or a narrow dark streak along each feather shaft. WINGS (dorsal).—Dark brown coverts are broadly edged with pale tawny. WINGS (ventral) and TAIL.—As described for Juveniles of Arctic. *Note:* A common type for either sex. Some may be the offspring of a pale type adult; however, since they are common, they may also be paler variants that occur throughout their range.

**JUVENILES OF "AMERICAN" (*F. p. anatum*):** HEAD.—**Malar mark is the wide type and blends with the dark brown cap; nape may be dark or have some pale patches.** Forehead is dark or has a narrow tawny patch. The tawny auriculars are unmarked. BODY (dorsal).—Dark brown with very narrow tawny-rufous tips on some scapular feathers or can be virtually all dark brown. BODY (ventral).—Rufous-tawny (more reddish than *tundrius* juveniles) with moderately wide dark brown streaking on the breast, belly, and lower belly. The flanks are marked as on "JUVENILES OF ARCTIC." Tawny leg feathers typically have moderately wide dark streaking, but can be narrow streaking. WINGS (dorsal).—Dark brown with very narrow tawny-rufous tips on the larger coverts. WINGS (ventral).—As on Juveniles of "Arctic." TAIL.—Dorsal surface either is unmarked dark brown or grayish brown or with partial, narrow rufous-tawny pale bands. *Note:* This type is found throughout this subspecies range in the West and on many released birds east of the Great Plains.

**JUVENILES OF "AMERICAN" (LIGHTLY MARKED TYPE *F. p. anatum*):** HEAD.—**Wide type or moderately wide type black malar mark. Both malar types connect with a broad black eyeline that is on the top half of the auriculars and behind the eyes above the auriculars.** The head is quite pale because the rear two-thirds or one-half of the crown is dark but the forward one-third or one-half and forehead are rufous-tawny. There is either a short pale tawny-rufous supercilium just above the eyes or the supercilium is wide and extends onto the nape. The nape is quite pale and mottled with dark brown. *Note:* Head can be as pale as on darker-headed *tundrius.* BODY (dorsal).—Dark brown with very narrow or narrow rufous-tawny edges on the lower part of the back and all scapular feathers. *Note:* Can be as lightly edged as on darker *tundrius.* BODY (ventral).—Rufous-tawny with narrow dark brown streaking on the breast, belly, and lower belly. *Note:* Similar to many *tundrius* but more reddish on the ventral base color. The forward flanks are streaked and the distal flanks are covered with large arrowhead-shaped markings and barring on the inner portion of each feather. The leg feathers are narrowly streaked. *Note:* Leg feathers are marked similarly to many *tundrius.* The undertail coverts are barred with dark brown. WINGS (dorsal).—Dark brown with narrow tawny-rufous edges on most coverts. WINGS (ventral).—Marked as described for Juvenile Traits. TAIL.—Dark brown or grayish brown on the dorsal surface with four or five narrow, pale rufous-tawny bands. Ventral surface is barred with narrow rufous-tawny bands.

*Note:* This type is regularly found throughout the Rocky Mts., particularly in arid regions. Some museum specimens of *anatum* from Arizona, Colorado, and Wyoming indicate that lightly marked types with pale heads existed historically long before there was any chance of possible intergrading from released stock (AMNH; C. M. White pers. comm.). This is a common type of *anatum* that has been released in e. Canada.

**JUVENILES OF "PEALE'S" (*F. p. pealei*):** Data based primarily on museum specimens and information from C. M. White (pers. comm.). Plumage varies considerably depending on geographic region. Overall plumage gets clinally darker to the north and west. There are three main types: (1) Heavily marked/Aleutian type generally is found on the Aleutian Islands, but sparingly found south of the Aleutians to Gulf of Alaska, and very sparingly to Washington. (2) Moderately marked/Queen Charlotte Islands type is found south of the Aleutian Islands and represents most birds in the Gulf of Alaska population and half of the population on the Queen Charlotte Islands. B.C. (3) Lightly marked/light Queen Charlotte Islands type make up one-fourth of the Queen Charlotte Island population and a few possibly north to the Gulf of Alaska. BODY (ventral).—All pale markings are pale yellowish olive (C. M. White pers. comm.) and lack the warmer tawny tone of *tundrius* or very warm reddish tone of *anatum.* For examples of the range of juvenile plumages, *see* Burnham (1997).

**JUVENILES OF "PEALE'S" (HEAVILY MARKED/ALEUTIAN TYPE *F. p. pealei*):** HEAD.—**Wide type of black malar mark, but is sometimes very wide type.** May have a small pale forehead or forehead is dark. Little if any pale markings on the head except possibly for a very small amount of pale mottling on the nape. Auriculars are heavily streaked with dark brown. BODY (dorsal).—Dark brown with perhaps a grayish tone (bloom) at times and lacks pale markings on the back and scapulars. BODY (ventral).—Very wide dark brown streaking. Underparts basically are dark brown with very narrow pale tawny edges on all breast, belly, flank, and leg feathers. At a distance, the underparts will appear virtually solid dark brown and are much darker than any other subspecies or variation of juvenile *pealei.* The undertail coverts are broadly marked with large dark brown arrowhead shapes. WINGS (dorsal).—All upperwing coverts are solid dark brown and lack pale edgings. WINGS (ventral).—Coverts are dark brown with a small amount of tawny and white spotting; median underwing coverts have more white spotting than the lesser underwing coverts. The underwing coverts are much darker than on all other races. **The underside of the dark gray remiges have smaller pale spotting and barring than on other races, and markings are whitish rather than pale pinkish tawny as on other races. The underwing appears somewhat two-toned, with dark coverts contrasting with paler gray remiges (uniformly marked on all other races).** TAIL.—Dorsal surface is dark brown or grayish brown and unmarked. The ventral surface either is unmarked or has a few faint, pale, partial tawny-rufous bands.

**JUVENILES OF "PEALE'S" (MODERATELY MARKED/QUEEN CHARLOTTE ISLANDS TYPE *F. p. pealei*):** HEAD.—**Wide type of black malar mark.** Auriculars are streaked with dark brown. Crown of head is dark brown except for a narrow white forehead. In fresh plumage, there may be some thin pale edges on the central crown feathers. There is a small tawny supercilium patch over the eyes and a small amount of pale tawny on the nape. BODY (dorsal).—Dark brown with perhaps a grayish tone (bloom) at times. In fresh plumage, there are very narrow tawny tips on the lower scapular feathers, but these wear off by late fall. BODY (ventral).—Wide dark brown streaking on all breast, belly, flanks, and leg feathers. The underparts are dark brown with narrow pale tawny edges on all feathers. *Note:* The pale feather edgings are a bit wider than on heavily marked type so they are not quite as dark; however, they are still quite dark on the ventral areas and still darker than on any other subspecies. WINGS (dorsal).—In fresh plumage, there is a small amount of thin pale edging on the larger upperwing coverts; however, these edgings wear off by late fall and the dorsal surface becomes uniformly dark brown. WINGS (ventral).—Dark brown underwing coverts have a small amount of whitish mottling and are darker than other races. **The underside of the remiges are marked as on heavily marked type juvenile and are darker than other races. The underwing appears somewhat two-toned, with the dark coverts contrasting with the paler gray remiges (uniformly marked on other races).** TAIL.—Dorsal surface may be brown or grayish and unmarked or may have partial, pale tawny-rufous bars. The ventral surface always has partial or nearly full tawny-rufous barring.

**JUVENILES OF "PEALE'S" (LIGHTLY MARKED/LIGHT QUEEN CHARLOTTE TYPE *F. p. pealei*):** HEAD.—**Black malar mark is the narrow type or moderate type.** The broad pale tawny or white forehead often extends part of the way up on the center of the crown and may connect with the front of the long, pale supercilium that extends onto the nape. Crown of head either is dark with narrow pale edgings on some feathers or is dark on the rear half. Auriculars may be unmarked or have a small amount of dark streaking. Nape is pale with some dark mottling. *Note:* Head can appear identical to a darker-headed juvenile *tundrius.* BODY (dorsal).—Dark brown and either like a darker *pealei* with minimal pale markings or distinctly tawny-scalloped on all feather edges and similar to *tundrius.* BODY (ventral).—Pale tawny or nearly pale olive-tawny with narrow dark brown streaking *Note:* Similarly streaked to an average *tundrius* and lightly marked type *anatum,* but ventral color is not as warm-colored as these two races. WINGS (dorsal).—Dark brown with minimal amount of pale feather edgings or may have wider edgings on the palest birds. WINGS (ventral).—Coverts are mottled and barred with tawny and white and remiges are spotted and barred as on most other races and appear uniformly marked underneath. TAIL.—Brown or grayish on the dorsal surface and often narrowly barred with tawny-rufous. Ventral surface is narrowly barred. *Note:* These are very pale types of *pealei* and mainly found in small numbers on the Queen Charlotte Islands, British Columbia, and sparingly north to the Gulf of Alaska. This type is nearly identical to many juvenile *tundrius.*

**JUVENILES OF "EASTERN" (no subspecies designation):** HEAD.—Black malar can vary: a narrow type, wide type, or very wide type. The crown of the head may be pale as in many *tundrius*; fairly dark as in many *anatum, pealei,* and *peregrinus*; or dark as in many *anatum* and in *brookei.* BODY.—Dorsal color and pattern vary from distinctly edged with tawny as in *tundrius* to darker as in many other races. Ventral region can vary from being lightly streaked as in *tundrius* to heavily streaked as in *pealei.*

**ABNORMAL PLUMAGES:** A few sight records of partial albinos with a few white feathers and sight records and captured juveniles in the cream-colored imperfect albino plumage (Clark and Wheeler 2001). *Note:* Very rare plumage types and are not depicted in accompanying photographs.

**HABITAT:** ***F. p. tundrius.* Summer.**—Low-elevation tundra biome of the Arctic. Breeding areas are along lakes, rivers, and sea coasts that provide embankments and cliffs for nesting sites. Peregrines are absent from montane and ice-pack regions, particularly interior areas of many Arctic islands. Climate is cool or cold and damp.

***F. p. tundrius.* Winter.**—In the U.S., found in mid-latitude and subtropical low elevations along coastal beaches, marshes, and tidal flats, and other riparian zones on coastal and southern interior locations. Urban areas and semi-open agricultural, grazed, and natural areas are also inhabited. South of the U.S., found in similar habitats but in a tropical environment. Occasionally found in montane areas, and have been seen at 14,000 ft. (4,300 m) in the Andes Mts. of South America. Cliffs are not required at this time of year but, if present, may be used for roosting. Climate is hot and dry or wet.

***F. p. tundrius.* Migration.**—Found throughout a wide variety of open to forested regions: coastal lowlands, plains, agricultural, and moderately high montane elevations. In interior locations, riparian habitat is extensively used because of high prey density of shorebirds and waterfowl. The largest numbers of falcons frequent the historical coastal migration location of Padre Island region of south Texas. The beaches, dunes, marshes, and tidal flats of these barrier islands and coastal zones offer high prey density for hungry migrant falcons. Climate is highly variable.

***F. p. anatum.* Summer.**—Natural areas are in rugged, semi-open and wooded, often montane regions with rocky cliffs, outcrops, and canyons that are at least 30 ft. (10 m) high, but may be well over 1,000 ft. (310m) high. Coastal habitat is on mainland and near-coastal islands and rock outcrops. Interior locations are adjacent to lakes, rivers, or streams. Sea level to montane elevations are inhabited. Montane breeding areas may be found up to nearly 10,000 ft. (3,100 m) in southern and central latitudes of the U.S., but are at lower elevations in the northern latitudes of Alaska and Canada. Urban areas are inhabited in some states and most provinces, either from released populations or natural colonization. Climate varies from cool to hot and wet or dry, depending on geographic area, elevation, and latitude.

***F. p. anatum.* Winter.**—In the contiguous U.S., many lower-elevation pairs remain in rugged breeding habitat. High-elevation and northern-latitude birds winter in southern montane regions with rugged terrain and cliffs or, as *tundrius,* in lowland habitat that often lack cliffs.

In w. Canada, released-era adults are wintering at more northern latitudes than were formerly known. An adult female has wintered in Winnipeg, Manitoba, for several years. In the winter of 2001–2002, an adult female overwintered in Edmonton, Alberta, the most northern wintering of *anatum* (formerly to Prince Rupert, B.C.).

***F. p. anatum.* Migration.**—Found in a variety of habitats, mainly between the Great Plains and Atlantic Coast. Individuals migrating south of the U.S. often pass through the historical Padre Island region of s. Texas with its beaches, dunes, and saltwater flats that are a haven to shorebirds.

***F. p. pealei.*** **All seasons.**—Coastal island marine habitat. South of the Aleutian Islands of Alaska, found on rugged, wooded coastal islands and headlands of coastal mainland with low to high cliffs. Cliffs heights in British Columbia range from 39 to 1,200 ft. (12–366 m); cliffs on islands in the Gulf of Alaska can be equally as high. On the treeless Aleutian Islands, they are in locations with similarly low to high cliffs. Climate in all areas is cool to cold and very wet.

***F. p. pealei.*** **Winter/migration.**—Similar to previously described breeding areas or on open or wooded coastal beaches and mainland headlands.

**"Eastern" Peregrine. All seasons.**—Predominantly year-round in moderate-sized and large urban centers with tall buildings, bridges, smokestacks, and other tall structures that provide ledges and cavities for nesting sites. Interior forested and semi-open areas along lakes and rivers with cliffs were historically used and are now used by a several pairs. Climate varies from hot to warm and is moderately wet or wet.

**"Eastern" Peregrine. Winter and migration.**—Similar to many *anatum* and some *tundrius.*

**HABITS:** Fairly tame to tame species. Solitary, but a few birds may loosely assemble in areas of high prey density during migration and winter. Peregrines frequently bathe and drink, and virtually all nest sites are near water. Peregrines are active at all times of the day, including dawn and dusk, and are sometimes nocturnal. Radio-tagged migrants have been documented flying at night.

As in all falcons, Peregrines regularly fluff the throat and forehead feathers when relaxed or in cool temperatures. In strong winds, the tail is braced against the perch for added stability.

Peregrines perch on any elevated natural or artificial structure and readily stand on the ground. In coastal areas, Peregrines will stand on sandy beaches, on debris, or on dunes.

A playful species; even adults have been seen picking up and dropping and catching plastic objects. Peregrines sometimes engage in vicious, often deadly battles for nesting territories.

**FEEDING:** A aerial and perch hunter; rarely engages in pirating, scavenging, or terrestrial hunting. Peregrines typically prey on a few select species within regional and local areas. Particularly with migrants, diet may vary seasonally. Being crepuscular, Peregrines regularly hunt during early morning and late evening. Avian prey is mainly captured in flight, but are also captured while on the ground or water.

Females generally hunt larger prey than do males. Peregrines mainly feed on birds ranging from small passerines to mid-size waterfowl. Unusually large avian prey such as cormorants, geese, and large herons are very rarely taken by females. Breeding pairs of Peregrine Falcons also cooperatively hunt. In mid- and lower latitudes, Peregrines feed on bats. Large flying insects, captured in flight, are mainly eaten by juveniles.

Prey of all sizes are typically grabbed and held by Peregrines. Small prey are carried to a safe eating location. Prey that is larger than a small duck (e.g., teal-size) is too heavy for even female Peregrines to carry. Large prey is latched onto and the Peregrine glides to the ground and begins to feed. Rarely, large or small prey are hit at high speeds and raked by the rear talon of a clenched foot; this attack injures or kills prey, which drops to the ground or water. The falcon then flies down and lands on the prey or picks it off the surface of the water or ground and takes it to another location to eat it. Avian prey evading capture will often dive to the ground, into dense vegetation, or onto or into water. Peregrines may veer around and try to snatch such prey before they can gain flight.

*F. p. pealei* typically hit prey and let it drop into the water or hit prey that is on the water.

Prey species use evasive techniques to escape capture. Some passerines and shorebirds form tight flocks, which deter Peregrines.

Peregrines immediately break the neck of captured prey. Neck-breaking occurs in flight or on the ground or perch. Prey is partially or fully plucked before eating. On large prey, often only the meat on the neck and breast is eaten. On small prey, the wings are torn off before the body is eaten; on large birds, the wings are left intact. Adults pluck and decapitate prey before bringing it back to feed nestlings. The head and wings are left intact on medium-sized and large avian prey at all times. During other seasons, large and some small prey are eaten at the point of capture; small prey may also be eaten while the Peregrine is flying or carried to another, often elevated location to be eaten. Prey that is eaten in flight is held by one foot when the falcon is gliding, kiting, or soaring. Falcons migrating over the ocean may use tall structures on ships as feeding and resting posts.

*Aerial hunting.*—Quarry is often targeted from long distances with the falcons engaging in moderate to long pursuits. Angled or vertical dives are initiated while gliding, kiting, or soaring at low to high altitudes. Prey is also captured by being tail-chased or intercepted on level flight from the side or front angles. Short dives are used when the previous methods fail. Peregrines will tail-chase, intercept from the side or front, or angle up underneath intended prey from low-level flight. Peregrines regularly use low-altitude, high-speed, surprise-and-flush forays.

*Perch hunting.*—From high perches, falcons engage in a high-speed, shallow-angled dive toward prey. From low perches, as on beaches, Peregrines will tail-chase, intercept from various angles, use short dives, or angle up underneath intended prey from low-level flight.

*Pirating.*—A biologist in Oregon has seen Peregrines pirate mammalian prey from Red-tailed Hawks and fish from Osprey. Peregrines have also been seen pirating prey from Merlin and Sharp-shinned Hawks and also can kill the smaller raptor.

*Scavenging.*—Juveniles and rarely adults eat carrion, primarily in winter, if live prey is difficult to capture.

*Terrestrial hunting.*—*F. p. anatum* in the West occasionally pursues prey on foot into dense vegetation, and *F. p. tundrius* walks or hops on the ground on the tundra, capturing voles, lemmings, and young birds. Young *anatum* may also feed on insects on the ground. "Eastern" juveniles will walk into pigeon traps in order to capture pigeons.

Peregrines are also robbed of prey by Red-tailed Hawks, Swainson's Hawks, Northern Harriers, and both eagle species.

*F. p. tundrius.*—The breeding season coincides with the nesting season of Arctic songbirds. In many regions, recently fledged, inexperienced and poorly flying Arctic songbirds make up an extensive part of this race's prey during the nesting season. The smaller, highly agile adult male Peregrines do all or most of the hunting during the early to mid-part of the nesting cycle, concentrating heavily on the small but common and easy-to-catch young songbirds.

At Rankin Inlet, Nunavut, Snow Buntings, Horned Larks, Lapland Longspurs, and American Pipits compose most of the Peregrine's diet. Small shorebirds, but especially Semipalmated Plovers and Semipalmated Sandpipers form a lesser but still substantial amount of their diet. Larger birds such as Rock Ptarmigan are regionally and seasonally important. Northern species of ducks, such as adult and young Northern Pintail and Long-tailed Ducks (Oldsquaw) and young Common and King eiders are also preyed upon. On coastal locations, Black Guillemots are hunted. Less common avian prey are Arctic Terns and Long-tailed Jaegers; gulls are also preyed upon to a lesser extent. Brown and Collared lemmings normally form a small part of the Peregrine's diet, but larger numbers are hunted when lemming numbers cyclically increase.

In the high Arctic, falcons also feed extensively on songbirds, which include the previously listed species, but also Northern Wheatear and Common and Hoary redpolls. Duck prey are mainly Long-tailed Duck and the two eider species. Alcids are important prey for coastal pairs. Gulls form a small part of their diet.

Shorebirds and songbirds form an important dietary component during fall and spring migration, and falcon movements coincide with prey movements. The historical fall migratory area on Padre Island, Texas hosts large numbers of shorebirds in fall and spring. In the spring, a smaller but still important staging area at Cheyenne Bottoms Wildlife Area in Barton Co., Kan., likewise coincides with peak shorebird migration.

Wintering Peregrines also feed extensively on North American shorebirds, songbirds, and bats. However, depending on geographic region, prey may encompass a variety of tropical bird species and bats.

*F. p. anatum.*—Diet is highly variable depending on geographic region and season. In cliff settings, since males do much of the hunting, they often specialize in capturing small, agile, but abundant swallows and swifts. However, they may also regularly prey on other passerines, waterfowl, and bats. In urban areas, they may feed extensively on House Sparrows, Rock Doves, and European Starlings; also on bats. Studies done of urban pairs in Edmonton, Alberta, show that nonurban type avian prey, such as shorebirds, grebes, gulls, terns, and passerines, form an important part of their diet.

During migration and winter, North American shorebirds, small- and mid-sized waterfowl, and songbirds become important prey items. In winter, tropical bird species and bats also become important food items.

*F. p. pealei.*—During the breeding season, *pealei* in all parts of their range feed nearly exclusively on alcids. Hunting is generally done in the immediate area of the eyrie and their diet is composed of alcid species that are common in a particular locale. In all parts of *pealei* range, Cassin's Auklet, Marbled Murrelet, Pigeon Guillemot, and Rhinoceros Auklet may be hunted. On the Queen Charlotte Islands, Gulf of Alaska, and eastern Aleutian Islands, Ancient Murrelets may be eaten. Kittlitz's Murrelets and Parakeet Auklets are major prey in the Gulf of Alaska and the Aleutian Islands. Crested, Least, and Whiskered auklets are resident in *pealei* range only on the Aleutian Islands, and they form the falcon's primary prey in this region. Large alcids, such as Horned and Tufted puffins, are taken in all of *pealei*'s range.

*F. p. pealei* that move out of breeding and/or natal areas feed extensively on shorebirds as well as alcids. This race, possibly more so than other races, is known for feeding on avian carrion that is washed up on shore.

"Eastern" Peregrine.—Feed on similar prey that the former eastern *F. p. anatum* did. However, in certain regions, the former Peregrine race probably fed extensively on Passenger Pigeons until they became rare in the late 1800s and extinct in the early 1900s. Urban Peregrines rely year-round on the abundant supply of Rock Doves, European Starlings, and House Sparrows. In many urban areas, and in rural and remote natural areas, Peregrines may feed on the previously listed urban species, but also capture Blue Jays, Common Snipe, Eastern and Western meadowlarks, Mourning Doves, Northern Flickers, Red-winged Blackbirds, several swallow species, and various waterbirds. Except for bats, mammals are rarely preyed upon.

Migrant and southern wintering "Eastern" Peregrines feed extensively on North American shorebirds and songbirds and bats, and on a variety of tropical species of birds and bats.

**FLIGHT:** Powered flight is an irregular sequence of powerful, moderately deep wingbeats. Peregrines may flap for considerable distances before gliding or alternate short sequences of flapping and gliding. Level flight with steadily beating wings has been documented to 75 mph. (121 km/hr), but more typically at 45–60 mph (72–97 km/hr) with intermittent flapping and gliding. Wings are held on a flat plane when gliding and soaring. Peregrines kite in strong winds and rarely hover.

Hunting flights: *Long-dive.*—High-altitude high-speed long, angled or vertical descent, with the wings partially or fully closed. The faster the dive, the more closed-winged and streamlined a Peregrine becomes. Vertical diving speeds of a juvenile female *F. p. anatum* used for falconry have been verified as exceeding 200 mph (322 km/hr) by sky-jumpers. It is believed that some birds may easily achieve 250 mph. (402 km/h) and possibly 300 mph (483 km/h), a speed that can be attained by some sky-jumpers. (A Gyrfalcon used for falconry has been clocked at 353 mph [568 km/hr].) Long dives are initiated while perching, soaring, or kiting. If soaring, they rise up to a high altitude above the intended prey, then proceed to make a long dive. *Short-dive.*—Often used when a long-dive was unsuccessful. The Peregrine swings upward above the intended prey and makes one or more short dives to intercept it. *Tail-chase.*—Direct, high-speed flight to overtake aerial prey at low or moderate altitudes. *Underside grab.*—May be initiated from a short-dive or tail-chase in which the falcon swings upward and grabs prey from underneath.

**VOICE:** Agitated vocalization at nest sites is a rapid, repeated, harsh *cack, cack, cack.* Vocalizations are rarely heard away from the breeding grounds. Courting birds utter several different sounds: (1) *eechip* (both sexes), (2) staccato chittering (both sexes), (3) whinning or begging *waik* (both sexes), (4) staccato chatter or chutter (males), (5) rapid *chips* (males), (6) *up-chip* (females), (7) *chup-chip* (females), and

(8) constant *kree, kree, kree* whine (females and young food-begging).

**STATUS AND DISTRIBUTION: *F. p. tundrius.***—*Very uncommon.* Their population was somewhat affected by the DDT era, but not as much as DDT affected *F. p. anatum.* This subspecies winters in areas of Central and South America where DDT and other organochlorine pesticides are still used. Numbers slowly increased over the last few decades and are possibly close to historical numbers.

***F. p. tundrius.* Summer.**—*Very uncommon.* Sporadically distributed in the Arctic. In Nunavut and e. Northwest Territories, found north of 60°N. In w. Northwest Territories, Yukon, and Alaska, primarily inhabits areas north of 68°N. There is a small population on the south side of the Seward Peninsula, Alaska, that is at a slightly lower latitude. At about 74°N, Banks Island of the Northwest Territories, is the northernmost inhabited location in the West. Permanent ice fields and lack of prey in extensive rocky areas prevent Peregrines from occupying many coastal and especially interior regions of many islands of the high Arctic. One of the highest breeding densities, and nearly equal to that of the high breeding density of *F. p. pealei* on the Queen Charlotte Is., B.C., is around Rankin Inlet, Nunavut.

***F. p. tundrius.* Winter.**—Very expansive winter range. In the w. U.S., mainly in s. California; however, *tundrius* adults are sometimes seen along the coast of Washington. Major winter range extends south along Baja California coasts and from n. coastal Sonora and southward throughout Mexico, all of the Caribbean, and the northern two-thirds of South America. Southernmost birds are found in n. Argentina, n. Chile, and Uruguay. There is a rare winter record of a juvenile banded at Rankin Inlet, Nunavut, that was found in Simpson Co., Ky. (in Dec. 1983). Winter density in all areas is very low.

***F. p. tundrius.* Movements.**—Highly migratory and engages in the longest migration of any North American raptor. Movements often entail long over-water crossings. According to telemetry data, some may migrate nocturnally. Telemetry data also illustrate that few birds make prolonged stops to feed and rest.

*Fall migration:* All birds must leave the Arctic prior to severe weather when prey becomes scarce or is not available. Adults may leave nesting grounds in the Arctic by mid- to late Aug. as soon as their young are independent. At Rankin Inlet, Nunavut, the last birds leave breeding areas by late Sep.; those of the high Arctic probably leave earlier. Adults form the bulk of the early migrants and juveniles the late migrants. The overall peak movement consists of both ages but with a large percentage of the adult population. Recently fledged juveniles may not leave natal areas until late Aug. to late Sep. Except on Padre Island, Tex., there is little noticeable movement across most of the West. Migration period in the West, as seen at Padre Island, mirrors that of the well-defined movements along the East Coast. Peak migration period occurs in early Oct. for much of the cen. and s. U.S. Numbers drop substantially after mid-Oct. with only stragglers by late Oct. In Veracruz, Mexico, the first migrants are seen in late Aug. and peak in early Oct.; the last are seen in mid-Nov.

Movement out of nesting grounds of n. Alaska and n. Canada spreads south in a broad band into s. Canada and the U.S. However, once in the U.S., most birds funnel south toward coastal Texas and through Padre Island. There is some movement along the Pacific Coast. Even birds nesting in the eastern high Arctic and Greenland may angle southwest and eventually through Padre Island. Alaskan birds have been documented taking a southeasterly diagonal transcontinental path in the fall and end up on Assateague Island, Md./Va., and on the Florida Keys. Winter grounds may be reached from mid-Sep. to Dec.

*Spring migration:* The largest number of spring migrants in North America are seen on Padre Island, Tex. Adults utilize the tidal flats and beaches, feeding on shorebirds, from Apr. to early May; juveniles mainly during May. From Padre Island, birds fan out northwards to nesting grounds from Alaska to Greenland. Adults arrive on their southernmost nesting grounds at Rankin Inlet, Nunavut by May 20, but may arrive as early as May 10. Those breeding in high-Arctic regions do not arrive at nesting areas until very late May and into Jun.

***F. p. anatum.* Summer/all seasons.**—*Rare.* The western population of *F. p. anatum* was greatly reduced and, in many areas, nearly extirpated during the organochlorine era of the late 1940s to early 1970s. Formerly, there was a small population of *anatum* that nested from Louisiana west to Kansas and north to Minnesota. Many were tree-nesting pairs that, at the time, nested high up in tall old-growth cottonwoods, cypress, and sycamore trees that had cavities or broken-off stumps. The largest number nested along the bluffs and cliffs on the Mississippi River, especially in n. Iowa and s. Minnesota. The last eastern *anatum* that nested east of the Rocky Mts. in the U.S. did so in 1967 on cliffs along the Mississippi River near Lansing, Iowa. In 1942, Louisiana had the last historical tree nesting in the w. U.S. near Tallulah, Madison Parish.

The remnant wild population was augmented by released juveniles from 1975 through the early 1990s in w. Canada and in the contiguous w. U.S. west of the Great Plains. Populations in all regions continue to steadily grow. The first nesting success of release-era Peregrines in the West occurred in 1977 in n. Alberta of a female that was fostered into one of the last nesting pairs of original *anatum*.

CANADA: All figures are for number of pairs in 2000 unless otherwise noted. Reproductive success is variable in the northern latitudes but has been especially poor in the last few years. *Alberta:* 45–55 pairs in 2001. There was one remaining pair in 1970. Nests in and around Calgary and Edmonton, Lake Athabasca, along the Red Deer River, and Wood Buffalo N.P. *British Columbia:* 17 pairs. Most are in southern and central parts of the province, including the Frazier River lowlands and se. Vancouver Island; also on the gulf islands. *Manitoba:* three pairs. Has nested or summered in Brandon, Gimli, Portage la Prairie, and Winnipeg. Iowa-reared bird of mixed lineage dispersed to Winnipeg and has nested. *Northwest Territories:* 77 pairs in the Mackenzie River Valley; this region is an intergrade zone with *F. p. tundrius. Nunavut:* Only *F. p. tundrius* breeds in the territory. *Saskatchewan:* four pairs breed in Regina and Saskatoon. All are from released stock. *Yukon:* 19 pairs on the Peel River, 26 pairs on the Porcupine River, and 43 pairs on the Yukon River. Reproductive success was low in 2000 and 2001.

*Note:* A few Canadian-released *anatum* migrated south and breed with mixed lineage birds from the U.S. in the n. Midwest.

UNITED STATES: Data are for known pairs as of 1998. More current data, if available are given in parentheses. *Alaska:* 301 pairs. Widespread south of the Brooks Range and below tundra elevation. Mainly found along major boreal forest drainages such as the Kuskokwim, Porcupine, Tanana, and Yukon Rivers. Based on juveniles, some birds on the Tanana and Yukon Rivers were formerly classic *anatum* in the pre-DDT era, but they are now are pale *tundrius*-like due to colonization by *tundrius* when *anatum* numbers dropped and territories were vacated during the pesticide era (*see* White and Kiff 1998). *Arizona:* 159 pairs (considered to have more than 200 pairs). This state has the highest number of breeding pairs in the U.S. Mainly resident throughout the state in canyon and montane areas, including "sky-island" mountains. The highest density and probably highest density for the subspecies is in the Grand Canyon. *California:* 167 pairs. Precipitous decline from 300 pairs before 1950 to two pairs by 1970. Resident along the northern and central coast, Cascade Range, Coast Range Mts., much of the Sierra Nevada, sporadically distributed in southern coastal areas and islands, and in cities (Los Angeles, Long Beach [6 pairs], San Diego). Population is increasing annually. Coastal and some interior pairs, however, are reproducing poorly due to chemical contamination. *Colorado:* 89 pairs (119 in 2001; 90 producing young). Population dropped to its lowest point in 1980 with four known pairs; rebounded thereafter, especially with reintroduction efforts. Found in most counties from the Front Range west to the Utah border. Increasing annually. Urban Peregrines, such as in Denver, did not do well because of high mortality. *Idaho:* 17 pairs. *Louisiana:* 0. The last pair of original *F. p. anatum* nested in a tree near Tallulah, Madison Parish, in 1942. *Montana:* 18 pairs. A 1973–1975 survey found only 23 known or suspected pairs. Only one territory was occupied from 1975 to 1977, and it did not produce young. *Nevada:* seven pairs. Breeds in the Highland, Jarbridge, Mormon, Ruby, and Spring Mts. A pair has also nested atop a hotel in Las Vegas. *New Mexico:* 32 pairs. Breeds in Catron, Dona Ana, Guadalupe, Hidalgo, and Lincoln Cos. *North Dakota:* historical cliff nesting near Medora in 1954. See "Eastern." *Ore-*

*gon:* 51 pairs. Resident along the northwest and southwest coasts, in the Cascade Mts., Hells Canyon along the Idaho border, Portland, east of Portland along the Columbia River, and in the Wallowa Mts. *Texas:* 11 pairs. Resident in Guadalupe Mts. N.P. and Big Bend N.P. Also found in adjacent canyon and montane areas in the Chisos Mts., and Santa Elena, Mariscal, and Boquillas Canyons, and smaller nearby canyons. *Utah:* 164 pairs. Resident in Salt Lake City and with a large number of pairs in the rugged canyonlands and montane regions of the southern part of the state. *Washington:* 45 (72 in 2001, but about 22 are in *pealei* range). Only four pairs remained in 1980. Released birds breed in Seattle, Spokane, and Tacoma. Also breeds in the Cascade Mts. and in several locations east of the Cascades. Highest density is in the San Juan Islands and Puget Sound. The population on the San Juan Islands may intergrade with *F. p. pealei*, but are mainly *anatum*; many are possible falconer escapees. Those breeding on the northwest coast are also intergrades with *pealei*, but are mainly *pealei*. *Wyoming:* 42 pairs (65 territories; not all used annually; increasing). Breeds near Devil's Tower, Flaming Gorge Reservoir, Glendo Reservoir, Teton Mts., Wind River Range, and in Yellowstone N.P.

MEXICO: *Baja California:* 42 pairs. Resident in San Pedro de Mártir N.P., Scammon's Lagoon, Ojo de Liebre Lagoon, Cedro Island, San Benitos Island, and San Roque Island. *Sonora:* Resident on the western coast, along the Rio Aros and Rio Yaqui, at Ajos-Bavispe Area de Proteccion de Los Recuros Naturales, Cajon del Diablo, and in the southeastern corner of the state. *Chihuahua:* A few eyries along the Sierra Madre Occidental and a few along the Rio Bravo and Rio Conchos near Texas. *Coahuila:* Sierra Carmen, Parque Nacional del Rio Bravo, and Sierra El Fueste. *Nuevo León:* Parque Nacional Cumbres de Monterey and mountains near Monterrey. *Durango:* scattered locations in the Sierra Madre Occidental in the northern, southwestern, and southern parts of the state. *Zacatecas:* Mountain and high plateau regions of the northwestern, northeastern, and south-central parts of the state. *Aguascalientes:* entire northern and eastern parts of the state. *Jalisco:* eastern part of the state. *Guanajuato:* resident along the northern, western, and southern borders of the state. *Michoachán:* Resident along the north-central border of the state. *Querétaro:* Resident along the joint border with Hidalgo and Estado de México. *Hidalgo:* Parque Nacional Los Marmoles. *Estado de México:* resident in the numerous volcanic mountains and national parks. *Puebla:* in the mountains south of the city of Puebla. *Guerrero:* in the national parks near Taxco.

***F. p. anatum*. Winter.**—Winter density is low. Winters on the West Coast from sw. British Columbia and south. Although most breeding areas in the w. U.S. are shown as purple for all-season occupancy, birds breeding in high montane regions depart breeding areas in the winter. Absent in winter from much of the Great Basin, cen. Rocky Mts., the Great Plains, and interior Canada. Winter range extends south of mapped area to n. Argentina. Based on banding data, *F. p. anatum* that breed in Alaska may winter as far south in South America as does *F. p. tundrius*. Peregrines released in the Rocky Mts. of the U.S. have wintered from n. and w. Mexico south to Panama; one went to s. California from Idaho. Telemetry-tracked adults from Edmonton and Wood Buffalo N.P., Alberta, wintered in the states of Sinaloa, Veracruz, and Yucatán in Mexico, and one male wintered in e. Brazil. One adult female tracked by telemetry from Edmonton, Alberta, wintered near Mazatlán, Sinaloa, for two consecutive winters. In pre-DDT era, wintered as far north as Prince Rupert, BC. Southern wintering grounds may be reached from late Sep. to mid-Nov.

***F. p. anatum*. Movements.**—Sedentary or a short- to long-distance migrant. All Alaskan and virtually all Canadian birds migrate. Long over-water crossings are sometimes made over the Gulf of Mexico and island-hopping across the Caribbean. Telemetry shows that birds may take less-than-straight paths to and from winter grounds and often curve around the Great Plains. Movements are typically swift with little tendency to stage to rest and feed. Northern populations may leapfrog over southern breeding populations and winter farther south.

*Fall migration:* All ages may disperse short distances in any direction prior to actually migrating south. Movement may be rapid or slow for portions of the journey and sometimes quite rapid for the duration. Based on telemetry data from the Canadian Wildlife Service (Environment Canada) on a few adults from Edmonton and Wood Buffalo N.P., Alberta, they may leave Edmonton as late as early Oct. and Wood Buffalo area from late Aug. to mid-Sep. Departure time for juveniles is probably similar. Alberta adults either arch southeast and skirt the Great Plains or angle west of the Rocky Mts. to also bypass the Plains. Birds that stay east of the Plains usually angle south into e. and s. Texas, e. Mexico, then through Central America and possibly into South America.

A female that crossed to the west side of the Rockies later angled southwest into nw. New Mexico, then south to Sinaloa, Mexico. She flew from Edmonton, Alberta, to near Mazatlán, Sinaloa, in less than 12 days, a journey of 2,170 miles (3,500 km). One female angled sharply southeast from ne. Alberta to s. Florida, then crossed onto islands of the Caribbean, but was thwarted twice from getting to n. South America by storms and perished in Hurricane Mitch while attempting to make her second over-water crossing. Banding data also shows many juveniles from e. Alaska making a transcontinental journey to the mid-Atlantic Coast and Florida, then probably crossing to the Caribbean islands or onto South America for winter.

*Spring migration:* Adults migrate before juveniles. Movement is typically rapid and direct. The Edmonton adult female made the journey from Mazatlán to Edmonton in less than nine days. The same route that was taken in the fall may be retraced again in the spring or a different route may be taken. Telemetry shows adults leaving winter grounds from early Mar. to early Apr. Juveniles leave in April. Juveniles are seen at hawkwatch sites in New Mexico in late Apr. and early May. As with *tundrius* race, a large number of *anatum* that winter in e. Mexico and Central and South America pass through Padre Island, Tex., in spring before continuing north and fanning out to breeding areas.

***F. p. pealei.*** **Summer/all seasons.**—*Uncommon.* Resident on the Commander and possibly Kuril Islands and Kamchatka region of e. Russia. Resident on the Aleutian Islands of Alaska, islands of the Gulf of Alaska, and southward mainly on the Queen Charlotte Islands, British Columbia. Lesser numbers found on gulf islands and mainland headlands on and around Vancouver Island. A few may interbreed with *anatum* on the San Juan Islands of Washington. A natural population on the Olympic Peninsula is rebounding from the pesticide era; some are intergrades with *anatum.*

CANADA: Data are for known pairs in 2000. Queen Charlotte Islands (including Langara I.): 63, has the highest breeding density of Peregrines in the world; nw. Vancouver Island and Scott Island: 21. There are several additional pairs on the central coastal headlands and islands of British Columbia.

UNITED STATES: Alaska: 600 pairs on the Aleutians and coastal areas that are deemed stable. Oregon: one bird paired with an *F. p. anatum* on the central coast in 2001 (data from J. Pagel). Washington: 17 in 1998 (about 22 in 2001; 31% of the 72 known pairs in the state are in *pealei* habitat regions). *F. p. pealei* may intergrade with *anatum* on the San Juan Islands.

***F. p. pealei.*** **Winter.**—Aleutian population are sedentary. The Gulf of Alaska population may be sedentary or some may move south. Some Queen Charlotte Island birds are known to winter south at least to sw. coastal Washington and probably farther south. An unknown number of all ages winter along the Pacific Coast of Oregon and California to at least San Diego, Calif., and possibly south along Baja California and coastal Sinaloa.

Data from D. Varland: An adult female banded in Mar. 1998 on the coast of sw. Washington bred that year on Langara Island, on the northern end of the Queen Charlotte Islands, B.C. She was observed the following winter in the same area of sw. Washington and in spring as breeder, again, on Langara Island. Another adult female from an unknown breeding location has also wintered in the same location on coastal sw. Washington for 6 years (as of winter of 2001–2002). The southwestern coast of Washington is a prime wintering area.

***F. p. pealei.*** **Movements.**—This subspecies was formerly considered sedentary. However, south of the Aleutian Islands, an unknown number of both ages regularly migrate south of breeding and natal areas for the winter. Movements are difficult to assess because migrants and wintering birds occur in prime coastal habitat at the same time.

*Fall migration:* Based on banding data from coastal sw. Washington, both ages may migrate through or begin their winter stint in this region by early Sep. Larger numbers occur from late Sep. through Oct. Movements may extend through Nov. and possibly later.

*Spring migration:* Information is based on banding data from sw. Washington. Breeding adults possibly move from late Feb. through Apr. with Mar. being the peak month. One-year-olds and nonbreeding adults often linger on wintering grounds on coastal beaches until late May and occasionally linger through Aug. Also, some individuals may wander throughout summer.

**"Eastern." Summer/all seasons in the U.S.**—*Rare.* This is a restoration population with an introduced lineage of Peregrines that replaced the extirpated eastern *F. p. anatum* that formerly nested east of the Great Plains in the U.S. The population east of the Rocky Mts. in the U.S. is from released birds that were part of an intense captive-release program that began in 1982 in the Midwest. The population east of the Rocky Mts. in Canada is *anatum* because Canada released only *anatum* subspecies (*see F. p. anatum.* Summer). However, Peregrines do not have international borders and there has been some genetic mixing in s. Manitoba (Winnipeg) and w. and cen. Ontario (along Lake Superior) of mixed-lineage Midwestern-released birds. Likewise, some Ontario-released *F. p. anatum* have migrated south into the n. Midwest.

UNITED STATES: Known pairs for 1998 are given, more current data, if known, are in parentheses. *Arkansas:* 0 (1 pair). Five were hacked on the White River, Independence Co., in 1993; three survived. Six were hacked in Little Rock in 1994. Only known pair is at Bayou Meto WMA, and is the only known pair in the U.S. to currently nest in a tree. *Iowa:* 2 pairs (5 in 2002). Introduced Peregrines breed in urban settings in Cedar Rapids (first nested in 1992), Davenport (first nested in 2002), Des Moines (first nested in 1992; failed in 2002), Lansing (on cliffs), and Louisa Power Plant in Louisa Co. (in a box on a smokestack). Falcons have also been released in Dubuque, Effigy Mounds National Monument in Allamakee Co., and Mason City. *Kansas:* 1 pair (2 in 2002). Introduced in Kansas City and Topeka. *Louisiana:* 0. *Minnesota:* 24 (23 in 1999). Intense release program produced urban nesters and pairs nesting on tall smokestacks; some pairs nest on natural cliffs. Nests or territorial pairs are in Bayport, Beaver Bay, Becker, Bloomington, Cohasset, Dakota Co. (Fort Snelling S.P.), Dresbach, Duluth (Bong Bridge), Eagan, Finn Church, near Grand Portage, Hastings, Hibbing, Lake City, Lake Co. (Beaver Bay, Kennedy Creek Cliff, Split Rock S.P., Tettegouche S.P.), Little Marais, Mendota, Minneapolis, Monticello, Prairie Island, Red Wing, Rochester, Sartell, St. Cloud, St. Paul, Tofte, and Winona Co. (John Latch S.P.). *Note:* Some Canadian-released *F. p. anatum* have immigrated to Minnesota and bred. *Missouri:* 4 pairs. Nested in Kansas City in 1997. Released in St. Louis, with breeding since 1991. Released in Springfield. Current pairs in Clayton, Kansas City, Springfield, and St. Louis. *Nebraska:* 1 pair. Released in Lincoln and Omaha. Has nested in Omaha. *North Dakota:* 0 (one pair). Pair in Fargo in 1999 and 2000 but did not nest; nested in 2001 and 2002. *Oklahoma:* 0. A pair was present in Tulsa in 1991 but did not nest. Juveniles have been seen in recent summers at Wichita Mts. NWR in Commanche Co. *South Dakota:* 0. Four were released in Sioux Falls in 1997.

**"Eastern." Winter.**—Most Minnesota birds leave the breeding grounds. However, a few breeding birds winter in and around Minneapolis/St. Paul, and Duluth often has a hardy wintering bird. Adults are often present year-round in urban breeding locations in Iowa, Missouri, and Nebraska. The extent of the winter range is unknown. It may extend, as it does for *anatum*, to cen. Brazil or even Uruguay.

**"Eastern." Movements.**—Little data. Some adults are sedentary. However, much of their breeding range is in areas with very severe winters, and birds are forced to move elsewhere to survive. Some birds may move only short distances and others may move long distances.

Released Peregrines do as their name implies: wander. Released birds in Iowa have dispersed to e. Kansas, s. Manitoba, Minnesota, and e. Nebraska. Released birds in n. Michigan and n. Minnesota have dispersed to s. Ontario and bred. Likewise, Ontario-released *F. p. anatum* have moved south and bred in the previous two states.

*Fall migration:* Little data. As with other mid-latitude Peregrines, they undoubtedly wander a great deal in late summer and early fall prior to any southward movements. Actual southward movements may begin, as it does for many eastern released birds, in Sep. or Oct. Some may not move far from nesting areas.

*Spring migration:* Little data. Even in northern areas such as Minnesota, many adults arrive at nest sites by early Mar. Movements of early nesters takes place during Feb. Those begin nesting activities in late Mar. or early Apr. migrate as do many *anatum*, during Mar. Juveniles migrate later in Apr. and May.

**NESTING:** *F. p. tundrius* begins nesting activities from late May through Jun. and ends from Aug. to late Sep. At Rankin Inlet, Nunavut, either sex may be the first to arrive on the breeding grounds. Pairs may arrive nearly simultaneously but there is no evidence that previously mated pairs arrive at the same time. Pair formation occurs quickly once on the breeding grounds. At Rankin Inlet, youngsters may fledge from mid-Aug. to the end of the first week of Sep. Late nesting birds of the high Arctic may not fledge young until mid- or late Sep.

*F. p. anatum* in Canada may begin nesting in Mar. and Apr., and possibly in May at their northern range limits, and end nesting activities from Jun. to Aug., occasionally later. Failed-nesters remain on territory until migration.

*F. p. pealei* begins nesting in Mar. and Apr. and ends in Jun. and Jul. Young may fledge by early Jun.

Depending on latitude, "Eastern" types may begin nesting from Feb. through Apr. and end from May to Jul., occasionally later. Pairs may remain in the general vicinity of their territory year-round, or with dispersing and migrating birds, males often establish territories.

Breeding typically does not occur until females are 2 years old and males are 3–5 years old; however, 1-year-old juvenile females regularly pair with older males and occasionally mate and rear young. Very rarely does a juvenile male pair with an older female, and it is extremely rare for them to mate and rear young.

**Courtship (perched).**—As with larger falcons, Peregrines engage in perched courting displays more than other raptors do. Courting activities are often more animated and vigorous than used by Gyrfalcons and Prairie Falcons. Both sexes have ledge displays with *eechip* calls: *male ledge display, female ledge display,* and *mutual ledge display.* Within these ledge displays, both sexes perform *billing, food transfer, horizontal head-low bow, vertical head-low bow,* and *tip-toe-walk.* In the *head-low bows,* the head is vigorously bobbed. Copulation displays involve the *curved-neck display* by males (male silent); females utter *eechip* and *upchip* and males the *chutter* call notes prior to copulation. Both sexes, but males, in particular, advertise territories with *high-perching display.*

**Courtship (flight).**—Both sexes perform *aerial food transfer, aerial-kissing, high-circling, mutual-floating display, passing-and-leading display,* and *talon-grappling.* Males perform *figure eights, flash roll, roll,* and *undulating roll* (variation of *sky-dancing*), and *z-flight.* Females perform the *flutter-flight.* Vocalization occurs with all but the most intense courting flights by males.

**Nest locations.**—*F. p. tundrius.*—Terrain varies from gentle, open slopes to low embankments, low or high rock outcrop, or tall sheer cliffs.

*F. p. anatum* and "Eastern."—Low cliffs that are at least 30 ft. (9 m) high to high cliffs that are several thousand feet high on coastal headlands and interior areas; large bridges; and tall buildings, smokestacks, and towers. Along northwest coastal areas from California to British Columbia, *anatum* also nest on "sea stacks," large rocky pinnacles protruding out of the water. Less common eyrie locations for *anatum* in Baja California are on channel markers, ground, old Osprey nests, and wrecked ships. In California, they have nested on top of a 55-gallon barrel.

*F. p. pealei.*—Very low cliffs that are at least 15 ft. (4.6 m) high to high cliffs that are up to 1,200 ft. (366 m) high (in British Columbia). All cliffs are either on coastal islands, "sea stack" rock formations, or coastal headlands. Lower cliffs may be heavily wooded adjacent to nest sites. *F. p. pealei* also nests in tall trees and uses vacant nests of other large birds.

**Nest sites.**—No nest is built. Generally nests in the same area in consecutive years but rarely uses the same eyrie site 2 years in a row. On cliffs, alternate eyrie sites are on the same cliff. After use and reuse, many eyries have a large amount of white excrement staining the area below the nest ledge.

*F. p. tundrius.*—Eyries are on a flat surface

on a slope or on a narrow or wide ledge on a cliff. The eyrie site may be totally exposed or is sometimes protected by an overhang. The actual site is a shallow scrape in the soil, on grasses or moss, or in an abandoned sheltered or unsheltered stick nest on a ledge of a Common Raven or Rough-legged Hawk; occasionally an old Golden Eagle nest. Eyries may be easily accessible to humans and predators or placed high on inaccessible cliffs.

*F. p. anatum* and "Eastern."—Eyries are in crevices and gutters on buildings, on ledges and gravel roofs on tall buildings; on ledges of tall smokestacks; in covered or on open wooden boxes placed on buildings and bridges; on covered wooden boxes placed on towers; and on top of girders, in hollow girders, or on concrete supports of bridges. Eyrie sites are generally high and are inaccessible to predators. Cliff sites are on narrow or wide ledges, and may be exposed or protected by an overhang. Wooden boxes were often placed on cliff areas for hacked birds to provide additional shelter. Some "Eastern" pairs in Minnesota and sw. Ontario may use abandoned stick nests of Common Ravens on cliffs. Old Golden Eagle nests have been used in n. Alberta. Formerly, eastern *anatum* regularly nested in broken-off trunks and cavities high up in tall trees, especially in cypress, cottonwoods, and sycamores.

*F. p. pealei.*—Eyries are either covered by protective rock, dirt, or tree root overhangs or are on exposed ledges. Many sites are near the top of an eroded cliff or hillside underneath exposed roots of spruce trees. Since all areas are very humid, the nest ledge is often matted with mosses and other vegetation. On the Queen Charlotte Islands, tree eyries are in large standing spruce with broken off limbs that form a partial cavity or on the top of tall broken off stumps. Occasionally, vacant stick nests of Bald Eagles and Pelagic Cormorants are used.

Three or 4 eggs are the common clutch size, 2 or 5 eggs are fairly common, and 1 or 6 eggs are uncommon to rare. The eggs are incubated for 33–35 days. Eggs are laid at 2-day intervals with incubation starting with the second or third egg. Both sexes incubate, but females perform the majority of the task, especially at night. Fledging time: males fledge in 39–46 days and females fledge in 41–49 days. With the short nesting season under the severe conditions of the Arctic, fledgling *F. p. tundrius* stay with their parents until the fall migration: a period of only 2–4 weeks, and rarely up to 6 weeks. With a longer nesting season in temperate areas, *F. p. anatum* and "Eastern" fledglings may stay with their parents for up to 8 weeks.

**CONSERVATION:** Peregrine Falcons needed a great deal of human assistance to regain their status after suffering from the deadly effects of pesticide contamination and human persecution. Greater public awareness and appreciation of raptors in general was one of the first accomplishments that was made. Population restoration programs were initiated throughout Canada and the U.S. Across s. Canada and from the Rocky Mts. westward in the U.S., programs were created to boost the nearly depleted wild population of *anatum.* East of the Great Plains, where *anatum* was extirpated, programs were also implemented to build a new Peregrine population. In Canada, *anatum* were used; in the U.S., various strains of subspecies were used as described in Subspecies.

**Pesticide bans.**—The first step that was taken to help not only Peregrine Falcons, but all wildlife and humankind, was to ban organochlorine pesticides, particularly DDT. This lethal pesticide was first used in 1946 in Canada and the U.S. This ban, however, was established too late to help retain the original native stock of eastern *anatum* Peregrines.

Canada took a series of steps to discontinue the sale and use of DDT that began in 1968 with a ban on spraying forests in national parks. The major Canadian ban came on Jan. 1, 1970 (announced Nov. 3, 1969), when DDT was permitted for insecticide use on only 12 of the 62 previously sprayed food crops. However, all registration for insecticide use on food crops was stopped by 1978. Canada, however, permitted DDT use for bat control and medicinal purposes until 1985. Canadian users and distributors were also allowed to use existing supplies of DDT until Dec. 31, 1990.

The U.S. also had a series of steps to ban DDT, but halted the overall sale and environmental use fairly quickly. In 1969, the USDA stopped the spraying on shade trees, and tobacco crops and at aquatic locations and in homes. The USDA placed further bans on its use on crops and commercial plants and for building purposes in 1970. The EPA banned all DDT sale and use on Dec. 31, 1972. However, limited use for military and medicinal purposes were permitted until Oct. 1989. In 1974, the U.S. banned the use of Aldrin and Dieldrin, both deadly pesticides that may have affected wildlife as much as DDT did.

Mexico was expected to discontinue government-sponsored DDT use for malaria control in 2002, and has planned a total ban of DDT by 2006.

As of 2000, possibly five other Latin American countries still use DDT and other organochlorine chemicals without restrictions. Peregrine Falcons and many of their prey species regularly winter in this region and are susceptible to contamination.

There are 122 countries, including the U.S. and Canada, that have signed a United Nations-sponsored treaty banning eight deadly organochlorine pesticides: Aldrin, Chlordane, DDT, Dieldrin, Endrin, Heptachlor, Mirex, and Toxaphene. There are also two industrial chemicals, Hexachlorobenzene (also a pesticide) and PCBs, and two by-products of industrial processes, dioxins and furans, that have also been banned.

**Protective laws.**—Legal protection assisted Peregrines to recover after suffering from organochlorine pesticides. The USFWS designated *F. p. tundrius* as an Endangered Species from Jun. 2, 1970 to Mar. 20, 1984, and as a Threatened Species from Mar. 20, 1984 to Oct. 5, 1994. In Canada, The committee on the Status of Endangered Wildlife in Canada (COSEWIC) listed *tundrius* as a Threatened Species in Apr. 1978, but downlisted it to a Vulnerable Species in Apr. 1992. (COSEWIC has week-long assessment meetings, thus no exact "dates" are available for rulings.)

The USFWS listed the *F. p. anatum*, which includes released "Eastern types," as an Endangered Species from Jun. 2, 1970 to Aug. 25, 1999. COSEWIC listed *anatum* as an Endangered Species in Apr. 1978 and downlisted them to a Threatened Species in Apr. 1999. Mexico redesignated *anatum* from an Endangered to a Threatened Species in 1999.

COSEWIC designated *F. p. pealei* as a Vulnerable Species in 1978 and as a Species of Concern in 1999. The USFWS never listed *pealei*, but it was protected under the "similarity of appearance" provision that took effect in Mar. 1984.

From the 1999 delisting of *anatum*, a 5-year period took effect in the U.S. and Canada to study the effects of being delisted as an Endangered Species under the Endangered Species Act of 1973.

In the U.S., both subspecies were first protected under the Endangered Species Conservation Act of 1969, then under the Endangered Species Act of 1973.

Provisions were also made in the 1973 act to accommodate the foreign races that were being used to boost the propagation programs.

**Population restoration (introduction) programs.**—Two main restoration programs were used in the West: (1) Captive-release programs designed to bolster the severely low wild *F. p. anatum* population in s. Canada, from the Rocky Mts. in the U.S. and westward; (2) captive-release programs to rebuild the defunct population of *anatum* east of the Rocky Mts. Only *anatum* stock was released from the Rocky Mts. and westward and in w. Canada (w. Ontario has mixed lineage of subspecies); east of the Rocky Mts. in the U.S., five subspecies were bred and released. *See* Subspecies for the five races used for captive breeding.

*Programs for the Rocky Mts. and westward.*—(1) *Hacking:* Placing typical-brood number (3–5) juvenile birds in sheltered wooden boxes on buildings, cliffs, and towers. Nestlings and fledglings were supplied food by human caretakers until they could fend for themselves. It was imperative during the hacking process that the young Peregrines did not bond with their caretakers so they would be truly wild when fledged. Hacking had a much higher success rate than other methods of releasing young birds. (2) *Fostering:* Placing captive-born nestlings into select wild eyries of remnant *anatum* that experienced egg failure and may have abandoned additional nesting attempts (egg failures were typically due to pesticides). (3) *Cross-fostering:* Placing Peregrine nestlings into active Prairie Falcon eyries. This was the least-used method.

*Programs east of the Rocky Mts.*—Only direct hacking in wooden boxes was used since there were no wild pairs of Peregrines left and Prairie Falcons breed only as far east as the western part of the Great Plains. Hacking locations were the same as previously described, but researchers also used boxes on smokestacks at power companies.

Released juveniles in Canada were produced from mated, captive pairs. In the U.S., offspring were from mated pairs and artificially inseminated adult females. With human manipulation, even mated pairs would lay more eggs than are typical in the wild, thus increasing the number of eggs and young available for release.

Another source of obtaining release birds that was used in California and Colorado was by taking eggs from wild pairs that had previous histories of eggs with pesticide-thinned eggshells and hatching them in an incubator. Dummy eggs replaced the thin-shelled eggs that would have otherwise broken if they had been left in the nest for the parents to incubate.

*Canada.*—A government-sponsored captive-breeding facility at Wainwright, Alberta, was established in 1972, and it produced birds for release in Canada from 1974 to 1996. As previously noted in Subspecies, all original stock came from native Canadian *F. p. anatum.* A small amount of "trading" of *anatum* occurred between Canada and the U.S. to reduce inbreeding of the captive population. At least four other Canadian nonprofit, private organizations also produced Peregrines. Some nonprofit organizations still raise and release Peregrines, especially after Wainwright closed its facilities. From 1974 to 1985, Wainwright raised 861 Peregrines, most for releases. When the Wainwright facility closed in 1996, 1,752 Peregrines were produced for release by Canadian facilities.

Releases continued to 2000 in s.-cen. British Columbia to restore breeding populations.

*United States.*—In its desire to get large numbers of Peregrines released into the wild as quickly as possible, the U.S. relied extensively on non-native North American races and foreign races. Virtually all breeding stock came from falconry birds, injured birds that could not be released back into the wild, and illegally possessed birds that were confiscated by authorities but also could not be returned back to the wild because they were imprinted on humans. The various subspecies that were used are listed in Subspecies. To retain genetics of the *F. p. anatum* race, Rocky Mt. releases were from *anatum* that were obtained from various parts of the Rocky Mts.

In the East, experimental captive breeding began in 1971 with *F. p. pealei* (from the Queen Charlotte Islands, B.C.) that were raised by Dr. Heinz Meng of New Paltz, N.Y. In 1970, once it was proven that captive breeding would succeed, a large captive breeding facility was constructed at Cornell University, Ithaca, N.Y. under the newly formed nonprofit organization, The Peregrine Fund, founded by Dr. Tom Cade. In 1973, 20 Peregrines were raised at Cornell; in 1974, 28 more were produced. Thereafter, production grew significantly.

In the West, James Enderson began the first captive-breeding in 1973 with *F. p. anatum* falconry birds in Colorado. In 1974, a captive breeding facility was built in Fort Collins, Colo., in association with the Colorado Division of Wildlife to assist large-scale propagation that was started at Cornell. In 1984, The BLM supplied land for The Peregrine Fund to create a large breeding facility for Peregrines and other endangered raptors and birds near Boise, Idaho. Subsequently, the captive-breeding facilities at Cornell and Fort Collins were closed.

The Santa Cruz Predatory Bird Research Group of the University of California at Santa Cruz bred and released nearly 800 *F. p. anatum* Peregrines in California from 1977 to 1992. They started with two nestlings that were obtained from the Peregrine Fund that were fostered by a wild pair of *anatum* on Morro Rock, San Luis Obispo Co. A few birds were released after 1992 to offset negative production due to human impact.

The Midwest Peregrine Restoration Group was formed in 1982 by Dr. Patrick Redig and Dr. Harrison Tordoff of the University of Minnesota. They supplied most of the released stock for Iowa, Kansas, Minnesota, Missouri, and Nebraska (and also for much of the Midwest region of the East).

Other small nonprofit groups and private breeders have also provided birds for releases. All release programs were assisted or directed by respective state wildlife agencies. Generous donations by corporations and the public also greatly helped, especially since all release programs were conducted by nonprofit organizations.

The first releases in the Rocky Mt. region took place in 1976 with fostering young in a wild eyrie in Colorado. In Alberta, the first experimental release occurred in 1975, also with fostering young at a wild eyrie in the northern part of the province. The Santa Cruz Predatory Bird Research Group began releases in 1977. By the late 1970s, large-scale releases were initiated throughout the w. U.S. and Canada. In the Rocky Mt. region, over half of the releases occurred in Colorado. Smaller numbers were released in Idaho. Large-scale releases continued into the early 1990s. Thereafter, small-scale programs were used.

The percentages of released subspecies and their breeding status have been carefully tracked by the Midwest Peregrine Restoration group for Iowa, Kansas, Minnesota, Missouri, and Nebraska (and in the Midwest in the East). Percentages of released/currently breeding subspecies: *anatum*, 57/54; *pealei*, 27/22; *peregrinus*, 6/9; *brookei*, 6/10; and *tundrius*, 4/5. Interestingly, in the very early stages of genetic ecological adaptation, the two foreign races, *peregrinus* of Europe and *brookei* of the Mediterranean region, have showed adaptability and have increased their status as breeders compared to the percentage that were released.

**Restoration in Canada.**—Over 1,700 Peregrines were released in Canada, but much of the large-scale releases ended in the early 1990s. *Alberta:* 45 Peregrines were released annually from 1992 to 2000. Urban releases occurred in Calgary and Edmonton, but also among remnant wild pairs in ne. Alberta and along rivers on the plains. *British Columbia:* Had minimal releases. From 1998 to 2000, releases took place near Kelowna, B.C. to re-establish the population in the Okanagan Valley region. There was a pair near Kamloops in 2001, the first pair in interior British Columbia in several decades. *Manitoba:* Released birds in larger cities as Brandon and Winnipeg. One Iowa-released bird dispersed to Winnipeg and has nested. *Saskatchewan:* Urban releases in Saskatoon and Regina.

**Restoration in the United States.**—*Alaska:* No programs were implemented. Population has been slowly recovering. *Arizona:* No release programs were implemented. Extant population was sufficient to rebuild itself. *Arkansas:* Five birds hacked on the White River in Independence Co. in 1993; three survived. Introduced in urban Little Rock in 1994. *California:* Nearly 800 *F. p. anatum* were released by Santa Cruz Predatory Bird Research Group from 1977 to 1992. *Colorado:* Over 500 captive-reared *F. p. anatum* were released from the late 1970s through the 1980s. *Idaho:* A strong reintroduction program. *Iowa:* Releases began in 1989 and continues today. Original releases occurred in Cedar Rapids, Des Moines, and Muscatine. Releases have recently occurred in Dubuque, Effigy Mounds National Monument in Allamakee Co., near Lansing (last nesting site in the U.S. east of the Rocky Mts. in 1967), and at the Louisa Power Plant in Louisa Co. Iowa-reared birds have dispersed widely north and west of original release sites. *Kansas:* Introduced in Kansas City and Topeka. Iowa-released bird dispersed to Topeka. *Louisiana:* No release programs and no current population. *Minnesota:* Intense restoration program (*see* "Eastern" Status and Distribution for all areas. *Missouri:* Introduction began in 1985 in St. Louis. Twenty-four birds released in Kansas City beginning in 1991. Also released in Springfield in 1997 and 1998. *Montana:* From 1981 to 1995, 448 were reintroduced at 23 sites. From 1996 to 2000, 30 more Peregrines were released, mainly along the Missouri River. A 1973–1975 survey revealed only 23 known or suspected territories; only one was occupied in 1975–1977, and it did not produce young. *Nebraska:* Released in Lincoln and Omaha. Iowa-released birds have dispersed to Omaha. *Nevada:* Unknown number of birds were released by the Santa Cruz Predatory Bird Research Group from 1977 to 1990. *New Mexico:* Unknown number of releases in the 1970s and 1980s in the northern part of the state and along the Canadian River. *North Dakota:* No releases. A pair residing at Fargo came from other areas. *Oklahoma:* No known release programs. Pair present in Tulsa in 1991. *Oregon:* Unknown number of birds were released by the Santa Cruz Predatory Research Group from 1977 to 1992. *South Dakota:* Four birds were released in 1997 in Sioux Falls. *Texas:* No release programs were initiated. Extant population was viable enough to increase on its own. *Utah:* Reintroduction of 80 birds from 1979 to 1989 in the greater Salt Lake City area. Large extant population in the rugged areas in the southern part of the state were not augmented with releases. *Washington:* Hacked birds from the Santa Cruz Predatory Research Group of California augmented a low extant population from at least 1977 to 1992 in some urban and natural areas. *Wyoming:* Reintroduction efforts ran from 1980 to 1994.

**Peregrine Falcon banding protocol for North America.**—To obtain scientific data on Peregrine movements and survival, wild-captured and captive-released falcons are fitted with an aluminum USFWS band with small numbers on the right leg. A colored band with large alpha-numeric designations is put on the left leg. On perched birds, the colored band can be read with a spotting scope from several hundred feet away. The colored bands reflect the subspecies and origin of a falcon and have a two- or three-digit alpha-numeric designation. COLOR DESIGNATIONS OF BANDS: Red: Captive-bred birds that are released in the e. U.S.; bicolor black/red or black/green (replaces the black/red), *F. p. tundrius* banded on breeding grounds; blue: *F. p. tundrius* or *F. p. anatum* captured during migration; black: *F. p. anatum* banded on breeding grounds; green: *F. p. pealei* banded in all seasons.

A few biologists are anodizing USFWS bands various colors to assist visual tracking of individuals.

**Mortality.**—All subspecies and released types are subject to illegal shooting in North America and on the winter grounds. Before legal protection, Peregrines were even shot for killing pigeons in cities. Poisoning, including that from organochlorine pesticides, will affect Peregrines to some extent for many years to come; however, levels have decreased in most breeding populations and should continue to slowly decrease. High levels of organochlorine chemicals are still prevalent on coastal California, but have shown up in nesting pairs in n. Alberta. Electrocution occurs from utility wires and poles. Adults often collide with utility wires and juveniles also collide with wires, cars, trains, and trees. Urban Peregrines, especially juveniles, suffer high mortality by striking windows. Urban juveniles often get caught in chimneys.

Natural mortality occurs with terrestrial predators such as Arctic Foxes, and aerial predators such as jaegers and gulls may kill young *F. p. tundrius.* Golden Eagles and Great Horned Owls were a problem to the population restoration efforts in the Rocky Mts., and the owls preyed on hacked juveniles east of the Great Plains. Captive-release locations were often in areas that had minimal depredation by owls. Migrant Peregrines may be affected by sudden, violent storms, especially hurricanes, when migrating over extensive open water of the Atlantic Ocean, Caribbean, and Gulf of Mexico. Mortality occasionally results from fierce territorial aggression between Peregrines during the breeding season.

Human disturbance at eyries by recreational rock climbing may cause nest failure. However, most popular climbing areas are closed to recreational use if nesting falcons are present.

**Falconry.**—A historically popular sport that is practiced by nearly 3,000 falconers in North America. Peregrine Falcons have been one of most favored species among falconers since medieval times. Prior to listing under the Endangered Species Act of 1973, Peregrines could be captured as migrants, taken from eyries as nestlings, or purchased from licensed breeders. After receiving protection from the act, only captive-bred birds could be used for falconry. However, the Aug. 1999 delisting permitted states west of the 100th meridian to individually regulate the taking of *F. p. anatum* eyasses from eyries for falconry use. In 2001, states were allowed to issue permits to licensed persons to take 5 percent of the state's known annual production of young Peregrines; but due to legal challenges, this was withdrawn in 2002. Federal statutes regulate that at least one eyass has to be left in an eyrie; however, some states, such as Colorado, dictate that at least two eyasses must be left in an eyrie. Except in Saskatchewan, which issued permits in 2001 (but no falcons were taken), Canada and states east of the 100th meridian do not allow the taking or possession of wild *F. p. anatum* or *F. p. tundrius* Peregrines.

*F. p. pealei* is on the Blue List in British Columbia, and the taking of wild birds is not permitted.

Falconry birds have been legally available from licensed facilities that raised captive birds during the period when Peregrines were protected from capture in the wild.

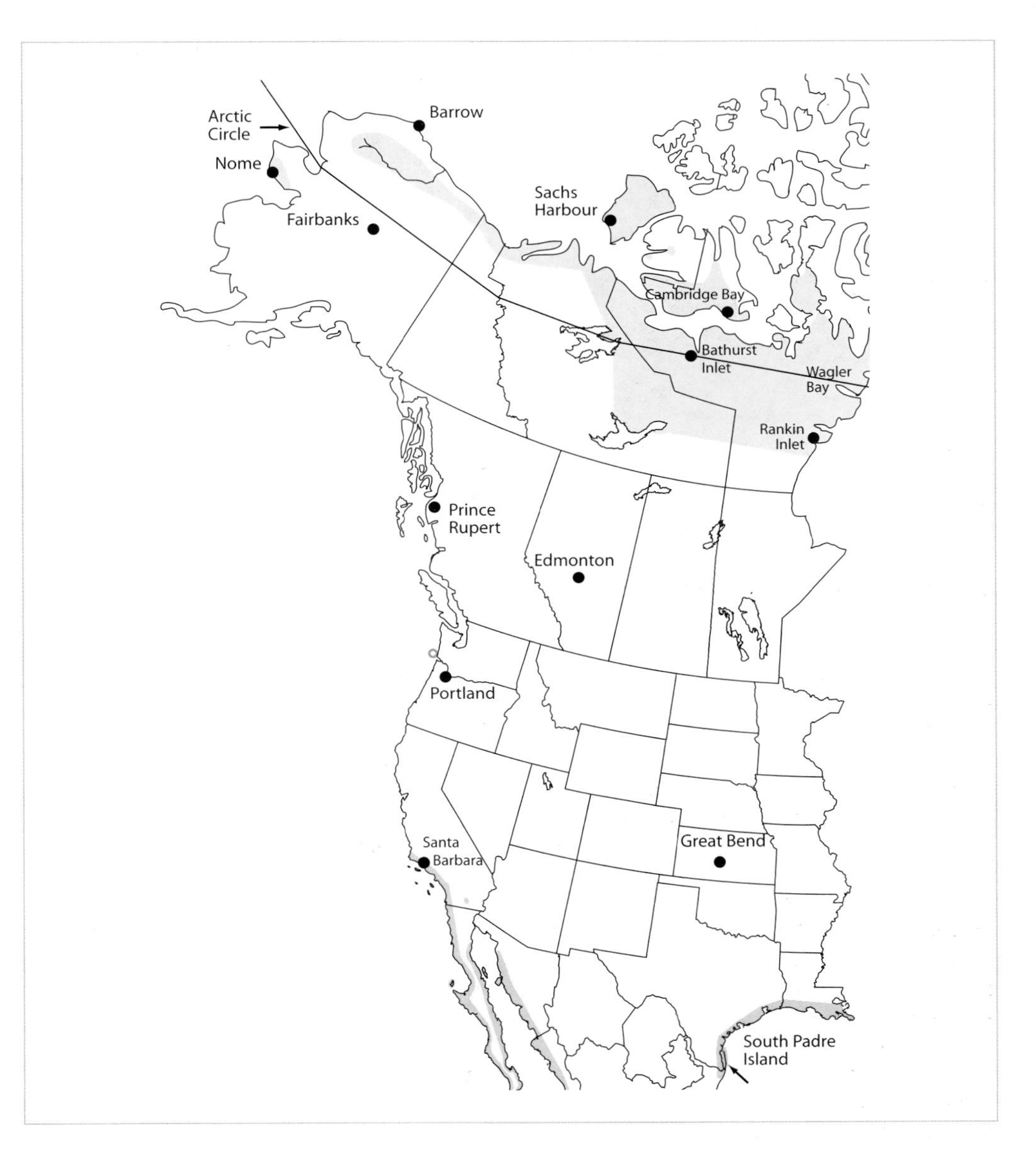


**"ARCTIC" PEREGRINE FALCON,** *Falco peregrinus tundrius:* Very uncommon. Winters south to s. South America. Exact North American winter range is unclear as races are often difficult to determine. Large fall and spring concentration of migrants on s. TX coast.

**SIMILAR SPECIES:** COMPARED TO ADULTS.—**(1) Gyrfalcon, intermediate (gray)-morph adults.**—PERCHED.—Crown and nape are gray, and dark malar mark is irregularly defined. Dorsal body similar and often very bluish gray. Ventral body can be similar on palest types, with a unmarked white breast and barred flanks and spotted belly. Wingtips are distinctly shorter than the tail tip. FLIGHT.—Underside of the remiges is pale and barred; secondaries often more heavily barred and darker than the rest of the remiges. From above in flight, all of upperparts, including the tail, are uniformly grayish: rump and uppertail coverts are same color as all other upperparts. **(2) Mississippi Kite, adults.**—

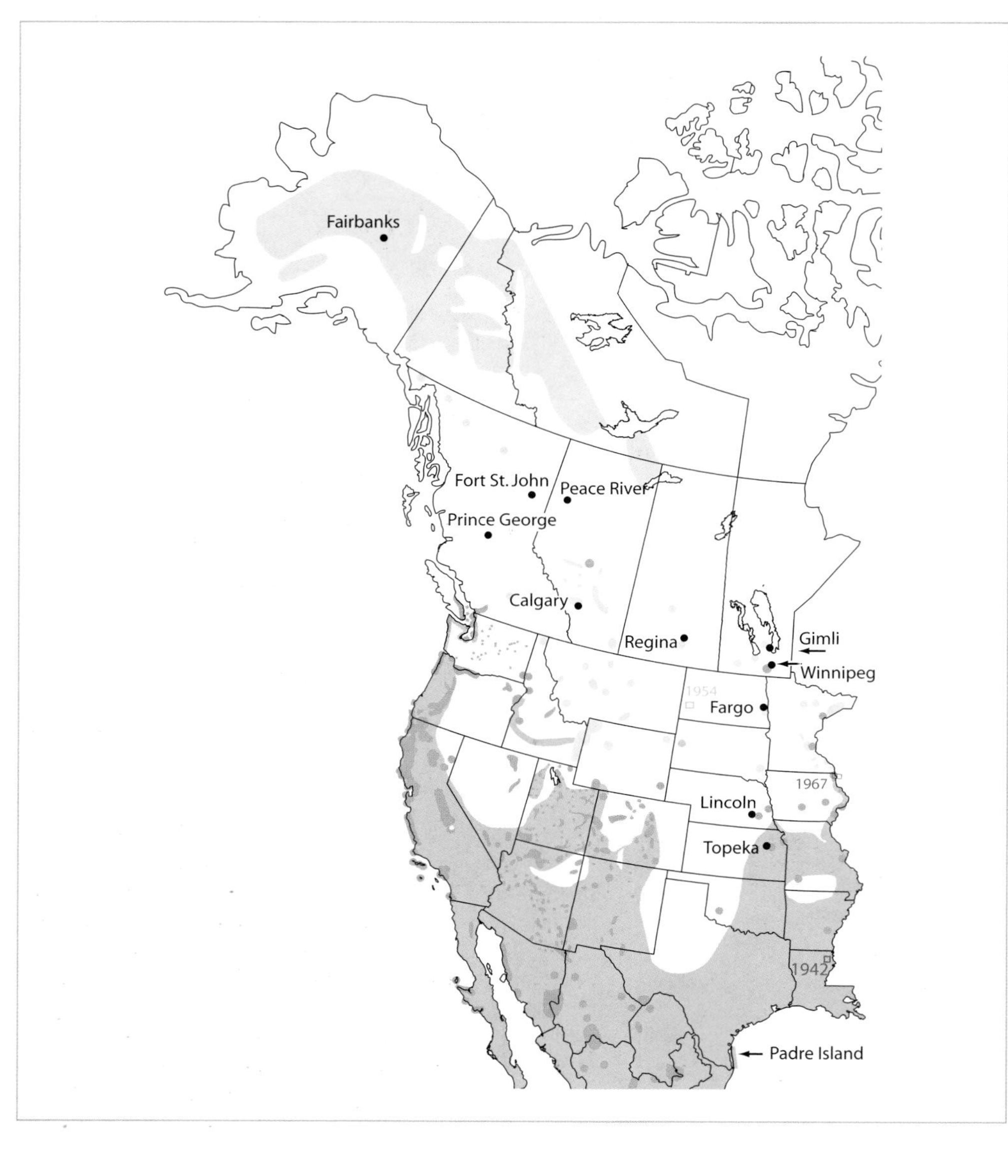


**"AMERICAN" PEREGRINE FALCON,** *Falco peregrinus anatum:* and "EASTERN" PEREGRINE FALCON (no assigned race): Rare. Was endangered. Rebounded due to release programs and a ban on DDT. The U.S. released "Eastern" stock east of the Great Plains from multiple, often foreign, races. Winters to South America.

FLIGHT.—Narrow wings and uniformly gray on underside with a white edge on the trailing edge of the secondaries. When soaring, the outermost primary is much shorter than the wingtip. Wingtips bend slightly upward when soaring. Dark tail is square-edged. COMPARED TO JUVENILES.—**(3) Gyrfalcon, intermediate (gray)-morph juveniles.**—PERCHED.—Head is similar but dark malar often not as distinct. Ventral areas are white not tawny. Wingtips much shorter than the tail tip. All pale banding on tail is white, not tawny. FLIGHT.—Underside of the remiges is not uniformly colored and marked as on Peregrine.

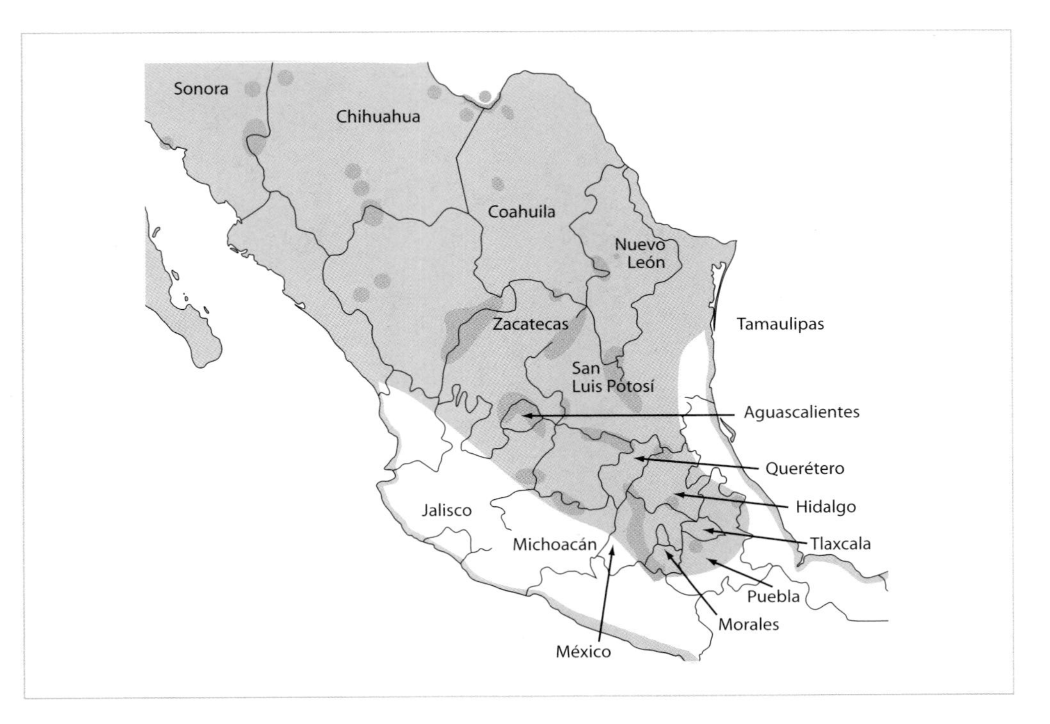


**"AMERICAN" PEREGRINE FALCON,** *Falco peregrinus anatum:* Mexico downgraded species from endangered to threatened in 1999.

The secondaries are more heavily barred or nearly solid gray and primaries are white and narrowly barred; coverts more heavily marked than remiges. Ventral areas as described in Perched. **(4) Gyrfalcon, dark-morph juveniles.**—Similar to *pealei* and released birds with *pealei* lineage. PERCHED.—Head often appears hooded as on darker-headed Peregrines; use caution. Ventral areas nearly solid dark brown with a minimal amount of pale whitish (not tawny) edgings. Wings much shorter than the tail tip. FLIGHT.—Underside of remiges is uniformly pale gray and unmarked. Dark brown underwing coverts are much darker than the remiges. **(5) Northern Harrier, adult females and juveniles.**—FLIGHT.—Shape is similar when gliding at high altitude. However, harriers have more distinct crook at the wrist of the wing, and the wrist area is farther out on the wing. Dark brown head and neck are hooded. Axillaries and median underwing coverts are darker than the rest of the underwing; secondaries darker than the primaries on juveniles. Wide dark tail bands. Wingbeats are loose and floppy. **(6) Merlin (*columbarius*) females.**—PERCHED.—Poorly defined dark malar mark. Wingtips are distinctly shorter than the tail tip. FLIGHT.—Very similar brown above and dark-streaked below. Underside of wings is nearly identically marked with pale spotting and barring on the dark feathers. Wingbeats are rapid; and rarely glide for long distances. **(7) Mississippi Kite, juveniles.**—PERCHED.—Lack a dark malar mark. Often have large white spots on scapulars. Toes are short. Wingtip-to-tail-tip ratio similar. FLIGHT.—Similar at higher altitudes with streaked ventral and often dark underside of the remiges. Underside of remiges, however, at close and moderate distances, differs as the lack of pale tawny-rufous spotting and barring is evident. When soaring, the short outermost primary is noticeable. Those lacking pale tail bands are similar at a distance. Primary tips bend upward when soaring. **(8) Prairie Falcon, juveniles.**—PERCHED.—White slash behind the eyes with the dark auricular patch. Tawny-colored ven-

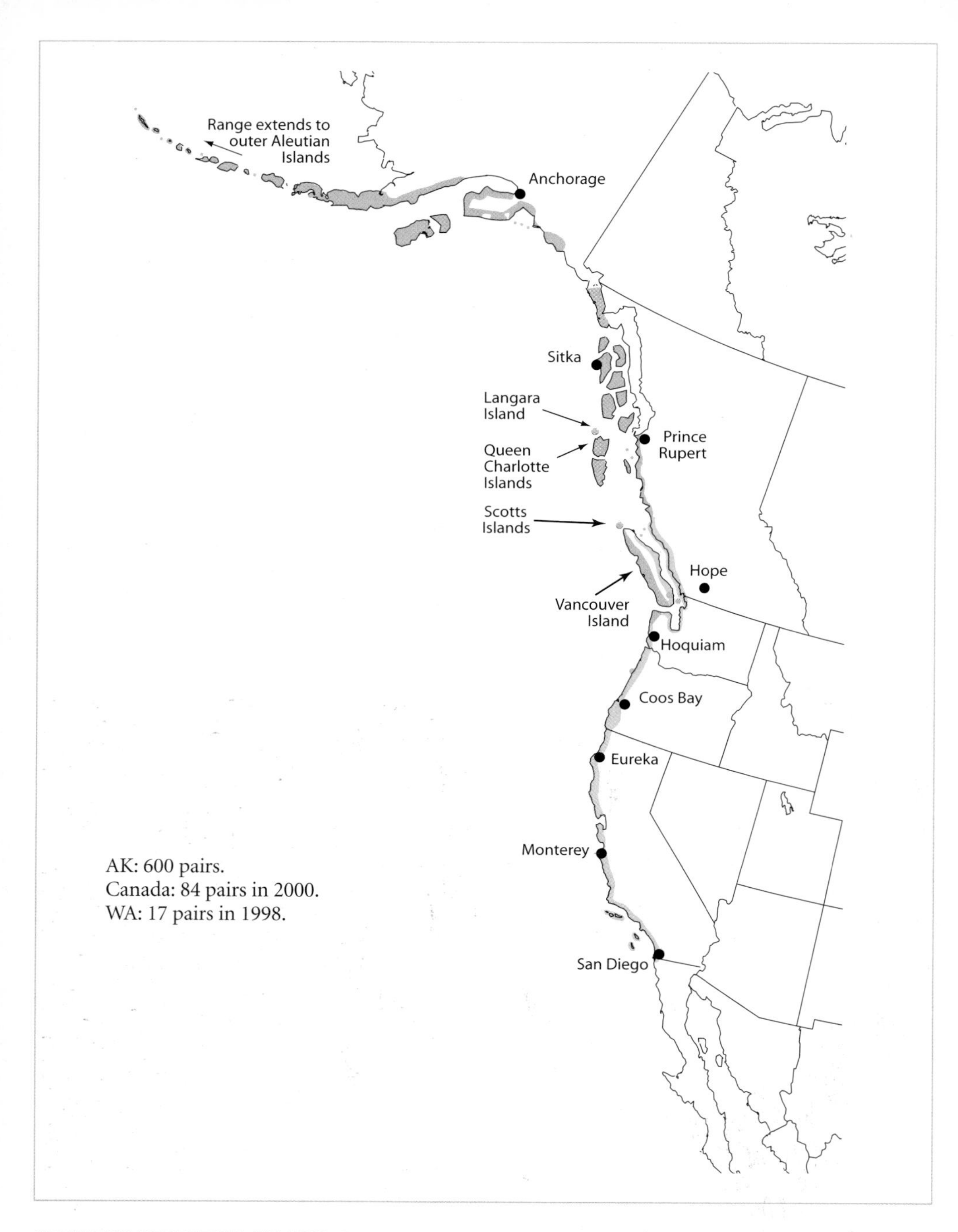


**"PEALE'S" PEREGRINE FALCON,** *Falco peregrinus pealei:* Densely populated on the Queen Charlotte Islands and some Aleutian islands. Breeds in nw. WA and intergrades with *F. p. anatum* there and at one known location on the OR coast. Some winter far south of breeding range.

tral region with the narrow dark brown streaking on recently fledged birds is similar to lightly marked type of *anatum* and many *tundrius*; use caution. Wingtips are distinctly shorter than the tail tip. FLIGHT.—Ventral surface of wing has black axillaries (armpit). Scalloped dorsal surface is similar to some recently fledged *anatum* and all *tundrius*.

**OTHER NAMES:** Peregrine. Formerly called Duck Hawk. Former eastern *F. p. anatum* was called the Rock Peregrine and Great-footed Hawk (by J. J. Audubon). Females are called "falcons" and males called "tiercels." Falconers call migrants "passage birds," and adults of all subspecies and types are called "haggards." *Spanish:* Halcón peregrino. *French:* Faucon pélerin.

---

REFERENCES: Arizona Breeding Bird Atlas 1993–1997; Beebe 1960; Bent 1961; Berger et al. 1968; Bond 1946; Burnham 1997; Bylan 1998, 1999; Byre 1990; Cade et al. 1988; Campbell et al. 1990; Canadian Peregrine Foundation 1999a, 2002; Castellanos et al. 1997; Center for Conservation, Research, and Technology 2000; Clark 1999; Clark and Wheeler 2001; Commission for Environmental Cooperation 1997; Court et al. 1988; Dekker 1980, 1987, 1988, 1999; Dodge 1988–1997, 1995; Environment Canada 1999, 2000a, 2000b, 2002; Ferguson-Lees and Christi 2001; Forsman 1999; Franklin 1999; Fyfe 1988; Godfrey 1986; Gross 1999; Gustafson and Hildenbrand 1998; Harris 1979; Hayes and Buchanan 2002; Hitchcock 1977; Howell and Webb 1995; Idaho Conservation Data Center 1999; Islands Protection Society 1984; Jacobs and Wilson 1997; Johnsgard 1990; Kaufman and Meng 1975; Kellogg 2000; Kent and Dinsmore 1996; Kessel 1989; Kingery 1998; Lott 1999; Montana Bird Distribution Committee 1996; Palmer 1988; Redig and Tordoff 1988, 2000; Rodriguez-Estrella and Brown 1990; Rowell 1991; Rowell and Holroyd 2001; Russell 1991; Santa Cruz Predatory Bird Research Group 1999c; Smith 1996; Struzik 1999; Tordoff 2000; Tordoff et al. 1997, 2000; Tucker et al. 1998; USFWS 1984, 1993, 1998c, 1999c, 2001a; Varland 1999, 2001, 2002; White 1968; White and Kiff 1998; White et al. 1995; Wildlife Management Advisory Council 2001.

**Plate 555. Peregrine Falcon, adult female "Arctic" (*F. p. tundrius*) [Jul.]** ▪ Moderately wide type black malar mark. Large white forehead and lores. ▪ Brownish gray upperparts with partial pale gray barring. Moderately marked white underparts are barred on the flanks and barred and spotted on the belly. Belly is usually pinkish tawny. ▪ *Note:* Photographed at Rankin Inlet, Nunavut, by Gordon Court.

**Plate 556. Peregrine Falcon, adult female "Arctic" (pale type *F. p. tundrius*) [Jul.]** ▪ Narrow type black malar mark. Center of crown and supercilium are white. Thin dark eyeline extends onto nape. ▪ White underparts are lightly marked. Leg feathers are unmarked. Brown upperparts (not shown). ▪ *Note:* Photographed at Rankin Inlet, Nunavut, by Gordon Court.

**Plate 557. Peregrine Falcon, adult male "American" (*F. p. anatum*) [Jun.]** ▪ Very wide type black malar mark that covers all of the side of the head; black lores and forehead. ▪ Bluish gray upperparts are barred with pale gray. Underparts are tawny-rufous except gray flanks and leg feathers. ▪ Black tail. ▪ *Note:* Mesa Co., Colo.

**Plate 558. Peregrine Falcon, adult female "Peale's" (moderately marked type *F. p. pealei*) [Mar.]** ▪ Wide type black malar mark; narrow white forehead and lores; spotted lower auriculars. ▪ White underparts are heavily marked. Blackish upperparts are edged with pale gray. ▪ Wingtips are a bit shorter than tail tip. ▪ *Note:* Grays Harbor Co., Wash.

**Plate 559. Peregrine Falcon, adult female "Peale's" (heavily marked type *F. p. pealei*) [Mar.]** ▪ Wide type black malar mark; black forehead and lores; spotted lower auriculars. ▪ White underparts are very heavily marked. ▪ *Note:* Grays Harbor Co., Wash.; typical plumage on the Aleutian Islands, Alaska, uncommon to rare elsewhere.

**Plate 560. Peregrine Falcon, adult female "Eastern" (no subspecies designation) [Apr.]** ▪ Wide type black malar mark. ▪ Breast is spotted/streaked, flanks are heavily barred, and tawny belly is barred. ▪ *Note:* Captive-released stock and a mix of several races. She appears much like a European Peregrine. Photographed in New Haven Co., Conn., by Jim Zipp.

**Plate 561. Peregrine Falcon, adult female "Eastern" (no subspecies) [Sep.]** ▪ Wide type black malar mark. ▪ Pale tawny-rufous breast is spotted/streaked, tawny-rufous flanks and belly are heavily barred. ▪ *Note:* Same bird as on plate 560, but in newly molted first-adult plumage. Photographed in New Haven Co., Conn., by Jim Zipp.

**Plate 562. Peregrine Falcon, adult male "Eastern" (no subspecies) [Apr.]** ▪ Very wide type black malar mark covers all of side of head; black forehead and lores. White breast; grayish flanks are narrowly barred; pale tawny belly is spotted and barred. ▪ Dark underside of tail. ▪ *Note:* Appears similar to Australian Peregrines. Photographed in New Haven Co., Conn., by Jim Zipp.

**Plate 563. Peregrine Falcon, adult female "Arctic" (*F. p. tundrius*) [Oct.]** ▪ Moderately wide type black malar mark. ▪ White underparts are moderately marked. ▪ Uniformly black and white marked underwing. Pointed wingtips. ▪ Blackish distal half of tail. ▪ *Note:* Cape May, N.J.

**Plate 564. Peregrine Falcon, adult female "Arctic" (pale type *F. p. tundrius*) [Jul.]** ▪ Narrow type black malar mark. ▪ White underparts are very lightly barred on flanks. ▪ Blackish tail. ▪ *Note:* Rare plumage type of the central Arctic. Photographed at Rankin Inlet, Nunavut, by Gordon Court.

**Plate 565. Peregrine Falcon, adult male "American" (*F. p. anatum*) [Jul.]** ▪ Very wide type black malar mark covers all of side of head. ▪ Tawny-rufous breast and belly, gray flanks. ▪ Uniformly marked underwing. Pointed wingtips. ▪ Blackish distal half of tail. ▪ *Note:* Park Co., Colo.

**Plate 566. Peregrine Falcon, adult female "American" (*F. p. anatum*) [Jun.]** ▪ Moderately wide type black malar is narrower and columnar-shaped (found on some adult females of this race). ▪ Tawny-rufous breast, belly, and undertail coverts; gray flanks. ▪ Blackish distal half of tail. ▪ *Note:* Mesa Co., Colo.

**Plate 567. Peregrine Falcon, adult male "American" (*F. p. anatum*) [Jun.]** ▪ Black head. ▪ Bluish gray upperparts: rump and uppertail coverts are paler bluish. ▪ Blackish tail contrasts with pale rump and uppertail coverts. ▪ *Note:* Adult females are more brownish or blackish than adult males but still distinctly paler rumps and uppertail coverts; tail is blackish. Mesa Co., Colo.

**Plate 568. Peregrine Falcon, juvenile female "Arctic" (*F. p. tundrius*) [Oct.]** ▪ Moderately wide type black malar mark connects with a moderately wide black eyeline. Pale tawny supercilium; pale tawny forehead and lores. Pale blue cere and orbital skin. ▪ Tawny edges on upperparts. ▪ Wingtips equal tail tip. ▪ *Note:* Assateague Island, Md.

**Plate 569. Peregrine Falcon, juvenile female "Arctic" (blonde type *F. p. tundrius*) [Jan.]** ▪ Narrow type black malar mark is broken at the gape. Pale tawny crown and supercilium with dark area on sides and rear of crown. Pale greenish cere and orbital skin (turns to yellow by spring) ▪ Lightly streaked tawny underparts. ▪ *Note:* Cameron Parish, La.

**Plate 570. Peregrine Falcon, juvenile female "American" (*F. p. anatum*) [Jul.]** ▪ Wide type black malar mark. Dark crown. Pale forehead, dark lores. ▪ Lightly streaked tawny rufous underparts, including leg feathers. ▪ *Note:* Pale type of ventral markings for *anatum*. Douglas Co., Colo.

**Plate 571. Peregrine Falcon, juvenile female "Peale's" (moderately marked type *F. p. pealei*) [Mar.]** ▪ "Wide type" black malar mark. Streaked auriculars. ▪ Upperparts lack distinct pale feather edges. Heavily streaked whitish underparts, including leg feathers. ▪ Wingtips nearly reach tail tip. ▪ *Note:* Scavenging on a Common Murre. Pacific Co., Wash.

**Plate 572. Peregrine Falcon, juvenile female "Arctic" (*F. p. tundrius*) [Oct.]** ▪ Moderately wide type black malar mark and black eyeline. ▪ Pale tawny underparts are lightly streaked with very narrow streaking on leg feathers. ▪ Uniformly marked dark underwing. Pale tawny-rufous markings on remiges. ▪ Distinctly banded tail. ▪ *Note:* Cape May, N.J.

**Plate 573. Peregrine Falcon, juvenile "American" (*F. p. anatum*) [Jun.]** ▪ Wide type black malar mark. ▪ Tawny-rufous underparts are moderately streaked. ▪ Uniformly dark underwing. Pale tawny-rufous markings on remiges. ▪ *Note:* Photographed in Moffat Co., Colo. by Mike Lanzone.

**Plate 574. Peregrine Falcon, juvenile male "American" (*F. p. anatum*) [Jul.]** ▪ Wide type black malar mark. Pale-headed type. ▪ Tawny-rufous underparts are lightly streaked (more rufous than other races). Arrowhead-shaped dark markings on leg feathers. ▪ *Note:* These pale-headed, lightly marked types are found throughout breeding range in the West.

**Plate 575. Peregrine Falcon, juvenile (1-year-old) female "Eastern" (no subspecies) [Jun.]** ▪ Wide type black malar mark. Pale yellow cere and orbital skin. ▪ Heavily streaked underparts. ▪ Dark, uniformly marked underwing. ▪ *Note:* Seen in adult plumage on plates 560 and 561. Appears much like a "Peale's" Peregrine. Photographed in New Haven Co., Conn., by Jim Zipp.

## GYRFALCON
### *(Falco rusticolus)*

**AGES:** Adult, 1 or 2 years in a subadult stage (basic I and basic II), and juvenile. Full adult plumage and color of the fleshy cere, orbital skin, and feet may not be obtained until they are 2–3 years old (Clum and Cade 1994). The exact plumage transition from juvenile to adult is unknown (Clum and Cade 1994). Plumage data for adults refer to birds that have replaced all juvenile feathering.

Subadults are mainly in an adult plumage but retain some old, worn juvenile feathering. Most retained juvenile feathers are on the upperwing coverts, and a lesser amount are on other dorsal regions. During this subadult period, the cere, orbital skin, and feet also slowly change from juvenile-like to adultlike color. The amount of retained juvenile feathering and degree of color change of fleshy areas varies individually. Such color change may be based not only on advancing age, but also on diet and other factors.

Juvenile age characters are worn the first year. Juveniles have slightly longer secondaries, which creates a broader wing, and longer rectrices, which creates a longer tail than of older ages.

**MOLT:** Falconidae wing and tail molt pattern (*see* chapter 4). As in most falcons, the first prebasic molt begins when somewhat less than 1 year old. The molt is partially completed as 1-year-olds during their second year of life. Most probably begin molt in a span from Mar. to May. As in most raptors, molt begins on the primaries, then begins somewhat later on the rectrices. The secondaries and body feathers begin molting after the remiges and rectrices have undergone a fair amount of molt. The wing and tail molt described below is based on Clum and Cade (1994). Primary molt sequence typically starts with p4; however, it may start with p3, 5, 6, or 7. Primary molt then continues in the following sequence: p5, 6, 3, 7, 2, 8, 9, 1, and 10. Rectrix molt begins with r1 (the deck set), then continues with r2, 3, 4, 6, and 5 or an alternate sequence of r2, 6, 3, 4, and 5. The outermost juvenile primary (p10) may not be molted in the first prebasic molt, but all secondaries, rectrices, and the majority of body plumage are molted. The first prebasic molt is not complete on the upperwing coverts and sometimes other dorsal body regions. A large amount of retained juvenile feathers are visible as a faded and frayed pale brown patch on the upperwing coverts. This molt would indicate birds that are 1-year-olds in their second year of life.

The second prebasic molt occurs on 2-year-old birds in their third year of life. This annual molt may also not be a complete molt on the upperwing coverts and possibly other dorsal areas. A small amount of juvenile plumage may still be retained as very old, frayed, and very pale brown feathers. Some may molt entirely out of any retained juvenile feathering and achieve full adult plumage at this time. The extent of the remix and rectrix molt is unknown.

Subsequent adult prebasic molts begin in Apr. or May and are completed from Sep. to Nov. (Clum and Cade 1994). Molt may not be a complete annual replacement. Females begin their annual molt during incubation and males begin toward the end of the nesting cycle. These molts would affect birds that are at least 3 years old, in their fourth year of life.

**SUBSPECIES:** Monotypic. Since color morphs are indicative of geographic regions, they were formerly considered subspecies.

**COLOR MORPHS:** Polymorphic, with three primary color morphs that have a full continuum of plumage variations among them. The primary color morphs are: light (white), intermediate (gray), and dark (black). There are two additional intermediate plumages that form the continuum among the three primary morphs. There is a great deal of individual variation within each color morph. Sexes are similarly marked. Color-morph distribution in the Arctic adheres to Gloger's Rule.

**SIZE:** A large raptor. Males average smaller than females, but there is overlap on all anatomical features. Juveniles average longer than adults and subadults because they have longer tails. Length: 19–24 in. (48–61 cm); wingspan: 43–51 in. (109–130 cm).

**SPECIES TRAITS:** HEAD.—The bill is pale bluish or pinkish on white morphs and pale bluish with a dark tip on all other color morphs. The orbital region is bare flesh. BODY.—The upper body region is very broad. Talons are whitish or pale pink on white morphs and black on all other color morphs. In flight, Gyrfalcon often shows a hump on the back. WINGS.—**When soaring, the secondaries are broad and taper to moderately pointed wingtips.** When perched, wingtips are distinctly shorter than the tail tip. TAIL.—When closed, the basal area is very broad. *Note:* **The upper body is very broad and tapers to the narrower wingtip and tail area.**

**ADULT TRAITS:** HEAD.—Cere, orbital skin, and feet are yellow. These fleshy regions may be brighter on males, especially the cere color. BODY.—Dorsal markings consist of a barred pattern. WINGS and TAIL.—When perched, the wingtips are moderately shorter than the tail tip.

**ADULT LIGHT (WHITE)-MORPH TRAITS (LIGHTLY MARKED TYPE):** HEAD.—**White with a few small dark specks on the nape or white and very thinly streaked with brown on the crown and nape. A very thin dark area may border the area under the eyes and in front of the orbital skin.** BODY (dorsal).—**Two main types of markings on the largely white upperparts: (1) A short horizontal brown bar on each back, scapular, and rump feather; or (2) a few have a small, narrow arrowhead-shaped or diamond-shaped mark on all dorsal feathers. The dark brown mark covers a small area of the white feather.** BODY (ventral).—Immaculate white breast, belly, leg feathers, and undertail coverts, but some birds may have a few small dark specks on the distal flanks. WINGS (dorsal).—**Two types of markings as described in dorsal body on the median and lesser coverts. Greater secondary and primary coverts are white with narrow brown barring. The white secondaries may be unmarked, partially barred with brown, or fully barred with brown. Inner primaries are either immaculate white with a few bars on the distal area of each feather or are moderately barred.** WINGS (ventral).—**White underwing coverts. Underwing coverts are unmarked, including the axillaries. The underwing coverts may have a few small dark spots that are visible only at close range. Underside of the remiges is white except for the dark brown tips of the outer six or seven primaries. The dark primary tips create a "dipped-in-ink" appearance. There is a narrow brown band on the trailing edge of the inner primaries and secondaries. On individuals that have distinct barring on the dorsal surface of the secondaries, the dark barring will also show on the inner portion of the secondaries in translucent light.** TAIL.—**There are three variations on the white dorsal surface of the tail: (1) is unmarked, (2) has a small amount of dark mottling on several rectrices, or (3) is partially or fully barred on only the deck rectrix set (r1). The ventral surface is white and unmarked on all types. White uppertail coverts have very small dark arrowhead-shaped marks on each feather.** *Note:* **Either sex can be this lightly marked. Many are older birds. Based on museum specimens, similarly marked juveniles (lightly marked type) may molt into an adultlike plumage that is identical to these very white adults.**

**ADULT LIGHT (WHITE)-MORPH TRAITS (HEAVILY MARKED TYPE):** HEAD.—**White with narrow brown streaking on the crown and nape, and most have fairly wide brown eyelines. There may be some faint streaking on the auriculars and even a few dark specks on the malar region. A large dark area is under the eyes and in front of the orbital skin.** BODY (dorsal).—**White, with one or two dark brown horizontal cross bars on each back, scapular, and rump feather. Dark brown bar on each feather covers about 50 percent of the white feather surface.** BODY (ventral).—Breast, belly, forward flanks, leg feathers, and undertail coverts are immaculate white. Some may have a few small dark specks on the white distal flank and leg feathers. WINGS (dorsal).—**White, but all coverts and remiges are either moderately or heavily barred with brown as on dorsal body.** WINGS (ventral).—**Underwing coverts are either immaculate white, including the axillaries, or are sparsely covered with small brown flecks. Remiges are either immaculate white or have partial or fully barred primaries and secondaries. With the dark tips on the outer six or seven feathers, the primaries appear as if they are dipped in ink.** TAIL.—**Dorsal surface is white with six to eight narrow, complete dark brown bars. However, barring may be confined mainly to**

**the deck set (r1), with dark mottling or partial barring on the outer webs of all other rectrices. Ventral surface is immaculate white or may have brown barring on the outer web on the outermost retrix set (r6).** Uppertail coverts are white and crossed with brown bars. ***Note:* Plumage type is on either sex. Similar type of juvenile (heavily marked type) molts into this type of plumage as a subadult.**

**ADULT LIGHT INTERMEDIATE (LIGHT GRAY/SILVER)-MORPH TRAITS:** HEAD.—White with a heavily streaked crown. The supercilium is paler but thinly streaked. The white nape is heavily streaked. There is a narrow or wide dark eyeline, which may extend downward as a dark stripe behind the auriculars. The white auriculars and cheeks may be narrowly streaked with brown. The region in front of the orbital skin and below the eyes is dark brown or blackish and typically extends onto the lower jaw as a defined dark malar mark. BODY (dorsal).—**Each back, scapular, and rump feather is dark brown with a broad white tip and a moderately wide mid-feather white bar. The dark brown area of each feather covers more than 50 percent of the feather surface.** BODY (ventral).—White with small dark spots on the sides of the neck and breast. The central area of the breast may also be covered with small spots or may be unmarked. The white flanks and belly are covered with small dark spots, the distal portion of the flanks are covered with large dark spots. Leg feathers are white with partial, narrow dark barring. Undertail coverts are white with a narrow dark bar on each feather. WINGS (dorsal).—Coverts are dark brown and narrowly or moderately barred with a white mid-feather bar on each feather as described on the dorsal body. All remiges are 50 percent light-dark barred except the dark barring is narrower on the outer half of the primaries. WINGS (ventral).—**White coverts, including axillaries, are covered with very small dark spots. The markings are most prevalent on the greater coverts and median coverts. Lesser coverts on the patagial region have very small dark markings. The primaries are white with narrow dark barring. The outer secondaries are white with moderately wide dark barring, and the inner secondaries are white with broad dark barring. Wingtips are dark brown on the outer six or seven primaries.** TAIL.—**Dorsal tail surface is very pale gray with six to eight complete dark bands that are somewhat narrower than the pale bands are.** Uppertail coverts are dark with white barring. Undertail is white with dark barring. ***Note:*** This plumage type has not been previously described or depicted. It is found on both sexes. Depicted herein on an adult from Nome, Alaska; also, a 7-year-old female that wintered each year in Edmonton, Alberta (G. Court pers. comm. and Unpubl. photographs).

**ADULT INTERMEDIATE (GRAY)-MORPH TRAITS:** HEAD.—*Pale type*: Pale gray or whitish with dark streaking on the crown and a white forehead. The supercilium is white and streaked with gray. There is a narrow dark eyeline above the auriculars. Auriculars are gray on the distal region but whitish on the forward area and cheeks. The dark malar stripe is either moderately distinct or ill-defined. Dark nape is mottled with white patches. *Dark type*: Dark gray crown with a white forehead and a narrow whitish supercilium that begins behind the eyes. Auriculars and cheeks are dark gray with a moderately distinct dark malar stripe. Dark nape is mottled with white. BODY (dorsal).—*Barred type* (either sex, but mainly on males): Base feather color is dark gray or brownish gray with a moderately wide or fairly wide pale gray mid-feather cross bar. The dark portion can cover 50 percent or more of the feather surface. Even on the palest birds, the pale mid-feather cross bar is gray, not white, as on light intermediate morphs. Pale cross bars are distinct on all of the scapulars and back feathers. *Partial-barred type* (either sex, but mainly on females): Dorsal color is dark gray or grayish brown with a very narrow, pale gray mid-feather cross bar. Pale cross bars are often ill-defined or nonexistent on the forward scapulars and back. Rump is grayish and barred on both types. BODY (ventral).—White with a variable amount of dark markings. *Lightly marked type* (mainly males, but may be on some females): White breast is either unmarked or very lightly marked with dark dashes. Forward flanks and belly are spotted; the rear portion of the flanks is covered with large dark spots or narrow dark bars. The white leg feathers are narrowly barred. *Heavily* *marked type* (both sexes): Breast is covered with moderately wide dark streaking or teardrop markings; belly, lower belly and forward flanks are covered with large dark spots; rear flanks are broadly barred. The white leg feathers are heavily barred. Both types have barred undertail coverts. WINGS (dorsal).—Marked like the two respective types of dorsal body markings as described for dorsal body. Barred types are likely to have pale gray barring on all coverts and remiges, including the inner portion of the outer primaries. Partially barred types have a reduced amount of pale barring on the coverts or many lack pale barring. Many have a reduced amount of pale barring on the remiges, including the outer primaries. **Both types, but especially barred types, have distinctly darker primaries and greater primary coverts that contrast sharply with the paler grayish secondaries and the rest of the upperwing coverts.** WINGS (ventral).—**White underwing coverts are covered with a light or moderate amount of dark markings. The median coverts may be more extensively streaked or barred and form a darker area on this feather tract. The outer primaries are white with narrow dark barring and forms a pale panel that contrasts with the more thickly barred, somewhat darker inner primaries and outer secondaries and the typically darker, solid gray inner secondaries. *Note:* As a rule, the underwing either appears uniformly pale with pale coverts and pale remiges or has a slightly darker band on the median coverts because of the coverts being more extensively marked.** TAIL.—**Dorsal tail is pale gray and banded and is the same color as the rest of the dorsal areas.** The dark bands may be chevron-shaped and up to 50 percent of the width of the pale area of each rectrix. There is a *possible* sexual difference in the width of the banding: Males may have narrow dark bands and females may have wide dark bands. Uppertail coverts are dark and crossed with pale gray bars.

**ADULT DARK INTERMEDIATE (DARK GRAY)-MORPH TRAITS:** HEAD.—Similar to dark type head of intermediate morph, including the pale forehead. Head is all dark, but is often paler on the cheeks. Dark malar mark is present. Pale supercilium is generally lacking, but may have a very small amount of pale mottling on the nape. Throat is white. BODY (dorsal).—Dark brownish gray or brownish black with very narrow pale gray edges on most feathers. The most distal scapulars may have a faint gray bar on each feather. Rump is dark and barred with gray. BODY (ventral).—White with moderately thick dark streaking or teardrop-shaped markings on the breast and belly; the flanks are broadly barred. Dark markings cover less than 50 percent of the surface of the underparts. White leg feathers are heavily barred. Undertail coverts are covered with fairly thick barring. WINGS (dorsal).—Upperwing coverts lack pale barring like the rest of the upperparts. Tertials and secondaries may have faint gray barring. Primaries are dark. WINGS (ventral).—**Underwing coverts are quite dark and extensively spotted and barred and contrast against the paler gray remiges. There is some pale barring on the outer primaries that blends into the solid gray inner primaries and secondaries. *Note:* Overall, the underwing will appear slightly two-toned with fairly dark coverts and pale gray remiges.** TAIL.—Uppertail coverts are barred with gray and blend as a uniform color unit with the grayish rump. Dorsal surface is dark with pale gray barring.

**ADULT DARK (BLACK)-MORPH TRAITS:** HEAD.—All of the head, including forehead and nape, is uniformly blackish brown. The throat is white. The area under and in front of the eyes and malar area are often more blackish. BODY (dorsal).—Uniformly dark blackish brown or with a very narrow partial gray crossbar on many feathers, particularly on the distal scapulars. Blackish-brown rump and uppertail coverts are barred with gray and somewhat paler than the rest of the dorsal body but similar to the tail. BODY (ventral).—**Breast is densely streaked and the belly is broadly streaked or has large dark arrowhead-shaped markings or spotting. The flanks are broadly barred. Dark markings cover 50–90 percent of the surface area of the underparts. Leg feathers are dark with narrow white barring. Undertail coverts are heavily barred.** WINGS (dorsal).—Virtually uniformly blackish brown. Occasionally, a few coverts may have a faint, partial gray bar or spot. Primaries are uniformly dark. WINGS (ventral).—**Underwing coverts are blackish brown and covered with**

**a small amount of white spotting, particularly on the median coverts. The dark coverts contrast sharply with the nearly uniformly pale gray remiges. The outer one to three primaries may have numerous, faint, narrow pale bars.** ***Note:*** **The underwing appears two-toned with the nearly solid dark coverts and pale gray remiges.** TAIL.—Dorsal surface is dark with numerous pale gray bands that are equal to or narrower than the dark bands. Ventral tail surface is medium gray and covered with narrow pale gray bands.

**SUBADULT TRAITS:** HEAD.—As on the respective color morph of adult except the cere and orbital skin are pale bluish in Gyrfalcon's first subadult year as a 1-year-old. Older subadults, which are subadult II or older, have pale yellow ceres and orbital skin. BODY.—As on the respective color morph of adult. Molt is incomplete and may retain a small amount of juvenile feathering on the dorsal regions, particularly on the rump. On white morphs, feet may be somewhat pinkish gray until changing to the yellow adult color. On all other color morphs, feet are grayish or grayish blue on younger subadults, then gradually change to pale yellow as they get older. WINGS.—Molt is incomplete, and 1-year-olds may retain a highly visible, large patch of juvenile feathers on the lesser upperwing coverts and sometimes a few feathers on the median covert tract. Retained feathers are faded brown in all color morphs, including white morphs. Older subadults molt into a full adult feathering or may retain a few very faded scattered juvenile feathers—now 3 years old—particularly on the upperwing coverts. These individuals may exhibit three ages of upperwing coverts. TAIL.—As on respective color morph of adult. All juvenile rectrices are fully replaced in the first prebasic molt.

**JUVENILE TRAITS:** HEAD.—Bluish or grayish cere and orbital skin. BODY.—Feet are pale pink on white morphs and grayish on all other color morphs. TAIL.—Long tail. Wingtips extend two-thirds of way down the tail when perched.

**JUVENILE LIGHT (WHITE)-MORPH TRAITS (LIGHTLY MARKED TYPE):** HEAD.—**Immaculate white with a few small dark specks on the nape or may have some dark streaking on the eyeline and nape.** BODY (dorsal).—**Scapular and back feathers are white with a narrow or moderately wide brown or dark brown vertical streak along the shaft region of each feather. Lower back and rump are white with very narrow dark brown streaking.** BODY (ventral).—White with a few dark specks on the sides of the breast and flanks or all of the breast, belly, and flanks may have small dark dashes. The white leg feathers are generally unmarked and the white undertail coverts are always unmarked. WINGS (dorsal).—**All lesser and median coverts have a narrow or moderately wide brown or dark brown streak along the shaft area of each feather. Greater coverts have partial brown barring. Secondaries may have partial brown barring or may be unmarked. Primaries have partial dark barring on the distal area of each primary and have a dark tip on all feathers.** WINGS (ventral).—**Immaculate white with dark brown "dipped-in-ink" tips on the outer six or seven primaries. There is a narrow brown subterminal band on the trailing edge of the remiges. In translucent light, partial barring is evident on the underside of the remiges.** TAIL.—Dorsal surface is white and is either unmarked or may have a very small amount of dark mottling or partial barring next to the dark shaft on the deck rectrix set.

**JUVENILE LIGHT (WHITE)-MORPH TRAITS (HEAVILY MARKED TYPE):** HEAD.—**White with thin brown streaking on the crown, auriculars, and nape.** BODY (dorsal).—**Back and scapulars are medium brown or dark brown with a medium wide white edge on each feather. The distal one-third of the scapulars may have a broad barred pattern. Lower back and rump also have brown feathering with broad white edges.** Uppertail coverts are white with a dark streak or partial narrow barring on each feather. BODY (ventral).—White with short medium brown or dark brown dashes on the breast, belly, and flanks. Lower belly, leg feathers, and undertail coverts are typically white and unmarked. WINGS (dorsal).—**All lesser and median coverts are medium brown or dark brown with broad white edges. Greater coverts and the secondaries are white with brown barring. The dark barring covers less than 50 percent of the feather surface. Greater primary coverts and most of the upper surface of the primaries are white with brown barring. The**

**tips of the outer six primaries are dark brown and forms a large, dark "dipped-in-ink" brown patch on the wingtips that is highly visible when perched or in flight.** WINGS (ventral).—**White underwing coverts are either unmarked or have small brown dashes. Underside of the remiges may have partial brown barring. The outer six or seven primaries have the "dipped-in-ink" dark brown wingtips.** TAIL.—Dorsal tail surface is white with several narrow dark brown bars. The ventral tail surface is white and unmarked, but in translucent light will show barring from the dorsal surface.

**JUVENILE LIGHT INTERMEDIATE (LIGHT GRAY)-MORPH TRAITS:** HEAD.—White with fairly narrow dark brown streaking on the crown, auriculars, and nape. The dark streaking may create an ill-defined malar mark. Head is similar to a pale-headed type of juvenile gray morph. Throat is white and unmarked. BODY (dorsal).—Dark brown with a moderately wide white edge on all back, scapular, and rump feathers. Additional white markings on the basal two-thirds of the scapular tract form a partially barred pattern and is similar to adult light intermediate morphs. The overall pattern of white edgings and other markings creates a darker dorsal surface than seen on a heavily marked type of white morph but paler than any gray morph. BODY (ventral).—White with narrow dark brown streaking on the breast, belly, and flanks. The white leg feathers are very narrowly streaked. The white undertail coverts have a thin dark streak along each feather shaft. WINGS (dorsal).—Lesser and median coverts are dark brown with moderately wide white edges and partial white barring on the median coverts. Greater coverts and secondaries are dark brown with white edges and two or three distinct narrow white bars on each feather. Upper surface of the primaries is also brown and barred throughout with white. WINGS (ventral).—**White underwing coverts are narrowly streaked on the lesser coverts and have a somewhat darker band on the median coverts and axillaries created by moderately wide broad dark barring. Remiges are whitish with narrow gray barring on all primaries, wide gray barring on the outer secondaries, and very wide gray barring on the darker inner secondaries. The primaries appear paler than the secondaries.** ***Note:*** **Overall appearance of the underwing is very pale with a somewhat darker barred area on the axillaries and median coverts. The wingtips are dark brown.** TAIL.—Dorsal surface is marked with equal-width white and dark brown bands. Uppertail coverts are dark brown with white edges and a white cross bar on each feather. *Note:* This plumage has not been previously described. It is darker than heavily marked juvenile light morphs but paler than juvenile intermediate morphs. Type specimens are in the Denver Museum of Nature and Science and the AMNH.

**JUVENILE INTERMEDIATE (GRAY)-MORPH TRAITS:** HEAD.—Forehead is white and the crown is white with narrow dark brown streaking, and the nape is very mottled with white. There is generally a broad dark eyeline and often a dark brown area on the distal region of the auriculars. Cheeks and forward part of the auriculars are white and streaked with dark brown. There is a moderately distinct or distinct dark malar mark. White nape is streaked with brown. BODY (dorsal).—Dark brown (not gray as in adults) with narrow white edgings on all feathers. By late winter, the pale edgings often wear off and the upperparts become uniformly dark brown. Very distal region of the scapulars may have a partial or narrow white mid-feather cross bar. BODY (ventral).—White with moderately wide dark brown streaking on the breast, belly, and flanks. Dark streaking covers 50 percent of each feather surface. White leg feathers are moderately streaked with dark brown. White undertail coverts have a thin dark brown streak along the feather shaft of some or all feathers. WINGS (dorsal).—Dark brown throughout with narrow white edgings on all lesser and median coverts. The greater coverts and secondaries have narrow white edges and two or three faint partial bars. Primaries are dark brown with small white spots on all feathers. WINGS (ventral).—**The secondaries are uniformly pale gray and are darker than the primaries, which are white and barred with gray. The axillaries and greater coverts are heavily barred and the lesser coverts are heavily streaked.** ***Note:*** **The underwing appears two-toned with darker streaked/barred coverts and pale grayish remiges.** TAIL.—Dorsal tail surface is dark brown with seven or eight narrow white bands. The white bands may be offset or chevron-shaped. Ventral surface is also narrowly banded. Uppertail coverts are dark brown with white edges and sometimes a partial white bar or white spot on each feather.

**JUVENILE DARK INTERMEDIATE (DARK GRAY)-MORPH TRAITS:** HEAD.—Forehead is white and crown is dark brown. There is a partial whitish supercilium and a small amount of white mottling on the nape. The cheeks and auriculars are dark brown, the dark malar mark is distinct. The throat is white. Very rarely, head is pale and streaked on intermediate morphs. BODY (dorsal).—Uniformly dark brown. BODY (ventral).—Dark brown streaking with narrow white edges on all of the breast, belly, flanks, and leg feathers. The dark area of each feather covers 50–75 percent of the feather surface. Undertail coverts are white with moderately wide dark streaks on the shaft area. WINGS (dorsal).—Uniformly dark brown on all coverts, secondaries, and primaries. WINGS (ventral).—**Dark brown with narrow white streaking on the lesser coverts and white spotting on the median coverts and axillaries. Primaries are whitish with narrow gray barring and are paler than the uniformly gray secondaries. Wingtips are dark.** *Note:* **The underwing appears two-toned with dark coverts and pale gray remiges.** TAIL.—Dorsal surface is uniformly dark brown or may have a few pale, partial bands. Ventral surface may be partially banded with narrow whitish bands or is fully banded with narrow bands. Uppertail coverts are either uniformly dark or have small white tips on each feather.

**JUVENILE DARK (BLACK)-MORPH TRAITS:** HEAD.—Head is uniformly dark brown except the pale forehead and whitish streaked throat. Very rarely, head is whitish and streaked as on intermediate morph. BODY (dorsal).—Uniformly dark brown. BODY (ventral).—Dark brown with very narrow white edges on the breast, belly, leg feathers, and forward portion of the flanks; rear flanks have a small white spot on each side of the feather shaft. Lower belly is dark. On the darkest individuals, the breast and belly are all dark, but there are very thin white edges on the upper belly, leg feathers, and forward flanks; rear flanks have small white spotting. Undertail coverts are white with broad dark barring. WINGS (dorsal).—Uniformly dark brown on all coverts and remiges. WINGS (ventral).—**Dark brown underwing coverts and axillaries have very narrow white edges on the lesser coverts and small white spots on the median coverts and axillaries. Remiges are uniformly pale gray with narrow white barring on the outer primaries. *Note:* Underwing appears two-toned with dark coverts and pale gray remiges.** TAIL.—Dorsal surface is uniformly dark brown. Ventral surface is dark with narrow partial white bands on the most distal part of each feather. Uppertail coverts are uniformly dark brown.

**ABNORMAL PLUMAGES:** None.

**HABITAT: Summer.**—Open, moist, and dry barren arctic and alpine tundra and fairly open areas along northern edge of the boreal forest. Breeding occurs from sea level to 5,350 ft. (1,630 m). Breeding pairs are found in areas with rocky outcrops, hard- or soft-substrate embankments and cliffs, and occasionally in small White Spruce groves. Most embankments and cliff areas are along seashores (adjacent to seabird colonies), rivers, lakes, and mountain ridges. Areas along major rivers support the highest breeding densities. Nonbreeding individuals may be found in open areas lacking cliffs and trees and include open tundra expanses, seashores, riverbeds, and lakeshores. **Winter.**—Identical to breeding habitat for most adults and some subadults and juveniles. Many subadults and juveniles winter in entirely different habitats at more southern latitudes. Open and semi-open areas include: fields, marshes, prairies, rocky and sandy seashores and lakeshores, reservoirs, airports, and harbors. Towns and cities are also frequented. **Migration.**—As in the above locations.

**HABITS:** Fairly tame to tame raptor. Gyrfalcons prefer to perch on elevated natural and human-made objects, but also readily perch on the ground. All types of artificial structures are used for perches, including grain elevators, bridges, on and in abandoned buildings, on occupied buildings, waterfowl hunting blinds, utility poles, fence posts, and hay bales.

As in all falcons, the throat and forehead feathers are often erected. In windy conditions, the long tail is pressed against a perch and used as a brace.

**FEEDING:** Perch and aerial hunter. Gyrfalcons primarily feed on birds up to the size of a goose, and, to a lesser extent, on mammals up to the size of a hare. Rock and Willow ptarmigan are the Gyrfalcon's major prey in most areas and seasons. Seasonally, Gyrfalcons feed on different ages, sexes, and species of ptarmigan. Displaying male ptarmigan are main prey in the early summer, young ptarmigan in late summer and fall; Gyrfalcon's diet often switches more toward Willow Ptarmigan in winter. Ptarmigan numbers fluctuate widely in possible cyclic patterns and often determine the breeding success of local falcon populations. Gyrfalcons nesting at seashore locations feed mainly on colonial nesting seabirds and waterfowl. Mammals are major prey at higher elevations and more northern altitudes. Arctic Ground Squirrels make up a large part of their summer diet in some of these areas. Even small prey like lemmings are extensively fed upon when they are at their peak cycles.

Avian prey are killed on or near the ground or may be pursued and captured at fairly high altitudes. Avian prey are typically hit by the hunting falcon, then fall to the ground; less frequently, may overtake and grab avian prey in flight. Ducks and geese will often try to land on the ground and hide to escape being captured.

During the nesting season, avian prey is generally decapitated, plucked, and the outer portion of the wing is broken off prior to being transported back to the nest. Larger mammals like Arctic Hare are cut into smaller segments before being taken back to the nest. During other seasons, prey is typically eaten at or near the point of capture. Juveniles and subadults may pirate food from conspecifics and other raptors.

**FLIGHT:** Powered flight is with moderately slow, stiff wingbeats interspersed with irregular glide sequences. Wings are held on a flat plane when soaring and gliding. Gyrfalcons kite on strong winds and may hover when searching for concealed prey.

Hunting methods include: (1) random, low-altitude flapping and gliding forays back and forth over an area, (2) moderate-altitude soaring and gliding stints, and (3) direct attack from a perch or while in flight.

Attack methods: (1) *Surprise-and-flush flight.*—Ground-hugging flight to surprise, panic, and capture prey. Gyrfalcons wintering in urban areas use buildings as cover to surprise prey. (2) *Pursuit flight.*—Prey that is in the open is (2A) tail-chased in level flight, often in erractic maneuvers; (2B) pursued to a higher altitude in a powerful, direct tail-chasing course, often in erratic maneuvers; or (2C) forced to the ground. (3) *Hover flight.*—Prey that is hidden in vegetation is flushed by hovering at a low altitude above the vegetation and making short dives toward the concealed prey. (4) *Diving.*—The Gyrfalcon rises above the intended prey by a rapid, direct-flapping ascent flight, then makes a short dive to capture it. (5) *Pendulum swoops.*—Shallow dives assisted by wing-flapping that begins above prey that is typically in flocks, then swings upward in a series of undulating angles from underneath into the flock to try to capture a bird.

Although Gyrfalcons typically do not engage in long vertical dives, a trained falconry bird attained the incredible speed of 353 mph (568 km/hr)! This is the fastest speed ever recorded for a bird.

**VOICE:** Alarm call during the breeding season is a repetitive, harsh *cack, cack, cack.* Also emitted by males during some aerial courting activities. A loud, punctuated *chup, chup, chup* is emitted by both sexes during perched courting displays, given by males during the perched food-transfer courting display, and as a feeding note by both sexes. A rapid *chi-chi-chi-chi* is given by females when males approach for copulation. Rarely vocalizes outside of the breeding season. The *waiiiik* call is a soft, protracted crescendo whine given by copulating females. The waiiik call can also be loud and vigorous and emitted by unmated males in some perched and aerial displays and by mated males when approaching the eyrie with prey. The high-pitched and harsh *scree, scree, scree* is a food-begging call given by females.

**STATUS AND DISTRIBUTION:** *Uncommon.* Population is stable. Local and regional fluctuations occur with ptarmigan cycles, but Gyrfalcon populations always remain healthy. Since their prey are mainly nonmigratory, the organochlorine era of the late 1940s to early 1970s that plagued southern latitudes had little effect on Gyfalcons.

The highest nesting densities are on major river drainages.

**Summer.**—*Alaska*: 375–635 pairs. Highest densities are on major drainages flowing out of the Brooks Range Mts. on the North Slope on the Colville, Kukpuk, Sagauanivktok, and Utokuk Rivers. *Northwest Territories and Nunavut* (Nunavut became a separate Territory in 1999): Estimated 1,300 pairs, with 5,000 individuals. High breeding numbers on the Anderson, Horton, and Thelon Rivers in the Northwest Territories. *Yukon*: Estimated 750 pairs, with 2,490–4,180 birds. High breeding densities are on the North Slope on the Firth and Anker Rivers. The estimated world population is 15,000–17,000 pairs.

**Winter.**—Virtually all adult males, most adult females, and most subadult and juvenile males remain in the harsh Arctic in the winter. Many subadult and juvenile females and a few adult females winter south of the Arctic. Most wintering birds in s. Canada and the n. U.S. are females. Gyrfalcons are regular in the mapped winter range. They are irregular to very irregular in most states south to the blue dashed line on the map. *California:*—Seven records in the following counties: (1) Late Oct. 1948 in Siskiyou, (2) mid-Jan. 1982 in Yolo and Solano, (3) late Oct. 1983 in Siskiyou, (4) late Jan. 1985 in Siskiyou, (5) late Dec. 1987 in Shasta, (6) early to late Nov. 1989 in Siskiyou and Modoc, and (7) late Oct. 1993 to mid-Feb. 1994 in Del Norte. *Kansas:*—Seven records in the following counties: Barton, Elk, Pawnee, Riley, Russell, and Sedgwick. Three records since 1990: (1) Early Nov. 1990 at Cheyenne Bottoms Wildlife Area, Barton Co.; (2) early Nov. 1991 at Wilson Reservoir, Russell Co.; and (3) late Nov. 2000 near Wichita, Sedgwick Co. *Missouri:*—(1) Early Dec. 1994 in Monroe Co. and (2) early Mar. 1996 in Linn Co. *Oklahoma:*—(1) Mid-Jan. 1974 near Grainola, Osage Co. (white morph); (2) early Nov. 1982 near Foraker, Osage Co. (gray morph); (3) early to mid-Feb. 1982 in Oklahoma City, Oklahoma Co. (juvenile dark morph); and (4) early Jan. 1991 near Foraker, Osage Co. (dark juvenile). *Texas:*—Late Jan. 2002 in Lubbock (juvenile gray-morph female). *Note:* Latter individual not reviewed by state bird records committee at time.

Winter territories may be established by adults and subadults. Juveniles are prone to wanderlust in the winter and rarely establish territories. In rare cases, a mated pair may winter together south of the breeding grounds.

**Movements.**—Southward movements from Arctic breeding areas are mainly by females.

*Fall migration*: Southward journeys out of the Arctic may begin in late Aug. Individuals may be seen in s. Canada as early as late Sep. and in the n. U.S. in late Oct. Southward movements, especially by nomadic juveniles, may continue throughout mid-winter. It is unknown if Gyrfalcons breeding in the high-Arctic latitudes of Nunavut move south of breeding areas during winter.

*Spring migration*: Adult females may begin heading north in late winter. There are numerous Gyrfalcon records in Mar. and Apr. for the U.S. There are May records for s. Canada.

**NESTING:** Breeding season depends on latitude and geography. Pairs may remain together year-round in some areas. New pair formation may begin in Feb. Pairs may not breed every year: Breeding is dependent on food supply. Eggs are laid from late Mar. to late May. Most young fledge between Jun. and late Aug. Breeding first occurs when birds are 2–4 years old.

**Courtship (perched).**—Either sex performs the following displays on the nest ledge: *vertical head-low bow*, *horizontal head-low bow*, *mutual ledge display*, and *billing*. Individual courtship behavior includes *male ledge display* and *female ledge display* (emitting *chup* and *waiiiik* calls), *wail-pluck display* by the male (emitting *waiiiik* call). Copulation display by male is the *curved-neck display*.

**Courtship (flight).**—Male displays: *roll*, *undulating roll*, *flash roll*, and *eyrie-flyby*. Female displays: *flutter-flight* (emitting *scree* call). Mutual displays: *aerial food transfer* (with *chup* and *waiiiik* calls), *high-circling*, *mutual-floating*, and *passing-and-leading display*.

Nest sites are on cliff ledges and usually in a location with an overhang to protect the incubating and brooding female and nestlings from the harsh weather. Gyrfalcons may create a shallow scrape in the soil on a ledge or occupy old stick nests on ledges of Golden Eagles, Common Ravens, or Rough-legged Hawks. On ledge nests, both sexes scrape a shallow bowl by laying on their bellies and kicking the dirt out behind them. Gyrfalcons occasionally use old stick nests of Common Ravens and Rough-legged Hawks that are placed in White Spruce. Nest sites may be occupied by a pair or several pairs for many years. Nests that are reused often accumulate a buildup of excrement below the eyrie. Stick nests on ledges are typically destroyed by active nestlings.

One to 5 eggs are laid. Females do most of the incubating. Eggs can be left unattended for long periods in freezing temperatures without suffering ill effects. The eggs hatch in about 35 days. Nestlings fledge in 46–50 days.

**CONSERVATION:** In 2000, the USFWS proposed to transfer the Gyrfalcon from Appendix I to Appendix II of the Convention on International Trade of Endangered Species of Wild Fauna and Flora (CITES). CITES rejected the proposal and the falcon was retained on Appendix I, the optimum protection typically granted for an endangered species. Gyrfalcons, particularly white morphs, have been popular for falconry for centuries. Canada and Alaska allow the take of Gyrfalcons for private falconry use but not for commercial use. Very few nestlings are taken from nest sites for falconry. Most falconry birds are obtained from licensed breeders of falcons raised in captivity. About 150 Gyrfalcons are raised by licensed breeders each year.

**Mortality.**—Birds die of natural causes in the Arctic. Those wintering south of the Arctic may become victims of illegal shooting, electrocution from power lines, and collisions with fence and power lines.

**SIMILAR SPECIES:** COMPARED TO ADULT INTERMEDIATE (GRAY) MORPH.—**(1) Northern Goshawk, adults.**—PERCHED.—Orange or red irises. Black and white head pattern is distinct; never has a hint of dark malar mark. Blue-gray dorsal plumage is unbarred. Tail is unbanded or has 3 or 4 wide black bands. FLIGHT.—Body and wing shape are similar, especially when gliding. Distinct black barring on the underside of outer primaries. Underparts are uniformly barred. On the dorsal surface, the remiges are darker than the coverts, including the greater primary coverts. **(2) Peregrine Falcon, adults (all races).**—PERCHED.—Head is black with more sharply defined malar. *F. p. tundrius* is most similar because head is often grayish and mottled on supercilium and nape. Wingtips equal the tail tip. FLIGHT.—On the ventral surface, remiges are uniformly spotted or barred. Tail is dark. On the dorsal surface, the lower back and rump are paler than the rest of the upper body and the tail is very dark. COMPARED TO JUVENILE INTERMEDIATE (GRAY) AND DARK MORPHS.—**(3) Northern Goshawk, juveniles.**—Very similar overall appearance; use caution! PERCHED.—Yellow irises. Ventral streaking and wingtip-to-tail-tip ratio are similar. Tail is distinctly banded with 3 or 4 dark bands. FLIGHT.—Underside of the remiges is thickly barred. Tail is prominently banded. **(4) Swainson's Hawk, subadult intermediate and dark-morph juveniles.**—Overlap in spring and fall. PERCHED.—Head is similar but lacks fleshy orbital region. Ventral areas similarly streaked: Intermediate morph like intermediate-morph Gyrfalcon and dark morph like dark-morph Gyrfalcon. Tarsi and feet are yellow. Wingtips equal or nearly equal tail tip. FLIGHT.—Pointed wingtips are similar. Underside of remiges is darker than the coverts or equally as dark. **(5) Ferruginous Hawk, light-morph juveniles.**—Similar to juvenile gray morph, especially dorsal views. PERCHED.—Pale irises. Shares similar thick dark eyeline and lack of a dark malar. Underparts are white and spotted, not streaked. Wingtips are nearly equal to the tail tip. FLIGHT.—Similar thick-chested look to body. Wings deceptively pointed when gliding. Whitish panels on upper surface of the primaries. Dusky tail bands. White uppertail coverts and basal region of tail. **(6) Ferruginous Hawk, rufous/dark-morph juveniles.**—Similar to dark morph from all angles of view. PERCHED.—Pale irises. Pronounced yellow gape. Underparts lack pale markings. Wingtips nearly equal to the tail tip. FLIGHT.—Similar thick-chested look to body. Wings deceptively pointed when gliding. Underside of remiges is white and nearly unmarked. Whitish panel on dorsal surface of primaries. Tail has dark subterminal smudge. **(7) Peregrine Falcon, juveniles (all races).**—*F. p. tundrius* similar to intermediate morph; *F. p. anatum* and *F. p. pealei* similar to dark morph. PERCHED.—Head features similar but dark malar typically more defined. Ventral areas tawny colored. Wingtips equal to or barely shorter than the tail tip. FLIGHT.—Underwing is uniformly dark with tawny-rufous spotting on all remiges. **(8) Prairie Falcon,**

**GYRFALCON,** *Falco rusticolus:* Estimated population: AK: 375–635 pairs, YK: 750 pairs, NT: 1,300 pairs. (Data for NT includes new territory of Nunavut that was created from NT in 1999.) Uncommon. Irregular to very irregular south to dashed line. Wintering birds in s. Canada and U.S. are mainly female gray morphs.

**adults and juveniles.**—Similar to gray morph. PERCHED.—Distinct white area behind the eyes. Underparts of juveniles are similarly streaked, especially on females; flanks are more solid dark. Wingtip-to-tail-tip similar. Dorsal areas similar with pale feather edgings. FLIGHT.—Axillaries are solid dark. Underwing is otherwise very similar: Darker area on the median covert tract on paler marked Prairies is identical to gray morph; those with all-dark median coverts are easily separable.

**OTHER NAMES:** Gyr (female), jerkin (male). *Spanish:* Halcón Gerifalte. *French:* Faucon Gerfaut.

REFERENCES: Alcorn 1988; Andrews and Righter 1992; Burt and Grossenheider 1976; Clum and Cade 1994; Dorn and Dorn 1990; Dekker and Lange 2001; Dotson and Mindell 1979; Ducey 1988; Ferguson-Lees and Christi 2001; Forsman 1999; Gilligan et al. 1994; Harris 1979; Janssen 1987; Kent and Dinsmore 1996; Godfrey 1986; Johnsgard 1990; McCaskie and San Miguel 1999; Montana Bird Distribution Committee 1996; Palmer 1988; Robbins and Easterla 1990; Rottenborn and Morlan 2000; Sanchez 1993, 1994; Seyffert 2001; Small 1994; Terres 1980; Tucker et al. 1998; USFWS 2000b; Wildlife Management Advisory Council (North Slope) 2001.

**Plate 576. Gyrfalcon, adult white morph (unknown sex) [May]** ▪ White head. Yellow cere and orbital skin. ▪ White back and scapulars are sparsely marked with arrowhead shapes and bars. Underparts are white. ▪ Sparsely barred white wing coverts. Black wingtips. ▪ Tail is barred on central rectrices. ▪ *Note:* Photograph by Robert Fyfe.

**Plate 577. Gyrfalcon, adult (unknown sex) light intermediate morph [Jun.]** ▪ Whitish head with a blackish malar mark. Thin dark eyeline. Yellow cere and orbital skin. ▪ Brownish black upperparts are distinctly barred with white. White underparts are lightly spotted; leg feathers are lightly barred. ▪ Wingtips are shorter than tail tip. ▪ *Note:* Photographed in Nome, Alaska, by Brian Small.

**Plate 578. Gyrfalcon, adult female intermediate (gray) morph [Jun.]** ▪ Dark type head. Yellow cere and orbital skin. ▪ Dark gray upperparts are the partial-barred type that lack distinct pale cross bars. White underparts are lightly spotted/streaked; leg feathers are heavily barred. ▪ *Note:* Photographed in Nome, Alaska, by Jim Zipp.

**Plate 579. Gyrfalcon, adult female intermediate (gray) morph [Feb.]** ▪ Dark type head. Yellow cere and orbital skin. ▪ Dark gray upperparts are the partial-barred type that lack distinct pale cross bars and are nearly uniformly dark. White underparts are moderately spotted/streaked with thick barring on flanks. ▪ Wingtips are distinctly shorter than tail tip. ▪ *Note:* Boulder Co., Colo.

**Plate 580. Gyrfalcon, subadult female dark morph [Feb.]** ▪ Blackish brown head. Pale yellow cere and orbital skin. ▪ Blackish brown underparts are speckled with white on breast and belly and barred on flanks. ▪ A few faded, pale brown juvenile feathers on upperwing coverts. Wingtips are distinctly shorter than tail tip. ▪ *Note:* Photographed in Ottawa, Ont., by Tony Beck.

**Plate 581. Gyrfalcon, adult (unknown sex) white morph** ▪ White underparts. ▪ White underwing with small black wingtips. ▪ White tail may be unmarked or variably banded. ▪ *Note:* Photographed in Thule, Greenland, by Jack Stephens.

**Plate 582. Gyrfalcon, adult (unknown sex) light intermediate morph [Jun.]** ▪ Lightly spotted white underparts. ▪ Pale underwings: gray and white barred secondaries are darker than lightly barred white primaries. Lightly marked white coverts. ▪ Tail is distinctly banded. ▪ *Note:* Photographed in Nome, Alaska, by Brian Small.

**Plate 583. Gyrfalcon, subadult female intermediate (gray) morph [Feb.]** ▪ Heavily barred/spotted white underparts; barred flanks. ▪ Gray secondaries are darker than white barred primaries. ▪ *Note:* Subadult features on upperwing coverts, with a few, faded brown juvenile feathers. Photographed in Duluth, Minn., by Frank Nicoletti.

**Plate 584. Gyrfalcon, subadult female dark morph [Feb.]** ▪ Blackish brown head. ▪ Underparts are speckled with white on breast and belly and barred on flanks. ▪ Gray remiges are paler than dark, spotted coverts. ▪ *Note:* Same bird as on plate 580. Photographed in Ottawa, Ont., by Tony Beck.

**Plate 585. Gyrfalcon, juvenile (sex unknown) white morph (lightly marked type) [Jul.]** ▪ Pale blue bill, cere, and orbital skin. Mainly white head. ▪ White upperparts and wing coverts are streaked. ▪ Black wingtips. ▪ White tail. ▪ *Note:* Recently fledged. Photographed on Ellesmere Island, Nunavut, by Wayne Lynch.

**Plate 586. Gyrfalcon, juvenile female white morph (heavily marked type) [Feb.]** ▪ Pale blue cere and orbital skin. ▪ Brown upperparts are edged and spotted with white. ▪ Dark brown wing coverts are edged and spotted with white and black wingtips. ▪ White tail is banded. ▪ *Note:* Photographed in Newfoundland by Bruce Mactavish.

**Plate 587. Gyrfalcon, juvenile female intermediate (gray) morph [Dec.]** ▪ Pale blue cere and orbital skin. Brown head with a pale supercilium and indistinct dark malar mark. ▪ White underparts, including leg feathers, are moderately streaked with dark brown. ▪ *Note:* Weld Co., Colo.

**Plate 588. Gyrfalcon, juvenile (unknown sex) dark intermediate (dark gray) morph [Jul.]** ▪ Pale blue cere and orbital skin. Dark brown head with a faint, pale supercilium. ▪ Dark brown upperparts. Heavily streaked white underparts, including leg feathers. ▪ *Note:* Recently fledged. Denali N.P., Alaska.

**Plate 589. Gyrfalcon, juvenile female dark morph [Feb.]** ▪ Blackish brown head. Pale blue cere and orbital skin. ▪ Nearly uniformly blackish brown underparts with a few thin white streaks and spots. Uniformly blackish brown upperparts. ▪ *Note:* A very heavily marked bird. Photographed in Ottawa, Ont. by Tony Beck.

**Plate 590. Gyrfalcon, juvenile female intermediate (gray) morph [Feb.]** ▪ White underparts are thinly streaked. ▪ Secondaries are barred gray and are darker than white primaries. Heavily barred axillaries and greater coverts and streaked lesser coverts are darker than remiges. ▪ *Note:* Pale extreme of gray morph. Photographed in Ottawa, Ont. by Tony Beck.

**Plate 591. Gyrfalcon, juvenile female intermediate (gray) morph [Jul.]** ▪ White underparts are moderately streaked with brown. ▪ Heavily barred axillaries and greater coverts and streaked lesser coverts are darker than gray remiges. ▪ *Note:* Typical gray morph. Photographed at Rankin Inlet, Nunavut, by Gordon Court.

---

# PRAIRIE FALCON
## *(Falco mexicanus)*

**AGES:** Adult and juvenile. Adult plumage is acquired when 1 year old. Adult plumage is somewhat sexually dimorphic, but not readily separable at long field distances. Juvenile plumage worn the first year. Juveniles have somewhat broader wings (longer secondaries) and longer tails than adults do.

**MOLT:** Falconidae wing and tail molt pattern (*see* chapter 4). First prebasic molt from juvenile to adult plumage occurs Apr.–Nov. Molt begins on the head, then expands onto the breast, back, and forward scapulars, and then proceeds to rear of the body. Remiges and rectrices molt after most contour plumage is acquired.

Subsequent annual adult prebasic molts begin Apr. to early May at lower elevations and southern latitudes (K. Steenhof pers. comm.); birds inhabiting high elevations and northern latitudes begin from late May through Jun. Females begin molt prior to males. Body, wings, and tail molt in unison. Remix molt, based on Palmer (1988), begins on the mid-primaries (p4) then in sequence of p5, 6, 3, 7, 2, 1, 9, and 10. There is a variance in p8: it could be before p2 or after p9. Shortly after remix molt is initiated, rectrix molt begins on the deck set (r1), then in sequence of r2, 6, 3, 4, and 5. Annual molt is fairly complete.

**SUBSPECIES:** Monotypic.

**COLOR MORPHS:** None.

**SIZE:** A medium-sized raptor. Sexually dimorphic with little or no overlap, but sexes are not always readily separable in the field. Males average smaller than females. Juveniles are longer than adults because of having longer rectrices and may be on the larger end of the size spectrum for each respective sex. MALE.—Length: 14–16 in. (35–41 cm); wingspan: 36–38 in.

(91–96 cm). FEMALE.—Length: 16–18 in. (35–46 cm); wingspan: 41–44 in. (104–112 cm).

**SPECIES TRAITS:** HEAD.—Large eyes with dark brown irises. Whitish supercilium either connects with the white forehead or begins over the eyes and may connect on the nape as a "V" shape. **Sharply defined white cheek area behind each eye is bordered by a brown auricular patch.** Narrow dark "mustache." Falcons have bare, fleshy orbital skin. WINGS.—**Solid dark brown axillaries with moderately dark or solid dark median underwing coverts.** When perched, wingtips are moderately shorter than the tail tip. **In flight, long wings taper to pointed wingtips.**

**ADULT TRAITS:** HEAD.—Yellow cere, orbital skin, and feet are brighter on males, but may be the same color for both sexes (K. Steenhof pers. comm.). Partially barred flanks and leg feathers. **Brown upperparts have a variably distinct pale tawny or gray cross bar on the mid-section of most feathers.** TAIL.—Dorsal surface is brown and may have numerous, very narrow pale bands. **Dorsal surface of tail is generally much paler than the rest of the upperparts.**

**Adult male:** HEAD.—White supercilium is very distinct and often extends onto the nape. BODY (dorsal).—**Back, scapulars, rump, and upperwing coverts are medium brown or pale brown and always has a distinct, wide, very pale tawny or gray outer edge and a mid-feather cross bar on each feather.** BODY (ventral).—Dark markings may be small and sparse or absent on the breast. Belly is marked with small or medium-sized spots or dash marks and may overlap with markings of some adult females. WINGS.—**Dark median underwing coverts are extensively spotted or barred with white and appear only slightly darker than rest of the underwing (but accentuates the solid dark brown axillaries). Upperwing coverts are distinctly barred with a tawny or grayish bar on each feather.** TAIL.—Deck rectrices are pale brown and are either unmarked or possess very pale, narrow banding; all other rectrices are covered with very pale, narrow bands. Moderately wide white terminal band.

**Adult female:** HEAD.—White supercilium may be pronounced and extend onto the nape or is only a partial white stripe above the eyes. BODY (dorsal).—*Unbarred type*: **Pale feather edges and cross bars are indistinct or absent and upperparts appear almost uniformly brown.** *Note:* Majority of females are this type. *Barred type*: **Medium brown upperparts have a distinct pale outer edge and a mid-feather cross bar on each feather (malelike).** *Note:* Only a small number of females are this type. BODY (ventral).—Breast and belly are moderately or heavily marked with large or very large dark spots. Some appear to be almost streaked because the spotting is clumped together. WINGS.—**Median underwing coverts either (1) are solid dark brown or (2) have a small amount of white spotting (but have a lesser amount of spotting than on most males; rare overlap with males). Combined with solid dark axillaries, the very dark brown underwing coverts form a broad dark band on the underwing and are much darker than on males.** Upperwing coverts are nearly solid brown on the unbarred type or have pale cross-barring on the barred type. TAIL.—**Dorsal surface is consistently paler than rest of the upperparts and exhibits little if any pale banding.**

**JUVENILE TRAITS:** HEAD.—Orbital skin and cere are pale yellow, but gradually turn a brighter yellow by spring. BODY (dorsal).—Medium brown upperparts have very narrow pale edges on all feathers, but lack the pale cross-barring of most adults. By late winter, pale edgings wear off and the upperparts become uniformly medium brown. By late winter and spring, upperparts also fade and wear to pale brown. BODY (ventral).—Underparts, including leg feathers and forward half of the flanks, are white with very narrow to moderately wide dark brown streaking; rear half of the flanks are mostly a solid brown patch. Males tend to have very narrow or narrow streaking, females have narrow or moderate streaking. Feet are pale yellow, but gradually turn to a brighter yellow by spring. WINGS.—**Solid dark brown axillaries are distinct.** Somewhat sexually dimorphic on the underwing pattern, but is not a consistent sexual trait as seen in adults. *Male*: (1) An extensive amount of white spotting on the dark coverts as on virtually all adult males, (2) a minimal amount of white spotting on the dark coverts, or (3) uniformly dark brown coverts as on many females. *Female:* (1) Uniformly dark brown coverts that blend with the uniformly dark axillaries or (2) a minimal or moderate amount of white spotting on the dark coverts. TAIL.—Uniformly brown on the dorsal surface in fresh plumage, but fades and becomes paler than the rest of the upperparts by mid-winter, particularly on the deck rectrices. Moderately wide terminal tail band.

**JUVENILE (RECENTLY FLEDGED):** HEAD.—Pale bluish cere, orbital skin, and feet. *Note:* Pale cheek area is sometimes ill-defined on a few individuals soon after fledging, but becomes more apparent after having been fledged for a few weeks and becomes a diagnostic field mark. BODY.—Dark brown upperparts with all feathers being edged with tawny-rufous; on some, the edgings are very distinctly marked with broad tawny-rufous edgings. Rich tawny-colored underparts are marked with dark brown streaking as described above. WINGS.—Pale spotting on dark median underwing coverts, if present, is either pale tawny or white. TAIL.—Uppertail is uniformly dark brown and the same color as the rest of the dorsal areas. White terminal band is very wide. (Juveniles can be separated from adults, which have narrow white terminal bands at this time of year, by width of white tail tip.) *Note:* This fresh and new plumage occurs from fledging (early Jun. to mid-Jul.) to at least Sep. Thereafter, plumage gradually fades and wears to a medium brown color as noted in Juvenile Traits.

**ABNORMAL PLUMAGES: Partial albino.**—Several records (Clark and Wheeler 1987; Steenhof 1998).

**Imperfect albino (dilute plumage).**—Several records (Clark and Wheeler 1987; Steenhof 1998). Specimens with pale "frosting" on most contour feathers (C. M. White pers. comm.). *Note:* These rare plumages are not depicted in accompanying photographs.

**HABITAT: Summer.**—Arid, very open regions of short grass or scrub vegetation with cliff formations that are at least 20 ft. (6 m) high. Nest cliffs are on buttes, hillside escarpments, or canyons and may be formed of soft or hard substrate. Breeding cliffs are sometimes in semi-open regions with scattered conifer trees and occasionally amidst dense conifer woodlands several miles from open foraging habitat. Although some nest cliffs may be near riparian areas, Prairie Falcons do not require water. Most areas are in undisturbed regions, but the falcons may inhabit areas interspersed with light agriculture or grazing lands. Pairs may also successfully breed in the proximity of human activity if they are not disturbed. Pairs primarily breed below timberline elevation, but are known to breed up to 12,000 ft. (3,700 m) in n. Rocky Mts. of Colorado. Semi-open woodlands, subalpine meadows, and alpine meadows up to 14,000 ft. (4,300 m) are inhabited in mid- to late summer in most montane regions of western states.

**Winter.**—Habitat is often the same as breeding areas but without a requirement for cliffs. Winter months are generally spent at much lower elevations than in summer. Prairie Falcons inhabit open or semi-open mountain valleys; open arid prairies and deserts; open or semi-open agricultural and rural areas; open and semi-open moderately humid areas; and, occasionally, semi-open humid coastal meadows, pastures, and salt flats.

**Migration.**—Same as for summer and winter.

**HABITS:** Generally solitary. Two to four juveniles (perhaps siblings) often associate with each other during the postfledging and fall migration periods and sometimes in winter. Juveniles often play games with each other, including talon-grappling. Several falcons may loosely accompany large passerine flocks during the nonbreeding season.

Falcons often erect throat and forehead feathers. Typical of large falcons, the tail may be braced against a perch for added stability in windy conditions. Quite wary in nonbreeding season, but tame individuals are sometimes encountered. Very aggressive in breeding season when defending territories and exhibit little fear of humans.

Exposed, elevated structures of any type are used for perches; Prairie Falcons also perch on the ground. If on the ground, they prefer to perch atop chunks of dirt. Utility poles and fence posts are commonly used perches in the nonbreeding season. When perching in trees, outermost and uppermost exposed branches are preferred. Prairie Falcons are highly tolerant of climate extremes.

**FEEDING:** Perch and aerial hunter. Perch hunting is a direct, low-altitude attack at intended prey. Aerial hunting consists of random surprise-and-flush forays. The falcons generally hunt singly, but possible cooperative hunting has been observed, with one bird flushing prey and the other chasing it.

Small and medium-sized rodents, small rabbits, small hares, and small and medium-sized birds are primary prey. Occasionally, larger-sized rodents, rabbits, hares, and birds are preyed upon. Ground squirrels are major food sources during the breeding season in many regions, especially *Spermophilus* spp. (e.g., Townsend's, Belding's, and Richardson's). Avian species become the primary food source in the nonbreeding season. Reptiles and large flying insects are also preyed upon in certain locales and seasons.

Hunting styles: (1) *Dive-chasing.*—Short, acrobatic, shallow-angled dives that begin at moderate or low altitudes, with the falcon swinging sharply upward and grabbing the prey from underneath. (2) *Tail-chasing.*—Direct, low-level aerial pursuit of many avian and all mammalian prey. (3) *Hover flight.*—Intermittent hovering and slow flight performed by inexperienced juveniles. (4) *Passive hunting.*—Mainly a perch-hunting method in which a Prairie Falcon of any age sits and waits for a foraging Northern Harrier to flush avian prey. (5) *Pirating.*—Used against raptors such as Northern Harriers or species as large as Red-tailed Hawks. (6) *Scavenging.*—Very rarely practiced form of obtaining food.

The Prairie Falcon's notched bill is used to break the prey's neck, then, with avian prey, it plucks the feathers and mammalian prey it plucks fur prior to eating.

**FLIGHT:** Powered flight is rapid and direct with moderately fast, shallow, stiff wingbeats. In powered flight, Prairie Falcons may flap for long distances or intersperse flapping with irregular gliding sequences. High-speed ground-skimming altitude often is used when hunting in direct pursuit and in random surprise-and-flush flights. High-altitude flights may occur at other times. Wings held on a flat plane when soaring and gliding. If wind velocity is sufficient, Prairie Falcons will kite for short periods along cliff faces. Juveniles may briefly hover at low altitudes when searching for hidden prey and may fly at slow speeds at low altitude between hovering stints (*see* Feeding).

**VOICE:** Highly vocal in the breeding season during courtship activities or when disturbed. A repetitive, harsh, and rapid *caack-caack-caack-caack* is the most typically heard vocalization. This call is often nasal in quality, especially in females. Males have a discernibly higher-pitched call. A high-pitched, wavering whine *kree, kree, kree* by food-begging adult females, nestlings, and fledglings. Females may emit a *cherk* or other short notes at nest sites. When courting and around nest sites, an *eechip* or *eechup* note is uttered. Silent away from the breeding grounds.

**STATUS AND DISTRIBUTION: Summer.**—*Uncommon.* There are an estimated 4,300–6,000 pairs in the U.S., Canada, and Mexico. The largest numbers are in Nevada (1,200 pairs) and Wyoming (820 pairs). Population has always been historically low but seemingly stable. In Canada, listed as a Species at Risk in British Columbia and as a Sensitive Species in Alberta and Saskatchewan.

Because of remote habitat preferences, Prairie Falcons have been subjected to only minimal human pressure; however, along the Front Range in Colorado, some canyons of Utah, Montana, Wyoming, s. British Columbia, and probably other western states and provinces, Prairie Falcons have suffered localized disturbance and habitat alteration due to urbanization and recreational activities. Except along western portions of the Snake River in sw. Idaho, at the Snake River Birds of Prey Natural Conservation Area, where nesting density is extremely high with about 200 pairs, Prairie Falcons are distributed in very low density. In many regions, cyclic ground squirrel populations periodically affect regional nesting success and status.

Rare west of Cascade Mts.; one breeding pair known in Jackson Co., Ore.

In Mexico, sparsely distributed breeder in arid highlands to cen. Baja California and Chihuahua, south to s.-cen. Durango and s. Coahuila, and an isolated area in ne. Zacatecas.

**Winter.**—Mainly winter south and east of breeding and natal areas. A few individuals nesting in Wyoming and sw. Idaho have wintered northeast of breeding areas in s. Manitoba and Montana, respectively. Solitary except in areas of abundant prey. A few individuals may loosely assemble in areas with ample prey.

Some pairs may remain together or form on the winter grounds: a pair of adults were present in Tarrant Co. in n.-cen. Texas for two consecutive years. Generally a highly nomadic species. Many Prairie Falcons follow large flocks of prairie-dwelling passerines, particularly Horned Larks and Eastern and Western Meadowlarks. Horned Larks, in particular, often mass in extraordinarily large numbers in agricultural areas, particularly wheat fields. Cattle feedlots and urban areas, especially near grain elevators, are also popular wintering locales because of large densities of European Starlings, House Sparrows, Rock Doves, Brown-headed Cowbirds, Common Grackles, or Great-tailed Grackles; however, territories may be established and the falcons may remain sedentary if prey base is ample and stable. Some winter territories may be used for consecutive years.

Major winter areas are in e. Colorado, w. Kansas, Texas panhandle, s. Arizona, and cen. and s. California. Regular in small numbers in cen. Iowa and w. and e.-cen. Missouri (mainly along the Missouri River basin). Irregular farther east to Scott Co., Iowa. Casual in w. Minnesota (one female wintered in St. Paul for several years). Regular, but uncommon on coastal areas of Skagit Co., Wash. Rare in winter west of the Cascade Mts. in Oregon. Uncommon west of the Coast Range Mts. in California. There are at least eight records for Arkansas: (1) Prairie Co. in early Feb. 1951, (2) Marion Co. in late Nov. 1981, (3) Miller Co. in late Dec. 1982, (4) Cleburne Co. in late Nov. 1984, (5) Crittenden Co. in early Sep. 1986, (6) Perry Co. in late Jan. 1991, (7) Cleburne Co. in late Jan. 2001, and (8) Cleburne Co. in early Feb. 2002.

In Mexico, some winter to s. Baja California and as far south as the state of México. In Toluca, Mexico, they have been seen at 9,500 ft. (2,900 m).

**Movements.**—Mid- to late summer dispersal occurs from mid-Jun. through Aug. Juveniles disperse immediately after becoming independent, typically from mid-June through mid-July, depending on elevation and latitude. Banding data indicates most head north, east, or southeast of natal areas. Juveniles disperse to the Great Plains, mountain valleys, or subalpine and alpine meadows. Juveniles from California have dispersed to regions east of the Continental Divide. Adult females nesting in sw. Idaho may leave breeding areas in Jul., often several weeks after nesting is completed, and head northeast to prairie regions in e. Montana, se. Alberta, and sw. Saskatchewan.

Mid- to late summer dispersal allows the falcons to hunt easy-to-catch juvenile ground squirrels, which remain active later in the summer at higher elevations and northern latitudes. Falcons moving onto the Great Plains also feed on the late-summer abundance of meadowlarks, Horned Larks, and Lark Buntings. Falcons that disperse onto the Great Plains may remain in favored areas for varying lengths of time, and a few may remain all winter. Juveniles may disperse eastward to e. North Dakota by late Jul. and into w. Minnesota by early to mid-Aug. They are regularly found in cen. Iowa by mid- to late fall and, rarely, as far east as e.-cen. Iowa. There is little information on dispersal movements by adult males.

*Fall migration*: Actual migration is often an extension of the mid- to late-summer dispersal. Movements may continue to at least early Nov. for adult males. However, late-season sightings may also include nomadic movements by adult females and juveniles. Migration is short to moderate in distance. Movements are generally in a southeasterly direction but can be lateral or even northeasterly. Falcons that have dispersed in a northeasterly direction, especially those that headed for northern prairies, often take a more southerly direction Aug.–Oct. Juveniles and adult females move simultaneously. Peak numbers of adult females and juveniles on the cen. and s. Great Plains occur from mid-Sep. to early Oct. Adult males are generally not seen on cen. Great Plains until Oct. Migrants passing through w. Minnesota generally peak in early Oct.

*Spring migration*: mid-Jan. to May. Adults are very early migrants. A notable movement of adult males occurs on the cen. Great Plains from mid-Jan. to early Feb. Adult females follow shortly thereafter. Most adults are on breeding territories by mid-Feb. to mid-Mar. Juveniles generally peak in Mar., and with some moving until May.

**Falconry.**—Prairie Falcons are a favored falconry species. Escaped individuals may be encountered anywhere and may sport jesses or telemetry antenna attachments on their backs or tails. Escaped individuals also may be artificial hybrids with Gyrfalcons or Peregrine Falcons. Natural hybridization rarely occurs with Peregrines.

**NESTING:** Feb.–Jul., depending on latitude and elevation. Birds in southern latitudes may be on territories by Dec., but pairs in northern latitudes and high elevations usually do not occupy territories until Feb. or mid-Mar. Males are generally the first to arrive on breeding grounds. Pairs may rarely remain together or unite on the winter grounds. Prairie Falcons normally breed when 2 years old. One-year-old females, in juvenile plumage, may occasionally breed with adult males.

**Courtship (flight).**—*Sky-dancing* and *eyrie fly-by* are performed by both sexes, but more vigorously by males, and *flight-play.*

**Courtship (perched).**—From *eyrie fly-by* sexes may also engage in *perch-and-fly. See* chapter 5 for courtship descriptions.

No nest is built. Nest sites are scraped depressions in cavities and potholes on cliffs with protective overhangs. The overhangs give protection against inclement late-winter and early-spring weather. Eyries may be as low as 15 ft. (5 m), but are typically at least 30 ft. (9 m) high. Eyries may be used annually for many years, and often for decades. Long-term nest sites may have obvious whitewash stains below them. Vacant stick nests of Common Ravens, Red-tailed Hawks, and Golden Eagles are often used. Abandoned tree nests and ledges of abandoned buildings are rarely used for nest sites. In prime locales, nest sites may be 200 ft. (61 m) apart, providing nests are out of sight of each other. Females assume nest duties and males hunt. Females hunt when nestlings are older. Four or 5 eggs are laid in Mar. and Apr. and are incubated for 29–34 days. Youngsters fledge in 36–41 days and are independent in as little as 60 days. Double clutches are rare.

Breeding is extremely dependent and timed according to life cycles of local ground-squirrel species in many regions. Many ground-squirrel species begin estivation in May–Jul., thus forcing Prairie Falcons to complete nesting activities prior to losing an abundant and easy-to-catch food source. Ground squirrels emerge in Jan. or Feb. at low elevations and latitudes, but later at higher elevations and northern latitudes.

**CONSERVATION:** The Snake River Birds of Prey Natural Conservation Area in Ada Co., Idaho, was established in 1993 to preserve nesting and foraging habitat for the largest known density of nesting Prairie Falcons. This is a relatively stable species since much of the falcon's habitat is in very remote regions. Increasing urban sprawl, as noted in Status and Distribution, is having localized negative impact on breeding pairs. Falconers locally harvest nestlings from easily reached eyries.

**Mortality.**—Illegal shooting, electrocution from utility wires, and collisions with fence lines and sometimes vehicles. Although quite tolerant of human disturbance in nesting territories, intense, prolonged disturbance forces adults to abandon nest sites. Unlike Peregrine Falcons, Prairie Falcons suffered only a minimal amount from the organochlorine pesticide era of the late 1940s to early 1970s. Prairie Falcons do not feed as extensively on avian prey as do Peregrines and thus did not accumulated large doses of deadly chemicals. Organophosphate poisoning has occurred in local farming areas.

The population in the Snake River Birds of Prey Natural Conservation Area has experienced a reduced breeding population in the last decade. The lower number of breeding pairs in the Snake River area may be partially attributed to the substantial reduction of the now-endangered Southern Idaho Ground Squirrel. The ground squirrel has lost over half of its historical habitat due to range fires and agriculture.

**SIMILAR SPECIES:** **(1) Merlin, adult females/juveniles.**—PERCHED.—Yellow cere, orbital skin, and legs in all seasons. Nondescript pattern behind eyes. Tail same color as upperparts with three or four narrow pale bands; except some *F. c. columbarius* and *F. c. suckleyi*, which lacks bands on dorsal surface of tail. *F. c. columbarius* and *suckleyi* have fairly distinct dark "mustaches." *F. c. richardsonii* lacks a "mustache." FLIGHT.—Uniformly dark underwings on *columbarius* and *suckleyi*, uniformly pale on *richardsonii.* Tail patterns as in Perched. *Note:* Large female *richardsonii* are deceptively similar in size and coloration to male Prairie Falcons. **(2) Peregrine Falcon, juve-**

**niles.**—PERCHED.—*F. p. tundrius* may have very narrow "mustache," crown area usually pale. *F. p. anatum* and *F. p. pealei* have wide, dark "mustaches." *F. p. tundrius* have very narrow streaking on underparts; upperparts pale, tawny edging on all feathers. Wingtips nearly reach or are equal to the tail tip. FLIGHT.—Underwings on all races are uniformly dark and have pale tawny spotting and barring. Uppertail surface is the same color as upperparts in all seasons. *Note:* Use caution in fall with *tundrius;* similar to recently fledged Prairie Falcons with the pale feather edgings. **(3) Gyrfalcon, gray morphs.**—PERCHED.—Nondescript head pattern: all dark or all pale behind eyes; some "mustache" definition. Similar wingtip-to-tail-tip ratio. FLIGHT.—Caution on underwing pattern: Many gray morphs have darker, mottled median coverts, identical to male Prairie Falcons; however, axillaries are mottled white and *are not* solid dark brown, as in Prairie Falcons. **(3A) Adult gray morphs.**—Upperparts very grayish barred. Uppertail distinctly barred with gray. The spotted or barred median underwings coverts are sometimes fairly dark, but the axillaries are also spotted or barred. **(3B) Juvenile gray morphs.**—Brown upperparts as on juvenile/adult female Prairie Falcons. Median underwing coverts and axillaries may be particularly dark but are always spotted or barred on the axillaries. Uppertail is same color as rest of upperparts in all seasons and has numerous, narrow whitish bands.

**OTHER NAMES:** None regularly used. Males are called "tiercels," females are called "falcons." *Spanish*: Halcón Mexicano, Halcón Pradeno, or Halcón Cafe. *French*: Faucon des Prairies.

---

REFERENCES: Adamus et al. 2001; Andrews and Righter 1992; Burt and Grossenheider 1976; Call 1978; Cecil 1997, 1999; Clark and Wheeler 1987; del Hoyo et al. 1994; Dorn and Dorn 1990; Ducey 1988; Fesnock 1997; Gilligan et al. 1994; Herron et al. 1985; Hitchcock 1977; Howell and Webb 1995; James and Neal 1986; Kent 1996, 1997; Kent and Dinsmore 1996; Kingery 1998; Lanning and Hitchcock 1991; Lanning and Lawson 1977; Millard 1993; Montana Bird Distribution Committee 1996; Oberholser 1974; Palmer 1988; Peterson 1995; Pulich 1988; Robbins and Easterla 1992; Russell and Monson 1997; Semenchuk 1992; Sharpe et al. 2001; Small 1994; Smith 1996; Steenhof 1998; Steenhof et al. 1984; Stewart 1975; Sutton 1967; Thompson and Ely 1989; USFWS 2001b; White 1962.

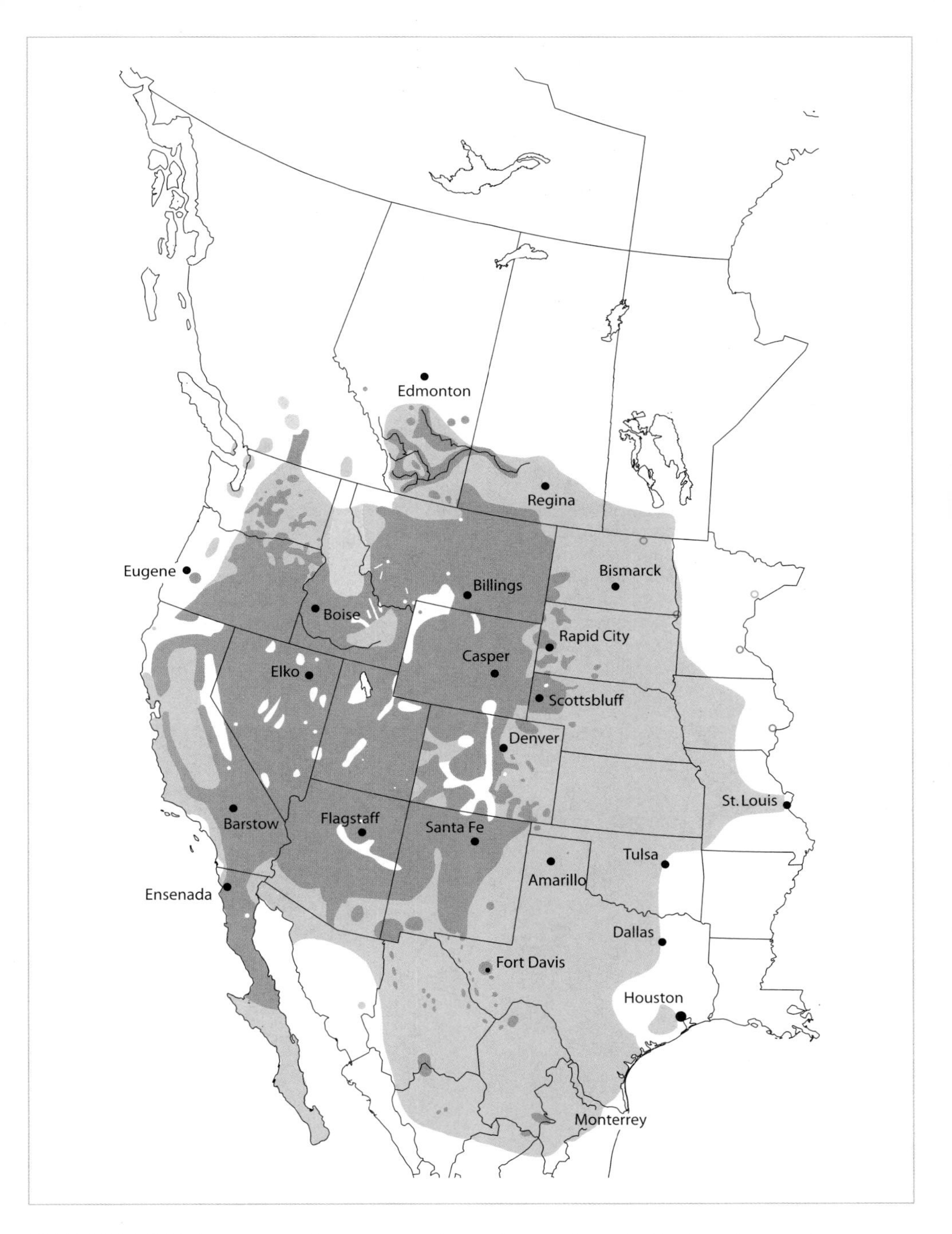


**PRAIRIE FALCON,** *Falco mexicanus:* Uncommon. 4,300–6,000 pairs. Sparsely distributed except along w. Snake River, ID. Disperses mid-Jun. to Aug. to Great Plains and east to w. MN and up to alpine elevations (blank white areas) in mountains of U.S.

**Plate 592. Prairie Falcon, adult male [Jun.]** ▪ White cheek extends up behind eye; brown auricular patch; narrow dark malar mark; thin dark eyeline; full white supercilium. ▪ Medium brown upperparts are always distinctly barred with pale brown or gray. ▪ Wingtips are shorter than tail tip. ▪ Pale brown, faintly banded tail.

**Plate 593. Prairie Falcon, adult female [Jun.]** ▪ White cheek extends up behind the eye; brown auricular patch; narrow dark malar mark; thin dark eyeline; short, white supercilium. ▪ Medium brown upperparts are often partially barred or unbarred and nearly uniformly brown. ▪ Wingtips are shorter than tail tip. ▪ Pale brown, often unbanded tail.

**Plate 594. Prairie Falcon, adult pair (male on left) [Jun.]** ▪ Example showing size difference between sexes that is typical of larger falcons. Males are much smaller than females.

**Plate 595. Prairie Falcon, adult male [Jun.]** ▪ Lightly to moderately spotted white underparts. ▪ Black axillaries. Black median underwing coverts are spotted with white and lesser coverts are lightly streaked with black. Remiges are white and uniformly barred with gray. ▪ Pale tail is banded on underside.

**Plate 596. Prairie Falcon, adult female [Jun.]** ▪ Moderately to heavily spotted white underparts. ▪ Black axillaries. All-black median underwing coverts blend with axillaries; lesser coverts are heavily streaked with black. Median coverts are sometimes lightly spotted with white. Remiges are white and uniformly barred with gray. ▪ Pale tail is banded on underside.

**Plate 597. Prairie Falcon, adult male [Jun.]** ▪ Medium brown upperparts are barred with pale brown or gray. ▪ Pale brown tail contrasts to darker upperparts. Narrow white terminal band. ▪ *Note:* Wing molt has begun on the wings: p4 is a new feather and p5 is dropped and ready to molt a new feather.

**Plate 598. Prairie Falcon, adult female [Jun.]** ▪ Nearly uniformly medium brown upperparts. ▪ Pale brown tail contrasts to darker upperparts. White terminal band is worn off. ▪ *Note:* Molting remiges. Primary molt with new p4 and 5 (darker, grayish feathers) and dropped p3 and 6; new middle secondaries s5 and 6.

**Plate 599. Prairie Falcon, juvenile female (recently fledged) [Jul.]** ▪ White cheek extends up behind eye; brown auricular patch; thin dark malar mark. Pale blue cere and orbital skin. ▪ Dark brown upperpart feathers are broadly edged with pale brown. ▪ Wingtips are shorter than tail tip. ▪ Tail is same color as upperparts.

**Plate 600. Prairie Falcon, juvenile female (worn plumage) [Jan.]** ▪ White cheek extends up behind the eye; brown auricular patch; thin dark malar mark. Pale yellow cere and orbital skin. ▪ Brown upperparts are narrowly edged with pale brown. Underparts fade to white by mid-fall. ▪ Wingtips are shorter than tail tip. ▪ Tail has faded to pale brown.

**Plate 601. Prairie Falcon, juvenile male (recently fledged) [Jul.]** ▪ Tawny underparts are thinly streaked with dark brown. ▪ Black axillaries. Black median coverts are spotted with white; lesser coverts are moderately streaked with black. Remiges are white and uniformly barred with gray. ▪ *Note:* Juvenile females may also have similar white spotting on median coverts.

**Plate 602. Prairie Falcon, juvenile female (recently fledged) [Jul.]** ▪ Tawny underparts are moderately streaked with dark brown. ▪ Black axillaries blend with all-black median coverts and heavily streaked lesser coverts. ▪ *Note:* Juvenile males may have similar all-black underwing markings.

**Plate 603. Prairie Falcon, juvenile female (recently fledged) [Jul.]** ▪ Uniformly dark brown upperparts. ▪ Dark brown tail is same color as upperparts. Broad white terminal band is wider than adult's in summer.

# BIBLIOGRAPHY


Adamus, P. R., K. Larson, G. Gillson, and C. Miller. 2001. *Oregon Breeding Bird Atlas.* CD-ROM. Oregon Field Ornithol., Eugene.

Alaska Natural Heritage Program. 1998. Status Report on the Queen Charlotte Goshawk (*Accipiter gentilis laingi*) in Alaska. Unpubl. rep. <http://www.uag.alaska.edu/enri/aknhp_web/index.htlm> [Mar. 10, 1999].

Alcorn, J. R. 1988. *The Birds of Nevada.* Fairview West Publ., Fallon, NV.

American Ornithologists' Union. 1997. Forty-first Supplement to the American Ornithologists' Union *Check-list of North American Birds. Auk* 114 (3): 542–552.

American Ornithologists' Union. 1998. *Check-list of North American Birds.* 7th ed. Am. Ornithol. Union, Washington, D.C.

American Ornithologists' Union. 2000. Forty-second Supplement to the American Ornithologists' Union *Check-list of North American Birds. Auk* 117 (3): 847–858.

Anderson, C. M., D. G. Roseneau, B. J. Walton, and P. J. Bente. 1988. New Evidence of a Peregrine Migration on the West Coast of North America. Pp. 507–516 *in Peregrine Falcon Populations* (T. J. Cade, J. H. Enderson, C. G. Thelander, and C. M. White, eds.). The Peregrine Fund, Boise, ID.

Andrews, R., and R. Righter. 1992. *Colorado Birds: A Reference to Their Distribution and Habitat.* Denver Mus. Nat. History, Denver, CO.

Arizona Breeding Bird Atlas 1993–1999. Arizona Game Fish Dep., Nongame Branch, Phoenix. Unpubl.

Arnold, K. A. 2001. Red-shouldered Hawk. The Texas Breeding Bird Atlas. Version 1.10. Texas A & M Univ. System, College Station and Corpus Christi, TX, <http://tbba.cbi.tamuc.edu> [Jan. 26, 2002].

Baicich, P. J., and C.J.O. Harrison. 1997. *A Guide to Nests, Eggs, and Nestlings of North American Birds.* 2nd ed. Academic Press, San Diego, CA.

Barker, K. 1996. Recent Records for Harris' Hawks in Oklahoma. *Bull. Oklahoma Ornithol. Soc.* 29: 21–22.

Baumgartner, F. M., and A. M. Baumgartner. 1992. *Oklahoma Bird Life.* Univ. of Oklahoma Press, Norman.

Bechard, M. J. 1997. Breeding Biology and Status of the Ferruginous Hawk. Paper read at the Hawk Migr. Assoc. N. Am. Conference VIII, Snowbird, UT, Jun. 12–15, 1997.

Bechard, M. J., and J. K. Schmutz. 1995. Ferruginous Hawk (*Buteo regalis*). *In* The Birds of North America, No. 172 (A. Poole and F. Gill, eds.). Acad. Nat. Sci., Philadelphia, PA, and Am. Ornithol. Union, Washington, D.C.

Becker, D. M. 1984. Reproductive Ecology and Habitat Utilization of Richardson's Merlin in Southeastern Montana. M.S. thesis, Univ. of Montana, Missoula.

Bednarz, J. C. 1983. Status of the Harris' Hawk in New Mexico. Unpubl. rep. New Mexico Dep. Game Fish, Albuquerque.

Bednarz, J. C. 1987. Pair and Group Reproductive Success, Polyandry and Cooperative Breeding in Harris' Hawks. *Auk* 104: 393–401.

Bednarz, J. C. 1988. Harris' Hawk Subspecies: Is *superior* Larger or Different Than *harrisi?* Pp. 294–300 *in Proc. Southwest Raptor Manage. Symp. Workshop* May 21–24, 1986. Univ. of Arizona, Tucson; Natl. Wildl. Fed., Washington, D.C.; Natl. Wildl. Fed. Sci. Tech. Ser. No. 11.

Bednarz, J. C. 1995. Harris's Hawk (*Parabuteo unicinctus*). *In* The Birds of North America, No. 146 (A. Poole and F. Gill, eds.). Acad. Nat. Sci., Philadelphia, PA, and Am. Ornithol. Union, Washington, D.C.

Beebe, F. L. 1960. The Marine Peregrine of the Northwest Pacific Coast. *Condor* 62 (3): 145–189.

Benson, K.L.P., and K. A. Arnold. 2001. The Texas Breeding Bird Atlas. Version 1.10. Texas A & M Univ. System, College Station and Corpus Christi, TX, <http://tbba.cbi.tamucc.edu> [Jan. 26, 2002].

Bent, A. C. 1961. *Life Histories of North American Birds of Prey.* Pts. 1 and 2. Dover Publ., New York, NY.

Berger, D. D., C. R. Sindelar, Jr., and K. E. Gamble. 1968. The Status of Breeding Peregrines in the Eastern United States. Pp. 165–173 *in Peregrine Falcon Populations: Their Biology and Decline* (J. J. Hickey, ed.). Univ. of Wisconsin Press, Madison.

Bildstein, K. L., and M. W. Collopy. 1985. Escorting Flight and Agonistic Interactions in Wintering Northern Harriers. *Condor* 87: 398–401.

Blanchard, M. 1999. White-tailed Kites in

Thurston County. *Washington Ornithol. Soc. News* 58 (Jan).

Block, W. M., M. L. Morrison, and M. Hildegard Reiser (eds.). 1994. The Northern Goshawk: Ecology and Management. *Stud. Avian Biol.*, No. 6.

Bloom, P. H. 1980. The Status of the Swainson's Hawk in California, 1979. California Dep. Fish Wildl. and U.S. Dep. Int. Bur. Land Manage. Final rep. (Nov.).

Bloom, P. H., and M. D. McCrary. 1996. The Urban Buteo: Red-shouldered Hawks in Southern California. Pp. 31–39 *in Raptors in Human Landscapes*, Academic Press, London, U.K.

Boal, C. W., D. E. Anderson, and P. L. Kennedy. 2001. Home Range and Habitat Use of Northern Goshawks (*Accipiter gentilis*) in Minnesota. Final rep. Minn. Coop. Fish Wildl. Res. Unit, St. Paul.

Bolen, E. G., and D. Flores. 1993. *The Mississippi Kite.* Univ. of Texas Press, Austin.

Bond, R. M. 1946. The Peregrine Populations of Western North America. *Condor* 48 (3): 101–116.

Breen, T. F., and J. W. Parrish. 1996. First Evidence for Double-brooding in Southeastern Kestrels in Georgia. *Oriole* 60: 81–83.

Breen, T. F., and J. W. Parrish. 1997. Distribution and Use of Nest Boxes by American Kestrels in the Coastal Plains of Georgia. *Fla. Field Nat.* 25: 129–138.

Breining, G. 1994. *Return of the Eagle: How America Saved Its National Symbol.* Falcon Press, Helena, MT.

Bridges, A. 2002. Captive Condors From U.S. to Be Released in Baja Park. World Wide Web Site for the *San Diego Tribune* (Jul. 24, 2002). <www.signonsandiego.com/news/uniontrib/wed/news_1ncondor.html> [Jul. 28, 2002].

Brockman, F. C. 1968. *Trees of North America.* Golden Press, New York, NY.

Brown, B. T. 1988. Additional Bald Eagle Nest Records From Sonora, Mexico. *J. Raptor Res.* 22 (1): 30–32.

Brown, B. T., P. L. Warren, and L. S. Anderson. 1987. First Bald Eagle Nesting Record From Sonora, Mexico. *Wilson Bull.* 99 (2): 279–280.

Brown, R. E., J. H. Williamson, and D. B. Boone. 1997. Swallow-tailed Kite Nesting in Texas: Past and Present. *Southwest. Nat.* 42 (1): 103–105.

Buchanan, J. B. 1991. Two Cases of Carrion Feeding by Peregrine Falcons in Western Washington. *Northwest. Nat.* 72: 28–29.

Burnham, B. 1997. *A Fascination With Falcons.* Hancock House, Surrey, BC, and Blaine, WA.

Burnham, W. A., W. Heinrich, C. Sandfort, E. Levine, D. O'Brien, and D. Kunkel. 1988. Recovery Effort for the Peregrine Falcon in the Rocky Mountains. Pp. 565–574 *in Peregrine Falcon Populations* (T. J. Cade, J. H. Enderson, C. G. Thelander, and C. M. White, eds.). The Peregrine Fund, Boise, ID.

Burnside, F. L. 1983. The Status and Distribution of the Red-cockaded Woodpecker in Arkansas. *Am. Birds* 37 (2): 142–145.

Burt, W. H., and R. P. Grossenheider. 1976. *A Field Guide to the Mammals.* Houghton Mifflin, Boston, MA.

Busby, W. H., and J. H. Zimmerman. 2001. *Kansas Breeding Bird Atlas.* Univ. Press of Kansas, Lawrence.

Bylan, S. (ed.). 1998. *Hawk Migr. Stud.* 24, No. 1.

Bylan, S. (ed.). 1998. *Hawk Migr. Stud.* 25, No. 1.

Byre, V. 1990. A Group of Young Peregrine Falcons Prey on Migrating Bats. *Wilson Bull.* 102: (4): 728–730.

Cade, T. J., J. H. Enderson, C. G. Thelander, and C. M. White (eds.). 1988. *Peregrine Falcon Populations.* The Peregrine Fund, Boise, ID.

Call, M. W. 1978. Nesting Habitats and Surveying Techniques for Common Western Raptors. U.S. Dep. Int. Bur. Land Manage., Denver, CO.

Campbell, R. W., N. K. Dawe, I. McTaggart-Cowan, J. M. Cooper, G. W. Kaiser, and M. C. E. McNall. 1990. *The Birds of British Columbia.* Vol. 2, Diurnal Birds of Prey Through Woodpeckers. Univ. of British Columbia Press, Vancouver, BC.

Canadian Peregrine Foundation. 1999a. Richmond Hill home page. World Wide Web Site <http://www.peregrine-foundation.ca/tops/rh.html> [Aug. 24, 2000].

Canadian Peregrine Foundation. 1999b. Guelph home page. World Wide Web Site <http://www.peregrine-foundation.ca/tops/gutop.html> [Sep. 9, 2000].

Canadian Peregrine Foundation. 2002. Project Track 'em. World Wide Web Site http://www.peregrine-foundation.ca/programs/tracem/track2002.html [May 16, 2000].

Cardiff, S. W. 1999. Central South Region (Hawks Through Cranes). *N. Am. Birds* 53 (3): 289.

Cartron, J. 2001. How the Population Status of the Prairie Dog May Be Essential to Nesting Ferruginous Hawk. *Aloft* 7 (3): 20.

Castellanos, A., F. Jaramillo, F. Salina, A. Ortega-Rubio, and C. Arquelles. 1997. Peregrine Falcon Recovery Along the West-Central Coast of the Baja California Peninsula, Mexico. *J. Raptor Res.* 31 (1): 1–6.

Cecil, R. I. 1997. Field Reports—Winter 1996–97. *Iowa Bird Life* 67 (2): 59–60.

Cecil, R. I. 1999. Field Reports—Winter 1989–99. *Iowa Bird Life* 69 (2): 67–68.

Center for Conservation, Research, and Technology. 2000. Current Projects: Broad-winged Hawk. World Wide Web Home Page for the Center for Conservation, Research, and Technology, Univ. of Maryland, Baltimore, and Boise State Univ. Boise, ID <http://www.ccrt.org> [Dec. 8, 2000].

Clark, W. S. 1998. First North American Record of a Melanistic Osprey. *Wilson Bull.* 110: 289–290.

Clark, W. S. 1999. *A Field Guide to the Raptors of Europe, the Middle East, and North Africa.* Oxford Univ. Press, Oxford, U.K.

Clark, W. S. 2001. Aging Bald Eagles. *Birding* 33 (1): 18–28.

Clark, W. S. 2002. First Breeding Record of a Dark Morph Hook-billed Kite in the U.S. *N. Am. Birds* 56 (3): 260–262.

Clark, W. S., and R. C. Banks. 1992. The Taxonomic Status of the White-tailed Kite. *Wilson Bull.* 104: 571–579.

Clark, W. S., and B. K. Wheeler. 1989. Unusual Roost Site Selection and Staging Behavior of Black-shouldered Kites. *J. Raptor Res.* 23: 116–117.

Clark, W. S., and B. K. Wheeler. 1998. 'Dark-morph' Sharp-shinned Hawk Reported From California Is a Normal Juvenile Female of Race *perobscurus. Bull. B.O.C.* 118 (3): 191–193.

Clark, W. S., and B. K. Wheeler. 2001. *A Field Guide to Hawks of North America.* 2nd ed. Houghton Mifflin, Boston, MA.

Clum, N. J., and T. J. Cade. 1994. Gyrfalcon (*Falco rusticolus*). *In* The Birds of North America, No. 114 (A. Poole and F. Gill, eds.). Acad. Nat. Sci., Philadelphia, PA, and Am. Ornithol. Union, Washington, D.C.

Commission for Environmental Cooperation. 1997. North American Regional Action Plan on DDT: North American Working Group for the Sound Management of Chemicals Task Force on DDT and Chlordane. World Wide Web Site <http://www.cec.org/programs_projects/pollutants_health/smoc/ddt.cfm?varlan=english.

Conrads, D. J. 1997. The Nesting Ecology of the Cooper's Hawk in Iowa. *Iowa Bird Life* 67: 33–42.

Conrads, D. J., M. Phelps, and T. H. Kent. 1989. Mississippi Kite at Dudgeon Lake. *Iowa Bird Life* 59: 118–119.

Contreras-Balderas, A. J., and F. Montiel-de la Garza. 1999. Swainson's Hawks in Nuevo Leon, Mexico. *J. Raptor Res.* 33 (2): 176–177.

Corman, T. E. 1998. Broad-winged Hawk. Pp. 89–91 *in The Raptors of Arizona* (R. L. Glinski, ed.). Univ. of Arizona Press, Tucson, and Arizona Game Fish Dep., Phoenix.

Court, G. S., C. C. Gates., and D. A. Boag. 1988. Natural History of the Peregrine Falcon in the Keewatin District of the Northwest Territories. *Arctic* 41 (1): 17–30.

Craig, T. H., E. H. Craig, and J. S. Marks. 1982. Aerial Talon-grappling in Northern Harriers. *Condor* 84: 239.

Crocoll, S. T. 1994. Red-shouldered Hawk (*Buteo lineatus*). *In* The Birds of North America, No. 107 (A. Poole and F. Gill, eds.). Acad. Nat. Sci., Philadelphia, PA, and Am. Ornithol. Union, Washington, D.C.

Davey, B. (ed.). 1999a. *Condor News*, U.S. Fish Wildl. Serv., Hopper Mountain NWR Complex, Ventura, CA. Vol. 3 (15).

Davey, B. (ed.). 1999b. *Condor News*, U.S. Fish Wildl. Serv., Hopper Mountain NWR Complex, Ventura, CA. Vol. 9 (17).

Dawson, J. W., and R. W. Mannan. 1995. Abstract: Electrocution as a Mortality Factor in an Urban Population of Harris' Hawks. *J. Raptor Res.* 29 (1): 55.

Debus, S. 1998. *The Birds of Prey of Australia.* Oxford Univ. Press, Melbourne, Australia.

Dechant, J. A., M. L. Sondreal, D. H. Johnson, L D., Igl, C. M. Goldale, et al. 1998 (rev. 1999). Effects of Management Practices on Grassland Birds: Northern Harrier. World Wide Web Site for the Northern Prairie Wildl. Res. Center, <http://www.npwrc.usgs.gov/resources/literatr/rasbird/harrier/harrier.htm> [Jun. 3, 2000].

Dekker, D. 1980. Hunting Success Rates, Foraging Habitats, and Prey Selection of Peregrine Falcons Migrating Through Central Alberta. *Can. Field-Nat.* 94 (4): 371–382.

Dekker, D. 1987. Peregrine Falcon Predation on Ducks in Alberta and British Columbia. *J. Wildl. Manage.* 51 (1): 156–159.

Dekker, D. 1988. Peregrine Falcon and Merlin Predation on Small Shorebirds and Passerines in Alberta. *Can. J. Zool.* 66: 925–928.

Dekker, D. 1999. *Bolt From the Blue: Wild Peregrines on the Hunt.* Hancock House, Surrey, BC, and Blaine, WA.

Dekker, D., and J. Lange. 2001. Hunting Methods and Success Rates for Gyrfalcon, *Falco rusticolus*, and Prairie Falcon, *Falco mexicanus*, Preying on Feral Pigeons (Rock Doves), *Columba*

*livia*, in Edmonton, Alberta. *Can. Field-Nat.* 15 (3): 395–401.
del Hoyo, J. A. Elliott, and J. Sargatal (eds.). 1994. *Handbook of the Birds of the World.* Vol. 2. New World Vultures to Guineafowl. Lynx Edicions, Barcelona, Spain.
Dickerman, R. W. 1994. Undescribed Subspecies of Red-tailed Hawk From Baja California. *Southwest. Nat.* 39 (4): 375–395.
Dickerman, R. W., and F. C. Parkes. 1987. Subspecies of the Red-tailed Hawk in the Northeast. *Kingbird* 37: 57–64.
Dinsmore, J. J. 1996. Field Reports—Summer 1996. *Iowa Bird Life* 66 (4): 129–132.
Dinsmore, J. J. 1997. Field Reports—Summer 1997. *Iowa Bird Life* 67 (4): 120–123.
Dinsmore, J. J. 1999. Field Reports—Summer 1999. *Iowa Bird Life* 69 (4): 123–130.
Dinsmore, J. J. 2000. Field Reports—Summer 2000. *Iowa Bird Life* 70 (4): 173.
Dinsmore, J. J. 2002. Field Reports—Summer 2001. *Iowa Bird Life* 71 (4): 163–180.
Dodd, N. L., and J. R. Vahle. 1998. Osprey. Pp. 37–41 *in The Raptors of Arizona* (R. L. Glinski, ed.). Univ. of Arizona Press, Tucson, and Arizona Game Fish Dep., Phoenix.
Dodge, J. (ed.). 1988–1997. *Hawk Migr. Stud.* Vols. 14–23.
Dodge, J. (ed.). 1995. Peregrine Falcons Tracked by Satellite. *Hawk Flights* 1: 2.
Dold, C. 1998. Making Room for Prairie Dogs. *Smithsonian* 28: 60–68.
Dorn, J. L., and R. D. Dorn. 1990. *Wyoming Birds.* Mountain West Publ., Cheyenne, WY.
Dotson, R. A., and D. P. Mindell. 1979. Raptor Surveys and River Profiles in the Kuskokwim, Unalakleet, and Yukon River Drainages, Alaska. Prepared for Bur. Land Manage. Anchorage Dist. Office, Anchorage, AK.
Dove, C. J., and R. C. Banks. 1999. A Taxonomic Study of Crested Caracaras (Falconidae). *Wilson Bull.* 111 (3): 330–339.
Ducey, J. E. 1988. *Nebraska Birds: Breeding Status and Distribution.* Simmons-Boardman, Omaha, NE.
Duncan, B. 1986. The Occurrence and Identification of Swainson's Hawks in Ontario. *Ont. Birds* 4 (2): 43–61.
Duncan, P., and D. A. Kirk. 1994. Status Report on the Queen Charlotte Goshawk, *Accipiter gentilis laingi*, in Canada. Comm. Status of Endangered Wildl. in Canada.
Dunk, J. R. 1995. White-tailed Kite (*Elanus leucurus*). *In* The Birds of North America, No. 178 (A. Poole and F. Gill, eds.). Acad. Nat. Sci., Philadelphia, PA, and Am. Ornithol. Union, Washington, D.C.
Dunn, J. L., and K. L. Garrett. 1997. A Field Guide to Warblers of North America. Houghton Mifflin Co., Boston, MA.
Dunne, P., D. Sibley, and C. Sutton. 1998. *Hawks in Flight: The Flight Identification of North American Migrant Raptors.* Houghton Mifflin, Boston, MA.
Eberly, C. 1999. Conservation From Outer Space: Boldly Tracking Migration Patterns. *Hawk Migr. Stud.* Aug.: 17–19.
Edelstam, C. 1984. Patterns of Molt in Large Birds of Prey. *Ann. Zool. Fenn.* 21: 271–276.
Ehresman, B. L. 1999. The Recovery of the Bald Eagle as an Iowa Nesting Species. *Iowa Bird Life* 69 (1): 1–12.
England, A. S., M. J. Bechard, and C. S. Houston. 1997. Swainson's Hawk (*Buteo swainsoni*). *In* The Birds of North America, No. 265 (A. Poole and F. Gill, eds.). Acad. Nat. Sci., Philadelphia, PA, and Am. Ornithol. Union, Washington, D.C.
Environment Canada. 1999. Species at Risk: Peregrine Falcon. World Wide Web Site for the Green Lane, <http://www.nais.ccrs.nrcan.gc.ca/schoolnet/issues/risk/birds/ebirds/prgfalcon/html> [Aug. 9, 2000].
Environment Canada. 2000a. Species at Risk: American Peregrine Falcon. World Wide Web Site for the Green Lane, <http://www.speciesatrisk.gc.ca/Species/English/SearchDetail.cfm?SpeciesID=29> [Jun. 13, 2002].
Environment Canada. 2000b. Species at Risk: Peale's Peregrine Falcon. World Wide Web Site for the Green Lane, <http://www.speciesatrisk.gc.ca/Species/English/SearchDetail.cfm?SpeciesID=54> [Jun. 13, 2002].
Environment Canada. 2001. Waiting for the Fiddler: Pesticides and the Environment in the Atlantic Region. World Wide Web Site for the Green Lane, <http://www.ns.ec.gc.ca/epb/fiddle/insectic.html> [Apr. 11, 2002].
Environment Canada. 2002. Peregrine Falcon: Previous Migrations. World Wide Web Site for the Green Lane, <http://www.pnr-rpn.ec.gc.ca/nature/endspecies/peregrine/db02s02.en.html> [May 28, 2002].
Evans, D. L., and R. N. Rosenfield. 1985. *Migration and Mortality of Sharp-shinned Hawks Ringed at Duluth, Minn.*, ICBP Tech. Publ. No. 5.
Ewins, P. J. 1995. Recovery of Osprey populations in Canada. Pp. 14–16 *in Bird Trends.* Vol. 4 (C. Hyslope, ed.). Can. Wildl. Serv., ON.
Ewins, P. J., and C. S. Houston. 1992. Recovery

Patterns of Ospreys, *Pandion haliaetus*, Banded in Canada Up to 1989. *Can. Field-Nat.* 106: 361–365.

Faanes, C. A., and G. R. Lingle. 1995. Breeding Birds of the Platte River Valley of Nebraska. World Wide Web Site for the Northern Prairie Wildlife Research Center, <http://www.npwrcs.usgs.gov/resource/distr/birds/platte/platte.htm> [Mar. 20, 1999].

Farquhar, C. C. 1992. White-tailed Hawk (*Buteo albicaudatus*). *In* The Birds of North America, No. 30 (A. Poole and F. Gill, eds.). Acad. Nat. Sci., Philadelphia, PA, and Am. Ornithol. Union, Washington, D.C.

Ferguson-Lees, J., and D. A. Christie. 2001. *Raptors of the World.* Christopher Helm–A & C Black, London, U.K.

Fesnock, A. L. 1997. Reproductive Status of Prairie Falcons at Pinnacles Nat'l. Monument, Calif. 1988–1997. Paper read at the Hawk Migr. Assoc. N. Am. Conference VIII, Snowbird, UT, Jun. 12–15, 1997.

Finn, S. P., J. M. Marzluff, and D. E. Varland. 1998. Northern Goshawk Occupancy and Productivity in Managed Forests of Western Washington. Final Draft. Second Annu. Rep., Mar. 17.

Fish, A., and B. Hull. 1996. Season Summary 1996. Golden Gate Raptor Observ., San Francisco, CA.

Forsman, D. 1999. *The Raptors of Europe and the Middle East.* T. & A. D. Poyser, London, U.K.

Franklin, K. 1999. Vertical Flight. *N. Am. Falconers Assoc. J.* 38: 68–72.

Friedman, H. 1950. The Birds of North and Middle America. *U.S. Natl. Mus. Bull.* 50., Pt. XI. U.S. Gov. Printing Office, Washington, D.C.

Fuller, J. L. 2001. Field Reports—Fall 2000. *Iowa Bird Life* 71 (1): 34.

Fuller, M. R., W. S. Seegar, and P. W. Howey. 1995. The Use of Satellite Systems for the Study of Bird Migration. *Israel J. Zool.* 41: 243–252.

Fuller, M. R., W. S. Seegar, and L. S. Schveck. 1998. Routes and Travel Rates of Migrating Peregrine Falcons, *Falco peregrinus*, and Swainson's Hawks, *Buteo swainsoni*, in the Western Hemisphere. *J. Avian Biol.* 29: 433–440.

Fung, K. 1999. *Atlas of Saskatchewan.* Univ. of Saskatchewan, Saskatoon.

Fyfe, R. W. 1988. The Canadian Peregrine Falcon Recovery Program, 1967–1985. Pp. 599–610 *in Peregrine Falcon Populations* (T. J. Cade, J. H. Enderson, C. G. Thelander, and C. M. White, eds.). The Peregrine Fund, Boise, ID.

Garber, G. (ed.). 2000. Nesting, Productivity, and Food Habits of Ferruginous Hawks as a Function at Prairie Dog Towns in Central, Western, and Northwestern New Mexico. Submitted to Bur. Land Manage; Socorro and Farmington District Offices, New Mexico Dep. Game Fish; Turner Found.; Hawks Aloft, Albuquerque, NM.

Garner, H. 1999. Distribution and Habitat Use of Sharp-shinned and Cooper's Hawks in Arkansas. *J. Raptor Res.* 33 (4): 329–332.

Garrison, B. A., and P. H. Bloom. 1993. Natal Origins and Winter Site Fidelity of Rough-legged Hawks Wintering in California. *J. Raptor Res.* 27 (2): 116–118.

Gatz, T. A. 1998. White-tailed Kite. Pp. 42–45 *in The Raptors of Arizona* (R. L. Glinski, ed.). Univ. of Arizona Press, Tucson, and Arizona Game Fish Dep., Phoenix.

Gibson, D. D., and B. Kessel. 1997. Inventory of the Species and Subspecies of Alaskan Birds. *West. Birds* 28: 45–95.

Gilligan, J. M. Smith, D. Rogers, and A. Contreras (eds.). 1994. *Birds of Oregon.* Cinclus Publ., McMinnville, OR.

Glinski, R. L. 1982. The Red-shouldered Hawk (*Buteo lineatus*) in Arizona. *Am. Birds* 36 (5): 801–803.

Glinski, R. L. 1998a. Gray Hawk. Pp. 82–85 *in The Raptors of Arizona* (R. L. Glinski, ed.). Univ. of Arizona Press, Tucson, and Arizona Game Fish Dep., Phoenix.

Glinski, R. L. 1998b. Mississippi Kite. Pp. 46–49 *in The Raptors of Arizona* (R. L. Glinski, ed.). Univ. of Arizona Press, Tucson, and Arizona Game Fish Dep., Phoenix.

Gloger, C. W. L. 1833. *The Variation of Birds Under the Influence of Climate.* August Schulz, Breslau, Germany.

Godfrey, E. W. 1986. *The Birds of Canada.* Natl. Mus. Nat. Sci., Ottawa, ON.

Goodrich, L. J., S. C. Crocoll, and S. E. Senner. 1996. Broad-winged Hawk (*Buteo platypterus*). *In* The Birds of North America, No. 218 (A. Poole and F. Gill, eds.). Acad. Nat. Sci., Philadelphia, PA, and Am. Ornithol. Union, Washington, D.C.

Granlund, J. 1999. Western Great Lakes Region (Raptors Through Shorebirds). *N. Am. Birds* 53 (3): 282.

Griffin, C. R. 1976. Preliminary Comparison of Texas and Arizona Harris' Hawk (*Parabuteo unicinctus*). *J. Raptor Res.* 10: 50–54.

Griffiths, C. S. 1994. Monophyly of the Falconiformes Based on Syringeal Morphology. *Auk* 111 (4): 787–805.

Gross, H. 1999. Hawkwatch International Band

Encounters and Recaptures, 1997–1999. *Raptor Watch* 13 (3): 8–9.

Grzybowski, J. A. 2000. Southern Great Plains Region (Raptors Through Terns). *N. Am. Birds* 54 (1): 69.

Gullion, G. 1984. *Grouse of the North Shore.* Willow Creek Press, Oshkosh, WI.

Gustafson, M. E., and J. Hildenbrand. 1998. Banding Protocol for Peregrine Falcons. World Wide Web Home Page for the Bird Banding Laboratory, <http://www.pwrc.usgs.gov/bbl/homepage/pefaprot.htm> [Aug. 9, 2001].

Hamerstrom, F. 1968. Aging and Sexing Harriers. *Inland Bird Banding News* 40: 43–46.

Hamerstrom, F., and F. Hamerstrom. 1978. External sex characters of Harris' Hawks in Winter. J. *Raptor Res.* 12: 1–14.

Haney, D. L., and C. M. White. 1999. Habitat Use and Subspecific Status of Merlins, *Falco columbarius,* Wintering in Central Utah. *Great Basin Nat.* 59 (3): 266–271.

Harris, J. T. 1979. *The Peregrine Falcon in Greenland: Observing an Endangered Species.* Univ. of Missouri Press, Columbia.

Harrison, H. H. 1979. *A Field Guide to Western Birds' Nests.* Houghton Mifflin, Boston, MA.

Hayes, G. E., and J. B. Buchanan. 2002. Washington State Status Report for the Peregrine Falcon. Washington Dep. Fish Wildl., Wildl. Prog., Olympia.

Hector, D. P. 1987. The Decline of the Aplomado Falcon in the United States. *Am. Birds* 41: 381–389.

Henny, C. J., and D. W. Anderson. 1979. Osprey Distribution, Abundance, and Status in Western North America: The Baja California and Gulf of California Population. *Bull. Southern California Acad. Sci.* 78 (2): 89–106.

Henny, C. J., B. Conant, and D. W. Anderson. 1993. Recent Distribution and Status of Nesting Bald Eagles in Baja California, Mexico. *J. Raptor Res.* 27: 203–209.

Herron, G. B., C. A. Mortimore, and M. S. Rawlings. 1985. Nevada Raptors. *Biol. Bull.* No. 8, Nevada Dep. Wildl.

Hiller, I. 1976. Rare Visitor. *Texas Parks Wildl.* Oct: 16–17.

Hitchcock, M. A. 1977. A Survey of the Peregrine Falcon Population in Northwestern Mexico, 1976–77. Unpubl. rep. Chihuahuan Desert Res. Inst., Alpine, TX, Contrib. No. 40.

Hoffman, M. L., and M. W. Collopy. 1988. Historical Status of the American Kestrel (*Falco sparverius paulus*) in Florida. *Wilson Bull.* 100: 91–107.

Holt, W. 2000. The Mystery of California's Wintering Swainson's Hawks. *West. Tanager* 66 (6): 1–3.

Houghton, L. M., and L. M. Rymon. 1997. Nesting, Distribution, and Population Status of U.S. Osprey in 1994. *J. Raptor Res.* 1 (1): 44–53.

Houston, C. S., 1967. Recoveries of Red-tailed Hawks Banded in Saskatchewan. *Blue Jay* Sep.: 109–111.

Houston, C. S. 1974. South American Recoveries of Franklin's Gulls and Swainson's Hawks Banded in Saskatchewan. *Blue Jay* 32: 156–157.

Houston, C. S. 1981. History of Richardson's Merlin in Saskatchewan. *Blue Jay* 39: 30–37.

Houston, C. S. 1982. Artificial Nesting Platforms for Ferruginous Hawks. *Blue Jay* 40 (4): 208–213.

Houston, C. S. 1985. Ferruginous Hawk Nest Platforms—Progress Report. *Blue Jay* 43 (4): 243–246.

Houston, C. S. 1995a. Swainson's Hawk Banding in North America. *N. Am. Bird Bander* 20: 120–127.

Houston, C. S. 1995b. Thirty-two Consecutive Years of Reproductive Success at a Ferruginous Hawk Nest. *J. Raptor Res.* 29 (4): 282–283.

Houston, C. S. 1998. Swainson's Hawk Productivity and Five-young Nest. *Blue Jay* 56 (3): 151–155.

Houston, C. S., and M. J. Bechard. 1983. Trees and the Red-tailed Hawk in Southern Saskatchewan. *Blue Jay* 41: 99–109.

Houston, C. S., and M. J. Bechard. 1984. Decline of the Ferruginous Hawk in Saskatchewan. *Am. Birds* 40: 166–170.

Houston, C. S., and K. I. Fung. 1999. Saskatchewan's First Swainson's Hawk With Satellite Radio. *Blue Jay* 57 (2): 69–72.

Houston, C. S., W. C. Harris, and A. Schmit. 1998. Ferruginous Hawk Banding in Saskatchewan. *Blue Jay* 56 (2): 92–94.

Houston, C. S., and K. A. Hodson. 1997. Resurgence of Breeding Merlins, *Falco columbarius richardsonii,* in Saskatchewan Grasslands. *Can. Field-Nat.* 3: 243–248.

Houston, C. S., and J. K. Schmutz. 1995. Declining Reproduction Among Swainson's Hawks in Prairie Canada. *J. Raptor Res.* 29: 198–201.

Houston, C. S., and F. Scott. 2001. Power Poles Assist Range Expansion of Ospreys in Saskatchewan. *Blue Jay* 59: 182–188.

Howell, S.N.G., and S. Webb. 1995. *A Guide to the Birds of Mexico and Northern Central America.* Oxford Univ. Press, New York, NY.

Howell, S.N.G., S. Webb, D. A. Sibley, and L. J. Prairie. 1992. First Record of a Melanistic Northern Harrier in North America. *West. Birds* 23: 79–80.

Humphrey, P. S., and K. C. Parkes. 1959. An Approach to the Study of Molts and Plumages. *Auk* 76: 1–31.

Hunt, W. G. 1983. Rare Aplomado May Return to Texas. *Texas Parks Wildl.* Jul.: 11–13.

Hunt, W. G. 1998. Bald Eagle. Pp. 50–54 *in The Raptors of Arizona* (R. L. Glinski, ed.). Univ. of Arizona, Tucson, and Arizona Game Fish Dep., Phoenix.

Hunt, W. G., R. R. Rogers, and D. J. Slowe. 1975. Migratory and Foraging Behavior of Peregrine Falcons on the Texas Coast. *Can. Field-Nat.* 89 (2): 111–123.

Idaho Conservation Data Center. 1999. Distribution of Special Status Vertebrate Species by County: Ferruginous Hawk (*Buteo regalis*). World Wide Web Site for the Idaho Dep. Fish Game, http://www.state.id.us./fishgame/ vert.htm>.

Ingle, D. 1999. Hare Decline: Extraterrestrial? World Wide Web Home Page for *All Outdoors* Magazine, <Alloutdoors.comHome>.

Islands Protection Society. 1984. *Islands at the Edge: Preserving the Queen Charlotte Islands Wilderness.* Douglas & McIntyre Ltd., Vancouver, BC.

Isley, L. D., and J. W. Lish. 1984. Nesting and Summer Records for Ospreys in Oklahoma. *Bull. Oklahoma. Ornithol. Soc.* 19: 2–3.

Jacobs, B., and J. D. Wilson. 1997. *Missouri Breeding Bird Atlas.* Missouri Dep. Conserv., Jefferson City.

James, D. A., and J. C. Neal. 1986. *Arkansas Birds: Their Distribution and Abundance.* Univ. of Arkansas Press, Fayetteville.

James, P. C. 1994. Urban-nesting of Swainson's Hawks in Saskatchewan. *Condor* 94: 773–774.

Janssen, R. B. 1987. *Birds in Minnesota.* Univ. of Minn. Press, Minneapolis.

Jeffers, R. D. 2000. The Mystery of the Dying Eagles. *Endangered Species Bull.*, U.S. Fish Wildl. Serv. 25 (5): 4–5.

Johnsgard, P. A. 1973. *Grouse and Quails of North America.* Univ. of Nebraska Press, Lincoln.

Johnsgard, P. A. 1990. *Hawks, Eagles, and Falcons of North America.* Smithson. Inst. Press, Washington, D.C.

Justus, K. 1997. Mississippi Kite: Purple Martins Being Included in the Aerial Hunter's Diet. *Purple Martin Update* 7 (3): 8–9.

Kaufman, J., and H. Meng. 1975. *Falcons Return: Restoring an Endangered Species.* William Morrow, New York, NY.

Kaufman, K. 1996. *Lives of North American Birds.* Houghton Mifflin, Boston, MA.

Keddy-Hector, D. P. 1998. Aplomado Falcon. Pp. 124–127 *in The Raptors of Arizona* (R. L. Glinski, ed.). Univ. of Arizona Press, Tucson, and Arizona Game Fish Dep., Phoenix.

Kellogg, S. (ed.). 2000. Fall 1999 Flyway Reports. *Hawk Migr. Stud.* 26 (1).

Kenne, M. C. 2000. Field Reports—Spring 2000. *Iowa Bird Life* 70 (3): 135.

Kent, T. H. 1996. Field Reports—Fall 1995. *Iowa Bird Life* 66 (2): 19.

Kent, T. H. 1997. Field Reports—Fall 1996. *Iowa Bird Life* 67 (1): 20–21.

Kent, T. H. 1998. Field Reports—Fall 1997. *Iowa Bird Life* 68 (1): 11–12.

Kent, T. H., and J. J. Dinsmore. 1996. *Birds in Iowa.* Publ. by the authors, Iowa City and Ames.

Kerlinger, P., and S. A. Gauthreaux, Jr. 1985. Seasonal Timing, Geographic Distribution, and Flight Behavior of Broad-winged Hawks During Spring Migration in South Texas: A Radar and Visual Study. *Auk* 102: 735–743.

Kessel, B. 1989. *Birds of the Seward Peninsula, Alaska: Their Biogeography, Seasonality, and Natural History.* Univ. of Alaska Press, Fairbanks.

Kingery, H. E. (ed.) 1998. *Colorado Breeding Bird Atlas.* Colorado Bird Atlas Partnership and Colorado Div. Wildl., Denver.

Kirk, D. A., and M. J. Mossman. 1998. Turkey Vulture (*Cathartes aura*). *In* The Birds of North America, No. 339 (A. Poole and F. Gill, eds.). Acad. Nat. Sci., Philadelphia, PA, and Am. Ornithol. Union, Washington, D.C.

Kochert, M. N., and M. Q. Moritsch. 1984. Dispersal and Migration of Southwestern Idaho Raptors. *J. Field Ornithol.* 55 (3): 357–368.

Kochert, M. N., K. Steenhof, J. M. Marzluff, and L. B. Carpenter. 1999. Effects of Fire on Golden Eagle Occupancy and Reproductive Success. *J. Wildl. Manage.* 63: 773–780.

Koford, C. B. 1953. *The California Condor.* Natl. Audubon Soc. Res. Rep. No. 4.

Krebs, C. J., S. Boutin, and R. Boonstra. 2001. *Ecosystem Dynamics of the Boreal Forest: The Kluane Project.* Oxford Univ. Press, Oxford, U.K.

Kricher, J. 1999. Rediscovering Condors. *Birders World* 13 (1): 40–45.

Laack, L. 1995. Return of the Aplomado. *Texas Parks Wildl.* Feb.: 12–15.

Lammertink, J. M., J. A. Rojas-Tome, F. M. Casillas-Orona, and R. L. Otto. 1996. *Status and Conservation of Old-growth Forests and Endemic Birds in the Pine-Oak Zone of the Sierra Madre Occidental, Mexico.* No. 69, Oct. Inst. Systematics and Pop. Biol., Univ. of Amsterdam, Amsterdam, The Netherlands.

Lane, J. J., and R. A. Fischer. 1997. Species Profile: Southeastern American Kestrel (*Falco sparverius paulus*) on Military Installations in the Southeastern United States. Tech. Rep. SERDP-97-4. U.S. Army Engineer Waterways Exp. Sta., Vicksburg, MS.

Lanning, D. V., and M. A. Hitchcock. 1991. Breeding Distribution and Habitat of Prairie Falcons in Northern Mexico. *Condor* 93: 762–765.

Lanning, D. V., and P. W. Lawson. 1977. Ecology of the Peregrine Falcon in Northeastern Mexico, 1977. Chihuahuan Desert Res. Inst., Alpine, TX. Contrib. No. 41.

Lemieux, J. 1995. Falcon Nest Finds Common Ground Among Ranchers, Government. *Caller Times*, Corpus Christi, May 29, 1995.

Levy, S. H. 1998. Crested Caracara. Pp. 115–117 *in The Raptors of Arizona* (R. L. Glinski, ed.). Univ. of Arizona Press, Tucson, and Arizona Game Fish Dep., Phoenix.

Ligon, J. S. 1961. *New Mexico Birds.* Univ. of New Mexico Press, Albuquerque.

Liguori, J. 2001. Pitfalls of Classifying Light Morph Red-tailed Hawks to Subspecies. *Birding* 53 (5): 436–446.

Line, L. 1996. Accord Is Reached to Recall Pesticide Devastating Hawk. *New York Times*, Science, Oct. 15, 1996.

Long, M. 1998. The Vanishing Prairie Dog. *Natl. Geogr.* 193 (4): 116–131.

Lott, C. 1999. Florida Keys Autumn Raptor Migration Census. Unpubl. rep. Hawkwatch Int. Proj. no. NG98-103.

Lowery, G. H. 1974. *Louisiana Birds.* Kingsport Press, Kingsport, TN.

Machtans, C. S. 2000. Extra-limital Observations of Broad-winged Hawk, *Buteo platypterus*, Connecticut Warbler, *Oporornis agilis*, and Other Bird Observations From the Liard Valley, Northwest Territories. *Can. Field-Nat.* 114 (4): 671–679.

MacWhirter, R. B., and K. L. Bildstein. 1996. Northern Harrier (*Circus cyaneus*). *In* The Birds of North America, No. 210 (A. Poole and F. Gill, eds.). Acad. Nat. Sci., Philadelphia, PA, and Am. Ornithol. Union, Washington, D.C.

Marchant, S. (ed.). 1994. *Handbook of Australian, New Zealand, and Antarctica Birds.* Vol. 2. Oxford Univ. Press, Oxford, U.K.

Martell, M., S. Willey, and J. Schladweiler. 1998. Nesting and Migration of Swainson's Hawks in Minnesota. *Loon* 70: 72–81.

Martin, R. 2000. Northern Great Plains (Kites Through Shorebirds). *N. Am. Birds* 54 (4): 396–397.

Maxwell, R. C., and M. S. Husak. 1999. Common Black-Hawk Nesting in West-Central Texas. *J. Raptor Res.* 33 (3): 270–271.

McCaskie, G., and K. L. Garrett. 2000. Southern Pacific Coast (Boobies Through Ptarmigan). *N. Am. Birds* 54 (4): 382.

McCaskie, G., and M. San Miguel. 1999. Report of the California Birds Record Committee: 1996 Records. *West. Birds* 30: 57–85.

McCollough, M. A. 1989. Molting Sequence and Aging of Bald Eagles. *Wilson Bull.* 101 (1): 1–10.

McKinley, J. O., and W. G. Mattox. 2001. A Brood of Five Swainson's Hawks in Southwestern Idaho. *J. Raptor Res.* 35 (2): 169.

Meehan, T. D., P. Ginrod, and S. Hoffman. 1997. Foreign Encountered Raptors Banded in the Goshute Mts., Nev. 1980–1996: Breeding and Winter Ranges, Mortality, and Migration Routes. Paper read at the Hawk Migr. Assoc. N. Am. Conference VIII, Snowbird, UT, Jun. 12–15, 1997.

Meretsky, V. J., and N.F.R. Snyder. 1992. Range Use and Movements of California Condors. *Condor* 94: 313–335.

Meretsky, V. J., N.F.R. Snyder, S. R. Beissinger, D. A. Clendenen, and J. W. Wiley. 2000. Demography of the California Condor: Implications for Reestablishment. *Conserv. Biol.* 14 (3): 63–66.

Meyer, K. 1994. Species Profile: American Swallow-tailed Kite. *Wildbird* Jan.: 44–49.

Meyer, K. D. 1995. Swallow-tailed Kite (*Elanoides forficatus*). *In* The Birds of North America, No. 138 (A. Poole and F. Gill, eds.). Acad. Nat. Sci., Philadelphia, PA, and Am. Ornithol. Union, Washington, D.C.

Meyer, K. D., J. D. Arnett, and A. Washburn. 1997. Abstract: Migration Routes and Winter Range or the Swallow-tailed Kite (*Elanoides forficatus*) Based on Satellite and VHF Telemetry. Paper presented at the Raptor Res. Found. 1997 Annu. Meeting, Savannah, GA, Oct. 30–Nov. 1, 1997.

Millar, J. G. 2002. The Protection of Eagles and the Bald and Golden Eagle Protection Act. *J. Raptor Res.* 36 (1): 29–31.

Millard, S. 1993. Prairie Falcon Movements Into

Western Minnesota. *Hawk Migr. Stud.* 19 (1): 16.

Miller, E. K., and J. A. Smallwood. 1997. Juvenal Plumage Characteristics of Male Southeastern American Kestrels. *J. Raptor Res.* 31 (3): 273–274.

Mindell, D. P. 1985. Plumage Variation and Winter Range of Harlan's Hawk (*Buteo jamaicensis harlani*). *Am. Birds* 39: 127–133.

Mlodinow, S., and B. Tweit. 2001. Oregon-Washington. *N. Am. Birds* 55 (2): 220.

Monson, G. 1998. White-tailed Hawk. Pp. 96–98 *in The Raptors of Arizona.* (R. L. Glinski, ed.). Univ. of Arizona Press, Tucson, and Arizona Game Fish Dep., Phoenix.

Montana Bird Distribution Committee. 1996. *P. D. Skaar's Montana Bird Distribution.* 5th ed. Montana Nat. Heritage Prog. Spec. Publ. no. 3, Helena.

Montaperto, I. 1988. First Northern Harrier Nest in Southwestern Oklahoma. *Bull. Okla. Ornithol. Soc.* 21: 12–13.

Montiel de la Garza, F., and A. J. Contreras-Balderas. 1990. First Hook-billed Kite Specimen From Nuevo Leon, Mexico. *Southwest. Nat.* 35 (3).

Montoya, A. B., P. J. Zwank, and M. Cardena. 1997. Breeding Biology of Aplomado Falcons in Desert Grasslands of Chihuahua, Mexico. *J. Field Ornithol.* 68: 135–143.

Moore, C. A., and A. L. Fesnock. 1997. Golden Eagle Plumage in the Western U.S. Paper read at the Hawk Migr. Assoc. N. Am. Conference VIII, Snowbird, UT, Jun. 12–15, 1997.

Moore, K. R., and C. J. Henny. 1983. Nestsite Characteristics of Three Coexisting Accipiter Hawks in Northeastern Oregon. *J. Raptor Res.* 17 (3): 65–76.

Morrison, J. L. 1996. Crested Caracara (*Caracara plancus*). *In* The Birds of North America, No. 249 (A. Poole and F. Gill, eds.). Acad. Nat. Sci., Philadelphia, PA, and Am. Ornithol. Union, Washington, D.C.

Mueller, H. C. 1972. Zone-tailed Hawk and Turkey Vulture: Mimicry or Aerodynamics? *Condor* 74: 221–222.

Munro, H. L., and D. A. Reid. 1982. Swainson's Hawks, *Buteo swainsoni*, Nesting Near Winnipeg. *Can. Field-Nat.* 96: 206–208.

New Mexico Natural Heritage Database. 1996. List of Species of New Mexico With NHP "Tracked" Status. World Wide Web Site <www.fw.vt.edu/fishex/nmex_main/species/040805.htm>.

New Mexico State Game Commission 2002. Agenda Item No. 17: Aplomado Falcon. Minutes of 17 July 2002 Meeting. World Wide Web Site <http://www.gm.fsh.state.nm.us/PageMill_Text/Commission/minutes.html> [Aug. 8, 2002].

Nicoletti, F. J. 1997. American Kestrel Migration Correlated With Green Darner Movements at Hawk Ridge in Duluth, Minn. Paper read at the Hawk Migr. Assoc. N. Am. Conference VIII, Snowbird, UT, Jun. 12–15, 1997.

Nicoletti, F. J., S. Millard, and B. Yokel. 1998. First Description of Albinism in a Dark-morph Red-tailed Hawk in North America. *Loon* 70: 117–118.

North American Bird Information Web Site. 2003. Turkey Vulture (*Cathartes aura*): Noteworthy Observations of Turkey Vulture in British Columbia. World Wide Web Site <http://www.birdinfo.com/index.html> for the North American Bird Information Web Site. British Columbia links <http:/www.birdinfo.com/TurkeyVulture_bc.html> and <http://www.birdinfo.com/TurkeyVulture_MAP.html> [Feb. 20, 2003].

Nye, P. E. 1988. A Review of Bald Eagle Hacking Projects and Early Results in North America. Pp. 95–112 *in Proc. Int. Symp. Raptor Reintroduction, 1985* (D. K. Garcelon and G. W. Roemer, eds.). Inst. Wildl. Stud., Arcata, CA.

Oberholser, H. C. 1974. *The Bird Life of Texas.* Vol. 1. Univ. of Texas Press, Austin.

Olendorff, R. R. 1993. Status, Biology, and Management of Ferruginous Hawks: A Review. Raptor Res. and Tech. Asst. Cen., Spec. Rep., U.S. Dep. Int., Bur. Land Manage., Boise, ID.

Olivo, C. 2001. Bolivia: Studying Migrating Raptors at Four Hawkwatch Sites. *Hawk Migr. Stud.* 26 (2): 32–38.

Olivo, C. 2002. First Full-season Autumn Raptor Migration Count in Concepcion, Bolivia. *Raptor Watch* 16 (3): 12–13.

Olson, C. V., and D. P. Arsenault. 2000. Differential Winter Distribution of Rough-legged Hawks (*Buteo lagopus*) by Sex in Western North America. *J. Raptor Res.* 34 (3): 157–166.

Palmer, P. 1989. A Melanistic American Kestrel. *Hawk Migr. Stud.* 14 (2): 41.

Palmer R. S. (ed.). 1988. Diurnal Raptors. *In Handbook of North American Birds.* Vols. 4 and 5. Yale Univ. Press, New Haven, CT.

Parker, J. W. 1999. Mississippi Kite. *In* The Birds of North America, No. 402 (A. Poole and F. Gill, eds.). Acad. Nat. Sci., Philadelphia, PA, and Am. Ornithol. Union, Washington, D.C.

Parvin, B. 1988. The Disappearing Wildlands of

the Rio Grande Valley. *Texas Parks Wildl.* Mar.: 3–15, 45.
Patten, M. A., and R. A. Erickson. 2000. Population Fluctuations of the Harris' Hawk (*Parabuteo unicinctus*) and Its Appearance in California. *J. Raptor Res.* 34 (3): 187–195.
Peregrine Fund, The. 1999a. Recovery of the Aplomado Falcons, 1999 Report. World Wide Web Home Page <www.peregrinefund.org/> [Apr. 1999].
Peregrine Fund, The. 1999b. California Condors Released Near Grand Canyon. World Wide Web Home Page <www.peregrinefund.org./> [May 1999].
Peregrine Fund, The. 1999c. Notes From the Field: California Condor. World Wide Web Home Page <http://www.peregrinefund.org/> [Dec. 1999].
Peregrine Fund, The. 2001. Wild California Condor Lays Egg. World Wide Web Home Page <http://www.peregrinefund.org> [May 2, 2001].
Perez, C. J., P. J. Zwank, and D. W. Smith. 1996. Survival, Movements, and Habitat Use of Aplomado Falcons Released in Southern Texas. *J. Raptor Res.* 30: 175–182.
Peterson, R. A. 1995. *The South Dakota Breeding Bird Atlas.* South Dakota Ornithol. Union, Aberdeen.
Pettingill, O. S., Jr. 1970. *Ornithology in Laboratory and Field.* 4th ed. Burgess Publ. Minneapolis, MN.
Pierce, A. 1998. Winter Months Organophosphate Poisoning in Raptors, Case Report and Overview. Colorado State Univ. Unpubl. rep.
Pittaway, R. 1993. Subspecies and Color Morphs of the Red-tailed Hawk. *Ont. Birds* 2: 23–29.
Poole, K. G. 1994. Lynx-Snowshoe Hare Cycle in Canada. *Cat News* 20 (Spring 1994).
Preston, C. R., and R. D. Beane. 1993. Red-tailed Hawk (*Buteo jamaicensis*). *In* The Birds of North America, No. 52 (A. Poole and F. Gill, eds.). Acad. Nat. Sci., Philadelphia, PA, and Am. Ornithol. Union, Washington, D.C.
Pulich, W. M. 1988. *The Birds of North-Central Texas.* Texas A & M Univ. Press, College Station.
Purrington, R. D. 1998. Central Southern Region (Diurnal Raptors). *Field Notes* 52 (4): 464–465.
Purrington, R. D. 2000. Central Southern Region (Raptors Through Rails). *N. Am. Birds* 54 (4): 392.
Quinn, M. S. 1991. Nest Site and Prey of a Pair of Sharp-shinned Hawks in Alberta. *J. Raptor Res.* 25 (1): 18.
Raptor Center, The. 1999a. Highway to the Tropics/Prairie Partners: Osprey (Tracking Osprey Via Satellite on the Internet). World Wide Web Home Page <http://www.raptor.cvm.umn.edu> [Jan. 12, 2001].
Raptor Center, The. 1999b. Highway to the Tropics/Prairie Partners: The Raptor Center Tracks Swainson's Hawks by Satellite. World Wide Web Home Page <http://www.raptor.cvm.umn.edu> [Jan. 12, 2001].
Rea, A. M. 1983. *Once a River: Bird Life and Habitat Changes on the Middle Gila.* Univ. of Arizona Press, Tucson.
Rea, A. M. 1998. Turkey Vulture. Pp. 27–31 *in The Raptors of Arizona.* (R. L. Glinski, ed.). Univ. of Arizona Press, Tucson, and Arizona Game Fish Dep., Phoenix.
Redig, P. T., and H. B. Tordoff. 1988. Peregrine Falcon Reintroduction in the Upper Mississippi Valley and Western Great Lakes Region. Pp. 559–563 *in Peregrine Falcon Populations* (T. J. Cade, J. H. Enderson, C. G. Thelander, and C. M. White, eds.). The Peregrine Fund, Boise, ID.
Reynolds, R. T. 1983. Management of Western Coniferous Forest Habitat for Nesting Accipiter Hawks. Gen. Tech. Rep. RM-102. U.S. Dep. Agric., Fort Collins, CO.
Reynolds, R. T., and E. C. Meslow. 1984. Partitioning of Food and Niche Characteristics of Coexisting Accipiters During Breeding. *Auk* 101: 76–79.
Reynolds, R. T., and H. M. Wight. 1978. Distribution, Density, and Productivity of Accipiter Hawks Breeding in Oregon. *Wilson Bull.* 90 (2): 182–196.
Rideout, D. W., and D. A. Swepston. 1984. Where the Golden Eagles Nest. *Texas Parks Wildl.* Jan.: 18–23.
Ripple, J. 2000. Kites of Fancy. *Birders World* 14 (3): 63–66.
Risenbrough, R., W. Schorloff, P. H. Bloom, and E. E. Littrell. 1989. Investigations of the Decline of Swainson's Hawk Populations in California. *J. Raptor Res.* 23 (3): 63–71.
Robbins, M. B., and D. A. Easterla. 1992. *Missouri Birds: Their Abundance and Distribution.* Univ. of Missouri Press, Columbia.
Rodriguez-Estrella, R., and B. T. Brown. 1990. Density and Habitat Use of Raptors Along the Rio Bavispe and Rio Yaqui, Sonora, Mexico. *J. Raptor Res.* 24 (3): 47–51.
Rodriguez-Estrella, R., J. Llinas-Gutierrez, and J. Cancino. 1991. New Golden Eagle Records From Baja California. *J. Raptor Res.* 25 (3): 68–71.

Rosenfield, R. N., and J. Bielefeldt. 1993. Cooper's Hawk (*Accipiter cooperii*). *In* The Birds of North America, No. 75 (A. Poole and F. Gill, eds.). Acad. Nat. Sci., Philadelphia, PA, and Am. Ornithol. Union, Washington, D.C.

Rosenfield, R. N., and J. Bielefeldt. 1997. Reanalysis of Relationships Among Eye Color, Age and Sex in Cooper's Hawk. *J. Raptor Res.* 31 (4): 313–316.

Rosenfield, R. N., and D. L. Evans. 1980. Migration Incidence and Sequence of Age and Sex Classes of the Sharp-shinned Hawk. *Loon* 52: 66–69.

Rottenborn, S. C., and J. Morlan. 2000. Report of the California Birds Record Committee: 1997 Records. *West. Birds* 31 (1): 1–37.

Rowell, G. 1991. Falcon Rescue. *Natl. Geogr.* 179 (4): 106–114.

Rowell, P., and G. L. Holroyd. 2001. *Results of the 2000 Canadian Peregrine Falcon* (Falco peregrinus) *Survey*. Can. Wildl. Serv., Environ. Canada, Edmonton, AB.

Russell, R. W. 1991. Nocturnal Flight by Migrant "Diurnal" Raptors. *J. Field Ornithol.* 62 (4): 505–508.

Russell, S. M., and G. Monson. 1997. *The Birds of Sonora*. Univ. of Arizona Press, Tucson.

Sailer, J. E. 1987. Adult Pair of Merlins in Southern Utah in June. *J. Raptor Res.* 21 (1): 38–39.

Salter, R., W. J. Richardson, and C. Holdworth. 1974. Spring Migration of Birds Through the Mackenzie Valley, N.W.T. April–May, 1973. *In* Arctic Gas Biological Report Series, Chapter II, Vol. 28: Ornithological Studies in the Mackenzie Valley, 1973 (W.W.H. Gunn, W. J. Richardson, R. E. Schweinsburg, and T. D. Wright, eds.). L.G.L. Limited, Environ. Res. Assoc.

Sanchez, G. 1993. Ecology of Wintering Gyrfalcons in Central South Dakota. M.S. thesis, Boise State Univ., Boise, ID.

Sanchez, G. 1994. Arctic Visitor. *South Dakota Conserv. Digest* 61: 6–9.

Santa Cruz Predatory Bird Research Group. 1995. A Pilot Golden Eagle Population Study in the Altamont Pass Wind Resource Area, California. World Wide Web Site for the Santa Cruz Predatory Research Group, Univ. of California, Long Marine Lab., Santa Cruz, <http://www.nrel.gov/wind/geagles.html> [Dec. 10, 2001].

Santa Cruz Predatory Bird Research Group. 1999a. Eagle Migration: Telemetry Maps for Wintering Adult, Subadult, and Juvenile Bald Eagles. World Wide Web Site for the Santa Cruz Predatory Research Group, Univ. of California, Long Marine Lab., Santa Cruz, <http://www2.ucsc.edu/scpbrg/migrationhtm> [Jan. 19, 2001].

Santa Cruz Predatory Bird Research Group. 1999b. Golden Eagles in a Changing Landscape. World Wide Web Site for the Santa Cruz Predatory Research Group, Univ. of California, Long Marine Lab., Santa Cruz, <http://www2.ucsc.edu/scpbrg/eagles.htm> [Dec. 10, 2001].

Santa Cruz Predatory Bird Research Group. 1999c. Peregrines: Peregrines Recovery. World Wide Web Site for the Santa Cruz Predatory Research Group, Univ. of California, Long Marine Lab., Santa Cruz, <http://www2.ucsc.edu/scpbrg/peregrines.htm> [Apr. 23, 2001].

Schmutz, J. K. 1996. Southward Migration of Swainson's Hawks: Over 10,000 Km in 54 Days. *Blue Jay* 54: 70–76.

Schmutz, J. K., and R. W. Fyfe. 1987. Migration and Mortality of Alberta Ferruginous Hawks. *Condor* 89: 169–174.

Schmutz, J. K., C. S. Houston, and S. J. Barry. 2001. Prey and Reproduction in a Meta Population Decline Among Swainson's Hawks, *Buteo swainsoni*. *Can. Field-Nat.* 115 (2): 257–273.

Schnell, J. H. 1994. Common Black-Hawk (*Buteogallus anthracinus*). *In* The Birds of North America, No.122 (A. Poole and F. Gill, eds.). Acad. Nat. Sci., Philadelphia, PA, and Am. Ornithol. Union, Washington, D.C.

Schnell, J. H. 1998. Common Black-Hawk. Pp. 73–76 *in The Raptors of Arizona* (R. L. Glinski, ed.). Univ. of Arizona Press, Tucson, and Arizona Game Fish Dep., Phoenix.

Scott, F., and C. S. Houston. 1983. Osprey Nesting Success in West-Central Saskatchewan. *Blue Jay* 41 (1): 27–31.

Seibold, I., and A. J. Helbig. 1995. Evolutionary History of New and Old World Vultures Inferred From Nucleotide Sequences of the Mitchondrial Cytochrome *b* Gene. *Phil. Trans. R. Soc. Lond.* 350: 163–178.

Semenchuk, G. P. 1992. *The Atlas of Breeding Birds of Alberta*. Fed. Alberta Nat., Edmonton, AB.

Sexton, C. 2001. Texas. *N. Am. Birds* 55 (2): 194.

Seyffert, K. 2001. *Birds of the Texas Panhandle*. Texas A & M Univ. Press, College Station.

Shackelford, C. E., D. Saenz, and R. R. Schaefer. 1996. Sharp-shinned Hawks Nesting in the Pineywoods of Eastern Texas and Western Louisiana. *Bull. Texas Ornithol. Soc.* 29: 23–25.

Shackelford, C. E., and G. G. Simons. 1999. An Annual Report of the Swallow-tailed Kite in

Texas: A Survey and Monitoring Project for 1998. *Texas Parks Wildl.* PWDBKW7000-496 (3/99).

Sharp, C. S. 1902. Nesting of Swainson's Hawk. *Condor* 4: 116–118.

Sharpe, R. S., W. R. Silcock, and J. G. Jorgensen. 2001. *The Birds of Nebraska: Their Ecology and Distribution.* Univ. of Nebraska Press, Lincoln.

Shefferly, N. 1996. *Lepus americanus.* World Wide Web Site for the Univ. of Michigan Mus. Zool., <animaldiversity.ummz.umich.edu/accounts/lepus/1._americansnarrative.html> [Jan. 18].

Shepard, M. G. 1999. British Columbia–Yukon Region. *N. Am. Birds* 53 (3): 318.

Sherrington, P. 1993. Golden Eagle Migration in the Front Range of the Alberta Rocky Mountains. *Birders J.* 2 (4): 195–204.

Shipman, M. S., and M. J. Bechard. 1998. Ecology of the Northern Goshawk in the Shrub Steppe of Northeastern Nevada: Breeding Biology and the Post-Fledging Dependency Period. M.S. thesis, Boise State Univ., Boise, ID.

Sieg, C. H., and D. M. Becker. 1990. Nest-site Habitat Selected by Merlins in Southeastern Montana. *Condor* 92: 688–694.

Sirois, J., and D. McRae. 1996. *The Birds of the Northwest Territories.* 2nd ed. Environ. Canada, Can. Wildl. Serv., Yellowknife, NT.

Small, A. 1994. *California Birds: Their Status and Distribution.* Ibis Publ., Vista, CA.

Smallwood, J. A. 1990. American Kestrel and Merlin. Pp. 29–37 *in Proc. Southeast Raptor Manage. Symp. Workshop,* Natl. Wildl. Fed., Washington, D.C. (B. G. Pendleton et al., eds.). World Wide Web Site <http://www.tnc.org/wings/wingresource.SMAInd.htm> [Dec. 12, 2000].

Smallwood, J. A., C. Natale, K. Steenhof, M. Meetz, C. D. Marti, et al. 1999. Clinal Variation in the Juvenal Plumage of American Kestrels. *J. Field Ornithol.* 70: 425–435.

Smith, A. R. 1996. *Atlas of Saskatchewan Birds.* Saskatchewan Nat. Hist. Soc., Regina.

Smith, D. W., and J. Ireland. 1992. First Record of the Black-shouldered Kite for Canada. *West. Birds* 23: 177–178.

Smith, M. R., P. W. Mattocks, Jr., and K. M. Cassidy. 1997. Breeding Birds of Washington State. *In Washington State Gap Analysis—Final Report.* Vol. 4 (K. M. Cassidy, C. E. Grue, M. R. Smith, and K. M. Dvornich, eds.). Seattle Audubon Soc. Publ. in Zool. No. 1, Seattle, WA.

Smith, R. 2002. Manzanos Study Solving Mysteries of Sharp-shinned Hawks. *Raptor Watch* 6 (3): 6–7.

Smith, V. J., J. A. Jenks, C. R. Berry, Jr., D. M. Fecske and C. J. Kopplin. 2002. *The South Dakota Gap Analysis Project Final Report.* U.S. Geol., Surv., Gap Analysis Prog.

Snyder, H. A. 1998a. Northern Harrier. Pp. 56–57 *in The Raptors of Arizona* (R. L. Glinski, ed.). Univ. of Arizona Press, Tucson, and Arizona Game Fish Dep., Phoenix.

Snyder, H. A. 1998b. Zone-tailed Hawk. Pp. 99–101 *in The Raptors of Arizona* (R. L. Glinski, ed.). Univ. of Arizona Press, Tucson, and Arizona Game Fish Dep., Phoenix.

Snyder, N.F.R., and A. M. Rea. 1998. California Condor. Pp. 32–36 *in The Raptors of Arizona* (R. L. Glinski, ed.). Univ. of Arizona Press, Tucson, and Arizona Game and Fish Dep., Phoenix.

Snyder, N.F.R., and H. A. Snyder. 1991. *Birds of Prey: Natural History and Conservation of North American Raptors.* Voyageur Press, Stillwater, MN.

Snyder, N.F.R., and H. A. Snyder. 1998a. Sharp-shinned Hawk. Pp. 58–62 *in The Raptors of Arizona* (R. L. Glinski, ed.). Univ. of Arizona Press, Tucson, and Arizona Game and Fish Dep., Phoenix.

Snyder, N.F.R., and H. A. Snyder. 1998b. Northern Goshawk. Pp. 68–72 *in The Raptors of Arizona* (R. L. Glinski, ed.). Univ. of Arizona Press, Tucson, and Arizona Game and Fish Dep., Phoenix.

Sodhi, N. S., and L. W. Oliphant. 1993. Prey Selection by Urban-breeding Merlins. *Auk* 110 (4): 727–735.

Sodhi, N. S., L. W. Oliphant, P. C. James, and I. G. Warkentin. 1993. Merlin (*Falco columbarius*). *In* The Birds of North America, No. 44 (A. Poole and F. Gill, eds.). Acad. Nat. Sci., Philadelphia, PA, and Am. Ornithol. Union, Washington, D.C.

Solensky, M. 2000. Merlins Nesting in Minneapolis. *Loon* 72: 72–75.

Solensky, M. J. 1997. Distribution, Productivity, and Nest-site Habitat of Taiga Merlin (*Falco columbarius columbarius*) in North-Central Wisconsin. M.S. thesis, Univ. of Wisconsin, Eau Claire.

Solensky, M. J. 2001. Merlin Nestsite Reoccupancy in the Twin Cities and North-central Wisconsin. 2001 Report. The Raptor Center, Univ. of Minnesota, St. Paul.

Stalmaster, M. V. 1987. *The Bald Eagle.* Universe Books, New York, NY.

Steenhof, K. 1998. Prairie Falcon (*Falco mexicanus*). *In* The Birds of North America, No. 346 (A. Poole and F. Gill, eds.). Acad. Nat. Sci.,

Philadelphia, PA, and Am. Ornithol. Union, Washington, D.C.
Steenhof, K., M. N. Kochert, and M. Q. Moritsch. 1984. Dispersal and Migration of Southwestern Idaho Raptors. *J. Field Ornithol.* 55 (3): 357–368.
Stevenson, H. M., and B. H. Anderson. 1994. *The Birdlife of Florida.* Univ. Press of Florida, Gainesville.
Stewart, G. R. 1979. Re-establishing the Harris' Hawk on the Lower Colorado River. Unpubl. rep. Bur. Land Manage., Yuma, AZ.
Stewart, R. E. 1975. *Breeding Birds of North Dakota.* Tri-College Center for Environmental Studies, Fargo, ND.: Northern Prairie Wildlife Research Center Home Page <www.npwrc.usgs.gov>, World Wide Web Site <http://www.npwrc.org/resource/distr/birds/bb_of_nd/bb_of_nd.htm>.
Struzik, E. 1999. Recovery Effort Takes Wing. World Wide Web Site for the *Edmonton Journal,* Edmonton, AB, <http://raysweb.net/specialplaces/pages-species-ej/peregrinefalcon-ej.html> [Mar. 21, 2002].
Sutton, G. M. 1967. *Oklahoma Birds.* Univ. of Oklahoma Press, Norman.
Svingen, P. H. 2000. First Minnesota Breeding Record of the Richardson's Merlin. *Loon* 72: 66.
Swepston, D. A., and D. W. Rideout. 1984. Golden Eagle Nest Survey. Texas Parks Wildl. Dep., Austin. Fin. rep. Job No. 34; Fed. Aid Proj. No. W-103-R-13.
Taverner, P. A. 1940. Variation in the American Goshawk. *Condor* 42: 157–160.
Terres, J. K. 1980. *The Audubon Society Encyclopedia of North American Birds.* Alfred A. Knopf, New York, NY.
Thompson, M. C., and C. Ely. 1989. *Birds in Kansas.* Vol. I. Univ. of Kansas Mus. Nat. Hist. Public Ed. Ser. no. 11.
Todd, W.E.C. 1950. A Northern Race of Red-tailed Hawk. *Ann. Carnegie Mus.* 31: 289–297.
Todd, W.E.C. 1963. *Birds of the Labrador Peninsula and Adjacent Areas.* Carnegie Mus., Pittsburgh, PA, and Univ. of Toronto Press, Toronto, ON.
Tordoff, H. B. 2000. Percentage of Subspecies of Peregrines Released in the Midwest and Their Success Surviving to Breeding. *Conserv. Biol.* 15: 528–532.
Tordoff, H. B., M. S. Martell, and P. T. Redig. 1997. *Midwest Peregrine Restoration, 1997 Report.* The Raptor Center, Univ. of Minnesota, St. Paul.
Tordoff, H. B., M. S. Martell, P. T. Redig, and M. J. Solensky. 2000. *Midwest Peregrine Falcon Restoration, 1999 Report.* The Raptor Center, Univ. of Minnesota, St. Paul.
Toups, J. A., J. A. Jackson, and E. Johnson. 1985. Black shouldered Kite: Range Expansion Into Mississippi. *Am. Birds* 39: 865–867.
Tucker, K. R. 1999. Good News for Swainson's Hawks. *Bird Calls* 3 (2): 9.
Tucker, V. A., T. J. Cade, and A. E. Tucker. 1998. Diving Speeds and Angles of a Gyrfalcon (*Falco rusticolus*). *J. Exp. Biol.* 201: 2061–2070.
Tufts, R. W. 1986. *Birds of Nova Scotia.* Nimbus Publ. and Nova Scotia Mus., Halifax, NS.
Tyler, J. D., S. J. Orr, and J. K. Banta. 1989. The Red-shouldered Hawk in Southwestern Oklahoma. *Bull. Oklahoma Ornithol. Soc.* 22 (3): 17–21.
U.S. Department of the Interior. 2000. Survival, Dispersal, and Long-range Movements of Prairie Falcons. World Wide Web Site for the ISDI and the For. and Rangeland Ecosystem Sci. Center, Snake River Field Sta., Idaho, <http://www.ris.idbsu.edu/PrairieFalcon.htm> [Feb. 8, 2000].
U.S. Fish and Wildlife Service. 1984. Reclassification of the Arctic Peregrine Falcon and Clarification of Its Status in Washington and Elsewhere in the Coterminous United States. Final Rule. *Fed. Reg.* 49 (55): 10520–10526.
U.S. Fish and Wildlife Service. 1992. Notice on Finding Petition to List the Ferruginous Hawk. *Fed. Reg.* 57: 37507–37513 [Aug. 19, 1992].
U.S. Fish and Wildlife Service. 1993. Draft Addendum to the Pacific Coast and Rocky Mountain/Southwest American Peregrine Falcon Recovery Plans. U.S. Fish Wildl. Serv., Portland, OR.
U.S. Fish and Wildlife Service. 1994. Final Rule. Endangered Status for Puerto Rican Sharp-shinned Hawk. *Fed. Reg.* 59: 46710.
U.S. Fish and Wildlife Service. 1995. Final Rule to Reclassify the Bald Eagle From Endangered to Threatened in All of the Lower 48 States. *Fed. Reg.* 60: 35999 [Jul. 2, 1995].
U.S. Fish and Wildlife Service. 1997a. *Return to the Wild, California Condor Recovery Program.* U.S. Fish Wildl. Serv., Ventura, CA.
U.S. Fish and Wildlife Service. 1997b. 90-Day Finding for a Petition to List the Northern Goshawk in the Contiguous United States West of the 100th Meridian. *Fed. Reg.* 62: 50892 [Sep. 29, 1997].
U.S. Fish and Wildlife Service. 1998a. *Condor News* (J. Hendron, ed.), Vol. 2. Hopper Mountain, NWR, Ventura, CA.
U.S. Fish and Wildlife Service. 1998b. Northern Goshawk Status Review. Office of Tech. Support, Portland, OR. Jun. 1998.

U.S. Fish and Wildlife Service. 1998c. Proposed Rule to Remove the Peregrine Falcon in North America From the List of Endangered and Threatened Wildlife. *Fed. Reg.* 63 (165): 45446–45463 [Aug. 26, 1998].

U.S. Fish and Wildlife Service. 1999a. Ninety-day Finding for a Petition to List the Black-tailed Prairie Dog as Threatened. *Fed. Reg.* 64 (57): 14424–14428 [Mar. 25, 1999].

U.S. Fish and Wildlife Service. 1999b. Proposed Rule to Remove Bald Eagle in the Lower 48 States From the List of Endangered and Threatened Wildlife. *Fed. Reg.* 64: 36453 [Jul. 6, 1999].

U.S. Fish and Wildlife Service. 1999c. Final Rule to Remove the American Peregrine Falcon From the Federal List of Endangered and Threatened Wildlife. *Fed. Reg.* 64: 46541–46558 [Aug. 25, 1999].

U.S. Fish and Wildlife Service. 2000a. Twelve-month Finding for a Petition to List the Black-tailed Prairie Dog as Threatened. *Fed. Reg.* 65 (24): 5476–5488 [Feb. 4, 2000].

U.S. Fish and Wildlife Service. 2000b. Proposal to Transfer the North American Population of *Falco rusticolus* (Gyrfalcon) From Appendix I to Appendix II of the Convention on International Trade in Endangered Species of Wild Fauna and Flora (CITES), submitted Nov. 12, 1999, for consideration at the eleventh meeting of the conference of the parties to CITES.

U.S. Fish and Wildlife Service. 2001a. Availability of Final Environmental Assessment of Take of Nestling American Peregrine Falcons in the Contiguous United States and Alaska for Falconry. *Fed. Reg.* 64 (92): 24149–24150 [May 11, 2001].

U.S. Fish and Wildlife Service. 2001b. Review of Plant and Animal Species That Are Candidates or Proposed for Listing as Endangered or Threatened Species. *Fed. Reg.* 66: 54808–54810.

U.S. Fish and Wildlife Service. 2002. Notice of Receipt of Applications for Import/Export Permit: Applicant Zool. Soc. San Diego/San Diego Zoo, San Diego, CA, PRT 056991 and 657398. *Fed. Reg.* 67 (107): 38516–17.

van Rossem, A. J. 1938. A Mexican Race of the Goshawk (*A. gentilis*) [Linnaeus]. Proc. Biol. Soc. Wash. 51: 99–100.

Varland, D. E. 1999. Peregrine Falcon Banding in Western Washington, 1998 Annual Report. Submitted to U.S. Fish Wildl. Serv., Portland, OR, Jan. 29. Unpubl. rep.

Varland, D. E. 2001. Raptor Surveys and Banding on Coastal Beaches of Western Washington: 2000 Annual Report. Submitted to Washington State Parks Recreation Comm. Environ. Prog., Olympia.

Varland, D. E., 2002. Coastal Raptor Surveys 2001 Annual Report. Report to the Washington State Parks Recreation Comm., Environ. Prog., Olympia, Apr. 30, 2002.

Varland, D. E., E. E. Klaas, and T. M. Loughin. 1991. Development of Foraging Behavior in the American Kestrel. *J. Raptor Res.* (1): 9–17.

Varland, D. E., E. E. Klaas, and T. M. Loughin. 1993. Use of Habitat and Perches, Causes of Mortality, and Time Until Dispersal in Post-fledging American Kestrels. *J. Field Ornithol.* 64 (2): 169–178.

Varland, D. E., and T. M. Loughin. 1992. Social Hunting in Broods of Two and Five American Kestrels After Fledging. *J. Raptor Res.* 26 (2): 74–80.

Varland, D. E., and T. M. Loughin. 1993. Reproductive Success of American Kestrels Nesting Along an Interstate Highway in Central Iowa. *Wilson Bull.* 105 (3): 465–474.

Ventana Wilderness Society. 2000a. Bald Eagle Restoration: Bald Eagles Return to Central Coast. World Wide Web Site <http://www.ventanaws.org/eagles.htm> [Jan. 22, 2000].

Ventana Wilderness Society. 2000b. Condor Reintroduction: Reintroducing the California Condor to Big Sur. World Wide Web Site <http://www.ventanaws.org/condors.htm> [Jul. 13, 2001].

Walsh, P. J. 1996. Notes on a Mississippi Kite Nest in Central Iowa. *Iowa Bird Life* 66 (1): 1–10.

Walton, B. J., J. Linthicum, and G. Stewart. 1988. Release and Re-establishment Techniques Developed for Harris' Hawks—Colorado River 1979–1986. Pp. 318–320 *in Proc. Southwest Raptor Manage. Symp. Workshop*, May 21–24, 1986 (R. L. Glinski, B. G. Pendleton, M. B. Moss, M. N. Le Franc, Jr., B. A. Millsap, and S. W. Hoffman, eds.)., Univ. of Arizona, Tucson; Natl. Wildl. Fed., Washington, D.C.; Natl. Wildl. Fed. Sci. and Tech. Ser. No. 11.

Walton, B. J., and C. G. Thelander. 1988. Peregrine Falcon Management Efforts in California, Oregon, Washington, and Nevada. Pp. 587–597 *in Peregrine Falcon Populations* (T. J. Cade, J. H. Enderson, C. G. Thelander, and C. M. White, eds.). The Peregrine Fund, Boise, ID.

Warkentin, I. G., and P. C. James. 1990. Winter Roost Site Selection by Urban Merlins. *J. Raptor Res.* 24 (1–2): 5–11.

Warkentin, I. G., P. C. James, and L. W. Oliphant.

1992. Use of a Plumage Criterion for Aging Female Merlins. *J. Field Ornithol.* 63 (4): 473–475.

Watson, J. W. 1999. Telemetry-tracking of Four Adult Ferruginous Hawks (*Buteo regalis*) Banded in South-Central Washington in 1999. Unpubl. rep. Washington Dep. Fish Wildl.

Watson, J. W., D. W. Hays, S. P. Finn, and P. Meehan-Martin. 1998. Prey of Breeding Northern Goshawks in Washington. *J. Raptor Res.* 32 (4): 297–305.

Wauer, R. 1992. *A Naturalist's Mexico.* Texas A & M Univ. Press, College Station.

Whaley, W. H., and C. M. White. 1994. *Trend in Geographic Variation of Cooper's Hawk and Northern Goshawk in North America: A Multivariate Analysis.* Vol. 5. West. Found. Vertebr. Zool.

Wheeler, B. K., and W. S. Clark. 1995. *A Photographic Guide to North American Raptors.* Academic Press, San Diego, CA.

White, C. M. 1962. Prairie Falcon Displays Accipitrine and Circinine Hunting Methods. *Condor* 64 (5): 439–440.

White, C. M. 1965. Goshawk Nesting in the Upper Sonoran in Colorado and Utah. *Condor* 67 (3): 269.

White, C. M. 1968. Diagnosis and Relationships of the North American Tundra-inhabiting Peregrine Falcons. *Auk* 85: 179–191.

White, C. M., R. E. Ambrose, and J. L. Longmire. 1995. Remarks on Systematics and the Sources of Variation in *Falco rusticolus*: The Reference to the Reintroduction of Falcons to Poland. *Acta Ornithol.* 30 (1): 31–41.

White, C. M., and L. F. Kiff. 1998. Holarctic Birds of Prey. *Proc. Int. Conference.* Badajoz, Extremadura (Spain), Apr. 17–22, 1995.

Wildlife Management Advisory Council. 2001. Gyrfalcon (*Falco rusticolus*) and Peregrine Falcon (*Falco peregrinus tundrius*): Yukon North Slope Long-Term Research and Monitoring Plan. World Wide Web Site <www.taiga.net/wmac/researchplan/reports/falcon.html> [Mar. 19, 2002].

Williams, D. 2002. Falcons Return to Their Native Land. World Wide Web Site for the *El Paso Times*, Special Report: Borderland Environment <http://www.elpasotimes.com/stores/borderland/20020706-127187.shtml> [Jul. 6, 2002].

Williams, S. O., III. 1997. Recent Occurrences of Aplomado Falcons in New Mexico: Is Natural Recolonization of Historic Range Underway? Paper presented at the 35th annual meeting of the New Mexico Ornithol. Soc., New Mexico State Univ., Las Cruces, Apr. 5, 1997.

Williams, S. O., III. 1998. Nesting Season, New Mexico Region. *Field Notes* 52 (4): 487.

Williams, S. O., III. 2000. New Mexico Regional Reports (Loons Through Raptors). *N. Am. Birds* 54 (1): 87.

Willis, E. V. 1963. Is the Zone-tailed Hawk a Mimic of the Turkey Vulture? *Condor* 65 (4): 313–317.

Wood, D. S., and G. D. Schnell. 1984. *Distributions of Oklahoma Birds.* Univ. of Oklahoma Press, Norman.

Wood, N. A. 1932. Harlan's Hawk. *Wilson Bull.* Jun.: 78–87.

Woodbridge, B. 1997. Tracking the Migration of Swainson's Hawks: Conservation Lessons in a Global Classroom. Paper read at the Hawk Migr. Assoc. N. Am. Conference VIII, Snowbird, UT, Jun. 12–15, 1997.

Yates, M. A., K. E. Riddle, and F. P. Ward. 1988. Recoveries of Peregrine Falcons Migrating Through the Eastern and Central United States, 1955–1985. Pp. 471–483 *in Peregrine Falcon Populations* (T. J. Cade, J. H. Enderson, C. G. Thelander, and C. M. White, eds.). The Peregrine Fund, Boise, ID.

Young, K. E., B. C. Thompson, and R. Valdez. 1999. Project Progress Report (Sep. 29, 1997–Dec. 1998), Determination of Habitat Suitability for Aplomado Falcons on Public Lands in Southern New Mexico. World Wide Web Site <http://leopold.nmsu.edu/kyoung/progress/jan_progress99.htm> [Nov. 16, 2000].

Yukon Department of Renewable Resources. 1990. Yukon Mammal Series: Snowshoe Hare. World Wide Web Site <http://www.renres.gov.yk.ca/wildlife/snhare.html> [May 11, 2000].

# INDEX

English names are printed in roman type; scientific names are in italic. An (m) before a number indicates a map.